AF776407

Applied Biocatalysis

Applied Biocatalysis

The Chemist's Enzyme Toolbox

Edited by

JOHN WHITTALL

University of Manchester
Manchester
UK

PETER W. SUTTON

Glycoscience S.L.
Barcelona
ES

WILEY

This edition first published 2021
© 2021 John Wiley & Sons Ltd

All rights reserved. No part of this publication may be reproduced, stored in a retrieval system, or transmitted, in any form or by any means, electronic, mechanical, photocopying, recording or otherwise, except as permitted by law. Advice on how to obtain permission to reuse material from this title is available at http://www.wiley.com/go/permissions.

The right of John Whittall and Peter W. Sutton to be identified as the authors of the editorial material in this work has been asserted in accordance with law.

Registered Offices
John Wiley & Sons, Inc., 111 River Street, Hoboken, NJ 07030, USA
John Wiley & Sons Ltd, The Atrium, Southern Gate, Chichester, West Sussex, PO19 8SQ, UK

Editorial Office
The Atrium, Southern Gate, Chichester, West Sussex, PO19 8SQ, UK

For details of our global editorial offices, customer services, and more information about Wiley products visit us at www.wiley.com.

Wiley also publishes its books in a variety of electronic formats and by print-on-demand. Some content that appears in standard print versions of this book may not be available in other formats.

Limit of Liability/Disclaimer of Warranty
In view of ongoing research, equipment modifications, changes in governmental regulations, and the constant flow of information relating to the use of experimental reagents, equipment, and devices, the reader is urged to review and evaluate the information provided in the package insert or instructions for each chemical, piece of equipment, reagent, or device for, among other things, any changes in the instructions or indication of usage and for added warnings and precautions. While the publisher and authors have used their best efforts in preparing this work, they make no representations or warranties with respect to the accuracy or completeness of the contents of this work and specifically disclaim all warranties, including without limitation any implied warranties of merchantability or fitness for a particular purpose. No warranty may be created or extended by sales representatives, written sales materials or promotional statements for this work. The fact that an organisation, website, or product is referred to in this work as a citation and/or potential source of further information does not mean that the publisher and authors endorse the information or services the organisation, website, or product may provide or recommendations it may make. This work is sold with the understanding that the publisher is not engaged in rendering professional services. The advice and strategies contained herein may not be suitable for your situation. You should consult with a specialist where appropriate. Further, readers should be aware that websites listed in this work may have changed or disappeared between when this work was written and when it is read. Neither the publisher nor authors shall be liable for any loss of profit or any other commercial damages, including but not limited to special, incidental, consequential, or other damages.

Library of Congress Cataloging-in-Publication Data

Names: Whittall, John, editor. | Sutton, Peter (Peter W.), editor.
Title: Applied biocatalysis : the chemist's enzyme toolbox / edited by Dr John Whittall, Dr Peter W. Sutton.
Description: First edition. | Hoboken, NJ : Wiley, 2021. | Includes bibliographical references and index.
Identifiers: LCCN 2020015374 (print) | LCCN 2020015375 (ebook) | ISBN 9781119487012 (cloth) | ISBN 9781119487029 (adobe pdf) | ISBN 9781119487036 (epub)
Subjects: LCSH: Biocatalysis.
Classification: LCC TP248.65.E59 A677 2021 (print) | LCC TP248.65.E59 (ebook) | DDC 660.6/34–dc23
LC record available at https://lccn.loc.gov/2020015374
LC ebook record available at https://lccn.loc.gov/2020015375

Cover Design: Wiley
Cover Images: (Background) © Angelatriks/Shutterstock, (Inset diagram) Courtesy of Gheorghe-Doru Roiban

Set in 10/12pt TimesLTStd by SPi Global, Chennai, India

Printed and bound by CPI Group (UK) Ltd, Croydon, CR0 4YY

10 9 8 7 6 5 4 3 2 1

Contents

Abbreviations

a/a	Area/area
AA	Amino acid
AADH	Amino acid dehydrogenase
ABTS	2,2′-Azino-bis(3-ethylbenzthiazoline-6-sulfonic acid
ABS	Amide bond synthetase
ACN	Acetonitrile (Also MECN)
AcOEt	Ethyl acetate (Also EtOAc)
ADH	Alcohol dehydrogenase (Also KRED)
ADP	Adenosine diphosphate
AEAA	(*S*)-*N*-(2-Aminoethyl)aspartic acid
AHAS	Acetohydroxyacid synthase
AHTC	Anhydrotetracycline
AMDase	Aryl malonate decarboxylase
AmDH	Amine dehydrogenase
AP	Area percent
API	Active pharmaceutical ingredient
ArgK	Arginine kinase
ArgK-LP	ArgK from *Limulus Polyphemus*
AroL	*E. coli* K12 shikimate kinase
AST	Arylsulfotransferase
ATA	Amine transaminase (Also TA)
ATase	Acetyl transferase
atm	Atmosphere (pressure)
ATP	Adenosine triphosphate
BCA	Bicinchoninic acid assay
BCL	Agarose beads crosslinked
BA	Benzaldehyde
BaSP	Sucrose phosphorylase from *Bifidobacterium adolescentis*
Bis-tris	2-[Bis(2-hydroxyethyl)amino]-2-(hydroxymethyl)propane-1,3-diol
BL21	Competent *Escherichia coli* BL21
BME	β-Mercaptoethanol
BMGH	Buffered minimal glycerol medium
BMGY	Buffered glycerol-complex medium
BMMH	Buffered minimal methanol medium
BMMY	Buffered methanol-complex medium
$BnONH_2$	*O*-Benzylhydroxylamine

Boc	t-Butoxycarbonyl
BSA	Bovine serum albumin
BSM	Basal salt medium
BVMO	Baeyer–Villiger monooxygenase
BVO	Baeyer–Villiger oxidation
c	Conversion
CA	Cholic acid
CAD	Charged aerosol detection
CALB	*Candida antarctica* lipase B
CAR	Carboxylic acid reductase
Cbz-OSu	N (Benzyloxycarbonyloxy)succinimide
CDCA	Chenodeoxycholic acid
CDW	Cell dry weight
CDMO	Cyclododecanone monooxygenase
CFE	Cell free extract
c.f.u	Colony-forming unit
CHMO	Baeyer–Villiger monooxygenase (cyclohexanone monoxygenase)
CLR	Controlled laboratory reactor
CoG	Cost of goods
CP	Citrate phosphate (buffer)
CPD	Cyclopentadecanone
CPME	Cyclopentyl methyl ether
CSU	Catalytic subunit
CSU-GST	Catalytic subunit – glutathione S-transferase tag
CV	Column volume
DAAO	D-Amino acid oxidase
DAAT	D-Amino acid transferase
DAD	Diode-array detection
DAPG	2,4-Diacetylphloroglucinol
DCC	Dicyclohexylcarbodiimide
DCM	Dichloromethane
ddH_2O	Double distilled water
DEA	Diethyl amine
DEAE	Diethylaminoethanol (group in ion-exchange resin)
DERA	2-Deoxyribose-5-phosphate aldolase
dH_2O	Distilled water
DHA	Dihydroxyacetone
DHAK	Dihydroxyacetone kinase
DHAP	Dihydroxyacetone phosphate
DHIQ	Dihydroisoquinoline
dI	Deoxyinosine
DI	Deionised
DIBAL-H	Diisobutyl aluminium hydride
DK(R)	Dynamic kinetic (resolution)
DMAP	4-Dimethylaminopyridine
DMF	Dimethylformamide

DMSO	Dimethyl sulfoxide
DNA	Deoxyribonucleic acid
DoE	Design of experiments
DSP	Downstream processing
DPPA	3,3-Diphenylpropionic acid
DTT	Dithiothreitol
E	Enantioselectivity
EDDS	Ethylenediamine-N,N′-disuccinic acid
EDTA	Ethylenediaminetetraacetic acid
ee	Enantiomeric excess
EH	Epoxide hydrolase
ELSD	Evaporative light scattering detector
ER	Ene-reductase (often same as ERED)
ERED	Enoate reductase (often same as ER)
Et_2O	Diethyl ether
EtOAc	Ethyl acetate (Also AcOEt)
EUGO	Eugenol oxidase
EV	Expansion vessel
EWG	Electron-withdrawing group
FA	Formic acid
FAD	Flavin adenine dinucleotide
$FADH_2$	Flavin adenine dinucleotide reduced form
FDH	Formate dehydrogenase
F&F	Flavour and Fragrance
FID	Flame ionisation detector
FMN	Flavin mononucleotide
FMO	Flavin monooxygenase enzyme
Fre	Flavin reductase
FSA	D-Fructose-6-phosphate aldolase
*Fs*DAAO	D-Amino acid oxidase from *Fusarium solani*
FTIR	Fourier-transform infrared spectroscopy
g	Gram (× g is centrifuge unit)
G6P	Glucose-6-phosphate
G6PDH	Glucose-6-phosphate dehydrogenase
GC	Gas chromatography
GC-FID	Gas chromatography/flame ionisation detection
GC-MS	Gas chromatography/mass spectrometry
GDH	Glucose dehydrogenase
GHMP	Galactokinase, homoserine kinase, mevalonate kinase
GMP	Good manufacturing practice
Gox	Galactose oxidase
GPDH	α-Glycerophosphate dehydrogenase
GPR40	G-protein-coupled receptor 40
GRAS	Generally recognised as safe
GST	Glutathione S-transferase
GTP	Guanosine-5′-triphosphate

hr	hour
HA	Hydroxyacid
HEPES	4-(2-Hydroxyethyl)piperazine-1-ethanesulfonic acid
HEWT	Amine transaminase from *Halomonas elongata* DSM 2581
HFA	Hydroxy fatty acid
HIC-DH	Hydroxyisocaproate dehydrogenase
HMM	Hidden Markov model
HMU	Worldwide PDB entry aminotransferase from *Silicibacter pomeroyi*
HPAEC	High-performance anion exchange chromatography
HPLC	High-performance liquid chromatography
HPLC-DAD	High-performance liquid chromatography with diode-array detection
HPLC-PAD	High-performance liquid chromatography with photodiode-array detection
HPLC-RI	High performance liquid chromatography with refractive index
HRMS	High-resolution mass spectrometry
HRP	Horse radish peroxidase
HSDH	Hydroxysteroid dehydrogenases
HTP	High throughput
HWE	Horner–Wadsworth–Emmons reaction
iBAT	Ileal bile acid transport
IBX	o-Iodoxybenzoic acid
ID	Internal diameter
IMAC	Immobilised metal affinity chromatography
IMB	Immobilised
IP	Intellectual property
IPA	Isopropyl alcohol
IPA	Isopropyl amine
IPAc	Isopropyl acetate
IPEA	Isopropenyl acetate
IPTG	Isopropyl β-D-1-thiogalactopyranoside
IR	Imine reductase
IRED	Imine reductase
IS	Internal standard
ISM	Iterative saturation mutagenesis
ISPC	*In situ* product crystallisation
ISPR	*In situ* product removal
kDA	Kilodalton
KHK	Ketohexokinase
KIRED	Ketimine reductase
KPB	Potassium phosphate buffer
KPi	Potassium phosphate buffer
KR	Kinetic resolution (Also ADH)
KRED	Ketoreductase
LAAD	L-Amino acid deaminase
LacC	D-Tagatose 6-phosphate kinase
LB	Lysogenic broth (also known as Luria–Bertani medium)
LC-MS	Liquid chromatography/mass spectrometry

LDH	Lactate dehydrogenase
LE-AmDH	Lysine amine dehydrogenase
LSD	Lysine specific histone demethylase
MALDI	Matrix-assisted laser desorption/ionisation
MAO-N	Monoamine oxidase from *Aspergillus niger*
MAP	Methoxyacetophenone
MeCN	Acetonitrile (Also ACN)
MenD	2-Succinyl-5-enolpyruvyl-6-hydroxy-3-cyclohexene-1-carboxylate synthase
MeOH	Methanol
MES	2-(*N*-Morpholino)ethanesulfonic acid
MeTHF	2-Methyl tetrahydrofuran
MDH	Malate dehydrogenase
MFC	Mass flow controller
min	Minute
MOPS	4-Morpholinepropanesulfonic acid
MPLC	Medium-pressure liquid chromatography
m.pt	Melting point
MRS	de Man, Rogosa and Sharpe broth
MS	Mass spectroscopy
$mS.cm^{-1}$	Millisiemens per centimeter
MTBE	Methyl *tert*-butyl ether
MWCO	Molecular weight cut-off
m/z	Mass-to-charge ratio
NA	Nutrient agar
NAC	*N*-Acetyl-L-cysteine
NAD^+	β-Nicotinamide adenine dinucleotide
NADH	β-Nicotinamide adenine dinucleotide, reduced form
NADPH	β-Nicotinamide adenine dinucleotide 2′-phosphate, reduced form
$NADP^+$	β-Nicotinamide adenine dinucleotide 2′-phosphate
$NH_3{\cdot}BH_3$	Ammonia-borane complex
Ni-NTA	Nickel-nitrilotriacetic acid
nm	Nanometre
NMAADH	*N*-Methylamino acid dehydrogenase
NMP	Nucleoside monophosphate
NMR	Nuclear magnetic resonance spectroscopy
NOX	NAD(P)H oxidase
NP	Nitrophenol
NP	Normal phase (chromatography)
NPS	Nitrophenyl sulfate
NR	Nitroreductase
nr	Nonredundant
NTP	Nucleoside triphosphate
OD	Optical density
Omd	Orotidine-5′-monophosphate decarboxylase
OMP	Orotidine-5′-monophosphate

OPA	o-Phthalaldehyde
Opt	Orotate phosphoribosyl transferase
OYE	Old yellow enzyme
OUR	Oxygen uptake rate
P	*ortho*-Phosphate
P450	Cytochrome P450
P5CR	Δ^1-Pyrroline-5-carboxylate reductase
PA	Phenyl acetate
PAD	Pulsed amperometric detection
PAGE	Polyacrylamide gel electrophoresis
PAL	Phenylalanine ammonia lyase
PAPS	Adenosine-3′-phospho-5′-phosphosulfate
PBS	Phosphate buffered saline
PCR	Polymerase chain reaction
PD	Potato-dextrose
PDa	Potato dextrose agar
PDA	Photodiode array
PDb	Potato dextrose broth
PDB	Protein data bank
PDC	Pyruvate decarboxylase
PDL	Pentadecanolide
PE	Petroleum ether
PEP	Phosphoenolpyruvic acid
PFAM	Protein family
Pf-TA	Transaminase from *Pseudomonas fluorescens*
Phe-DH	Phenylalanine dehydrogenase
PK	Pyruvate kinase
Pk	Porcine kidney
PLIF	Protein–ligand interaction fingerprint
PLP	Pyridoxal 5′-phosphate
PMP	Pyridoxamine 5-phosphate
PMSF	Phenylmethylsulfonyl fluoride
PNPase	Purine nucleoside phosphorylase
polyP	Polyphosphate
PP	Pyrophosphate
PpATaseCH	Acylase from *Pseudomonas protegens*
PPB	Potassium phosphate buffer
PPK	Polyphosphate kinase
PPO	4-(Hydroxy(methyl)phosphoryl)-2-oxobutanoic acid
Pps	PRPP synthetase
PPT	Phosphinothricin
PRPP	Phosphoribosyl pyrophosphate
PTM	*Pichia* trace metal salt supplement
PTV	Programmed temperature vaporisation
PWM	Position weight matrix
PYR	Pyruvate

QbD	Quality by design
rac	Racemic
R&D	Research and Development
RCF	Relative centrifugal force
Rdc2	Halogenase from *Pochonia chlamydosporia*
Rf	Retention factor
RhuA	Rhamnulose-1-phosphate aldolase
ROH	Generic alcohol
ROP	Ring-opening polymerisation
RP	Reverse phase
R-PAC	(*R*)-Phenylacetyl carbinol
rpm	Revolutions per minute
RSU	Regulatory subunit
rt	Room temperature
Rt	Retention time (also t_R)
SASA	Solvent accessible surface area
SDS	Sodium dodecyl sulfate
SDS-PAGE	SDS–polyacrylamide gels
sec	Seconds
SeH	Soluble epoxide hydrolase
SFC	Supercritical fluid chromatography
SLM	Supported liquid membrane
sLpm	Standard litre per minute
SOC	Super optimal broth with catabolite repression
STRAP	Structure based sequences alignment program
TA	Transaminase
TA-CV	ξ-Transaminase from *Chromobacterium violaceum*
TATP	Tri-acetone-triperoxide
TB	Terrific broth
TBAP	Tetrabutylammonium phosphate
TBDAc	2,3-Di-*O*-acetyl-6-*O*-t-butyldimethylsilyl)-β-cyclodextrin
TEA	Triethylamine
TEoA	Triethanolamine
TEV	Tobacco etch virus
TFA	Trifluoroacetic acid
ThDP	Thiamine diphosphate
THIQ	Tetrahydroisoquinoline
TLC	Thin layer chromatography
TMS	Tetramethyl silane
$TMSCHN_2$	Trimethylsilyl diazomethane
TOF	Time-of-flight
TPI	Triosephosphate Isomerase
t_R	Retention time (also Rt)
Tris	Tris(hydroxymethyl)aminomethane
TrpS	Tryptophan synthase
TY	Tryptone yeast extract broth

U	Units
UAB	Universitat Autònoma de Barcelona
UDCA	Ursodeoxycholic acid
UFA	Unsaturated fatty acid
UHPLC	Ultra-high-performance liquid chromatography
UPLC	Ultra-performance liquid chromatography
UTP	Uridine-5′-triphosphate
UV	Ultraviolet
UV-Vis	Ultraviolet/visible
VAO	Vanillyl alcohol oxidase
v/v	Volume/volume
vvm	Gas volume flow per unit of liquid volume per minute (vessel volume per minute)
wcp	Wet cell pellet
wcw	Wet cell weight
wrt	With respect to
wt	Wild type
w/v	Weight/volume
w/w	Weight/weight
x g	Centrifugal force (relative centrifugal force more precise than rpm)
YNB	Yeast nitrogen base
YPD	Yeast extract peptone dextrose medium
YT	Yeast extract tryptone
ZmPDC	*Zymomonas mobilis* pyruvate decarboxylase

1

Directed Evolution of Enzymes Driving Innovation in API Manufacturing at GSK

Jonathan Latham,[1] Anne A. Ollis,[2] Chris MacDermaid,[3] Katherine Honicker,[2] Douglas Fuerst[2] and Gheorghe-Doru Roiban,[1]*

[1]Synthetic Biochemistry, Medicinal Science & Technology, GlaxoSmithKline Medicines Research Centre, Stevenage, Hertfordshire, UK

[2]Synthetic Biochemistry, Medicinal Science & Technology, GlaxoSmithKline, Collegeville, PA, USA

[3]Molecular Design, Computational and Modelling Sciences, GlaxoSmithKline, Collegeville, PA, USA

1.1 Introduction

Biocatalysis has had a significant impact on the synthesis of active pharmaceutical ingredients (APIs) in recent years. The main driver for this is the ability to harness the regio- and stereoselectivity of enzymes to improve the efficiency of synthetic routes. For example, enzymes can offer direct access to enantiopure products, where traditional organic synthesis would require either resolution or the use of auxiliary groups [1], whilst enzymes applied in manufacture have improved syntheses or generated molecules that would otherwise be either impossible or impractical to synthesise. Other factors supporting the adoption of enzymes in the synthesis of APIs include:

- **Reduced manufacturing costs:** The selectivity of biocatalytic processes often results in fewer overall processing and purification steps, reducing labour and material requirements.

Applied Biocatalysis: The Chemist's Enzyme Toolbox, First Edition. Edited by John Whittall and Peter W. Sutton.
© 2021 John Wiley & Sons Ltd. Published 2021 by John Wiley & Sons Ltd.

- **Environmental sustainability:** Traditional organic solvents and reagents raise environmental concerns. Enzymes often function in aqueous solutions using natural biochemical co-factors as reagents, providing environmentally friendly and operationally safe alternatives to traditional chemical transformations.
- **Sustainable supply:** Enzymes fulfil sustainability principals [2]. They are produced from microbial fermentation – a renewable feedstock requiring simple sugars to grow – and are themselves biodegradable.
- **Simplified drug manufacturing:** The increasing complexity of APIs and the need to implement new products has ramped up the pressure to find new synthetic strategies to simplify the way drugs are made. Additionally, GlaxoSmithKline's (GSK) focus on making medicines broadly accessible across the globe requires economical manufacture of APIs and contributes to the drive for simplification of manufacturing processes.
- **Quality:** Regulatory pressures from government agencies to maintain high API quality standards whilst reducing carbon emissions drive the use of biocatalysis to harness the high regio-, chemo- and stereoselectivity offered.
- **Accessibility:** Technology advances such as next-generation sequencing and directed evolution have simplified biocatalysis adoption.

Early attempts to embrace biocatalysis relied on identification of native microorganisms and enzymes capable of catalysing the desired transformation with exquisite selectivity under the required process conditions. This represented a significant barrier, as an enzyme's natural sensitivity and substrate specificity often were not compatible with manufacturing process conditions, where solvent concentration, temperature and pH, for example, are often out of a typical physiological range. Early protein engineering attempts to overcome these challenges largely involved generating structure-guided, rational mutations to an enzyme's primary sequence through site-directed mutagenesis.

Advances in the directed evolution of proteins, however, have facilitated, greatly accelerated and enabled the wider-scale implementation of biocatalysis by providing an accessible means of producing fit-for-purpose enzymes and increasing the overall speed of enzyme engineering [3]. Directed evolution can be used to tailor multiple enzyme properties that historically challenged the uptake of biocatalysis, rapidly alleviating problems with properties such as activity, specificity, expression, thermostability and tolerance to process conditions. This ability to engineer biocatalytic enzymes has been a boon to all fields of chemical manufacture, including pharmaceuticals.

Despite all the technological advances and positive trends, adoption of industrial-scale biocatalytic processes has been generally slower than expected. This is mainly due to the fact that small-molecule drug manufacturing processes are far from simple and efficient; development of a small-molecule screening hit to a commercial product takes around 10–15 years [4]. Recently, in GSK, there has been a focus on adopting biocatalysis by first intent, rather than as a second- or third-generation process. This aids in the reduction of costs associated with filing a new process post-approval, as well as reducing the resource requirement over the lifecycle of the product [5]. As such, within GSK, we employ a unified technology platform to deliver our portfolio [6].

In addition to flow chemistry (continuous primary) and chemical catalysis, biocatalysis was identified as step-changer technology that would boost GSK's ability to manufacture APIs. To accelerate internal biocatalysis and enzyme engineering capability, in 2014,

GSK in-licensed the CodeEvolver® platform from Codexis, a California, USA-based biotechnology company. Internal investment in a team of scientists with expertise in biocatalysis, directed evolution, molecular biology, sequencing and computational chemistry provided dedicated support and operation of this platform. In addition, specialised theoretical and practical biocatalysis courses and workshops were organised for medicinal, organic and process chemists, to increase awareness of biocatalysis and directed evolution for synthetic chemistry applications. This training enabled scientists to identify potential enzymatic opportunities in chemistry routes and creen available biotransformations.

This chapter highlights how the directed evolution of biocatalysts has delivered impact for GSK, and contextualises this in the API manufacturing and drug development environment, enabling understanding of the timelines required to deliver a robust and manufacturable biocatalytic process. The chapter also provides examples of success stories in implementing directed evolution at GSK, and discusses the hurdles currently associated with embedding biocatalysis and how the process may be further accelerated.

1.2 Drug Development Stages

To understand how biocatalysis fits into the pipeline, knowledge of the drug development process is required. The journey from small-molecule screening to commercialisation of a medicine starts with *target selection* and *validation* (Figure 1.1). During this process, scientists gather evidence to support the role of a target (e.g. an enzyme in a biochemical pathway or a receptor) in a given disease, and the potential therapeutic benefit of modulating its function. The second stage is *lead discovery*, where the scientists seek molecules capable of interacting with the target (hits), which can then be used as a starting template for further optimisation. These hits are derivatised into synthetic small substrates called lead molecules, which interact with the target and have additional qualities that give the team confidence that they can be optimised to deliver a medicine (e.g. favourable target binding strength and selectivity). This process, from target selection to identification of lead molecules, can take between 4 and 24 months.

During the lead discovery and optimisation stages, the attrition rate is very high, with most compounds being discarded because of poor biochemical or biophysical properties. The focus, therefore, is on the quick delivery of a large number of diverse compounds using whichever chemistry works – making the adoption of biocatalysis more difficult at this stage. As only small quantities of compound are required (mgs), techniques like chromatography and chiral resolution are considered acceptable, meaning that the selectivity and process advantages offered by biocatalysis are less likely to be harnessed. In the identification of compounds with the desired pharmacological properties, however, biocatalysis can still be a powerful tool at this stage, as it enables the synthesis of drug entities inaccessible by other chemistries.

Once several pre-candidate molecules have been selected, medicinal chemists can begin to focus their effort on identifying biocatalysts that may provide a more efficient synthetic step. Quick read-outs and fast delivery are required, meaning that enzyme hits (those identified as capable of performing the desired chemistry) need to be easily scaled to deliver grams of material and allow delivery of product for further toxicological studies. Reactions utilising non process-ready biocatalysts (having low activity, poor stability or suboptimal

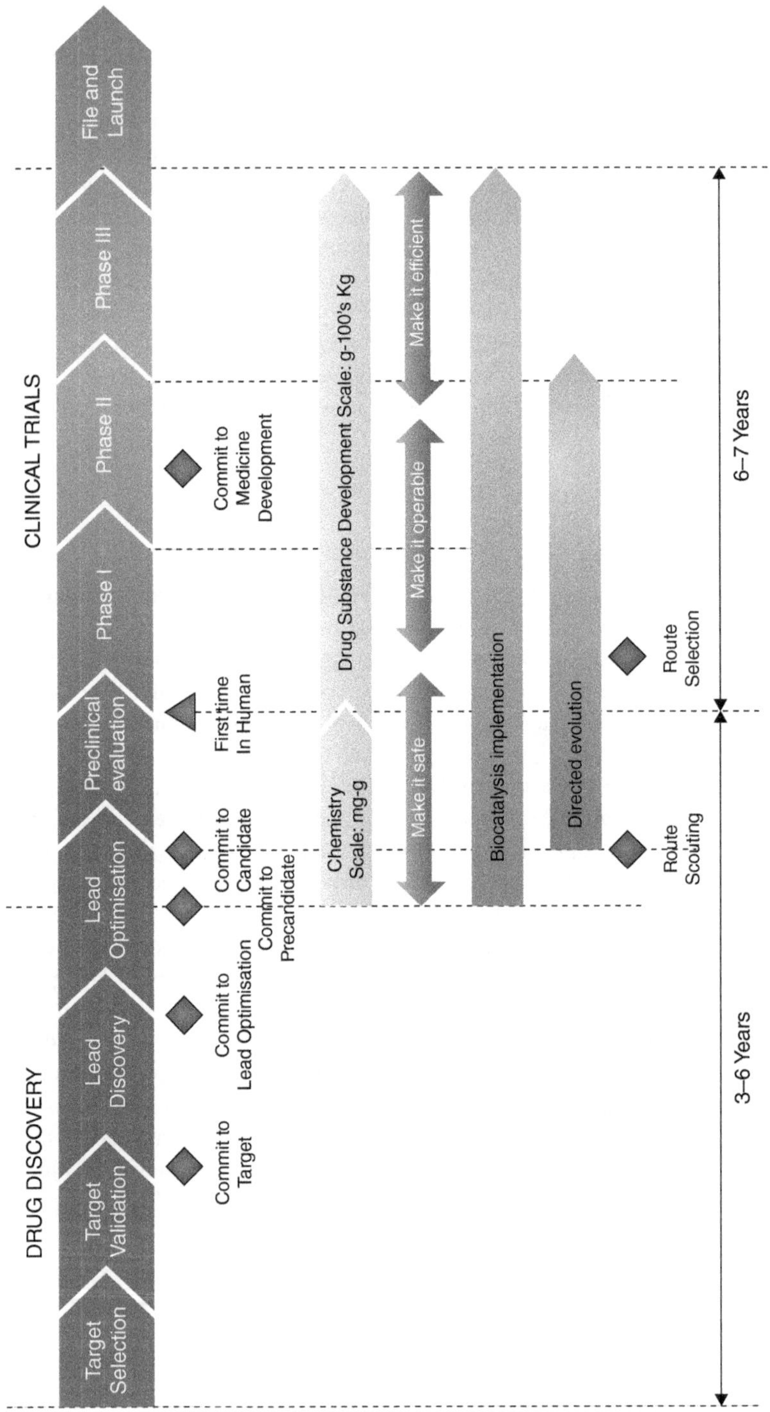

Figure 1.1 *A general drug discovery path overlapped with biocatalytic opportunities.*

selectivity) do not pose a significant problem at this stage, provided that the biotransformation affords product that meets the minimal quality criteria (e.g. chiral purity), since the quantity of product that must be prepared is small. Due to the relatively small scale, several of the downsides of non-process-ready enzymes can be easily mitigated (e.g. purification to meet quality requirements or centrifugation to deal with high enzyme loadings), affording some flexibility around the biocatalyst properties. However, if the catalyst presents challenges at this small scale (e.g. very low activity that cannot be reproduced on gram scale, or poor selectivity) that would require engineering to overcome, it is unlikely that the enzymatic step will be pursued as the high attrition of compounds means the resource commitment to enzyme evolution is difficult to justify. One solution to this problem is to increase the quality of the enzyme panels (collections of enzymes from various transformation classes that are initially screened for desired biotransformation), and therefore the likelihood of success on scale-up, by expanding the number and diversity of enzymes within the panels – either by acquisition of new enzymes or through panel expansion with engineered enzyme variants.

Once the few promising candidates have been selected, the drug journey continues with preclinical evaluation, where compounds are assessed for toxicity and efficacy using a combination of *in vitro* and *in vivo* animal models. With this data in hand, a decision is made as to whether or not a compound will progress to phase 1 clinical trials – also called 'commit to first time in human'. During phase 1, which typically takes between 12 and 18 months, batches of API are prepared for later dosing. The challenges that occur during API preparation can have a knock-on effect on clinical trials, causing them to slow or halt if drug supply is inadequate. Keeping these trials on time is key to timely assessment of drug candidates and ultimately to the delivery of approved medicines to patients. At this point, an ideal process for delivery of API is not required, provided that product quality is maintained. For assets in phase 1, the route employed by the medicinal chemistry team is usually scaled up to provide API for toxicology studies. Although these processes are usually not suitable or scalable for commercial-scale manufacturing, the time-critical nature of compound delivery and the small quantities required mean significant time isn't usually invested in process development at this stage.

Whilst API is being supplied using a non-ideal manufacturing process, route scouting activities are undertaken concurrently to identify more appropriate long-term syntheses. This is often where biocatalysis opportunities are identified and screened. At this point, the hits from enzyme panel screening must provide a significant advantage over the previous chemistry and provide API compound on a reasonable scale. Enzyme engineering becomes feasible at this stage and can play an important part in assessing a route's feasibility, impacting route scouting and selection.

As development progresses and the asset heads towards phase 2, delivery of API for clinical studies must occur concurrent with other process development activities. These include:

- Discovery of new routes of synthesis and chemistries to facilitate this (route scouting)
- Optimisation and understanding of the chemical process
- Thorough understanding of parameters impacting drug substance and intermediates quality

- Consideration of supply chain security for the process – that is, availability of starting materials and reagents on the required scale
- Transfer of the process to a full-scale manufacturing plant
- Preparation of regulatory documentation to support file and launch

If found to be effective during phase 1, an asset reaches phase 2 trials – a process that can take 2–3 years before commitment to medicine development. In addition to drug production for future trials, it is also important that scale-up process strategy is considered during this time. By now, the number of candidates has been reduced and route improvements driven by introducing biocatalysis have been linked directly with API production. The biocatalyst has been optimised to deliver a manufacturable process, but there is still room for further improvement and fine tuning if certain criteria (such as cost of goods) have not yet been met.

In phase 3, extensive work is undertaken to identify the logistics surrounding the distribution of clinical supply to the investigator sites and to develop a robust commercial, end-to-end supply chain to ensure continuity of both launch and long-term drug supply to patients. By this time, in addition to optimisation of enzymes through engineering and biocatalytic process development, a fermentation process for enzyme supply and the manufacturing chain must be established.

Throughout the drug development phases, the aim is to generate a process that is safe, operable and ultimately efficient. In this context, the decision to engineer an enzyme for use in an API manufacturing route is complex, due to the time and resource commitment required for protein engineering. It is often the case that either wild-type or panel enzymes will satisfy the requirements of early, small-scale manufacturing campaigns of nascent assets. However, once an asset is significantly advanced along the drug development pipeline, it becomes increasingly difficult for enzymes to provide a manufacturable process for larger-scale clinical supply campaigns. This is where directed evolution has a large impact. As strategies and technologies develop to expedite the evolution process, it becomes increasingly possible to adopt directed evolution earlier in the drug development cycle. In the next section, we discuss how recent developments have helped with this aim, and what future work is required to fully realise this vision.

1.3 Enzyme Panels

GSK has a significant number of panels, produced both internally and acquired from external sources, which are continuously enhanced through evolution and addition of new enzyme classes (Scheme 1.1).

A portfolio analysis reveals which transformations account for the most frequently used chemistries within GSK [6]. Heteroatom alkylation and arylations, together with aromatic heterocycle formations, make up approximately 40% of the portfolio. Functional group interconversion, C-C bond formation and reductions, oxidations and protections are other types of transformations frequently encountered. Focusing on enzymatic alternatives for these transformations would have the most impact.

New enzyme panels are assembled using a diverse set of enzymes which have been identified through previously demonstrated activities or using predictive tools (Figure 1.2). In some instances (e.g. lipases), a significant number of enzymes are already commercially available, and therefore these panels comprise mostly enzymes from a commercial source.

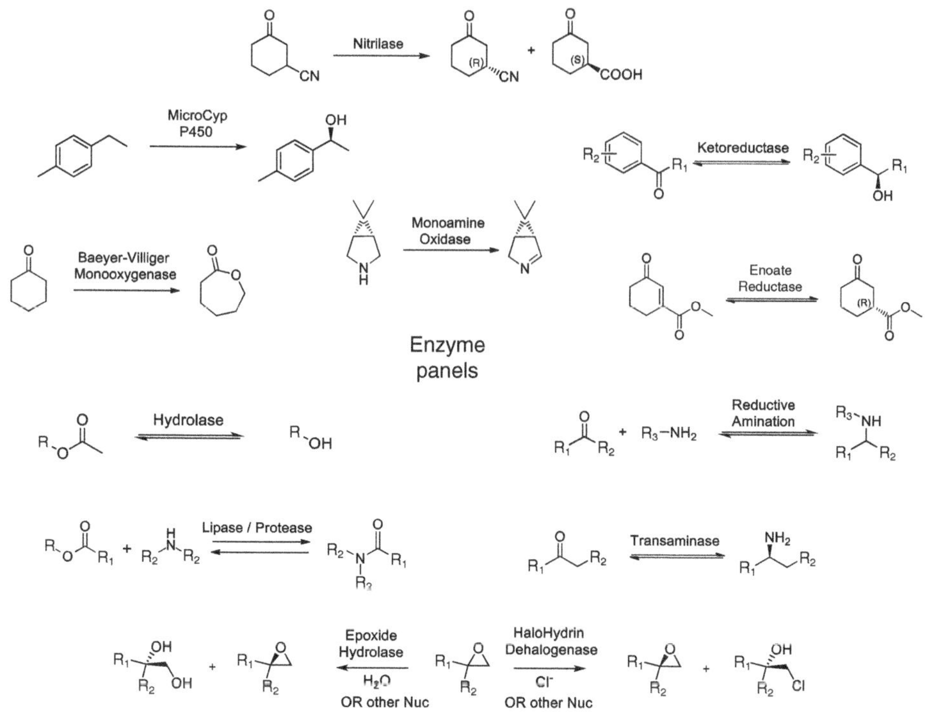

Scheme 1.1 *Selected enzyme classes currently in GSK's collection.*

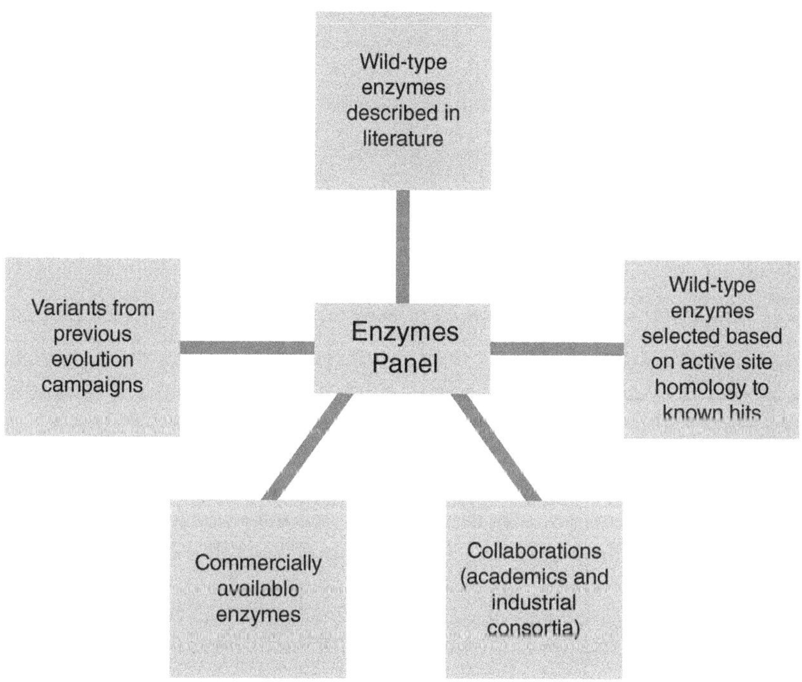

Figure 1.2 *Strategies for generating an enzyme panel.*

Although it is usually trivial to acquire larger quantities of commercial enzymes to facilitate scale-up of any hits for a particular transformation, commercial enzymes can carry intellectual property (IP) restrictions around their use for commercial-scale manufacture and can cause supply-chain concerns as most are single-source. Additionally, it is not usually possible to evolve a commercially-available enzyme without licensing agreements with the source company.

For most other enzyme classes, there are very few, if any, commercially-available enzymes available to screen. Although significant efforts are underway to change the situation, this often means that enzymes of the desired class must be identified by other means. Panels are often assembled by acquisition of the genes encoding wild-type enzymes reported to catalyse the reaction class of interest in the literature. In some cases, mutational studies have also been conducted, and variants of these enzymes may also be acquired.

Bioinformatic tools are the key to the assembly of an enzyme panel, allowing identification of additional putative enzymes from sequence databases based on similarity to known enzymes of the desired class. The goal of a search is straightforward: to identify naturally occurring enzymes that perform the same transformations but maximise their sequence and structural diversity in hopes of maximising the substrate scope, as well as the pliability of the enzymes to be evolved.

As with hit expansion for small molecules, enzymes identified in the literature are often used as seeds for subsequent similarity searching against large annotated sequence datasets, including those from Interpro [7], Uniprot [8], NCBI, PDB [9], CATH [10] and some metagenomics collections [11]. In a standard approach, homologous sequences are identified and clustered, and each cluster's functional annotations are examined. Homologous protein structures are often included in the clustering step as they help in the identification of the relevant clusters from which to sample. Selecting exemplars from the various clusters is often the most challenging part of the process. In a typical scenario, candidates are prioritised based on the availability of experimental annotations, starting with those with experimental data or structures or coming from extremophile organisms. In the absence of additional data, the remaining exemplars are chosen from a diverse set of clades that maximise coverage of each cluster's diversity. The HH-Suite toolset [12], maintained by the Soeding group, provides great tools for identifying homologues, as well as Mmseqs2 [11] and CLANS [13] for clustering. Phylogenetic reconstruction can be performed using MEGA [14] or ETE3 [15] and their corresponding tree-reconstruction and evolutionary-analysis workflows.

One way to improve this approach is to select for enzymes that exhibit similar active sites, and thereby maintain activity and selectivity, by developing 'fingerprints' or protein-based pharmacophores, sometimes referred to as PLIFs, which describe the residue composition and positionality within the active site and the interactions with bound ligands [16]. Once a fingerprint is created, it can be used to identify clusters of sequences that exhibit similar active site makeups, which can then be prioritised for acquisition. The goal of this approach is often to acquire enzymes that minimally perturb the active site but which sample diversity throughout the remainder of the enzyme. The method can also be extended to sample diversity at specific positions within the active site. The success of this approach hinges on the availability of experimentally determined structures from the structural families of interest. The approach also assumes that the active site can be unambiguously identified.

Beyond the acquisition of wild-type enzymes, it is also possible to leverage variants produced in previous enzyme evolution campaigns to enrich a panel. As these evolved enzymes are likely to have generally improved properties (e.g. activity, stability and expression) compared to wild-type enzymes, they can often provide hits more amenable to immediate scale-up. These enzymes can be the result of in-house evolution, in-licensing of other companies' enzyme panels or from collaborations.

As these panels are used for the initial assessment of a biocatalytic transformation, it is important to maintain a significant amount of sequence diversity within them, in order to increase the likelihood that enzymes with the desired activity and selectivity can be identified for a wide range of substrates.

One example is the GSK IRED panel, where 85 IREDs were assembled from different sources [17]. These wild-type variants were tested under industrially relevant conditions using equimolar loadings of amine. Screening a diverse set of substrates, selected to cover a broad chemical space, showed most enzymes were capable of driving reactions to completion (Scheme 1.2). Enzyme-dependent stereoselectivity was observed for many products and successful scale-up was performed for several enzymes and substrates.

Reductive amination of keto acids represents another area where we focused our efforts and successfully generated a small collection of enzymes that are very promising for the synthesis of *N*-alkylated amino acids [18, 19]. Enzyme such as *N*-methylamino acid dehydrogenases (NMAADHs, EC 1.5.1.1 and EC 1.5.1.21), ketimine reductases (KIREDs, EC 1.5.1.25) and Δ^1-pyrroline-5-carboxylate reductases (P5CR, EC 1.5.1.2) have been shown to deliver highly enantioselective alkylated products, further extending the scope of these substrates (Scheme 1.3).

Following several successful in-house enzyme evolution campaigns, we have also constructed enzyme panels incorporating the best variants produced therein. One of the drivers

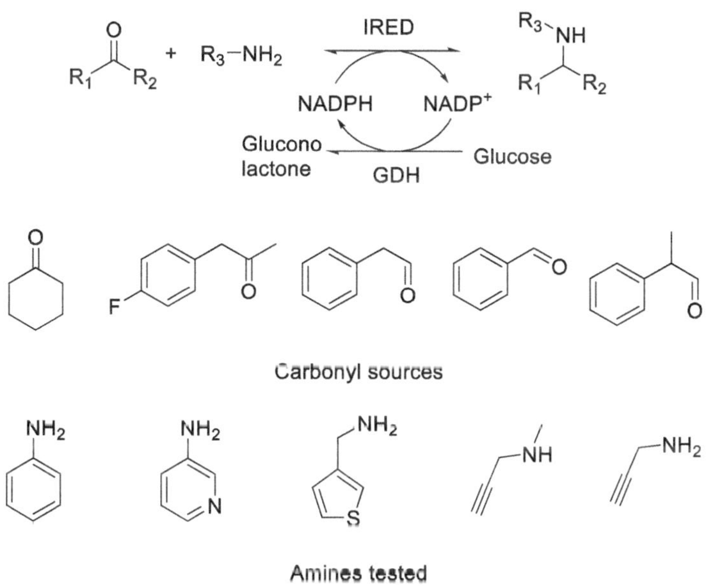

Scheme 1.2 *Reductive amination of ketones performed by imine reductases.*

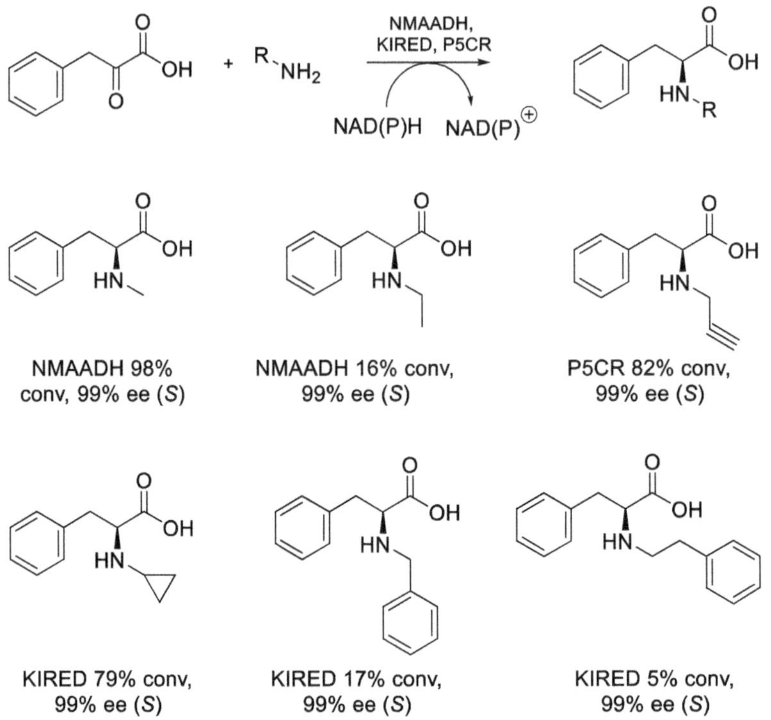

Scheme 1.3 *Successful synthesis of alkylation products via enzymatic reductive amination of keto acids.*

for this practice is that, by creating panels of robust enzymes with diverse activities, it becomes more likely that a panel can be used directly without the need for further enzyme engineering – greatly increasing the speed at which the biocatalytic process can be implemented, and therefore increasing uptake across the portfolio.

GSK is also collaborating with various academics in the field with the purpose of continuously expanding its portfolio of enzymes (e.g. halogenases, nitration enzymes, unspecific peroxygenases, etc).

1.4 Enzyme Engineering

With high-quality enzyme panels in hand, it becomes possible to develop a manufacturable process for an API using the hit from panel screening without the need for further enzyme optimisation by either engineering or evolution. The definition of a manufacturable process changes throughout the drug development cycle and becomes more stringent as the asset matures and the quantity of API required increases – making it more likely that engineering of a panel enzyme will be necessary to meet these needs. The specific process parameters required for assets at each stage of development are complex, but the following list summarises some key aspects that must be considered before committing to the evolution of an enzyme:

- **Product quality:** This is always a key factor in considering any manufacturing process. Depending upon the quantity of product required, more laborious purification steps (e.g. chromatography) may be considered in order to supply a compound in a timely manner. For large-scale synthesis of later-stage assets, processes capable of delivering the required purity directly are required, and enzyme engineering is often employed to achieve this.
- **Enzyme supply:** For small-scale synthesis of early-stage assets, it is likely that laboratory-scale fermentation can provide sufficient quantities of enzyme for a reaction – even if activity is low and high enzyme loading is required. For large-scale synthesis of later-stage assets, enzyme loading needs to be reduced as low as is practicable for biocatalyst supply to be economical, and enzyme engineering is often employed to improve enzyme activity and reduce the loading of enzyme required.
- **Physical properties:** High enzyme loadings, as the result of low activity with a particular substrate, can lead to the formation of emulsions upon extractive workup and therefore long, protracted filtration steps for enzyme removal. For early-phase fast-moving assets, long filtration times can often be accepted, provided product of the required quality is delivered. For small-scale synthesis, centrifugation is often a viable option for enzyme removal and can be significantly faster than traditional filtration. For later-stage assets, however, enzyme loading needs to be reduced as low as practicable in order to afford a tractable workup, often by improving enzyme activity through directed evolution.
- **Process intensity:** This refers to the quantity of product afforded per reaction volume and depends upon the quantity of material that must be supplied. For highly potent, low-volume or early-phase compounds, a high-volume process may be acceptable. When larger quantities of product are required, process intensity can be a key determinant, particularly once the synthesis is conducted on plant scale, as physical limitations on reactor size can mean multiple batches of the synthesis are required, significantly increasing the cost of a manufacturing campaign. Enzyme engineering is often used to alleviate substrate or product inhibition, increase enzyme activity and improve enzyme stability with respect to temperature or organic solvent, to allow for process intensification.

Enzyme variants resulting from each round of evolution are tested under representative process conditions to validate the hits and assess their fitness (Figure 1.3). Should an enzyme variant fail to meet any of the listed criteria, subsequent rounds of directed evolution will usually be conducted until a fit-for-purpose enzyme is obtained. In most cases, reaction engineering goes hand in hand with protein engineering. Continuous engineering of the enzyme expands the catalyst operating space and brings the process towards a feasible, manufacturable and cost-effective process. Nevertheless, a close interaction between chemists, engineers, process chemists and biochemists is key to ensuring the design of a robust enzymatic process. Real-time feedback of process improvements is necessary to drive the evolution trajectory and determine which protein properties still need to be improved. Because of this feedback between the enzyme engineering and process chemistry workstreams, protein engineering needs to fit in the overall picture of process design and must tie in with process development activities. Once the final enzyme variant is generated, design of experiments (DoE), quality by design (QbD) studies, systematic risk based approaches to drug development, manufacturing and lifecycle management [20] are all performed in order to scale up

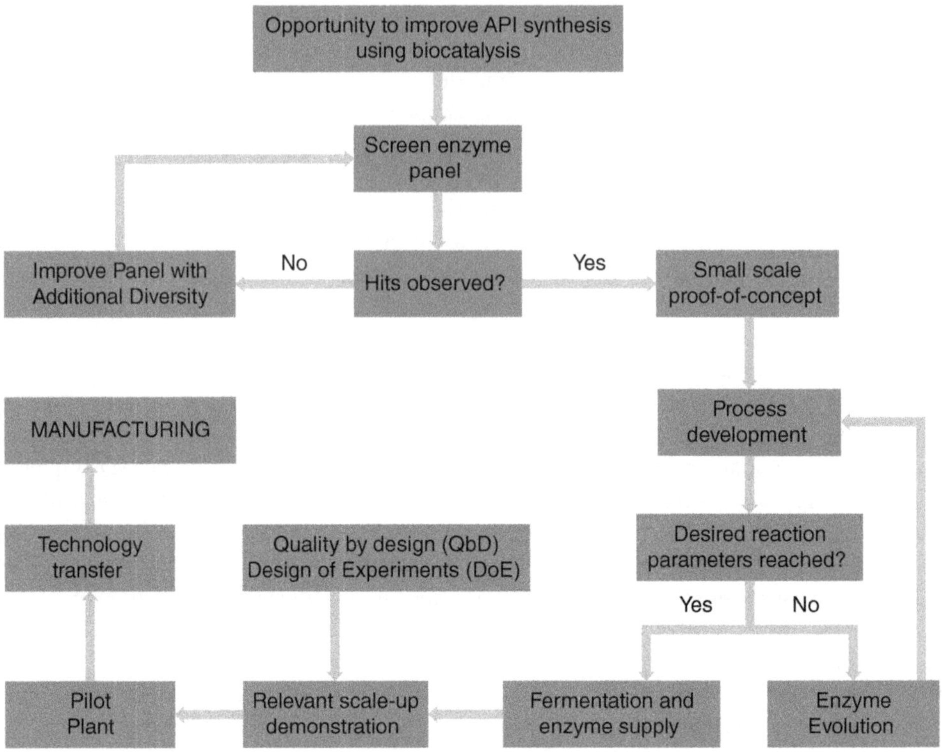

Figure 1.3 *Enzyme evolution and biocatalysis adoption.*

the process and demonstrate its feasibility on pilot plant scale prior to technology transfer and manufacturing.

Depending on the desired manufacturing process, enzyme formulation (e.g. immobilisation) is another important parameter that may impact evolution timelines. Ideally, an evolved enzyme would be tractable to use as a free lyophilised powder in order to reduce the costs and time associated with developing an immobilisation process, but the formulation of an enzyme is ultimately dictated by its performance in the desired manufacturing process. An important aspect to keep in mind is that enzymes are considered reagents and therefore their manufacture does not have to be performed under Good Manufacturing Practice (GMP) conditions, which has a positive effect on the cost of goods (CoG).

Broadly speaking, enzyme engineering involves exploring the sequence space of an enzyme to determine how amino acid substitutions influence its fitness in the desired transformation. Early attempts largely relied on making structure-guided, rational modifications through site-directed mutagenesis. Although this approach has yielded significant results, the complexity of enzyme structures, combined with the common need to improve several parameters simultaneously, slowed the engineering process significantly. Directed evolution – an iterative process whereby enzyme variants are rapidly generated and screened using a selective pressure – allows this space to be explored significantly faster.

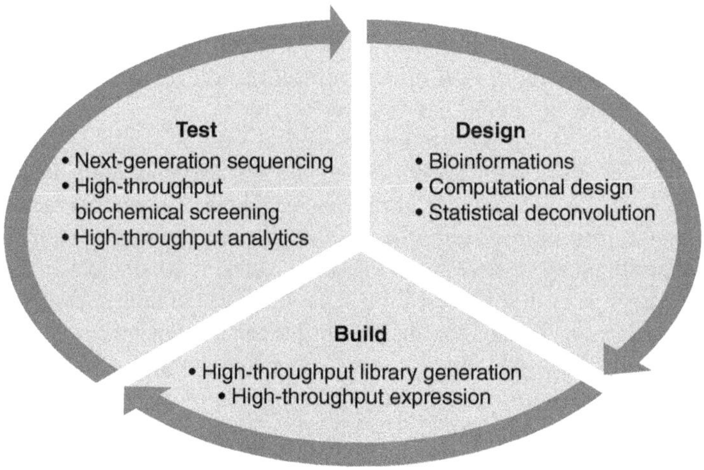

Figure 1.4 Design–build–test cycle for an evolution round.

Modern directed evolution occurs via iterative rounds of the design–build–test cycle, whereby positions believed to affect an enzyme's performance in the desired process are identified using computational tools. Diversity is then generated at these positions using molecular biology tools, and the resulting library of variants are screened using high-throughput (HTP) biochemical assays (Figure 1.4). The sequence and activity data obtained in each round is fed back into subsequent cycles until an enzyme suitable for the desired process is obtained. Directed evolution campaigns therefore involve multidisciplinary teams of computational chemists and biologists, molecular biologists and biochemists in addition to the other disciplines required to ensure implementation of the enzyme on the manufacturing scale, as already described.

Depending on the depth of sequence space explored in each round and the specific techniques employed, each turn of the design–build–test cycle (i.e. each evolution round) can take up to 12 weeks. As many enzymes will require several rounds of evolution before they can afford a manufacturable biocatalytic process, it is important to include the time required for enzyme evolution in project timelines and to allow process development activities to occur concurrently with the creation of improved enzyme variants. The time required to evolve an enzyme can be a significant pressure on the adoption of biocatalysis. Process chemists must therefore have confidence that a workable process can be achieved using the biocatalytic reaction and that the final enzyme will provide a manufacturable process. Efforts to accelerate the directed evolution cycle will continue to build confidence in applying biocatalytic processes.

1.4.1 Library Design (Design Phase)

Prioritisation of positions or mutations for experimental investigation ultimately depends on the goals of the evolution campaign. Even the most thoughtfully designed libraries can falter if they are not tailored to the hypothesis of the round or campaign. Library design draws on knowledge of the sequence, structure and function of the enzyme class of interest. Coupled with knowledge derived from literature and experiment, a strategic library design approach can significantly reduce the number of rounds of evolution needed to reach the campaign's goals, saving both time and money.

Classical examples of directed evolution involved the use of methods to introduce random DNA replication errors, such as error-prone polymerase chain reaction (PCR), HNO_2 or mutator strains, all of which generate libraries of enzymes with random mutations. Given that most proteins consist of ~300 amino acids, there are an overwhelming number of possible combinations that can be introduced using the standard 20 amino acids, if addressed completely randomly. In addition to creating capacity issues in the generation and screening of such large numbers of enzymes, these libraries can often be of low quality, with few active or improved enzymes identified – because the random nature of diversity generation means it is likely that catalytically active residues, or residues that are essential for protein expression and folding, will be mutated. The introduction of several mutations at once also creates an analytical challenge when attempting to deconvolute the effect of each individual mutation on performance. The capacity problem with random approaches to diversity generation can be overcome by using a selection pressure at the colony level – in such a way that only colonies expressing an active protein grow [21], or produce a coloured compound [22], allowing colonies expressing active protein to be readily identified and the corresponding variant enzymes further analysed. Whilst this approach can allow randomly generated libraries to be effectively traversed, a lack of general selection approaches means it is not readily applied to many reaction classes. Additionally, the effective engineering of enzymes relies upon screening under a condition that is representative of the desired operating condition (see later), which can be difficult to replicate on an agar plate.

A so-called 'fourth wave' of biocatalysis has therefore emerged, in which enzymes are evolved through rational directed evolution. This involves the identification of positions thought to affect the property to be modified using computational chemistry and bioinformatics, followed by the generation of diversity at those positions. Several computational tools have arisen that facilitate the design of these smart libraries, including CASTER, B-FITTER and 3DM [23]. An example is CAST-ISM, in which positions within and adjacent to the active site are subjected to iterative saturation mutagenesis (ISM) in order to assess the effect of substitutions on activity [24]. Any beneficial mutations identified during this process can be combined with one another in subsequent evolution rounds, in the hope of producing additive effects that further improve the desired properties. Herein, we highlight some of the typical approaches for smart library design.

The first steps often involve gathering and curating sequence and structural information. Sequence statistics, like per-position amino acid diversity, are often calculated from profile alignments generated via a pairwise BLAST [25] or iterative PSIBLAST [26] search. Hits below a particular expectation value are then aligned, and insertion states are removed. If the chosen method also generates a hidden Markov profile [12a, 27] or position weight matrix (PWM) [26], the amino acid diversity can conveniently be determined directly from back-calculation of the pseudo counts or interpretation of the PWM, respectively. Once the distribution of mutations is known, the conservation, probabilities and entropies are easily calculated and can be used to identify sequence positions that are amenable to mutation: that is, those with a low conservation, which are different from the consensus residue identity or which have a favourable weighting in the PWM. The amino acid distributions can also be used to evaluate statistical potentials like secondary structure propensities [28] or hydrophobicites [29], which help zero in on positions that deviate significantly from the homologous distribution's consensus.

Structural information also plays a crucial role in library design, but oftentimes crystal structures of the enzymes of interest are not available; thus, homology models must be constructed from close homologues. Identifying template structures is straightforward. Hhsearch [12a], RaptorX [30] or one of the many other homology detection and structure prediction tools available [31] can be used to search across a database of crystal structure sequences, like those contained in the PDB [9]. However, the selection of templates to use for model construction is often a challenge [32]; priority should be given to the quality of the alignment, followed by structures with high completeness and resolution, the presence of bound ligands or cofactors and, when possible, the sampling of multiple conformations. Supporting experimental evidence for the native oligomerisation state or so-called biological unit is also desirable.

Depending on the per cent identity and number and size of the indels in the alignment, the homology modelling procedure employed varies [32a, 33]. In cases where there is to be minimal perturbation to the structure (no large indels), the generation of 10–20 models is often enough to identify structural features and determine the ensemble average values of the desired descriptors of interest. However, hundreds of models or more may be required if there is poor homology or the insertion or deletion of large loops or secondary structural elements, or where there is evidence of large conformational changes. For simple structural perturbations where the per cent identity between the target sequence and template is high, most homology modelling software tools will perform similarly. For the most challenging cases, though, advanced comparative modelling tools and workflows present in packages like Rosetta (Robetta [32b, 34]), I-TASSER [35] or Modeller [36] can sometimes provide additional insight. The Continuous Automated Model EvaluatiOn (CAMEO) [31, 37] website (www.cameo3d.org) is an excellent resource for up-to-date tools for use in identifying templates and performing comparative or *de novo* modelling tasks.

Structural models are useful in several cases. The first and most obvious is identification of the active site and the residues within it. For campaigns where the objectives hinge on modulating the activity or selectivity of the enzyme, libraries designed to impact on the active site are a must. In cases where the location of the active site is ambiguous, pocket detection methods such as CCG's site finder [38] or Schrodinger's site map [39] can be employed.

Identification of buried or solvent-exposed residues, or those that are buried at a protein–protein interface, can also be useful when designing libraries. Tools that calculate the solvent-accessible surface area (SASA) include DSSP [40], MSMS and AREAIMOL [41], as well as those found in popular molecular modelling suites. The rationale behind targeting exposed residues centres on the notion that residues confined to the enzyme surface will be more amenable to mutation, whilst those that are buried may significantly impact the hydrophobic packing and likely folding of the enzyme and should only be mutated after careful consideration of the library or evolution project goals. In some cases, repacking the hydrophobic core or residues near a protein–protein interface may lead to increased stabilisation of the enzyme and therefore might be desirable. In cases where nonhomogeneous solvents are to be employed, targeting only those positions that are solvent-exposed might be appropriate.

A structural model also enables the use of advanced modelling tools and applications, such as calculation of the folding free energy change for a mutation or for groups of mutations. The so-called delta–delta G (ΔΔG) calculation has become ubiquitous in modern protein modelling packages and platforms [42], but its applicability is limited to those campaigns where there is a clear relationship between the change in folding free energy and the properties being optimised; that is, an established structure–activity relationship. Instead, a more qualitative structural modelling approach is typically applied during library design, especially when using a homology model. Thus, more routine calculations are performed, such as modelling structural perturbations of a combination of mutations present in a library or identifying residues that disrupt hydrophobic patches on the enzyme's surface.

Site-saturation libraries are often the starting point for a directed evolution campaign or are called upon when beneficial diversity has been exhausted after multiple rounds of recombination. These libraries are the most straightforward to design and take advantage of the sequence and structural properties discussed previously. Examples include libraries targeting positions lining the active site, which are often the most fruitful when an increase in activity or selectivity is desired. Others include those that target sites with low conservation, a different consensus amino acid based on the alignment of close homologues, inconsistencies with secondary structure preferences or solvent exposure. Mutations of conserved or buried amino acids may be beneficial, for example, when attempting to increase stability, but care must be taken not to target positions that are functionally important. In general, targeting roughly 80–90% of the enzyme for site saturation is common, but in cases where time or resources are constrained, prioritising libraries for impact based on their sequence and structural properties can work well.

The design of combinatorial libraries is a more challenging task as specific mutations must be chosen to be combined. Often, the designer does not know how particular mutations will behave when recombined as their effects are typically non-additive. Some basic recombination strategies include: grouping mutations based on family homology (i.e. a homology combi), combining the best beneficial diversity to create an 'elite' library and randomisation of beneficial mutations into a predefined number of libraries pursuant to a desired statistical distribution. These libraries are quick to design but typically miss combinations of mutations that maximise beneficial residue–residue interactions. In the simplest notion, maximising the pairwise interactions between mutations involves determining via experiment or approximating through computation as many of the $N \times N$ pairwise interactions as possible, where N is the number of beneficial mutations. In some cases, these pairwise interactions can be as simple as Cartesian distances between positions in space on a structural model, coefficients determined from statistical models used to fit the experimental data (Codexis' MOSAIC) or even energetic values determined *in silico* (ΔΔGs). Recombination of the beneficial mutations then becomes a straightforward optimisation problem wherein one seeks to maximise the pairwise effects between each of the mutations in the library, subject to external constraints imposed by the designer or platform. Suitable algorithms for this task include simulated annealing Monte Carlo, integer-linear programming and genetic algorithms. The number of libraries designed is then dependent on the desired coverage of the beneficial mutation space, as well as on whether specific mutations are permitted in multiple libraries.

1.4.2 Library Generation (Build Phase)

The build phase of this cycle requires the rapid generation and expression of the desired libraries of variants. Review of contemporary molecular biology methods for generating diverse libraries of variants has been covered in other works and is beyond the scope of this chapter. Methods for HTP expression of library variants can vary from expression and screening steps carried out in 96- or 384-well plate formats to batch expression and sorting to agar-based selections. Industrial approaches typically take advantage of advances in automation to facilitate plate-based expression and screening, with liquid-handling robots and automated colony pickers enabling HTP expression efforts. Molecular foundries are also becoming more prevalent, increasing accessibility to automated HTP molecular biology approaches [43].

Speed is a key focus for industrial directed evolution, and efforts to accelerate directed-evolution molecular biology workflows include generating smarter and smaller libraries, reducing the number of variants to be generated and screened [44]. Outsourcing library generation with ultra-HTP gene synthesis could also reduce the bench time required for a directed-evolution campaign and ensure that all of the mutations targeted are synthesised – something that is not guaranteed using the methods described herein [44]. However, the cost of such libraries is currently too high for most lab budgets. Efforts to speed up the typical multi-day bacterial expression protocols are also a key factor in accelerating directed evolution. This includes cell-free protein-expression efforts [45] and identification of molecular biology-amenable bacterial strains with faster doubling times [46].

1.4.3 Screening (Test Phase)

Even with small, smart libraries, the number of variants that must be tested is usually on the order of 1000s. In order to keep to the rapid timelines required for delivery, these variants must be rapidly assessed under the appropriate selection pressures. To achieve this laboratories need to be equipped with HTP platforms for reaction setup and data analysis. As many industrial processes involve the use of enzymes in organic solvents, screening platforms compatible with both organic and aqueous systems must be in place.

In order to assess the enzyme variants at the pace required to fit with directed-evolution timelines, screening predominantly occurs in microtitre plate format. Plates (96- and 384-well) must be available which are compatible with a broad range of solvents (e.g. methylcyclohexane, toluene, methyl *tert*-butyl ether, methyl isobutyl ketone) and with higher temperatures (up to 80 °C), because of the range of conditions used in industrial processes. These plates must also be compatible with the automation used for reaction setup, such as plate sealers and liquid-handling robots. Automated platforms have greatly facilitated the speed, throughput, accuracy and reliability of screening enzyme libraries. Screening format is also a critical parameter. Screening directly with clarified cell lysate yields more rapid results, but the requirements of the final process and end-user preferences typically dictate screening formulations. It is common to screen enzymes as lyophilised powders, as this is usually the preparation favoured for the final process, meaning access to a lyophiliser capable of drying dozens of enzyme plates at once is required.

Dedicated plate-ready analysis is a must: ultra-performance liquid chromatography (UPLC), gas chromatography (GC), liquid chromatography–mass spectrometry (LCMS), collision-activated dissociation (CAD), RapidFire® or matrix-assisted laser desorption/ionisation (MALDI). Considering the number of variants that need to be assessed, the shortest analytical method possible is required. UPLC is commonly used for such screening campaigns because of its broad applicability and amenability to short sample run-times. The development of short (<3 min per sample) and robust (>1000 samples per column) UPLC methods ensures screening reactions can be rapidly analysed. Software packages capable of efficiently processing the outputs from these instruments are the key to ensuring rapid analysis of screening data. Alternative ways of analysing reactions such as RapidFire® and MALDI can drastically reduce analysis times (<10 sec per sample), but these platforms are yet to find widespread use in directed evolution workflows. Visualisation tools (e.g. Spotfire) are simplifying the interpretation of results.

Once enzyme variants have been screened and hits identified, it is important to have access to chemistry facilities (e.g. controlled laboratory reactors, parallel synthesis stations) to allow scale-up and verification. The ability to rapidly scale up expression of any hit enzymes from microtitre plates to shake flask or fermenter is also key, as this can allow improved enzyme variants to be used immediately for product delivery.

1.5 Case Studies

Protein engineering has been a significant enabler to embedding biocatalysis across the pharmaceutical industry as it allows enzyme panel hits to be tailored towards particular process requirements. There are several reported cases of successful biocatalysis implementations enabled by directed evolution from an array of companies in the sector [47]. A seminal example, reported by Merck and Codexis [47a] employed substrate walking and a total of 11 rounds of directed evolution to afford a process-ready transaminase for the selective installation of a chiral amine moiety in the antidiabetic drug Sitagliptin. Codexis have also reported the directed evolution of a Baeyer–Villiger monooxygenase (BVMO) for chiral sulfide oxidation to the corresponding sulfone Esomeprazole (a proton pump inhibitor, PPI) [47d]. The wild-type BMVO showed barely detectable activity for the desired transformation, but, after 19 rounds of directed evolution, a highly selective and process-ready enzyme was afforded. Evolution of a ketoreductase (KRED) for improved activity in the manufacturing-scale synthesis of Vibegron (a treatment for overactive bladder syndrome) has also been reported [47e]. In addition to improving KRED activity and selectivity for the desired substrate, this evolution campaign was focused on improving KRED tolerance to high-pH reaction buffer in order to effect racemisation of the starting ketone and therefore afford a dynamic kinetic resolution of the neighbouring amine centre. In another important example, transaminase evolution variants have been used to catalyse a dynamic asymmetric transamination in order to set up the two contiguous chiral centres in a cyclohexane ring [48]. Finally, monoamine oxidase-N (MAO-N) from *Aspergillus niger* variants have been used to deliver an efficient synthesis for Telaprevir [49].

Occasionally, with enzyme panels of sufficiently high quality (well-expressed, stable enzymes with high retention of activity under a range of process conditions), enzymes identified as hits can be directly implemented without the need for engineering through directed evolution. This section highlights a number of examples from GSK where biocatalysis has been successfully employed in the manufacture of APIs – both through the use of 'ready-now' panel enzymes and via directed evolution.

1.5.1 Implementation of Process-Ready Panel Hits

1.5.1.1 Epoxide Hydrolases (EHs)

Enzymes which are process-ready without the requirement for additional or extensive directed evolution constitute an ideal scenario. One example involves the synthesis of compound GSK2330672, an ileal bile acid transport (iBAT) inhibitor which targets cholestatic pruritus [50]. Means of accessing the enantioenriched epoxide (*R*)-2-butyl-2-ethyloxirane were scouted in order to dramatically simplify the process and reduce CoG (Scheme 1.4) [51]. To this end, a range of genes encoding epoxide hydrolases were identified and acquired, and the resulting enzymes were tested for the desired biotransformation. Pleasantly, a few enzymes showed selectivity and delivered the (*R*)-enantiomer, with EH5 from *Agromyces mediolanus* being progressed to process development. The wild-type EH5 showed very good activity and stability with high substrate loading and delivered a three-volume process (substrate loadings >330 g.L^{-1}) at 10% w/w lyophilised powder loading – yielding (*R*)-2-butyl-2-ethyloxirane with >95% ee (Scheme 1.4).

To further improve the process, several mutants of EH5 [52], previously developed for improved enantioselectivity towards epichlorohydrin [53], were tested for further evaluation. Variants N240D, S207V, W128F, S207V/N240D, W128F/S207V and S207V/N240D/W182F were generated and screened. Results showed variant N240D (EH15) was slightly more enantioselective and active compared to EH5. Further experiments and reaction optimisation showed that by using EH15, enzyme loading could be reduced to 5–8% w/w of lyophilised lysate. This approach is a rare example where a wild-type enzyme and several purely rationally designed enzyme variants could lead to the implementation of a sustainable process without the need for enzyme evolution.

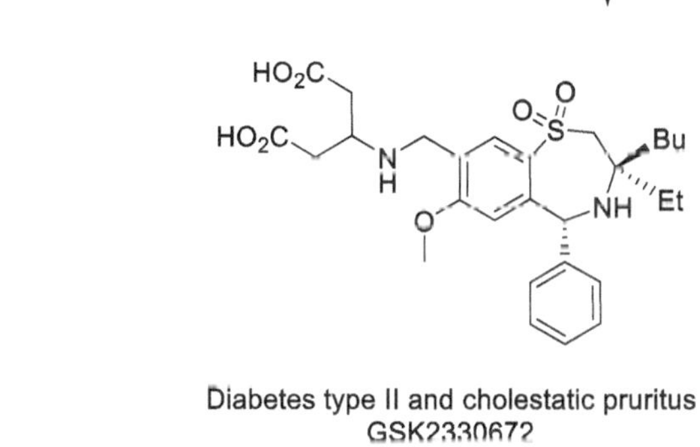

Scheme 1.4 *Epoxide hydrolase (EH5)-catalysed resolution of rac-2-butyl-2-ethyloxirane.*

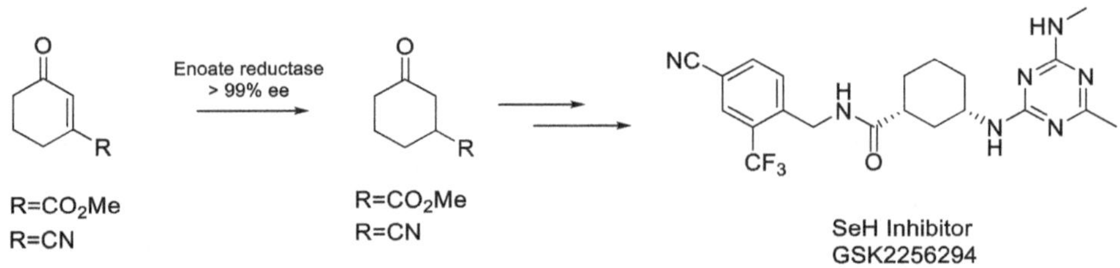

Scheme 1.5 *Synthesis of chiral synthons for the preparation of SeH inhibitor GSK2256294.*

1.5.1.2 Enoate Reductases (EREDs)

A robust and diverse panel of optimised enzyme variants derived from previous directed evolution efforts [54] allows quick identification of hits and can potentially provide access to hundreds of grams of product in a short period of time. This was the case when an ERED panel was tested against methyl 3-oxocyclohex-1-ene-1-carboxylate to deliver the desired (*R*)- and (*S*)- configurated products (Scheme 1.5). The hits observed were easily scaled up to hundreds of grams – an exercise which significantly increased process chemists' confidence that biocatalysis can rapidly deliver scalable processes.

1.5.2 Implementation of Panel Hits that Required Evolution

1.5.2.1 Purine Nucleoside Phosphorylases (PNPases)

Directed evolution can successfully deliver process-ready enzymes with relatively small libraries of variants. An early GSK project involved the evolution of a PNPase for utilisation in the preparation of Nelarabine (Scheme 1.6) [5, 55]. PNPase immobilised on a resin catalysed the phosphorolysis of uridine, leading to the formation of Nelarabine. The yield was significantly reduced, however, by product binding to the resin. It was envisioned that directed evolution could be used to increase the activity of the PNPase, thereby reducing the resin loading and increasing the API yield.

Using rational design, 20 amino acids were selected for random diversity generation. Screening of 25 000 clones using Rapidfire® MS identified variants with up to ninefold activity improvement compared to the wild-type enzyme. Several of the most active variants were then investigated under process conditions, which allowed identification of the best

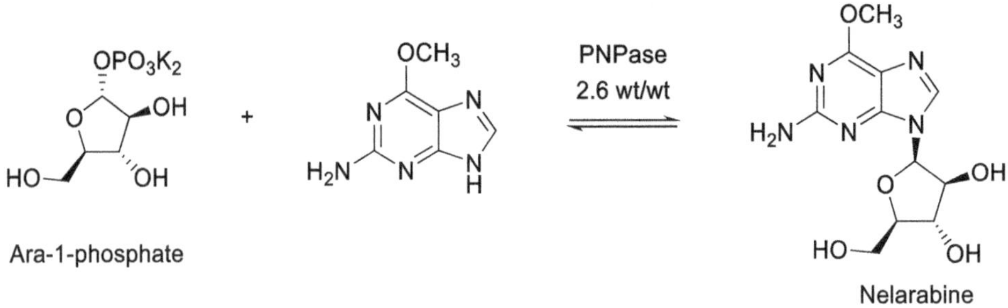

Scheme 1.6 *Nelarabine formation catalysed by PNPase.*

variant for the process. A cost-analysis exercise showed a fivefold improvement over the parent. Scaled-up reactions performed using this immobilised variant were successful, and the process was ultimately demonstrated in a pilot-plant campaign on 50 L scale.

1.5.2.2 Imine Reductases (IREDs)

IREDs are an attractive option for the biocatalytic synthesis of amines, as they can generate secondary and even tertiary amines [56]. Recent efforts to expand the biocatalytic toolbox at GSK have led to the delivery of a panel containing 85 IREDs – a number of which are capable of accepting a wide range of equimolar quantities of amine [17]. Construction of the GSK IRED panel coincided with the opportunity to synthesise the secondary amine intermediate shown in Scheme 1.7, which is used in the synthesis of the LSD1 inhibitor GSK2879552. This inhibitor is currently under clinical trial for use in treating small-cell lung cancer [57].

The earlier manufacturing route for API GSK2879552 required cryogenic reaction conditions, workup with multiple extractions and the use of several nonpreferred and highly flammable solvents [57]. It also involved a separate step for the resolution of an amine intermediate, which increased manufacturing costs. Given the problems with this route, there was an opportunity to incorporate an IRED-catalysed reductive amination (Scheme 1.8) [58]. Scoping experiments led to the identification of IR-46, which proved to be enantioselective towards the desired enantiomer but did not meet minimal process conditions, meaning that it required enzyme engineering.

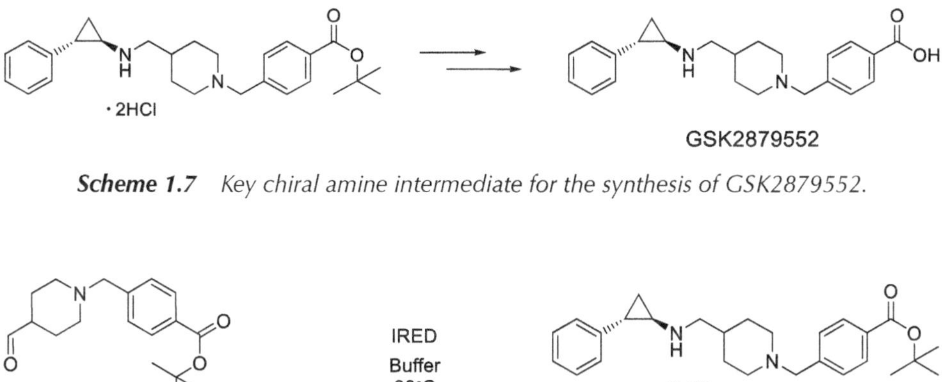

Scheme 1.7 *Key chiral amine intermediate for the synthesis of GSK2879552.*

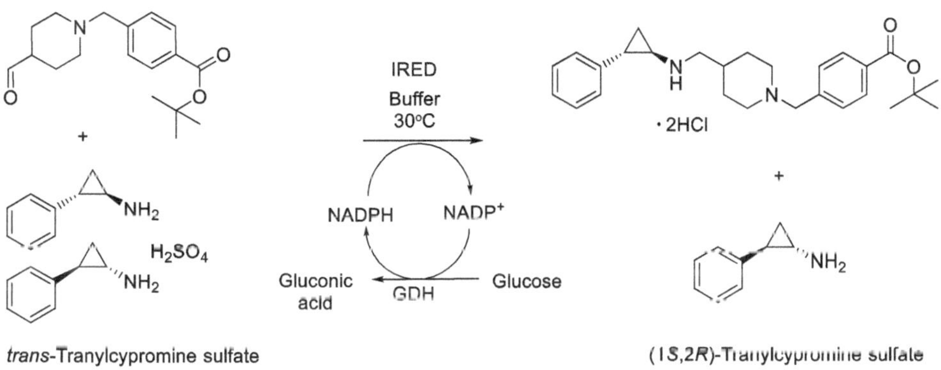

Scheme 1.8 *Modified conditions reflecting phosphate buffer synthesis of tert-butyl 4-((4-((((1R,2S)-2-phenylcyclopropyl)amino)methyl)piperidin-1-yl)methyl)benzoate.*

Three rounds of evolution were undertaken, screening approximately 10 000 different mutants under process-relevant conditions. Initially, enzyme lysate was used, but as rounds progressed, enzyme stability was increased sufficiently that lyophilised powder could be screened. Buffer pH was also decreased across the rounds, to achieve function at pH 4.2; this was necessitated by the manufacturing process. Different stressful conditions were applied during screening to allow identification of the optimal variants. As a result, enzyme loadings were reduced from 450 to ~1% w/w.

In 2017, this new enzymatic process for GSK2879552 delivered API on the kilogram scale that met the target specification (99.7% purity and 99.7% ee). The simplified route removed a synthetic step, reduced the number of solvents and the cycle time and generated better green metrics. The success of this work has helped promote enzyme-catalysed reductive amination as a cheaper and more sustainable alternative to traditional approaches, with other IRED-catalysed reactions following it.

1.6 Outlook

The directed evolution of enzymes has had a significant impact on the way in which APIs are synthesised at GSK – both improving the sustainability of manufacture and reducing CoG. The work described herein, as well as other examples, has shown that panels of evolved variants can lead to the implementation of efficient and environmentally sound processes, replacing traditional chemical approaches across the portfolio and supporting GSK's leading role in the field of industrial biocatalysis. In addition to the significant scientific advances in that field in recent years, success has also been facilitated by regular communication across several global teams, continuous interaction with key stakeholders and the flexibility to adapt to changing project needs.

These achievements have increased the confidence in biocatalysis, both internally and externally, demonstrating that panels of enzymes can be used to identify efficient enzymatic reactions and that enzymes can be evolved where necessary to yield fit-for-purpose biocatalysts. This confidence supported the use of biocatalytic reactions in ~30% of new synthetic chemical routes in GSK in 2018. Encouraged by these successes in small-molecule manufacture, there is increasing focus on the application of directed evolution in other areas of the pharmaceutical sector (e.g. vaccines, cell and gene therapies).

Despite these advances, there are still challenges ahead, with the constant need for enzyme evolution to deliver in significantly shorter timeframes and to affect assets earlier in the drug development pipeline. Accelerating the timelines of directed evolution would increase the adoption of biocatalysis. Alternative directed evolution methods are therefore emerging in the hope of realising this ambition, such as:

- Leveraging growing sequence databases (along with bioinformatic tools) and the falling cost of gene synthesis to identify more suitable wild-type enzymes without the need for directed evolution.
- Bolstering the current computational chemistry toolkit to allow smaller – and smarter – libraries of variants to be screened. By creating more reliable models to identify which mutations will most improve an enzyme's performance in the desired process, more targeted variant libraries can be designed – reducing the time and resources required

to achieve an enzyme with the desired properties. This includes *in silico* evolution methods, which would obviate much of the wet work required in directed evolution.
- Cell-free protein synthesis to accelerate library generation by removing the need for microbial cultivation during protein expression.
- Microfluidics, which can allow faster screening of variants reduces the amount of compound needed.

Acknowledgements

The authors would like to acknowledge all current and former members of the Synthetic Biochemistry team across the United Kingdom and United States and all GSK stakeholders integral to the successful implementation of directed evolution and biocatalysis in small-molecule API manufacturing. We thank K. Tiffin for proofreading the manuscript and making valuable suggestions.

References

1. Gröger, H. and Asano, Y. (2012) Introduction – principles and historical landmark of enzyme catalysis in organic synthesis, in *Enzyme Catalysis in Organic Synthesis* (eds K. Drauz, H. Gröger and O. May), Wiley-VCH Verlag, pp. 1–42.
2. Graedel, T.E. (2007) Green chemistry and sustainable development, in *Handbook of Green Chemistry and Technology* (eds J. Clark and D. Macquarrie), Blackwell Science, pp. 56–61.
3. Reetz, M.T. (2016) *Directed Evolution of Selective Enzymes*, Wiley-VCH Verlag.
4. Mahmoudian, M. (2006) A decade of biocatalysis at Glaxo Wellcome, in *Biocatalysis in the Pharmaceutical and Biotechnology Industries* (ed. R.N. Patel), Taylor & Francis.
5. Kern, M., Roiban, G.-D., Fosberry, A. and Snajdrova, R. (2018) GSK: biocatalyst discovery and optimisation, in *Biocatalysis: An Industrial Perspective* (eds G. de Gonzalo and P.D. de María), The Royal Society of Chemistry, pp. 257–275.
6. Sutton, P.W., Adams, J.P., Wade, C. and Wheelhouse, K. (2018) New technologies in process development, in *Early Drug Development* (ed. F.G. Giordanetto), Wiley-VCH Verlag, pp. 73–124.
7. Finn, R.D., Attwood, T.K., Babbitt, P.C. *et al.* (2017) *Nucleic Acids Research*, **45**, D190–D199.
8. UniProt Consortium (2019) *Nucleic Acids Research*, **47**, D506–D515.
9. Berman, H.M., Westbrook, J., Feng, Z. *et al.* (2000) *Nucleic Acids Research*, **28**, 235–242.
10. Dawson, N.L., Lewis, T.E., Das, S. *et al.* (2017) *Nucleic Acids Research*, **45**, D289–D295.
11. Steinegger, M. and Soding, J. (2018) *Nature Communications*, **9**, 2542.
12. (a) Remmert, M., Biegert, A., Hauser, A. and Soding, J. (2011) *Nature Methods*, **9**, 173–175; (b) Soding, J. (2005) *Bioinformatics*, **21**, 951–960.
13. Frickey, T. and Lupas, A. (2004) *Bioinformatics*, **20**, 3702–3704.
14. Kumar, S., Stecher, G., Li, M. *et al.* (2018) *Molecular Biology and Evolution*, **35**, 1547–1549.
15. Huerta Cepas, J., Serra, F. and Bork, P. (2016) *Molecular Biology and Evolution*, **33**, 1635–1638.
16. (a) Da, C. and Kireev, D. (2014) *Journal of Chemical Information and Modeling*, **54**, 2555–2561; (b) Hu, B. and Lill, M.A. (2012) *Journal of Chemical Information and Modeling*, **52**, 1046–1060.
17. Roiban, G.-D., Kern, M., Liu, Z. *et al.* (2017) *ChemCatChem*, **9**, 4475–4479.
18. Hyslop, J.F., Lovelock, S.L., Sutton, P.W. *et al.* (2018) *Angewandte Chemie International Edition*, **57**, 13 821–13 824.

19. Hyslop, J.F., Lovelock, S.L., Watson, A.J.B. *et al.* (2019) *Journal of Biotechnology*, **293**, 56–65.
20. Popkin, M.E., Borman, P.J., Omer, B.A. *et al.* (2018) *Journal of Pharmaceutical Innovation*, **13**, 367–372.
21. Fernández-Álvaro, E., Snajdrova, R., Jochens, H. *et al.* (2011) *Angewandte Chemie International Edition*, **50**, 8584–8587.
22. Slabu, I., Galman, J.L., Lloyd, R.C. and Turner, N.J. (2017) *ACS Catalysis*, **7**, 8263–8284.
23. Reetz, M.T. and Carballeira, J.D. (2007) *Nature Protocols*, **2**, 891–903.
24. Qu, G., Li, A., Sun, Z. *et al.* (2019) *Angewandte Chemie International Edition*, doi:10.1002/anie .201901491.
25. (a) Altschul, S.F., Gish, W., Miller, W. *et al.* (1990) *Journal of Molecular Biology*, **215**, 403–410; (b) Camacho, C., Coulouris, G., Avagyan, V. *et al.* (2009) *BMC Bioinformatics*, **10**, 421.
26. Altschul, S.F., Madden, T.L., Schaffer, A.A. *et al.* (1997) *Nucleic Acids Research*, **25**, 3389–3402.
27. Eddy, S.R. (2011) *PLoS Computational Biology*, **7**, e1002195.
28. (a) Smith, C.K., Withka, J.M. and Regan, L. (1994) *Biochemistry*, **33**, 5510–5517; (b) Engel, D.E. and DeGrado, W.F. (2004) *Journal of Molecular Biology*, **337**, 1195–1205.
29. (a) Wimley, W.C., Creamer, T.P. and White, S.H. (1996) *Biochemistry*, **35**, 5109–5124; (b) Wimley, W.C. and White, S.H. (1996) *Nature Structural Biology*, **3**, 842–848.
30. Kallberg, M., Wang, H., Wang, S. *et al.* (2012) *Nature Protocols*, **7**, 1511–1522.
31. Haas, J., Barbato, A., Behringer, D. *et al.* (2018) *Proteins*, **86**(Suppl. 1), 387–398.
32. (a) Meier, A. and Soding, J. (2015) *PLoS Computational Biology*, **11**, e1004343; (b) H. Park, D.E. Kim, S. Ovchinnikov *et al.*, *Proteins* 2018, **86**(Suppl. 1), 283–291.
33. (a) Muhammed, M.T. and Aki-Yalcin, E. (2019) *Chemical Biology & Drug Design*, **93**, 12–20; (b) Bordner, A.J. (2012) *Methods in Molecular Biology*, **857**, 83–106; (c) Dolan, M.A., Noah, J.W. and Hurt, D. (2012) *Methods in Molecular Biology*, **857**, 399–414; (d) Joo, K., Lee, J. and Lee, J. (2012) *Methods in Molecular Biology*, **857**, 175–188.
34. Park, H., Bradley, P., Greisen, P. Jr. *et al.* (2016) *Journal of Chemical Theory and Computation*, **12**, 6201–6212.
35. Zhang, Y. (2008) *BMC Bioinformatics*, **9**, 40.
36. Webb, B. and Sali, A. (2014) *Current Protocols in Bioinformatics*, **47**, 5.6.1–5.6.32.
37. Haas, J., Roth, S., Arnold, K. *et al.* (2013) *Database (Oxford)*, **2013**, bat031.
38. (a) Soga, S., Shirai, H., Kobori, M. and Hirayama, N. (2007) *Journal of Chemical Information and Modeling*, **47**, 400–406; (b) Volkamer, A., Griewel, A., Grombacher, T. and Rarey, M. (2010) *Journal of Chemical Information and Modeling*, **50**, 2041–2052.
39. (a) Halgren, T. (2007) *Chemical Biology & Drug Design*, **69**, 146–148; (b) Halgren, T.A. (2009) *Journal of Chemical Information and Modeling*, **49**, 377–389.
40. Kabsch, W. and Sander, C. (1983) *Biopolymers*, **22**, 2577–2637.
41. Winn, M.D., Ballard, C.C., Cowtan, K.D. *et al.* (2011) *Acta Crystallographica Section D: Structural Biology*, **67**, 235–242.
42. (a) Beard, H., Cholleti, A., Pearlman, D. *et al.* (2013) *PLoS One*, **8**, e82849; (b) Dehouck, Y., Grosfils, A., Folch, B. *et al.* (2009) *Bioinformatics*, **25**, 2537–2543; (c) Kellogg, E.H., Leaver-Fay, A. and Baker, D. (2011) *Proteins*, **79**, 830–838; (d) Schrödinger (2019) Release 2019-2, Schrödinger LLC, New York, NY; (e) Chemical Computing Group (2018) Molecular Operating Environment (MOE) 2013.08, Chemical Computing Group ULC, Montreal, QB; (f) Dehouck, Y., Grosfils, A., Folch, B. *et al.* (2019) *Bioinformatics*, **25**, 2537–2543; (g) Goldstein, R.F. (1994) *Biophysical Journal*, **66**, 1335–1340; (h) Guerois, R., Nielsen, J.E. and Serrano, L. (2002) *Journal of Molecular Biology*, **320**, 369–387; (i) Katz, B.A. and Kossiakoff, A. (1986) *Journal of Biological Chemistry*, **261**, 15 480–15 485; (j) Labute, P. (2009) *Proteins*, **75**, 187–205.
43. (a) Chao, R., Mishra, S., Si, T. and Zhao, H. (2017) *Metabolic Engineering*, **42**, 98–108; (b) Hillson, N., Caddick, M., Cai, Y. *et al.* (2019) *Nature Communications*, **10**, 2040.

44. Acevedo-Rocha, C.G., Ferla, M. and Reetz, M.T. (2018) Directed evolution of proteins based on mutational scanning, in *Protein Engineering: Methods and Protocols* (eds U.T. Bornscheuer and M. Höhne), Springer, pp. 87–128.
45. (a) Perez, J.G., Stark, J.C. and Jewett, M.C. (2016) *Cold Spring Harbor Perspectives in Biology*, **8**, a023853; (b) Karim, A.S. and Jewett, M.C. (2018) Cell-free synthetic biology for pathway prototyping, in *Methods in Enzymology*, Vol. **608** (ed. N. Scrutton), Academic Press, pp. 31–57.
46. Weinstock, M.T., Hesek, E.D., Wilson, C.M. and Gibson, D.G. (2016) *Nature Methods*, **13**, 849.
47. (a) Savile, C.K., Janey, J.M., Mundorff, E.C. *et al.* (2010) *Science*, **329**, 305; (b) Bong, Y.K., Clay, M.D., Collier, S.J. *et al.* (2011) Patent WO2011071982; (c) Prier, C.K. and Kosjek, B. (2019) *Current Opinion in Chemical Biology*, **49**, 105–112; (d) Bong, Y.K., Song, S., Nazor, J. *et al.* (2018) *Journal of Physical Chemistry*, **83**, 7453–7458; (e) Xu, F., Kosjek, B., Cabirol, F.L. *et al.* (2018) *Angewandte Chemie International Edition*, **57**, 6863–6867.
48. Limanto, J., Ashley, E.R., Yin, J. *et al.* (2014) *Organic Letters*, **16**, 2716–2719.
49. Znabet, A., Polak, M.M., Janssen, E. *et al.* (2010) *Chemical Communications*, **46**, 7918–7920.
50. Barcan, G., Guo, J., Morgan, C. *et al.* (2018) Patent WO2018002827A1.
51. Roiban, G.-D., Sutton, P.W., Splain, R. *et al.* (2017) *Organic Process Research & Development*, **21**, 1302–1310.
52. Xue, F., Liu, Z.-Q., Zou, S.-P. *et al.* (2014) *Process Biochemistry*, **49**, 409–417.
53. Xue, F., Liu, Z.-Q., Wan, N.-W. *et al.* (2015) *RSC Advances*, **5**, 31 525–31 532.
54. Bougioukou, D.J., Kille, S., Taglieber, A. and Reetz, M.T. (2009) *Advanced Synthesis & Catalysis*, **351**, 3287–3305.
55. Cohen, M.H., Johnson, J.R., Justice, R. and Pazdur, R. (2008) *The Oncologist*, **13**, 709–714.
56. Mangas-Sanchez, J., France, S.P., Montgomery, S.L. *et al.* (2017) *Current Opinion in Chemical Biology*, **37**, 19–25.
57. Johnson, N.W., Kasparec, J. (2014) Patent 8853408 B2.
58. Schober, M., MacDermaid, C., Ollis, A.A. *et al.* (2019) *Nature Catalysis*, **2**, 909–915.

2

Survey of Current Commercial Enzyme and Bioprocess Service Providers

Simon Charnock,[1] Andrea Martínez Bernardini,[2] Emanuele Monza,[2] Maria Fatima Lucas[2] and Peter W. Sutton,[3]*
[1]Prozomix Ltd, Haltwhistle, Northumberland, UK
[2]Zymvol Biomodeling SL, Barcelona, Spain
[3]Glycoscience SL, Bizkaia, Spain

In 2012, David Rozzell published an excellent and concise survey of commercial enzyme suppliers [1]. Given that the field of biocatalysis is moving rapidly, we felt that it would be timely to perform another such survey (Table 2.1), but to also include details of companies that provide one or more of the many services focused towards bioprocess development (Table 2.2).

We have tried to be as comprehensive as possible. Regarding Table 2.1, we have attempted to contact all of the companies included so that they can contribute their own information and provide a simple 'yes' or 'no' answer to each of our column headings. We have not been able to get a response in all cases, but we felt that the information on those companies we could not contact would still be of value, based on data available on their websites. We apologise to any suppliers that we were unaware of and that have been omitted.

The data gathered in this limited space is very general, but we hope that sufficient detail is provided to allow the reader to select appropriate companies to contact directly for more information.

This survey is primarily aimed at the industrial and academic synthetic chemist who needs to source enzymes or services for biocatalytic process development. We therefore also include Table 2.3, which lists the chemical transformations of a selection of enzymes taken from Table 2.1.

Applied Biocatalysis: The Chemist's Enzyme Toolbox, First Edition. Edited by John Whittall and Peter W. Sutton.
© 2021 John Wiley & Sons Ltd. Published 2021 by John Wiley & Sons Ltd.

2.1 Commercial Enzyme Suppliers/Distributors

Table 2.1 *Commercial enzyme suppliers/distributors.*

Enzyme name	Supplier	Company contact details	Number of enzymes in panel	Screening quantities available (i.e. between 1 mg and 1 g of each available on immediate basis for sale)	Larger quantities available (i.e. between 1 g and 1 kg available within 1 month of request)	Attributes																
						Recombinant	Homologously expressed	Metagenomic	Engineered panel	Evolved panel	Sequence diverse members	Unique proprietary members	Nagoya compliant	Psychrophilic members	Mesophilic members	Thermophilic members	NAD+-dependent members	NAD(P)+-dependent	Animal-free	BSE-TSE-certified	GMO-free	GMP grade
Endo-beta-N-acetylglucosaminidase	NZYTech	www.nzytech.com	19	Y	N	Y	N	N	N	N	N	N	N	N	Y	Y	N	N	N	N	N	N
	Merck KGgA/ MilliporeSigma	www.sigmaaldrich.com	3	Y	N	Y	N	N	N	N	N	N	N	N	N	N	N	N	N	N	N	N
	Prozomix Ltd	www.prozomix.com	1	Y	Y	Y	N	N	N	N	Y	Y	Y	N	Y	N	N	N	Y	Y	Y	N
Aconitase	Merck KGgA/ MilliporeSigma	www.sigmaaldrich.com	1	Y	N	N	N	N	N	N	N	N	N	N	N	N	N	N	N	N	N	N
Acylase (amino)	Amano Enzyme	www.amano-enzyme.co .jp	2																			
	Enzymicals AG	www.enzymicals.com	6	N	Y	Y	N	N	Y	N	Y	Y	Y	N	Y	Y	N	N	Y	Y	Y	N
	EnzymeWorks	www.enzymeworking .com	10	Y	Y	Y	Y	N	Y	Y	Y	Y	Y	N	Y	Y	N	N	Y	Y	Y	N
	Merck KGgA/ MilliporeSigma	www.sigmaaldrich.com	1	Y	Y	N	N	N	N	N	N	N	N	N	N	N	N	N	N	N	N	N
	Protéus	www.proteus.fr	9	Y	Y	Y	N	N	Y	Y	Y	Y	Y	N	Y	Y	N	N	Y	Y	Y	N
Acylase (penicillin)	ASA Spezialenzyme GmbH	www.asa-enzyme.de	1	Y	Y	Y	N	N	N	N	N	Y	N	N	Y	N	N	N	Y	Y	N	N
	Codexis Inc.	www.codexis.com	90	Custom item	>1 month	Y	N	N	Y	Y	Y	Y	Y				-	-	N	Y	Y	N
	Enzagen	www.enzagen.com	1																			

	Gecco Biotech	www.gecco-biotech.com	1	Y	N	Y	N	Y	N	N	N	N	N	N	Y	N	N	N	Y	N	N	N
	Merck KGgA/ MilliporeSigma	www.sigmaaldrich.com	1	Y	N	N	N	N	N	N	N	N	N	N	N	N	N	N	N	N	N	N
Adenosine 5'-triphosphatase	Merck KGgA/ MilliporeSigma	www.sigmaaldrich.com	2	Y	N	N	N	N	N	N	N	N	N	N	N	N	N	N	N	N	N	N
Agarase	Merck KGgA/ MilliporeSigma	www.sigmaaldrich.com	1	Y	N	N	N	N	N	N	N	N	N	N	N	N	N	N	N	N	N	N
	NZYTech	www.nzytech.com	24																			
Aldehyde reductase	Almac	www.almacgroup.com	16	Y	Y	Y	N	N	N	N	Y	Y	Y	N	Y	Y	Y	Y	Y	Y	Y	N
	EnzymeWorks	www.enzymeworking .com	40	Y	Y	Y	Y	N	Y	Y	Y	Y	Y	N	Y	Y	Y	Y	Y	Y	Y	N
	Proteus	www.proteus.fr	9	Y	Y	Y	N	N	Y	Y	Y	Y	Y	N	Y	Y	Y	Y	Y	Y	Y	N
	Prozomix Ltd	www.prozomix.com	1	Y	Y	Y	N	N	N	N	Y	Y	Y	N	Y	N	N	Y	Y	Y	Y	N
Aldolase	Almac	www.almacgroup.com	96	Y	Y	Y	N	N	N	N	Y	Y	Y	N	Y	Y	N	N	Y	Y	Y	N
	Enzymaster	www.enzymaster.de	162	N	Y	Y	N	N	Y	Y	Y	Y	Y	N	Y	Y	N	N	Y	Y	Y	N
	EnzymeWorks	www.enzymeworking .com	25	Y	Y	Y	Y	N	Y	Y	Y	Y	Y	N	Y	Y	N	N	Y	Y	Y	N
Aldolase A	Merck KGgA/ MilliporeSigma	www.sigmaaldrich.com	1	Y	N	Y	N	N	N	N	N	N	N	N	N	N	N	N	N	N	N	N
Aldolase (N-acetylneuraminic acid)	Codexis Inc.	www.codexis.com	Custom item 1	Custom item	>1 month	Y	N	N	N	N	N	N	Y				-	-	N	Y	Y	N
	Evoxx technologies	www.evoxx.com	2	N	Y	Y	Y	Y	Y	Y	Y	Y	Y						Y	Y	Y	N
	Merck KGgA/ MilliporeSigma	www.sigmaaldrich.com	2	Y	N	Y	N	N	N	N	N	N	N	N	N	N	N	N	N	N	N	N
Aldolase (deoxyribose-phosphate)	Almac	www.almacgroup.com	96	Y	Y	Y	N	N	N	N	Y	Y	Y	N	Y	Y	N	N	Y	Y	Y	N
	Codexis Inc.	www.codexis.com	Custom items >100	Custom items	>1 month	Y	N	N	Y	Y	Y	Y	Y				-	-	N	Y	Y	N
	Enzymicals AG	www.enzymicals.com	3	N	Y	Y	N	N	N	N	Y	Y	Y	N	Y	Y	N	N	Y	Y	Y	N

Table 2.1 (Continued)

Enzyme name	Supplier	Company contact details	Number of enzymes in panel	Screening quantities available (i.e. between 1 mg and 1 g of each available on immediate basis for sale)	Larger quantities available (i.e. between 1 g and 1 kg available within 1 month of request)	Attributes: Recombinant	Homologously expressed	Metagenomic	Engineered panel	Evolved panel	Sequence diverse members	Unique proprietary members	Nagoya compliant	Psychrophilic members	Mesophilic members	Thermophilic members	NAD+-dependent members	NAD(P)+-dependent	Animal-free	BSE-TSE-certified	GMO-free	GMP grade
	EnzymeWorks	www.enzymeworking.com	15	Y	Y	Y	Y	N	Y	Y	Y	Y	Y	N	Y	Y	N	N	Y	Y	Y	Y
	Evoxx technologies	www.evoxx.com	5	N	Y	Y	Y	Y	Y	Y	Y	Y	Y		Y	Y	Y	Y	Y	Y	Y	N
Aldolase (deoxyribose-phosphate) (DERA)	Merck KGgA/ MilliporeSigma	www.sigmaaldrich.com	1	Y	N	Y	N	N	N	N	N	N	N	N	N	N	N	N	N	N	N	N
	Prozomix Ltd	www.prozomix.com	210	Y	Y	Y	N	Y	N	N	Y	Y	Y	N	Y	N	N	N	Y	Y	Y	N
	Syncozymes	en.syncozymes.com	8	Y	Y	Y	Y	N	Y	N	Y	N	N	N	N	N	N	N	Y	N	N	N
Aldolase (fructose-1,6-diphosphate)	Merck KGgA/ MilliporeSigma	www.sigmaaldrich.com	2	Y	N	N	N	N	N	N	N	N	N	N	N	N	N	N	N	N	N	N
Aldolase (FSA)	Prozomix Ltd	www.prozomix.com	96	Y	Y	Y	N	Y	N	N	Y	Y	Y	N	Y	N	N	N	Y	Y	Y	N
Aldolase (HBPA)	Prozomix Ltd	www.prozomix.com	50	Y	Y	Y	N	Y	N	N	Y	Y	Y	N	Y	N	N	N	Y	Y	Y	N
Aldolase (4-oxalocrotonate tautomerase)	Prozomix Ltd	www.prozomix.com	96	Y	Y	Y	N	Y	N	N	Y	Y	Y	N	Y	N	N	N	Y	Y	Y	N
Aldolase (spingosine-1-phosphate)	Prozomix Ltd	www.prozomix.com	25	Y	Y	Y	N	Y	N	N	Y	Y	Y	N	Y	N	N	N	Y	Y	Y	N
Aldolase (threonine)	Almac	www.almacgroup.com	50	Y	Y	Y	N	Y	Y	N	Y	Y	Y	N	Y	Y	N	N	Y	Y	Y	N

	Codexis Inc.	www.codexis.com	Custom items 10	Custom items	>1 month	Y	N	N	Y	Y	Y	N	Y				-	-	N	Y	Y	N
	EnzymeWorks	www.enzymeworking .com	15	Y	Y	Y	Y	N	Y	Y	Y	Y	Y	N	Y	Y	N	N	Y	Y	Y	Y
	Protéus	www.proteus.fr	3	Y	N	Y	N	N	Y	Y	Y	Y	Y	N	Y	Y	N	Y	Y	Y	Y	N
Aldolase (Yfau)	Prozomix Ltd	www.prozomix.com	35	Y	Y	Y	N	Y	N	N	Y	Y	Y	N	Y	N	N	N	Y	Y	Y	N
Alkyltransferase	Prozomix Ltd (CoEBio3)	www.prozomix.com, www.coebio3. manchester.ac.uk	21	Y	Y	Y	N									N	N	Y	Y	Y	Y	N
Amidase	Almac	www.almacgroup.com	25	Y	Y	Y	N	N	N	N	Y	Y	Y	N	Y	Y	N	N	Y	Y	Y	N
	EnzymeWorks	www.enzymeworking .com	30	Y	Y	Y			Y	Y	Y	Y							Y	Y		
	EUCODIS	www.eucodis.com	2	Y	Y	Y	N	N	N	N	Y	Y	Y	N	Y	Y	N	N	Y	Y	Y	N
	Merck KGgA/ MilliporeSigma	www.sigmaaldrich.com	1	Y	N	Y	N	N	N	N	N	N	N	N	N	N	N	N	N	N	N	N
	Protéus	www.proteus.fr	9	Y	N	Y	N	N	Y	Y	Y	Y	Y	N	Y	Y	N	N	Y	Y	Y	N
	Syncozymes	en.syncozymes.com	18	Y	Y	Y	Y	N	Y	N	Y	N	N	N	N	N	N	N	Y	N	N	N
Aminopeptidase	Merck KGgA/ MilliporeSigma	www.sigmaaldrich.com	5	Y	N	N	N	N	N	N	N	N	N	N	N	N	N	N	N	N	N	N
Ammonia mutase	Almac	www.almacgroup.com	1	Y	Y	Y	N	N	N	N	Y	Y	Y	N	Y	Y	N	N	Y	Y	Y	N
	EnzymeWorks	www.enzymeworking .com	5	Y	Y	Y	N	N	Y	Y	Y	Y	Y	N	Y	Y	N	N	Y	Y	Y	N
	GECCO Biotech	www.gecco-biotech.com	12	Y	N	Y	N	N	N	N	Y	N	N	N	Y	N	N	N	N	N	N	N
Amylase	Amano Enzyme	www.amano-enzyme.co .jp	1																			
	Asahi Kasei Pharma Corporation	www.asahi-kasei.co.jp/ shindan/en/	2																			
	ASA Spezialenzyme GmbH	www.asa-enzyme.de	3			N	Y	N	N	N	N	N	N	N	Y	N	N	N	Y	Y	Y	N
	Biocatalysts Ltd	www.biocatalysts.com	2	Y	Y	N	N	N	N	N	Y	N	Y	N	Y	Y	N	N	Y	Y	Y	N

Table 2.1 *(Continued)*

Enzyme name	Supplier	Company contact details	Number of enzymes in panel	Screening quantities available (i.e. between 1 mg and 1 g of each available on immediate basis for sale)	Larger quantities available (i.e. between 1 g and 1 kg available within 1 month of request)	Attributes																
						Recombinant	Homologously expressed	Metagenomic	Engineered panel	Evolved panel	Sequence diverse members	Unique proprietary members	Nagoya compliant	Psychrophilic members	Mesophilic members	Thermophilic members	NAD+-dependent members	NAD(P)+-dependent	Animal-free	BSE-TSE-certified	GMO-free	GMP grade
	GECCO Biotech	www.gecco-biotech.com	1	Y	N	Y	N	N	N	N	N	N	N	N	N	Y	N	N	Y	N	N	N
	Megazyme	www.megazyme.com	14	N	Y	Y/N	N	N	N	N	Y	Y	N	N	Y	Y	N	N	N	N	N	N
	Merck KGgA/ MilliporeSigma	www.sigmaaldrich.com	24	Y	Y	N	N	N	N	N	N	N	N	N	N	N	N	N	N	N	N	N
	Novozymes A/S	pharmaceuticals@ novozymes.com	1	N	Y	Y																
	NZYTech	www.nzytech.com	15	Y	N	Y	N	N	N	N	N	N	N	N	Y	Y	N	N	N	N	N	N
	Prozomix Ltd	www.prozomix.com	3																			
Amyloglucosidase	Megazyme	www.megazyme.com	6	N	Y	Y/N	N	N	N	N	Y	Y	N	N	Y	N	N	N	Y	N	N	N
	Merck KGgA/ MilliporeSigma	www.sigmaaldrich.com	9	Y	Y	N	N	N	N	N	N	N	N	N	N	N	N	N	N	N	N	N
Amylomaltase	NZYTech	www.nzytech.com	7	Y	N	Y	N	N	N	N	N	N	N	N	Y	Y	N	N	N	N	N	N
Amylosucrase	Prozomix Ltd	www.prozomix.com	5	Y	Y	Y	N	N	N	N	Y	Y	Y	N	Y	N	N	N	Y	Y	Y	N
Apotryptophanase	Merck KGgA/ MilliporeSigma	www.sigmaaldrich.com	1	Y	N	N	N	N	N	N	N	N	N	N	N	N	N	N	N	N	N	N
Apyrase	Merck KGgA/ MilliporeSigma	www.sigmaaldrich.com	5	Y	N	N	N	N	N	N	N	N	N	N	N	N	N	N	N	N	N	N

Arabinanase	Megazyme	www.megazyme.com	9	N	Y	Y/N	N	N	N	N	Y	Y	N	N	Y	N	N	N	Y	N	N	N
	NZYTech	www.nzytech.com	8	Y	N	Y	N	N	N	N	N	N	N	N	Y	Y	N	N	N	N	N	N
	Prozomix Ltd	www.prozomix.com	3	Y	Y	Y	N	N	N	N	Y	Y	Y	N	Y	N	N	N	Y	Y	Y	N
Arabinofuranosidase	Megazyme	www.megazyme.com	6	N	Y	Y/N	N	N	N	N	Y	Y	N	N	Y	N	N	N	Y	N	N	N
	NZYTech	www.nzytech.com	35	Y	N	Y	N	N	N	N	N	N	N	N	Y	Y	N	N	N	N	N	N
	Prozomix Ltd	www.prozomix.com	28	Y	Y	Y	N	N	N	N	Y	Y	Y	N	Y	N	N	N	Y	Y	Y	N
Arabinoxylanase	NZYTech	www.nzytech.com	2	Y	N	Y	N	N	N	N	N	N	N	N	Y	Y	N	N	N	N	N	N
Arabinoxylan arabinofuranohydrolase	Prozomix Ltd	www.prozomix.com	1	Y	Y	Y	N	N	N	N	Y	Y	Y	N	Y	N	N	N	Y	Y	Y	N
L-Arginase	Merck KGgA/ MilliporeSigma	www.sigmaaldrich.com	1	Y	N	N	N	N	N	N	N	N	N	N	N	N	N	N	N	N	N	N
Asparaginase	EUCODIS	www.eucodis.com	1	Y	Y	Y	N	N	N	N	N	Y	Y	N	Y	N	N	N	Y	Y	Y	N
	Merck KGgA/ MilliporeSigma	www.sigmaaldrich.com	1	Y	N	N	N	N	N	N	N	N	N	N	N	N	N	N	N	N	N	N
Bromelain	Merck KGgA/ MilliporeSigma	www.sigmaaldrich.com	2	Y	N	N	N	N	N	N	N	N	N	N	N	N	N	N	N	N	N	N
Carbamoylase	Almac	www.almacgroup.com	20	Y	Y	Y	N	N	Y	N	Y	Y	Y	N	Y	Y	N	N	Y	Y	Y	N
Carbamoyltransferase	Almac	www.almacgroup.com	5	Y	Y	Y	N	N	N	N	Y	Y	Y	N	Y	Y	N	N	Y	Y	Y	N
Carbonic anhydrase	Merck KGgA/ MilliporeSigma	www.sigmaaldrich.com	13	Y	N	Y	N	N	N	N	N	N	N	N	N	N	N	N	N	N	N	N
Carboxylase (acetyl-coenzyme A)	Merck KGgA/ MilliporeSigma	www.sigmaaldrich.com	2	Y	N	N	N	N	N	N	N	N	N	N	N	N	N	N	N	N	N	N
Carboxylase (phosphoenolpyruvate)	Toyobo	www.toyobo-global.com	1																			
Carboxylase (pyruvate)	Merck KGgA/ MilliporeSigma	www.sigmaaldrich.com	2	Y	N	N	N	N	N	N	N	N	N	N	N	N	N	N	N	N	N	N
Carboxylase (D-ribulose-1,5-diphosp	Merck KGgA/ MilliporeSigma	www.sigmaaldrich.com	1	Y	N	N	N	N	N	N	N	N	N	N	N	N	N	N	N	N	N	N
Carboxylic acid reductase	Almac	www.almacgroup.com	35	Y	Y	Y	N	N	N	N	Y	Y	Y	N	Y	Y	N	Y	Y	Y	Y	N
	EnzymeWorks	www.enzymeworking.com	5	Y	Y	Y	N	N	Y	Y	Y	Y	Y	N	Y	Y	N	Y	Y	Y	Y	N

Table 2.1 *(Continued)*

Enzyme name	Supplier	Company contact details	Number of enzymes in panel	Screening quantities available (i.e. between 1 mg and 1 g of each available on immediate basis for sale)	Larger quantities available (i.e. between 1 g and 1 kg available within 1 month of request)	Attributes																
						Recombinant	Homologously expressed	Metagenomic	Engineered panel	Evolved panel	Sequence diverse members	Unique proprietary members	Nagoya compliant	Psychrophilic members	Mesophilic members	Thermophilic members	NAD+-dependent members	NAD(P)+-dependent	Animal-free	BSE-TSE-certified	GMO-free	GMP grade
	Protéus	www.proteus.fr	3	Y	N	Y	N	N	Y	Y	Y	Y	Y	N	Y	Y	N	Y	Y	Y	Y	N
	Prozomix Ltd	www.prozomix.com	20	Y	Y	Y	N	N	N	N	Y	Y	Y	N	Y	N	N	N	Y	Y	Y	N
Carboxypeptidase	Merck KGgA/ MilliporeSigma	www.sigmaaldrich.com	6	Y	N	N	N	N	N	N	N	N	N	N	N	N	N	N	N	N	N	N
Carnitine acetyltransferase	Merck KGgA/ MilliporeSigma	www.sigmaaldrich.com	1	Y	N	N	N	N	N	N	N	N	N	N	N	N	N	N	N	N	N	N
Carrageenase	NZYTech	www.nzytech.com	2	Y	N	Y	N	N	N	N	N	N	N	N	Y	N	N	N	N	N	N	N
Catalase	Biocatalysts Ltd	www.biocatalysts.com	1	Y	Y	N	Y	N	N	N	N	N	Y	N	Y	N	N	N	Y	Y	Y	N
	EUCODIS	www.eucodis.com	1	Y	Y	Y	N	N	Y	N	N	N	Y	N	N	Y	N	N	Y	Y	Y	N
	Gecco Biotech	www.gecco-biotech.com	1	y	N	Y	N	N	N	N	N	N	N	N	Y	N	N	N	Y	N	N	N
	Megazyme	www.megazyme.com	1	N	Y	N	N	N	N	N	Y	Y	N	N	Y	N	N	N	Y	N	N	N
	Merck KGgA/ MilliporeSigma	www.sigmaaldrich.com	12	Y	Y	N	N	N	N	N	N	N	N	N	N	N	N	N	N	N	N	N
	Novozymes A/S	pharmaceuticals@ novozymes.com	1	N	Y	Y																
	Toyobo	www.toyobo-global.com	1																			
Cellobiohydrolase	Megazyme	www.megazyme.com	2	N	Y	Y/ N	N	N	N	N	Y	Y	N	N	Y	N	N	N	Y	N	N	N
	Merck KGgA/ MilliporeSigma	www.sigmaaldrich.com	1	Y	N	Y	N	N	N	N	N	N	N	N	N	N	N	N	N	N	N	N
	NZYTech	www.nzytech.com	10	Y	N	Y	N	N	N	N	N	N	N	N	Y	Y	N	N	N	N	N	N

Cellodextrinase	NZYTech	www.nzytech.com	2	Y	N	Y	N	N	N	N	N	N	N	N	Y	Y	N	N	N	N	N	N
Cellulase	Amano Enzyme	www.amano-enzyme.co.jp	1																			
	ASA Spezialenzyme GmbH	www.asa-enzyme.de	1	Y	Y	N	Y	N	N	N	N	N	N	N	Y	N	N	N	Y	Y	Y	N
	Biocatalysts Ltd	www.biocatalysts.com	1	Y	Y	N	N	N	N	N	N	N	Y	N	Y	N	N	N	Y	Y	Y	N
	Codexis Inc.	www.codexis.com	Custom item	Custom item	>1 month	Y	N	N	N	N	Y	Y	Y				-	-	N	Y	N	N
	Megazyme	www.megazyme.com	6	N	Y	Y/N	N	N	N	N	Y	Y	N	N	Y	N	N	N	Y	N	N	N
	Merck KGgA/MilliporeSigma	www.sigmaaldrich.com	10	Y	Y	N	N	N	N	N	N	N	N	N	N	N	N	N	N	N	N	N
	NZYTech	www.nzytech.com	81	Y	N	Y	N	N	N	N	N	N	N	N	Y	Y	N	N	N	N	N	N
	Protéus	www.proteus.fr	6	Y	N	Y	N	N	Y	Y	Y	Y	Y	N	Y	Y	N	N	Y	Y	Y	N
	Prozomix Ltd	www.prozomix.com	7	Y	Y	Y	N	N	N	N	Y	Y	Y	N	Y	N	N	N	Y	Y	Y	N
Chitinase	ASA Spezialenzyme GmbH	www.asa-enzyme.de	1	Y	Y	N	Y	N	N	N	N	N	N	N	Y	N	N	N	N	N	Y	N
	Biocatalysts Ltd	www.biocatalysts.com	1	Y	Y	Y	N	Y	N	N	N	N	Y	N	Y	N	N	N	Y	Y	Y	N
	EUCODIS	www.eucodis.com	3	Y	Y	Y	N	N	Y	N	Y	Y	Y	N	Y	N	N	N	Y	Y	Y	N
	Megazyme	www.megazyme.com	1	N	Y	Y	N	N	N	N	Y	Y	N	N	Y	N	N	N	Y	N	N	N
	Merck KGgA/MilliporeSigma	www.sigmaaldrich.com	2	Y	Y	N	N	N	N	N	N	N	N	N	N	N	N	N	N	N	N	N
	NZYTech	www.nzytech.com	7	Y	N	Y	N	N	N	N	N	N	N	N	Y	Y	N	N	N	N	N	N
	Prozomix Ltd	www.prozomix.com	1																			
Chitin deaminase	EUCODIS	www.eucodis.com	1	Y	Y	Y	N	N	N	N	N	Y	Y	N	Y	N	N	N	Y	Y	Y	N
Chitosanase	EUCODIS	www.eucodis.com	1	Y	Y	Y	N	N	N	N	N	N	Y	N	Y	N	N	N	Y	Y	Y	N
	Merck KGgA/MilliporeSigma	www.sigmaaldrich.com	2	Y	N	N	N	N	N	N	N	N	N	N	N	N	N	N	N	N	N	N
	NZYTech	www.nzytech.com	3	Y	N	Y	N	N	N	N	N	N	N	N	Y	Y	N	N	N	N	N	N
Chloramphenicol acetyltransferase	Merck KGgA/MilliporeSigma	www.sigmaaldrich.com	1	Y	N	N	N	N	N	N	N	N	N	N	N	N	N	N	N	N	N	N

Table 2.1 (Continued)

Enzyme name	Supplier	Company contact details	Number of enzymes in panel	Screening quantities available (i.e. between 1 mg and 1 g of each available on immediate basis for sale)	Larger quantities available (i.e. between 1 g and 1 kg available within 1 month of request)	Attributes																
						Recombinant	Homologously expressed	Metagenomic	Engineered panel	Evolved panel	Sequence diverse members	Unique proprietary members	Nagoya compliant	Psychrophilic members	Mesophilic members	Thermophilic members	NAD+-dependent members	NAD(P)+-dependent	Animal-free	BSE-TSE-certified	GMO-free	GMP grade
Chloroperoxidase	ASA Spezialenzyme GmbH	www.asa-enzyme.de	1	Y	Y	N	Y	N	N	N	N	N	N	N	Y	N	N	N	Y	Y	Y	N
	Enzymicals AG		1	N	Y	Y	N	N	N	N	Y	Y	Y	N	Y	Y	N	N	Y	Y	Y	N
	EUCODIS	www.eucodis.com	2	Y	Y	Y	Y	N	Y	N	Y	Y	Y	N	Y	N	N	N	Y	Y	Y	N
	Merck KGgA/ MilliporeSigma	www.sigmaaldrich.com	3	Y	N	N	N	N	N	N	N	N	N	N	N	N	N	N	N	N	N	N
Chondroitinase	Merck KGgA/ MilliporeSigma	www.sigmaaldrich.com	4	Y	N	N	N	N	N	N	N	N	N	N	N	N	N	N	N	N	N	N
Clostripain	Merck KGgA/ MilliporeSigma	www.sigmaaldrich.com	1	Y	N	N	N	N	N	N	N	N	N	N	N	N	N	N	N	N	N	N
Collagenase	Amano Enzyme	www.amano-enzyme.co .jp	1																			
	Merck KGgA/ MilliporeSigma	www.sigmaaldrich.com	29	N	Y	N	N	N	N	N	N	N	N	N	N	N	N	N	N	N	N	N
Creatine amidinohydrolase	Merck KGgA/ MilliporeSigma	www.sigmaaldrich.com	2	Y	N	Y	N	N	N	N	N	N	N	N	N	N	N	N	N	N	N	N
	Toyobo	www.toyobo-global.com	3																			
Creatine phosphokinase	Merck KGgA/ MilliporeSigma	www.sigmaaldrich.com	2	Y	N	N	N	N	N	N	N	N	N	N	N	N	N	N	N	N	N	N

Creatinine amidohydrolase	Merck KGgA/ MilliporeSigma	www.sigmaaldrich.com	1	Y	N	Y	N	N	N	N	N	N	N	N	N	N	N	N	N	N	N	N
	Toyobo	www.toyobo-global.com	2																			
Cutinase	ASA Spezialenzyme GmbH	www.asa-enzyme.de	2	Y	Y	Y	N	N	N	N	N	Y	N	N	Y	N	N	N	Y	N	N	N
	Enzymicals AG	www.enzymicals.com	1	Y	Y	Y	N	N	N	N	Y	Y	Y	N	Y	Y	N	N	Y	Y	Y	N
	Evoxx technologies	www.evoxx.com	10	Y	Y	Y	N	Y	Y	Y	Y	Y	Y				N	N	Y	Y	Y	N
Cyclodextrin glucanotransferase	Amano Enzyme	www.amano-enzyme.co .jp	1																			
	Novozymes A/S	pharmaceuticals@ novozymes.com	1	N	Y	Y																
Cyclooxygenase	Almac	www.almacgroup.com	11	Y	Y	Y	N	N	N	N	Y	Y	Y	N	Y	Y	N	N	Y	Y	Y	N
	Merck KGgA/ MilliporeSigma	www.sigmaaldrich.com	1	Y	N	N	N	N	N	N	N	N	N	N	N	N	N	N	N	N	N	N
Cyclophilin	Merck KGgA/ MilliporeSigma	www.sigmaaldrich.com	1	Y	N	N	N	N	N	N	N	N	N	N	N	N	N	N	N	N	N	N
Cytochrome P450 reductase	Merck KGgA/ MilliporeSigma	www.sigmaaldrich.com	1	Y	Y	Y	N	N	N	N	N	N	N	N	N	N	N	N	N	N	N	N
Deaminase	Biocatalysts Ltd	www.biocatalysts.com	1	Y	Y	N	Y	N	N	N	N	N	Y	N	Y	N	N	N	Y	Y	Y	N
Deaminase (adenosine)	Merck KGgA/ MilliporeSigma	www.sigmaaldrich.com	5	Y	N	N	N	N	N	N	N	N	N	N	N	N	N	N	N	N	N	N
Creatinine deiminase	Toyobo	www.toyobo-global.com	1																			
Decarboxylase	Almac	www.almacgroup.com	30	Y	Y	Y	N	N	N	N	Y	Y	Y	N	Y	Y	N	N	Y	Y	Y	N
	Enzymaster	www.enzymaster.de	40	N	Y	Y	N	N	Y	Y	Y	Y	Y	N	Y	Y	N	N	Y	Y	Y	N
	EnzymeWorks	www.enzymeworking .com	10	Y	Y	Y	N	N	Y	Y	Y	Y	Y	N	Y	Y	N	N	Y	Y	Y	N
Decarboxylase (acetoacetate)	Prozomix Ltd	www.prozomix.com	20	Y	Y	Y	N	N	N	N	Y	Y	Y	N	Y	N	N	N	Y	Y	Y	N
Decarboxylase (aspartate)	Almac	www.almacgroup.com	9	Y	Y	Y	N	N	N	N	Y	Y	Y	N	Y	Y	N	N	Y	Y	Y	N
	Codexis Inc.	www.codexis.com	Custom items 10	Custom items	>1 month	Y	N	N	N	N	Y	N	Y				-	-	N	Y	Y	N

Table 2.1 *(Continued)*

Enzyme name	Supplier	Company contact details	Number of enzymes in panel	Screening quantities available (i.e. between 1 mg and 1 g of each available on immediate basis for sale)	Larger quantities available (i.e. between 1 g and 1 kg available within 1 month of request)	Attributes: Recombinant	Homologously expressed	Metagenomic	Engineered panel	Evolved panel	Sequence diverse members	Unique proprietary members	Nagoya compliant	Psychrophilic members	Mesophilic members	Thermophilic members	NAD+-dependent members	NAD(P)+-dependent	Animal-free	BSE-TSE-certified	GMO-free	GMP grade
Decarboxylase (malonate)	Enzymicals AG	www.enzymicals.com	1	N	Y	Y	N	N	N	N	Y	Y	Y	N	Y	Y	N	N	Y	Y	Y	N
Decarboxylase (oxalate)	Merck KGgA/ MilliporeSigma	www.sigmaaldrich.com	1	Y	N	Y	N	N	N	N	N	N	N	N	N	N	N	N	N	N	N	N
Decarboxylase (oxaloacetate)	Asahi Kasei Pharma Corporation	www.asahi-kasei.co.jp/ shindan/en/	1																			
	Merck KGgA/ MilliporeSigma	www.sigmaaldrich.com	1	Y	N	N	N	N	N	N	N	N	N	N	N	N	N	N	N	N	N	N
Decarboxylase (pyruvate)	Merck KGgA/ MilliporeSigma	www.sigmaaldrich.com	1	Y	N	N	N	N	N	N	N	N	N	N	N	N	N	N	N	N	N	N
	Prozomix Ltd	www.prozomix.com	50	Y	Y	Y	N	N	N	N	Y	Y	Y	N	Y	N	N	N	Y	Y	Y	N
Decarboxylase (L-tyrosine)	Merck KGgA/ MilliporeSigma	www.sigmaaldrich.com	2	Y	N	N	N	N	N	N	N	N	N	N	N	N	N	N	N	N	N	N
Dehalogenase	Almac	www.almacgroup.com	9	Y	Y	Y	N	N	N	N	Y	Y	Y	N	Y	Y	N	N	Y	Y	Y	N
	EnzymeWorks	www.enzymeworking .com	5	Y	Y	Y	N	N	Y	Y	Y	Y	N	N	Y	Y	N	N	Y	Y	Y	N
	Gecco Biotech	www.gecco-biotech.com	2	y	N	Y	N	N	Y	N	N	N	N	N	Y	Y	N	N	Y	N	N	N
	Protéus	www.proteus.fr	4	Y	N	Y	N	N	Y	Y	Y	Y	Y	N	Y	Y	N	N	Y	Y	Y	N

Dehalogenase (halohydrin)	Prozomix Ltd	www.prozomix.com	21	Y	Y	Y	N	N	N	N	Y	Y	Y	N	Y	N	N	N	Y	Y	Y	N
Dehydratase	Almac	www.almacgroup.com	34	Y	Y	Y	N	N	N	N	Y	Y	Y	N	Y	Y	N	N	Y	Y	Y	N
	EnzymeWorks	www.enzymeworking .com	4	Y	Y	Y	N	N	Y	Y	Y	Y	Y	N	Y	Y	N	N	Y	Y	Y	N
Dehydratase (cyanide)	Almac	www.almacgroup.com	8	Y	Y	Y	N	N	Y	N	Y	Y	Y	N	Y	Y	N	N	Y	Y	Y	N
	Enzymicals AG	www.enzymicals.com	3	N	Y	Y	N	N	N	N	Y	Y	Y	N	Y	Y	N	N	Y	Y	Y	N
Dehydrogenase (alanine)	EnzymeWorks	www.enzymeworking .com	5	Y	Y	Y	N	N	Y	Y	Y	Y	Y	N		Y	Y	Y	Y	Y	Y	N
	Evoxx technologies	www.evoxx.com	1	Y	Y	Y	N	Y	Y	Y	Y	Y	Y						Y	Y	Y	N
	Johnson Matthey	www.matthey.com/ biocatalysts/	10	Y	Y	Y	N	N	N	N	Y	Y	Y	Y	Y	Y	Y	N	Y	Y	Y	Y
	Merck KGgA/ MilliporeSigma	www.sigmaaldrich.com	3	Y	N	Y	N	N	N	N	N	N	N	N	N	N	N	N	N	N	N	N
	Prozomix Ltd	www.prozomix.com	1	Y	Y	Y	N	N	N	N	Y	Y	Y	N	Y	N	N	N	Y	Y	Y	N
Dehydrogenase (alcohol)/ketoreductase/ carbonyl reductase	Almac	www.almacgroup.com	>200	Y	Y	Y	N	N	N	N	Y	Y	Y	N	Y	Y	Y	Y	Y	Y	Y	N
	c-LEcta GmbH	www.c-lecta.com	>200	Y	Y	Y	N	Y	Y	Y/ N	Y	Y	Y	N	Y	N	Y	Y	Y	Y	Y	N
	Codexis Inc.	www.codexis.com	24 (kit), 260 (panel) +custom items	Y from kit Lysate panel Custom items	Up to 100 g from kit Others are custom items	Y	N	N	Y	Y	Y	Y	Y				Y	Y	N	Y	Y	N
	Enzymaster	www.enzymaster.de	177	N	Y	Y	N	Y	Y	Y	Y	Y	Y	N	Y	Y	Y	Y	Y	Y	Y	N
	EnzymeWorks	www.enzymeworking .com	200	Y	Y	Y			Y	Y	Y	Y	Y	N	Y	Y	Y	Y	Y	Y	Y	Y
	Evoxx technologies GmbH	www. evoxx.com	25	Y	Y	Y	Y	Y	Y	Y	Y	Y	Y		Y	Y	Y	Y	Y	Y	Y	N

Table 2.1 *(Continued)*

Enzyme name	Supplier	Company contact details	Number of enzymes in panel	Screening quantities available (i.e. between 1 mg and 1 g of each available on immediate basis for sale)	Larger quantities available (i.e. between 1 g and 1 kg available within 1 month of request)	Attributes: Recombinant	Homologously expressed	Metagenomic	Engineered panel	Evolved panel	Sequence diverse members	Unique proprietary members	Nagoya compliant	Psychrophilic members	Mesophilic members	Thermophilic members	NAD+-dependent members	NAD(P)+-dependent	Animal-free	BSE-TSE-certified	GMO-free	GMP grade
	Gecco Biotech	www.gecco-biotech.com	8	Y	Y	Y	N	N	N	N	Y	N	N	N	Y	N	Y	Y	Y	N	N	N
	InnoSyn BV	www.innosyn.com	88	Y	Y	Y	N	N	N	N	Y	N	Y	Y	Y	Y	Y	Y	Y	Y	Y	N
	Johnson Matthey	www.matthey.com/ biocatalysts/	180	Y	Y	Y	N	Y	Y	Y	Y	Y	Y	N	Y	Y	Y	Y	Y	Y	Y	Y
	Megazyme	www.megazyme.com	1	N	Y	Y	N	N	N	N	Y	Y	N	N	Y	N	Y	N	Y	N	N	N
	Merck KGgA/ MilliporeSigma	www.sigmaaldrich.com	5	Y	N	Y	N	N	N	N	N	N	N	N	N	N	N	N	N	N	N	N
	Protéus	www.proteus.fr	15	Y	Y	Y	N	N	Y	Y	Y	Y	Y	N	Y	Y	Y	Y	Y	Y	Y	N
	Prozomix Ltd	www.prozomix.com	768	Y	Y	Y	N	Y	N	N	Y	Y	Y	Y	Y	Y	Y	Y	Y	Y	Y	N
	Syncozymes	en.syncozymes.com	174			Y	Y	N	Y	Y	Y	N	N	N	N	N	Y	Y	Y	N	N	N
Dehydrogenase (alcohol, F420 dependent)	Gecco Biotech	www.gecco-biotech.com	1	Y	N	Y	N	N	N	N	N	N	Y	N	N	Y	N	N	Y	N	N	N
Dehydrogenase (aldehyde)	Merck KGgA/ MilliporeSigma	www.sigmaaldrich.com	4	Y	N	N	N	N	N	N	N	N	N	N	N	N	N	N	N	N	N	N
	Prozomix Ltd	www.prozomix.com	48	Y	Y	Y	N	Y	N	N	Y	Y	Y	Y	Y	Y	Y	Y	Y	Y	Y	N
Dehydrogenase (amine)	Almac	www.almacgroup.com	50	Y	Y	Y	N	N	N	N	Y	Y	Y	N	Y	Y	Y	N	Y	Y	Y	N
	Codexis Inc.	www.codexis.com	24 (beta kit)	Y	>1 month	Y	N	N	N	Y		Y	Y					Y	N	Y	Y	N

	EnzymeWorks	www.enzymeworking .com	5	Y	Y	Y	N	N	Y	Y	Y	Y	Y	Y	Y	Y	N	Y	Y	Y	Y	N
	Evoxx technologies	www.evoxx.com	>20	Y	Y	Y	N	Y	Y	Y	Y	Y	Y	Y	Y	Y			Y	Y	Y	N
	Johnson Matthey	www.matthey.com/ biocatalysts/	15	Y	Y	Y	N	N	Y	Y	Y	Y	Y	N	Y	Y	Y	N	Y	Y	Y	Y
Dehydrogenase (D- and/or L-amino acid)	Almac	www.almacgroup.com	25	Y	Y	Y	N	N	Y	N	Y	Y	Y	N	Y	Y	Y	Y	Y	Y	Y	N
	ASA Spezialenzyme GmbH	www.asa-enzyme.de	2	Y	Y	N	Y	N	N	N	N	N	N	N	Y	N	Y	N	Y	N	Y	N
	EnzymeWorks	www.enzymeworking .com	40	Y	Y	Y	Y	N	Y	Y	Y	Y	N	N	Y	N	Y	N	Y	Y	Y	N
	Evoxx technologies	www.evoxx.com	>10	N	Y	Y	N	Y	Y	Y	Y	Y	Y						Y	Y	Y	N
	Johnson Matthey	www.matthey.com/ biocatalysts/	20	N	N	Y	N	N	N	N	Y	Y	Y	N	Y	Y	Y	N	Y	Y	Y	Y
	Syncozymes	en.syncozymes.com	17	Y	Y	Y	Y	N	Y	Y	Y	N	N	N	N	N	N	Y	Y	N	N	N
Dehydrogenase (1,5-anhydroglucitol-6-phosphate)	Asahi Kasei Pharma Corporation	www.asahi-kasei.co.jp/ shindan/en/	1																			
Dehydrogenase (cellobiose)	NZYTech	www.nzytech.com	1	Y	N	Y	N	N	N	N	N	N	N	N	Y	N	N	N	N	N	N	N
Dehydrogenase (cholesterol)	Amano Enzyme	www.amano-enzyme.co .jp	1																			
Dehydrogenase (formaldehyde)	Merck KGgA/ MilliporeSigma	www.sigmaaldrich.com	1	Y	N	N	N	N	N	N	N	N	N	N	N	N	N	N	N	N	N	N
	Toyobo	www.toyobo-global.com	1																			
Dehydrogenase (formate)	Codexis Inc.	www.codexis.com	1	Y	Y	Y	N	N	N	N	N	N	Y				Y	N	N	Y	Y	N
	EnzymeWorks	www.enzymeworking .com	5	Y	Y	Y	N	Y	Y	Y	Y	Y	Y	N	Y	Y	Y	N	Y	Y	Y	Y
	Evoxx technologies	www.evoxx.com	2	Y	Y	Y	N	Y	Y	Y	Y	Y	Y	N	Y	Y	Y	N	Y	Y	Y	N
	Johnson Matthey	www.matthey.com/ biocatalysts/	10	Y	Y	Y	N	N	Y	N	Y	Y	Y	N	Y	Y	Y	N	Y	Y	Y	Y

Table 2.1 (*Continued*)

Enzyme name	Supplier	Company contact details	Number of enzymes in panel	Screening quantities available (i.e. between 1 mg and 1 g of each available on immediate basis for sale)	Larger quantities available (i.e. between 1 g and 1 kg available within 1 month of request)	Attributes																
						Recombinant	Homologously expressed	Metagenomic	Engineered panel	Evolved panel	Sequence diverse members	Unique proprietary members	Nagoya compliant	Psychrophilic members	Mesophilic members	Thermophilic members	NAD+-dependent members	NAD(P)+-dependent	Animal-free	BSE-TSE-certified	GMO-free	GMP grade
	Megazyme	www.megazyme.com	1	N	Y	Y	N	N	N	N	Y	Y	N	N	Y	N	Y	N	Y	N	N	N
	Merck KGgA/ MilliporeSigma	www.sigmaaldrich.com	2	Y	N	N	N	N	N	N	N	N	N	N	N	N	N	N	N	N	N	N
	Protéus	www.proteus.fr	4	Y	N	Y	N	N	Y	Y	Y	Y	Y	N	Y	Y	Y	Y	Y	Y	Y	N
	Prozomix Ltd	www.prozomix.com	12	Y	Y	Y	N	Y	N	N	Y	Y	Y	N	Y	N	N	N	Y	Y	Y	N
	Syncozymes	en.syncozymes.com	4	Y	Y	Y	Y	N	Y	N	Y	N	N	N	N	N	Y	N	Y	N	N	N
Dehydrogenase (D-fructose)	Toyobo	www.toyobo-global.com	1																			
Dehydrogenase (galactose)	ASA Spezialenzyme GmbH	www.asa-enzyme.de	1	Y	N	Y	N	N	N	N	N	Y	N	N	Y	N	Y	N	Y	N	N	N
	Megazyme	www.megazyme.com	2	N	Y	Y	N	N	N	N	Y	Y	N	N	Y	N	Y	N	Y	N	N	N
	Merck KGgA/ MilliporeSigma	www.sigmaaldrich.com	3	Y	N	Y	N	N	N	N	N	N	N	N	N	N	N	N	N	N	N	N
Dehydrogenase (glucose)	Almac	www.almacgroup.com	2	Y	Y	Y	N	N	Y	N	Y	Y	Y	N	Y	Y	Y	Y	Y	Y	Y	N
	Amano Enzyme	www.amano-enzyme.co.jp	3																			
	c-LEcta GmbH	www.c-lecta.com	1	Y	Y	Y	N	N	Y	N	N	Y	Y	N	Y	N	Y	Y	Y	Y	Y	N

	Codexis Inc.	www.codexis.com	2	Y	Y	Y	N	N	Y	Y	N	Y	Y				Y	Y	N	Y	Y	N
	EnzymeWorks	www.enzymeworking .com	5	Y	Y	Y	N	N	Y	Y	Y	Y	Y	N	Y	Y	Y	Y	Y	Y	Y	N
	Enzymicals AG	www.enzymicals.com	2	N	Y	Y	N	N	Y	N	Y	Y	Y	N	Y	Y	Y	Y	Y	Y	Y	N
	Evoxx technologies	www.evoxx.com	2	Y	Y	Y	N	Y	Y	Y	Y	Y	Y	N	Y	Y	Y	Y	Y	Y	Y	N
	Gecco Biotech	www.gecco-biotech.com	1	Y	N	Y	N	N	N	N	N	N	Y	N	N	Y	Y	Y	Y	N	N	N
	InnoSyn BV	www.innosyn.com	2	Y	Y	Y	N	N	N	N	N	N	Y	N	Y	Y	Y	Y	Y	Y	Y	N
	Johnson Matthey	www.matthey.com/ biocatalysts/	20	Y	Y	Y	N	N	Y	N	Y	Y	Y	N	Y	Y	Y	Y	Y	Y	Y	Y
	Merck KGgA/ MilliporeSigma	www.sigmaaldrich.com	1	Y	N	N	N	N	N	N	N	N	N	N	N	N	N	N	N	N	N	N
	Protéus	www.proteus.fr	2	Y	N	Y	N	N	Y	Y	Y	Y	Y	N	Y	Y	Y	Y	Y	Y	Y	N
	Syncozymes	en.syncozymes.com	9	Y	Y	Y	Y	N	Y	Y	Y	N	N	N	N	N	Y	Y	Y	N	N	N
	Toyobo	www.toyobo-global.com	2																			
Dehydrogenase (glucose-6-phosphate)	Asahi Kasei Pharma Corporation	www.asahi-kasei.co.jp/ shindan/en/	1																			
	Megazyme	www.megazyme.com	1	N	Y	Y/ N	N	N	N	N	Y	Y	N	N	Y	N	N	Y	Y	N	N	N
	Merck KGgA/ MilliporeSigma	www.sigmaaldrich.com	8	Y	N	N	N	N	N	N	N	N	N	N	N	N	N	N	N	N	N	N
	Toyobo	www.toyobo-global.com	2																			
Dehydrogenase (glucose-6-phosphate) F-420-dependent	Gecco Biotech	www.gecco-biotech.com	5	Y	N	Y	N	N	N	N	N	N	Y	N	N	Y	N	N	Y	N	N	N
Dehydrogenase (glutamate)	Evoxx technologies	www.evoxx.com		Y	Y	Y	N	Y	Y	Y	Y	Y	Y	N	Y	Y			Y	Y	Y	N
	Merck KGgA/ MilliporeSigma	www.sigmaaldrich.com	5	Y	N	N	N	N	N	N	N	N	N	N	N	N	N	N	N	N	N	N
	Toyobo	www.toyobo-global.com	1																			
Dehydrogenase (glycerol)	Evoxx technologies	www.evoxx.com	1	N	Y	Y	N	Y	Y	Y	Y	Y	Y	N	Y	Y			Y	Y	Y	N
	Merck KGgA/ MilliporeSigma	www.sigmaaldrich.com	1	Y	N	N	N	N	N	N	N	N	N	N	N	N	N	N	N	N	N	N
	Toyobo	www.toyobo-global.com	1																			

Table 2.1 *(Continued)*

Enzyme name	Supplier	Company contact details	Number of enzymes in panel	Screening quantities available (i.e. between 1 mg and 1 g of each available on immediate basis for sale)	Larger quantities available (i.e. between 1 g and 1 kg available within 1 month of request)	Attributes																
						Recombinant	Homologously expressed	Metagenomic	Engineered panel	Evolved panel	Sequence diverse members	Unique proprietary members	Nagoya compliant	Psychrophilic members	Mesophilic members	Thermophilic members	NAD+-dependent members	NAD(P)+-dependent	Animal-free	BSE-TSE-certified	GMO-free	GMP grade
Dehydrogenase (glyceraldeyhyde 3-phosphate)	Merck KGgA/ MilliporeSigma	www.sigmaaldrich.com	4	Y	N	N	N	N	N	N	N	N	N	N	N	N	N	N	N	N	N	N
Dehydrogenase (α-glycerophosphate)	Merck KGgA/ MilliporeSigma	www.sigmaaldrich.com	3	Y	N	N	N	N	N	N	N	N	N	N	N	N	N	N	N	N	N	N
Dehydrogenase (haloalkane)	Prozomix Ltd	www.prozomix.com	13	Y	Y	Y	N	Y	N	N	Y	Y	Y	N	Y	N	N	N	Y	Y	Y	N
Dehydrogenase (D-3-hydroxybutyrate)	Almac	www.almacgroup.com	4	Y	Y	Y	N	N	N	N	Y	N	Y	N	N	N	Y	Y	Y	Y	Y	N
	Amano Enzyme	www.amano-enzyme.co .jp	1																			
	Asahi Kasei Pharma Corporation	www.asahi-kasei.co.jp/ shindan/en/	1																			
	Evoxx technologies	www.evoxx.com		Y	Y	Y	N	Y	Y	Y	Y	Y	Y	N	Y	Y			Y	Y	Y	N
	Megazyme	www.megazyme.com	1	N	Y	Y	N	N	N	N	Y	Y	N	N	Y	N	Y	N	Y	N	N	N
	Merck KGgA/ MilliporeSigma	www.sigmaaldrich.com	2	Y	N	N	N	N	N	N	N	N	N	N	N	N	N	N	N	N	N	N
	Toyobo	www.toyobo-global.com	1																			

Dehycrogenase (3-a pha-hydroxysteroid)	Almac	www.almacgroup.com	3	Y	Y	Y	N	N	N	N	Y	N	Y	N	N	N	Y	Y	Y	Y	Y	N
	Asahi Kasei Pharma Corporation	www.asahi-kasei.co.jp/shindan/en/	1	Y	N	Y	N	N	N	N	N	Y	N	N	Y	N	Y	N	Y	Y	N	N
	ASA Spezialenzyme GmbH	www.asa-enzyme.de	1																			
	Merck KGgA/ MilliporeSigma	www.sigmaaldrich.com	1	Y	N	N	N	N	N	N	N	N	N	N	N	N	N	N	N	N	N	N
Dehydrogenase (12-alpha-hydroxysteroid)	ASA Spezialenzyme GmbH	www.asa-enzyme.de	1	Y	N	N	Y	N	N	N	N	N	N	N	Y	N	N	Y	Y	Y	Y	N
	Iris Biotech GmbH	www.iris-biotech.de/de/	1																			
Dehydrogenase (inosine monophosphate)	Merck KGgA/ MilliporeSigma	www.sigmaaldrich.com	1	Y	N	Y	N	N	N	N	N	N	N	N	N	N	N	N	N	N	N	N
Dehydrogenase (myo-inositol)	Megazyme	www.megazyme.com	1	N	Y	Y	N	N	N	N	Y	Y	N	N	Y	N	Y	N	Y	N	N	N
	Merck KGgA/ MilliporeSigma	www.sigmaaldrich.com	1	Y	N	N	N	N	N	N	N	N	N	N	N	N	N	N	N	N	N	N
Dehydrogenase (isocytrate)	Evoxx technologies	www.evoxx.com	1	Y	Y	Y	N	Y	Y	Y	Y	Y	Y	N	Y	Y			Y	Y	Y	N
	Megazyme	www.megazyme.com/	1	N	Y	Y	N	N	N	N	Y	Y	N	N	Y	N	N	Y	Y	N	N	N
	Merck KGgA/ MilliporeSigma	www.sigmaaldrich.com	2	Y	N	Y	N	N	N	N	N	N	N	N	N	N	N	N	N	N	N	N
Dehydrogenase (α-ketoglutarate)	Merck KGgA/ MilliporeSigma	www.sigmaaldrich.com	1	Y	N	Y	N	N	N	N	N	N	N	N	N	N	N	N	N	N	N	N
Dehydrogenase (lactate)	Amano Enzyme	www.amano-enzyme.co.jp	1																			
	Asahi Kasei Pharma Corporation	www.asahi-kasei.co.jp/shindan/en/	1																			
	Codexis Inc.	www.codexis.com	1	Y	Y	Y	N	N	N	N	N	N	Y				N	Y	N	Y	Y	N
	EnzymeWorks	www.enzymeworking.com	5	Y	Y	Y	N	Y	Y	Y	Y	Y	Y	N	Y	Y	Y	Y	Y	Y	Y	N
	Evoxx technologies	www.evoxx.com	1	Y	Y	Y	N	Y	Y	Y	Y	Y	Y	N	Y	Y			Y	Y	Y	N
	Johnson Matthey	www.matthey.com/biocatalysts/	20	Y	Y	Y	N	N	N	N	Y	Y	Y	N	Y	Y	Y	N	Y	Y	Y	Y

Table 2.1 (*Continued*)

Enzyme name	Supplier	Company contact details	Number of enzymes in panel	Screening quantities available (i.e. between 1 mg and 1 g of each available on immediate basis for sale)	Larger quantities available (i.e. between 1 g and 1 kg available within 1 month of request)	Attributes: Recombinant	Homologously expressed	Metagenomic	Engineered panel	Evolved panel	Sequence diverse members	Unique proprietary members	Nagoya compliant	Psychrophilic members	Mesophilic members	Thermophilic members	NAD+-dependent members	NAD(P)+-dependent	Animal-free	BSE-TSE-certified	GMO-free	GMP grade
	Megazyme	www.megazyme.com	2	N	Y	Y	N	N	N	N	Y	Y	N	N	Y	N	Y	N	N	N	N	N
	Merck KGgA/ MilliporeSigma	www.sigmaaldrich.com	7	Y	N	N	N	N	N	N	N	N	N	N	N	N	N	N	N	N	N	N
	Syncozymes	en.syncozymes.com	1	Y	Y	Y	Y	N	Y	N	N	N	N	N	N	N	Y	N	Y	N	N	N
	Toyobo	www.toyobo-global.com	3																			
Dehydrogenase (D-lactate)	Merck KGgA/ MilliporeSigma	www.sigmaaldrich.com	4	Y	N	N	N	N	N	N	N	N	N	N	N	N	N	N	N	N	N	N
Dehydrogenase (leucine)	Codexis Inc.	www.codexis.com	1 (CDX-012)	Y	Y	Y	N	N	N	N	N	Y	Y				N	Y	N	Y	Y	N
	EnzymeWorks	www.enzymeworking.com	5	Y	Y	Y	Y	Y	Y	Y	Y	Y	Y	N	Y	Y	Y	Y	Y	Y	Y	N
	Evoxx technologies	www.evoxx.com		Y	Y	Y	N	Y	Y	Y	Y	Y	Y	N	Y	Y			Y	Y	Y	N
	Merck KGgA/ MilliporeSigma	www.sigmaaldrich.com	1	Y	N	N	N	N	N	N	N	N	N	N	N	N	N	N	N	N	N	N
	Syncozymes	en.syncozymes.com	1	Y	Y	Y	Y	N	Y	N	N	N	N	N	N	N	Y	N	Y	N	N	N
	Toyobo	www.toyobo-global.com	1																			
Dehydrogenase (L-malate)	Amano Enzyme	www.amano-enzyme.co.jp	1																			
	Evoxx technologies	www.evoxx.com	1	Y	Y	Y	N	Y	Y	Y	Y	Y	Y	N	Y	Y			Y	Y	Y	N

	Megazyme	www.megazyme.com	1	N	Y	Y	N	N	N	N	Y	Y	N	N	Y	N	Y	N	Y	N	N	N
	Merck KGgA/ MilliporeSigma	www.sigmaaldrich.com	5	Y	N	N	N	N	N	N	N	N	N	N	N	N	N	N	N	N	N	N
	Prozomix Ltd	www.prozomix.com	1																			
	Toyobo	www.toyobo-global.com	1																			
Dehydrogenase (mannitol)	Biocatalysts Ltd	www.biocatalysts.com	1	Y	Y	N	Y	N	N	N	N	Y	Y	N	Y	N	N	Y	Y	Y	Y	N
	Megazyme	www.megazyme.com	1	N	Y	Y	N	N	N	N	Y	Y	N	N	Y	N	Y	N	Y	N	N	N
	Merck KGgA/ MilliporeSigma	www.sigmaaldrich.com	1	Y	N	N	N	N	N	N	N	N	N	N	N	N	N	N	N	N	N	N
Dehydrogenase (phenylalanine)	Almac	www.almacgroup.com	5	Y	Y	Y	N	N	N	N	Y	Y	Y	N	Y	Y	Y	N	Y	Y	Y	N
	Biocatalysts Ltd	www.biocatalysts.com	2	Y	Y	N	Y	N	N	N	Y	Y	Y	N	Y	N	N	Y	Y	Y	Y	N
	EnzymeWorks	www.enzymeworking .com	5	Y	Y	Y	Y	Y	Y	Y	Y	Y	Y	N	Y	Y	Y	N	Y	Y	Y	NY
	Evoxx technologies	www.evoxx.com	1	Y	Y	Y	N	Y	Y	Y	Y	Y	Y	N	Y	Y			Y	Y	Y	N
	Merck KGgA/ MilliporeSigma	www.sigmaaldrich.com	1	Y	N	N	N	N	N	N	N	N	N	N	N	N	N	N	N	N	N	N
	Syncozymes	en.syncozymes.com	8	Y	Y	Y	Y	N	Y	N	N	N	N	N	N	N	Y	N	Y	N	N	N
Dehydrogenase (phosphite/NAD(P)H)	Gecco Biotech	www.gecco-biotech.com	1	Y	N	Y	N	N	N	N	N	N	Y	N	N	Y	Y	Y	Y	N	N	N
Dehydrogenase (6-phosphogluconate)	Megazyme	www.megazyme.com	1	N	Y	Y	N	N	N	N	Y	Y	N	N	Y	N	N	Y	Y	N	N	N
	Merck KGgA/ MilliporeSigma	www.sigmaaldrich.com	1	Y	N	N	N	N	N	N	N	N	N	N	N	N	N	N	N	N	N	N
Dehydrogenase (threonine)	EnzymeWorks	www.enzymeworking .com	2	Y	Y	Y	Y	N	Y	Y	Y	Y	N	N	N	N	Y	N	Y	Y	Y	N
	Syncozymes	en.syncozymes.com	1	Y	Y	Y	Y	N	Y	N	N	N	N	N	N	N	Y	N	Y	N	N	N
Dehydrogenase (xanthine)	Asahi Kasei Pharma Corporation	www.asahi-kasei.co.jp/ shindan/en/	1																			
Dehydrogenase (xylose)	Megazyme	www.megazyme.com	1	N	Y	Y	N	N	N	N	Y	Y	N	N	Y	N	Y	N	Y	N	N	N

Table 2.1 *(Continued)*

Enzyme name	Supplier	Company contact details	Number of enzymes in panel	Screening quantities available (i.e. between 1 mg and 1 g of each available on immediate basis for sale)	Larger quantities available (i.e. between 1 g and 1 kg available within 1 month of request)	Attributes: Recombinant	Homologously expressed	Metagenomic	Engineered panel	Evolved panel	Sequence diverse members	Unique proprietary members	Nagoya compliant	Psychrophilic members	Mesophilic members	Thermophilic members	NAD+-dependent members	NAD(P)+-dependent	Animal-free	BSE-TSE-certified	GMO-free	GMP grade
Dextranase	ASA Spezialenzyme GmbH	www.asa-enzyme.de	1	Y	Y	N	Y	N	N	N	N	N	N	N	Y	N	N	N	Y	Y	Y	N
	Megazyme	www.megazyme.com	1	N	Y	N	N	N	N	N	Y	Y	N	N	Y	N	N	N	Y	N	N	N
	Merck KGgA/ MilliporeSigma	www.sigmaaldrich.com	4	Y	N	N	N	N	N	N	N	N	N	N	N	N	N	N	N	N	N	N
	NZYTech	www.nzytech.com	5	Y	N	Y	N	N	N	N	N	N	N	N	Y	Y	N	N	N	N	N	N
Dextran sucrase	Merck KGgA/ MilliporeSigma	www.sigmaaldrich.com	1	Y	N	N	N	N	N	N	N	N	N	N	N	N	N	N	N	N	N	N
Diaphorase	Megazyme	www.megazyme.com	1	N	Y	Y	N	N	N	N	Y	Y	N	N	Y	N	N	N	Y	N	N	N
	Merck KGgA/ MilliporeSigma	www.sigmaaldrich.com	1	Y	Y	Y																
	Toyobo	www.toyobo-global.com	2																			
Diaphorase (NADH)	Asahi Kasei Pharma Corporation	www.asahi-kasei.co.jp/ shindan/en/	1																			
Diaphorase (NADPH)	Asahi Kasei Pharma Corporation	www.asahi-kasei.co.jp/ shindan/en/	1																			
Diastase	Amano Enzyme	www.amano-enzyme.co .jp	1																			

Dihydrofolate reductase	Merck KGgA/ MilliporeSigma	www.sigmaaldrich.com	2	Y	N	Y	N	N	N	N	N	N	N	N	N	N	N	N	N	N	N	N
Dioxygenase (biphenyl)	Almac	www.almacgroup.com	4	Y	Y	Y	N	N	Y	N	Y	Y	Y	N	Y	Y	Y	N	Y	Y	Y	N
Dioxygenase (4-hydroxyphenyl-pyruvate)	Merck KGgA/ MilliporeSigma	www.sigmaaldrich.com	1	Y	N	Y	N	N	N	N	N	N	N	N	N	N	N	N	N	N	N	N
Dioxygenase (protocatechuate)	Merck KGgA/ MilliporeSigma	www.sigmaaldrich.com	1	Y	N	N	N	N	N	N	N	N	N	N	N	N	N	N	N	N	N	N
	Toyobo	www.toyobo-global.com	1																			
DNA gyrase	Merck KGgA/ MilliporeSigma	www.sigmaaldrich.com	1	Y	N	N	N	N	N	N	N	N	N	N	N	N	N	N	N	N	N	N
DNA ligase	Merck KGgA/ MilliporeSigma	www.sigmaaldrich.com	1	Y	N	N	N	N	N	N	N	N	N	N	N	N	N	N	N	N	N	N
Endoglycoceramidase	Merck KGgA/ MilliporeSigma	www.sigmaaldrich.com	1	Y	N	N	N	N	N	N	N	N	N	N	N	N	N	N	N	N	N	N
Endoglycosidase	Merck KGgA/ MilliporeSigma	www.sigmaaldrich.com	6	Y	N	N	N	N	N	N	N	N	N	N	N	N	N	N	N	N	N	N
Ene reductase	Almac	www.almacgroup.com	>200	Y	Y	Y	N	Y	Y	Y	Y	Y	Y	N	Y	Y	Y	Y	Y	Y	Y	N
	Codexis Inc.	www.codexis.com	7 (kit) + custom items	Y from kit	Up to 100 g from kit	Y	N	N	Y	Y	Y	Y	Y				N	Y	N	Y	Y	N
	EnzymeWorks	www.enzymeworking.com	15	Y	Y	Y	N	Y	Y	Y	Y	Y	Y	N	Y	Y	Y	Y	Y	Y	Y	N
	Enzymicals AG	www.enzymicals.com	3	N	Y	Y	N	N	N	N	Y	Y	Y	N	Y	Y	N	N	Y	Y	Y	N
	Gecco Biotech	www.gecco-biotech.com	2	Y	N	Y	N	N	N	N	N	N	Y	N	Y	N	Y	Y	Y	N	N	N
	Johnson Matthey	www.matthey.com/biocatalysts/	110	Y	Y	Y	N	Y	Y	N	Y	Y	Y	N	Y	Y	Y	Y	Y	Y	Y	Y
	Protéus	www.proteus.fr	6	Y	N	Y	N	N	Y	Y	Y	Y	Y	N	Y	Y	N	N	Y	Y	Y	N
	Prozomix Ltd	www.prozomix.com	144	Y	Y	Y	N	Y	N	N	Y	Y	Y	N	Y	N	Y	Y	Y	Y	Y	N
	Syncozymes	en.syncozymes.com	29	Y	Y	Y	Y	N	Y	N	N	N	N	N	N	N	Y	Y	Y	N	N	N

Table 2.1 *(Continued)*

Enzyme name	Supplier	Company contact details	Number of enzymes in panel	Screening quantities available (i.e. between 1 mg and 1 g of each available on immediate basis for sale)	Larger quantities available (i.e. between 1 g and 1 kg available within 1 month of request)	Attributes: Recombinant	Homologously expressed	Metagenomic	Engineered panel	Evolved panel	Sequence diverse members	Unique proprietary members	Nagoya compliant	Psychrophilic members	Mesophilic members	Thermophilic members	NAD+-dependent members	NAD(P)+-dependent	Animal-free	BSE-TSE-certified	GMO-free	GMP grade
Enolase	Merck KGgA/ MilliporeSigma	www.sigmaaldrich.com	2	Y	N	N	N	N	N	N	N	N	N	N	N	N	N	N	N	N	N	N
Enterokinase	Merck KGgA/ MilliporeSigma	www.sigmaaldrich.com	5	Y	N	N	N	N	N	N	N	N	N	N	N	N	N	N	N	N	N	N
Epimerase	Almac	www.almacgroup.com	12	Y	Y	Y	N	N	N	N	Y	Y	Y	N	Y	Y	N	N	Y	Y	Y	N
Epoxide hydrolase	Almac	www.almacgroup.com	12	Y	Y	Y	N	N	Y	Y	Y	Y	Y	N	Y	Y	N	N	Y	Y	Y	N
	Codexis Inc.	www.codexis.com	10 custom items	Custom items	>1 month	Y	N	N	N	N	N	N	Y				-	-	N	Y	Y	N
	EnzymeWorks	www.enzymeworking .com	20	Y	Y	Y	N	N	Y	Y	Y	Y	Y	N	Y	Y	N	N	Y	Y	Y	N
	Enzymicals AG	www.enzymicals.com	1	N	Y	Y	N	N	N	N	Y	Y	Y	N	Y	Y	N	N	Y	Y	Y	N
	Gecco Biotech	www.gecco-biotech.com	2	Y	N	Y	N	N	N	N	N	N	N	N	Y	Y	N	N	Y	N	N	N
	Protéus	www.proteus.fr	4	Y	N	Y	N	N	Y	Y	Y	Y	Y	N	Y	Y	N	N	Y	Y	Y	N
	Prozomix Ltd	www.prozomix.com	96	Y	Y	Y	N	N	N	N	Y	Y	Y	N	Y	N	N	N	Y	Y	Y	N
	Syncozymes	en.syncozymes.com	2	Y	Y	Y	Y	N	Y	N	N	N	N	N	N	N	N	N	Y	N	N	N

Esterase (acetylcholine)	Merck KGgA/ MilliporeSigma	www.sigmaaldrich.com	4	Y	N	Y	N	N	N	N	N	N	N	N	N	N	N	N	N	N	N	N
Esterase (acetylxylan)	Megazyme	www.megazyme.com	1	N	Y	Y	N	N	N	N	Y	Y	N	N	Y	N	N	N	N	N	N	N
Esterase (butyrylcholine)	Merck KGgA/ MilliporeSigma	www.sigmaaldrich.com	4	Y	N	N	N	N	N	N	N	N	N	N	N	N	N	N	N	N	N	N
Esterase (carbohydrate)	NZYTech	www.nzytech.com	131	Y	N	Y	N	N	N	N	N	N	N	N	Y	Y	N	N	N	N	N	N
Esterase (carboxyl)	Almac	www.almacgroup.com	96	Y	Y	Y	N	N	N	N	Y	Y	Y	N	Y	Y	N	N	Y	Y	Y	N
	Amano Enzyme	www.amano-enzyme.co .jp	1																			
	ASA Spezialenzyme GmbH	www.asa-enzyme.de	3	Y	Y	Y/ N	Y/ N	N	N	N	N	Y	N	N	Y	N	N	N	Y	Y	Y/ N	N
	Biocatalysts Ltd	www.biocatalysts.com	2																			
	c-LEcta GmbH	www.c-lecta.com	63	Y	Y	Y	N	Y	Y	N	Y	Y	Y	N	Y	N	N	N	Y	Y	Y	N
	Codexis Inc.	www.codexis.com	85	Custom items	>1 month	Y	N	N	Y	Y	Y	Y	Y				-	-	N	Y	Y	N
	EnzymeWorks	www.enzymeworking .com	15	Y	Y	Y	N	N	Y	Y	Y	Y	Y	N	Y	Y	N	N	Y	Y	Y	N
	Enzymicals AG	www.enzymicals.com	14	Y	Y	Y	N	N	Y	N	Y	Y	Y	N	Y	Y	N	N	Y	Y	Y	N
	EUCODIS	www.eucodis.com	25	Y	Y	Y	N	N	Y	N	Y	Y	Y	Y	Y	Y	N	N	Y	Y	Y	N
	Evoxx technologies	www.evoxx.com	>20	N	Y	Y	N	Y	Y	Y	Y	Y	Y	N	Y	Y			Y	Y	Y	N
	Johnson Matthey	www.matthey.com/ biocatalysts/	80	Y	Y	Y	N	N	N	Y	Y	Y	Y	N	Y	Y	Y	Y	Y	Y	Y	Y
	Merck KGgA/ MilliporeSigma	www.sigmaaldrich.com	17	Y	N	N	N	N	N	N	N	N	N	N	N	N	N	N	N	N	N	N
	Protéus	www.proteus.fr	2	Y	N	Y	N	N	Y	Y	Y	Y	Y	N	Y	Y	N	N	Y	Y	Y	N
	Prozomix Ltd	www.prozomix.com	22	Y	Y	Y	N	N	N	N	Y	Y	Y	N	Y	N	N	N	Y	Y	Y	N
	Synccozymes	en.syncozymes.com	14	Y	Y	Y	Y	N	Y	Y	N	N	N	N	N	N	N	N	Y	N	N	N
Esterase (cholesterol)	Amano Enzyme	www.amano-enzyme.co .jp	1																			
	Asahi Kasei Pharma Corporation	www.asahi-kasei.co.jp/ shindan/en/	2																			

Table 2.1 *(Continued)*

Enzyme name	Supplier	Company contact details	Number of enzymes in panel	Screening quantities available (i.e. between 1 mg and 1 g of each available on immediate basis for sale)	Larger quantities available (i.e. between 1 g and 1 kg available within 1 month of request)	Attributes: Recombinant	Homologously expressed	Metagenomic	Engineered panel	Evolved panel	Sequence diverse members	Unique proprietary members	Nagoya compliant	Psychrophilic members	Mesophilic members	Thermophilic members	NAD+-dependent members	NAD(P)+-dependent	Animal-free	BSE-TSE-certified	GMO-free	GMP grade
	EUCODIS	www.eucodis.com	1	Y	Y	Y	N	N	N	N	Y	Y	Y	N	Y	Y	N	N	Y	Y	Y	N
	Merck KGgA/ MilliporeSigma	www.sigmaaldrich.com	5	Y	N	N	N	N	N	N	N	N	N	N	N	N	N	N	N	N	N	N
	Toyobo	www.toyobo-global.com	4																			
Esterase (feruloyl)	Megazyme	www.megazyme.com	2	N	Y	Y	N	N	N	N	Y	Y	N	N	Y	N	N	N	Y	N	N	N
Esterase (PLE)	Enzymicals AG	www.enzymicals.com	6	Y	Y	Y	N	N	Y	N	Y	Y	Y	N	Y	Y	N	N	Y	Y	Y	N
	InnoSyn BV	www.innosyn.com	7	Y	Y	Y	N	N	Y	N	N	Y	Y	N	Y	N	N	N	Y	Y	Y	N
Ficin	Merck KGgA/ MilliporeSigma	www.sigmaaldrich.com	3	Y	N	N	N	N	N	N	N	N	N	N	N	N	N	N	N	N	N	N
Flavin-reductase	GECCO Biotech	www.gecco-biotech.com	2	Y	N	Y	N	N	N	N	N	N	N	N	Y	Y	Y	Y	Y	N	N	N
Fructanase/Fructosidase	Megazyme	www.megazyme.com	8	N	Y	Y/N	N	N	N	N	Y	Y	N	N	Y	N	N	N	Y	N	N	N
	NZYTech	www.nzytech.com	3	Y	N	Y	N	N	N	N	N	N	N	N	Y	N	N	N	N	N	N	N
beta-Fructofuranosidase (Invertase)	ASA Spezialenzyme GmbH	www.asa-enzyme.de	1	Y	Y	N	Y	N	N	N	N	N	N	N	Y	N	N	N	Y	Y	Y	N
	Merck KGgA/ MilliporeSigma	www.sigmaaldrich.com	1	Y	N	N	N	N	N	N	N	N	N	N	N	N	N	N	N	N	N	N
	NZYTech	www.nzytech.com	4	Y	N	Y	N	N	N	N	N	N	N	N	Y	N	N	N	N	N	N	N
	Toyobo	www.toyobo-global.com	1																			

Fucosidase	Megazyme	www.megazyme.com	3	N	Y	Y	N	N	N	N	Y	Y	N	N	Y	Y	N	N	Y	N	N	N
	Merck KGgA/ MilliporeSigma	www.sigmaaldrich.com	1	Y	N	N	N	N	N	N	N	N	N	N	N	N	N	N	N	N	N	N
	NZYTech	www.nzytech.com	8	Y	N	Y	N	N	N	N	N	N	N	N	Y	Y	N	N	N	N	N	N
Fumarase	Merck KGgA/ MilliporeSigma	www.sigmaaldrich.com	2	Y	N	N	N	N	N	N	N	N	N	N	N	N	N	N	N	N	N	N
Galactanase	Megazyme	www.megazyme.com	3	N	Y	Y/ N	N	N	N	N	Y	Y	N	N	Y	N	N	N	Y	N	N	N
	NZYTech	www.nzytech.com	6	Y	N	Y	N	N	N	N	N	N	N	N	Y	Y	N	N	N	N	N	N
	Prozomix Ltd	www.prozomix.com	4	Y	Y	Y	N	N	N	N	Y	Y	Y	N	Y	N	N	N	Y	Y	Y	N
Galactose-1-phosphate uridyltransferase	Merck KGgA/ MilliporeSigma	www.sigmaaldrich.com	1	Y	N	N	N	N	N	N	N	N	N	N	N	N	N	N	N	N	N	N
Galactan 1,3-beta-galactosidase	Prozomix Ltd	www.prozomix.com	1	Y	Y	Y	N	N	N	N	Y	Y	Y	N	Y	N	N	N	Y	Y	Y	N
alpha-Galactosidase	Amano Enzyme	www.amano-enzyme.co .jp	1																			
	EnzymeWorks	www.enzymeworking .com	4	Y	Y	Y	N	N	Y	Y	Y	Y	N	N	Y	N	N	N	Y	Y	Y	N
	Megazyme	www.megazyme.com	4	N	Y	Y/ N	N	N	N	N	Y	Y	N	N	Y	N	N	N	Y	N	N	N
	Merck KGgA/ MilliporeSigma	www.sigmaaldrich.com	3	Y	N	Y	N	N	N	N	N	N	N	N	N	N	N	N	N	N	N	N
	NZYTech	www.nzytech.com	13	Y	N	Y	N	N	N	N	N	N	N	N	Y	N	N	N	N	N	N	N
	Prozomix Ltd	www.prozomix.com	2	Y	Y	Y	N	N	N	N	Y	Y	Y	N	Y	N	N	N	Y	Y	Y	N
beta-Galactosidase	ASA Spezialenzyme GmbH	www.asa-enzyme.de	2	Y	Y	N	Y	N	N	N	N	N	N	N	Y	N	N	N	Y	Y	Y	N
	Codexis Inc.	www.codexis.com	Custom items 12	Custom items	>1 month	Y	N	N	N	N	N	N	Y				-	-	N	Y	Y	N
	EnzymeWorks	www.enzymeworking .com	5	Y	Y	Y	N	N	Y	Y	Y	Y	N	N	N	N	N	N	Y	Y	Y	N

Table 2.1 *(Continued)*

Enzyme name	Supplier	Company contact details	Number of enzymes in panel	Screening quantities available (i.e. between 1 mg and 1 g of each available on immediate basis for sale)	Larger quantities available (i.e. between 1 g and 1 kg available within 1 month of request)	Attributes: Recombinant	Homologously expressed	Metagenomic	Engineered panel	Evolved panel	Sequence diverse members	Unique proprietary members	Nagoya compliant	Psychrophilic members	Mesophilic members	Thermophilic members	NAD+-dependent members	NAD(P)+-dependent	Animal-free	BSE-TSE-certified	GMO-free	GMP grade
	Megazyme	www.megazyme.com	1	Y	N	Y/N	N	N	N	N	Y	Y	N	N	Y	N	N	N	Y	N	N	N
	Merck KGgA/ MilliporeSigma	www.sigmaaldrich.com	9	Y	N	N	N	N	N	N	N	N	N	N	N	N	N	N	N	N	N	N
	NZYTech	www.nzytech.com	23	Y	N	Y	N	N	N	N	N	N	N	N	Y	Y	N	N	N	N	N	N
	Prozomix Ltd	www.prozomix.com	4	Y	Y	Y	N	N	N	N	Y	Y	Y	N	Y	N	N	N	Y	Y	Y	N
	Toyobo	www.toyobo-global.com	1																			
Galactosyltransferase	Merck KGgA/ MilliporeSigma	www.sigmaaldrich.com	1	Y	N	N	N	N	N	N	N	N	N	N	N	N	N	N	N	N	N	N
Galacturonidase	NZYTech	www.nzytech.com	7	Y	N	Y	N	N	N	N	N	N	N	N	Y	N	N	N	N	N	N	N
Glucanase	Biocatalysts Ltd	www.biocatalysts.com	1	Y	Y	N	N	N	N	N	N	N	Y	N	Y	N	N	N	Y	Y	Y	N
	Megazyme	www.megazyme.com	8	N	Y	Y/N	N	N	N	N	Y	Y	N	N	Y	N	N	N	Y	N	N	N
	Merck KGgA/ MilliporeSigma	www.sigmaaldrich.com	3	Y	N	N	N	N	N	N	N	N	N	N	N	N	N	N	N	N	N	N
	Prozomix Ltd	www.prozomix.com	1	Y	Y	Y	N	N	N	N	Y	Y	Y	N	Y	N	N	N	Y	Y	Y	N

alpha-Glucosidase	ASA Spezialenzyme GmbH	www.asa-enzyme.de	1	Y	Y	N	Y	N	N	N	N	N	N	N	Y	N	N	N	Y	Y	Y	N
	Megazyme	www.megazyme.com	5	Y	N	Y/N	N	N	N	N	Y	Y	N	N	Y	N	N	N	Y	N	N	N
	Merck KGgA/ MilliporeSigma	www.sigmaaldrich.com	5	Y	N	N	N	N	N	N	N	N	N	N	N	N	N	N	N	N	N	N
	NZYTech	www.nzytech.com	13	Y	N	Y	N	N	N	N	N	N	N	N	Y	Y	N	N	N	N	N	N
	Prozomix Ltd	www.prozomix.com	3	Y	Y	Y	N	N	N	N	Y	Y	Y	N	Y	N	N	N	Y	Y	Y	N
	Toyobo	www.toyobo-global.com	1																			
beta-Glucosidase	Biocatalysts Ltd	www.biocatalysts.com	1	Y	Y	N	N	N	N	N	N	Y	Y	N	Y	N	N	N	Y	Y	Y	N
	EnzymeWorks	www.enzymeworking .com	3	Y	Y	Y	N	N	Y	Y	Y	Y	Y	N	Y	Y	N	N	Y	Y	Y	N
	Megazyme	www.megazyme.com	4	N	Y	Y/N	N	N	N	N	Y	Y	N	N	Y	Y	N	N	Y	N	N	N
	Merck KGgA/ MilliporeSigma	www.sigmaaldrich.com	7	Y	N	N	N	N	N	N	N	N	N	N	N	N	N	N	N	N	N	N
	NZYTech	www.nzytech.com	25	Y	N	Y	N	N	N	N	N	N	N	N	Y	Y	N	N	N	N	N	N
	Proteus	www.proteus.fr	8	Y	N	Y	N	N	Y	Y	Y	Y	Y	N	Y	Y	N	N	Y	Y	Y	N
	Prozomix Ltd	www.prozomix.com	80	Y	Y	Y	N	Y/N	N	N	Y	Y	Y	N	Y	N	N	N	Y	Y	Y	N
	Toyobo	www.toyobo-global.com	1																			
alpha-Glucuronidase	Megazyme	www.megazyme.com	1	N	Y	Y	N	N	N	N	Y	Y	N	N	Y	N	N	N	Y	N	N	N
	Prozomix Ltd	www.prozomix.com	2	Y	Y	Y	N	N	N	N	Y	Y	Y	N	Y	N	N	N	Y	Y	Y	N
beta-Glucuronidase	Megazyme	www.megazyme.com	1	N	Y	Y	N	N	N	N	Y	Y	N	N	Y	N	N	N	Y	N	N	N
	Merck KGgA/ MilliporeSigma	www.sigmaaldrich.com	16	Y	Y	Y	N	N	N	N	N	N	N	N	N	N	N	N	N	N	N	N
Glucoamylase	Merck KGgA/ MilliporeSigma	www.sigmaaldrich.com	6	Y	Y	N	N	N	N	N	N	N	N	N	N	N	N	N	N	N	N	N
	Toyobo	www.toyobo-global.com	1																			

Table 2.1 *(Continued)*

Enzyme name	Supplier	Company contact details	Number of enzymes in panel	Screening quantities available (i.e. between 1 mg and 1 g of each available on immediate basis for sale)	Larger quantities available (i.e. between 1 g and 1 kg available within 1 month of request)	Attributes																
						Recombinant	Homologously expressed	Metagenomic	Engineered panel	Evolved panel	Sequence diverse members	Unique proprietary members	Nagoya compliant	Psychrophilic members	Mesophilic members	Thermophilic members	NAD+-dependent members	NAD(P)+-dependent	Animal-free	BSE-TSE-certified	GMO-free	GMP grade
Glutaminase	Amano Enzyme	www.amano-enzyme.co.jp	1																			
	Merck KGgA/ MilliporeSigma	www.sigmaaldrich.com	1	Y	N	N	N	N	N	N	N	N	N	N	N	N	N	N	N	N	N	N
gamma-Glutamyl-hydrolase	ASA Spezialenzyme GmbH	www.asa-enzyme.de	1	Y	Y	N	N	N	N	N	N	N	N	N	Y	N	N	N	N	N	N	N
	Merck KGgA/ MilliporeSigma	www.sigmaaldrich.com	1	Y	N	N	N	N	N	N	N	N	N	N	N	N	N	N	N	N	N	N
gamma-Glutamyltransferase	Asahi Kasei Pharma Corporation	www.asahi-kasei.co.jp/ shindan/en/	1																			
	Biocatalysts Ltd	www.biocatalysts.com	1	Y	Y	N	N	N	N	N	N	N	Y	N	Y	N	N	N	Y	Y	Y	N
	Merck KGgA/ MilliporeSigma	www.sigmaaldrich.com	2	Y	N	N	N	N	N	N	N	N	N	N	N	N	N	N	N	N	N	N
Glutathione peroxidase	Merck KGgA/ MilliporeSigma	www.sigmaaldrich.com	1	Y	N	N	N	N	N	N	N	N	N	N	N	N	N	N	N	N	N	N
Glutathione reductase	Merck KGgA/ MilliporeSigma	www.sigmaaldrich.com	2	Y	N	Y	N	N	N	N	N	N	N	N	N	N	N	N	N	N	N	N
Glutathione S-transferase	Merck KGgA/ MilliporeSigma	www.sigmaaldrich.com	2	Y	N	N	N	N	N	N	N	N	N	N	N	N	N	N	N	N	N	N

Glycogen branching enzyme	Prozomix Ltd	www.prozomix.com	3	Y	Y	Y	N	N	N	N	Y	Y	Y	N	Y	N	N	N	Y	Y	Y	N
Glyoxalase	Merck KGgA/ MilliporeSigma	www.sigmaaldrich.com	1	Y	N	N	N	N	N	N	N	N	N	N	N	N	N	N	N	N	N	N
Haloalkane dehalogenase	Enantis s.r.o.	www.enantis.com	20	Y	N	Y	Y	Y	Y	Y	N	Y	Y	Y	Y	Y	N	N	Y	N	N	N
	EnzymeWorks	www.enzymeworking .com	5	Y	Y	Y	N	N	Y	Y	Y	Y	Y	N	Y	y	N	N	Y	Y	Y	N
	Gecco Biotech	www.gecco-biotech.com	5	Y	N	Y	N	N	N	N	N	N	Y	N	Y	Y	N	N	Y	N	N	N
Halogenase	Almac	www.almacgroup.com	30	Y	Y	Y	N	N	Y	Y	Y	Y	Y	N	Y	Y	N	N	Y	Y	Y	N
	Prozomix Ltd (CoEBio3)	www.prozomix.com, www.coebio3. manchester.ac.uk	51	Y	Y	Y	N									N	N	Y	Y	Y	Y	N
Halogenase (flavin-dependent)	GECCO Biotech	www.gecco-biotech.com	2	Y	N	Y	N	N	N	N	N	N	N	N	Y	Y	N	N	Y	N	N	N
Halohydrin dehalogenase	Almac	www.almacgroup.com	25	Y	Y	Y	N	N	N	N	Y	Y	N	N	Y	Y	N	N	Y	Y	Y	N
	Codexis Inc.	www.codexis.com	90 (kit)	Custom items	>1 month	Y	N	N	Y	Y	Y	Y	Y				-	-	N	Y	Y	N
	EnzymeWorks	www.enzymeworking .com	10	Y	Y	Y	N	N	Y	Y	Y	Y	Y	N	Y	Y	N	N	Y	Y	Y	N
	Enzymicals AG	www.enzymicals.com	30	Y	Y	Y	N	N	N	N	Y	Y	Y	N	Y	Y	N	N	Y	Y	Y	N
	GECCO Biotech	www.gecco-biotech.com	2	Y	N	Y	N	N	N	N	N	N	N	N	Y	Y	N	N	Y	N	N	N
Haloperoxidase (vanadium)	Enzymicals AG	www.enzymicals.com	2	N	Y	Y	N	N	N	N	Y	Y	Y	N	Y	Y	N	N	Y	Y	Y	N
Heparinase	Merck KGgA/ MilliporeSigma	www.sigmaaldrich.com	3	Y	N	N	N	N	N	N	N	N	N	N	N	N	N	N	N	N	N	N
Hexosaminidase	Megazyme	www.megazyme.com	4	N	Y	Y	N	N	N	N	Y	Y	N	N	Y	N	N	N	Y	N	N	N
	NZYTech	www.nzytech.com	15	Y	N	Y	N	N	N	N	N	N	N	N	Y	N	N	N	N	N	N	N
Hyaluronate lyase	Prozomix Ltd	www.prozomix.com	5	Y	Y	Y	N	N	N	N	Y	Y	Y	N	Y	N	N	N	Y	Y	Y	N
Hyaluronicase	Amano Enzyme	www.amano-enzyme.co .jp	1																			

Table 2.1 *(Continued)*

Enzyme name	Supplier	Company contact details	Number of enzymes in panel	Screening quantities available (i.e. between 1 mg and 1 g of each available on immediate basis for sale)	Larger quantities available (i.e. between 1 g and 1 kg available within 1 month of request)	Attributes																
						Recombinant	Homologously expressed	Metagenomic	Engineered panel	Evolved panel	Sequence diverse members	Unique proprietary members	Nagoya compliant	Psychrophilic members	Mesophilic members	Thermophilic members	NAD+-dependent members	NAD(P)+-dependent	Animal-free	BSE-TSE-certified	GMO-free	GMP grade
	Merck KGgA/ MilliporeSigma	www.sigmaaldrich.com	10	Y	N	N	N	N	N	N	N	N	N	N	N	N	N	N	N	N	N	N
	NZYTech	www.nzytech.com	2	Y	N	Y	N	N	N	N	N	N	N	N	Y	N	N	N	N	N	N	N
Hydantoinase	Almac	www.almacgroup.com	25	Y	Y	Y	N	N	Y	N	Y	Y	Y	N	Y	Y	N	N	Y	Y	Y	N
	EnzymeWorks	www.enzymeworking .com	4	Y	Y	Y	N	N	Y	Y	Y	Y	Y	N	Y	Y	N	N	Y	Y	Y	N
	Enzymicals AG	www.enzymicals.com	1	N	Y	Y	N	N	N	N	Y	Y	Y	N	Y	Y	N	N	Y	Y	Y	N
Hydratase	InnoSyn BV	www.innosyn.com	37	Y	Y	Y	N	Y/ N	Y/ N	N	Y	Y	Y	N	Y	N	N	N	Y	Y	Y	N
	Protéus	www.proteus.fr	5	Y	N	Y	N	N	Y	Y	Y	Y	Y	N	Y	Y	N	N	Y	Y	Y	N
Hydroxylase (p-hydroxybenzoate)	Gecco Biotech	www.gecco-biotech.com	1	Y	N	Y	N	N	N	N	N	N	Y	N	Y	N	N	Y	Y	N	N	N
	Toyobo	www.toyobo-global.com	1																			
Hydroxylase (phenol)	Gecco Biotech	www.gecco-biotech.com	1	Y	N	Y	N	N	N	N	N	N	Y	N	Y	N	N	Y	Y	N	N	N
Hydroxylase (proline)	Almac	www.almacgroup.com	1	Y	Y	Y	N	N	N	N	Y	N	N	N	N	N	N	N	Y	Y	Y	N
	Prozomix Ltd (CoEBio3)	www.prozomix.com, www.coebio3. manchester.ac.uk	7	Y	Y	Y	N									N	N	Y	Y	Y	Y	N

Iduronicase	Merck KGgA/ MilliporeSigma	www.sigmaaldrich.com	1	Y	N	N	N	N	N	N	N	N	N	N	N	N	N	N	N	N	N	N
Imine reductase	Almac	www.almacgroup.com	>100	Y	Y	Y	N	N	Y	N	Y	Y	Y	N	Y	Y	N	Y	Y	Y	Y	N
	Codexis Inc.	www.codexis.com	24 (kit) +custom items	Y from kit	Up to 100g from kit	Y	N	N	Y	Y	Y	Y	Y				N	Y	N	Y	Y	N
	EnzymeWorks	www.enzymeworking .com	20	Y	Y	Y	N	N	Y	Y	Y	Y	Y	N	Y	Y	Y	Y	Y	Y	Y	N
	Enzymicals AG	www.enzymicals.com	30	Y	Y	Y	N	N	N	N	Y	Y	Y	N	Y	Y	N	Y	Y	Y	Y	N
	Johnson Matthey	www.matthey.com/ biocatalysts/	85	Y	Y	Y	N	Y	Y	Y	Y	Y	Y	N	Y	Y	N	Y	Y	Y	Y	Y
	Protéus	www.proteus.fr	10	Y	N	Y	Y	N	Y	Y	Y	Y	Y	N	Y	Y	N	N	Y	Y	Y	N
	Prozomix Ltd	www.prozomix.com	576	Y	Y	Y	N	Y	N	N	Y	Y	Y	N	Y	N	Y	Y	Y	Y	Y	N
	Prozomix Ltd (CoEBio3)	www.prozomix.com, www.coebio3. manchester.ac.uk	46	Y	Y	Y	N									Y	Y	Y	Y	Y	Y	N
	Syncozymes	en.syncozymes.com	6	Y	Y	Y	Y	N	Y	N	Y	N	N	N	N	N	N	Y	Y	N	N	N
Inulinase	Merck KGgA/ MilliporeSigma	www.sigmaaldrich.com	1	Y	N	N	N	N	N	N	N	N	N	N	N	N	N	N	N	N	N	N
	NZYTech	www.nzytech.com	2	Y	N	Y	N	N	N	N	N	N	N	N	Y	Y	N	N	N	N	N	N
Invertase	Merck KGgA/ MilliporeSigma	www.sigmaaldrich.com	4	Y	N	N	N	N	N	N	N	N	N	N	N	N	N	N	N	N	N	N
Isoamylase	EUCODIS	www.eucodis.com	1	Y	Y	Y	N	N	N	N	N	Y	Y	N	Y	N	N	N	Y	Y	Y	N
	Merck KGgA/ MilliporeSigma	www.sigmaaldrich.com	1	Y	N	N	N	N	N	N	N	N	N	N	N	N	N	N	N	N	N	N
	Prozomix Ltd	www.prozomix.com	1	Y	Y	Y	N	N	N	N	Y	Y	Y	N	Y	N	N	N	Y	Y	Y	N
Isomerase (protein disulfide)	Merck KGgA/ MilliporeSigma	www.sigmaaldrich.com	1	Y	N	N	N	N	N	N	N	N	N	N	N	N	N	N	N	N	N	N
Isomerase (glucose)	Merck KGgA/ MilliporeSigma	www.sigmaaldrich.com	1	Y	N	N	N	N	N	N	N	N	N	N	N	N	N	N	N	N	N	N

Table 2.1 *(Continued)*

Enzyme name	Supplier	Company contact details	Number of enzymes in panel	Screening quantities available (i.e. between 1 mg and 1 g of each available on immediate basis for sale)	Larger quantities available (i.e. between 1 g and 1 kg available within 1 month of request)	Attributes: Recombinant	Homologously expressed	Metagenomic	Engineered panel	Evolved panel	Sequence diverse members	Unique proprietary members	Nagoya compliant	Psychrophilic members	Mesophilic members	Thermophilic members	NAD+-dependent members	NAD(P)+-dependent	Animal-free	BSE-TSE-certified	GMO-free	GMP grade
Isomerase (phosphoglucose)	Asahi Kasei Pharma Corporation	www.asahi-kasei.co.jp/shindan/en/	1																			
	Megazyme	www.megazyme.com	3	N	Y	Y	N	N	N	N	Y	Y	N	N	Y	N	N	N	Y	N	N	N
	Merck KGgA/ MilliporeSigma	www.sigmaaldrich.com	4	Y	N	N	N	N	N	N	N	N	N	N	N	N	N	N	N	N	N	N
Isomerase (triosephosphate)	Merck KGgA/ MilliporeSigma	www.sigmaaldrich.com	3	Y	N	N	N	N	N	N	N	N	N	N	N	N	N	N	N	N	N	N
Kinase (3-phosphoglyceric phospho)	Merck KGgA/ MilliporeSigma	www.sigmaaldrich.com	1	Y	N	N	N	N	N	N	N	N	N	N	N	N	N	N	N	N	N	N
Isomerase (phosphomannose)	Megazyme	www.megazyme.com	1	N	Y	Y	N	N	N	N	Y	Y	N	N	Y	N	N	N	Y	N	N	N
Isomerase (phosphoribose)	Merck KGgA/ MilliporeSigma	www.sigmaaldrich.com	1	Y	N	N	N	N	N	N	N	N	N	N	N	N	N	N	N	N	N	N
Isomerase (xylose)	NZYTech	www.nzytech.com	1	Y	N	Y	N	N	N	N	N	N	N	N	Y	N	N	N	N	N	N	N
Kinase	Almac	www.almacgroup.com	30	Y	Y	Y	N	N	N	N	Y	Y	Y	N	Y	Y	N	N	Y	Y	Y	N
Kinase (acetate)	Merck KGgA/ MilliporeSigma	www.sigmaaldrich.com	2	Y	N	N	N	N	N	N	N	N	N	N	N	N	N	N	N	N	N	N

Kinase (adenylate)	EnzymeWorks	www.enzymeworking.com	5	Y	Y	Y	N	N	Y	Y	Y	Y	N	N	Y	N	N	N	Y	Y	N	N
	Megazyme	www.megazyme.com	1	N	Y	Y	N	N	N	N	Y	Y	N	N	Y	N	N	N	Y	N	N	N
	Merck KGgA/ MilliporeSigma	www.sigmaaldrich.com	1	Y	N	N	N	N	N	N	N	N	N	N	N	N	N	N	N	N	N	N
Kinase (creatine)	Asahi Kasei Pharma Corporation	www.asahi-kasei.co.jp/shindan/en/	1																			
	Merck KGgA/ MilliporeSigma	www.sigmaaldrich.com	2	Y	N	N	N	N	N	N	N	N	N	N	N	N	N	N	N	N	N	N
Kinase (cytidylate)	Megazyme	www.megazyme.com	1	N	Y	Y	N	N	N	N	Y	Y	N	N	Y	N	N	N	Y	N	N	N
Kinase (fructose-6-phosphate)	Merck KGgA/ MilliporeSigma	www.sigmaaldrich.com	2	Y	N	N	N	N	N	N	N	N	N	N	N	N	N	N	N	N	N	N
Kinase (galacto)	Merck KGgA/ MilliporeSigma	www.sigmaaldrich.com	1	Y	N	Y	N	N	N	N	N	N	N	N	N	N	N	N	N	N	N	N
Kinase (guanylate)	Merck KGgA/ MilliporeSigma	www.sigmaaldrich.com	1	Y	N	N	N	N	N	N	N	N	N	N	N	N	N	N	N	N	N	N
Kinase (gluco)	Merck KGgA/ MilliporeSigma	www.sigmaaldrich.com	1	Y	N	N	N	N	N	N	N	N	N	N	N	N	N	N	N	N	N	N
Kinase (gluconate)	Megazyme	www.megazyme.com	1	N	Y	Y	N	N	N	N	Y	Y	N	N	Y	N	N	N	Y	N	N	N
Kinase (glycerol)	Amano Enzyme	www.amano-enzyme.co.jp	1																			
	Asahi Kasei Pharma Corporation	www.asahi-kasei.co.jp/shindan/en/	1																			
	Merck KGgA/ MilliporeSigma	www.sigmaaldrich.com	4	Y	N	N	N	N	N	N	N	N	N	N	N	N	N	N	N	N	N	N
	Toyobo	www.toyobo-global.com	1																			
Kinase (guanylate)	Megazyme	www.megazyme.com	1	N	Y	N	N	N	N	N	Y	Y	N	N	Y	N	N	N	Y	N	N	N
	Merck KGgA/ MilliporeSigma	www.sigmaaldrich.com	1	Y	N	N	N	N	N	N	N	N	N	N	N	N	N	N	N	N	N	N

Table 2.1 (Continued)

Enzyme name	Supplier	Company contact details	Number of enzymes in panel	Screening quantities available (i.e. between 1 mg and 1 g of each available on immediate basis for sale)	Larger quantities available (i.e. between 1 g and 1 kg available within 1 month of request)	Attributes: Recombinant	Homologously expressed	Metagenomic	Engineered panel	Evolved panel	Sequence diverse members	Unique proprietary members	Nagoya compliant	Psychrophilic members	Mesophilic members	Thermophilic members	NAD+-dependent members	NAD(P)+-dependent	Animal-free	BSE-TSE-certified	GMO-free	GMP grade
Kinase (hexo)	Asahi Kasei Pharma Corporation	www.asahi-kasei.co.jp/shindan/en/	2																			
	Megazyme	www.megazyme.com	2	N	Y	Y/N	N	N	N	N	Y	Y	N	N	Y	N	N	N	Y	N	N	N
	Merck KGgA/MilliporeSigma	www.sigmaaldrich.com	2	Y	N	N	N	N	N	N	N	N	N	N	N	N	N	N	N	N	N	N
	Toyobo	www.toyobo-global.com	1																			
Kinase (myo)	Merck KGgA/MilliporeSigma	www.sigmaaldrich.com	1	Y	N	N	N	N	N	N	N	N	N	N	N	N	N	N	N	N	N	N
Kinase (NAD)	Merck KGgA/MilliporeSigma	www.sigmaaldrich.com	3	Y	N	Y	N	N	N	N	N	N	N	N	N	N	N	N	N	N	N	N
Kinase (nucleoside 5′-diphosphate)	Merck KGgA/MilliporeSigma	www.sigmaaldrich.com	2	Y	N	N	N	N	N	N	N	N	N	N	N	N	N	N	N	N	N	N
Kinase (polynucleotide)	Merck KGgA/MilliporeSigma	www.sigmaaldrich.com	1	Y	N	N	N	N	N	N	N	N	N	N	N	N	N	N	N	N	N	N
Kinase (pyruvate)	Megazyme	www.megazyme.com	1	N	Y	N	N	N	N	N	Y	Y	N	N	Y	N	N	N	Y	N	N	N
	Merck KGgA/MilliporeSigma	www.sigmaaldrich.com	4	Y	N	N	N	N	N	N	N	N	N	N	N	N	N	N	N	N	N	N
Kinase (thymidylate)	Megazyme	www.megazyme.com	1	N	Y	Y	N	N	N	N	Y	Y	N	N	Y	N	N	N	Y	N	N	N

Kinase (uridylate)	Megazyme	www.megazyme.com	1	N	Y	Y	N	N	N	N	Y	Y	N	N	Y	N	N	N	Y	N	N	N
Laccase	Almac	www.almacgroup.com	96	Y	Y	Y	N	N	N	N	Y	Y	Y	N	Y	Y	N	N	Y	Y	Y	N
	Amano Enzyme	www.amano-enzyme.co.jp	1																			
	Asahi Kasei Pharma Corporation	www.asahi-kasei.co.jp/shindan/en/	4																			
	ASA Spezialenzyme GmbH	www.asa-enzyme.de	5	Y	Y	N	Y	N	N	N	N	Y	N	N	Y	N	N	N	Y	Y	Y	N
	GECCO Biotech	www.gecco-biotech.com	1	Y	N	Y	N	N	N	N	N	N	N	N	N	Y	N	N	Y	N	N	N
	Merck KGgA/MilliporeSigma	www.sigmaaldrich.com	3	Y	N	Y	N	N	N	N	N	N	N	N	N	N	N	N	N	N	N	N
	Novozymes A/S	pharmaceuticals@novozymes.com	1	N	Y	Y																
	NZYTech	www.nzytech.com	3	Y	N	Y	N	N	N	N	N	N	N	N	Y	Y	N	N	N	N	N	N
	Proteus	www.proteus.fr	19	Y	N	Y	N	N	Y	Y	Y	Y	Y	N	Y	Y	N	N	Y	Y	Y	N
Lactamase	Almac	www.almacgroup.com	6	Y	Y	Y	N	N	N	N	Y	Y	Y		Y	Y	N	N	Y	Y	Y	N
	EnzymeWorks	www.enzymeworking.com	8	Y	Y	Y	N	N	Y	Y	Y	Y	Y	N	Y	Y	N	N	Y	Y	Y	N
	Merck KGgA/MilliporeSigma	www.sigmaaldrich.com	3	Y	N	N	N	N	N	N	N	N	N	N	N	N	N	N	N	N	N	N
beta-Lactamase	EUCODIS	www.eucodis.com	6	Y	Y	Y	N	N	Y	N	Y	Y	Y	N	Y	N	N	N	Y	Y	Y	N
Lactase	Amano Enzyme	www.amano-enzyme.co.jp	1																			
	ASA Spezialenzyme GmbH	www.asa-enzyme.de	1	Y	Y	N	Y	N	N	N	N	Y	N	N	Y	N	N	N	Y	Y	Y	N
	Biocatalysts Ltd	www.biocatalysts.com	2	Y	Y	N	N	N	N	N	N	N	Y	N	Y	N	N	N	Y	Y	Y	N
Lactoperoxidase	Merck KGgA/MilliporeSigma	www.sigmaaldrich.com	2	Y	N	N	N	N	N	N	N	N	N	N	N	N	N	N	N	N	N	N
Laminarinase	NZYTech	www.nzytech.com	14	Y	N	Y	N	N	N	N	N	N	N	N	Y	Y	N	N	N	N	N	N
	Prozomix Ltd	www.prozomix.com	1	Y	Y	Y	N	N	N	N	Y	Y	Y	N	Y	N	N	N	Y	Y	Y	N

Table 2.1 *(Continued)*

Enzyme name	Supplier	Company contact details	Number of enzymes in panel	Screening quantities available (i.e. between 1 mg and 1 g of each available on immediate basis for sale)	Larger quantities available (i.e. between 1 g and 1 kg available within 1 month of request)	Attributes: Recombinant	Homologously expressed	Metagenomic	Engineered panel	Evolved panel	Sequence diverse members	Unique proprietary members	Nagoya compliant	Psychrophilic members	Mesophilic members	Thermophilic members	NAD+-dependent members	NAD(P)+-dependent	Animal-free	BSE-TSE-certified	GMO-free	GMP grade
Levanase	NZYTech	www.nzytech.com	4	Y	N	Y	N	N	N	N	N	N	N	N	Y	N	N	N	N	N	N	N
Levansucrase	NZYTech	www.nzytech.com	3	Y	N	Y	N	N	N	N	N	N	N	N	Y	N	N	N	N	N	N	N
Lichenase	NZYTech	www.nzytech.com	11	Y	N	Y	N	N	N	N	N	N	N	N	Y	Y	N	N	N	N	N	N
Licheninase	Prozomix Ltd	www.prozomix.com	1	Y	Y	Y	N	N	N	N	Y	Y	Y	N	Y	N	N	N	Y	Y	Y	N
Licheninase + cellulase	Prozomix Ltd	www.prozomix.com	2	Y	Y	Y	N	N	N	N	Y	Y	Y	N	Y	N	N	N	Y	Y	Y	N
Lipase	Almac	www.almacgroup.com	96	Y	Y	Y	N	N	Y	N	Y	Y	Y	Y	Y	Y	N	N	Y	Y	Y	N
	Amano Enzyme	www.amano-enzyme.co.jp	9																			
	Asahi Kasei Pharma Corporation	www.asahi-kasei.co.jp/shindan/en/	1																			
	ASA Spezialenzyme GmbH	www.asa-enzyme.de	1	Y	Y	N	Y	N	N	N	N	Y	N	N	N	N	N	N	Y	Y	Y	N
	Biocatalysts Ltd	www.biocatalysts.com	9	Y	Y	Y/N	Y/N	N	Y/N	N	Y	Y	Y	N	Y	N	N	N	Y	Y	Y	N
	ChiralVision	www.chiralvision.com	18	Y	Y	N	N	N	N	N	Y	Y	N	N	Y	Y	N	N	Y	Y	Y	N
	c-LEcta GmbH	www.c-lecta.com	3	Y	Y	Y	N	N	Y	N	Y	Y	Y	Y	N	N	N	N	Y	Y	Y	N
	Codexis Inc.	www.codexis.com	See esterases	See esterases	See esterases	Y	N	N	Y	Y	Y	Y	Y				-	-	N	Y	Y	N

	Enzagen	www.enzagen.com	25	Y	Y	y/N	Y/N	N	N	N	Y	N	Y	N	Y	Y/N	N	N	Y	Y	Y/N	N
	EnzymeWorks	www.enzymeworking.com	15	Y	Y	Y	N	N	Y	N	Y	Y	Y	N	Y	Y	N	N	Y	Y	Y	N
	Enzymicals AG	www.enzymicals.com	10	Y	Y	Y	N	N	N	N	Y	Y	Y	N	Y	Y	N	N	Y	Y	Y	N
	EUCODIS	www.eucodis.com	25	Y	Y	Y	N	N	Y	N	Y	Y	Y	Y	Y	Y	N	N	Y	Y	Y	N
	Evoxx technologies	www.evoxx.com	1	Y	Y	Y	N	Y	Y	Y	Y	Y	Y			Y			Y	Y	Y	N
	Merck KGgA/ MilliporeSigma	www.sigmaaldrich.com	26	Y	Y	N	N	N	N	N	N	N	N	N	N	N	N	N	N	N	N	N
	Novozymes A/S	pharmaceuticals@ novozymes.com	10	N	Y	Y																
	Nzomics	www .nzomicsbiocatalysis .co.uk																				
	Protéus	www.proteus.fr	9	Y	N	Y	N	N	Y	Y	Y	Y	Y	N	Y	Y	N	N	Y	Y	Y	N
	Purolite Life Sciences	www.purolite.com/life-sciences/	12	Y	Y	N	N	N	N	N	Y	Y	N	N	Y	Y	N	N	Y	Y	Y	N
	Syncozymes	en.syncozymes.com	1	Y	Y	Y	Y	N	Y	N	N	N	N	N	N	N	N	N	Y	N	N	N
Lipase (lipoprotein)	Amano Enzyme	www.amano-enzyme.co.jp	1																			
	Merck KGgA/ MilliporeSigma	www.sigmaaldrich.com	3	Y	Y	N	N	N	N	N	N	N	N	N	N	N	N	N	N	N	N	N
	Toyobo	www.toyobo-global.com	2																			
Lipoxygenase	Almac	www.almacgroup.com	5	Y	Y	Y	N	N	N	N	Y	Y	Y	N	Y	Y	N	N	Y	Y	Y	N
	EnzymeWorks	www.enzymeworking.com	5	Y	Y	Y	N	N	Y	Y	Y	Y	Y	N	Y	Y	N	N	Y	Y	Y	N
	Enzymicals AG	www.enzymicals.com	2	N	Y	Y	N	N	N	N	Y	Y	Y	N	Y	Y	N	N	Y	Y	Y	N
	Merck KGgA/ MilliporeSigma	www.sigmaaldrich.com	2	Y	N	N	N	N	N	N	N	N	N	N	N	N	N	N	N	N	N	N

Table 2.1 *(Continued)*

Enzyme name	Supplier	Company contact details	Number of enzymes in panel	Screening quantities available (i.e. between 1 mg and 1 g of each available on immediate basis for sale)	Larger quantities available (i.e. between 1 g and 1 kg available within 1 month of request)	Attributes: Recombinant	Homologously expressed	Metagenomic	Engineered panel	Evolved panel	Sequence diverse members	Unique proprietary members	Nagoya compliant	Psychrophilic members	Mesophilic members	Thermophilic members	NAD+-dependent members	NAD(P)+-dependent	Animal-free	BSE-TSE-certified	GMO-free	GMP grade
Luciferase	Merck KGgA/ MilliporeSigma	www.sigmaaldrich.com	2	Y	N	N	N	N	N	N	N	N	N	N	N	N	N	N	N	N	N	N
Lyase (alginate)	Megazyme	www.megazyme.com	1	Y	N	Y	N	N	N	N	Y	Y	N	N	Y	N	N	N	Y	N	N	N
	Merck KGgA/ MilliporeSigma	www.sigmaaldrich.com	1	Y	N	N	N	N	N	N	N	N	N	N	N	N	N	N	N	N	N	N
Lyase (ammonia)	Almac	www.almacgroup.com	50	Y	Y	Y	N	N	N	N	Y	Y	Y	N	Y	Y	N	N	Y	Y	Y	N
	Codexis Inc.	www.codexis.com	>100 custom items	Custom items	>1 month	Y	N	N	Y	Y	Y	Y	Y				-	-	N	Y	Y	N
	EnzymeWorks	www.enzymeworking.com	5	Y	Y	Y	N	N	Y	Y	Y	Y	N	N	Y	Y	N	N	Y	Y	Y	N
	GECCO Biotech	www.gecco-biotech.com	2	Y	N	Y	N	N	N	N	N	N	N	N	Y	Y	N	N	Y	N	N	N
	Prozomix Ltd	www.prozomix.com	48	Y	Y	Y	N	Y	N	N	Y	Y	Y	N	Y	N	N	N	Y	Y	Y	N
	Prozomix Ltd (CoEBio3)	www.prozomix.com, www.coebio3.manchester.ac.uk	20	Y	Y	Y	N									N	N	Y	Y	Y	Y	N
Lyase (aspartate ammonia)	Almac	www.almacgroup.com	9	Y	Y	Y	N				Y	Y	Y	N	Y	Y	N	N	Y	Y	Y	N

Lyase (benzaldehyde)	Almac	www.almacgroup.com	5	Y	Y	Y	N	N	N	N	Y	Y	Y	N	Y	Y	N	N	Y	Y	Y	N
	EnzymeWorks	www.enzymeworking .com	5	Y	Y	Y	N	N	Y	Y	Y	Y	Y	N	Y	Y	N	N	Y	Y	Y	N
	Evoxx technologies	www.evoxx.com	1	Y	Y	Y	N	Y	Y	Y	Y	Y	Y	N	Y	Y			Y	Y	Y	N
	GECCO Biotech	www.gecco-biotech.com	1	Y	N	Y	N	N	N	N	N	N	N	N	N	Y	N	N	Y	N	N	N
Lyase (citrate)	Merck KGgA/ MilliporeSigma	www.sigmaaldrich.com	1	Y	N	N	N	N	N	N	N	N	N	N	N	N	N	N	N	N	N	N
Lyase (hydroxyni- trile)/Oxynitrilase	Almac	www.almacgroup.com	24	Y	Y	Y	N	N	Y	N	Y	Y	Y	N	Y	Y	N	N	Y	Y	Y	N
	ASA Spezialenzyme GmbH	www.asa-enzyme.de	1	Y	Y	N	Y	N	N	N	N	N	N	N	Y	N	N	N	Y	Y	Y	N
	Enzymaster	www.enzymaster.de	2	N	Y	Y	N	N	N	N	N	N	Y	N	Y	Y	N	N	Y	Y	Y	N
	EnzymeWorks	www.enzymeworking .com	5	Y	Y	Y	N	N	Y	Y	Y	Y	Y	N	Y	Y	N	N	Y	Y	Y	N
	Evoxx technologies	www.evoxx.com	3	Y	Y	Y	N	Y	Y	Y	Y	Y	Y	N	Y	Y			Y	Y	Y	N
	Innosyn	www.innosyn.com	5																			
	InnoSyn BV	www.innosyn.com	5	Y	Y	Y	N	N	N	N	N	Y	Y	N	Y	N	N	N	Y	Y	Y	N
	Johnson Matthey	www.matthey.com/ biocatalysts/	10	N	N	Y	N	N	N	N	Y	Y	Y	N	Y	Y	N	N	Y	Y	Y	Y
	Syncozymes	en.syncozymes.com	29																			
Lyase (mandelonitrile)	Merck KGgA/ MilliporeSigma	www.sigmaaldrich.com	1	Y	N	N	N	N	N	N	N	N	N	N	N	N	N	N	N	N	N	N
Lyase (pectate)	Megazyme	www.megazyme.com	3	N	Y	Y/ N	N	N	N	N	Y	Y	N	N	Y	N	N	N	Y	N	N	N
Lyase (phenylalanine ammonia)	Merck KGgA/ MilliporeSigma	www.sigmaaldrich.com	1	Y	N	N	N	N	N	N	N	N	N	N	N	N	N	N	N	N	N	N
Lyase (polysaccharide)	NZYTech	www.nzytech.com	62	Y	N	Y	N	N	N	N	N	N	N	N	Y	Y	N	N	N	N	N	N
Lyase (tyrosine phenol)	Enzymaster	www.enzymaster.de	24	N	Y	Y	N	N	Y	Y	Y	Y	Y	N	Y	Y	N	N	Y	Y	Y	N
Lyase (xanthan)	Megazyme	www.megazyme.com	1	N	Y	Y	N	N	N	N	Y	Y	N	N	Y	N	N	N	Y	N	N	N

Table 2.1 *(Continued)*

Enzyme name	Supplier	Company contact details	Number of enzymes in panel	Screening quantities available (i.e. between 1 mg and 1 g of each available on immediate basis for sale)	Larger quantities available (i.e. between 1 g and 1 kg available within 1 month of request)	Attributes: Recombinant	Homologously expressed	Metagenomic	Engineered panel	Evolved panel	Sequence diverse members	Unique proprietary members	Nagoya compliant	Psychrophilic members	Mesophilic members	Thermophilic members	NAD+-dependent members	NAD(P)+-dependent	Animal-free	BSE-TSE-certified	GMO-free	GMP grade
Lysophospholipase	Asahi Kasei Pharma Corporation	www.asahi-kasei.co.jp/shindan/en/	1																			
Lysozyme	Merck KGgA/MilliporeSigma	www.sigmaaldrich.com	11	Y	Y	Y	N	N	N	N	N	N	N	N	N	N	N	N	N	N	N	N
	NZYTech	www.nzytech.com	4	Y	N	Y	N	N	N	N	N	N	N	N	Y	N	N	N	N	N	N	N
Macrolide glycosyltransferase	EnzymeWorks	www.enzymeworking.com	5	Y	Y	Y	N	N	Y	Y	Y	Y	N	N	Y	N	N	N	Y	Y	Y	N
	Prozomix Ltd	www.prozomix.com	1	Y	Y	Y	N	N	N	N	Y	Y	Y	N	Y	N	N	N	Y	Y	Y	N
Maltooligosyl trehalose trehalohydrolase	Prozomix Ltd	www.prozomix.com	1	Y	Y	Y	N	N	N	N	Y	Y	Y	N	Y	N	N	N	Y	Y	Y	N
alpha-Mannanase	NZYTech	www.nzytech.com	3	Y	N	Y	N	N	N	N	N	N	N	N	Y	N	N	N	N	N	N	N
beta-Mannanase	Megazyme	www.megazyme.com	5	N	Y	Y/N	N	N	N	N	Y	Y	N	N	Y	N	N	N	Y	N	N	N
	NZYTech	www.nzytech.com	25	Y	N	Y	N	N	N	N	N	N	N	N	Y	Y	N	N	N	N	N	N
	Prozomix Ltd	www.prozomix.com	3	Y	Y	Y	N	N	N	N	Y	Y	Y	N	Y	N	N	N	Y	Y	Y	N
alpha-Mannosidase	Megazyme	www.megazyme.com	2	N	Y	Y	N	N	N	N	Y	Y	N	N	Y	N	N	N	Y	N	N	N
	Merck KGgA/MilliporeSigma	www.sigmaaldrich.com	1	Y	N	N	N	N	N	N	N	N	N	N	N	N	N	N	N	N	N	N
	NZYTech	www.nzytech.com	23	Y	N	Y	N	N	N	N	N	N	N	N	Y	Y	N	N	N	N	N	N

beta-Mannosidase	Megazyme	www.megazyme.com	1	N	Y	Y	N	N	N	N	Y	Y	N	N	Y	N	N	N	Y	N	N	N
	Merck KGgA/ MilliporeSigma	www.sigmaaldrich.com	2	Y	N	N	N	N	N	N	N	N	N	N	N	N	N	N	N	N	N	N
	NZYTech	www.nzytech.com	5	Y	N	Y	N	N	N	N	N	N	N	N	Y	Y	N	N	N	N	N	N
	Prozomix Ltd	www.prozomix.com	15	Y	Y	Y	N	N	N	N	Y	Y	Y	N	Y	N	N	N	Y	Y	Y	N
Monooxygenase (Baeyer–Villiger)	Almac	www.almacgroup.com	50	Y	Y	Y	N	N	Y	N	Y	Y	Y	N	Y	Y	Y	Y	Y	Y	Y	N
	Codexis Inc.	www.codexis.com	90 custom items	Custom items	>1 month	Y	N	N	Y	Y	Y	Y	Y				N	Y	N	Y	Y	N
	EnzymeWorks	www.enzymeworking .com	20	Y	Y	Y	N	N	Y	Y	Y	Y	Y	N	Y	Y	N	Y	Y	Y	Y	N
	Enzymicals AG	www.enzymicals.com	6	Y	Y	Y	N	N	N	N	Y	Y	Y	N	Y	Y	N	Y	Y	Y	Y	N
	Gecco Biotech	www.gecco-biotech.com	65	Y	Y	Y	N	N	Y	N	Y	Y	N	N	Y	Y	Y	Y	Y	N	N	N
	InnoSyn BV	www.innosyn.com	16	Y	Y	Y	N	N	N	N	Y	N	Y	N	Y	N	N	Y	Y	Y	Y	N
	Merck KGgA/ MilliporeSigma	www.sigmaaldrich.com	1	Y	Y	Y	N	N	N	N	N	N	N	N	N	N	N	N	N	N	N	N
	Proteus	www.proteus.fr	10	Y	N	Y	N	N	Y	Y	Y	Y	Y	N	Y	Y	N	N	Y	Y	Y	N
	Prozomix Ltd (CoEBio3)	www.prozomix.com, www.coebio3. manchester.ac.uk	23	Y	Y	Y	N									Y	Y	Y	Y	Y	Y	N
Monooxygenase (flavin-containing FMOs) – bacterial	Gecco Biotech	www.gecco-biotech.com	16	Y	N	Y	N	N	Y	N	N	N	Y	N	Y	Y	N	Y	Y	N	N	N
Monooxygenase (flavin-containing FMOs) – mammalian	Gecco Biotech	www.gecco-biotech.com	4	Y	Y	Y	N	N	Y	N	N	N	Y	N	Y	Y	N	Y	Y	N	N	N
Monooxygenase (Human P450)	Almac	www.almacgroup.com	10	Y	Y	Y	N	N	N	N	Y	Y	Y	N	Y	Y	N	Y	Y	Y	Y	N
	EnzymeWorks	www.enzymeworking .com	15	Y	Y	Y	N	N	Y	Y	Y	Y	Y	N	Y	Y	N	Y	Y	Y	Y	N
Monooxygenase (kynurenine)	Merck KGgA/ MilliporeSigma	www.sigmaaldrich.com	1	Y	N	Y	N	N	N	N	N	N	N	N	N	N	N	N	N	N	N	N

Table 2.1 *(Continued)*

Enzyme name	Supplier	Company contact details	Number of enzymes in panel	Screening quantities available (i.e. between 1 mg and 1 g of each available on immediate basis for sale)	Larger quantities available (i.e. between 1 g and 1 kg available within 1 month of request)	Attributes: Recombinant	Homologously expressed	Metagenomic	Engineered panel	Evolved panel	Sequence diverse members	Unique proprietary members	Nagoya compliant	Psychrophilic members	Mesophilic members	Thermophilic members	NAD+-dependent members	NAD(P)+-dependent	Animal-free	BSE-TSE-certified	GMO-free	GMP grade
Monooxygenase (Lytic polysaccharide/ LPMO)	EUCODIS	www.eucodis.com	2	Y	Y	Y	N	N	Y	N	Y	Y	Y	N	Y	N	N	N	Y	Y	Y	N
Monooxygenase (Microbial P450)	Almac	www.almacgroup.com	96	Y	Y	Y	N	N	Y	N	Y	Y	Y	N	Y	Y	N	Y	Y	Y	Y	N
	Codexis Inc.	www.codexis.com	11 (kit) + custom items	Y from kit	Up to 100 g from kit	Y	N	N	Y	Y	Y	Y	Y				N	Y	N	Y	N	N
	EnzymeWorks	www.enzymeworking.com	30	Y	Y	Y	N	N	Y	Y	Y	Y	Y	N	Y	Y	N	Y	Y	Y	Y	N
	Enzymicals AG	www.enzymicals.com	6	N	Y	Y	N	N	N	N	Y	Y	Y	N	Y	Y	N	N	Y	Y	Y	N
	Gecco Biotech	www.gecco-biotech.com	4	Y	N	Y	N	N	N	N	N	N	N	N	Y	Y	Y	Y	Y	N	N	N
	InnoSyn BV	www.innosyn.com	150	Y	Y	Y	N	N	Y	N	Y	Y	Y	N	Y	N	Y	Y	Y	Y	Y	N
	Johnson Matthey	www.matthey.com/biocatalysts/	10	N	N	Y	N	N	N	N	Y	Y	Y	N	N	N	N	Y	Y	Y	Y	Y
	Protéus	www.proteus.fr	5	Y	N	Y	N	N	Y	Y	Y	Y	Y	N	Y	Y	N	N	Y	Y	Y	N

Monooxygenase (styrene)	Prozomix Ltd	www.prozomix.com	161	Y	Y	Y	N	Y	N	N	Y	Y	Y	N	Y	N	Y	Y	Y	Y	Y	N
	Prozomix Ltd (CoEBio3)	www.prozomix.com/ www.coebio3 .manchester.ac.uk	43	Y	Y	Y	N									Y	Y	Y	Y	Y	Y	N
	Syncozymes	en.syncozymes.com	8	Y	Y	Y	Y	N	Y	N	Y	N	N	N	N	N	N	Y	Y	N	N	N
	EnzymeWorks	www.enzymeworking .com	5	Y	Y	Y	N	N	Y	Y	Y	Y	Y	N	Y	Y	N	Y	Y	Y	Y	N
	Enzymicals AG	www.enzymicals.com	1	N	Y	Y	N	N	N	N	Y	Y	Y	N	Y	Y	N	N	Y	Y	Y	N
	Gecco Biotech	www.gecco-biotech.com	1	Y	N	Y	N	N	N	N	N	N	Y	N	N	N	N	Y	Y	N	N	N
	Johnson Matthey	www.matthey.com/ biocatalysts/	5	N	N	Y	N	N	N	N	Y	Y	Y	N	N	N	N	Y	Y	Y	Y	Y
NADase	Merck KGgA/ MilliporeSigma	www.sigmaaldrich.com	1	Y	N	N	N	N	N	N	N	N	N	N	N	N	N	N	N	N	N	N
Neuraminidase	Merck KGgA/ MilliporeSigma	www.sigmaaldrich.com	11	Y	N	N	N	N	N	N	N	N	N	N	N	N	N	N	N	N	N	N
Nitrate reductase	Merck KGgA/ MilliporeSigma	www.sigmaaldrich.com	2	Y	N	N	N	N	N	N	N	N	N	N	N	N	N	N	N	N	N	N
Nitrilase	Almac	www.almacgroup.com	96	Y	Y	Y	N	N	Y	N	Y	Y	Y	N	Y	Y	N	N	Y	Y	Y	N
	c-LEcta GmbH	www.c-lecta.com	94	Y	Y	Y	N	N	Y	Y	N	Y	Y	N	Y	N	N	N	Y	Y	Y	N
	Codexis Inc.	www.codexis.com	12 (kit)	Y from kit	Up to 100 g	Y	N	N	Y	Y	Y	Y	Y				-	-	N	Y	Y	N
	EnzymeWorks	www.enzymeworking .com	20	Y	Y	Y	N	N	Y	Y	Y	Y	Y	N	N		N	N	Y	Y	Y	N
	Merck KGgA/ MilliporeSigma	www.sigmaaldrich.com	1	Y	N	Y	N	N	N	N	N	N	N	N	N	N	N	N	N	N	N	N
	Proteus	www.proteus.fr	10	Y	N	Y	N	N	Y	Y	Y	Y	Y	N	Y	Y	N	N	Y	Y	Y	N
	Prozomix Ltd	www.prozomix.com	96	Y	Y	Y	N	Y	N	N	Y	Y	Y	N	Y	N	N	N	Y	Y	Y	N
	Syncozymes	en.syncozymes.com	40	Y	Y	Y	Y	N	Y	N	Y	N	N	N	N	N	N	N	Y	N	N	N

Table 2.1 (Continued)

Enzyme name	Supplier	Company contact details	Number of enzymes in panel	Screening quantities available (i.e. between 1 mg and 1 g of each available on immediate basis for sale)	Larger quantities available (i.e. between 1 g and 1 kg available within 1 month of request)	Attributes																
						Recombinant	Homologously expressed	Metagenomic	Engineered panel	Evolved panel	Sequence diverse members	Unique proprietary members	Nagoya compliant	Psychrophilic members	Mesophilic members	Thermophilic members	NAD+-dependent members	NAD(P)+-dependent	Animal-free	BSE-TSE-certified	GMO-free	GMP grade
Nitrile hydratase	Almac	www.almacgroup.com	9	Y	Y	Y	N	N	N	N	Y	Y	Y	N	Y	Y	N	N	Y	Y	Y	N
	EnzymeWorks	www.enzymeworking .com	15	Y	Y	Y	N	N	Y	Y	Y	Y	Y	N	Y	Y	N	N	Y	Y	Y	N
	EUCODIS	www.eucodis.com	10	Y	y	Y	N	N	Y	Y	Y	Y	Y	N	Y	N	N	N	Y	Y	Y	N
	InnoSyn BV	www.innosyn.com	11	Y	Y	Y	N	N	N	N	Y	N	Y	N	Y	N	N	N	Y	Y	Y	N
	Protéus	www.proteus.fr	10	Y	N	Y	N	N	Y	Y	Y	Y	Y	N	Y	Y	N	N	Y	Y	Y	N
	Prozomix Ltd	www.prozomix.com	18	Y	Y	Y	N	N	N	N	Y	Y	Y	N	Y	N	N	N	Y	Y	Y	N
Nitroreductase	EnzymeWorks	www.enzymeworking .com	20	Y	Y	Y	N	Y	Y	Y	Y	Y	Y	N	Y	Y	Y	Y	Y	Y	Y	N
	Johnson Matthey	www.matthey.com/ biocatalysts/	60	Y	Y	Y	N	Y	Y	N	Y	Y	Y	N	Y	Y	Y	Y	Y	Y	Y	Y
	Merck KGgA/ MilliporeSigma	www.sigmaaldrich.com	1	Y	N	Y	N	N	N	N	N	N	N	N	N	N	N	N	N	N	N	N
	Prozomix Ltd	www.prozomix.com	96	Y	Y	Y	N	Y	N	N	Y	Y	Y	N	Y	N	N	N	Y	Y	Y	N
	Syncozymes	en.syncozymes.com	12	Y	Y	Y	Y	N	Y	N	Y	N	N	N	N	N	Y	Y	Y	N	N	N
Nuclease	c-LEcta GmbH	www.c-lecta.com	1	Y	Y	Y	N	N	Y	N	N	N	Y	N	Y	N	N	N	Y	Y	Y	Y/ N
	Merck KGgA/ MilliporeSigma	www.sigmaaldrich.com	6	Y	N	Y	N	N	N	N	N	N	N	N	N	N	N	N	N	N	N	N

Nucleoside deoxy-ribosyltransferase	Merck KGgA/ MilliporeSigma	www.sigmaaldrich.com	1	Y	N	Y	N	N	N	N	N	N	N	N	N	N	N	N	N	N	N	N
5′-Nucleoticase	Merck KGgA/ MilliporeSigma	www.sigmaaldrich.com	1	Y	N	N	N	N	N	N	N	N	N	N	N	N	N	N	N	N	N	N
Oligosaccharide reducing-end xylanase	NZYTech	www.nzytech.com	2	Y	N	Y	N	N	N	N	N	N	N	N	Y	N	N	N	N	N	N	N
Oxidase (Acyl-CoA)	Asahi Kasei Pharma Corporation	www.asahi-kasei.co.jp/ shindan/en/	1																			
Oxidase (alcohol)	Asahi Kasei Pharma Corporation	www.asahi-kasei.co.jp/ shindan/en/	1																			
	ASA Spezialenzyme GmbH	www.asa-enzyme.de	1	Y	Y	N	Y	N	N	N	N	N	N	N	Y	N	N	N	Y	Y	Y	N
	EnzymeWorks	www.enzymeworking .com	30	Y	Y	Y	N	N	Y	Y	Y	Y	Y	N	Y	Y	Y	Y	Y	Y	Y	N
	Enzymicals AG	www.enzymicals.com	3	N	Y	Y	N	N	N	N	Y	Y	Y	N	Y	Y	Y	Y	Y	Y	Y	N
	Gecco Biotech	www.gecco-biotech.com	8	Y	Y	Y	N	N	Y	N	N	Y	N	N	Y	Y	N	N	Y	N	N	N
	Merck KGgA/ MilliporeSigma	www.sigmaaldrich.com	3	Y	N	N	N	N	N	N	N	N	N	N	N	N	N	N	N	N	N	N
	Syncozymes	en.syncozymes.com	5	Y	Y	Y	Y	N	Y	N	Y	N	N	N	N	N	N	N	Y	N	N	N
Oxidase (amine)/monoamine oxidase (MAO)	Almac	www.almacgroup.com	96	Y	Y	Y	N	N	N	N	Y	Y	Y	N	Y	Y	N	N	Y	Y	Y	N
	Codexis Inc.	www.codexis.com	90 custom items	Custom items	>1 month	Y	N	N	Y	Y	Y	Y	Y				-	-	N	Y	Y	N
	EnzymeWorks	www.enzymeworking .com	15	Y	Y	Y	N	N	Y	Y	Y	Y	N	N	N	N	N	Y	Y	Y	Y	N
	Gecco Biotech	www.gecco-biotech.com	1	Y	N	Y	N	N	N	N	N	N	N	N	N	N	N	N	Y	N	N	N
	Merck KGgA/ MilliporeSigma	www.sigmaaldrich.com	2	Y	N	Y	N	N	N	N	N	N	N	N	N	N	N	N	N	N	N	N
	Syncozymes	en.syncozymes.com	5	Y	Y	Y	Y	N	Y	N	Y	N	N	N	N	N	N	N	Y	N	N	N
Oxidase (R-selective monoamine)/MAO	Prozomix Ltd (CoEBio3)	www.prozomix.com, www.coebio3. manchester.ac.uk	36	Y	Y	Y	N									N	N	Y	Y	Y	Y	N

Table 2.1 *(Continued)*

Enzyme name	Supplier	Company contact details	Number of enzymes in panel	Screening quantities available (i.e. between 1 mg and 1 g of each available on immediate basis for sale)	Larger quantities available (i.e. between 1 g and 1 kg available within 1 month of request)	Attributes																
						Recombinant	Homologously expressed	Metagenomic	Engineered panel	Evolved panel	Sequence diverse members	Unique proprietary members	Nagoya compliant	Psychrophilic members	Mesophilic members	Thermophilic members	NAD+-dependent members	NAD(P)+-dependent	Animal-free	BSE-TSE-certified	GMO-free	GMP grade
Oxidase (S-selective monoamine)/MAO	Prozomix Ltd (CoEBio3)	www.prozomix.com, www.coebio3.manchester.ac.uk	96	Y	Y	Y	N									N	N	Y	Y	Y	Y	N
Oxidase (D-Amino acid)	Merck KGgA/ MilliporeSigma	www.sigmaaldrich.com	1	Y	N	N	N	N	N	N	N	N	N	N	N	N	N	N	N	N	N	N
	Prozomix Ltd	www.prozomix.com	40	Y	Y	Y	N	Y	N	N	Y	Y	Y	N	Y	N	N	N	Y	Y	Y	N
Oxidase (L-amino acid)	Gecco Biotech	www.gecco-biotech.com	1	Y	N	Y	N	N	N	N	N	N	Y	N	Y	Y	N	N	Y	N	N	N
	Merck KGgA/ MilliporeSigma	www.sigmaaldrich.com	3	Y	N	N	N	N	N	N	N	N	N	N	N	N	N	N	N	N	N	N
Oxidase (ascorbate)	Amano Enzyme	www.amano-enzyme.co.jp	1																			
	Asahi Kasei Pharma Corporation	www.asahi-kasei.co.jp/shindan/en/	1																			
	Merck KGgA/ MilliporeSigma	www.sigmaaldrich.com	2	Y	N	N	N	N	N	N	N	N	N	N	N	N	N	N	N	N	N	N
	Toyobo	www.toyobo-global.com	3																			
Oxidase (bilirubin)	Amano Enzyme	www.amano-enzyme.co.jp	1																			

	Asahi Kasei Pharma Corporation	www.asahi-kasei.co.jp/shindan/en/	1																			
	Merck KGgA/MilliporeSigma	www.sigmaaldrich.com	1	Y	N	N	N	N	N	N	N	N	N	N	N	N	N	N	N	N	N	N
Oxidase (cholesterol)	Amano Enzyme	www.amano-enzyme.co.jp	1																			
	Asahi Kasei Pharma Corporation	www.asahi-kasei.co.jp/shindan/en/	2																			
	Merck KGgA/MilliporeSigma	www.sigmaaldrich.com	6	Y	N	N	N	N	N	N	N	N	N	N	N	N	N	N	N	N	N	N
Oxidase (choline)	Asahi Kasei Pharma Corporation	www.asahi-kasei.co.jp/shindan/en/	1																			
	Merck KGgA/MilliporeSigma	www.sigmaaldrich.com	3	Y	Y	Y	N	N	N	N	N	N	N	N	N	N	N	N	N	N	N	N
	Prozomix Ltd/MIB	www.prozomix.com	39	Y	Y	Y	N	N	N	Y	Y	Y	Y	N	Y	N	N	N	Y	Y	Y	N
	Toyobo	www.toyobo-global.com	1																			
Oxidase (diamine)	Merck KGgA/MilliporeSigma	www.sigmaaldrich.com	1	Y	N	N	N	N	N	N	N	N	N	N	N	N	N	N	N	N	N	N
Oxidase (eugenol/vanillyl alcohol)	Gecco Biotech	www.gecco-biotech.com	2	Y	Y	Y	N	N	N	N	N	N	N	N	Y	N	N	N	Y	N	N	N
	InnoSyn BV	www.innosyn.com	1	Y	Y	Y	N	N	N	N	N	N	Y	N	Y	N	N	N	Y	Y	Y	N
Oxidase (fructosyl-amino acid)	Merck KGgA/MilliporeSigma	www.sigmaaldrich.com	1	Y	N	Y	N	N	N	N	N	N	N	N	N	N	N	N	N	N	N	N
Oxidase (galactose)	Codexis Inc.	www.codexis.com	>100 cus-tom items	Custom items	>1 month	Y	N	N	Y	Y	Y	Y	Y				-	-	N	Y	Y	N
	Merck KGgA/MilliporeSigma	www.sigmaaldrich.com	2	Y	N	N	N	N	N	N	N	N	N	N	N	N	N	N	N	N	N	N
	Prozomix Ltd/MIB	www.prozomix.com	3	Y	Y	Y	N	N	N	Y	Y	Y	Y	N	Y	N	N	N	Y	Y	Y	N

Table 2.1 *(Continued)*

Enzyme name	Supplier	Company contact details	Number of enzymes in panel	Screening quantities available (i.e. between 1 mg and 1 g of each available on immediate basis for sale)	Larger quantities available (i.e. between 1 g and 1 kg available within 1 month of request)	Attributes: Recombinant	Homologously expressed	Metagenomic	Engineered panel	Evolved panel	Sequence diverse members	Unique proprietary members	Nagoya compliant	Psychrophilic members	Mesophilic members	Thermophilic members	NAD+-dependent members	NAD(P)+-dependent	Animal-free	BSE-TSE-certified	GMO-free	GMP grade
Oxidase (glucose)	Almac	www.almacgroup.com	25	Y	Y	Y	N	N	N	N	Y	Y	Y	Y	Y	Y	N	N	Y	Y	Y	N
	Amano Enzyme	www.amano-enzyme.co.jp	4																			
	ASA Spezialenzyme GmbH	www.asa-enzyme.de	1	Y	Y	N	Y	N	N	N	N	N	N	N	Y	N	N	N	Y	Y	Y	N
	Biocatalysts Ltd	www.biocatalysts.com	1	Y	Y	N	N	N	N	N	N	N	Y	N	Y	N	N	N	Y	Y	Y	N
	EnzymeWorks	www.enzymeworking.com	5	Y	Y	Y	N	N	Y	Y	Y	Y	Y	N	Y	N	N	Y	Y	Y	Y	N
	Megazyme	www.megazyme.com	1	N	Y	N	N	N	N	N	Y	Y	N	N	Y	N	N	N	Y	N	N	N
	Merck KGgA/ MilliporeSigma	www.sigmaaldrich.com	4	Y	Y	N	N	N	N	N	N	N	N	N	N	N	N	N	N	N	N	N
	Toyobo	www.toyobo-global.com	1																			
Oxidase (L-glutamate)	Merck KGgA/ MilliporeSigma	www.sigmaaldrich.com	2	Y	N	N	N	N	N	N	N	N	N	N	N	N	N	N	N	N	N	N
Oxidase (glycerol)	Gecco Biotech	www.gecco-biotech.com	1	Y	N	Y	N	N	N	N	N	Y	Y	N	Y	Y	N	N	Y	N	N	N
Oxidase (L-alpha-glycerophosphate)	Amano Enzyme	www.amano-enzyme.co.jp	1																			
	Asahi Kasei Pharma Corporation	www.asahi-kasei.co.jp/shindan/en/	2																			
	Toyobo	www.toyobo-global.com	1																			

Oxidase (HMF)	Gecco Biotech	www.gecco-biotech.com	4	Y	N	Y	N	N	N	N	N	N	Y	N	Y	Y	N	N	Y	N	N	N
Oxidase (lactate)	Asahi Kasei Pharma Corporation	www.asahi-kasei.co.jp/shindan/en/	1																			
	Toyobo	www.toyobo-global.com	1																			
Oxidase (lysine)	Merck KGgA/MilliporeSigma	www.sigmaaldrich.com	1	Y	N	N	N	N	N	N	N	N	N	N	N	N	N	N	N	N	N	N
Oxidase (methanol)	Gecco Biotech	www.gecco-biotech.com	1	Y	N	Y	N	N	N	N	N	Y	Y	N	Y	Y	N	N	Y	N	N	N
Oxidase (NAD(P)H)	Almac	www.almacgroup.com	5	Y	Y	Y	N	N	N	N	Y	Y	Y	N	Y	Y	Y	Y	Y	Y	Y	N
	c-LEcta	www.c-LEcta.com	1	Y	Y	Y	N	Y	Y	Y	N	Y	Y	N	Y	N	Y	Y	Y	Y	Y	N
	Codexis Inc.	www.codexis.com	2 custom items	Custom items	>1 month	Y	N	N	Y	N	N	N	Y				Y	Y	N	Y	Y	N
	EnzymeWorks	www.enzymeworking.com	3	Y	Y	Y	N	N	Y	Y	Y	Y	Y	N	Y	Y	N	Y	Y	Y	Y	N
	Enzymicals AG	www.enzymicals.com	2	N	Y	Y	N	N	N	N	Y	Y	Y	N	Y	Y	N	N	Y	Y	Y	N
	GECCO Biotech	www.gecco-biotech.com	1	Y	N	Y	N	N	N	N	N	N	N	N	N	Y	Y	Y	Y	N	N	N
	InnoSyn BV	www.innosyn.com	2	Y	Y	Y	N	N	Y	N	N	N	Y	N	Y	N	Y	Y	Y	Y	Y	N
	Johnson Matthey	matthey.com/biocatalysts	2	Y	Y	Y	N	N	N	N	Y	Y	Y	N	Y	Y	Y	Y	Y	Y	Y	Y
	Prozomix Ltd	www.prozomix.com	17	Y	Y	Y	N	Y	N	N	Y	Y	Y	N	Y	N	Y	Y	Y	Y	Y	N
Oxidase (oligosaccharide)	Gecco Biotech	www.gecco-biotech.com	2	Y	N	Y	N	N	N	N	N	N	N	N	Y	Y	N	N	Y	N	N	N
Oxidase (putrescine)	Gecco Biotech	www.gecco-biotech.com	2	Y	N	Y	N	N	N	N	N	N	Y	N	Y	Y	N	N	Y	N	N	N
Oxidase (pyruvate)	Asahi Kasei Pharma Corporation	www.asahi-kasei.co.jp/shindan/en/	1																			
	Merck KGgA/MilliporeSigma	www.sigmaaldrich.com	2	Y	N	N	N	N	N	N	N	N	N	N	N	N	N	N	N	N	N	N
	Toyobo	www.toyobo-global.com	1																			

Table 2.1 *(Continued)*

Enzyme name	Supplier	Company contact details	Number of enzymes in panel	Screening quantities available (i.e. between 1 mg and 1 g of each available on immediate basis for sale)	Larger quantities available (i.e. between 1 g and 1 kg available within 1 month of request)	Attributes: Recombinant	Homologously expressed	Metagenomic	Engineered panel	Evolved panel	Sequence diverse members	Unique proprietary members	Nagoya compliant	Psychrophilic members	Mesophilic members	Thermophilic members	NAD+-dependent members	NAD(P)+-dependent	Animal-free	BSE-TSE-certified	GMO-free	GMP grade
Oxidase (sarcosine)	Asahi Kasei Pharma Corporation	www.asahi-kasei.co.jp/shindan/en/	1																			
	Merck KGgA/MilliporeSigma	www.sigmaaldrich.com	1	Y	N	N	N	N	N	N	N	N	N	N	N	N	N	N	N	N	N	N
	Toyobo	www.toyobo-global.com	1																			
Oxidase (tyramine)	Asahi Kasei Pharma Corporation	www.asahi-kasei.co.jp/shindan/en/	1																			
Oxidase (urea)	ASA Spezialenzyme GmbH	www.asa-enzyme.de	1	Y	N	Y	N	N	N	N	N	Y	N	N	Y	N	N	N	Y	Y	N	N
Oxidase (xanthine)	Merck KGgA/MilliporeSigma	www.sigmaaldrich.com	5	Y	Y	N	N	N	N	N	N	N	N	N	N	N	N	N	N	N	N	N
	Toyobo	www.toyobo-global.com	1																			
Papain	Merck KGgA/MilliporeSigma	www.sigmaaldrich.com	7	Y	N	N	N	N	N	N	N	N	N	N	N	N	N	N	N	N	N	N
Pectate lyase	Prozomix Ltd	www.prozomix.com	1	Y	Y	Y	N	N	N	N	Y	Y	Y	N	Y	N	N	N	Y	Y	Y	N
Pectinase	ASA Spezialenzyme GmbH	www.asa-enzyme.de	1	Y	Y	N	Y	N	N	N	N	N	N	N	Y	N	N	N	Y	Y	Y	N
	Biocatalysts Ltd	www.biocatalysts.com	4	Y	Y	N	N	N	N	N	N	Y	Y	N	Y	N	N	N	Y	Y	N	N
	Merck KGgA/MilliporeSigma	www.sigmaaldrich.com	5	Y	N	N	N	N	N	N	N	N	N	N	N	N	N	N	N	N	N	N

Pectolyase	Merck KGgA/ MilliporeSigma	www.sigmaaldrich.com	2	Y	N	N	N	N	N	N	N	N	N	N	N	N	N	N	N	N	N	N
Pectin methylesterase	Merck KGgA/ MilliporeSigma	www.sigmaaldrich.com	1	Y	N	N	N	N	N	N	N	N	N	N	N	N	N	N	N	N	N	N
Penicillinase	Merck KGgA/ MilliporeSigma	www.sigmaaldrich.com	2	Y	N	N	N	N	N	N	N	N	N	N	N	N	N	N	N	N	N	N
Peptidase (Carboxy)	ASA Spezialenzyme GmbH	www.asa-enzyme.de	1	Y	Y	Y	N	N	Y	N	N	Y	N	N	Y	N	N	N	Y	Y	N	N
	Merck KGgA/ MilliporeSigma	www.sigmaaldrich.com	6	Y	Y	N	N	N	N	N	N	N	N	N	N	N	N	N	N	N	N	N
Peptidase/Protease	Amano Enzyme	www.amano-enzyme.co.jp	10																			
	ASA Spezialenzyme GmbH	www.asa-enzyme.de	4	Y	Y	Y	N	N	N	N	N	Y	N	N	Y	N	N	N	Y	Y	N	N
	Biocatalysts Ltd	www.biocatalysts.com	16	Y	Y	Y/N	Y/N	Y/N	N	N	Y	Y	Y	N	Y	Y	N	N	Y	Y	Y/N	N
	ChiralVision	www.chiralvision.com	22	Y	Y	N	N	N	N	N	Y	Y	N	N	Y	Y	N	N	Y	Y	Y	N
	Enzagen	www.enzagen.com	10	Y	Y	y/N	Y/N	N	N	N	Y	N	Y	N	Y	Y/N	N	N	Y	Y	Y/N	N
	EUCODIS	www.eucodis.com	3	Y	Y	Y	N	N	Y	N	Y	Y	Y	N	Y	N	N	N	Y	Y	Y	N
	Megazyme	www.megazyme.com	3	N	Y	Y/N	N	N	N	N	Y	Y	N	N	Y	N	N	N	Y	N	N	N
	Novozymes A/S	pharmaceuticals@novozyr.com	8	N	Y	Y																
	Toyobo	www.toyobo-global.com	1																			
Peptidoglycan lytic exotransglycosylase	NZYTech	www.nzytech.com	4	Y	N	Y	N	N	N	N	N	N	N	N	Y	N	N	N	N	N	N	N
Peroxidase	Amano Enzyme	www.amano-enzyme.co.jp	1																			
	Merck KGgA/ MilliporeSigma	www.sigmaaldrich.com	9	Y	Y	N	N	N	N	N	N	N	N	N	N	N	N	N	N	N	N	N
	Toyobo	www.toyobo-global.com	1																			

Table 2.1 *(Continued)*

Enzyme name	Supplier	Company contact details	Number of enzymes in panel	Screening quantities available (i.e. between 1 mg and 1 g of each available on immediate basis for sale)	Larger quantities available (i.e. between 1 g and 1 kg available within 1 month of request)	Attributes																
						Recombinant	Homologously expressed	Metagenomic	Engineered panel	Evolved panel	Sequence diverse members	Unique proprietary members	Nagoya compliant	Psychrophilic members	Mesophilic members	Thermophilic members	NAD+-dependent members	NAD(P)+-dependent	Animal-free	BSE-TSE-certified	GMO-free	GMP grade
Peroxidase (dye-decolorizing – DyP)	Gecco Biotech	www.gecco-biotech.com	10	Y	Y	Y	N	N	Y	Y	N	N	Y	N	Y	Y	N	N	Y	Y	Y	Y
Peroxidase (manganese)	Merck KGgA/ MilliporeSigma	www.sigmaaldrich.com	1	Y	N	N	N	N	N	N	N	N	N	N	N	N	N	N	N	N	N	N
Peroxidase (myelo)	Merck KGgA/ MilliporeSigma	www.sigmaaldrich.com	1	Y	N	N	N	N	N	N	N	N	N	N	N	N	N	N	N	N	N	N
Peroxygenase	EUCODIS	www.eucodis.com	6	Y	Y	Y	N	N	Y	N	Y	Y	Y	N	Y	N	N	N	Y	Y	Y	N
	Johnson Matthey	matthey.com/biocatalysts	5	N	N	Y	N	N	N	N	Y	Y	Y	N	Y	Y	N	N	Y	Y	Y	Y
Phosphatase	Almac	www.almacgroup.com	96	Y	Y	Y	N	N	N	N	Y	Y	Y	N	Y	Y	N	N	Y	Y	Y	N
	Asahi Kasei Pharma Corporation	www.asahi-kasei.co.jp/ shindan/en/	2																			
	EnzymeWorks	www.enzymeworking .com	5	Y	Y	Y	N	N	Y	Y	Y	Y	Y	N	Y	Y	N	N	Y	Y	Y	N
	Enzymicals AG	www.enzymicals.com	1	N	Y	Y	N	N	N	N	Y	Y	Y	N	Y	Y	N	N	Y	Y	Y	N
	Megazyme	www.megazyme.com	2	N	Y	Y	N	N	N	N	Y	Y	N	N	Y	N	N	N	Y	N	N	N
	Merck KGgA/ MilliporeSigma	www.sigmaaldrich.com	18	Y	N	Y	N	N	N	N	N	N	N	N	N	N	N	N	N	N	N	N
	Prozomix Ltd	www.prozomix.com	23	Y	Y	Y	N	N	N	N	Y	Y	Y	N	Y	N	N	N	Y	Y	Y	N
	Toyobo	www.toyobo-global.com	1																			

Phosphatase (diisopropyl-fluoro)	Merck KGgA/ MilliporeSigma	www.sigmaaldrich.com	1	Y	N	N	N	N	N	N	N	N	N	N	N	N	N	N	N	N	N	N
Phosphatase (glucose-6-	Merck KGgA/ MilliporeSigma	www.sigmaaldrich.com	2	Y	N	N	N	N	N	N	N	N	N	N	N	N	N	N	N	N	N	N
Phosphodiesterase	Asahi Kasei Pharma Corporation	www.asahi-kasei.co.jp/ shindan/en/	1																			
	Biocatalysts Ltd	www.biocatalysts.com	1	Y	Y	N	Y	N	N	N	N	N	Y	N	Y	N	N	N	Y	Y	Y	N
	Merck KGgA/ MilliporeSigma	www.sigmaaldrich.com	11	Y	N	Y	N	N	N	N	N	N	N	N	N	N	N	N	N	N	N	N
	Prozomix Ltd	www.prozomix.com	17	Y	Y	Y	N	Y/ N	N	N	Y	Y	Y	N	Y	N	N	N	Y	Y	Y	N
Phosphofructokinase	Asahi Kasei Pharma Corporation	www.asahi-kasei.co.jp/ shindan/en/	1																			
	Merck KGgA/ MilliporeSigma	www.sigmaaldrich.com	2	Y	N	N																
Phospho-β-galactosidase	NZYTech	www.nzytech.com	4	Y	N	Y	N	N	N	N	N	N	N	N	Y	N	N	N	N	N	N	N
alpha-Phosphoglucomutase	Megazyme	www.megazyme.com	1	N	Y	Y	N	N	N	N	Y	Y	N	N	Y	N	N	N	Y	N	N	N
	Merck KGgA/ MilliporeSigma	www.sigmaaldrich.com	5	Y	Y	N	N	N	N	N	N	N	N	N	N	N	N	N	N	N	N	N
Phospholipase	Asahi Kasei Pharma Corporation	www.asahi-kasei.co.jp/ shindan/en/	5																			
	Biocatalysts Ltd	www.biocatalysts.com	2	Y	Y	Y/ N	Y/ N	N	N	N	N	N	Y	N	Y	N	N	N	Y	Y	Y	N
	EnzymeWorks	www.enzymeworking .com	4	Y	Y	Y	Y	N	Y	Y	Y	Y	Y	N	Y	N	N	N	Y	Y	Y	N
	EUCODIS	www.eucodis.com	4	Y	Y	Y	N	N	Y	N	Y	Y	Y	N	Y	N	N	N	Y	Y	Y	N
	Evoxx technologies	www.evoxx.com	10	N	Y	Y	N	Y	Y	Y	Y	Y	Y	N	Y	Y			Y	Y	Y	N
	Merck KGgA/ MilliporeSigma	www.sigmaaldrich.com	15	Y	N	N	N	N	N	N	N	N	N	N	N	N	N	N	N	N	N	N
Phosphorylase a	Merck KGgA/ MilliporeSigma	www.sigmaaldrich.com	1	Y	N	N	N	N	N	N	N	N	N	N	N	N	N	N	N	N	N	N

Table 2.1 *(Continued)*

Enzyme name	Supplier	Company contact details	Number of enzymes in panel	Screening quantities available (i.e. between 1 mg and 1 g of each available on immediate basis for sale)	Larger quantities available (i.e. between 1 g and 1 kg available within 1 month of request)	Attributes																
						Recombinant	Homologously expressed	Metagenomic	Engineered panel	Evolved panel	Sequence diverse members	Unique proprietary members	Nagoya compliant	Psychrophilic members	Mesophilic members	Thermophilic members	NAD+-dependent members	NAD(P)+-dependent	Animal-free	BSE-TSE-certified	GMO-free	GMP grade
Phosphorylase b	Merck KGgA/ MilliporeSigma	www.sigmaaldrich.com	2	Y	N	N	N	N	N	N	N	N	N	N	N	N	N	N	N	N	N	N
Phosphorylase (galactosyl-N-acetylhexosamine)	NZYTech	www.nzytech.com	2	Y	N	Y	N	N	N	N	N	N	N	N	Y	N	N	N	N	N	N	N
Phosphorylase (maltose)	Merck KGgA/ MilliporeSigma	www.sigmaaldrich.com	1	Y	N	Y	N	N	N	N	N	N	N	N	N	N	N	N	N	N	N	N
Phosphorylase (mannosylglucose)	NZYTech	www.nzytech.com	2	Y	N	Y	N	N	N	N	N	N	N	N	Y	N	N	N	N	N	N	N
Phosphorylase (S-methyl-5′-thioadenosine)	Merck KGgA/ MilliporeSigma	www.sigmaaldrich.com	1	Y	N	N	N	N	N	N	N	N	N	N	N	N	N	N	N	N	N	N
Phosphorylase (nucleoside)	Merck KGgA/ MilliporeSigma	www.sigmaaldrich.com	2	Y	N	N	N	N	N	N	N	N	N	N	N	N	N	N	N	N	N	N
Phosphorylase (polynucleotide)	Merck KGgA/ MilliporeSigma	www.sigmaaldrich.com	2	Y	N	Y	N	N	N	N	N	N	N	N	N	N	N	N	N	N	N	N
Phosphorylase (purine nucleoside)	Asahi Kasei Pharma Corporation	www.asahi-kasei.co.jp/shindan/en/	1																			
	Codexis Inc.	www.codexis.com	>100 custom items	Custom items	>1 month	Y	N	N	Y	Y	Y	Y	Y				-	-	N	Y	Y	N

	EnzymeWorks	www.enzymeworking .com	8	Y	Y	Y	N	N	Y	Y	Y	Y	Y	N	Y	Y	N	N	Y	Y	Y	N
	Evoxx technologies	www.evoxx.com	2	Y	Y	Y	N	Y	Y	Y	Y	Y	Y	N	Y	Y			Y	Y	Y	N
	Merck KGgA/ MilliporeSigma	www.sigmaaldrich.com	1	Y	N	Y	N	N	N	N	N	N	N	N	N	N	N	N	N	N	N	N
	Toyobo	www.toyobo-global.com	1																			
Phosphorylase (pyrimidine nucleoside)	Merck KGgA/ MilliporeSigma	www.sigmaaldrich.com	1	Y	N	Y	N	N	N	N	N	N	N	N	N	N	N	N	N	N	N	N
Phosphorylase (thymidine)	Merck KGgA/ MilliporeSigma	www.sigmaaldrich.com	2	Y	N	Y	N	N	N	N	N	N	N	N	N	N	N	N	N	N	N	N
Phosphorylase (sucrose)	Merck KGgA/ MilliporeSigma	www.sigmaaldrich.com	2	Y	N	Y	N	N	N	N	N	N	N	N	N	N	N	N	N	N	N	N
	Prozomix Ltd	www.prozomix.com	8	Y	Y	Y	N	N	N	N	Y	Y	Y	N	Y	N	N	N	Y	Y	Y	N
Phosphotransacetylase	Megazyme	www.megazyme.com	1	N	Y	Y	N	N	N	N	Y	Y	N	N	Y	N	N	N	Y	N	N	N
	Merck KGgA/ MilliporeSigma	www.sigmaaldrich.com	1	Y	N	N	N	N	N	N	N	N	N	N	N	N	N	N	N	N	N	N
Phosphotransferase	EnzymeWorks	www.enzymeworking .com	5	Y	Y	Y	N	N	Y	Y	Y	Y	Y	N	Y	Y	N	N	Y	Y	Y	N
	Enzymicals AG	www.enzymicals.com	3	Y	Y	Y	N	N	N	N	Y	Y	Y	N	Y	Y	N	N	Y	Y	Y	N
Phosphotriesterase	Almac	www.almacgroup.com	8	Y	Y	Y	N	N	N	N	Y	Y	Y	N	Y	Y	N	N	Y	Y	Y	N
Phytase	Enzagen	www.enzagen.com	1																			
	Merck KGgA/ MilliporeSigma	www.sigmaaldrich.com	1	T	N	N	N	N	N	N	N	N	N	N	N	N	N	N	N	N	N	N
	Protéus	www.proteus.fr	3	Y	N	Y	N	N	Y	Y	Y	Y	Y	N	N	Y	N	N	Y	Y	Y	N
PNGase F	Merck KGgA/ MilliporeSigma	www.sigmaaldrich.com	5	Y	N	Y	N	N	N	N	N	N	N	N	N	N	N	N	N	N	N	N
Polygalacturonase	Megazyme	www.megazyme.com	3	N	Y	Y/ N	N	N	N	N	Y	Y	N	N	Y	N	N	N	Y	N	N	N
	Merck KGgA/ MilliporeSigma	www.sigmaaldrich.com	3	Y	N	N	N	N	N	N	N	N	N	N	N	N	N	N	N	N	N	N
	NZYTech	www.nzytech.com	6	Y	N	Y	N	N	N	N	N	N	N	N	Y	Y	N	N	N	N	N	N

Table 2.1 (*Continued*)

Enzyme name	Supplier	Company contact details	Number of enzymes in panel	Screening quantities available (i.e. between 1 mg and 1 g of each available on immediate basis for sale)	Larger quantities available (i.e. between 1 g and 1 kg available within 1 month of request)	Attributes																
						Recombinant	Homologously expressed	Metagenomic	Engineered panel	Evolved panel	Sequence diverse members	Unique proprietary members	Nagoya compliant	Psychrophilic members	Mesophilic members	Thermophilic members	NAD+-dependent members	NAD(P)+-dependent	Animal-free	BSE-TSE-certified	GMO-free	GMP grade
Porphyranase	NZYTech	www.nzytech.com	4	Y	N	Y	N	N	N	N	N	N	N	N	Y	N	N	N	N	N	N	N
Prolidase	Merck KGgA/ MilliporeSigma	www.sigmaaldrich.com	1	Y	N	N	N	N	N	N	N	N	N	N	N	N	N	N	N	N	N	N
Proteinase K	Merck KGgA/ MilliporeSigma	www.sigmaaldrich.com	6	Y	N	N	N	N	N	N	N	N	N	N	N	N	N	N	N	N	N	N
Pullulanase	GECCO Biotech	www.gecco-biotech.com	1	Y	N	Y	N	N	N	N	N	N	N	N	Y	N	N	N	Y	N	N	N
	Megazyme	www.megazyme.com	3	N	Y	Y/N	N	N	N	N	Y	Y	N	N	Y	N	N	N	Y	N	N	N
	Merck KGgA/ MilliporeSigma	www.sigmaaldrich.com	1	Y	N	N	N	N	N	N	N	N	N	N	N	N	N	N	N	N	N	N
	NZYTech	www.nzytech.com	3	Y	N	Y	N	N	N	N	N	N	N	N	Y	Y	N	N	N	N	N	N
	Prozomix Ltd	www.prozomix.com	4	Y	Y	Y	N	N	N	N	Y	Y	Y	N	Y	N	N	N	Y	Y	Y	N
Pyrophosphatase	Merck KGgA/ MilliporeSigma	www.sigmaaldrich.com	2	Y	N	N	N	N	N	N	N	N	N	N	N	N	N	N	N	N	N	N
Quinone reductase (F420-dependent)	Gecco Biotech	www.gecco-biotech.com	5	Y	N	Y	N	N	N	N	N	N	Y	N	Y	Y	N	N	Y	N	N	N
Quinone reductase	Gecco Biotech	www.gecco-biotech.com	5	Y	Y	Y	N	N	N	N	N	N	N	N	Y	N	Y	N	Y	N	N	N

Racemase	Prozomix Ltd	www.prozomix.com	96	Y	Y	Y	N	Y/N	N	N	Y	Y	Y	N	Y	N	N	N	Y	Y	Y	N
Reductase (F420)	Gecco Biotech	www.gecco-biotech.com	1	Y	N	Y	N	N	N	N	N	N	Y	N	Y	N	N	N	Y	N	N	N
Reductive aminase	Almac	www.almacgroup.com	50	Y	Y	Y	N	N	N	N	Y	Y	Y	N	Y	Y	N	Y	Y	Y	Y	N
	Codexis Inc.	www.codexis.com	24 (kit) + custom items	Y from kit	>1 month	Y	N	N	Y	Y	Y	Y	Y				Y	N	N	Y	Y	N
	EnzymeWorks	www.enzymeworking.com	10	Y	Y	Y	N	N	Y	Y	Y	Y	Y	N	Y	Y	Y	Y	Y	Y	Y	N
	Enzymicals AG	www.enzymicals.com	2	N	Y	Y	N	N	N	N	Y	Y	Y	N	Y	Y	N	Y	Y	Y	Y	N
	Prozomix Ltd	www.prozomix.com	576	Y	Y	Y	N	Y	N	N	Y	Y	Y	N	Y	N	Y	Y	Y	Y	Y	N
Rhamnogalacturonase	NZYTech	www.nzytech.com	13	Y	N	Y	N	N	N	N	N	N	N	N	Y	N	N	N	N	N	N	N
alpha-L-Rhamnosidase	Megazyme	www.megazyme.com	1	N	Y	Y	N	N	N	N	Y	Y	N	N	Y	N	N	N	Y	N	N	N
	Prozomix Ltd	www.prozomix.com	2	Y	Y	Y	N	N	N	N	Y	Y	Y	N	Y	N	N	N	Y	Y	Y	N
Rhodanese	Merck KGgA/ MilliporeSigma	www.sigmaaldrich.com	1	Y	N	N	N	N	N	N	N	N	N	N	N	N	N	N	N	N	N	N
Sialidase	Megazyme	www.megazyme.com	2	N	Y	Y	N	N	N	N	Y	Y	N	N	Y	N	N	N	Y	N	N	N
	Merck KGgA/ MilliporeSigma	www.sigmaaldrich.com	6	Y	N	N	N	N	N	N	N	N	N	N	N	N	N	N	N	N	N	N
	NZYTech	www.nzytech.com	6	Y	N	Y	N	N	N	N	N	N	N	N	Y	N	N	N	N	N	N	N
Sialyltransferase	Merck KGgA/ MilliporeSigma	www.sigmaaldrich.com	2	Y	N	Y	N	N	N	N	N	N	N	N	N	N	N	N	N	N	N	N
	Prozomix Ltd	www.prozomix.com	10	Y	Y	Y	N	N	N	N	Y	Y	Y	N	Y	N	N	N	Y	Y	Y	N
Sphingomyelinase	Asahi Kasei Pharma Corporation	www.asahi-kasei.co.jp/shindan/en/	1																			
	Merck KGgA/ MilliporeSigma	www.sigmaaldrich.com	3	Y	Y	N	N	N	N	N	N	N	N	N	N	N	N	N	N	N	N	N

Table 2.1 *(Continued)*

Enzyme name	Supplier	Company contact details	Number of enzymes in panel	Screening quantities available (i.e. between 1 mg and 1 g of each available on immediate basis for sale)	Larger quantities available (i.e. between 1 g and 1 kg available within 1 month of request)	Attributes: Recombinant	Homologously expressed	Metagenomic	Engineered panel	Evolved panel	Sequence diverse members	Unique proprietary members	Nagoya compliant	Psychrophilic members	Mesophilic members	Thermophilic members	NAD+-dependent members	NAD(P)+-dependent	Animal-free	BSE-TSE-certified	GMO-free	GMP grade
Sulfatase	Merck KGgA/ MilliporeSigma	www.sigmaaldrich.com	5	Y	N	N	N	N	N	N	N	N	N	N	N	N	N	N	N	N	N	N
Superoxide dismutase	Merck KGgA/ MilliporeSigma	www.sigmaaldrich.com	8	Y	Y	N	N	N	N	N	N	N	N	N	N	N	N	N	N	N	N	N
	Toyobo	www.toyobo-global.com	1																			
Synthase (allene Oxide)	Merck KGgA/ MilliporeSigma	www.sigmaaldrich.com	1	Y	N	N	N	N	N	N	N	N	N	N	N	N	N	N	N	N	N	N
Synthase (citrate)	Megazyme	www.megazyme.com	1	N	Y	Y	N	N	N	N	Y	Y	N	N	Y	N	N	N	Y	N	N	N
	Merck KGgA/ MilliporeSigma	www.sigmaaldrich.com	1	Y	N	N	N	N	N	N	N	N	N	N	N	N	N	N	N	N	N	N
Synthase (mannosylglycerate)	Prozomix Ltd	www.prozomix.com	1	Y	Y	Y	N	N	N	N	Y	Y	Y	N	Y	N	N	N	Y	Y	Y	N
Synthase (nitric oxide)	Merck KGgA/ MilliporeSigma	www.sigmaaldrich.com	1	Y	N	Y	N	N	N	N	N	N	N	N	N	N	N	N	N	N	N	N
Synthetase (acyl-CoA)	Asahi Kasei Pharma Corporation	www.asahi-kasei.co.jp/ shindan/en/	1																			
	Merck KGgA/ MilliporeSigma	www.sigmaaldrich.com	2	Y	N	N	N	N	N	N	N	N	N	N	N	N	N	N	N	N	N	N
	Megazyme	www.megazyme.com	1	N	Y	Y	N	N	N	N	Y	Y	N	N	Y	N	N	N	Y	N	N	N

Synthetase (CMP-Sialic acid)	Merck/MilliporeSigma KGgA	www.sigmaaldrich.com	1	Y	N	Y	N	N	N	N	N	N	N	N	N	N	N	N	N	N	N	N
Synthetase (L-glutamine)	Asahi Kasei Pharma Corporation	www.asahi-kasei.co.jp/shindan/en/	1																			
	Merck KGgA/MilliporeSigma	www.sigmaaldrich.com	2	Y	N	N	N	N	N	N	N	N	N	N	N	N	N	N	N	N	N	N
Synthetase (NAD)	Asahi Kasei Pharma Corporation	www.asahi-kasei.co.jp/shindan/en/	1																			
	EnzymeWorks	www.enzymeworking.com	2	Y	Y	Y	N	n	Y	Y	Y	Y	N	N	N	N	N	N	Y	Y	Y	N
Synthetase (succinyl-CoA)	Megazyme	www.megazyme.com	1	N	Y	Y	N	N	N	N	Y	Y	N	N	Y	N	N	N	Y	N	N	N
Tannase	Almac	www.almacgroup.com	50	Y	Y	Y	N	N	N	N	Y	Y	Y	N	Y	Y	N	N	Y	Y	Y	N
Tautomerase	Merck KGgA/MilliporeSigma	www.sigmaaldrich.com	1	Y	N	N	N	N	N	N	N	N	N	N	N	N	N	N	N	N	N	N
Thioglucosidase	Merck KGgA/MilliporeSigma	www.sigmaaldrich.com	2	Y	N	N	N	N	N	N	N	N	N	N	N	N	N	N	N	N	N	N
Thioredoxin reductase	Merck KGgA/MilliporeSigma	www.sigmaaldrich.com	2	Y	N	N	N	N	N	N	N	N	N	N	N	N	N	N	N	N	N	N
Topoisomerase	Merck KGgA/MilliporeSigma	www.sigmaaldrich.com	3	Y	N	Y	N	N	N	N	N	N	N	N	N	N	N	N	N	N	N	N
Transaminase (aminotransferase)	Almac	www.almacgroup.com	>300	Y	Y	Y	N	Y	Y	Y	Y	Y	Y	N	Y	Y	N	N	Y	Y	Y	N
	Asahi Kasei Pharma Corporation	www.asahi-kasei.co.jp/shindan/en/	2																			
	c-LEcta GmbH	www.c-lecta.com	15	Y	Y	Y	N	Y	Y	Y	Y	Y	Y	N	Y	N	N	N	Y	Y	Y	N
	Codexis Inc.	www.codexis.com	24 (kit) + custom items	Y from kit	Up to 100 g from kit	Y	N	N	Y	Y	Y	Y	Y				-	-	N	Y	Y	N
	Enzymaster	www.enzymaster.de	83	N	Y	Y	N	N	Y	Y	Y	Y	Y	N	Y	Y	N	N	Y	Y	Y	N
	EnzymeWorks	www.enzymeworking.com	100	Y	Y	Y	N	N	Y	Y	Y	Y	Y	N	Y	Y	N	N	Y	Y	Y	N

Table 2.1 *(Continued)*

Enzyme name	Supplier	Company contact details	Number of enzymes in panel	Screening quantities available (i.e. between 1 mg and 1 g of each available on immediate basis for sale)	Larger quantities available (i.e. between 1 g and 1 kg available within 1 month of request)	Attributes: Recombinant	Homologously expressed	Metagenomic	Engineered panel	Evolved panel	Sequence diverse members	Unique proprietary members	Nagoya compliant	Psychrophilic members	Mesophilic members	Thermophilic members	NAD+-dependent members	NAD(P)+-dependent	Animal-free	BSE-TSE-certified	GMO-free	GMP grade
	Enzymicals AG	www.enzymicals.com	20	Y	Y	Y	N	N	N	N	Y	Y	Y	N	Y	Y	N	N	Y	Y	Y	N
	Evoxx technologies GmbH	www.evoxx.com	18	Y	Y	Y	Y	Y	Y	Y	Y	Y	N				Y	Y	Y	Y	Y	N
	Gecco Biotech	www.gecco-biotech.com	4	Y	N	Y	N	N	N	N	N	N	N	N	Y	N	N	N	Y	N	N	N
	InnoSyn BV	www.innosyn.com	74	Y	Y	Y	N	Y/N	Y/N	Y/N	Y	N	Y	Y	Y	Y	N	N	Y	Y	Y	N
	Megazyme	www.megazyme.com	3	N	Y	Y	N	N	N	N	Y	Y	N	N	Y	N	N	N	Y	N	N	N
	Merck KGgA/ MilliporeSigma	www.sigmaaldrich.com	3	Y	N	N	N	N	N	N	N	N	N	N	N	N	N	N	N	N	N	N
	Protéus	www.proteus.fr	9	Y	N	Y	N	N	Y	Y	Y	Y	Y	N	Y	Y	N	N	Y	Y	Y	N
	Prozomix Ltd	www.prozomix.com	384	Y	Y	Y	N	Y	N	N	Y	Y	Y	N	Y	N	N	N	Y	Y	Y	N
	Syncozymes	en.syncozymes.com	58	Y	Y	Y	Y	N	Y	N	Y	N	N	N	N	N	N	N	Y	N	N	N
R-Transaminase	Almac	www.almacgroup.com	>200	Y	Y	Y	N	Y	Y	Y	Y	Y	Y	N	Y	Y	N	N	Y	Y	Y	N
	Codexis Inc.	www.codexis.com	12 out of 24 (kit) + cus-tom items	Y from kit	Up to 100 g from kit	Y	N	N	Y	Y	Y	Y	Y				-	-	N	Y	Y	N

	EnzymeWorks	www.enzymeworking.com	40	Y	Y	Y	N	NY	Y	Y	Y	Y	Y	N	Y	Y	N	N	Y	Y	Y	N
S-Transaminase	Johnson Matthey	www.matthey.com/biocatalysts/	100	Y	Y	Y	N	Y	Y	N	Y	Y	Y	N	Y	Y	N	N	Y	Y	Y	Y
	Almac	www.almacgroup.com	>100	Y	Y	Y	N	Y	Y	Y	Y	Y	Y	N	Y	Y	N	N	Y	Y	Y	N
	Codexis Inc.	www.codexis.com	12 out of 24 (kit) custom items	Y from kit	Up to 100 g from kit	Y	N	N	Y	Y	Y	Y	Y				-	-	N	Y	Y	N
Transglutaminase	EnzymeWorks	www.enzymeworking.com	50	Y	Y	Y	N	N	Y	Y	Y	Y	Y	N	N	N	N	N	Y	Y	Y	N
	Johnson Matthey	www.matthey.com/biocatalysts/	170	Y	Y	Y	N	Y	Y	N	Y	Y	Y	N	Y	Y	N	N	Y	Y	Y	Y
	ASA Spezialenzyme GmbH	www.asa-enzyme.de	1	Y	Y	N	Y	N	N	N	N	Y	N	N	Y	N	N	N	Y	Y	N	N
	Codexis Inc.	www.codexis.com	Custom items	Custom items	>1 month	Y	N	N	Y	Y	N	Y	Y				-	-	N	Y	Y	N
	EnzymeWorks	www.enzymeworking.com	2	Y	Y	Y	N	N	Y	Y	Y	Y	N	N	N	N	N	N	Y	Y	Y	N
	Merck KGgA/MilliporeSigma	www.sigmaaldrich.com	1	Y	N	N	N	N	N	N	N	N	N	N	N	N	N	N	N	N	N	N
Transketolase	Almac	www.almacgroup.com	50	Y	Y	Y	N	N	N	N	Y	Y	Y	N	Y	Y	N	N	Y	Y	Y	N
	Merck KGgA/MilliporeSigma	www.sigmaaldrich.com	1	Y	N	N	N	N	N	N	N	N	N	N	N	N	N	N	N	N	N	N
Trehalase	Megazyme	www.megazyme.com	1	N	Y	Y	N	N	N	N	Y	Y	N	N	Y	N	N	N	Y	N	N	N
	Merck KGgA/MilliporeSigma	www.sigmaaldrich.com	1	Y	N	N	N	N	N	N	N	N	N	N	N	N	N	N	N	N	N	N
	NZYTech	www.nzytech.com	13	Y	N	Y	N	N	N	N	N	N	N	N	Y	N	N	N	N	N	N	N

Table 2.1 *(Continued)*

Enzyme name	Supplier	Company contact details	Number of enzymes in panel	Screening quantities available (i.e. between 1 mg and 1 g of each available on immediate basis for sale)	Larger quantities available (i.e. between 1 g and 1 kg available within 1 month of request)	Attributes: Recombinant	Homologously expressed	Metagenomic	Engineered panel	Evolved panel	Sequence diverse members	Unique proprietary members	Nagoya compliant	Psychrophilic members	Mesophilic members	Thermophilic members	NAD+-dependent members	NAD(P)+-dependent	Animal-free	BSE-TSE-certified	GMO-free	GMP grade
Trehalose synthase	Codexis Inc.	www.codexis.com	6 custom items	Custom items	>1 month	Y	N	N	N	N	Y	N	Y				-	-	N	Y	Y	N
	Prozomix Ltd	www.prozomix.com	2	Y	Y	Y	N	N	N	N	Y	Y	Y	N	Y	N	N	N	Y	Y	Y	N
Tyrosinase	Almac	www.almacgroup.com	5	Y	Y	Y	N	N	N	N	Y	Y	Y	N	Y	Y	N	N	Y	Y	Y	N
	Merck KGgA/ MilliporeSigma	www.sigmaaldrich.com	1	Y	Y	N	N	N	N	N	N	N	N	N	N	N	N	N	N	N	N	N
Urease	Almac	www.almacgroup.com	96	Y	Y	Y	N	N	Y	N	Y	Y	Y	N	Y	Y	N	N	Y	Y	Y	N
	Merck KGgA/ MilliporeSigma	www.sigmaaldrich.com	7	Y	N	N	N	N	N	N	N	N	N	N	N	N	N	N	N	N	N	N
	Toyobo	www.toyobo-global.com	1																			
Uricase	Amano Enzyme	www.amano-enzyme.co.jp	1																			
	Asahi Kasei Pharma Corporation	www.asahi-kasei.co.jp/shindan/en/	1																			
	Merck KGgA/ MilliporeSigma	www.sigmaaldrich.com	3	Y	N	N	N	N	N	N	N	N	N	N	N	N	N	N	N	N	N	N
	Toyobo	www.toyobo-global.com	2																			

Uridine-5′-diphosphogalactose 4-Epimerase	Merck KGgA/ MilliporeSigma	www.sigmaaldrich.com	1	Y	N	N	N	N	N	N	N	N	N	N	N	N	N	N	N	N	N	N
Uridine-5′-diphosphoglucose Pyrophosphorylase	Merck KGgA/ MilliporeSigma	www.sigmaaldrich.com	2	Y	N	N	N	N	N	N	N	N	N	N	N	N	N	N	N	N	N	N
Urokinase	Merck KGgA/ MilliporeSigma	www.sigmaaldrich.com	1	Y	N	N	N	N	N	N	N	N	N	N	N	N	N	N	N	N	N	N
Xylanase	ASA Spezialenzyme GmbH	www.asa-enzyme.de	1	Y	Y	N	Y	N	N	N	N	N	N	N	Y	N	N	N	Y	Y	Y	N
	Biocatalysts Ltd	www.biocatalysts.com	3	Y	Y	Y/ N	Y/ N	N	N	N	N	N	Y	N	Y	N	N	N	Y	Y	Y	N
	EUCODIS	www.eucodis.com	2	Y	Y	Y	N	N	Y	N	Y	Y	Y	N	Y	Y	N	N	Y	Y	Y	N
	Megazyme	www.megazyme.com	11	N	Y	Y/ N	N	N	N	N	Y	Y	N	N	Y	N	N	N	Y	N	N	N
	Merck KGgA/ MilliporeSigma	www.sigmaaldrich.com	2	Y	N	N	N	N	N	N	N	N	N	N	N	N	N	N	N	N	N	N
	NZYTech	www.nzytech.com	40	Y	N	Y	N	N	N	N	N	N	N	N	Y	Y	N	N	N	N	N	N
	Proteus	www.proteus.fr	6	Y	N	Y	N	N	Y	Y	Y	Y	Y	N	Y	Y	N	N	Y	Y	Y	N
	Prozomix Ltd	www.prozomix.com	18	Y	Y	Y	N	N	N	N	Y	Y	Y	N	Y	N	N	N	Y	Y	Y	N
Xyloglucanase	Megazyme	www.megazyme.com	2	N	Y	Y	N	N	N	N	Y	Y	N	N	Y	N	N	N	Y	N	N	N
	NZYTech	www.nzytech.com	7	Y	N	Y	N	N	N	N	N	N	N	N	Y	Y	N	N	N	N	N	N
Xylosidase	Biocatalysts Ltd	www.biocatalysts.com	1	Y	N	Y	Y	Y	N	N	N	Y	Y	N	Y	N	N	N	Y	Y	Y	N
	EUCODIS	www.eucodis.com	6	Y	Y	Y	N	N	Y	N	Y	Y	Y	N	Y	Y	N	N	Y	Y	Y	N
	Megazyme	www.megazyme.com	3	N	Y	Y	N	N	N	N	Y	Y	N	N	Y	N	N	N	Y	N	N	N
	NZYTech	www.nzytech.com	18	Y	N	Y	N	N	N	N	N	N	N	N	Y	Y	N	N	N	N	N	N
	Prozomix Ltd	www.prozomix.com	4	Y	Y	Y	N	N	N	N	Y	Y	Y	N	Y	N	N	N	Y	Y	Y	N

2.2 Bioprocess Service Providers

Table 2.2 *Bioprocess service providers.*

<table>
<tr><th>Company Name</th><th>Website</th><th>Enzyme/biocatalyst scale-up service</th><th>Bespoke enzyme preparation service</th><th>Enzyme immobilisation service</th><th>Directed evolution service</th><th>In silico enzyme engineering service</th><th>Enzyme screening service</th><th>Microbial screening service</th><th>Metagenomic screening service</th><th>Biotrans process development service</th><th>Non-GMP Biotrans manufacturing serv</th><th>GMP Biotrans manufacturing service</th></tr>
<tr><td>Almac</td><td>www.almacgroup.com</td><td>Y</td><td>Y</td><td>Y</td><td>Y</td><td>Y</td><td>Y</td><td>Y</td><td></td><td>Y</td><td>Y</td><td>N</td></tr>
<tr><td colspan="13">Almac Biocatalysis is a leader in the discovery and application of enzymes in chemical processes from a few kg to commercial manufacture. Almac has a proven track record of >15 years in the field of biotransformation process development and scale-up and integrates both chemistry and biology to deliver processes for their customers. Almac Biocatalysis can also increase enzyme performance through enzyme engineering and enzyme immobilisation and subsequent supply. Additional services include enzyme discovery, fermentation development and analytical development to support the process. Almac's goal is to meet customer needs for a rapid, cost-effective and integrated offering for supply of enzyme and/or chemical product. For more information, see https://www.almacgroup.com/api-chemical-development/biocatalysis-solutions/.</td></tr>
<tr><td>Arzeda</td><td>www.arzeda.com</td><td></td><td></td><td></td><td></td><td></td><td></td><td></td><td></td><td></td><td></td><td></td></tr>
<tr><td colspan="13">Arzeda, the Protein Design Company™, pairs advanced, proprietary computational and machine learning algorithms with DNA synthesis and high-throughput laboratory testing to design and develop high-performance proteins with unique functional properties and uses them to improve bioprocesses and bioproducts with industry partners across many verticals, including agriculture, chemicals, pharmaceuticals and diagnostics and food.</td></tr>
<tr><td>ASA Spezialenzyme GmbH</td><td>www.asa-enzyme.com</td><td>Y</td><td>Y</td><td>Y/N</td><td>Y</td><td>N</td><td>Y</td><td>Y</td><td>N</td><td>Y</td><td>Y</td><td>N</td></tr>
<tr><td colspan="13">Currently, there are many enzymes in development. ASA Spezialenzyme GmbH offers contract research for the development of biotechnological processes or the large-scale production of enzymes. Ongoing establishing of GMP. ASA Spezialenzyme GmbH is specialised in the field of industrial used enzymes.</td></tr>
<tr><td>Biocatalysts Ltd</td><td>www.biocatalysts.com</td><td>Y</td><td>Y</td><td>Y</td><td>Y</td><td>Y</td><td>Y</td><td>N</td><td>Y</td><td>N</td><td>Y</td><td>N</td></tr>
<tr><td colspan="13">Biocatalyst Ltd (and our parent company BRAIN AG) discover, develop and manufacture speciality enzymes for various applications in many sectors. We produce more than 300 enzymes at kg–tonne scale for general sale. We produce a large and diverse set of enzymes exclusively for individual customers including companies producing APIs, diagnostic test kits, food chemistry, flavour and fragrance molecules, genome sequencing kits, agrichemicals and nutraceuticals. Our own manufacturing assets allow us to manufacture enzymes at multiple scales: <5 L, 50 L, 750 L and 10 000 L. Our largest manufacturing processes currently run at 18 000 L fermentation volume. We meet all our customers' regulatory needs.</td></tr>
</table>

Table 2.2 *(Continued)*

Company Name	Website	Enzyme/biocatalyst scale-up service	Bespoke enzyme preparation service	Enzyme immobilisation service	Directed evolution service	*In silico* enzyme engineering service	Enzyme screening service	Microbial screening service	Metagenomic screening service	Biotrans process development service	Non-GMP Biotrans manufacturing serv	GMP Biotrans manufacturing service
Bio-Prodict B.V.	www.bio-prodict.com	N	N	N	N	Y	N	N	N	N	N	N
Bio-Prodict B.V. is a bio-informatics service company that is specialised in the prediction of mutations using data from protein superfamilies. Bio-Prodict has developed 3DM, a proprietary software suite for guiding many types of protein-related R&D including protein engineering. 3DM contains tools for hotspot prediction, construct design, patent analysis, literature analysis and a powerful AI pipeline for accurate mutation prediction.												
Candidum	www.candidum.bio	N	N	N	Y	Y	Y	Y	Y	Y	Y	N
Candidum is an enzyme development service provider optimising custom enzyme solutions from the conceptual stage, the discovery of optimal engineering templates, to the engineering of optimal variants or libraries. Our platform integrates virtual prototyping and laboratory processes to streamline development. Beyond the optimisation of the enzyme, we also improve expression, resolve IP concerns and ready for scale-up with trusted strategic partners.												
ChiralVision	www.chiralvision.com	Y	N	Y	N	N	Y	N	N	Y	Y	N
Finding the right enzyme takes you only halfway. Designing the production process around it and applying immobilisation to allow for more extreme conditions, recycling and full removal takes you all the way. ChiralVision designs custom polymers and has developed tonne-scale immobilised enzyme processes for chiral pharma, chiral intermediates, life sciences, cosmetics, food, feed, biofuels, environment and more, meeting all technical and economical demands.												
c-LEcta GmbH	www.c-lecta.com	Y	Y	Y	Y	Y	Y	Y	Y	Y	Y	N
c-LEcta's mission is to develop customised industrial solutions for all relevant enzyme classes in strategic partnerships. The company covers the value chain from early-stage development down to commercial-scale production and product approval. A key technology is a highly efficient and flexible enzyme engineering platform. Enzyme libraries are designed by proprietary bioinformatics and molecular biology toolboxes designed for maximum speed.												

(Continued)

Table 2.2 (*Continued*)

Company Name	Website	Enzyme/biocatalyst scale-up service	Bespoke enzyme preparation service	Enzyme immobilisation service	Directed evolution service	*In silico* enzyme engineering service	Enzyme screening service	Microbial screening service	Metagenomic screening service	Biotrans process development service	Non-GMP Biotrans manufacturing serv	GMP Biotrans manufacturing service
Codexis Inc.	www.codexis.com	Y	Y	N	Y	Y	Y	N	N	Y	Y	With CMO
Codexis is a leading protein engineering company that applies its proprietary CodeEvolver® technology to develop proteins for a variety of applications, including as biocatalysts for the commercial manufacture of pharmaceuticals, fine chemicals and industrial enzymes, and enzymes as biotherapeutics and for use in molecular diagnostics. Codexis' proven technology enables improvements in protein performance, meeting customer needs for rapid, cost-effective and sustainable manufacturing in multiple commercial-scale implementations of biocatalytic processes.												
CPI	www.uk-cpi.com	Y	Y	N	N	N	Y	Y	N	Y	Y	N
As a member of the UK Catapult network, CPI is an open-access CRO/CMO that has extensive expertise in scaling biotechnology processes for multiple markets and industry sectors, providing a broad service offering to efficiently support commercialisation of biotechnological innovations, whilst minimising both risk and cost to your process, enabling you to transfer the final process to a CMO of your choice. CPI has the expertise and facilities to support enzyme production, biotransformation and downstream processing from small-scale high-throughput systems (1 mL) through to lab- (1–20 L), pilot- and demonstration-scale (10 000 L) vessels.												
Cyrus Biotech	www.cyrusbio.com	N	N	N	N	Y	N	N	N	N	N	N
Cyrus Biotech offers software and collaborative services for cutting-edge enzyme design using the Rosetta computational platform along with associated cheminformatics and bioinformatics software and proprietary in-house methods.												
Dezyme	www.dezyme.com	N	N	N	N	Y	N	N	N	N	N	N
Dezyme has developed a series of efficient bioinformatics tools to optimise proteins in terms of stability, thermal resistance or solubility through targeted amino acid substitutions. A license to access this software suite and *in silico* optimisation services are offered.												

Table 2.2 (*Continued*)

Company Name	Website	Enzyme/biocatalyst scale-up service	Bespoke enzyme preparation service	Enzyme immobilisation service	Directed evolution service	*In silico* enzyme engineering service	Enzyme screening service	Microbial screening service	Metagenomic screening service	Biotrans process development service	Non-GMP Biotrans manufacturing serv	GMP Biotrans manufacturing service
Enantis	www.enantis.com	N	Y	Y/N	Y	Y	N	N	N	N	N	N
Enantis is an enzyme and protein engineering company that develops and sells its own products globally as well as offering tailored contract research services in the same area. Enantis has a panel of haloalkane dehalogenases for sale, as well as thermostable fibroblast growth factors for stem-cell research. We offer contract research services to companies in all fields and we use our unique approach of *in silico* design based on energy and evolution towards the prediction of stabilising mutations. The enzymes and proteins are more thermostable or have longer lifetimes than their wild-type counterparts.												
EnginZyme AB	www.enginzyme.com	Y	Y	Y	N	N	Y	N	N	Y	N	N
EnginZyme has a general immobilisation platform with high activity retention and high enzyme loading, enabling enzymatic reactions in neat organic solvents and continuous-flow operation. EnginZyme offers R&D services to convert an enzyme into an immobilised biocatalyst ready for commercial operation.												
EntreChem	www.entrechem.com	N	N	N	N	N	Y	Y	N	Y	Y	N
Discovery and development of new drugs, mainly for oncology, from bacterial sources applying metabolic pathway genetic engineering and biocatalysis technologies. EntreChem also offers enantiomerically pure chiral building blocks, specifically high-optical- and chemical-purity aminoalcohols and diamines for medicinal chemistry or pharmaceutical intermediates.												
Enzagen	www.enzagen.com	N	N	N	N	N	Y	Y	N	Y	N	N
Enzyme screening kits and consultancy available with 25+ years' experience in industrial biocatalysis.												
Enzymaster	www.enzymaster.de	Y	Y	Y	Y	Y	Y	N	N	Y	Y	(Y)
Enzymaster offers a broad range of R&D services, comprising enzyme identification, directed evolution and customised scale-up for commercialisation. Our proprietary R&D platform, BioEngine®, is strongly guided by *in silico* enzyme engineering and additionally provides biocatalytic process and enzyme fermentation process development. Furthermore, enzyme and biocatalytic manufacturing services are offered with strategic partners.												

(*Continued*)

Table 2.2 *(Continued)*

Company Name	Website	Enzyme/biocatalyst scale-up service	Bespoke enzyme preparation service	Enzyme immobilisation service	Directed evolution service	*In silico* enzyme engineering service	Enzyme screening service	Microbial screening service	Metagenomic screening service	Biotrans process development service	Non-GMP Biotrans manufacturing serv	GMP Biotrans manufacturing service
Enzymicals AG	www.enzymicals.com	Y	Y	Y	Y	Y	Y	Y	Y	Y	Y	(Y)
The focus of Enzymicals is tailor-made enzymes, customised chemicals and individual process solutions with a recognised expertise up to industrial scale. Contract research, development and manufacturing services extend also to enzymes classes not present in the current portfolio.												
EnzymeWorks	www.enzymeworking .com	Y	Y	Y	Y	Y	Y	Y	Y	Y	Y	Y
EnzymeWorks offers over 55 classes of enzymes and CRO and CDMO services, including for (i) designing chemoenzymatic processes, (ii) enzyme discovery and engineering, (iii) contract biomanufacturing of enzymes and co-factors and (iv) contract production of downstream biochemicals, which are audited by a number of global companies with various certificates, including ISO, Kosher, etc.												
EUCODIS Bioscience GmbH	www.eucodis.com	Y	Y	Y	Y	Y	Y	Y	Y	Y	Y	(Y)
EUCODIS Bioscience is a provider of more than 50 enzymes, which can be delivered from screening scale to industrial scale for applications in pharma, food & feed, fine chemistry and cosmetics, and offers custom services including enzyme and strain identification, enzyme engineering, scale-up and industrial production of enzymes and proteins under ISO9001:2015. Food-grade and GMP-level enzymes as well as large-scale immobilised enzymes are produced with well-established, selected industrial partners. Whole-cell biocatalysts, cell factories and biocatalytic conversion processes are individually developed for our customers.												
Evoxx technologies GmbH	www-evoxx.com	Y	Y	Y	Y	y	y	Y	Y	Y	Y	N
Evoxx technologies, the European headquarter of the global enzyme manufacturer Advanced Enzymes, is focusing on research, development and production (lab to industrial scale) of industrial enzymes for a wide range of industries. Apart from own developments, Evoxx is operating a CDMO model, supporting customers at any stage in enzyme and biocatalytic process development.												

Table 2.2 (*Continued*)

Company Name	Website	Enzyme/biocatalyst scale-up service	Bespoke enzyme preparation service	Enzyme immobilisation service	Directed evolution service	*In silico* enzyme engineering service	Enzyme screening service	Microbial screening service	Metagenomic screening service	Biotrans process development service	Non-GMP Biotrans manufacturing serv	GMP Biotrans manufacturing service
Fermentation Pilot Plant (UAB)	ppf.uab.cat	Y	Y	Y	N	N	N	N	N	Y	Y	N
Contract manufacturing services (CMO) for companies and public institutions. Bioprocess development from lab to pilot scale based on cell factories: bacterial and yeast (2, 5, 50, 250 L scale) and mammalian cells (single-use wave reactor up to 20 L). Training, technical advice and consultancy services.												
Gecco Biotech	www.gecco-biotech.com	Y	Y	Y	Y	Y	Y	N	N	Y	N	N
GECCO Biotech offers novel enzyme discovery, engineering and application services, with a focus on flavin-containing redox enzymes and computational enzyme stabilisation based on FRESCO. GECCO operates as CRO and has capabilities to provide mg to g amounts of enzymes.												
Ingenza	www.ingenza.com	N	Y	N	Y	N	Y	Y	Y	Y	Y	Y
Ingenza develops and deploys industrial biotechnology for cell, metabolic pathway and protein engineering/optimisation. Our expertise spans bioinformatics, combinatorial genetics, expression system design, enzyme evolution, solid/liquid phase screening, fermentation, bioprocessing and synthetic/analytical chemistry. Our expression/host platforms in bacteria, yeast and mammalian cells are used worldwide in fermentation-based manufacturing of chemicals, GMP biologics, pharmaceuticals and biofuels. Downstream capabilities from cell harvest to enzyme purification and formulation enable us to introduce and improve diverse commercial bioprocesses for many industrial sectors. Our synthetic and analytical chemists prepare substrates and reference materials, determine enzyme catalyst parameters and conduct retro-biosynthetic analyses that inform novel biological routes to high value industrial molecules.												

(*Continued*)

Table 2.2 *(Continued)*

Company Name	Website	Enzyme/biocatalyst scale-up service	Bespoke enzyme preparation service	Enzyme immobilisation service	Directed evolution service	*In silico* enzyme engineering service	Enzyme screening service	Microbial screening service	Metagenomic screening service	Biotrans process development service	Non-GMP Biotrans manufacturing serv	GMP Biotrans manufacturing service
Innophore	www.innophore.com	N	N	N	Y (*in silico*)	Y	Y (*in silico*)	N	N	N	N	N
Innophore uses the patented 3D point cloud technology (so-called Catalophores) to identify enzymes for biocatalytic industrial processes. The search in a proprietary database of active-site cavities allows customers to find new enzymes that are otherwise unpredictable. The platform is a digital search engine for partners in the fields of pharma, chemistry, biotechnology, food, cosmetics, agriculture, textiles or diagnostics. Artificial intelligence is applied to refine searches and get the most out of our proprietary database.												
InnoSyn BV	www.innosyn.com	Y	Y	N	N	Y[a]	Y	N	N	Y	Y	N
InnoSyn BV is a young SME originating from a spin-out of DSM's biocatalysis, chemocatalysis and organic chemistry R&D department, Innovative Synthesis. InnoSyn BV provides services for all phases of technology and process development, from idea generation all the way to running a business. Over 25 years of experience and history as the biocatalysis R&D group at DSM has resulted in a broad biocatalysis competence base ranging all the way from enzyme identification and engineering, enzyme production in bacteria and yeast hosts, reaction engineering and scale-up to full-scale process development and implementation, which is now made available to customers from any chemical or biotech business segment.												
Johnson Matthey	www.matthey.com/biocatalysts/	Y	Y	Y	Y	Y	Y	Y	Y	Y	Y	Y
Enzyme services extend to enzyme classes not present in the current portfolio.												
Megazyme	www.megazyme.com	N	N	N	N	N	N	N	N	N	N	N
Megazyme is a biotech company primarily focused on the analytical and diagnostic markets. Our primary expertise is in the area of carbohydrate-acting enzymes (CAZymes). We hold >10 years' supply of all catalogue enzymes and produce bulk quantities on request.												

Table 2.2 (Continued)

Company Name	Website	Enzyme/biocatalyst scale-up service	Bespoke enzyme preparation service	Enzyme immobilisation service	Directed evolution service	*In silico* enzyme engineering service	Enzyme screening service	Microbial screening service	Metagenomic screening service	Biotrans process development service	Non-GMP Biotrans manufacturing serv	GMP Biotrans manufacturing service
Merck KGgA/ MilliporeSigma	www.sigmaaldrich .com	Y	N	N	N	N	N	N	N	N	N	N
For general sale globally, Merck KGgA/MilliporeSigma manufactures from small to industrial large scale an enormous variety of enzymes, which can be searched with the Enzyme Explorer. The Enzyme Explorer provides various search tools, such as Application Index, Enzyme Index, Substrate Index, Inhibitor Index, Cofactor Index, Lectin Index, Diagnostic and Elite Enzyme ISO 13485 grade, Industrial Enzymes, Recombinant Enzymes as animal-free alternatives, Enzyme for Glycobiology and Molecular Biology, Drug Discovery and Cell Signalling Enzymes, Protease Finder, Assay and Protocol Library. More details on the enzyme products and the Enzyme Explorer search tool can be found at https://www.sigmaaldrich.com/life-science/metabolomics/enzyme-explorer.html. At Sigma-Aldrich® and SAFC®, many complex manufacturing processes are used in the course of manufacturing 46 000 chemicals and biologicals. These capabilities are also available within a number of services ranging from Strategic Services, Laboratory Support & Research Services, Product and Process Development, Manufacturing & Production Services such as Contract Manufacturing and OEM Services to Regulatory Compliance Services and Facility Operations Services & Support. More details on these services can be found at https://www.sigmaaldrich.com/customer-service/services.html. For specific and business-to-business sales globally, Merck KGgA/MilliporeSigma provides customer-specific process solutions and project management.												
Nostrum biodiscocvery S.L.	www.nostrum biodiscovery.com	N	N	N	Y (*in silico*)	N	Y (*in silico*)	N	N	N	N	N
Nostrum Biodiscovery (NBD) is an *in silico* company specialised in accelerating early drug discovery and enzyme engineering projects based on a combination of computational tools. It has successfully applied its proprietary PELE simulation platform in several industrial and academic projects, involving the discovery of active molecules against all sorts of targets and which enhance enzymatic properties boosting catalysis through computer simulations, saving time and resources.												

(*Continued*)

Table 2.2 *(Continued)*

Company Name	Website	Enzyme/biocatalyst scale-up service	Bespoke enzyme preparation service	Enzyme immobilisation service	Directed evolution service	*In silico* enzyme engineering service	Enzyme screening service	Microbial screening service	Metagenomic screening service	Biotrans process development service	Non-GMP Biotrans manufacturing serv	GMP Biotrans manufacturing service
Novozymes A/S	pharmaceuticals@ novozymes.com											
Novozymes is the world leader in biological solutions. Together with customers, partners and the global community, we improve industrial performance while preserving the planet's resources and helping to build better lives. As the world's largest provider of enzyme and microbial technologies, our bioinnovation enables cost savings and greater sustainability. Biocatalysis can also improve your productivity and the quality of your APIs and intermediates.												
NZYTech, Genes & Enzymes	www.nzytech.com	Y	Y	N	N	N	N	N	N	N	N	N
NZYTech offers protein expression, production and purification services. In addition, NZYTech offers a service for production of difficult-to-express proteins. This includes expression of recombinant proteins in fusion with several tags known to enhance solubility and/or promote correct folding.												
Peaccel	www.peaccel.com	N	N	N	Y	Y	Y	N	N	Y	N	N
PEACCEL's proprietary innov'SAR machine learning and AI platform is an industry-proven tool which enables the rapid identification and efficient selection of improved protein variants *in silico*. The capacity of the innov'SAR algorithms to predict protein improvements replaces extensive wet lab mutation/screening cycles (generation, expression and testing of protein mutants, etc.). Hence, our technology allows a reduction of the cost, time and risks associated with the discovery and engineering of proteins such as industrial enzymes, antibodies or even entire synthetic pathways. Together with our partner companies, Peaccel covers the value chain from enzyme discovery to directed evolution and small-scale production.												

Table 2.2 *(Continued)*

Company Name	Website	Enzyme/biocatalyst scale-up service	Bespoke enzyme preparation service	Enzyme immobilisation service	Directed evolution service	*In silico* enzyme engineering service	Enzyme screening service	Microbial screening service	Metagenomic screening service	Biotrans process development service	Non-GMP Biotrans manufacturing serv	GMP Biotrans manufacturing service
Protéus	www.proteus.fr	Y	Y	Y	Y	N	Y	Y	Y	Y	Y	Y
Protéus, an affiliate of SEQENS, has been active for 20 years in the development of biocatalytic processes in pharmaceuticals, food, flavours and fragrances, agriscience, nutraceuticals, cosmetics and chemicals. Protéus provides its customers with specific development strategies according to their needs: (i) screening of proprietary enzymes with a portfolio of more than 400 enzymes, 90 which are readily available in kits for screening; (ii) discovery of new enzymes from a proprietary strain library (including 1300 sequenced strains); and (iii) directed evolution of enzymes, including the proprietary L-shuffling method for large recombination of proteins. Protéus offers process optimisation, immobilisation of enzymes and upscaling of biocatatlytic processes to commercial scale within the SEQENS group.												
Prozomix Ltd	www.prozomix.com	Y	Y	Y	N	N	Y	Y	Y	N	N	N
GMP-grade bulk biocatalysis enzyme service from 2020 onwards. Customer/third-party biocatalyst panel production service available (presentation in vials and/or 96- and 384-well microtitre plate format). Enzyme services extend to enzyme classes not present in the current portfolio.												
Purolite	www.purolite.com /life-sciences/	Y	Y	Y	N	N	Y	N	N	Y	Y	N
SeSaM-Biotech GmbH	www.sesam-biotech.com	N	Y	Y	Y	Y	Y	N	Y	Y	N	N
You have an enzyme-related question? We have the answer. Located in Germany, SeSaM-Biotech is your go-to fee-for-service provider for enzyme discovery, expression and especially evolution. With more than 10 years of R&D expertise, we are active worldwide in all Industries with enzyme-based products and processes.												

(*Continued*)

Table 2.2 (*Continued*)

Company Name	Website	Enzyme/biocatalyst scale-up service	Bespoke enzyme preparation service	Enzyme immobilisation service	Directed evolution service	*In silico* enzyme engineering service	Enzyme screening service	Microbial screening service	Metagenomic screening service	Biotrans process development service	Non-GMP Biotrans manufacturing serv	GMP Biotrans manufacturing service
Syncozymes	en.syncozymes.com	Y	N	Y	Y	Y	Y	N	N	Y	Y	N
SyncoZymes is a leading company specialising in green process chemistry and biotransformation for API and intermediate manufacture. We are mainly engaged in directed evolution, enzyme modification, fermentation, immobilisation, process development and application of enzymes. SyncoZymes is also the largest manufacturer of coenzymes NMN, NAD, NADP, NADH and NADPH in China.												
Zymtronix Catalytic Systems Inc.	www.zymtronix.com	N	N	Y	N	N	N	N	N	Y	Y	N
Customised enzyme immobilisation and process development for batch, packed-bed and flow manufacturing with a tiered approach from small-scale screening and optimisation to large-scale processes. Complex multienzyme systems including co-factor regeneration systems, cascade reactions and modular enzymatic syntheses.												
Zymvol	www.zymvol.com	N	N	N	Y (*in silico*)	Y	Y (*in silico*)	N	N	N	N	N
Service provision firm specialised in computer-driven biocatalyst discovery and development for the pharmaceutical and fine chemicals industries. The use of high-performance computing to identify and optimise custom-made biocatalysts allows the accurate simulation of the entire biochemical and biophysical enzymatic process and the testing of more than 50 000 virtual enzymes per day.												

[a]With two strategic partner companies within projects.

2.3 Chemical Transformations of Selected Commercially Available Enzymes

Table 2.3 Chemical transformations of selected commercially available enzymes

Enzyme Name	Chemical Transformation	Supplier	Company contact details	Number of enzymes in panel
Endo-beta-N-acetylglucosaminidase	Cleaves all nonreducing terminal β-linked N-acetylglucosamine residues from complex carbohydrates and glycoproteins	NZYTech Merck KGgA/ MilliporeSigma Prozomix Ltd	www.nzytech.com www.sigmaaldrich.com www.prozomix.com	19 1 1
Acylase (amino)	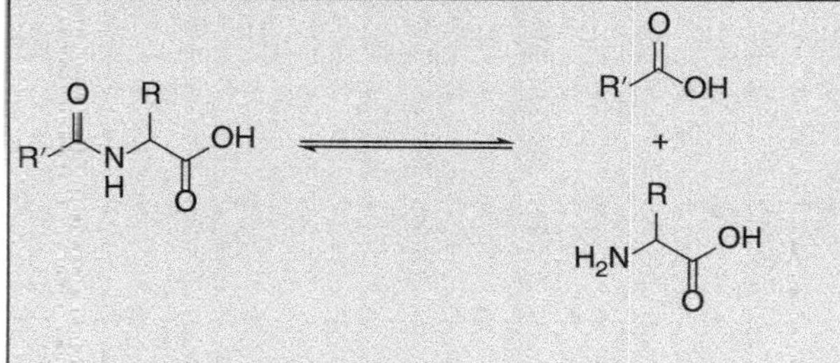	Amano Enzyme Enzymicals AG EnzymeWorks Merck KGgA/ MilliporeSigma Protéus	www.amano-enzyme.co.jp www.enzymicals.com www.enzymeworking.com www.sigmaaldrich.com www.proteus.fr	2 6 10 1 9
Acylase (penicillin)	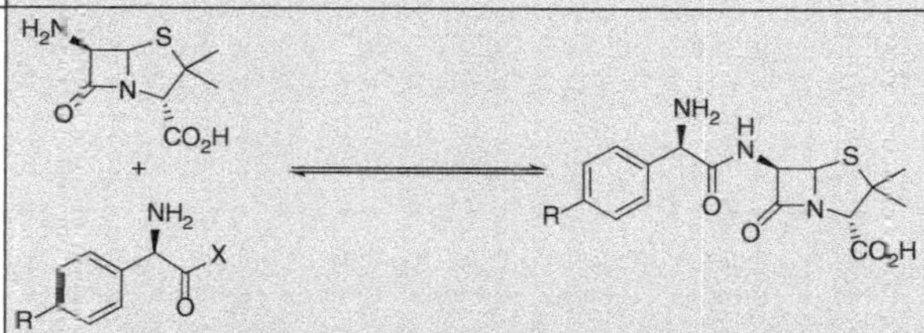 X = OMe, NH2, ... R = H, OH	ASA Spezialenzyme GmbH Codexis Inc. Enzagen Gecco Biotech Merck KGgA/ MilliporeSigma	www.asa-enzyme.de www.codexis.com www.enzagen.com www.gecco-biotech.com www.sigmaaldrich.com	1 1 90 1 1

(*Continued*)

Table 2.3 (*Continued*)

Enzyme Name	Chemical Transformation	Supplier	Company contact details	Number of enzymes in panel
Agarase	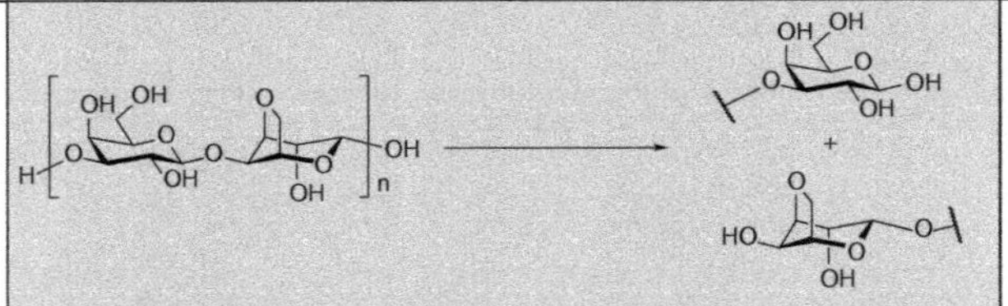	Merck KGgA/ MilliporeSigma NZYTech	www.sigmaaldrich.com www.nzytech.com	1 24
Aldehyde reductase		Almac EnzymeWorks Protéus Prozomix Ltd	www.almacgroup.com www.enzymeworking.com www.proteus.fr www.prozomix.com	16 40 9 1
Aldolase	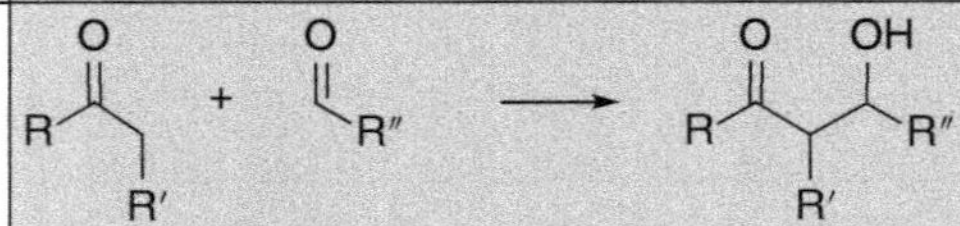	Enzymaster EnzymeWorks	www.enzymaster.de www.enzymeworking.com	162 25
Aldolase (N-acetylneuraminic acid)	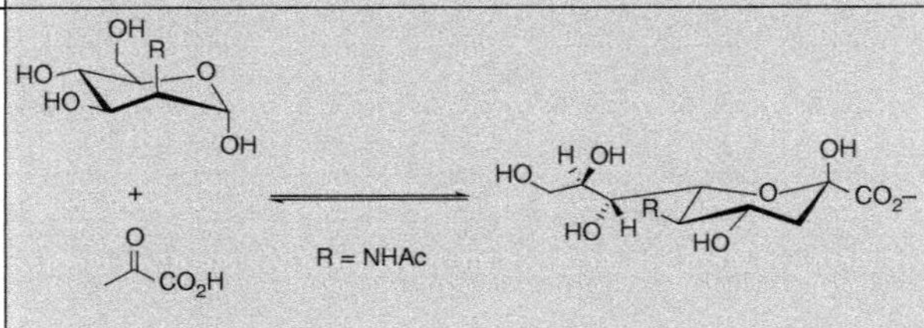	Codexis Inc. Evoxx technologies Merck KGgA/ MilliporeSigma	www.codexis.com www.evoxx.com www.sigmaaldrich.com	Custom item 1 2 2

Aldolase (deoxyribose-phosphate)	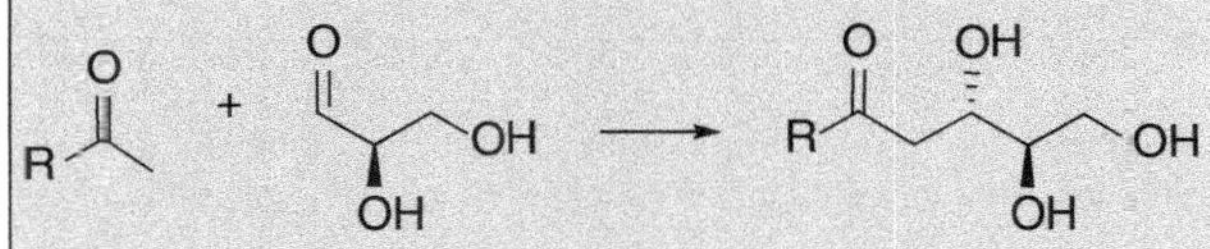	Almac Codexis Inc. Enzymicals AG EnzymeWorks Evoxx technologies	www.almacgroup.com www.codexis.com www.enzymicals.com www.enzymeworking.com www.evoxx.com	96 Custom items >100 3 15 5
Aldolase (deoxyribose-phosphate) (DERA)	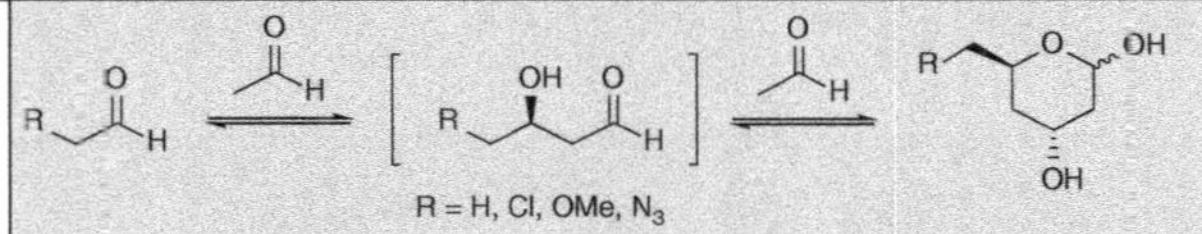	Merck KGgA/ MilliporeSigma Prozomix Ltd Syncozymes	www.sigmaaldrich.com www.prozomix.com en.syncozymes.com	1 210 8
Aldolase (threonine)	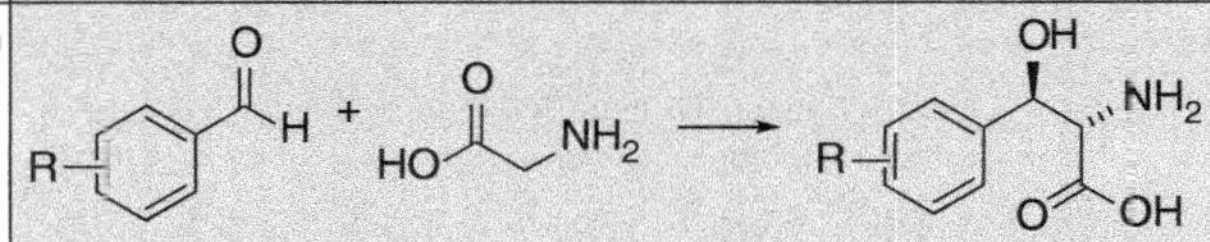	Almac Codexis Inc. EnzymeWorks Protéus	www.almacgroup.com www.codexis.com www.enzymeworking.com www.proteus.fr	50 Custom items 10 15 3
Amidase	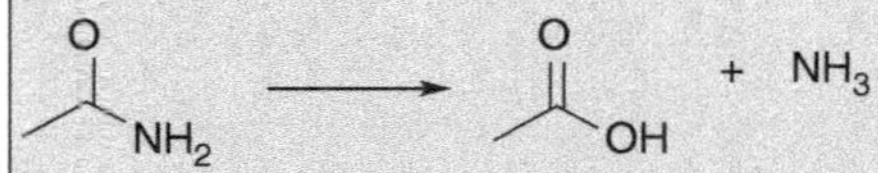	Almac EnzymeWorks EUCODIS Merck KGgA/ MilliporeSigma Protéus Syncozymes	www.almacgroup.com www.enzymeworking.com www.eucodis.com www.sigmaaldrich.com www.proteus.fr en.syncozymes.com	6 30 2 1 9 18

(*Continued*)

Table 2.3 *(Continued)*

Enzyme Name	Chemical Transformation	Supplier	Company contact details	Number of enzymes in panel
Ammonia mutase	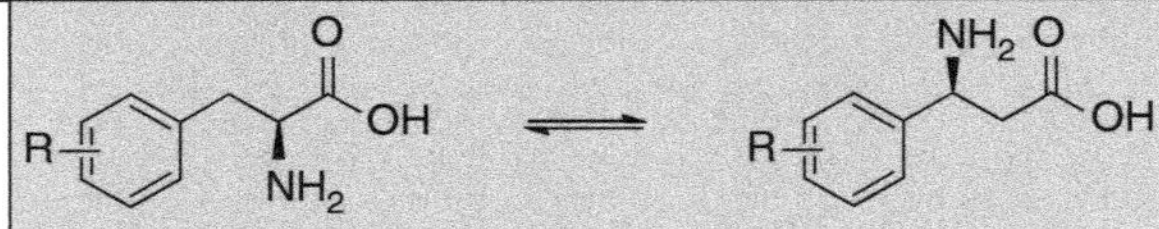	Almac EnzymeWorks GECCO Biotech	www.almacgroup.com www.enzymeworking.com www.gecco-biotech.com	1 5 12
Amylase	Cleaves the α-1,4-glycosidic bonds of starch, where alpha-amylase cleaves at random positions, beta-amylase cleaves selectively at the second glycosidic position from the reducing end and gamma-amylase cleavages selectively at the first glycosidic position from the reducing end	Amano Enzyme Asahi Kasei Pharma Corporation ASA Spezialenzyme GmbH Biocatalysts Ltd GECCO Biotech Megazyme Merck KGgA/ MilliporeSigma Novozymes A/S NZYTech Prozomix Ltd	www.amano-enzyme.co.jp www.asahi-kasei.co.jp/shindan/en/ www.asa-enzyme.de www.biocatalysts.com www.gecco-biotech.com www.megazyme.com www.sigmaaldrich.com pharmaceuticals@novozymes.com www.nzytech.com www.prozomix.com	1 2 3 2 1 14 24 1 15 3
Amyloglucosidase	Hydrolyses terminal α-1,4 and α-1,6 D-glucose residues successively from nonreducing ends of oligosaccharides	Megazyme Merck KGgA/ MilliporeSigma	www.megazyme.com www.sigmaaldrich.com	6 9

Amylomaltase	Catalyses the reversible intermolecular transfer of a segment of a α -1,4-D-glucan to a new 4-position in another α -1,4-D-glucan	NZYTech	www.nzytech.com	7
Amylosucrase	Transfers a D-glucose residue, typically obtained from sucrose as the donor, to acceptor molecules such as D-glucose itself, D-fructose or α-1,4-glycosidically linked glucose oligomers and polymers, particularly glycogen, with the resultant formation of a new α-1,4-glycosidic bond	Prozomix Ltd	www.prozomix.com	5
Arabinanase	Catalyses the endohydrolysis of α-1,5-arabinofuranosidic linkages in (1→5)-arabinans	Megazyme NZYTech Prozomix Ltd	www.megazyme.com www.nzytech.com www.prozomix.com	9 8 3
Arabinofuranosidase	Hydrolyses terminal nonreducing alpha-L-arabinofuranoside residues in alpha-L-arabinosides	Megazyme NZYTech Prozomix Ltd	www.megazyme.com www.nzytech.com www.prozomix.com	6 35 28
Arabinoxylanase	Hydrolyses arabinose from arabinoxylans	NZYTech	www.nzytech.com	2
Asparaginase		EUCODIS Merck KGgA/ MilliporeSigma	www.eucodis.com www.sigmaaldrich.com	1 1
Carbamoylase	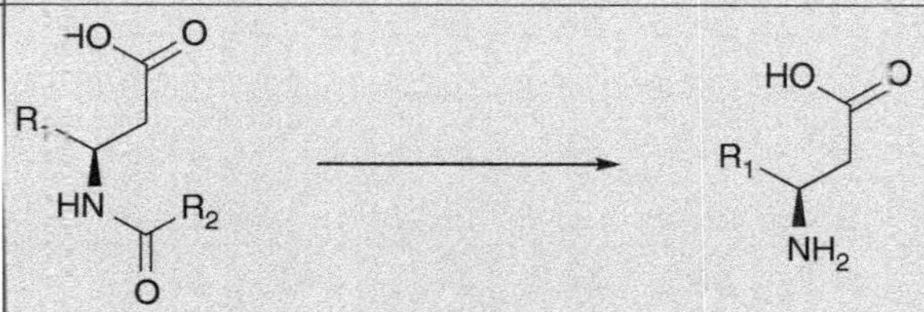	Almac	www.almacgroup.com	20

(Continued)

Table 2.3 *(Continued)*

Enzyme Name	Chemical Transformation	Supplier	Company contact details	Number of enzymes in panel
Carbamoyl-transferase	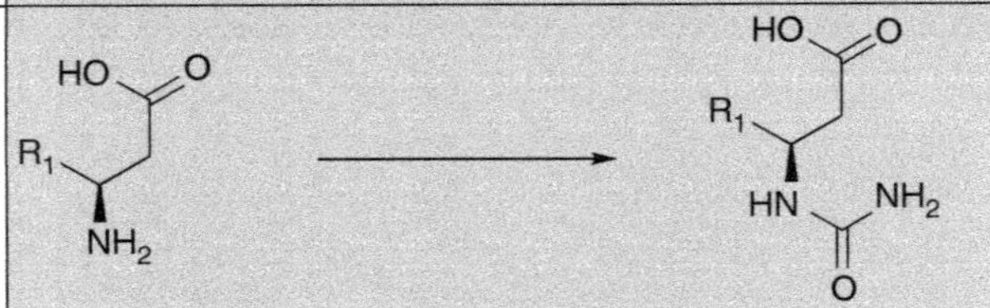	Almac	www.almacgroup.com	5
Carboxylase (phosphoenolpyruvate)	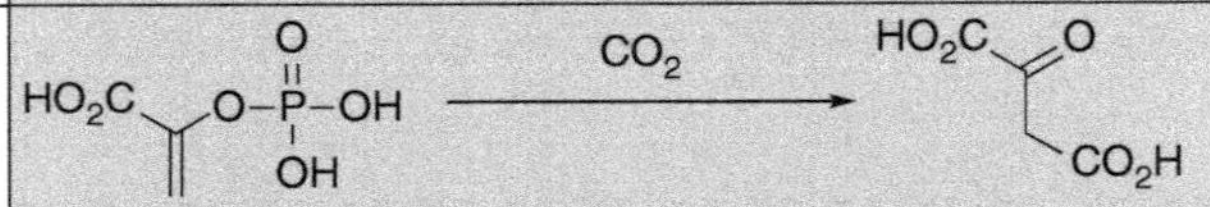	Toyobo	www.toyobo-global.com	1
Carboxylic acid reductase		Almac EnzymeWorks Protéus Prozomix Ltd	www.almacgroup.com www.enzymeworking.com www.proteus.fr www.prozomix.com	35 5 3 20
Carrageenase	Cleaves the internal β-(1–4) linkages of k-, ι- and λ-carrageenans, yielding products of the oligo-carrageenans	NZYTech	www.nzytech.com	2
Catalase	$2H_2O_2 \longrightarrow 2H_2O + O_2$	Biocatalysts Ltd EUCODIS Gecco Biotech Megazyme Merck KGgA /MilliporeSigma Novozymes A/S Toyobo	www.biocatalysts.com www.eucodis.com www.gecco-biotech.com www.megazyme.com www.sigmaaldrich.com pharmaceuticals@novozymes.com www.toyobo-global.com	1 1 1 1 12 1 1

Cellobiohydrolase	Participates in the hydrolysis of both crystalline and amorphous cellulose	Megazyme	www.megazyme.com	2
		Merck KGgA/ MilliporeSigma	www.sigmaaldrich.com	1
		NZYTech	www.nzytech.com	10
Cellodextrinase	Catalyses the hydrolysis of glycocidic linkages in β-1,4-D-glucans, removing successive glucose units	NZYTech	www.nzytech.com	2
Cellulase	Endohydrolyses (1→4)-β-D-glucosidic linkages in cellulose, lichenin and cereal β-D-glucans	Amano Enzyme	www.amano-enzyme.co.jp	1
		ASA Spezialenzyme GmbH	www.asa-enzyme.de	1
		Biocatalysts Ltd	www.biocatalysts.com	1
		Codexis Inc.	www.codexis.com	Custom item
		Megazyme	www.megazyme.com	6
		Merck KGgA/ MilliporeSigma	www.sigmaaldrich.com	10
		NZYTech	www.nzytech.com	81
		Protéus	www.proteus.fr	6
		Prozomix Ltd	www.prozomix.com	7
Chitinase	Randomly endohydrolyses N-acetyl-β-D-glucosaminide (1→4)-β-linkages in chitin and chitodextrins	ASA Spezialenzyme GmbH	www.asa-enzyme.de	1
		Biocatalysts Ltd	www.biocatalysts.com	1
		EUCODIS	www.eucodis.com	3
		Megazyme	www.megazyme.com	1
		Merck KGgA/ MilliporeSigma	www.sigmaaldrich.com	2
		NZYTech	www.nzytech.com	7
		Prozomix Ltd	www.prozomix.com	1

(Continued)

Table 2.3 (*Continued*)

Enzyme Name	Chemical Transformation	Supplier	Company contact details	Number of enzymes in panel
Chitosanase	Endohydrolyses β-(1→4)-linkages between D-glucosamine residues in a partly acetylated chitosan	EUCODIS Merck KGgA/ MilliporeSigma NZYTech	www.eucodis.com www.sigmaaldrich.com www.nzytech.com	1 2 3
Chloroperoxidase	Halogenation of electron rich centres in the presence of halide. Oxidation of a diverse range of substrates including unactivated olefins, primary alcohols, thioethers, often enantioselectively.	ASA Spezialenzyme GmbH Enzymicals AG EUCODIS Merck KGgA/ MilliporeSigma	www.asa-enzyme.de www.eucodis.com www.sigmaaldrich.com	1 1 2 3
Cyclodextrin glucanotransferase	Cyclises part of a (1→4)-α-D-glucan chain by formation of a (1→4)-α-D-glucosidic bond	Amano Enzyme Novozymes A/S	www.amano-enzyme.co.jp pharmaceuticals@ novozymes.com	1 1
Cyclooxygenase	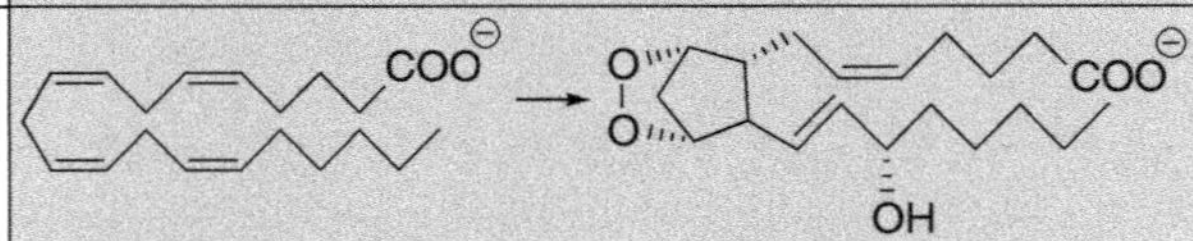	Almac Merck KGgA/ MilliporeSigma	www.almacgroup.com www.sigmaaldrich.com	11 1
Decarboxylase	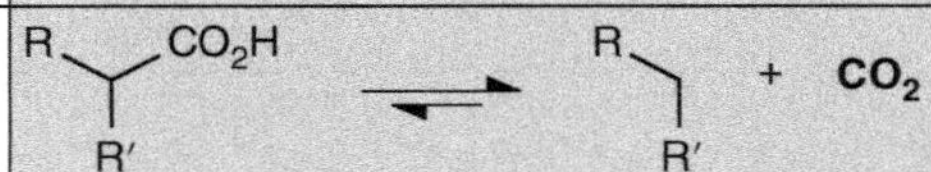	Enzymaster EnzymeWorks	www.enzymaster.de www.enzymeworking.com	40 10

Decarboxylase (aspartate)	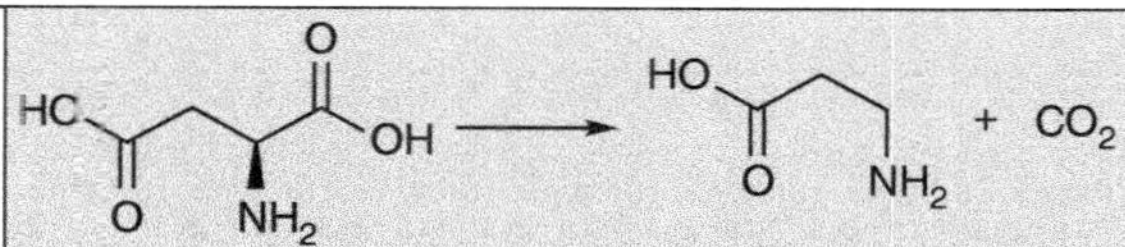	Almac Codexis Inc.	www.almacgroup.com www.codexis.com	9 Custom items 10
Dehydratase (cyanide)	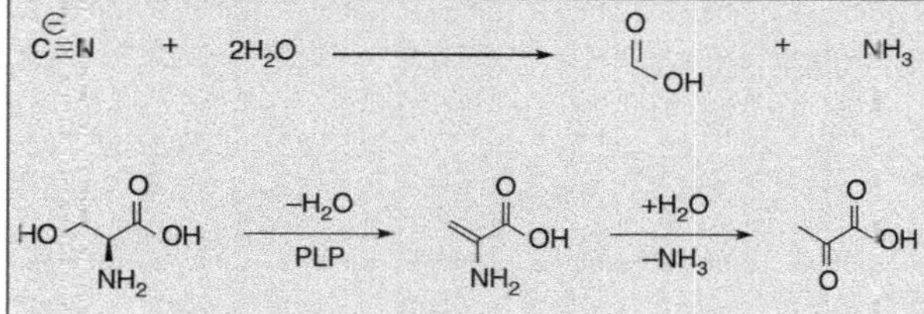	Almac Enzymicals AG	www.almacgroup.com www.enzymicals.com	8 3
Dehydrogenase (alcohol)/ ketoreductase/ carbonyl reductase	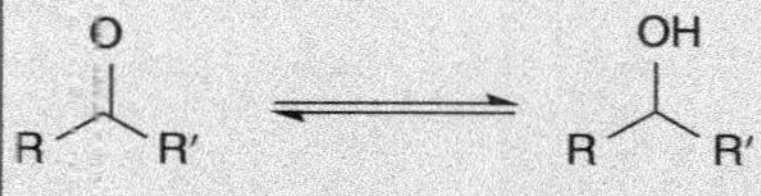 R = alkyl or aryl; R′ = H, alkyl or aryl	Almac c-LEcta GmbH Codexis Inc. Enzymaster EnzymeWorks Evoxx technologies GmbH	www.almacgroup.com www.c-lecta.com www.codexis.com www.enzymaster.de www.enzymeworking.com www.evoxx.com	96 >200 24 (kit) 260 (panel) + custom items 177 200 25

(*Continued*)

Table 2.3 (*Continued*)

Enzyme Name	Chemical Transformation	Supplier	Company contact details	Number of enzymes in panel
		Gecco Biotech	www.gecco-biotech.com	8
		InnoSyn BV	www.innosyn.com	88
		Johnson Matthey	www.matthey.com/biocatalysts/	180
		Megazyme	www.megazyme.com	1
		Merck KGgA/MilliporeSigma	www.sigmaaldrich.com	5
		Protéus	www.proteus.fr	15
		Prozomix Ltd	www.prozomix.com	768
		Syncozymes	en.syncozymes.com/	174
Dehydrogenase (amine)	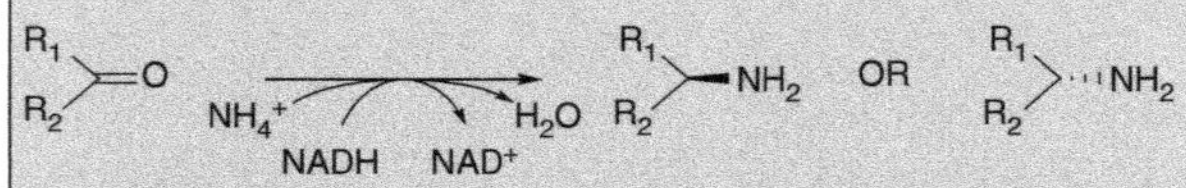	Almac	www.almacgroup.com	5
		Codexis Inc.	www.codexis.com	24 (beta kit)
		EnzymeWorks	www.enzymeworking.com	5
		Evoxx technologies	www.evoxx.com	>20
		Johnson Matthey	www.matthey.com	15
Dehydrogenase (D- and/or L-amino acid)	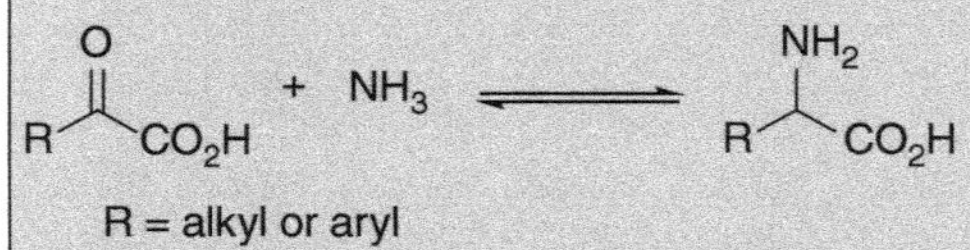	Almac	www.almacgroup.com	25
		ASA Spezialenzyme GmbH	www.asa-enzyme.de	2
		EnzymeWorks	www.enzymeworking.com	40
		Evoxx technologies	www.evoxx.com	>10
		Johnson Matthey	www.matthey.com/biocatalysts/	20
		Syncozymes	en.syncozymes.com	17

Dehydrogenase (cellobiose)	Cellobiose + acceptor = cellobiono-1,5-lactone + recuced acceptor	NZYTech	www.nzytech.com	1
Dehydrogenase (formate)	$HCO_2^{\ominus} \longrightarrow CO_2$	Codexis Inc.	www.codexis.com	1
		EnzymeWorks	www.enzymeworking.com	5
		Evoxx technologies	www.evoxx.com	2
		Johnson Matthey	www.matthey.com	10
		Megazyme	www.megazyme.com	1
		Merck KGgA/ MilliporeSigma	www.sigmaaldrich.com	2
		Protéus	www.proteus.fr	4
		Prozomix Ltd	www.prozomix.com	12
		Syncozymes	en.syncozymes.com	4
Dehydrogenase (glucose)	Glucose ⟶ Gluconic acid	Almac	www.almacgroup.com	2
		Amano Enzyme	www.amano-enzyme.co.jp	3
		c-LEcta GmbH	www.c-lecta.com	1
		Codexis Inc.	www.codexis.com	2
		EnzymeWorks	www.enzymeworking.com	5
		Enzymicals AG	www.enzymicals.com	2
		Evoxx technologies	www.evoxx.com	2
		InnoSyn BV	www.innosyn.com	2
		Johnson Matthey	www.matthey.com/ biocatalysts/	20
		Merck KGgA/ MilliporeSigma	www.sigmaaldrich.com	1
		Protéus	www.proteus.fr	2
		Syncozymes	en.syncozymes.com	9
		Toyobo	www.toyobo-global.com	2

(Continued)

Table 2.3 *(Continued)*

Enzyme Name	Chemical Transformation	Supplier	Company contact details	Number of enzymes in panel
Dehydrogenase (lactate)		Almac	www.almacgroup.com	1
		Amano Enzyme	www.amano-enzyme.co.jp	1
		Asahi Kasei Pharma Corporation	www.asahi-kasei.co.jp/shindan/en/	1
		Codexis Inc.	www.codexis.com	1
		EnzymeWorks	www.enzymeworking.com	5
		Evoxx technologies	www.evoxx.com	1
		Johnson Matthey	www.matthey.com/biocatalysts/	20
		Megazyme	www.megazyme.com	2
		Merck KGgA/MilliporeSigma	www.sigmaaldrich.com	7
		Syncozymes	en.syncozymes.com	1
		Toyobo	www.toyobo-global.com	3
Dehydrogenase (phenylalanine)	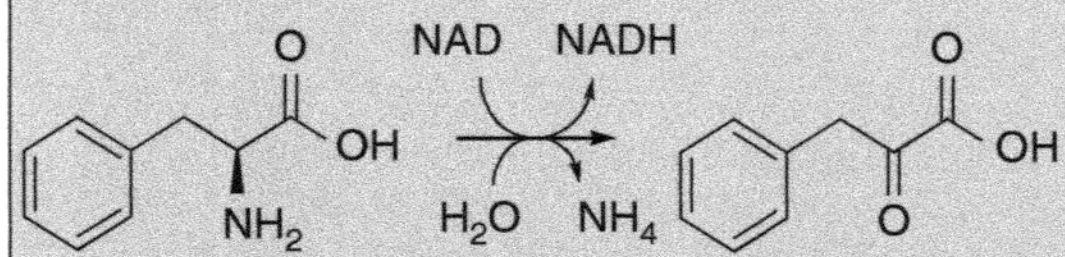	Almac	www.almacgroup.com	5
		Biocatalysts Ltd	www.biocatalysts.com	2
		EnzymeWorks	www.enzymeworking.com	5
		Evoxx technologies	www.evoxx.com	1
		Merck KGgA/MilliporeSigma	www.sigmaaldrich.com	1
		Syncozymes	en.syncozymes.com	8

Dextranase	Endohydrolyses (1→6)-α-D-glucosidic linkages in dextran	ASA Spezialenzyme GmbH	www.asa-enzyme.de	1
		Megazyme	www.megazyme.com	1
		Merck KGgA/ MilliporeSigma	www.sigmaaldrich.com	4
		NZYTech	www.nzytech.com	5
Diastase	Transforms starch into maltose, then glucose	Amano Enzyme	www.amano-enzyme.co.jp	1
Dioxygenase (biphenyl)	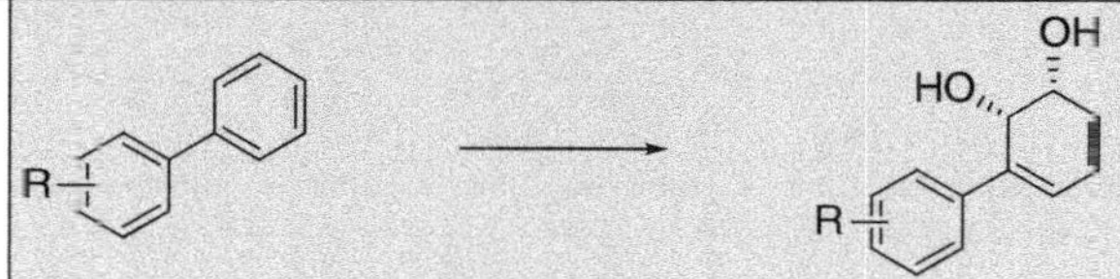	Almac	www.almacgroup.com	4
Ene reductase	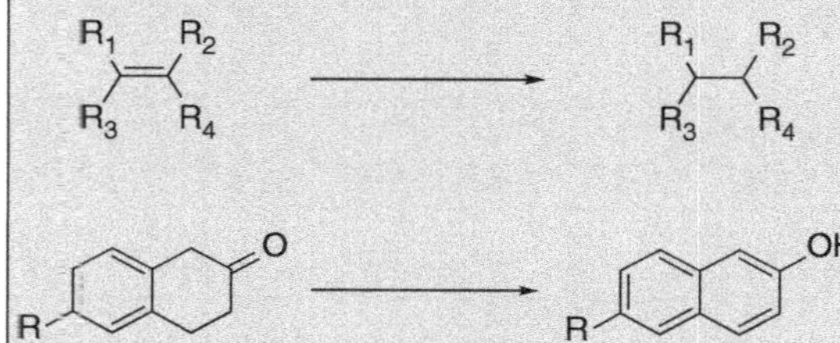	Almac	www.almacgroup.com	>200
		Codexis Inc.	www.codexis.com	7 (kit) + custom items
		EnzymeWorks	www.enzymeworking.com	15
		Enzymicals AG	www.enzymicals.com	3
		Johnson Matthey	www.matthey.com/ biocatalysts/	110
		Protéus	www.proteus.fr	6
		Prozomix Ltd	www.prozomix.com	144
		Syncozymes	en.syncozymes.com	29
Epimerase	UDP-Galactose → UDP-Glucose	Almac	www.almacgroup.com	12

(*Continued*)

Table 2.3 *(Continued)*

Enzyme Name	Chemical Transformation	Supplier	Company contact details	Number of enzymes in panel
Epoxide hydrolase		Almac Codexis Inc. EnzymeWorks Enzymicals AG Gecco Biotech Protéus Prozomix Ltd Syncozymes	www.almacgroup.com www.codexis.com www.enzymeworking.com www.enzymicals.com www.gecco-biotech.com www.proteus.fr www.prozomix.com en.syncozymes.com	12 10 Custom items 20 1 2 4 96 2
Esterase (carboxyl)	 R^1, R^2 = alkyl or aryl	Almac Amano Enzyme ASA Spezialenzyme GmbH Biocatalysts Ltd c-LEcta GmbH Codexis Inc. EnzymeWorks Enzymicals AG EUCODIS Evoxx technologies	www.almacgroup.com www.amano-enzyme.co.jp www.asa-enzyme.de www.biocatalysts.com www.c-lecta.com www.codexis.com www.enzymeworking.com www.enzymicals.com www.eucodis.com www.evoxx.com	96 1 3 2 63 85 15 20 25 >20

		InnoSyn BV	www.innosyn.com	7
		Johnson Matthey	www.matthey.com/biocatalysts/	80
		Merck KGgA/MilliporeSigma	www.sigmaaldrich.com	17
		Protéus	www.proteus.fr	2
		Prozomix Ltd	www.prozomix.com	22
		Syncozymes	en.syncozymes.com	14
Fructanase/fructosidase	Hydrolyses terminal, nonreducing (2→1)- and (2→6)-linked β-D-fructofuranose residues in fructans	Megazyme	www.megazyme.com	8
		NZYTech	www.nzytech.com	3
beta-Fructo-furanosidase (invertase)	Hydrolyses terminal nonreducing β-D-fructofuranoside residues in β-D-fructofuranosides	ASA Spezialenzyme GmbH	www.asa-enzyme.de	1
		Merck KGgA/MilliporeSigma	www.sigmaaldrich.com	1
		NZYTech	www.nzytech.com	4
		Toyobo	www.toyobo-global.com	1
Fucosidase	α-L-fucoside + H_2O to L-fucose + an alcohol	Megazyme	www.megazyme.com	3
		Merck KGgA/MilliporeSigma	www.sigmaaldrich.com	1
		NZYTech	www.nzytech.com	8
Galactanase	The enzyme specifically hydrolyses β-1,3-galactan and β-1,3-galactooligosaccharides	Megazyme	www.megazyme.com	3
		NZYTech	www.nzytech.com	6
		Prozomix Ltd	www.prozomix.com	4

(Continued)

Table 2.3 (*Continued*)

Enzyme Name	Chemical Transformation	Supplier	Company contact details	Number of enzymes in panel
Galactan 1,3-beta-galactosidase	Hydrolyses terminal, nonreducing β-D-galactose residues in (1→3)-β-D-galactopyranans	Prozomix Ltd	www.prozomix.com	1
alpha-Galactosidase	Hydrolyses terminal, nonreducing α-D-galactose residues in α-D-galactosides, including galactose oligosaccharides, galactomannans and galactolipids	Amano Enzyme	www.amano-enzyme.co.jp	1
		EnzymeWorks	www.enzymeworking.com	4
		Megazyme	www.megazyme.com	4
		Merck KGgA/ MilliporeSigma	www.sigmaaldrich.com	3
		NZYTech	www.nzytech.com	13
		Prozomix Ltd	www.prozomix.com	2
beta-Galactosidase	Hydrolyses terminal nonreducing β-D-galactose residues in β-D-galactosides	ASA Spezialenzyme GmbH	www.asa-enzyme.de	2
		Codexis Inc.	www.codexis.com	Custom items 12
		EnzymeWorks	www.enzymeworking.com	5
		Megazyme	www.megazyme.com	1
		Merck KGgA/ MilliporeSigma	www.sigmaaldrich.com	9
		NZYTech	www.nzytech.com	23
		Prozomix Ltd	www.prozomix.com	4
		Toyobo	www.toyobo-global.com	1

Galacturonidase	$[(1\rightarrow4)$-α-D-galacturonide$]_n + H_2O =$ $[(1\rightarrow4)$-α-D-galacturonide$]_{n-1}$ + D-galacturonate	NZYTech	www.nzytech.com	7
alpha-Glucosidase	Hydrolyses terminal, nonreducing (1→4)-linked α-D-glucose residues, with release of α-D-glucose	ASA Spezialenzyme GmbH	www.asa-enzyme.de	1
		Megazyme	www.megazyme.com	5
		Merck KGgA/ MilliporeSigma	www.sigmaaldrich.com	5
		NZYTech	www.nzytech.com	13
		Prozomix Ltd	www.prozomix.com	3
		Toyobo	www.toyobo-global.com	1
beta-Glucosidase	Hydrolyses terminal, nonreducing β-D-glucosyl residues, with release of β-D-glucose	Biocatalysts Ltd	www.biocatalysts.com	1
		EnzymeWorks	www.enzymeworking.com	3
		Megazyme	www.megazyme.com	4
		Merck KGgA/ MilliporeSigma	www.sigmaaldrich.com	7
		NZYTech	www.nzytech.com	25
		Protéus	www.proteus.fr	8
		Prozomix Ltd	www.prozomix.com	80
		Toyobo	www.toyobo-global.com	1
alpha-Glucuronidase	α-D-glucuronoside + H_2O = an alcohol + D-glucuronate	Megazyme	www.megazyme.com	1
		Prozomix Ltd	www.prozomix.com	2
beta-Glucuronidase	β-D-glucuronoside + H_2O = D-glucuronate + an alcohol	Megazyme	www.megazyme.com	1
		Merck KGgA/ MilliporeSigma	www.sigmaaldrich.com	16
Glucoamylase	Hydrolyses terminal (1→4)-linked α-D-glucose residues successively from nonreducing ends of the chains, with release of β-D-glucose	Merck KGgA/ MilliporeSigma	www.sigmaaldrich.com	6

(Continued)

Table 2.3 (*Continued*)

Enzyme Name	Chemical Transformation	Supplier	Company contact details	Number of enzymes in panel
Haloalkane dehalogenase	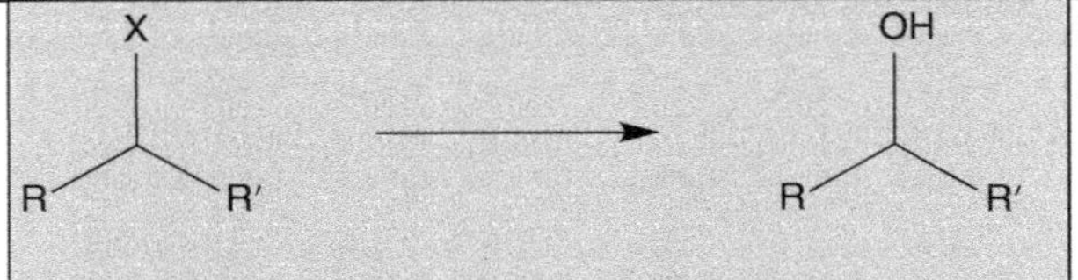	Enantis s.r.o. EnzymeWorks	www.enantis.com www.enzymeworking.com	20 5
Hexosaminidase	Hydrolyses terminal nonreducing *N*-acetyl-D-hexosamine residues in *N*-acetyl-β-D-hexosaminides	Megazyme NZYTech	www.megazyme.com www.nzytech.com	4 15
Hyaluronate lyase	Cleaves hyaluronate chains at a β-D-GlcNAc-(1→4)-β-D-GlcA bond, ultimately breaking the polysaccharide down to 3-(4-deoxy-β-D-gluc-4-enuronosyl)-*N*-acetyl-D-glucosamine	Prozomix Ltd	www.prozomix.com	5
Hyaluronidase	Randomly hydrolyses (1→4)-linkages between *N*-acetyl-β-D-glucosamine and D-glucuronate residues in hyaluronate	Amano Enzyme Merck KGgA/ MilliporeSigma NZYTech	www.amano-enzyme.co.jp www.sigmaaldrich.com www.nzytech.com	1 10 2
Hydantoinase	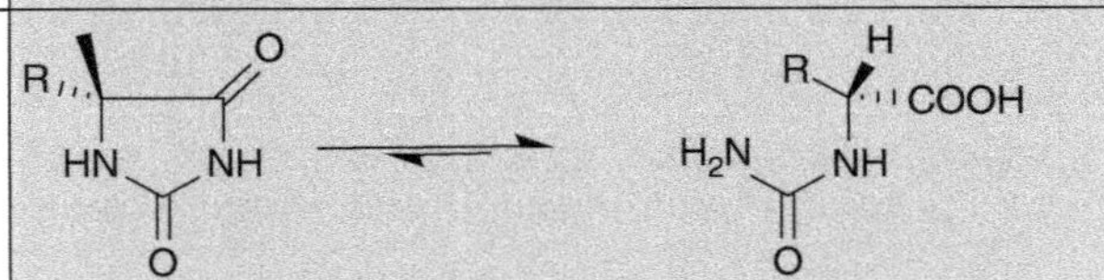	Almac EnzymeWorks Enzymicals AG	www.almacgroup.com www.enzymeworking.com www.enzymicals.com	25 4 1

Hydratase	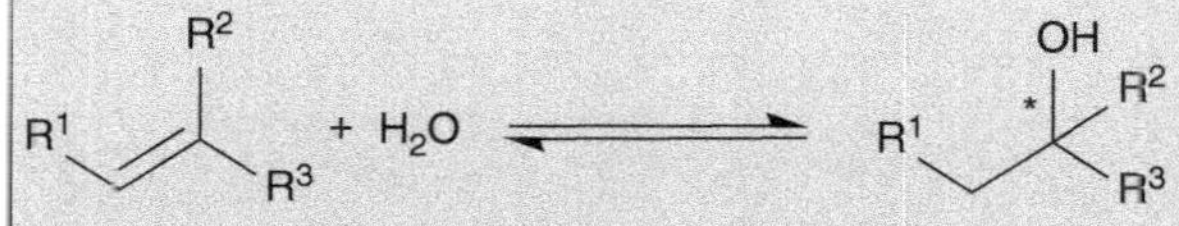	InnoSyn BV	www.innosyn.com	37
		Protéus	www.proteus.fr	5
Imine reductase	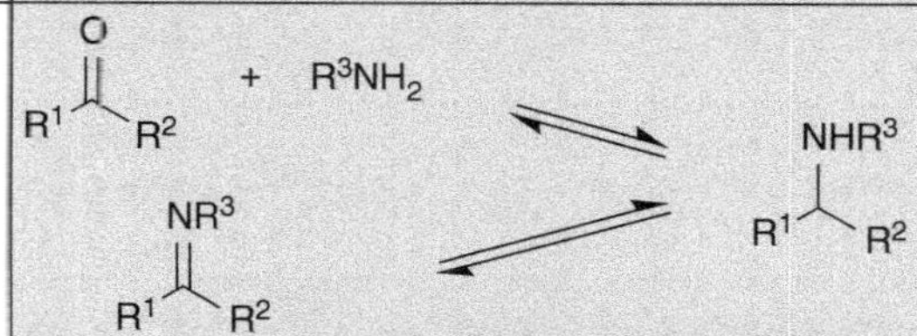 R^1, R^2,R^3 = H, alkyl or aryl	Almac	www.almacgroup.com	50
		Codexis Inc.	www.codexis.com	24 (kit) + custom items
		EnzymeWorks	www.enzymeworking.com	20
		Enzymicals AG	www.enzymicals.com	30
		Johnson Matthey	www.matthey.com/biocatalysts/	85
		Protéus	www.proteus.fr	10
		Prozomix Ltd	www.prozomix.com	576
		Prozomix Ltd (CoEBio3)	www.prozomix.com, www.coebio3.manchester.ac.uk	46
		Syncozymes	en.syncozymes.com	6
Inulase	Endohydrolyses (2→1)-β-D-fructosidic linkages in inulin	Merck KGgA/MilliporeSigma	www.sigmaaldrich.com	1
		NZYTech	www.nzytech.com	2
Isoamylase	Hydrolyses (1→6)-α-D-glucosidic branch linkages in glycogen, amylopectin and their β-limit dextrins	EUCODIS	www.eucodis.com	1

(*Continued*)

Table 2.3 (*Continued*)

Enzyme Name	Chemical Transformation	Supplier	Company contact details	Number of enzymes in panel
		Merck KGgA/ MilliporeSigma	www.sigmaaldrich.com	1
		Prozomix Ltd	www.prozomix.com	1
Laccase	  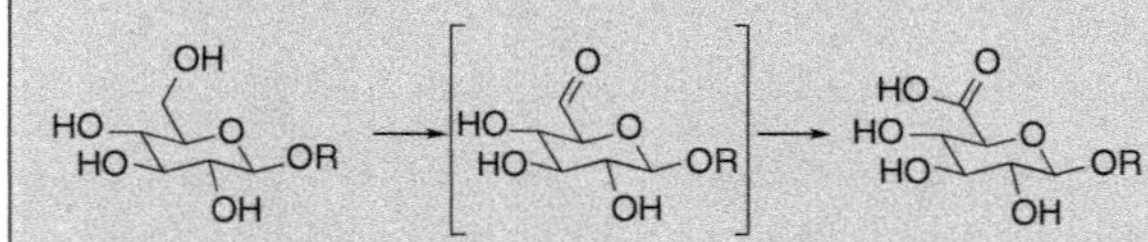	Almac	www.almacgroup.com	96
		Amano Enzyme	www.amano-enzyme.co.jp	1
		Asahi Kasei Pharma Corporation	www.asahi-kasei.co.jp/ shindan/en/	4
		ASA Spezialenzyme GmbH	www.asa-enzyme.de	5
		GECCO Biotech	www.gecco-biotech.com	1
		Merck KGgA/ MilliporeSigma	www.sigmaaldrich.com	3
		Novozymes A/S	pharmaceuticals@ novozymes.com	1
		NZYTech	www.nzytech.com	3
		Protéus	www.proteus.fr	19
Lactamase	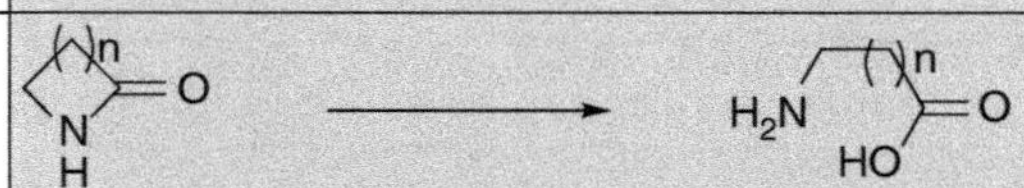	Almac	www.almacgroup.com	6
		EnzymeWorks	www.enzymeworking.com	8
		Merck KGgA/ MilliporeSigma	www.sigmaaldrich.com	3

Lipase	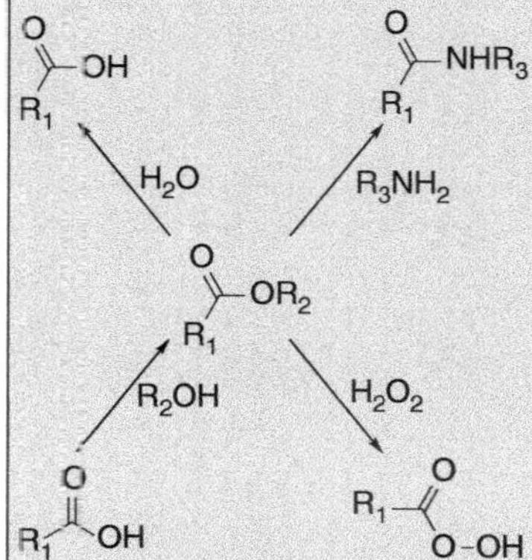	Almac	www.almacgroup.com	96
		Amano Enzyme	www.amano-enzyme.co.jp	9
		Asahi Kasei Pharma Corporation	www.asahi-kasei.co.jp/shindan/en/	1
		ASA Spezialenzyme GmbH	www.asa-enzyme.de	1
		Biocatalysts Ltd	www.biocatalysts.com	9
		ChiralVision	www.chiralvision.com	18
		c-LEcta GmbH	www.c-lecta.com	3
		Codexis Inc.	www.codexis.com	See esterase
		Enzagen	www.enzagen.com	25
		EnzymeWorks	www.enzymeworking.com	15
		Enzymicals AG	www.enzymicals.com	10
		EUCODIS	www.eucodis.com	25
		Evoxx technologies	www.evoxx.com	1
		Merck KGgA/MilliporeSigma	www.sigmaaldrich.com	26
		Novozymes A/S	pharmaceuticals@novozymes.com	10
		Nzomics	www.nzomicsbiocatalysis.co.uk	
		Protéus	www.proteus.fr	9
		Purolite Life Sciences	www.purolite.com/life-sciences/	12
		Syncozymes	en.syncozymes.com	1

(*Continued*)

Table 2.3 (*Continued*)

Enzyme Name	Chemical Transformation	Supplier	Company contact details	Number of enzymes in panel
Lipoxygenase	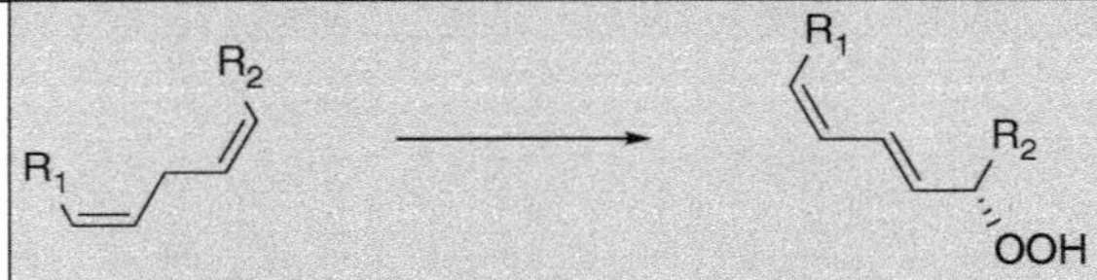	Almac EnzymeWorks Enzymicals AG Merck KGgA/ MilliporeSigma	www.almacgroup.com www.enzymeworking.com www.enzymicals.com www.sigmaaldrich.com	5 5 2 2
Lyase (ammonia)	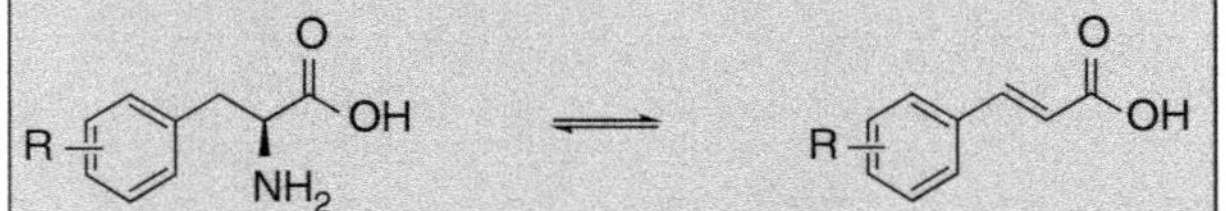	Almac Codexis Inc. EnzymeWorks GECCO Biotech Prozomix Ltd Prozomix Ltd (CoEBio3)	www.almacgroup.com www.codexis.com www.enzymeworking.com www.gecco-biotech.com www.prozomix.com www.prozomix.com, www.coebio3 .manchester.ac.uk	50 >100 cus- tom items 5 2 48 20
Lyase (aspartate ammonia)	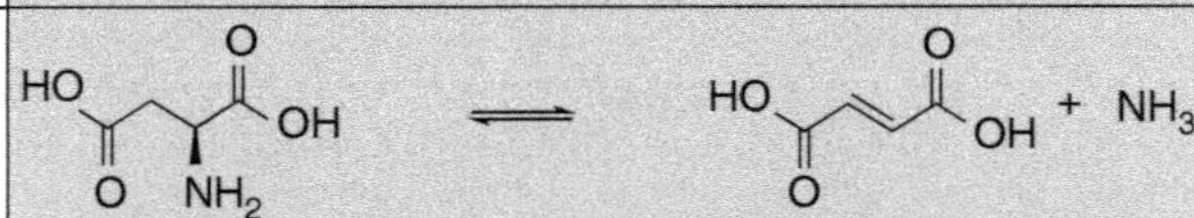	Almac	www.almacgroup.com	9
Lyase (benzaldehyde)	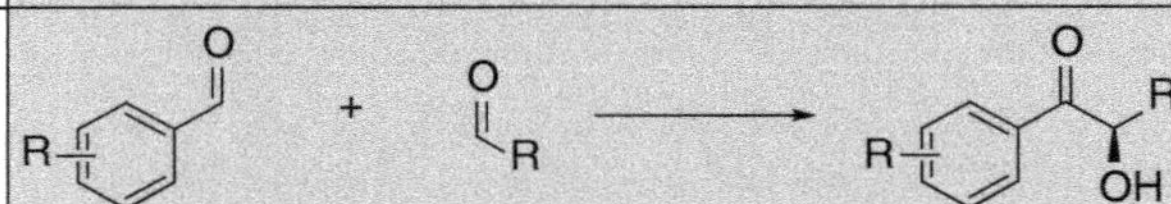	Almac EnzymeWorks Evoxx technologies GECCO Biotech	www.almacgroup.com www.enzymeworking.com www.evoxx.com www.gecco-biotech.com	5 5 1 1

Lyase (hydroxynitrile)/oxynitrilase		Almac	www.almacgroup.com	24
		ASA Spezialenzyme GmbH	www.asa-enzyme.de	1
		Enzymaster	www.enzymaster.de	2
		EnzymeWorks	www.enzymeworking.com	5
		Evoxx technologies	www.evoxx.com	3
		InnoSyn BV	www.innosyn.com	5
		Johnson Matthey	www.matthey.com/biocatalysts/	10
		Syncozymes	en.syncozymes.com	29
Monooxygenase (Baeyer–Villiger)		Almac	www.almacgroup.com	50
		Codexis Inc.	www.codexis.com	90 custom items
		EnzymeWorks	www.enzymeworking.com	20
		Enzymicals AG	www.enzymicals.com	6
		Gecco Biotech	www.gecco-biotech.com	65
		InnoSyn BV	www.innosyn.com	16
		Merck KGgA/MilliporeSigma	www.sigmaaldrich.com	1
		Protéus	www.proteus.fr	10
		Prozomix Ltd (CoEBio3)	www.prozomix.com, www.coebio3.manchester.ac.uk	23

(*Continued*)

Table 2.3 *(Continued)*

Enzyme Name	Chemical Transformation	Supplier	Company contact details	Number of enzymes in panel
Monooxygenase (human P450)	 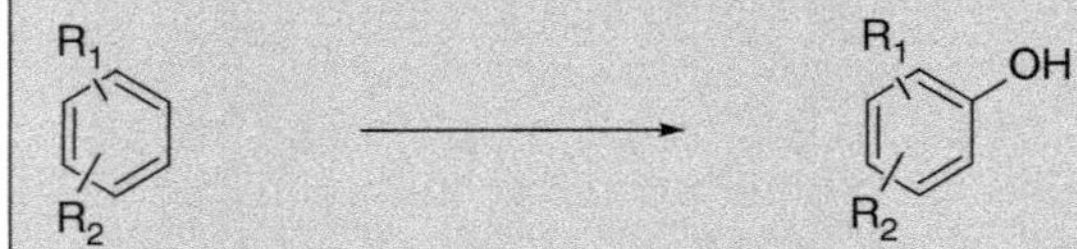  	Almac EnzymeWorks	www.almacgroup.com www.enzymeworking.com	10 15

Monooxygenase (microbial P450)	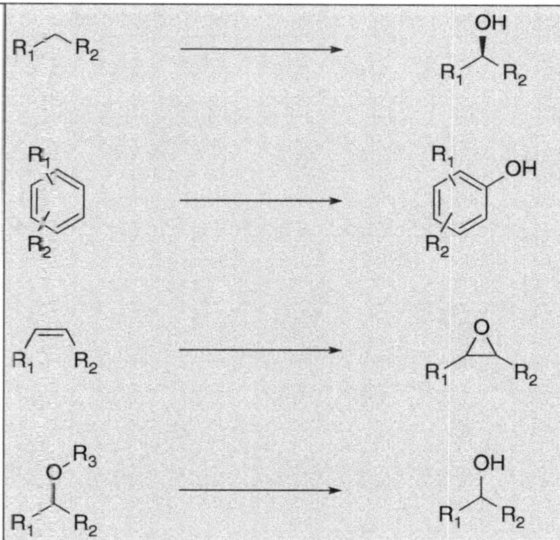	Almac Codexis Inc. EnzymeWorks Enzymicals AG Gecco Biotech InnoSyn BV Johnson Matthey Protéus Prozomix Ltd Prozomix Ltd (CoEBio3) Syncozymes	www.almacgroup.com www.codexis.com www.enzymeworking.com www.enzymicals.com www.gecco-biotech.com www.innosyn.com www.matthey.com/biocatalysts/ www.proteus.fr www.prozomix.com www.prozomix.com, www.coebio3.manchester.ac.uk en.syncozymes.com	96 11 (kit) + custom items 30 6 4 150 10 5 161 43 8
Nitrilase	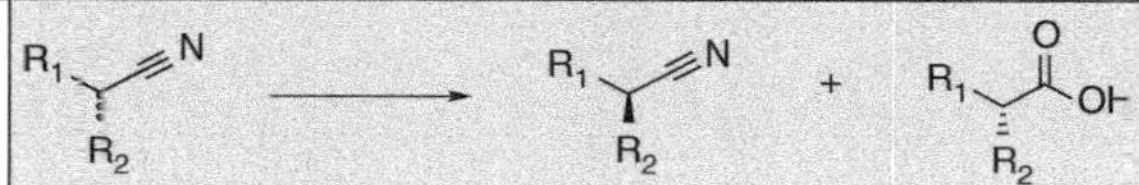	Almac c-LEcta GmbH Codexis Inc. EnzymeWorks Merck KGgA/MilliporeSigma Protéus Prozomix Ltd Syncozymes	www.almacgroup.com www.c-lecta.com www.codexis.com www.enzymeworking.com www.sigmaaldrich.com www.proteus.fr www.prozomix.com en.syncozymes.com	96 94 12 (kit) 20 1 10 96 40

(*Continued*)

Table 2.3 *(Continued)*

Enzyme Name	Chemical Transformation	Supplier	Company contact details	Number of enzymes in panel
Nitrile hydratase	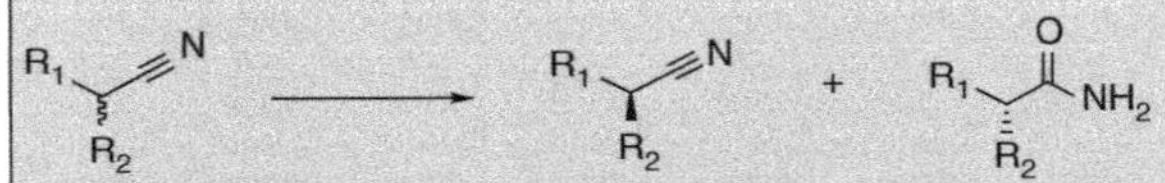	Almac	www.almacgroup.com	9
		EnzymeWorks	www.enzymeworking.com	15
		EUCODIS	www.eucodis.com	10
		InnoSyn BV	www.innosyn.com	11
		Protéus	www.proteus.fr	10
		Prozomix Ltd	www.prozomix.com	18
Nitroreductase	$Ar-NO_2 \longrightarrow Ar-NH_2$ Ar = aryl	EnzymeWorks	www.enzymeworking.com	20
		Johnson Matthey	www.matthey.com/biocatalysts/	60
		Merck KGgA/MilliporeSigma	www.sigmaaldrich.com	1
		Prozomix Ltd	www.prozomix.com	96
		Syncozymes	en.syncozymes.com	12
Nuclease		c-LEcta GmbH	www.c-lecta.com	1
		Merck KGgA/MilliporeSigma	www.sigmaaldrich.com	6
Oxidase (amine)/monoamine oxidase (MAO)	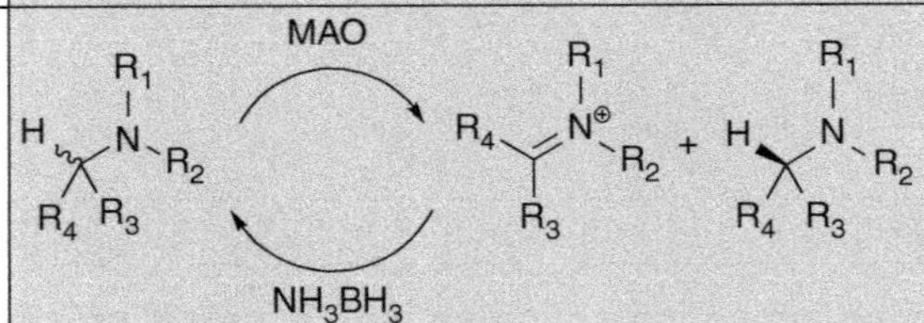	Almac	www.almacgroup.com	96
		Codexis Inc.	www.codexis.com	90 custom items
		EnzymeWorks	www.enzymeworking.com	15
		Gecco Biotech	www.gecco-biotech.com	1
		Merck KGgA/MilliporeSigma	www.sigmaaldrich.com	2
		Syncozymes	en.syncozymes.com	5

Oxidase (glucose)	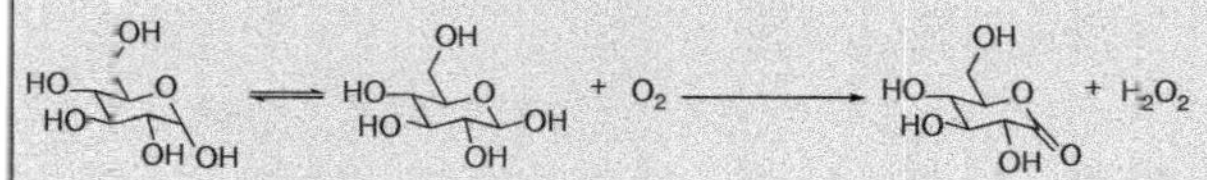	Almac	www.almacgroup.com	25
		Amano Enzyme	www.amano-enzyme.co.jp	4
		ASA Spezialenzyme GmbH	www.asa-enzyme.de	1
		Biocatalysts Ltd	www.biocatalysts.com	1
		EnzymeWorks	www.enzymeworking.com	5
		Megazyme	www.megazyme.com	1
		Merck KGgA/ MilliporeSigma	www.sigmaaldrich.com	4
		Toyobo	www.toyobo-global.com	1
Phosphatase	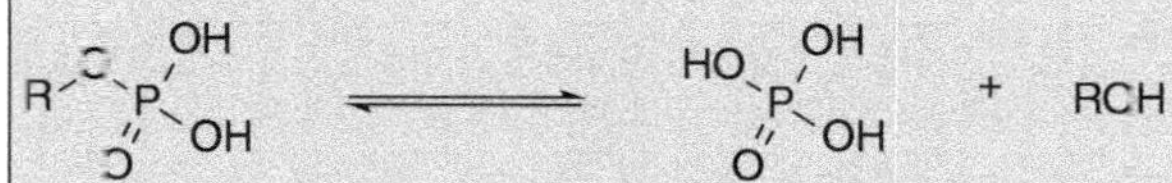	Almac	www.almacgroup.com	96
		Asahi Kasei Pharma Corporation	www.asahi-kasei.co.jp/ shindan/en/	2
		EnzymeWorks	www.enzymeworking.com	5
		Enzymicals AG	www.enzymicals.com	1
		Megazyme	www.megazyme.com	2
		Merck KGgA/ MilliporeSigma	www.sigmaaldrich.com	18
		Prozomix Ltd	www.prozomix.com	23
		Toyobo	www.toyobo-global.com	1
Phosphotriesterase	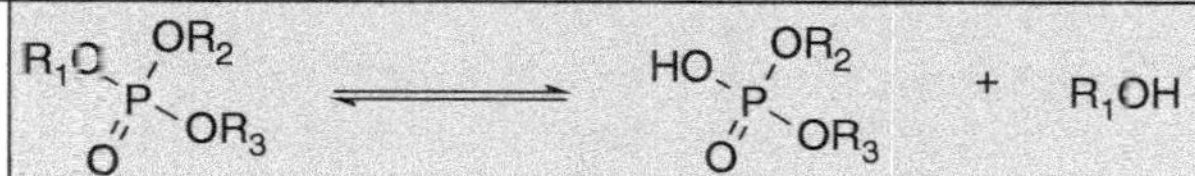	Almac	www.almacgroup.com	8

(*Continued*)

Table 2.3 (*Continued*)

Enzyme Name	Chemical Transformation	Supplier	Company contact details	Number of enzymes in panel
Reductive aminase	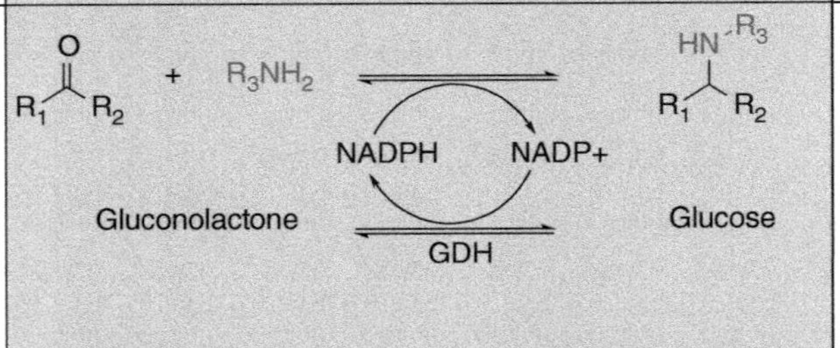	Almac Codexis Inc. EnzymeWorks Enzymicals AG Prozomix Ltd	www.almacgroup.com www.codexis.com www.enzymeworking.com www.enzymicals.com www.prozomix.com	50 24 (kit) + custom items 10 2 576
Tannase	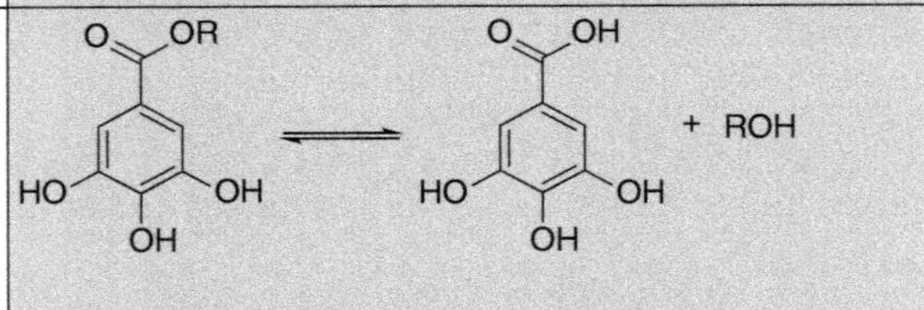	Almac	www.almacgroup.com	50
Transaminase (aminotransferase)	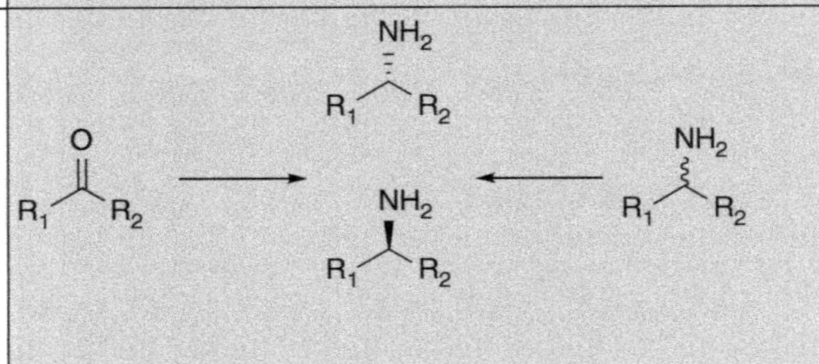	Almac Asahi Kasei Pharma Corporation c-LEcta GmbH Codexis Inc.	www.almacgroup.com www.asahi-kasei.co.jp/shindan/en/ www.c-lecta.com www.codexis.com	>300 2 15 24 (kit) + custom items

		Enzymaster	www.enzymaster.de	83
		EnzymeWorks	www.enzymeworking.com	100
		Enzymicals AG	www.enzymicals.com	20
		Evoxx technologies GmbH	www.evoxx.com	18
		Gecco Biotech	www.gecco-biotech.com	4
		InnoSyn BV	www.innosyn.com	74
		Megazyme	www.megazyme.com	3
		Protéus	www.proteus.fr	9
		Prozomix Ltd	www.prozomix.com	384
		Syncozymes	en.syncozymes.com	58
R-Transaminase	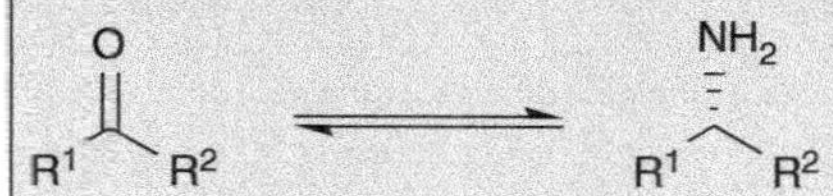	Almac	www.almacgroup.com	>200
		Codexis Inc.	www.codexis.com	12 out of 24 (kit) + custom items
		EnzymeWorks	www.enzymeworking.com	40
		Johnson Matthey	www.matthey.com/biocatalysts/	100
S-Transaminase	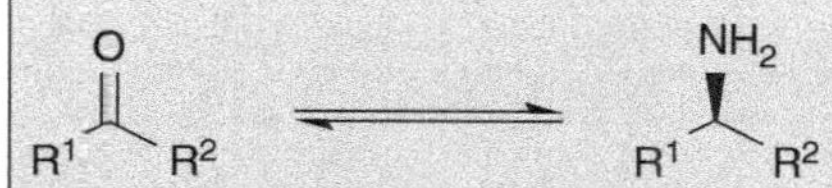 R^1, R^2 = H, alkyl or aryl	Almac	www.almacgroup.com	>100
		Codexis Inc.	www.codexis.com	12 out of 24 (kit) custom items

(*Continued*)

Table 2.3 (*Continued*)

Enzyme Name	Chemical Transformation	Supplier	Company contact details	Number of enzymes in panel
		EnzymeWorks Johnson Matthey	www.enzymeworking.com matthey.com/biocatalysts	50 170
Transketolase	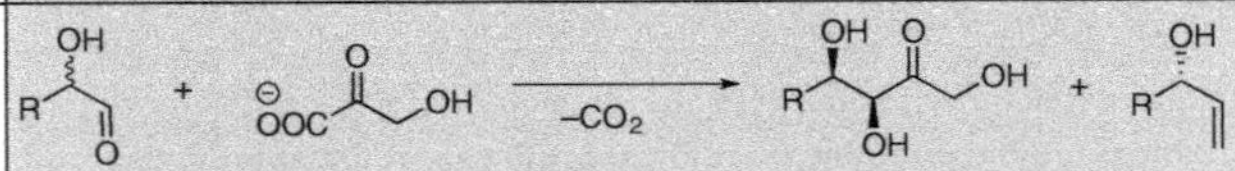	Almac Merck KGgA/ MilliporeSigma	www.almacgroup.com www.sigmaaldrich.com	50 1
Trehalose synthase	Maltose → trehalose	Codexis Inc. Prozomix Ltd	www.codexis.com www.prozomix.com	6 custom items 2
Tyrosinase		Almac Merck KGgA/ MilliporeSigma	www.almacgroup.com www.sigmaaldrich.com	5 1
Urease	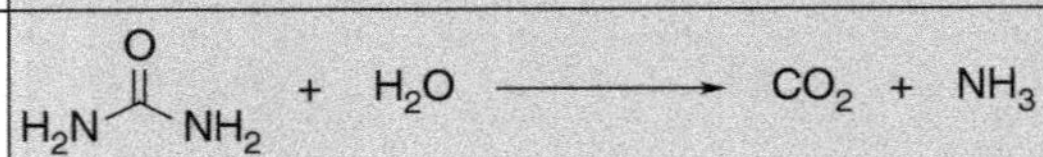	Almac Merck KGgA/ MilliporeSigma Toyobo	www.almacgroup.com www.sigmaaldrich.com www.toyobo-global.com	12 7 1

[a]Reaction descriptions adapted from http://www.qmul.ac.uk/sbcs/iubmb/.

The definitions used in some examples came from International Union of Biochemistry and Molecular Biology – Recommendations on Biochemical & Organic Nomenclature, Symbols & Terminology etc. as found on School of Biological and Chemical Sciences, Queen Mary, University of London website.

Acknowledgements

We would like to thank Eric Althoff, Alison Arnold, Alessandra Basso, Christina Baumann, Sven Bension, Joana Bras, Fraser Brown, Gareth Brown, Alzbeta Cardova, Reuben Carr, Jill Caswell, Arno Cordes, Stephane Corgie, Thomas Daussmann, Beatriz Dominguez, Martina Doring, Andrew Ellis, Beatrix Ellis, Franck Escalletes, Iwona Kaluzna, Karim Engelmark Cassimjee, David Entwistle, Ian Fotheringham, Jane Gao, Robert Gates, Gloria Gonzalez, Javier Gonzalez Sabin, Alexander Hoepker, Henk-Jan Joosten, Jonathan Kennedy, Grzegorz Kubik, Martin Lindegaard, Rio Luo Han, Nikola Loncar, David Mangan, Juliette Martin, Jan Modregger, Tom Moody, Francisco Morís, Rudy Pandjaitan, David Pearlman, Yuyin Qi, Derek Quinn, Marianne Rooman, Gerard Santiago, Martin Schürmann, Simona Serban, Alex Tao, Steve Taylor, Christopher Trummer, Merdan Urcanli, Hugo van Beek, Andreas Vogel, Rainer Wardenga, Roland Wohlgemuth and George Yeh, whose rapid responses made it possible to compile the data found in this chapter.

Reference

1. Rozzell, D. (2012) Tabular survey of available enzymes in enzyme catalysis, in *Organic Synthesis*, 3rd edn (eds K. Drauz, H. Gröger and O. May), Wiley-VCH Verlag, pp. 1847–1938.

3

Imine Reductases

3.1 Imine Reductase-Catalysed Enantioselective Reductive Amination for the Preparation of a Key Intermediate to Lysine-Specific Histone Demethylase 1 (LSD1) Inhibitor GSK2879552

Diluar Khan,[1] Joseph Hosford,[2] Katie Rufell,[1] Markus Schober,[2] Batool Ahmed Omer,[1] Gheorghe-Doru Roiban,*[2] and Mahesh J. Sanganee,*[1]

[1]*Chemical Development, Medicinal Science & Technology, GlaxoSmithKline, Stevenage, Hertfordshire, UK*

[2]*Synthetic Biochemistry, Medicinal Science & Technology, GlaxoSmithKline, Stevenage, Hertfordshire, UK*

tert-Butyl 4-((4-((((1*R*,2*S*)-2-phenylcyclopropyl)amino)methyl)piperidin-1-yl)methyl) benzoate **3** is produced at the key convergence point in the synthesis of GSK2879552, indicated as an inhibitor of LSD1 [1]. The initial route towards the active pharmaceutical ingredient (API) employed a classical resolution of *rac-trans*-**2** followed by reductive amination by borohydride. Directed evolution of imine reductase IR46 [2] from *Saccharothrix espanaensis* yields a biocatalyst (M3), which enables a one-pot resolution and reductive amination to produce **3** with high enantioselectivity (Scheme 3.1) [3].

3.1.1 Procedure 1: Enantioselective Reductive Amination of *tert*-Butyl 4-((4-formylpiperidin-1-yl)methyl)benzoate (1) with *rac*-(*trans*)-2-Phenylcyclopropanamine (2) Using Imine Reductase M3

3.1.1.1 Materials and Equipment

- *tert*-Butyl-4-((4-formylpiperidin-1-yl)methyl)benzoate **1** (329.2 g)
- *trans*-2-Phenylcyclopropanamine (*rac-trans*-**2**; 476.0 g)
- M3 lyophilised clarified cell lysate (4.0 g) [3]
- Glucose dehydrogenase (GDH-CDX-901) from Codexis Inc. (4.0 g)

Applied Biocatalysis: The Chemist's Enzyme Toolbox, First Edition. Edited by John Whittall and Peter W. Sutton.
© 2021 John Wiley & Sons Ltd. Published 2021 by John Wiley & Sons Ltd.

- DMSO (2.34 L)
- Sodium acetate trihydrate (93.2 g)
- Acetic acid (40.8 mL)
- D-glucose (280.0 g)
- Nicotinamide adenine dinucleotide phosphate disodium salt ($NADP^+$; 20.5 g)
- Water (16 L)
- Acetone (6 L)
- Celite 545 filtering agent (200.2 g)
- Sodium chloride solution (25% wt/v)
- Controlled laboratory reactors (CLR, 20 L)
- PTFE mini filter (with glass-fibre filter paper and Whatman No. 113 filter paper)
- Buchner flask (20 L)
- Peristaltic pump (with a 1 μm inline filter)
- V2 filter dryer (with 10 μm Hastelloy C22 poroplate filter)
- High-performance liquid chromatograph (HPLC) equipped with an Agilent Bonus RP column (4.6 × 150 mm, 3.5 μm) column.
- HPLC equipped with a ChiralPak AD-RH (2.1 × 150 mm, 5 μm)

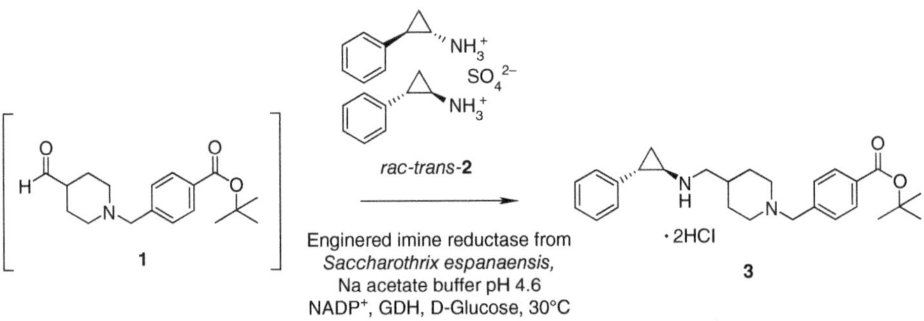

Scheme 3.1 *Imine reductase M3-catalysed enantioselective reductive amination of **1**.*

3.1.1.2 Procedure

3.1.1.2.1 Preparation of the Sodium Acetate Buffer.

1. Distilled water (4 L), sodium acetate trihydrate (93.2 g) and acetic acid (40.8 mL) were charged to a 20 L CLR. Additional water (10 L) was then charged.
2. Stirring was started at 200 rpm until all the solid components dissolved.
3. The pH of the solution at this point was 4.6.
4. Reagents were charged to the CLR: *rac-trans-***2** (476.0 g, 2.4 equiv.), D-glucose (280.0 g, 1.43 equiv.), $NADP^+$ (20.5 g).
5. Stirring was continued and the CLR jacket temperature set to 30 ± 2 °C.
6. Complete dissolution of reagents, clear solution attained for at least 45 min.

3.1.1.2.2 Reductive Amination.

1. Biocatalysts were charged to the CLR, lyophilised clarified cell lysate of imine reductase M3 (4.0 g, 0.012 wt) and GDH-CDX-901 (4.0 g, 0.012 wt).

2. Stirring for 10 min resulted in total resuspension of biocatalysts. Stirring was then continued for 30 min.
3. *tert*-Butyl-4-((4-formylpiperidin-1-yl)methyl)benzoate **1** (329.2 g, 1 equiv) in DMSO (2.34 L) was charged over 1 hr, whilst heating and stirring continued.
4. 4 hr after the addition of aldehyde **1**, the reaction was sampled and analysed by achiral HPLC, and conversion was determined (typical reaction conversion was >95% a/a (area/area)).
5. Reaction was quenched by the addition of acetic acid (1.2 L) over 10 min and stirred for a further 10 min.
6. A filtering agent, Celite 545 (200.2 g), was added and the suspension was stirred for 1 hr, to allow for the total precipitation of the biocatalysts.
7. The slurry was then filtered using a 24 cm PTFE mini filter fitted with glass-fibre filter paper and Whatman No. 113 wet-strengthened filter paper. A 20 L Buchner flask was used to receive the filtrate. Filtration of the biocatalysts took approximately 8 min.
8. The reaction filtrate was transferred to a clean 20 L CLR via peristaltic pump with a 1 μm inline filter.
9. CLR jacket temperature was set to 6–7 °C. Once the temperature had been reached, sodium chloride solution (25% w/v; 880 mL) was added over 60 min.
10. The suspension was filtered on a V2 filter dryer equipped with 10 μm Hastelloy C22 poroplate filter.
11. To achieve maximum recovery of product, the CLR was washed with water (2 × 1 L) and acetone (1 × 4 L, 2 × 1 L), each time discharging on to the filter.
12. The product **3** was dried under vacuum (50 mbar), first at 20 °C for 2 hr and then at 40 °C for a further 12 hr.
13. *tert*-Butyl 4-((4-((((1*R*,2*S*)-2-phenylcyclopropyl)amino)methyl)piperidin-1-yl)methyl) benzoate **3** was afforded as a white solid (452.1 g, yield 84.4%, 100.0% purity, 99.7% ee).

3.1.2 Analytical Method

3.1.2.1 Sampling Procedure

Sampling was performed whilst stirring:

1. Reaction mixture (100 μL) was added to acetonitrile : 1 M acetic acid (1 mL), mixed by inversion, and then a 0.45 μm PTFE filter was used to remove the biocatalyst debris.
2. The supernatant of the prepared sample was injected on achiral HPLC and chiral HPLC as described later.
3. Enantiomeric excess (ee) and conversion were calculated.

3.1.2.2 Achiral HPLC

HPLC was equipped with an Agilent Bonus RP column (4.6 × 150 mm, 3.5 μm). Trifluoroacetic acid (0.05%) in water (mobile phase A) and in acetonitrile (mobile phase B) was set at a flow rate of 1 mL.min^{-1} at a column temperature of 40 °C. The method was set up starting from a 95 : 5 ratio of A : B, followed by a 12 min linear gradient to 77 : 23 ratio, then a 6 min linear gradient to 5 : 95, then a 2 min hold at 5 : 95 and then an immediate change to 95 : 5. Elution of compounds was monitored at 220 nm. *Rac*-(*trans*)-(**2**) elutes at ~4.5 min, aldehyde (**1**) at ~12.8 min and product *tert*-butyl 4-((4-((((1*R*,2*S*)-2-phenylcyclopropyl)amino)methyl)piperidin-1-yl)methyl) benzoate (**3**) at ~15.0 min.

3.1.2.3 Chiral HPLC

HPLC equipped with a ChiralPak AD-RH (2.1 × 150 mm, 5 μm) and guard column using 20 mM sodium tetraborate decahydrate pH 9.0 (mobile phase C) and methanol (mobile phase D) was set up at a 10 : 90 ratio of C : D, flow rate 0.45 mL.min^{-1} and column temperature 45 °C. Elution of compounds was monitored at 230 nm. 4-((4-((((1*S*,2*R*)-2-Phenylcyclopropyl)amino)methyl)piperidin-1-yl)methyl) benzoate (**3**) elutes at ~17.7 min and 4-((4-((((1*R*,2*S*)-2-phenylcyclopropyl)amino)methyl)piperidin-1-yl)methyl) benzoate (**3**) at ~22.3 min.

3.1.3 Conclusion

The use of M3-engineered imine reductase enabled the API to be produced with high enantioselectivity. A simple non-extractive workup aided the separation of biocatalyst and product, allowing for high product isolation. This protocol has been demonstrated in the production of over 1 kg of API.

References

1. Johnson, N.W. and Kasparec, J. (2014) Patent 8853408 B2.
2. Li, H., Luan, Z.-J., Zheng, G.-W. and Xu, J.-H. (2015) *Advanced Synthesis & Catalysis*, **357** (8), 1692–1696.
3. Schober, M., MacDermaid, C., Ollis, A.A. *et al.* (2019) *Nature Catalysis*, **2**, 909–915.

3.2 Expanding the Collection of Imine Reductases Towards a Stereoselective Reductive Amination

Elina Siirola,[1] Charlie Moore,[1] Robert Bruccoleri[2] and Radka Snajdrova[1]
[1]*Novartis Institute for Biomedical Reseach, Global Discovery Chemistry, Basel, Switzerland*
[2]*Congenomics LLC, Glastonbury, CT, USA*

Imine reductases or reductive aminases are biocatalysts with huge potential in the synthesis of pharmaceuticals, providing access to asymmetric reductive aminations that are difficult to achieve enantioselectively when applying conventional chemical methods [1]. These enzymes complement the applications of transaminases by offering a route not only to primary amines, but to secondary and tertiary ones as well [2]. However, access to imine reductases from commercial sources can be challenging, which has led institutions to create their own imine reductase panels using different selection criteria [3–5].

Here, we describe a method for the targeted selection of imine reductases towards a highly selective reductive amination of a prochiral ketone (Scheme 3.2). In three rounds of enzyme selection, the first based on published literature, the second on sequence similarity to an early hit and the third on refined selection, we were able to proceed from low ($c = 8\%$) to high ($c = 83\%$) conversion and near-perfect chiral purity (>99% ee) of the

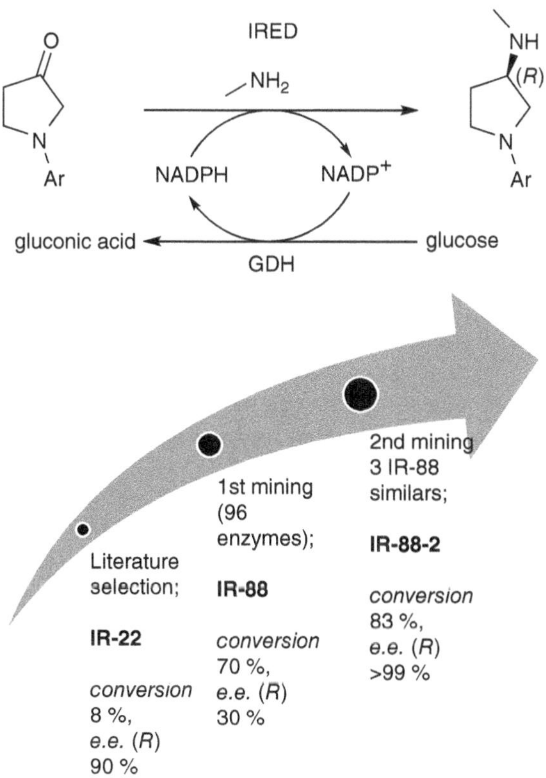

Scheme 3.2 *Reductive amination target and a summary of the genome mining approach.*

desired (*R*)-enantiomer. This approach demonstrates a relatively short timeline approach for building a diverse imine reductase panel and illustrates a method for starting point selection towards further improvement by enzyme engineering.

3.2.1 Procedure 1: Imine Reductase Sequence Selection from Public Databases

3.2.1.1 Materials and Equipment

- National Center for Biotechnology Information (NCBI) nonredundant protein sequence database (nr)
- Clustal Omega multiple sequence alignment algorithm (http://www.ebi.ac.uk/Tools/msa/clustalo/)
- HMMER 2.0 hidden Markov model generator (www.hmmer.org)
- PDB protein structure database and visualisation tools (www.rcsb.org)
- JCat codon optimisation package (www.jcat.de)

3.2.1.2 Procedure

1. A set of 28 known imine reductases were used as a training set to generate a protein multiple sequence alignment using Clustal Omega running default settings.

2. The alignment was used to generate an imine reductase hidden Markov model (HMM) using HMMER 2.0 to populate a database of 15 352 putative imine reductases from the NCBI nr.
3. Using two imine reductase protein structures with substrate bound (PDB ID: 5FWN and 5G6S), the approximate substrate binding region of imine reductases was defined and assumed to be constant across all imine reductases. The substrate region was defined in the PDB viewer simply by viewing the structure with substrate bound (tetrahydroisoquinoline, THIQ) and approximating the protein regions that might contribute to substrate recognition. This region was defined as amino acids 74–83, 101–107, 116–121, 163–181, 206–219 and 229–245 in imine reductase 5G6S numbering.
4. Sequences of these substrate binding regions were extracted from all imine reductases in the database and pairwise comparisons of these regions was performed. Levenshtein distances [6] (the sum of all amino changes required to transform one string being compared – amino acids in this case – into the other) between each of these sequences were computed to generate a pair list [7].
5. The Levenshtein distance pairs were used to cluster the imine reductases using single-link clustering; that is, each cluster was composed of imine reductases, where every imine reductase had a Levenshtein distance no bigger than a specific threshold to at least one other imine reductase in the cluster. By trying thresholds from 0 to 100, we found one where just 100 clusters were formed. Centroids from each cluster were selected manually by prioritising thermophilic organism sources, proprietary genome sources from our collection and overall phylogenetic diversity of sources, in that order. Genes were codon-optimised for *Escherichia coli* using JCat, with the subsequent genes synthesised and cloned into pET28a by Twist Bioscience (San Francisco, CA).

3.2.2 Procedure 2: Imine Reductase Expression in a 96-Well Deep Plate

3.2.2.1 Materials and Equipment

- Selected imine reductase genes cloned into pET28a using NdeI/XhoI restriction sites
- Chemically competent *E. coli* BL21(DE3) cells and SOC medium
- Lysogenic broth (LB) agar plates containing kanamycin (50 μg.mL^{-1})
- Static incubator at 37 °C and a rotary incubator with, preferably, a deep-well plate adapter
- TB medium supplemented with kanamycin (50 μg.mL^{-1}) and lactose (0.5% w/w)
- 96-well deep plates

3.2.2.2 Procedure

1. Plasmids (pET28a) containing the imine reductase genes of interest were transformed into chemically competent *E. coli* BL21(DE3) following the manufacturer's instructions, and the transformations were plated on LB agar plates with kanamycin as a selective antibiotic. The plates were incubated at 37°C overnight.
2. On the following day, a deep-well plate was filled with TB medium (1 mL.well^{-1}), and one colony from each agar plate was picked into an individual well. The culture plate was incubated at 25 °C for 24 hr.
3. Biomass was harvested by centrifuging the plate at 5000 rpm for 20 min. Supernatant was discarded and the plate was stored at −20 °C until screening (see Procedure 3).

3.2.3 Procedure 3: Reductive Amination in 96-Well Plates

3.2.3.1 Materials and Equipment

- Biomass plate from Procedure 1
- Lysis buffer (KPi buffer (94 mM K_2HPO_4 and 6 mM KH_2PO_4), pH 8 containing lysozyme (2 mg.mL^{-1}) and benzonase™ genetically-engineered endonuclease (0.1 μL.mL^{-1}))
- Reaction buffer (KPi buffer (94 mM K_2HPO_4 and 6 mM KH_2PO_4), pH 8 containing methylamine hydrochloride (400 mM), ketone (10 mM), D-glucose (60 mM), glucose dehydrogenase (0.4 mg.mL^{-1}) and $NADP^+$ (2 mM). Note that the final concentrations in each reaction were half of the concentrations in the reaction buffer.
- Acetonitrile (400 μL per well)
- Microtitre plate or vials compatible with an UPLC autosampler
- Aluminium foil plate seals

3.2.3.2 Procedure

1. The biomass in each well was resuspended into the lysis buffer (200 μL.well^{-1}) and the plate was sealed by an aluminium foil seal. The lysis reactions were incubated on a rotary incubator at 30 °C, 250 rpm for 30 min.
2. Reaction buffer (200 μL) was added to each well. The plate was sealed and incubated at 30 °C, 250 rpm for 16 hr.
3. The reactions were quenched with acetonitrile (400 μL) and the plate was centrifuged at 5000 rpm, 30 min. Supernatant was transferred into a microtitre plate for ultra-performance liquid chromatography (UPLC) analysis.

3.2.4 Procedure 4: Retesting of Reaction and Chiral Analysis

3.2.4.1 Materials and Equipment

- Culture tubes (50 mL)
- Shake flasks (500 mL)
- LB medium and TB medium containing kanamycin (50 μg.mL^{-1})
- Isopropyl-β-D-1-thiogalactopyranoside (IPTG) stock solution in water (1 M)
- KPi buffer (94 mM K_2HPO_4 and 6 mM KH_2PO_4), pH 8
- Sonicator
- Eppendorf tubes
- Eppendorf thermomixer
- Reaction buffer (KPi buffer (94 mM K_2HPO_4 and 6 mM KH_2PO_4), pH 8 containing methylamine hydrochloride (200 mM), ketone (5 mM), D-glucose (30 mM), glucose dehydrogenase (0.2 mg.mL^{-1}) and $NADP^+$ (1 mM)
- Water-immiscible organic solvent compatible with the chiral analysis method of choice (e.g. EtOAc, MTBE or Me-THF)
- Microtitre plate or vials compatible with the analytical instrument

3.2.4.2 Procedure

1. Single colonies from selected agar plates (see Procedure 2) were picked into LB medium (5 mL) and the pre-cultures were incubated at 37 °C, 160 rpm overnight

2. Pre-culture (2 mL) was used to inoculate TB main cultures (200 mL) in shake flasks. The cultures were incubated at 37 °C, 160 rpm until OD_{600} ~0.8. The protein expression was induced by the addition of IPTG (1 mM final concentration). The expression was carried out at 30 °C, 160 rpm overnight.
3. The cells were harvested at 6000 rpm for 10 min and resuspended into three aliquots of KPi buffer. The cells were disrupted by sonication and the lysates were clarified by centrifugation.
4. The clarified supernatants were lyophilised, yielding an enzyme powder, but the lysates could also be used freshly in the reactions.
5. The reactions were set up in Eppendorf tubes by dissolving 5 mg of each enzyme powder into 500 μL reaction buffer. After overnight incubation at 30 °C, 900 rpm, the pH of the reactions was adjusted to ~10 and the result was extracted into the solvent of choice. The organic layers were subjected to chiral analysis.

3.2.5 Analytical Method

Chiral HPLC method: Chiralpak AD-H column (4.6 × 252 mm, Shimadzu, Lot. No. ADH0CE-QK128). Heptane : 2-propanol : ethanol 70 : 5 : 25 with 0.2% diethylamine, flow 1 mL.min^{-1}, 30 °C, detection at 275 nm. Typical retention times: (*S*)-amine 12.9 min; (*R*)-amine 14.5 min; (*S*)-alcohol 9.8 min; (*R*)-alcohol 11.4 min (Figure 3.1).

3.2.6 Conclusion

Gene mining offers an interesting alternative to protein engineering when looking for improved variants of a particular enzyme. Our approach, outlined here, relies on simple concepts and free software. We believe that this should be a more effective way of building enzyme libraries compared to using simple BLAST algorithms because it focuses on the active site, which is where diversity is most wanted. Significantly shorter timelines and low investment were required in order to obtain an active and enantioselective imine reductase for reductive amination of a particular target molecule in a single screen. We have expanded our in-house enzyme collection and have a new powerful set of biocatalysts to address needs within our drug-discovery portfolio.

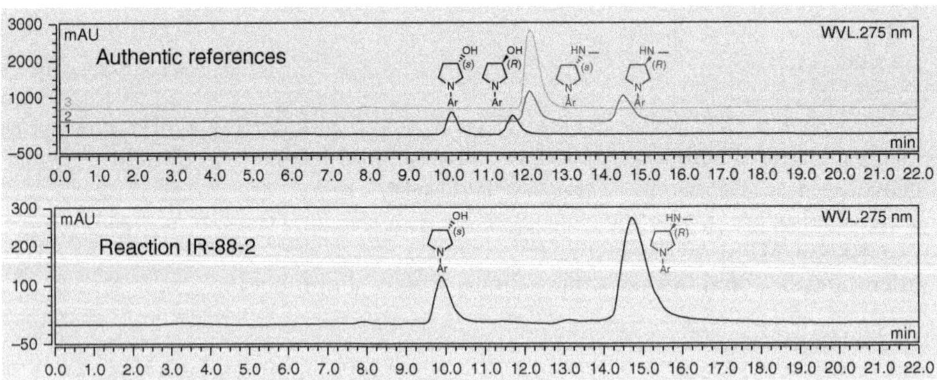

Figure 3.1 Chiral analysis of the reductive amination catalysed by IR88-2.

References

1. Schrittwieser, J.H., Velikogne, S. and Kroutil, W. (2015) *Advanced Synthesis & Catalysis*, **357**, 1655–1685.
2. Mangas-Sanchez, J., France, S.P., Montgomery, S.L. *et al.* (2017) *Current Opinion in Chemical Biology*, **37**, 19–25.
3. Wetzl, D., Berrera, M., Sandon, N. *et al.* (2015) *ChemBioChem*, **16**, 1749–1756.
4. Roiban, G.-D., Kern, M., Liu, Z. *et al.* (2017) *ChemCatChem*, **9**, 4475–4479.
5. France, S.P., Howard, R.M., Steflik, J. *et al.* (2018) *ChemCatChem*, **10**, 510–514.
6. Navarro, G. (2001) *ACM Computing Surveys*, **33** (1), 31–88.
7. Code reference: https://metacpan.org/pod/Text::Levenshtein.

3.3 Asymmetric Synthesis of the Key Intermediate of Dextromethorphan Catalysed by an Imine Reductase

Peiyuan Yao, Zefei Xu, Shanshan Yu, Qiaqing Wu* and Dunming Zhu*
National Engineering Laboratory for Industrial Enzymes, Tianjin Engineering Research Center of Biocatalytic Technology, Tianjin Institute of Industrial Biotechnology, Chinese Academy of Sciences, Tianjin, People's Republic of China

Dextromethorphan **1**, which has been widely used as a non-opioid antitussive for over 50 years and also has anticonvulsant and neuroprotective properties [1], can be prepared from (*S*)-1-(4-methoxybenzyl)-1,2,3,4,5,6,7,8-octahydroisoquinoline (*S*)-**3** by formylation, Grewe cyclisation and reduction of the *N*-formyl group (Scheme 3.3) [2]. As an alternative to the traditional chemical methods, biocatalysis deserves special attention because of its mild reaction conditions, nontoxic reagents and high stereoselectivity [3]. For this reason, we identified a highly hindrance-tolerant imine reductase that could efficiently convert 1-(4-methoxybenzyl)-3,4,5,6,7,8-hexahydroisoquinoline **2** into (*S*)-**3** with high enantioselectivity and conversion (Scheme 3.3).

The co-expression plasmid pRSFDuet-GDH-IR30 harbouring the imine reductase IR30 from *Sciscionella marina* and glucose dehydrogenase (GDH) from *Bacillus subtilis* was constructed and transformed into *E. coli* BL21 (DE3) for overexpression. With the co-expression strains (*S*)-**3**, the key intermediate of **1** was produced on a preparative scale [4].

3.3.1 Procedure 1: Construction of Recombinant Bacteria *E. coli* BL21(DE3) Harbouring Plasmid pRSFDuet-GDH-IR30

3.3.1.1 Materials and Equipment

- pRSFDuet™-1 vector (Novagen)
- Recombinant plasmid pET28a-IR30 containing the IR30 gene from *Sciscionella marina* (Protein identifier: WP_020496004.1)
- Recombinant plasmid pET15b-GDH containing the GDH gene from *Bacillus subtilis* (Protein identifier: WP_003246720.1)
- *FastPfu* DNA polymerase (TransGen Biotech)
- Restriction enzyme *Nco* I, *Hind* III, *Nde* I and *Xho* I (Thermo Scientific)
- T4 DNA ligase (Thermo Scientific)
- Distilled water (dH2O)
- *E. coli* BL21(DE3) chemically competent cell (TransGen Biotech)

- Cycle-pure kit (Omega Bio-Tek)
- Plasmid mini kit (Omega Bio-Tek)
- Polymerase chain reaction (PCR) instrument (Eppendorf)
- Lysogenic broth (LB) agar plate with 50 μg.mL^{-1} kanamycin
- 30% w/v glycerol in water (pressure steam-sterilised)
- Test tube (12 × 100 mm) with rubber stopper
- Constant-temperature incubator shaker

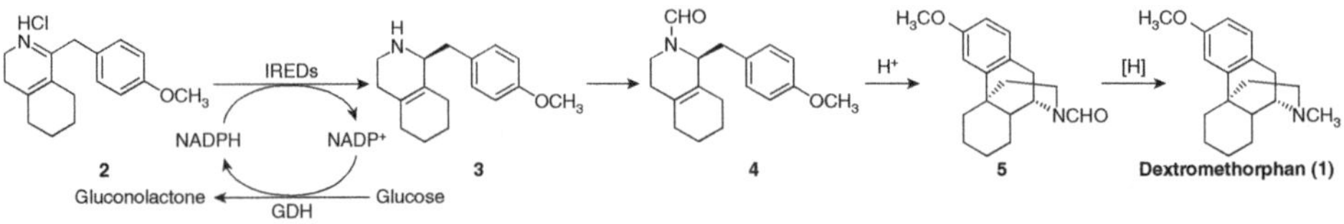

Scheme 3.3 *Synthesis of dextromethorphan **1** from 1-(4-methoxybenzyl)-3,4,5,6,7,8-hexahydroisoquinoline **2**.*

3.3.1.2 Procedure

3.3.1.2.1 Construction of Plasmid pRSFDuet-GDH. The plasmid pRSFDuet-GDH was constructed by inserting the GDH gene between the *Nco* I and *Hind* III sites of pRSFDuet-1.

1. The GDH gene was amplified by PCR from pET15b-GDH using primers GDH-F/GDH-R (Table 3.1). 50 μL reaction mixtures typically contained 30 μL dH_2O, 10 μL FastPfu DNA polymerase buffer (5×), 5 μL 2 mM dNTPs, 1 μL (50 ng) template DNA, 0.2 μM primers (1 μL of each) and 1 μL FastPfu DNA polymerase. The following amplification profile was followed: 95 °C for 2 min, followed by 30 cycles at 95 °C for 10 sec, 60 °C for 20 sec and 72 °C for 25 sec, and finally 72 °C for 5 min. The PCR products were analysed on agarose gel by electrophoresis and purified using a cycle-pure kit.
2. The PCR-amplified product and plasmid pRSFDuet-1 were digested with *Nco* I and *Hind* III at 37 °C for 1 hr and purified using the cycle-pure kit. These two fragments were linked by T4 DNA ligase at 22 °C for 1 hr. 10 μL of ligation mixture typically contained 6.9 μL dH_2O, 1 μL T4 DNA ligase buffer (10×), 1 μL pRSFDuet fragment (4.6 kb, 50 ng), 1 μL GDH gene fragment (0.8 kb, 10–50 ng) and 0.1 μL T4 DNA ligase.
3. The ligation mixture was transformed into *E. coli* BL21 (DE3). 10 μL mixture was added into 100 μL BL21(DE3) chemically competent cell and ice-bathed for 30 min, then heat-shocked at 42 °C for 45 sec and ice-bathed for 3 min. 900 μL sterile LB medium (no antibiotics) was added and shaken for 1 hr at 37 °C, before being coated on an LB agar plate containing 50 μg.mL^{-1} kanamycin. After overnight incubation at 37 °C, two or three single colonies were selected from the LB plate and placed in the LB medium (4 mL) for further culture. After confirmation of the sequence by DNA sequencing, the plasmid pRSFDuet-GDH was extracted from culture containing positive clone using the plasmid mini kit and handled further as described next.

3.3.1.2.2 Construction of Plasmid pRSFDuet-GDH-IR30. The plasmid pRSFDuet-GDH-30 was constructed by inserting IR30 gene between the *Nde* I and *Xho* I sites of pRSFDuet-GDH.

1. The IR30 gene was amplified by PCR from template DNA pET28a-IR30 using primer IR30-F/IR30-R (Table 3.1). The operation was as described for the construction of plasmid pRSFDuet-GDH.
2. The PCR-amplified product and plasmid pRSFDuet-GDH were digested with *Nde* I and *Xho* I at 37 °C for 1 hr and purified using the cycle-pure kit. These two fragments were linked by T4 DNA ligase at 22 °C for 1 hr. 10 μL of ligation mixture typically contained 6.9 μL dH_2O, 1 μL T4 DNA ligase buffer (10×), 1 μL pRSFDuet-GDH fragment (5.5 kb, 50 ng), 1 μL IR30 gene fragment (0.9 kb, 10–50 ng) and 0.1 μL T4 DNA ligase.
3. The ligation mixture was transformed into *E. coli* BL21 (DE3) and coated on an LB agar plate containing 50 μg.mL^{-1} kanamycin. After overnight incubation at 37 °C, two or three single colonies were selected from the LB plate and placed in the LB medium for further culture. After confirmation of the sequence by DNA sequencing, the *E. coli* BL21(DE3) strain harbouring the co-expression plasmid pRSFDuet-GDH-IR30 was mixed with the same amount of 30% w/v glycerol in water (sterilised) and stored at −80 °C.

3.3.2 Procedure 2: Co-expression of GDH and IR30 in *E. coli* BL21(DE3)

3.3.2.1 Materials and Equipment

- Tryptone (20 g)
- Yeast extract (10 g)
- NaCl (20 g)
- Distilled water (dH_2O)
- Kanamycin (50 mg.mL^{-1} in dH_2O, filter-sterilised)
- Isopropyl-β-D-1-thiogalactopyranoside (IPTG; 1 M in dH_2O, filter-sterilised)
- Glycerol stock of *E. coli* BL21(DE3) harbouring plasmid pRSFDuet-GDH-IR30 from Procedure 1
- Tris HCl buffer (50 mM, pH 9.0)
- 0.1 L Erlenmeyer flask with seal film
- 2 L Erlenmeyer flask with seal film
- Vertical autoclave
- Constant-temperature incubator shaker
- High-speed freezing centrifuge (min. 4500× g)

Table 3.1 *Primers used for amplification.*

Primer	Sequence (5′~3′)
GDH-F	CCCATGG (*Nco* I) GTATGGGTTATCCGGATTTAAAAGGAAAAGTCG
GDH-R	CAAGCTT (*Hind* III) TTAACCGCGGCCTGCCTGG
IR30-F	GGAATTCCATATG (*Nde* I) ACCGATAAACCGCCGGTGAC
IR30-R	GACGGAGCTCCTCGAG (*Xho* I) TTATGCATTGGCACC

3.3.2.2 Procedure

1. Tryptone (20 g), yeast extract (10 g) and NaCl (20 g) were dissolved in dH2O (2L) to give LB medium. The solution was distributed to some 0.1 L Erlenmeyer flasks (20 mL medium per flask) and two 2 L Erlenmeyer flasks (800 mL medium per flask) with seal film, and then autoclaved (20 min, 121 °C) to give sterile LB medium.
2. To prepare the pre-culture, kanamycin (20 μL) stock solutions were added to a sterile 0.1 L Erlenmeyer flask with 20 mL LB medium to reach final concentrations of 50 μg.mL^{-1}. The solution was inoculated with a little of the glycerol stock of *E. coli* BL21 (DE3) harbouring co-expression plasmid pRSFDuet-GDH-IR30 and shaken overnight at 37 °C, 200 rpm.
3. The next day, kanamycin (800 μL) stock solution was added to each 2 L Erlenmeyer flask with LB medium (800 mL) to reach final concentrations of 50 μg.mL^{-1}. The main culture was inoculated with the pre-culture to a final concentration of 1%.
4. The cells were grown at 37 °C, 200 rpm until an OD_{600} of 0.6–0.8 was reached. Then, the cultivation was cooled to 25 °C and expression of the recombinant enzymes was induced by the addition of 80 μL IPTG stock solution to a 0.1 mM final concentration. The expression was performed for 8 hr at 25 °C. The final OD_{600} was approximately 4.0.
5. The cells were harvested by centrifugation for 10 min at 4500× g and 4 °C. The supernatant was discarded. Cell pellets were washed with 50 mL Tris HCl buffer (50 mM, pH 9.0) and stored at −20 °C for further use. Total weight of cell paste was approximately 7.0 g.

3.3.3 Procedure 3: Biocatalytic Conversion of 1-(4-Methoxybenzyl)-3,4,5,6,7,8-Hexahydroisoquinoline (2)

3.3.3.1 Materials and Equipment

- 1-(4-Methoxybenzyl)-3,4,5,6,7,8-hexahydroisoquinoline **2** stock solution (0.5 M in DMSO)
- Glucose (0.4g)
- $NADP^+$ (6 mg)
- Tris HCl buffer (50 mM, pH 9.0)
- Cell pellets from Procedure 2
- 0.25 L round-bottom flask
- Flask stopper with nitrogen balloon
- Orbital shaker
- Aqueous NaOH (5 M)
- pH paper
- Ethyl acetate
- 50 mL Eppendorf tube
- Na_2SO_4, dry
- Rotary evaporator
- High-performance liquid chromatography (HPLC) system with a DAD detector (Agilent 1200 series)
- Chiral HPLC column (Daicel CHIRALPAK OJ-H: 250 × 4.6 mm, 5 μm)

- High-speed freezing centrifuge (min. 6000× g)
- TLC silica gel plates
- UV lamp (254 nm)

3.3.3.2 Procedure

1. 1-(4-Methoxybenzyl)-3,4,5,6,7,8-hexahydroisoquinoline **2** was chemically synthesised, as recently reported elsewhere [4].
2. Cells obtained in Procedure 2 (5.0 g) were resuspended in 50 mM Tris HCl buffer, pH 9.0 (100 mL) with glucose (0.4 g, 20 mM final conc.) and NADP (6 mg) in a 0.25 L round-bottom flask.
3. 1-(4-Methoxybenzyl)-3,4,5,6,7,8-hexahydroisoquinoline **2** stock solution (2 mL) was added to the round-bottom flask (10 mM final concentration). The mixture was shaken at 200 rpm, 25 °C under an atmosphere of nitrogen gas on an orbital shaker for 18 hr.
4. The reaction mixture was carefully adjusted to pH 10–11 with 5N NaOH (aq). The aqueous layer was extracted three times with ethyl acetate (3 × 100 mL) and the phase separation was facilitated by centrifugation (6000× g, 15 min).
5. The combined organic layers were dried over Na_2SO_4 and filtered, and the solvent was removed under reduced pressure.
6. The crude product was purified by chromatography over silica gel using dichloromethane/methanol (98 : 2) as eluent. Fractions were collected and checked for the desired substances via thin-layer chromatography (TLC) using dichloromethane/methanol (95 : 5) as eluent and via visualisation using a UV lamp. Fractions containing the desired product were combined, and the solvent was removed by distillation under reduced pressure.
7. Samples were analysed by HPLC with a CHIRALPAK OJ-H column (see Table 3.2 for the HPLC method and Table 3.3 for retention times). The desired (*S*)-**3** was obtained in 190 mg as a viscous oil (74% yield, 98% ee).

3.3.4 Analytical Method

$[\alpha]^{25}{}_D = -117.8°$ (c = 1.0, CH_3OH, l = 50 mm); literature values for (*S*)-**3**: $[\alpha]^{24}{}_D = -175.8$ (c = 1.7, ether) [2b]. ^1H NMR (600 MHz, methanol-d_4) $\delta = 7.15$ (d, $J = 8.4$ Hz, 2 H), 6.87 (d, $J = 8.4$ Hz, 2 H), 3.76 (s, 3 H), 3.35 (d, $J = 9.5$ Hz, 1 H), 3.05–3.00 (m, 2 H), 2.75–2.69 (m, 1 H), 2.53 (dd, $J = 10.3$, 13.9 Hz, 1 H), 2.19–2.09 (m, 1 H), 2.05–1.85 (m, 5 H), 1.79–1.67 (m, 2 H), 1.60–1.51 (m, 2 H). ^{13}C NMR (150MHz, methanol-d_4) $\delta = 158.59$, 130.40, 129.87, 128.73, 128.28, 113.78, 58.17, 54.29, 39.56, 36.98, 29.90, 29.32, 26.68, 22.83, 22.49 [4].

Table 3.2 *HPLC method.*

Mobile phase	Flow rate	Absorbance	Column temperature	Duration
n hexane/2 propanol/ethanolamine (90 : 10 : 0.05 v/v/v)	0.5 mL.min^{-1}	230 nm	30 °C	15 min

Table 3.3 Retention times for HPLC analysis.

Product	Retention (min)
(*S*)-**3**	9.4
(*R*)-**3**	10.7

3.3.5 Conclusion

The described procedure enabled the chemo- and stereoselective production of the key intermediate for the synthesis of dextromethorphan **1** in high yield and high enantioselectivity using standard chemical and microbiological laboratory equipment and methods.

References

1. (a) Taylor, C.P., Traynelis, S.F., Siffert, J. *et al.* (2016) *Pharmacology & Therapeutics*, **164**, 170–182; (b) Nguyen, L., Thomas, K.L., Lucke-Wold, B.P. *et al.* (2016) *Pharmacology & Therapeutics*, **159**, 1–22.
2. (a) Kumaraguru, T. and Fadnavis, N.W. (2014) *Organic Process Research & Development*, **18** (1), 174–178; (b) Meyers, A.I. and Bailey, T.R. (1986) *Journal of Organic Chemistry*, **51** (6), 872–875.
3. (a) Bornscheuer, U.T., Huisman, G.W., Kazlauskas, R.J. *et al.* (2012) *Nature*, **485** (7397), 185–194; (b) Reetz, M.T. (2013) *Journal of the American Chemical Society*, **135** (34), 12 480–12 496; (c) Hönig, M., Sondermann, P., Turner, N.J. and Carreira, E.M. (2017) *Angewandte Chemie International Edition*, **56** (31), 8942–8973.
4. Yao, P., Xu, Z., Yu, S. *et al.* (2019) *Advanced Synthesis & Catalysis*, **361** (3), 556–561.

3.4 Identification of Imine Reductases for Asymmetric Synthesis of 1-Aryl-tetrahydroisoquinolines

Jinmei Zhu,[1] Zixin Deng[1] and Xudong Qu*[1,2]

[1]*Key Laboratory of Combinatorial Biosynthesis and Drug Discovery, Ministry of Education, School of Pharmaceutical Sciences, Wuhan University, Wuhan, China*

[2]*State Key Laboratory of Microbial Metabolism and School of Life Sciences and Biotechnology, Shanghai Jiao Tong University, Shanghai, China*

Tetrahydroisoquinolines (THIQs) with a C1-aryl-substituted group are common in many natural and synthetic compounds of biological importance [1, 2]. Enzymatic synthesis of these molecules using imine reductases is attractive because of their cost effectiveness, high catalytic efficiency and enantioselectivity [3]. However, the C1-aryl substituents make the imine bond very sterically hindered, which poses a tremendous challenge for enzyme reduction. To address this bottleneck, our group recently identified a few steric-hindrance-tolerant imine reductases and successfully applied them to asymmetric synthesis of 1-aryl THIQs [4].

In this method, we will describe in detail how to identify these unusual imine reductases through database mining and apply them for synthesis of C1-aryl THIQs (Scheme 3.4). These enzymes can accept highly bulky C1-aryl dihydroisoquinolines (DHIQs) and convert

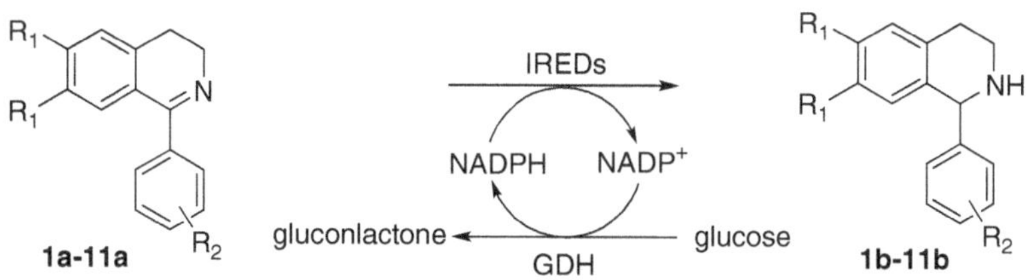

R_1: H or OMe; R_2: *para-*, *meta-*, *ortho-* H, CH_3, Cl, OCH_3

Scheme 3.4 *Asymmetric reduction of DHIQs by imine reductases.*

them into the corresponding (*R*)- or (*S*)-THIQs with very high enantioselectivity and conversion. The availability of these enzymes provides a solid basis for unravelling the unusual substrate specificity of imine reductases in order to further generate more efficient enzymes via protein engineering.

3.4.1 Procedure 1: Identification of Candidate Imine Reductases from a Database

3.4.1.1 Materials and Equipment

- Internet-connected computer
- Imine reductase engineering database [5]
- Interactive structure-based sequences alignment program (STRAP) [6]

3.4.1.2 Procedure

1. Imine reductases (IR4, IR26, IR71, IR82, IR97) are known to be able to reduce 1-methyl DHIQ to yield 1-methyl THIQ. Amongst them, IR71 was confirmed to be able to convert 1-phenyl DHIQ into 1-phenyl THIQ with a modest conversion (7%) and enantioselectivity (*R*-selective, 30% ee). Details of their activities towards 1-aryl DHIQ can be found in our previous paper [4].
2. Homologous proteins of IR71 in the imine reductase engineering database (https://ired.biocatnet.de/) were searched for using the Blast module.
3. The top 150 homologous sequences were collected and aligned using STRAP (http://www.bioinformatics.org/strap/).
4. Proteins with high amino acid identity to one another (identity more than 97%) were excluded, leaving a total of 95 novel imine reductase sequences (identity 32–82% to IR71), including 68 *R*-type (IR1–IR3, IR5–IR25, IR27–IR70) and 27 *S*-type (IR72–IR81, IR83–IR96, IR98–IR100) imine reductases (classified based on the residue in position 187, according to reference [5]).
5. The imine reductase sequences were codon-optimised for *Escherichia coli* and synthesised by GENEWIZ (Genewiz Biotech Co. Ltd, China). Accession numbers of the synthetic imine reductases IR1–IR100 are MF540776–MF540874.
6. The synthetic genes were cloned into the *Nde*I and *Hin*dIII sites of pET28a by GENEWIZ (Genewiz Biotech Co. Ltd, China) for overexpression in *E. coli*.

3.4.2 Procedure 2: Expression of Imine Reductases and GDH in *E. coli* BL21(DE3)

3.4.2.1 Materials and Equipment

- Tryptone (5 g)
- Yeast extract (2.5 g)
- NaCl (5 g)
- Distilled water (dH_2O)
- Kanamycin (50 mg.mL^{-1} in dH_2O, filter-sterilised)
- Isopropyl-β-D-1-thiogalactopyranoside (IPTG; 1 M in dH_2O, filter-sterilised)
- LB agar plate with colonies of *E. coli* BL21(DE3) harbouring the expression vectors pWHUIR1–3, 5–25, 27–70, 72–81, 83–96 and 98–100 and pWHUGDH containing genes encoding imine reductases and glucose dehydrogenase (GDH) (see our previous paper [4]).
- 1 L Schott bottles
- 50 mL Erlenmeyer flask
- Orbital shaker (Zhicheng ZWY-2112D)
- Table autoclave (Hirayama HVA-85)
- Cooling centrifuge (Beckman Coulter Avanti-J-26 XP)

3.4.2.2 Procedure

1. Tryptone (5 g), yeast extract (2.5 g) and NaCl (5 g) were dissolved in dH_2O (500 mL) and autoclaved (20 min, 121 °C) in a 1 L Schott bottle to give sterile lysogenic broth (LB) medium.
2. To prepare the pre-culture, sterile LB medium (10 mL) was placed into a sterile 50 mL Erlenmeyer flask. Afterwards, kanamycin stock solutions (10 μL) were added to reach final concentrations of 50 μg.mL^{-1}. The solution was inoculated with single colony of each *E. coli* BL21(DE_3) separately harbouring each plasmid and shaken at 37 °C and 200 rpm for 12 hr with OD_{600} 1.8–2.0.
3. Sterile LB medium (500 mL) in 2 L Erlenmeyer flasks was supplemented with kanamycin (500 μL) stock solutions to a final concentration of 50 μg.mL^{-1} and then was inoculated with 5 mL LB pre-cultures.
4. The cells were grown at 37 °C, 200 rpm until an OD_{600} of 0.6–0.8 was reached.
5. The cultivation was cooled to 18 °C and IPTG stock solution (50 μL) was added to reach the final concentration of 0.1 mM. The expression was performed at 18 °C, 200 rpm for 16 hr.
6. The cells were harvested by centrifugation (10 min, 4500 rpm, 4 °C) and stored at −80 °C or handled further as described in Procedure 3.

3.4.3 Procedure 3: Purification of Imine Reductases

3.4.3.1 Materials and Equipment

- 4-(2-Hydroxyethyl)piperazine-1-ethanesulfonic acid (HEPES; 4.7662 g)
- NaCl (17.532 g)
- Glycerol (100 mL)
- Buffer A (25 mM HEPES, 300 mM NaCl, 10% glycerol, pH 7.5)

- Buffer B (25 mM HEPES, 50 mM NaCl, 10% glycerol, pH 7.5)
- Imidazole (10 g)
- Double-distilled water (ddH_2O)
- Kanamycin (50 mg.mL^{-1} in dH_2O, filter-sterilised)
- pH meter (Mettler Toledo Five Plus)
- Nano homogenize machine (ATS Engineering, AH100B)
- Cooling centrifuge (BECKMAN Optima L)
- HisPur Ni-IDA Resin (Smart-Lifesciences)
- Ultrafiltration tube (Amicon Ultra-4, 10 kDa)
- Desalting column (GE Disposable PD-10)
- NanoDrop 2000 spectrophotometer (Thermo Scientific)

3.4.3.2 Procedure

1. The cells harvested in Procedure 2 were resuspended in lysis buffer (30 mL) and lysed by nano homogenize machine (500 mpa, 2 min, 4 °C). The cell lysates were centrifuged (12 000 rpm, 60 min, 4 °C) to separate the supernatants. Ni-NTA agarose resin (1 mL) was added to the supernatant (2 mL.L^{-1}) and shaken at 4 °C for 30 min.
2. The protein resin mixtures were loaded on to a gravity-flow column, and proteins were eluted with increasing concentrations of imidazole (25, 50, 100, 300 mM) in buffer A. The elution volume of each imidazole buffer was 20 mL. The eluents were collected and analysed by 12% acrylamide SDS-PAGE.
3. The eluents containing purified imine reductases were concentrated by centrifugation using an Amicon Ultra-4 (10 kDa, GE Healthcare) to 2.5 mL, loaded into PD-10 desalting columns and desalted using 3.5 mL buffer B, and then concentrated by centrifugation using an Amicon Ultra-4 (10 kDa, GE Healthcare). The proteins' purity was evaluated again by 12% acrylamide SDS-PAGE and concentrations were determined by NanoDrop 2000 Spectrophotometer (Thermo Scientific) based on the absorbance of 280 nm. Using these processes, a total of 88 novel imine reductases were successfully expressed and purified (IR1–IR3, IR5–IR9, IR11–IR22, IR24–IR25, IR27–IR66, IR68–IR70, IR72, IR74–IR81, IR83, IR86–IR96, IR98–IR100), and were stored at −80 °C for further use (see our previous paper [4]).

3.4.4 Procedure 4: Biocatalytic Reduction of 1-Aryl DHIQs

3.4.4.1 Materials and Equipment

- 1-Aryl DHIQs (**1a–11a**, Scheme 3.5) [4] stock solution (50 mM in dimethyl sulfoxide, DMSO)
- Glucose (1 M in ddH_2O)
- $NADP^+$ (50 mM in ddH_2O)
- GDH (synthesised and cloned into the *Nde*I and *Hin*dIII sites of pET28a by GENEWIZ (Genewiz Biotech Co. Ltd, China) and expressed in *E. coli* BL21(DE3) through the same method as the imine reductases described in Procedure 2; for details, see our previous paper [7])
- NaOH (10 M in ddH_2O)
- Ethyl acetate

- High-performance liquid chromatography (HPLC)-grade MeCN
- Dichloromethane
- KPB buffer (100 mM)
- Double-distilled water (ddH_2O)
- $MgSO_4$, dry
- NaCl (saturated solution in ddH_2O)
- Purified imine reductaes and GDH from Procedure 2
- 2 mL Eppendorf tube
- Orbital shaker (InforsHT Multitron 2 Standard)
- pH meter (Mettler Toledo FiveEasy Plus)
- Centrifuge (Thermo Scientific Pico17)
- HPLC (Shimadzu, LC-20A)
- HPLC (Agilent 1260 Infinity Series)
- Freeze dryer (Boyikang Beijing)
- Rotary evaporator (Buchi R-210/R-220SE 220SE)

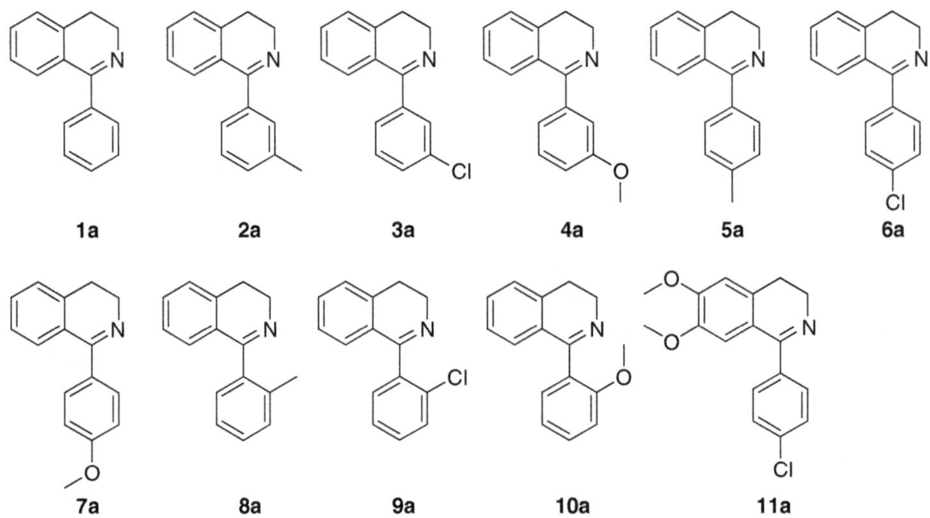

Scheme 3.5 *1-Aryl DHIQ substrates.*

3.4.4.2 Procedure

1. 11 representative DHIQ substrates (**1a–11a**) were synthesised, as previously reported [4].
2. A typical 500 μL reaction mixture contained 20 mM D-glucose, 0.2 mg.mL^{-1} GDH, 5 mM NADP+, 0.2 mg.mL^{-1} imine reductase, 2 mM DHIQ and potassium phosphate buffer (100 mM, pH 7.0).

3. Reactions were incubated at 30 °C with shaking at 200 rpm for 24 hr, and then quenched by the addition of 10 M NaOH (30 μL) and extracted with ethyl acetate (3 × 300 μL). The organic phase was combined and dried over $MgSO_4$ for 2 hr before filtration. Ethyl acetate was removed by vacuum and residual DMSO was removed by lyophilisation.
4. For analysis of the reaction, samples from Step 3 were redissolved in MeCN (500 μL) and analysed by reverse-phase HPLC (Shimadzu, LC-20A) with an Agela Promosil-C18 column (5 μm, 4.6 × 250 mm, Agela Technologies) at 254 nm. HPLC analysis was performed at a flow rate of 1 mL.min^{-1} over a 28 min gradient programme A (for **1a–10a**): T = 0 min, 70% B; T = 17 min, 100% B; T = 18 min, 100% B; T = 19 min, 70% B; T = 28 min, 70% B (A = H_2O with 0.01% TEA, B = MeCN with 0.01% TEA); gradient program B (for **11a**): T = 0 min, 20% B; T = 5 min, 100% B; T = 15 min, 100% B; T = 18 min, 20% B; T = 28 min, 20% B (A = H_2O with 0.01% TEA, B = MeCN with 0.01% TEA) (see Table 3.4 for retention times).

Table 3.4 *Retention times of DHIQs and THIQs in HPLC analysis.*

Compound	Retention time (min)	Compound	Retention time (min)	Compound	Retention time (min)
1a	8.53	**5a**	7.58	**9a**	9.24
1b	7.79	**5b**	7.10	**9b**	8.73
2a	7.78	**6a**	8.71	**10a**	6.78
2b	7.16	**6b**	7.80	**10b**	5.57
3a	8.87	**7a**	8.19	**11a**	16.85
3b	7.70	**7b**	7.49	**11b**	17.28
4a	8.54	**8a**	9.82		
4b	7.59	**8b**	8.58		

Table 3.5 *HPLC conditions used for the chiral analysis of THIQ products.*

Substance	Column	Wavelength (nm)	Flow (mL.min^{-1})	Solvent A:B	THIQs retention time (min) T^1	T^2
1a	Chiralpak® OJ-H	220	1.0	85:15	7.54 (*R*)	8.48 (*S*)
2a	Chiralpak® OJ-H	220	1.0	85:15	6.53 (*R*)	7.44 (*S*)
3a	Chiralpak® OJ-H	220	1.0	90:10	8.05 (*R*)	9.64 (*S*)
4a	Chiralpak® OJ-H	220	1.0	90:10	14.12 (*R*)	16.28 (*S*)
5a	Chiralpak® OJ-H	220	1.0	85:15	6.52 (*R*)	7.54 (*S*)
6a	Chiralpak® OJ-H	220	1.0	90:10	8.54 (*R*)	9.99 (*S*)
7a	Chiralpak® OJ-H	220	1.0	90:10	14.43 (*R*)	18.74 (*S*)
8a	Chiralpak® OJ-H	220	1.0	90:10	8.40 (*R*)	9.85 (*S*)
9a	Chiralpak® OJ-H	220	1.0	90:10	9.80 (*R*)	13.34 (*S*)
10a	Chiralpak® OJ-H	220	1.0	90:10	10.05 (*R*)	10.81 (*S*)
11a	Chiralpak®AS-H	254	0.8	70:30	14.24 (*R*)	18.61 (*S*)

Table 3.6 *Conversion and enantioselectivity of selected imine reductases towards* ***1a–11a***.

Substance	Conversion (%), enantiomeric excess									
	IR2	IR8	IR17	IR19	IR20	IR32	IR45	IR71	IR96	IR99
1a	100; $>96^R$	99; $>99^R$	100; $>99^R$	99; $>99^R$	99; $>99^R$	99; $>99^R$	99; $>99^S$	7; 30^R	78; 90^S	97, 97^S
2a	99; $>99^R$	99; $>99^R$	99; $>99^R$	100; $>99^R$	100; $>99^R$	100; $>99^R$	99; $>99^S$	4; 84^R	29; 47^R	24; 16^R
3a	100; $>99^R$	100; 97^R	100; $>99^R$	99; $>99^R$	100; $>99^R$	99; 98^R	99; 91^S	47; 58^R	74; 81^R	50; 21^S
4a	100; $>99^R$	100; $>99^R$	100; $>99^R$	99; $>99^R$	100; $>99^R$	100; $>99^R$	99; $>99^S$	26; 19^R	26; 8^S	16; 2^R
5a	100; $>99^R$	99; $>99^R$	99; $>99^R$	99; $>99^R$	99; $>99^R$	99; $>99^R$	100; 35^S	94; 72^R	46; 97^R	24; 91^R
6a	99; $>99^R$	99; $>99^R$	99; $>99^R$	100; $>99^R$	100; $>99^R$	99; $>99^R$	79; 36^R	82; $\mathbf{3^S}$	52; 95^R	27; 34^R
7a	100; $>99^R$	99; $>99^R$	100; $>99^R$	100; $>99^R$	99; $>99^R$	99; $>99^R$	5; 21^S	28; 93^R	4; 97^R	5; 93^R
8a	36; $>99^R$	0	0	0	0	0	7; 26^S	0	0	0
9a	23; 5^S	0	0	0	0	0	9; $>99^S$	0	0	0
10a	5; $>99^R$	0	0	0	0	0	8; 98^R	0	0	0
11a	62; $>99^S$	6; -	9; -	5; -	18; -	0	62; $>92^S$	14; -	1; -	4; -

-, not detected.

5. For determination of the enantioselectivity of imine reductases, the corresponding THIQ products need to be derivatised with acetic anhydride. Products from Step 3 were dissolved in DCM (100 μL) and mixed with acetic anhydride (10 μL) and TEA (2 μL). The reaction was shaken at 30 °C, 200 rpm for 2 hr. Solvents were further removed by evaporation under reduced pressure at room temperature, then isopropanol (200 μL) was added to dissolve the derivatised products. The derivatised samples were analysed on a chiral column by an Agilent 1260 Infinity Series HPLC System equipped with a DAD detector (for details, see Table 3.5). *R*- and *S*- annotation was according to the literature [8, 9]. For details, see [4].
6. The conversion and enantioselectivity of imine reductases towards **1a–11a** are revealed in Table 3.6 Based on the preceding procedures, we successfully identified 10 imine reductases capable of reducing 1-aryl DHIQs to THIQs. Amongst them, IR2, IR8, IR17, IR19, IR20 and IR32 can convert the *meta*- and *para*-substituted substrates (**2a–7a**) to 1-aryl THIQs in very high conversion (>99%) and *R*-specificity (>99% ee); IR45 is able to convert substrates **2a–7a** to corresponding THIQs in moderate-to-perfect conversion and *S*-specificity. Moreover, IR2 and IR45 are also able to convert the very steric-hindered *ortho*-substituted aryl DHIQs (**8a–10a**) and 6,7-dimethoxy chlorophenyl DHIQ (**11a**) into corresponding THIQ products.

3.4.5 Conclusion

The described procedure enabled us to identify a group of novel steric hindrance tolerant imine reductases. These enzymes can convert *meta*- and *para* substituted chloro-, methyl- and methoxyphenyl dihydroisoquinolines (DHIQs) into the corresponding (*R*)- or (*S*)-THIQs with very high enantioselectivity and conversion. In particular, the two most hindrance-tolerated enzymes, IR2 and IR45 (with different stereospecificities), are also able to convert *ortho*substituted chloro-, methyl- and methoxyphenyl DHIQs and 6,7-dimethoxy-DHIQ into THIQ products with good enantiomeric excess. This protocol not only provides an enzymatic solution to synthesise these important THIQs but also sets a solid basis for further of identification of more efficient and hindrance-tolerated imine reductases via database mining and protein engineering.

Acknowledgements

This work was financially supported by the NSFC (Nos 31570057). We thank Prof. Wenbo Liu for providing the chiral analysis.

References

1. Liu, W., Liu, S., Jin, R. *et al.* (2015) *Organic Chemistry Frontiers*, **2**, 288–299.
2. Chrzanowska, M., Grajewska, A. & Rozwadowska, M.D. (2016) *Chemical Reviews*, **116**, 12 369–12 465.
3. Grogan, G. & Turner, N.J. (2016) *Chemistry: A European Journal*, **22**, 1900–1907.
4. Zhu, J., Tan, H., Yang, L. *et al.* (2017) *ACS Catalysis*, **7**, 7003–7007.
5. Scheller, P.N., Fademrecht, S., Hofelzer, S. *et al.* (2014) *ChemBioChem*, **15**, 2201–2204.

6. Gille, C., Fähling, M., Weyand, B. *et al.* (2014) *Nucleic Acids Research*, **42**, W3–W6.
7. Peng, H., Wei, E., Wang, J. *et al.* (2016) *ACS Chemical Biology*, **11**, 3278–3283.
8. Chang, M., Li, W. and Zhang, X. (2011) *Angewandte Chemie International Edition*, **50**, 10 679–10 681.
9. Zhou, H., Liu, Y., Yang, S. *et al.* (2017) *Angewandte Chemie International Edition*, **56**, 2725–2729.

3.5 Preparation of Imine Reductases at 15 L Scale and Their Application in Asymmetric Piperazine Synthesis

Niels Borlinghaus,[1] Sebastian Gergel,[1] Bettina M. Nestl,*[1] Franziska Thol[2] and Shanshan Yu*[2]
[1]*Institute of Biochemistry and Technical Biochemistry, Department of Technical Biochemistry, University of Stuttgart, Stuttgart, Germany*
[2]*Sanofi-Aventis Deutschland GmbH, BioDev Microbial Development, Frankfurt am Main, Germany*

Nitrogen heterocycles are important structural elements present in natural products, biologically active synthetic compounds, agrochemicals and pharmaceuticals [1, 2]. Recent years have seen increasingly rapid advances in the field of identifying and developing imine reductases for the preparation of nitrogen heterocyclic products. In earlier works, we investigated the cultivation and purification of enantiocomplementary imine reductases at the 4 L cultivation scale [3]. Piperazines, thus, have attracted great attention in the light of recent studies that reported direct, selective synthesis of *C*- and *N*-substituted piperazines, homopiperazines and derivatives thereof from dicarbonyls and diamines [4]. In the respective reactions, isolated imine reductases are applied in a double-reductive amination process for the highly selective synthesis of (*R*)-*N*-methyl-3-phenylpiperazine from phenylglyoxal and *N*-methylethylenediamine (Scheme 3.6).

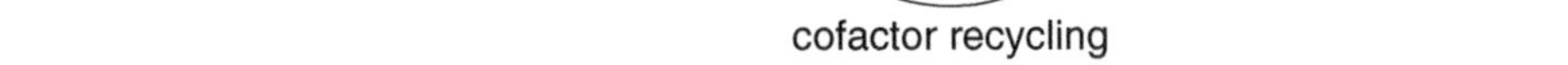

Scheme 3.6 *Imine reductase-catalysed preparation of enantiopure (R)-N-methyl-3-phenylpiperazine from phenylglyoxal and N-methylethylendiamine via double-reductive amination.*

3.5.1 Procedure 1: Recombinant Expression of Imine Reductases at 15 L Scale

3.5.1.1 Materials and Equipment

- Luri Bertani (LB) liquid medium (tryptone 10 g.L^{-1}, yeast extract 5 g.L^{-1}, NaCl 5 g.L^{-1}) and LB agar plates containing agar (15 g.L^{-1}) and kanamycin (50 μg.mL^{-1})

- 1000× stock solution of kanamycin (50 mg.mL^{-1}): kanamycin dissolved in sterilised bidest water and sterilised by filtering through a 0.20 μm filter
- Plasmid pET28a(+)-(*R*)-IRED containing *Streptosporangium roseum* DSM43021 imine reductase gene (NCBI accession number YP_003336672.1) inserted into the standard cloning vector pET28a(+) using NdeI and XhoI restriction sites, which carries an *N*-terminal His·Tag®/thrombin/T7·Tag
- Plasmid pET28a(+)-(*S*)-IRED containing *Paenibacillus elgii* imine reductase gene (NCBI accession number WP_010497949.1) inserted into the standard cloning vector pET28a(+) using NdeI and XhoI restriction sites, which carries an *N*-terminal His·Tag®/thrombin/T7·Tag [5]
- High cell-density medium modified from Korz et al. [6] (D-(+)-glucose monohydrate 11.0 g.L^{-1}, potassium phosphate monobasic 13.3 g.L^{-1}, ammonium sulfate 6.0 g.L^{-1}, citric acid monohydrate 6 g.L^{-1}, magnesium sulfate heptahydrate 2.0 g.L^{-1}, thiamine hydrochloride 0.0045 g.L^{-1}, manganese(II)-sulfate monohydrate 0.013 g.L^{-1}, copper(II)-sulfate pentahydrate 0.0022 g.L^{-1}, ammonium heptamolybdate tetrahydrate 0.0018 g.L^{-1}, zinc sulfate heptahydrate 0.017 g.L^{-1}, potassium iodide 0.004 g.L^{-1}, iron(III)-sulfate hydrate 0.5 g.L^{-1})
- *ortho*-Phosphoric acid (85%, as required)
- Antifoam, e.g. alkoxylated fatty acid esters (Schill + Seilacher 'J673')
- Feed solution consisting of D-(+)-glucose monohydrate 600 g.L^{-1}, citric acid monohydrate 2 g.L^{-1}, dissolved in distilled water and autoclaved
- Isopropyl-β-D-thiogalactopyranoside (IPTG, 1.8 g) dissolved in distilled water (50 mL) and sterilised by filtering through a 0.20 μm filter
- Rubidium chloride-competent cells of *Escherichia coli* strain BL21(DE3)
- Sodium phosphate buffer (50 mM, pH 7.0)
- 2 L Erlenmeyer flasks with cotton caps
- Orbital shaker (2.5 cm amplitude)
- Cooling centrifuge (5200× g and 17 000× g)
- 15 L Fermenter system (B.Braun, Biostat ED, Melsungen)
- Exhaust gas measurement (ProLine Process MS, AMETEK)
- Online glucose analyser (BioPAT® Trace, Sartorius Stedim)
- Thermostat
- Photometer
- Filter (e.g. Pall SUPOR EAV Mini KLEENPAK, 2.0 μm)

3.5.1.2 Procedure

1. To transform the *E. coli* cells, an aliquot (50 μL) of competent cells was mixed with plasmid DNA (1 μL of pET28a(+)-(*R*)-IRED or pET28a(+)-(*S*)-IRED). The mixture was kept on ice for 30 min, heated at 42 °C for 45 sec and placed back on ice for 1 min. After adding LB medium (500 μL), the cell suspension was incubated at 37 °C for 1 hr, and then dilutions were plated on LB agar plates containing kanamycin (50 μg.mL^{-1}). Incubation was overnight at 37 °C, yielding colonies of transformed *E. coli* cells.
2. Pre-cultures were prepared by inoculating LB sterile medium (5 mL) containing kanamycin (50 μg.mL^{-1}) with a single colony of the transformed *E. coli* cells. The pre-culture was incubated overnight under shaking (180 rpm) at 37 °C. 1 mL of

pre-culture was then diluted in sterile LB medium (500 mL) containing kanamycin (50 µg.mL^{-1}) in a 2 L Erlenmeyer shake flask and incubated at 37 °C with rotary shaking at 220 rpm until an OD_{600} of approximately 1 was reached.

3. The modified high cell-density medium was prepared using the quantities already given, based on final concentrations for a target fermentation volume of 7.0 L. First, trace elements were dissolved in distilled water (0.5 L). The potassium phosphate monobasic plus ammonium sulfate and citric acid monohydrate was dissolved in distilled water (6.0 L) and the trace-element solution and antifoam were added. Phosphoric acid was used to adjust the pH to 2.5. The complete solution was placed in a 15 L fermenter and sterilised (121 °C for 20 min). Glucose monohydrate, magnesium sulfate heptahydrate and citric acid monohydrate were dissolved in distilled water (0.5 L), autoclaved and added to the fermenter after medium sterilisation. Thiamine hydrochloride was dissolved in the kanamycin stock solution (7.0 mL; final concentration 50 mg.L^{-1}), filter-sterilised and also added to the sterilised medium. The fermenter contents were incubated at 30 °C and 1 bar overpressure. Dissolved oxygen was regulated to maintain 25% by first varying the stirrer speed from 500 to 1500 rpm and then varying the airflow from 7.5 to 20 standard litres per minute (sLpm).
4. The fermentation broth was inoculated with pre-culture (500 mL) and run in batch mode until glucose was consumed to 1 g.L^{-1} level. The online measured glucose level triggered the start of the feeding phase, which maintained a constant level of 1 g.L^{-1} glucose until the end of fermentation. An oxygen uptake rate (OUR) of 300 mmol.L^{-1}.h^{-1} initiated the cooling of the fermentation broth to 25 °C, preventing the formation of inclusion bodies upon induction. After reaching 25 °C, the IPTG solution was added to the culture for induction of imine reductase expression. After induction, the culture was run for a further 5 hr.
5. Cells were harvested by centrifugation for 20 min at 5200× g and 4 °C. The medium was removed and the cell pellet was resuspended in phosphate buffer solution (50 mM, pH 7.0, 1.5 mL.g^{-1}).

3.5.2 Procedure 2: Purification of Imine Reductase

3.5.2.1 Materials and Equipment

- Buffer A (affinity chromatography): sodium phosphate buffer (50 mM, pH 7.0), 300 mM KCl, 5% *v/v* glycerol and 1 mM mercaptoethanol
- Buffer B (affinity chromatography): sodium phosphate buffer (50 mM, pH 7.0), 300 mM KCl, 5% *v/v* glycerol, 1 mM mercaptoethanol and 500 mM imidazole
- Buffer A (anion exchange chromatography): Bis-Tris buffer (20 mM, pH 7.0) and 5% *v/v* glycerol (1.5 mS.cm^{-1})
- Buffer B (anion exchange chromatography): Bis-Tris buffer (20 mM, pH 7.0), 5% *v/v* glycerol and 1 M KCl (82 mS.cm^{-1})
- Sodium phosphate buffer (50 mM, pH 7.0)
- Äkta purifier, GE Healthcare
- Affinity chromatography: column material, Chelat Sepharose (GE Healthcare) loaded with Zn^{2+}/column dimension, height 11.6 cm and internal diameter 10 cm (ca. 910 mL)

- Anion exchange chromatography: column material, Source 30Q (GE Healthcare)/ column dimension, height 43 cm and internal diameter 5 cm (840 mL)
- Tangential flow filtration membrane cassette holder and filter: 30 kD Sartocon Hydrosart, low protein binding
- Photometer
- pH meter
- Conductivity meter
- Equipment and material for SDS-polyacrylamide gel electrophoresis
- High-pressure homogeniser (APV-1000)
- High-speed centrifuge with rotor and tubes

3.5.2.2 Procedure

1. A cell extract was prepared by treatment with a high-pressure homogenisation (3 cycles, about 1000 bar and intensive cooling) of the cell suspension, after which debris was removed by centrifugation (40 min, 17 000× g and 4°C) and filter sterilisation.
2. The supernatant was collected and applied to an affinity chromatography column, pre-equilibrated with buffer. The column was washed with buffer A affinity chromatography (5 column volumes (CV) and a flow rate of 79 cm.hr^{-1}). The purification of polyhistidine-tagged imine reductase was performed by linear-gradient 0–100% elution buffer B affinity chromatography over 10 CV at 280/340 nm wavelength. Fractions were collected by an automatic fraction collector with a fraction size of 0.5–1 L. Pooling of fraction was done based on SDS-PAGE analysis.
3. Fractions that contained desired proteins were concentrated and the buffer was exchanged by tangential flow filtration with Bis-Tris buffer (20 mM, pH 7.0, buffer A anion exchange chromatography). Desalted proteins were loaded on to the Source 30Q column using a flow rate of 150 cm.hr^{-1}. Elution was performed with a linear-gradient 0–100% buffer B anion exchange chromatography over 10 CV. Pooling of fractions was done based on SDS-PAGE analysis.
4. A final buffer exchange was carried out with the pooled fraction by tangential flow filtration with sodium phosphate buffer (50 mM, pH 7.0).
5. SDS-PAGE analysis with Coomassie Blue staining showed that the purity of the imine reductase was more than 95%.

3.5.2.2.1 General NADPH-Depletion Assay for the Determination of Enzyme Reduction Activity.

1. Purified imine reductase solution (0.1 mg.mL^{-1} (*R*)-selective and 0.5 mg.mL^{-1} (*S*)-selective imine reductase, respectively) in sodium phosphate buffer (50 mM, pH 7.0) was added to substrate solution (10 mM 2-methylpyrroline for (*R*)-selective and 10 mM 3,4-dihydroisoquinoline (DHIQ) for (*S*)-selective imine reductase, respectively) and 0.3 mM NADPH to a final volume of 1 mL.
2. The reaction was monitored at 30 °C by following the decrease of the signal of NADPH at 340 nm for the (*R*)-selective imine reductase and 370 nm for the (*S*)-selective imine reductase. (Due to background absorbance of 3,4-DHIQ at 340 nm, reduction activity

of (*S*)-selective imine reductases was determined at 370 nm.) A control experiment confirmed that no spontaneous reduction of substrate occurred.

3. Based on the following equation, specific activity ($U.mg^{-1} = \mu mol.min^{-1}.mg^{-1}$) was calculated from the quotient of volumetric activity by total protein concentration, determined by Bradford assay.

$$\text{Specific activity} = \frac{\Delta E \cdot V_{total}}{\varepsilon \cdot V_{protein} \cdot c_{protein} \cdot d}$$

where E is change in absorbance per minute ($1\,min^{-1}$), V_{total} is total assay volume (L), ε is NADPH extinction coefficient ($6200\,L.mol^{-1}.cm^{-1}$ at 340 nm and $2066\,L.mol^{-1}.cm^{-1}$ at 370 nm), $V_{protein}$ is volume of enzyme solution added (mL), $c_{protein}$ is concentration of protein solution added ($mg.mL^{-1}$) and d is optical path length (cm).

3.5.3 Procedure 3: Piperazine Synthesis Using Isolated Imine Reductases at 50 mM Scale

3.5.3.1 Materials and Equipment

- Phosphate buffer (0.1 M, pH 7.0)
- $MgCl_2$ hexahydrate
- *N*-Methylethylendiamine ($74.13\,g.mol^{-1}$, 0.001 mol)
- Phenylglyoxal hydrate ($134.13\,g.mol^{-1}$, 0.001 mol)
- Glucose-6-phosphate (G6P)
- NADPH tetrasodium salt
- Glucose-6-phosphate dehydrogenase (G6PDH) from *Leuconostoc mesenteroides*
- Purified imine reductase
- Syringe pump (LA-30, Landgraf Laborsysteme HLL GmbH)
- Magnetic stirrer
- Vortex mixer
- Aqueous NaOH (5 M)
- Dichloromethane
- Methyl-*tert*-butyl ether (MTBE)
- Acetic anhydride
- Rotary evaporator (Laborota 4000, Heidolph Instruments)
- Gas chromatography (GC) system with FID detection (Shimadzu GC-2010)
- GC column (HP-1ms UI, Agilent, 30 m × 0.25 mm × 0.25 μm)
- Chiral GC column (CP-Chirasil-DEX CB, Agilent, 30 m × 0.25 mm × 0.25 μm)

3.5.3.2 Procedure

1. Purified imine reductase ($1\,mg.mL^{-1}$ final concentration) was placed with phosphate buffer solution (0.1 M, pH 7.0) containing 50 mM *N*-methylethylendiamine (74 mg), 2.5 mM NADPH (42 mg), 110 mM G6P (572 mg), 10 $U.mL^{-1}$ G6PDH and 2.5 mM $MgCl_2$ (10 mg) in an appropriate glass vial (20 mL total volume). The reaction mixture was stirred with a magnetic stirrer.

2. Aqueous phenylglyoxal (1 equivalent, 134 mg dissolved in buffer) was added with a syringe pump to the reaction mixture over a time course of 5 hr and stirred at room temperature for 1 hr after feeding had been completed.
3. The transformation was stopped by adding 5 M NaOH to reach a final NaOH concentration of 0.5 M.
4. The product was extracted with dichloromethane (2 × 2 volumes). The combined organic layers were dried over $MgSO_4$ and dichloromethane solvent was removed under reduced pressure.
5. Alternatively, the product may be isolated in its acetylated form. Then, the biotransformation mixture is stopped by adding 5 M NaOH and derivatised by adding two volumes of derivatisation mix (dichloromethane with 50 µL.mL^{-1} acetic anhydride). The mixture is shaken for 2 min using a vortex mixer. The organic phase is transferred to a new vessel and the aqueous phase is again extracted with two volumes of dichloromethane. The combined organic layers are dried over $MgSO_4$ and the solvent is removed under reduced pressure.
6. Transformations were analysed by achiral GC to determine the product formation and by chiral GC to determine the enantiomeric excess. For achiral GC analysis, 100 µL of sample was extracted with 750 µL MTBE. The organic phase was diluted with MTBE (1 : 10) and analysed on a Shimadzu GC-2010 system using an HP-1ms UI column and an isothermal method at 150 °C for 10 min (split = 15). Retention time: *N*-methyl-3-phenylpiperazine 6.82 min. 2-Phenylpiperazine can be used as internal standard with a retention time of 7.08 min.
7. For chiral GC analysis, 100 µL of sample was derivatised directly in aqueous biotransformation solution by adding 250 µL of derivatisation mix (dichloromethane with 50 µL.mL^{-1} acetic anhydride). The mixture was vortexed for 30 sec. After centrifugation, the organic layer was diluted with dichloromethane (1 : 10) and analysed on a Shimadzu GC-2010 system using a CP-Chirasil-DEX CB column (split = 40). The following method was used to separate the product enantiomers: 2 min at 120 °C, temperature increased to 180 °C at 50 °C.min^{-1}, temperature increased to 200 °C at 5 °C.min^{-1} and held for 2 min. Retention times: (*R*)-*N*-methyl-3-phenylpiperazine, 7.68 min; (*S*)-*N*-methyl-3-phenylpiperazine, 7.78 min.

3.5.4 Conclusion

The described procedure enabled the regio- and stereoselective production of piperazine nitrogen heterocycles. The imine reductase-catalysed reaction afforded (*R*)-*N*-methyl-3-phenylpiperazine with excellent enantiomeric excess (>99%) and high yield (92% product formation, 87% isolated yield, 8.1 g.L^{-1}) in 6 hr reaction time. In principle, this procedure can be adapted to obtain a broad range of piperazines with versatile substitution patterns from the corresponding dicarbonyl and diamine substrates [4].

References

1. Lechner, H., Pressnitz, D. and Kroutil, W. (2015) *Biotechnology Advances*, **33**, 457–480.
2. Vo, C.V.T. and Bode, J.W. (2014) *Journal of Organic Chemistry*, **79**, 2809–2815.

3. Lenz, M., Scheller, P.N., Richter, S.M. *et al.* (2017) *Protein Expression and Purification*, **133**, 199–204.
4. Borlinghaus, N., Gergel, S. and Nestl, B.M. (2018) *ACS Catalysis*, **8**, 3727–3732.
5. (a) Scheller, P.N., Fademrecht, S., Hofelzer, S. *et al.* (2014) *ChemBioChem*, **15**, 2201–2204; (b) Scheller, P.N. and Nestl, B.M. (2016) *Applied Microbiology and Biotechnology*, **100**, 10 509–10 520.
6. Korz, D.J., Rinas, U., Hellmuth, K. *et al.* (1995) *Journal of Biotechnology*, **39**, 59–65.

3.6 Screening of Imine Reductases and Scale-Up of an Oxidative Deamination of an Amine for Ketone Synthesis

Gareth J. Brown[1] and Thomas S. Moody[1,2]
[1]*Almac Sciences, Almac House, County Armagh, Craigavon, UK*
[2]*Arran Chemical Company, Unit 1 Monksland Industrial Estate, Athlone, Co. Roscommon, Ireland*

Enzymatic oxidative deamination of an amine-containing compound via selectAZyme™ imine reductase technology (Scheme 3.7) was used to access a ketone functional group for further elaboration. Initial screening of a panel of 87 imine reductases with stoichiometric co-factor (NADP) identified 16 enzymes that showed the required activity, 5 of which were taken forward for further investigation. A key aspect of developing an economic process is the use of minimal co-factor. A co-factor recycle system involving the use of NADPH oxidase (NOX) enzymes was also investigated as part of this study.

3.6.1 Procedure 1: Screening Procedure

The following method describes the initial screening to identify imine reductases showing the desired activity.

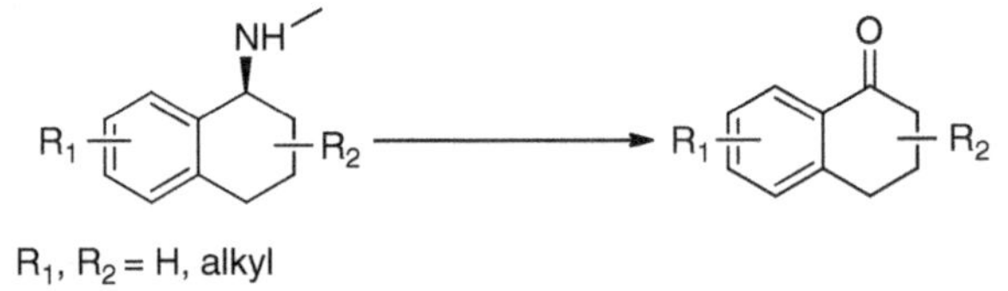

Scheme 3.7 *Imine reductase-mediated oxidative deamination.*

3.6.1.1 Materials and Equipment

- Almac selectAZyme™ IRESK-panel (lyophilised cell free extracts; 5–10 mg)
- 2 mL Eppendorf tubes
- 50 mM Glycine-NaOH buffer (pH 10.6, 0.95 mL)
- NADP (1 eq wrt substrate)
- Amine substrate (5 mg)
- Dimethyl sulfoxide (DMSO; 50 μL)
- Acetonitrile (1 mL)

3.6.1.2 Procedure

1. A set of 87 Eppendorf tubes was labelled with codes corresponding to the different imine reductases in the panel.
2. Imine reductase (5–10 mg) was added to each vial.
3. 50 mM glycine-NaOH buffer (pH 10.6, 0.95 mL, containing 1 eq NADP with respect to substrate) was added to each vial.
4. Amine substrate (5 mg) in DMSO (50 μL) was added to each vial.
5. The reaction vials were sealed and shaken overnight at 20 °C.
6. The next day, the screening reactions were quenched by addition of acetonitrile (1 mL).
7. Initial analysis of samples by thin-layer chromatography (TLC) identified ketone formation.
8. The reaction mixtures were centrifuged at 15 000 rpm for 2 min and the supernatant was separated and analysed by high-performance liquid chromatography (HPLC).

3.6.1.3 Analytical Method

Determination of percentage conversion for screening reactions was achieved by HPLC analysis, as shown in Table 3.7.

Table 3.7 *HPLC conditions for analysis of conversion of amine substrate to ketone product.*

		Conditions	
Column	Type/packing	Phenomenex Luna C18 (2) 150 × 4.6 mm, 3 μm P/No. 00F-4251-E0 S/No. 351720-12	
	Flow rate	1 mL.min^{-1}	
	Eluent	A – 0.1% TFA in water B – 0.1% TFA in ACN	
		Time (min)	%A
		0	70
		15	10
		16	70
		17	70
	Injection volume	10 μL	
	Temperature	30 °C	
Detector	Type	UV	
	Wavelength	220 nm	
	Run time	17 min	

3.6.2 Procedure 2: Preparative Biooxidation Procedure – Production of Ketone by Imine Reductase-Mediated Oxidative Deamination

A 20 g-scale biooxidation was carried out under optimised conditions for the imine reductase and NOX (for co-factor recycle).

3.6.2.1 Materials and Equipment

- Amine (20 g)
- IR20 crude lyophilised cell-free extract (57 g)
- NOX SF crude lyophilised cell-free extract (57 g)
- NADP co-factor (400 mg)
- DMSO (60 mL)
- 50 mM Glycine-NaOH buffer (pH 10.6, 510 mL)
- 1 L jacketed vessel
- Mechanical stirrer
- Baffle
- pH probe
- Pump
- Sintered funnel
- Celite
- Silica
- Rotary evaporator

3.6.2.2 Procedure

1. Buffer (50 mM glycine-NaOH, pH 10.6, 50 mL) was charged to a 1 L jacketed vessel containing a baffle and equipped with mechanical stirrer and pH probe.
2. NADP (400 mg) was charged.
3. Imine reductase 20 (57 g) and NOX (57 g) were added and the pH was readjusted to 10.6 with 5M NaOH.
4. Substrate (HCl salt, 20 g) in DMSO (60 mL) was added to the reaction and the pH was readjusted to 10.6 with 5M NaOH.
5. Air was introduced into the reaction mixture by means of a pump (maintaining flow at a suitable rate to avoid foaming).
6. The reaction was vigorously stirred at 20 °C and monitored by HPLC until complete.
7. Celite (240 g) was added to the reaction mixture with ethyl acetate (200 mL) and stirred vigorously, then the suspension was discharged and filtered through a sintered funnel.
8. The filtrate layers were separated.
9. The aqueous layer was extracted with ethyl acetate (2×100 mL).
10. The combined organics were concentrated to leave an oil, which was purified over a silica plug (100 g; heptane →80/20 heptane/EtOAc ~1 L) with 82.4% isolated yield.

For HPLC analysis, see Table 3.7 Substrate retention time = 6.9 min, product retention time = 13.9 min.

3.6.3 Conclusion

Imine reductase technology has proven to be a useful tool for the generation of an added-value ketone from an inexpensive amine substrate using imine reductase/NOX and air as an oxidation reagent. The approach can be applied generally to a range of amines.

4

Transaminases

4.1 A Practical Dynamic Kinetic Transamination for the Asymmetric Synthesis of the CGRP Receptor Antagonist Ubrogepant

Birgit Kosjek

Process Research and Development, Merck & Co. Inc., Rahway, NJ, USA

4.1.1 Enzymatic Dynamic Kinetic Transformations: Enabling the Asymmetric Synthesis of Intermediates with Multiple Stereocentres

The development of efficient and ecological manufacturing routes has become increasingly important as the chemical and pharmaceutical industry has moved towards more sustainable processing [1]. Setting multiple stereocentres simultaneously typically reduces the number of steps and separations involved in a synthesis, saving time, capital, labour and overhead, and improving the environmental footprint. Biocatalysis offers the advantage of supreme specificity and selectivity, and is thus ideally suited for the synthesis of complex intermediates with several chiral centres. Recent advances in protein engineering have enabled the use of biocatalytic dynamic kinetic transformations under conditions that allow for epimerisation of the starting material. These reactions challenge and surpass the known limitations of enzyme performance in a process environment of high solvent concentration, high pH and elevated temperature. Process development, in parallel with directed evolution efforts, has been critical to affording a set of novel enzymes tailored to mastering efficiency and stereoselectivity under reaction conditions set to fit substrate racemisation requirements. The developed processes expand across multiple enzyme classes and have proven robust during implementation from laboratory to manufacturing scale. Further examples of dynamic kinetic transformations are reported in Sections 7.1 and 7.2.

Applied Biocatalysis: The Chemist's Enzyme Toolbox, First Edition. Edited by John Whittall and Peter W. Sutton.
© 2021 John Wiley & Sons Ltd. Published 2021 by John Wiley & Sons Ltd.

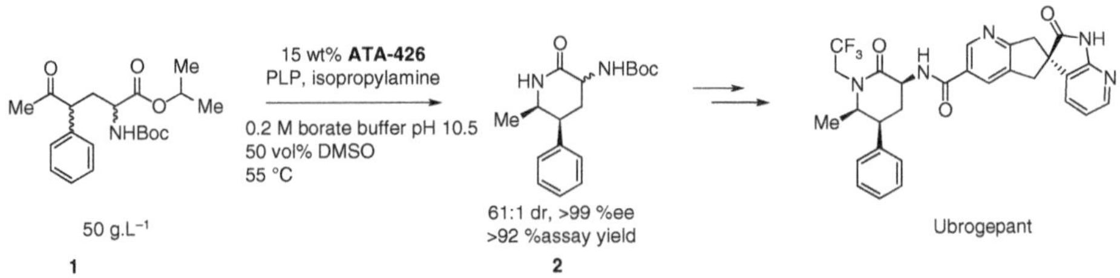

Scheme 4.1 *Industrial-scale production of tert-butyl((5S,6R)-6-methyl-2-oxo-5-phenyl piperidin-3-yl) carbamate **2** using a dynamic kinetic transamination including spontaneous lactamisation of the product under high-pH conditions.*

4.1.2 Ubrogepant

The total synthesis of Ubrogepant relies on an enzyme-catalysed dynamic kinetic transamination (DK-TA) for the production of a key chiral lactam fragment, setting two of its three stereocentres (Scheme 4.1) [2]. Both the epimerisation of the keto ester starting material **1** and the *in situ* lactamisation of the amine product **2** required high pH and elevated temperature. Screening of a transaminase library at pH 10.5 established proof of concept for a dynamic kinetic system and demonstrated complementary selectivity, allowing access to both *syn*- and *anti*-epimers. ATA-301 afforded the desired *syn*-lactam **2** at 7 : 1 diastereoselective ratio but low productivity. Directed evolution first improved the selectivity around ninefold to a final >60 : 1 dr, and – in tandem with process optimisation – went on to significantly increase the activity and stability under the set conditions. Final variant ATA-426 provided lactam **2** with a >92% assay yield in <24 hr. The optimised process takes advantage of product removal through *in situ* lactamisation of **2** over common engineering solutions to remove acetone and so drive the reaction equilibrium towards high conversion.

4.1.3 Procedure 1: Dynamic Kinetic Transamination of Isopropyl 2-((*tert*-Butoxycarbonyl) Amino)-4-Oxo-3-Phenylpentanoate 1

4.1.3.1 Materials and Equipment

- Sodium tetraborate decahydrate (26.2 g)
- Deionised water (1.4 L)
- Isopropylamine (IPA, 82.8 g, 1.4 mol)
- Aqueous HCl (6 N)
- Pyridoxal-5-phosphate hydrate (2.8 g, 11 mmol)
- Transaminase ATA-426 (21 g, Codexis Inc.)
- Isopropyl 2-((*tert*-butoxycarbonyl)amino)-4-oxo-3-phenylpentanoate **1** (140 g, 385 mmol)
- Dimethyl sulfoxide (DMSO, 1.4 L)
- Aqueous IPA solution (8 M)
- Isopropyl acetate (3.84 L)
- 2-Propanol (1.76 L)

4.1.3.2 Procedure

1. Isopropyl 2-((tert-butoxycarbonyl)amino)-4-oxo-3-phenylpentanoate **1** was chemically synthesised as reported in [2].

2. A 0.2 M borate buffer was prepared by dissolving sodium tetraborate decahydrate (26.2 g) in deionised water (1.4 L). IPA (82.8 g, 1.4 mol) was added and the pH was adjusted to 10.5 using 6 M HCl. The buffer solution was cooled to room temperature.
3. Pyridoxal-5-phosphate (2.8 g, 11 mmol) followed by transaminase (21 g, 15 wt%) was added and slowly dissolved at room temperature.
4. A solution of **1** (140 g, 385 mmol) in DMSO (1.4 L) was added to the reactor over 5–10 min.
5. The reaction was heated to 55 °C.
6. The pH was adjusted to 10.5 at 55 °C and the reaction aged for 24 hr. The pH was maintained via an automated pH loop using 8 M aqueous IPA.
7. At >95% conversion as determined by high-performance liquid chromatography (HPLC), a mixture of 2-propanol : isopropyl acetate (3 : 4, 2.8 L) was added, followed by stirring for 20 min at room temperature. The phases were separated and the aqueous layer back-extracted with 2-propanol : isopropyl acetate (2 : 8, 2.8 L). The organic layers were combined and washed with deionised water (0.5 L).
8. The extract was analysed by HPLC, showing a 114.6 g assay yield for the desired tert-butyl((5*S*,6*R*)-6-methyl-2-oxo-5-phenylpiperidin-3-yl) carbamate **2** (97.6% assay yield, 61 : 1 dr, >99% ee).
9. The extract was concentrated under vacuum and redissolved in dichloromethane, and the organic solution was washed with water and brine before crystallising **2** from methyl *tert*-butyl ether/n-hexane (2 : 3) at room temperature in 99.6 g (80%) yield.

4.1.4 Analytical Method

^{1}H and ^{13}C NMR data are available in [2].

Conversion and diastereomeric ratio were determined by reversed-phase HPLC using a Chiralcel OD-RH column (150 × 4.6 mm, 5 um) at a gradient from 20 : 80 acetonitrile : water (both 2mM ammonium formate, pH 3.5) to 80 : 20 acetonitrile : water (both 2mM ammonium formate, pH 3.5) over 45 min at 0.75 mL.min^{-1} and 25 °C. Detection at 210nm. Lactam epimers: 23, 24.5, 25.7, 27 min (desired). Keto ester epimers: 32.5, 33.2, 34.5, 35.3 min.

4.1.5 Conclusion

The development of this efficient DKR-TA presents a key asymmetric step to a lactam core, avoiding separations and shortening the overall step count. Significant productivity improvements were achieved through hand-in-hand process development and protein engineering, each part informing and challenging the other. This procedure can be adapted to related epimerisable starting materials [2–4].

References

1. Roschangar, F., Sheldon, R.A., Senanayake, C.H. (2015) *Green Chemistry*, **17**, 752.
2. Nobuyoshi Y., Cleator, E., Kosjek, B. *et al.* (2017) *Organic Process Research & Development*, **21**, 1851.
3. Limanto, J., Ashley, E.R., Yin, J., *et al.* (2014) *Organic Letters*, **16**, 2716.
4. Chung, C.K., Bulger, P.G., Kosjek, B. *et al.* (2014) *Organic Process Research & Development*, **18**, 215.

4.2 Asymmetric Biosynthesis of L-Phosphinothricin by Transaminase

Li-Qun Jin, Feng Peng, Feng Cheng, Ya-Ping Xue* and Yu-Guo Zheng
Key Laboratory of Bioorganic Synthesis of Zhejiang Province, College of Biotechnology and Bioengineering; Engineering Research Center of Bioconversion and Biopurification of Ministry of Education; and National and Local Joint Engineering Research Center for Biomanufacturing of Chiral Chemicals, Zhejiang University of Technology, Hangzhou, People's Republic of China

Asymmetric biosynthesis utilising enzymes is one of the most convenient and economically favoured routes for the synthesis of target chiral compounds under mild conditions. Transaminases have emerged as powerful catalysts for the production of chiral amines and amino acids, which are important intermediates for the preparation of a broad range of drugs and fine chemicals [1]. Starting from the prochiral ketones, transaminases can produce desired chiral amines and amino acids with a theoretical maximum yield of 100% [2]. For this reason, we cloned a novel transaminase from *Pseudomonas fluorescens* ZJB09-108 (*Pf*-TA) and expressed it in the *Escherichia coli* BL21 (DE3) for the production of L-phosphinothricin (L-PPT) [3].

This asymmetric biosynthesis route (Scheme 4.2) allowed us to produce L-PPT **2** from 4-(hydroxy(methyl)phosphoryl)-2-oxobutanoic acid (PPO) **1** in excellent yield [3]. The first step was to transfer the amino group from L-alanine **3** to the carbonyl group of pyridoxal 5'-phosphate (PLP) under the action of aminotransferase to form pyridoxamine 5-phosphate (PMP) and ketone **4** corresponding to the amino donor [4, 5]. The second part of the catalytic cycle was the binding of the acceptor ketone or aldehyde with the PMP; the combined compound acted as the amine donor to reconvert into PLP again, which was subsequently covalently bound by the active-site lysine [6].

4.2.1 Procedure 1: Gene Cloning of a Transaminase from *P. fluorescens* ZJB09-108

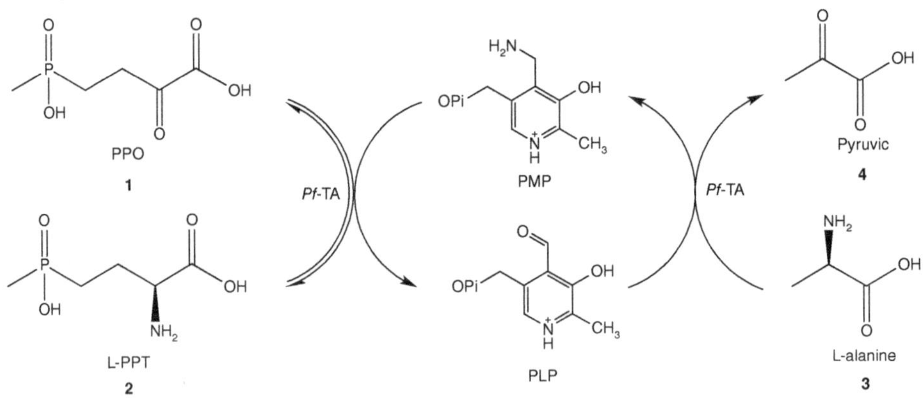

Scheme 4.2 *Enzymatic asymmetric synthesis of L-PPT **2** from PPO **1** by Pf-TA.*

4.2.1.1 Materials and Equipment

- *P. fluorescens* ZJB09-108 (Deposit number: CCTCC M 2012539)

- *E. coli* strains JM109 (Tiangen Biotech Co. Ltd)
- Plasmids pMD18-T (TaKaRa)
- Plasmids pET-28b (+) (Novagen)
- Distilled water (dH_2O)
- Kanamycin ($50\,mg.mL^{-1}$ in dH_2O, filter-sterilised)
- Peptone (Oxoid)
- Yeast extract (Oxoid)
- Glucose (Sinopharm Chemical Reagent Co. Ltd)
- Potato
- NaCl (Sinopharm Chemical Reagent Co. Ltd)
- *E. coli* BL21(DE3) (Invitrogen)
- 10 0.5 L Schott bottles with screw caps
- Orbital shaker ZWTR-2403 (Shanghai Zhicheng)
- Cooling centrifuge 5451R (Eppendorf)
- Phanta Max Super-Fidelity DNA Polymerase (Vazyme-innovation in Enzyme Technology)
- dNTP Mix (Vazyme-innovation in Enzyme Technology)
- XbaI (New England Biolabs)
- XhoI (New England Biolabs)
- T4 DNA ligase (Vazyme-innovation in Enzyme Technology)

4.2.1.2 Procedure

1. Potato (200 g) was placed in a saucepan with approximately 1.2 L of tap water and boiled on a stove for 1 hr. The liquid was then cooled until it could be comfortably handled (about 10 min). The liquid potato extract was poured through a strainer and collected in another container. Glucose (20 g) was added to 1 L of the extract water. The solution was autoclaved (20 min, 121 °C) in 0.5 L Schott bottles to give sterile PD medium.
2. *P. fluorescens* ZJB09-108 was cultivated and harvested at the late log phase in PD medium at 30 °C for 24 hr. The genomic DNA of *P. fluorescens* ZJB09-108 was then extracted by a FastDNA SPIN Kit for Siol (MP Biomedicals) according to its protocol, and the genomic DNA was used as a template for transaminase gene (*Pf*-TA) amplification by polymerase chain reaction (PCR).
3. The primers for transaminase gene amplification were P1: 5'-ATGTCTAAAAACGAAT CTCTGCTGCAGC-3' and P2: 5'-TTAAGCCAGTTCGTCGAAGCATTC-3'. The PCR reagents were: 2 × Phanta Max buffer (25 μL); dNTP mix (10 mM each, 1 μL), upstream primer P1 (10 mM each, 1 μL), downstream primer P2 (10 mM each, 1 μL); Phanta Max Super-Fidelity DNA Polymerase (1 μL); and template DNA (from *P. fluorescens* ZJB09-108) (0.5 μL). The PCR was operated using the following programme: 1 cycle at 94 °C for 5 min; 35 cycles at 94 °C for 45 sec; 55 °C for 45 sec; 72 °C for 2 min; and 72 °C for 10 min.
4. The amplified PCR products were double-digested with corresponding restriction endonucleases and inserted into the expression vector pET-28b. The gel recycling product of amplified PCR (30 μL), *Xba*I (2.5 μL), *Xho*I (2.5 μL), cut smart buffer (8 μL), dd H_2O (37 μL) was incubated at 37 °C for 1.5 hr. The vector pET-28b was

double-digested with *Xba*I and *Xho*I, and then the insert (the PCR product) (6.5 μL), vector (1.5 μL), T4 DNA ligase (1 μL), and T4 DNA ligase buffer 1 μL, dd H_2O (5 μL) were incubated at 37 °C for 45 min. The constructed plasmids were transformed into *Escherichia coli* BL21 (DE3) by the heat-shock method.

4.2.2 Procedure 2: Recombinant Expression of the *Pf*-TA and Preparation of Resting Cells

4.2.2.1 Materials and Equipment

- Peptone (10 g)
- Yeast extract (5 g)
- NaCl (10 g)
- Distilled water (dH_2O)
- Kanamycin (50 mg.mL^{-1} in dH_2O, filter-sterilised)
- 10 0.5 L Schott bottles with screw caps
- Isopropyl-β-D-1-thiogalactopyranoside (IPTG; 0.5 M in dH_2O, filter-sterilised)

4.2.2.2 Procedure

1. Peptone (10 g), yeast extract (5 g) and NaCl (10 g) were dissolved in dH_2O (1 L) and autoclaved (20 min, 121 °C) in 0.5 L Schott bottles to give sterile lysogenic broth (LB) medium.
2. Sterile LB medium (100 mL) was placed into a sterile 0.5 L Erlenmeyer flask. Afterwards, kanamycin (20 μL) stock solutions were added to reach final concentrations of 50 μg.mL^{-1}. The solution was inoculated with a single colony of *E. coli* BL21(DE3) harbouring pET28b (+)-*Pf*-TA and shaken at 37 °C and 200 rpm until the OD_{600} reached 0.6.
3. The cultivation was cooled to 28 °C and expression of the recombinant transaminase was induced by the addition of 20 μL IPTG (0.1 mM final conc.) to every flask. The expression was performed at 28 °C for 12 hr.
4. The cells were harvested by centrifugation for 10 min at 7656× g under 4 °C. About 1.0 g of resting cells from each flask was obtained and stored at −80 °C for further use.

4.2.3 Procedure 3: Biocatalytic Conversion of L-PPT

4.2.3.1 Materials and Equipment

- PPO (Yonon Inc.)
- PLP (J&K Chemical Technology)
- L-Alanine (J&K Chemical Technology)
- Tris (Aladdin)
- Aqueous HCl and NaOH (Xilong Chemical Co. Ltd)
- Resins 001 × 7 (Zhengguang Resin Co.)
- Resting cells (25 mg) from Procedure 2
- pH meter

4.2.3.2 Procedure

1. Preparation of 50 mM Tris HCl buffer (pH 8.5): 0.1 M aqueous solution of Tris (50 mL) was mixed with 0.1 M aqueous solution of hydrochloric acid (14.7 mL) and diluted to 100 mL with dH_2O.
2. Resting cells (25 mg) obtained from Procedure 2 were suspended in 2 mL 50 mM Tris HCl buffer (pH 8.5) and mixed with 5 mL substrate stock solution containing 200 mM PPO and 700 mM L-alanine. The mixed solution (about 7 mL) was then diluted to 10 mL with 50 mM Tris HCl buffer (pH 8.5). The biocatalytic conversion was performed at 35 °C, 180 rpm in an orbital shaker.
3. The concentration of L-PPT was checked every hour with the Ultimate 3000 high-performance liquid chromatography (HPLC) system. After 99% conversion, the bioreaction mixture was centrifuged (10 min, 7656× g, 4 °C) to remove cells.
4. The solution was concentrated by a rotary evaporator at 50 °C to a final volume of 5 mL. The pH value of the solution was adjusted to 4.0 by 6.0 M HCl and extracted twice with diethyl ether in a total ratio of 1.5 : 1 to remove pyruvic acid in the reaction solution.
5. The solution containing L-PPT and L-alanine was loaded on a cationic exchanger (001 × 7 Resin, 45 mL). The L-alanine was removed by absorbing onto the cationic exchanger (001 × 7 Resin). The effluent liquid containing L-PPT was collected, and its pH adjusted to 2.0 by 6 M HCl. The solution was then loaded on to the cationic exchanger (001×7 Resin, 45 mL) and eluted with NH_4OH solution (2 M). The eluate was collected and concentrated to a final volume of 1 mL by a rotary evaporator at 50 °C.
6. The L-phosphinothricin **2** was obtained in 148.3 mg (75% yield, >99.9% ee) by crystallisation under 4 °C.

4.2.4 Analytical Method

4.2.4.1 Materials and Equipment

- Absolute ethanol (Aladdin)
- Sodium borate (Aladdin)
- Double-distilled water (ddH_2O)
- *o*-Phthalaldehyde and *N*-acetyl-L-cysteine (J&K Chemical Technology)
- Phosphoric acid (aqueous 1 M) (Lingfeng Chemical Reagent Co. Ltd)
- Ultimate 3000 HPLC system (Dionex) equipped with an UltiMate fluorescence detector (Sunnyvale), a well-plate sampler (WPS-3000 SL) and a column oven (TCC-3000RS) and C18 column (Ultimate XB-C18, 4.6 × 250 mm, 5 μm).

4.2.4.2 Procedure

1. Preparation of boric acid buffer (0.1 M): 19 g sodium borate was dissolved to 470 mL ddH_2O. The pH value was adjusted to 9.8 by 1 M NaOH. ddH_2O was added to make 500 mL.
2. Preparation of derivatisation reagents: 0.1 g *o*-phthalaldehyde (OPA) and 0.12 g *N* acetyl L-cysteine (NAC) were dissolved in10 mL absolute ethanol. The prepared boric acid buffer (0.1 M) was added to make up 50 mL.

3. Preparation of ammonium acetate solution (50 mM): 3.854 g ammonium acetate was dissolved in 970 mL ddH_2O. The pH value was adjusted to 5.7 by 1.0 M phosphoric acid solution. ddH_2O was added to make 1 L. The mobile phase was composed of 50 mM ammonium acetate and methanol (90 : 10 v/v).
4. The samples were first derivatised with *o*-phthalaldehyde and *N*-acetyl-L-cysteine as chiral derivatisation reagent. Then a pair of enantiomer derivatives with fluorescence absorption was synthesised. The Ultimate 3000 HPLC system was used to analyse the samples. The column temperature and flow rate were set at 35 °C and $1.0\,mL.min^{-1}$, respectively. The excitation wavelength (λ_{ex}) and emission wavelength (λ_{em}) of the method were confirmed as 340 and 450 nm, respectively. The retention times of L-PPT and D-PPT were 10.6 and 12.6 min, respectively. The enantiomeric excess (ee) value of the products was calculated according to the equation ([L-PPT] − [D-PPT])/([L-PPT] + [D-PPT]) × 100%, where [L-PPT] and [D-PPT] are the concentrations of L-PPT and D-PPT, respectively.
5. The ^{13}C and ^{1}H nuclear magnetic resonance (NMR) spectra of L-PPT was obtained on an NMR spectrometer (AVANCE 500 MHz, Bruker) with D_2O as the solvent, using 500 and 125 MHz for carbon and proton determinations, respectively.
6. The product L-PPT was identified by ^{1}H-NMR (500 MHz, D_2O) δ 3.70–3.61 (m, 1H), 2.05–1.83 (m, 2H), 1.60–1.37 (m, 2H), 1.12 (d, 3H, J = 13.5 Hz)], ^{13}C-NMR (126 MHz; D2O) δ 174.09 (s), 55.21 (d, J = 14.4 Hz), 27.06 (d, J = 91.0 Hz), 24.18 (s), 14.95 (d, J = 93.1 Hz). The molecular mass of the product was confirmed by liquid chromatography–mass spectrometry (LCMS). The liquid phase detection method followed the same derivatisation method as HPLC. LCMS revealed the following result: 441 (m/z) (M^-, calculated value for $C_{18}H_{23}N_2O_7PS$, 442).

4.2.5 Conclusion

A novel transaminase from *P. fluorescens* ZJB09-108 was cloned and expressed in *E. coli* BL21 (DE3). Using whole cells of *E. coli* harbouring transaminase as biocatalyst, a high yield of L-PPT with 99.9% ee was attained. These results suggest the recombinant transaminase is a potential candidate for the biosynthesis of L-PPT from PPO.

4.2.6 Enzyme Source

The genomic DNA of *P. fluorescens* ZJB09-108 was extracted according to a standard protocol, which was used as a template for transaminase gene (*Pf*-TA) amplification by PCR. The *P. fluorescens* ZJB09-108 strain has been preserved in the China Center for Type Culture Collection, deposit number CCTCC M2012539. The nucleotide sequence of the *Pf*-TA has been submitted to the National Center for Biotechnology Information (NCBI) database, GenBank Accession No. MK840960.

References

1. Xue, Y.P., Cao, C.H. and Zheng, Y.G. (2018) *Chemical Society Reviews*, **47** (4), 1516–1561.
2. Lopez-Iglesias, M., Gonzalez-Martinez, D., Rodriguez-Mata, M. *et al.* (2017) *Advanced Synthesis & Catalysis*, **359**, 279–291.

3. Jin, L.Q., Peng, F., Liu, H.L. *et al.* (2019) *Process Biochemistry*, **85**, 60–67.
4. Shin, J.S. and Kim, B.G. (2002) *Journal of Organic Chemistry*, **67**, 2848–2853.
5. Franco, T.M.A., Favrot, L., Vergnolle, O. and Blanchard, J.S. (2017) *ACS Chemical Biology*, **12**, 1235–1244.
6. Schell, U., Wohlgemuth, R. and Ward, J.M. (2009) *Journal of Molecular Catalysis B: Enzymatic*, **59**, 279–285.

4.3 Application of *In Situ* Product Crystallisation in the Amine Transaminase from *Silicibacter pomeroyi*-Catalysed Synthesis of (*S*)-1-(3-Methoxyphenyl)ethylamine

Dennis Hülsewede, Jan Neuburger and Jan von Langermann*
Biocatalytic Synthesis Group, Institute of Chemistry, University of Rostock, Rostock, Germany

Over the past few years, transaminases have become a powerful tool for the synthesis of chiral amines, which are valuable building blocks in the preparation of active pharmaceutical ingredients (APIs) and agrochemicals [1]. Examples include applications at both laboratory and industrial scale, such as in the synthesis of Sitagliptin [2]. The major advantage of amine transaminases (ATAs) is their ability to synthesise highly selectively optically pure amines or amino acids [3]. This can be achieved by kinetic resolution from a racemic amine mixture, which is limited by design to a maximum yield of 50%, or by the preferred asymmetric synthesis with a theoretical maximum yield of 100%. The asymmetric synthesis reaction requires a primary amine, which serves as a donor for the amination of a prochiral carbonyl compound (amine acceptor), forming the chiral product amine and the deaminated co-product [4, 5]. Commonly used donor amines are isopropylamine (IPA) and alanine [6].

The asymmetric synthesis is unfortunately often limited by the unfavourable thermodynamic equilibrium towards the desired product amine. A conventional solution for this major limitation is the use of a significant excess of the amine donor over the amine acceptor to push the reaction to the product side [6, 7]. However, such nonphysiological conditions can cause secondary issues with the biocatalyst itself, which may easily become a significant problem in the synthesis process [5]. To circumvent this issue, (bio)catalytic cascade reactions were developed to remove the side product and thereby shift the equilibrium [8]. One of the most common examples includes the use of alanine as an amine donor, forming pyruvate as a byproduct, which is then further reduced to lactate by lactate dehydrogenase (LDH). The required regeneration of the co-factor is typically obtained with glucose dehydrogenase (GDH) and glucose as a secondary substrate [9]. In addition, a variety of tailor-made donor amines have been proposed, which include secondary reactions after deamination and result in a similar equilibrium shift [10]. The consequence of all these techniques is that the overall complexity of the transaminase-catalysed reaction is increased, involving further side products, which decreases atom efficiency and increases costs.

Alternative noncatalytic solutions for equilibrium shift include *in situ* product removal (ISPR) techniques such as supported liquid membrane (SLM) [11] and *in situ* product crystallisation (ISPC) of the product amine [12, 13], which is the technique we use here. The advantage of a crystallisation is that it provides direct removal of the product amine from the reaction solution and subsequent isolation by a simple filtration [13]. The following example shows the enzymatic transformation of 3-methoxyacetophenone (3MAP) **1** into (*S*)-1-(3-methoxyphenyl)ethylamine **2** using the conventional amine donor IPA and the ATAs from *Silicibacter pomeroyi*. The integrated *in situ* crystallisation is achieved with the counterion of 3,3-diphenylpropionic acid (3DPPA) **4**.

4.3.1 Procedure 1: Synthesis of IPA-3DPPA Salt (Donor Salt) 5

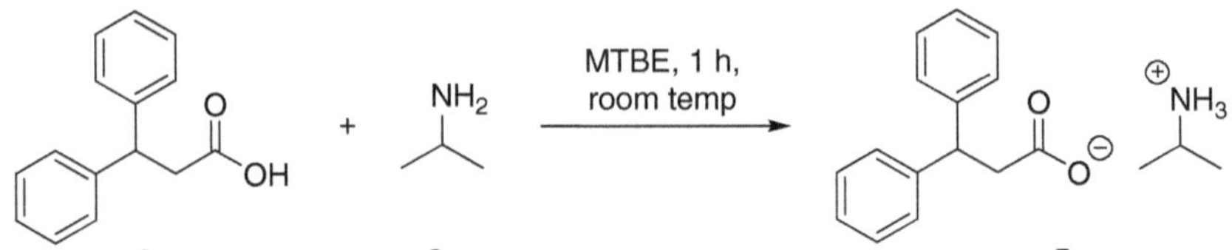

Scheme 4.3 *Synthesis of IPA-3DPPA salt (donor salt)* ***5***.

4.3.1.1 Materials and Equipment

- Methyl *tert*-butyl ether (MTBE)
- 3,3-Diphenylpropionic acid (3DPPA) **4** (5 g, 44.2 mmol)
- IPA **3**
- 250 mL round-bottom flask

4.3.1.2 Procedure

1. In a 250 mL round-bottom flask, 3DPPA (5 g, 22.1 mmol) was dissolved in MTBE (90 mL). IPA (1.9 mL, 22.1 mmol) was added within 2 min and the resulting mixture was stirred overnight at room temperature.
2. The resulting precipitate was filtered and washed with ice-cold MTBE (ca. 20 mL) and dried overnight at room temperature (in a fume hood) and for a few hours at 60 °C until all remaining MTBE was evaporated (see Scheme 4.3).

4.3.1.3 Isolated Yield

6.3 g (>99% yield) of a white powder.

^{1}H-NMR (500.13 MHz; DMSO-d6): δ 7.31–7.08 (m, 10H, Ar H), 6.22 (s, 3H, NH_3^+), 4.47 (t, J = 7.7 Hz, 1H, CH), 3.05 (m, J = 6.3 Hz, 1H, CH), 2.86 (d, J = 7.7 Hz, 2H, CH_2), 1.04 (d, J = 6.3 Hz, 6H, CH_3).

^{13}C-NMR (125.8 MHz; DMSO-d6): δ 172.8 (CO_2), 144.7 (Ar), 127.7 (Ar), 127.2 (Ar), 125.4 (Ar), 46.9 (CH), 41.8 (CH_2), 41.6 (CH), 23.6 (CH_3).

4.3.2 Procedure 2: ISPC-Based Synthesis of (*S*)-1-(3-Methoxyphenyl)ethylammonium 3,3-Diphenylpropionate 2

4.3.2.1 Materials and Equipment

- IPA **3**
- Isopropylammonium 3,3-diphenylpropionate (IPA-3DPPA) **5**

- Distilled water (dH_2O)
- Cyclopentyl methyl ether (CPME)
- conc. H_3PO_4 (>85%)
- sat. NaOH solution
- Internal standard solution: 9.7 μL n-decane in 1.99 mL CPME
- *N*-(2-Hydroxyethyl)piperazine-*N*-(2-ethanesulfonic acid) (HEPES)
- ATA from *Silicibacter pomeroyi* (commercially available as ECS ATA08 from Enzymicals AG)
- 3-Methoxyacetophenone (3MAP) **1**
- Pyridoxal 5′-phosphate hydrate (PLP)
- 50 mL Schlenk flask
- 1.5 mL Eppendorf tubes
- pH meter (WTW pH 320, WTW)
- Centrifuge (Mini Centrifuge C-1200, National Labnet)
- Vacuum pump (Diaphragm Pump, Vacuumbrand)
- Gas chromatograph (GC) (Trace 1310 Gas Chromatograph, Thermo Scientific)
- GC column CP-Chirasil-Dex CB; 25 m, 0.25 mm, 0.25 μm (Agilent)

4.3.2.2 *Procedure*

1. HEPES (119 mg, 0.5 mmol) was dissolved in dH_2O (20 mL) to obtain a 25 mM HEPES solution.
2. The HEPES solution (20 mL, 25 mM) was transferred into a 50 mL Schlenk flask. IPA (214 μL, 2.5 mmol) was added to obtain a concentration of 125 mM, and the pH was adjusted to 7.5 by adding conc. H_3PO_4.
3. IPA-3DPPA salt (713.5 mg, 2.5 mmol) (donor salt) was added to the reaction mixture as a continuous feed of IPA and 3DPPA for the ISPC of the product amine salt. Please note that IPA-3DPPA is only partially soluble and an excess of solid donor salt will be present.
4. PLP (24.7 mg, 0.1 mmol) and ATA (240 U) were added to the resulting suspension and the reaction was started by the addition of 3-methoxyacetophenone (3MAP) **1** (275 μL, 2.0 mmol). The reaction mixture was stirred at 30 °C for 23 hr.
5. The reaction vessel was held under reduced pressure (50–100 mbar) for 1 hr to partially remove acetone.
6. A 500 μL sample was taken, quenched with a sat. NaOH solution (50 μL), mixed with CPME (500 μL) and vortexed for 30 sec. The emulsion was centrifuged to enhance phase separation. 200 μL of the organic layer was added to CPME (800 μL) and a 25 mM n-decane solution in CPME (200 μL) (internal standard) for gas chromatography analysis (see later for details).
7. Substrate adjustment after each cycle (24 hr): Based on the analysis in Step 6, the 3MAP concentration was readjusted to 100 mM and IPA-3DPPA to 125 mM. The reaction was then allowed to proceed for another 23 hr.
8. Steps 5–7 were repeated until a product amine concentration of 180–190 mM was reached. During the last two repetitions, only 3MAP was readjusted to 100 mM, to allow the consumption of all solid IPA-3DPPA. This led to the presence of only product amine salt as the remaining solid phase, which was eventually filtered off and washed with small amounts of cold water and CPME (see Scheme 4.4).

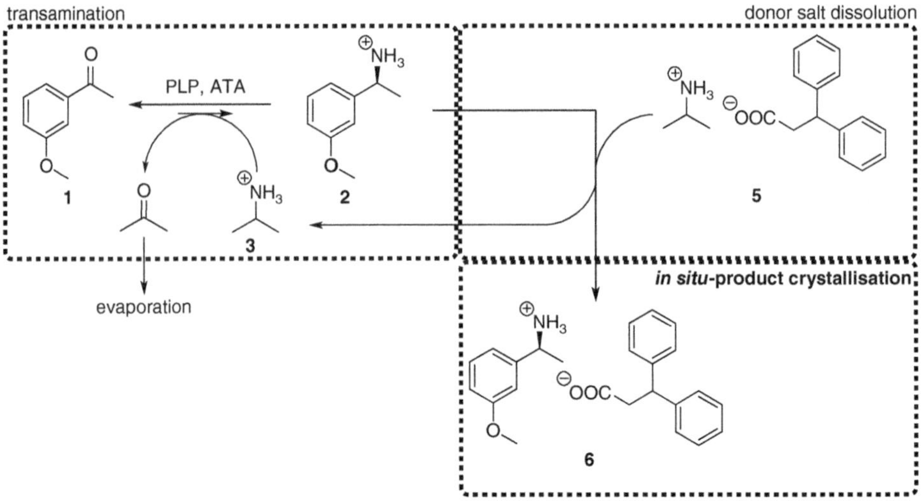

Scheme 4.4 *ISPC-based conversion of 3-methoxyacetophenone **1** into (S)-1-(3-methoxyphe nyl)ethylamine **2**. Equilibrium shift and product isolation result from a continuous product amine salt precipitation **6** with 3,3-diphenylpropionate, which originates from a constant dissolution of the donor amine salt isopropylammonium 3,3-diphenylpropionate **5**.*

4.3.2.3 Isolated Yield

1 g (92% yield, based on IPA-3DPPA consumption).

^{1}H-NMR (500.13 MHz; DMSO-d6): δ 7.33–6.73 (m, 14H, Ar H), 4.47 (t, J = 7.8 Hz, 1H, CH), 4.43 (s, 3H, NH_3^+), 4.00 (q, J = 6.7 Hz, 1H, CH), 3.77 (s, 3H, O-CH_3), 2.97 (d, J = 7.8 Hz, 2H, CH_2), 1.29 (d, J = 6.6 Hz, 3H, CH_3).

^{13}C-NMR (62.9 MHz; DMSO-d6): δ 172.0 (CO_2), 159.0 (Ar), 149.3 (Ar), 143.9 (Ar), 128.5 (Ar), 127.7 (Ar), 127.0 (Ar), 125.5 (Ar), 117.6 (Ar), 111.5 (Ar), 111.3 (Ar), 54.6 (O-CH_3), 50.0 (CH), 46.4 (CH), 40.1 (CH_2), 25.0 (CH_3).

The free amine can be obtained by dissolution of the product salt in alkaline solution (pH > 13), extraction into CPME and evaporation of CPME.

4.3.2.4 Analytical GC Method

See Tables 4.1 and 4.2.

Table 4.1 *GC method.*

Temperature program (r = (°C.min^{-1}))	Duration
90 °C – 2 r → 100 °C – 20 r → 130 °C– 2 r → 138 °C – 20 r → 160 °C	11.6 min

Table 4.2 Retention times for GC analysis.

Substance	Retention (min)
n-Decane (standard)	3.9
(*S*)-1-(3-Methoxyphenyl)ethylamine (3MPEA)	8.7
3-Methoxyacetophenone (3MAP)	9.0

4.3.3 Conclusion

This procedure describes the continuous synthesis of (*S*)-1-(3-methoxyphenyl)ethylamine **1** beyond the initial thermodynamic equilibrium in high purity. The continuous dissolution of the donor amine salt IPA-3DPPA provides the basis for a constant crystallisation of the product amine salt.

References

1. (a) Slabu, I., Galman, J.L., Lloyd, R.C. and Turner, N.J. (2017) *ACS Catalysis*, **7**, 8263–8284; (b) Fuchs, M., Farnberger, J.E. and Kroutil, W. (2015) *European Journal of Organic Chemistry*, **2015**, 6965–6982; (c) Kohls, H., Steffen-Munsberg, F. and Hohne, M. (2014) *Current Opinion in Chemical Biology*, **19**, 180–192.
2. (a) Savile, C.K., Janey, J.M., Mundorff, E.C. *et al.* (2010) *Science*, **329**, 305–309; (b) Narancic, T., Davis, R., Nikodinovic-Runic, J. and O' Connor, K. E. (2015) *Biotechnology Letters*, **37**, 943–954.
3. (a) Ferrandi, E.E. and Monti, D. (2017) *World Journal of Microbiology & Biotechnology*, **34**, 13; (b) Kelly, S.A., Pohle, S., Wharry, S. *et al.* (2018) *Chemical Reviews*, **118**, 349–367; (c) Desai, A.A. (2011) *Angewandte Chemie International Edition*, **50**, 1974–1976.
4. (a) Gomm, A. and O'Reilly, E. (2018) *Current Opinion in Chemical Biology*, **43**, 106–112; (b) Dold, S.-M., Syldatk, C. and Rudat, J. (2016) Transaminases and their applications, in *Green Biocatalysis* (ed. R.N. Patel), John Wiley and Sons, pp. 715–746.
5. Guo, F. and Berglund, P. (2017) *Green Chemistry*, **19**, 333–360.
6. Gundersen, M.T., Abu, R., Schürmann, M. and Woodley, J.M. (2015) *Tetrahedron: Asymmetry*, **26**, 567–570.
7. (a) Voges, M., Abu, R., Gundersen, M.T. *et al.* (2017) *Organic Process Research & Development*, **21**, 976–986; (b) Tufvesson, P., Lima-Ramos, J., Jensen, J.S. *et al.* (2011) *Biotechnology and Bioengineering*, **108**, 1479–1493; (c) Tufvesson, P., Nordblad, M., Krühne, U. *et al.* (2015) *Organic Process Research & Development*, **19**, 652–660.
8. (a) Cassimjee, K.E., Branneby, C., Abedi, V. *et al.* (2010) *Chemical Communications*, **46**, 5569–5571; (b) Simon, R.C., Richter, N., Busto, E. and Kroutil, W. (2014) *ACS Catalysis*, **4**, 129–143.
9. Koszelewski, D., Lavandera, I., Clay, D. *et al.* (2008) *Advanced Synthesis & Catalysis*, **350**, 2761–2766.
10. Gomm, A., Lewis, W., Green, A.P. and O'Reilly, E. (2016) *Chemistry: A European Journal*, **22**, 12 692–12 695.

11. (a) Rehn, G., Adlercreutz, P. and Grey, C. (2014) *Journal of Biotechnology*, **179**, 50–55; (b) Börner, T., Rehn, G., Grey, C. and Adlercreutz, P. (2015) *Organic Process Research & Development*, **19**, 793–799; (c) Rehn, G., Ayres, B., Adlercreutz, P. and Grey, C. (2016) *Journal of Molecular Catalysis B: Enzymatic*, **123**, 1–7; (d) Satyawali, Y., Ehimen, E., Cauwenberghs, L. *et al.* (2017) *Biochemical Engineering Journal*, **117**, 97–104.
12. (a) Hülsewede, D., Tänzler, M., Süss, P. *et al.* (2018) *European Journal of Organic Chemistry*, **18**, 2130–2133; (b) Hülsewede, D., Dohm, J.-N., von Langermann, J. (2019) *Advanced Synthesis & Catalysis*, **361**, 2727–2733.
13. Hülsewede, D., Meyer, L.-E. and von Langermann, J. (2019) *Chemistry: A European Journal*, **25**, 4871–4884.

4.4 Enantioselective Synthesis of Industrially Relevant Amines Using an Immobilised ω-Transaminase

Elisabetta Parodi,[1] Oreste Piccolo[2] and Antonella Petri[1]

[1]*Chemistry and Industrial Chemistry Department, University of Pisa, Pisa, Italy*
[2]*Studio di Consulenza Scientifica SCSOP, Sirtori, Italy*

The use of ω-transaminases has been identified as a greener and more sustainable method for the preparation of enantiomerically pure amines, which are important precursors of biologically active compounds with different industrial applications [1–4]. Kinetic resolution of racemic amines and asymmetric synthesis starting from the corresponding carbonyl compounds are the two procedures carried out in the transamination reaction. Asymmetric synthesis is generally a preferable method as it offers the possibility of obtaining the product with almost quantitative yields. Recent studies have favoured the application of transaminases at the industrial level, so it is not surprising that they are frequently found amongst the enzymes designed for the large-scale synthesis of enantioenriched amines [1, 3]. In this context, immobilisation of transaminases has been developed in order to obtain a more stable biocatalyst and to simplify the reaction workup and purification of the product [5–7]. Successful applications of immobilised transaminases have been reported thanks to enhanced stability to temperature and in reuse experiments both in batch and in continuous-mode reactions.

We have recently reported the asymmetric transamination of the selected substrate 1-Boc-3-piperidone **1** using several commercially available immobilised transaminases (TA-IMBs), which facilitate the preparation of 3-amino-1-Boc-piperidine **2**. The reaction was carried out whilst evaluating the catalytic activity and the enantioselectivity under different experimental conditions [8].

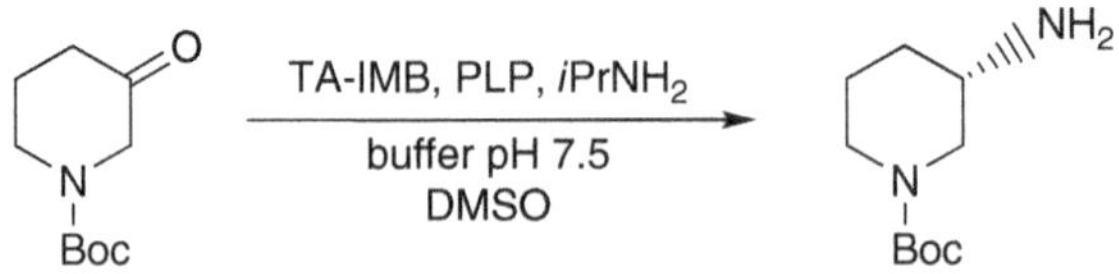

1

(*R*)–**2**
conv >99 %
ee >99 %

Scheme 4.5 *Synthesis of optically pure (R)-3-amino-1-Boc-piperidine* **2**.

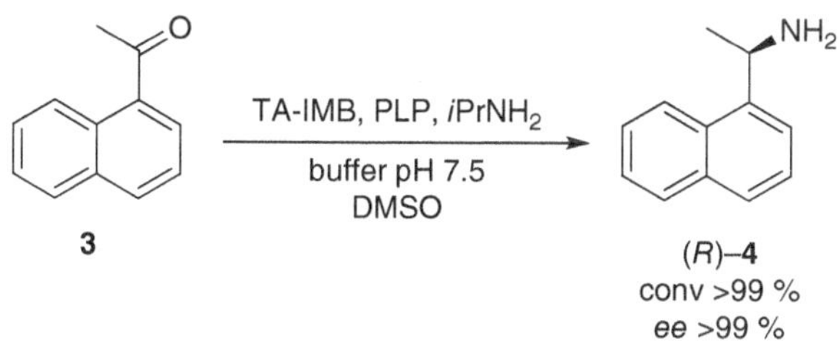

Scheme 4.6 *Synthesis of (R)-1-(1-naphthyl)ethylamine **4**.*

This procedure provides experimental details on the synthesis of the (*R*)-enantiomer of **2** (Scheme 4.5), which is a useful precursor of bioactive compounds [9]. A more efficient and lower-cost commercial TA-IMB enzyme was used.

The procedure was also applied to the synthesis of (*R*)-1-(1-naphthyl)ethylamine **4** (Scheme 4.6), the key chiral amine precursor of Cinacalcet, a calcimimetic agent [10].

4.4.1 Procedure 1: Enantioselective Synthesis of (*R*)-3-Amino-1-Boc-Piperidine 2

4.4.1.1 Materials and Equipment

- Immobilised transaminase enzyme ZAM 14151-1 (0.5 g; Enzymaster (Ningbo) Bio-Engineering Co. Ltd)
- Substrate **1** (0.5 g, 2.5 mmol) from commercial suppliers
- Triethanolamine buffer (25 mL, 100 mM, pH 7.5) containing isopropylamine (IPA; $iPrNH_2$, 1.1 M)
- Pyridoxal-5'-phosphate (PLP, 10.9 mg, 0.041 mmol)
- Dimethyl sulfoxide (DMSO, 7mL)
- Na_2SO_4, dry
- K_2CO_3
- Benzyl chloroformate
- CH_3CN
- Ethyl acetate
- 50 mL Falcon tubes
- ThermoMixer
- pH meter
- Glass-filter funnel with glass equipment
- Rotary evaporator
- Ultrasonic wave generator
- High-performance liquid chromatography (HPLC) system with UV detection

4.4.1.2 Procedure

1. Substrate **1** (0.5 g, 2.51 mmol) was dissolved in DMSO (3.5 mL) and preheated at 35 °C.
2. 100 mM triethanolamine buffer, pH 7.5 containing IPA (1.1 M) and PLP (1.3 mM) (25 mL) was added to a 50 mL Falcon tube.
3. TA-IMB (0.5 g) was added to the tube and the mixture was stirred in a ThermoMixer at 35 °C, 550 rpm for 10 min.

4. The reaction was started by addition of substrate solution and the resultant mixture was incubated at 50 °C, 550 rpm.
5. The progress of the reaction was monitored by HPLC. Samples (100 μL) were taken from the reaction mixture, adjusted to pH 13 by addition of 5 M NaOH (20 μL) and diluted with100 μL of mobile phase.
6. Upon complete conversion (~4 hr), the immobilised enzyme was filtered under vacuum through a glass-filter funnel with fritted disc (medium porosity) fitted with a Buchner flask and washed with buffer (3 × 5 mL).
7. The combined filtrate and washings were adjusted to pH 13 by addition of an aqueous solution of 5M NaOH and saturated with solid NaCl.
8. The aqueous layer was extracted with CH_2Cl_2 (3 × 15 mL) and the organic layers were combined and washed with saturated NaCl solution.
9. The organic portion was dried over Na_2SO_4 and filtered, and the solvent was removed by distillation under reduced pressure on a rotary evaporator to provide 458 mg (2.28 mmol, 91% yield) of amine **2** with >99% ee.
10. For determination of the enantiomeric excess (ee), a sample of (*R*)-**2** (15 mg) was added to an Eppendorf tube containing CH_3CN (300 μL), water (600 μL) and K_2CO_3 (15 mg). Benzyl chloroformate (15 μL) was slowly added dropwise and the solution was sonicated for 10 min. After the addition of ethyl acetate (500 μL), the solution was sonicated for a further 5 min and extracted with ethyl acetate (2 × 500 μL). The combined organic layer was dried over Na_2SO_4 and filtered. After evaporation of the solvent by distillation under reduced pressure, the residue was diluted with the mobile phase (485 μL).
11. The immobilised enzyme was reused for three subsequent reactions under the same experimental conditions, affording quantitative conversion of the substrate within 7 hr; the catalyst maintained high enantioselectivity (>99% ee) in all cases.

4.4.2 Procedure 2: Enantioselective Synthesis of (*R*)-1-(1-Naphthyl)ethylamine 4

Transamination of substrate **3** (0.1 g, 0.59 mmol) following the same procedure as for substrate **1** yielded the product with complete conversion after 24 hr (95 mg, 0.55 mmol, 93% yield), with >99% ee. This is the first example of preparation of this important intermediate employing an immobilised transaminase. (Transamination of **3** with free enzymes has recently been described [11].)

4.4.3 Analytical Method

4.4.3.1 Determination of Conversion

HPLC analyses were performed using a Kinetex EVO C18 column (150 × 4.6 mm). Elution was carried out at 1 mL.min^{-1} with detection at 220 nm and column temperature of 25 °C.

4.4.3.2 Amine 2

The eluent was water/acetonitrile (60 : 40 v/v). Retention times were 2.1 min for **2** and 3.5 min for **1**.

4.4.3.3 Amine 4

The eluent was water/acetonitrile. Samples were eluted with the following gradients: from 10 to 80% acetonitrile over 2 min; then from 80 to 90% acetonitrile over 4 min; and then from 90 to 10% acetonitrile over 1.5 min. Retention times were 5.8 min for **3** and 7.6 min for **4**.

4.4.3.4 Determination of Enantiomeric Excess

(*R*)-**2**: HPLC analyses were performed after derivatisation using a Lux-Cellulose 3 (250 × 4.6 mm) column with a flow rate of 0.5 mL.min^{-1}, detection at 216 nm and column temperature of 25 °C. The eluent was hexane/isopropyl alcohol (90 : 10 v/v). Racemic compound was used as reference. The absolute configuration was assigned by comparison of elution order of an authentic standard. Retention times were 14.2 min for the derivative of (*R*)-**2** and 20.8 min for the derivative of (*S*)-**2**.
(*R*)-**4**: HPLC analyses were performed without derivatisation using a Lux-Cellulose 1 (250 × 4.6 mm) column with a flow rate of 1 mL.min^{-1}, detection at 275 nm and column temperature of 25 °C. The eluent was hexane/isopropyl alcohol (90 : 10 v/v). Racemic compound was used as reference. Retention times were 14.2 min for (*S*)-**4** and 16.5 min for (*R*)-**4**.

4.4.3.5 Optical Rotation

(*R*)-**2**: $[\alpha]_D{}^{20}$ −31.93 c = 0.31, EtOH
(*R*)-**4**: $[\alpha]_D{}^{25}$ +47.7 c = 1, EtOH

4.4.3.6 NMR Spectroscopy

(*R*)-**2**: ^{1}H-NMR (400 MHz; $CDCl_3$) δ = 3.91 (bs, 1H), 3.81 (d, 1H), 2.81 (ddd, 1H), 2.78 (m, 1H), 2.57 (bs, 1H), 1.89 (m, 1H), 1.66 (m, 1H), 1.45 (m, 1H), 1.44 (s, 9H, (CH3)3), 1.23 (m, 1H). ^{13}C-NMR (101 MHz; $CDCl_3$) δ = 155, 79.4, 52.12, 47.6, 43.64, 33.96, 28.43, 23.74.
(*R*)-**4**: ^{1}H-NMR (400 MHz; $CDCl_3$) δ 8.15 (d, *J* = 7.5 Hz, 1H), 7.88 (d, *J* = 8.0 Hz, 1H), 7.76 (d, *J* = 8.2 Hz, 1H), 7.66 (d, *J* = 7.2 Hz, 1H), 7.57–7.45 (m, 3H), 4.97 (q, *J* = 6.6 Hz, 1H), 1.71 (bs, 3H), 1.56 (d, *J* = 6.6 Hz, 3H). ^{13}C-NMR (101 MHz; $CDCl_3$) δ 143.44, 134.00, 130.83, 129.08, 127.34, 126.06, 125.74, 125.53, 123.01, 121.47, 46.61, 24.94.

4.4.4 Conclusion

By performing the asymmetric synthesis with a TA-IMB, it was possible to obtain the amine (*R*)-**2** in quantitative yield and with 99% ee. The use of a highly efficient enzyme allowed the complete conversion of the substrate in 4 hr. Under these experimental conditions, the degradation of the substrate due to the low solubility of the substrate in aqueous media and the long reaction times previously observed by ourselves [8] and others [12] were avoided. The use of immobilised enzyme allows simple separation of the product and easy catalyst recycling. This method may be generally applicable and transferable to other substrates of industrial interest.

References

1. Kelly, S.A., Pohle, S., Wharry, S. *et al.* (2018) *Chemical Reviews*, **118**, 349–367.
2. Guo, F. and Berglund, P. (2017) *Green Chemistry*, **19**, 333–360.
3. Fuchs, M., Farnberger, J.E. and Kroutil, W. (2015) *European Journal of Organic Chemistry*, **2015** (32), 6965–6982.
4. Koszelewski, D., Tauber, K., Faber, K. and Kroutil, W. (2010) *Trends in Biotechnology*, **28**, 324–332.
5. Molnar, Z., Farkas, E., Lakó, A. *et al.* (2019) *Catalysts*, **9**, 438.
6. Mallin, H., Höhne, M. and Bornscheuer, U.T. (2014) *Journal of Biotechnology*, **191**, 32.
7. Päiviö, M. and Kanerva, L.T. (2013) *Process Biochemistry*, **48**, 1488–1494.
8. Petri, A., Colonna, V. and Piccolo, O. (2019) *Beilstein Journal of Organic Chemistry*, **15**, 60–66.
9. (a) Feng, J., Zhang, Z., Wallace, M.B. *et al.* (2007) *Journal of Medicinal Chemistry*, **50**, 2297–2300; (b) Eckhardt, M., Langkopf, E., Mark, M. *et al.* (2007) *Journal of Medicinal Chemistry*, **50**, 6450–6453.
10. Nemeth, E.F., Van Wagenen, B.C. and Balandrin, M.F. (1993) Patent WO9304373.
11. (a) Marx, L., Ríos-Lombardía, N., Farnberger, J.F. *et al.* (2018) *Advanced Synthesis & Catalysis*, **360**, 2157–2165; (b) Chen, H., Bong, Y.K., Wang, J. *et al.* (2019) Patent WO2109/128894 A1.
12. Gundersen, M.T., Tufvesson, P., Rackham, E.J. *et al.* (2016) *Organic Process Research & Development*, **20**, 602–608.

4.5 Amination of Sugars Using Transaminases

Fabiana Subrizi,[1] John M. Ward[2] and Helen C. Hailes*[1]

[1]*Department of Chemistry, Christopher Ingold Laboratories, University College London, London, UK*

[2]*Department of Biochemical Engineering, University College London, London, UK*

Carbohydrates, a major constituent of biomass, have attracted attention as a sustainable source of building blocks for chemicals, materials and biofuels due to their low cost, ready availability and stereochemical diversity [1]. Aminated carbohydrates, such as aminosugars and polyhydroxylated amines, are important carbohydrate mimetics for pharmaceutical applications [2] and can also be used as monomers or crosslinkers in biopolymers [2a, 2b, 3]. Transaminase enzymes represent a sustainable alternative to traditional synthetic methods for the amination of carbohydrates, which normally involves several steps in order to achieve regioselective control [2c] and impacts on the atom economy of any process [2a, 2b, 4]. Here, we show a sustainable biocatalytic approach to the upgrade of carbohydrates to value-added nitrogen-containing compounds such as aminopolyols. The direct transamination of a reducing sugar, either an aldose (Scheme 4.7) or a ketose (Scheme 4.8), with transaminases gave access to chiral acyclic aminopolyols from sugars such as D-xylose and D-fructose in high yields. Notably, the reaction of the ketose D-fructose was performed at a preparative enzymatic scale to give a chiral amine with high stereoselectivity, and the use of an enantiocomplementary transaminase can furthermore provide access to a diastereomeric aminosugar.

4.5.1 Procedure 1: Preparation of *Rh*-TA and Amination of D-Xylose

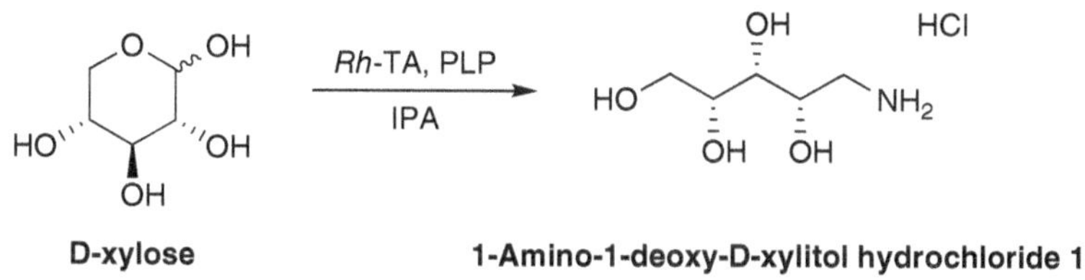

Scheme 4.7 Synthesis of 1-amino-1-deoxy-D-xylitol hydrochloride **1**.

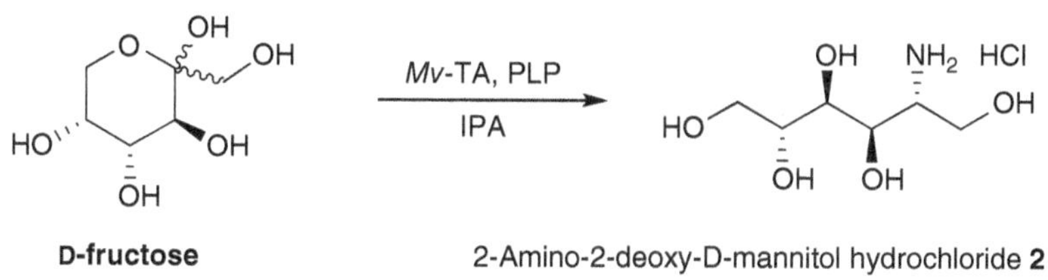

Scheme 4.8 Synthesis of 2-amino-2-deoxy-D-mannitol hydrochloride **2**.

4.5.1.1 Materials and Equipment

- Glycerol stock of *Escherichia coli* BL21(DE3)/*Rh*-TA [5]
- 2xTY (Tryptone yeast extract) broth (16 g.L^{-1} tryptone, 10 g.L^{-1} yeast extract, 5 g.L^{-1} NaCl)
- Kanamycin
- Isopropyl-β-D-1-thiogalactopyranoside (IPTG)
- 0.1 M phosphate buffer
- Pyridoxal 5'-phosphate (PLP)
- Lysis buffer (0.1 M potassium phosphate buffer, pH 8.0, containing PLP 0.5 mM)
- Glycerol
- D-Xylose
- Isopropylamine (IPA)
- Methanol (MeOH)
- 1 M aqueous HCl
- 1 M aqueous NaOH
- MilliQ water
- 28% aq NH_3
- Soniprep 150 sonicator (MSE Sanyo)
- Dowex 50WX8 ion exchange resin

4.5.1.2 Procedure

4.5.1.2.1 Culture and Expression.

1. An aliquot from a frozen glycerol stock of *E. coli* BL21(DE3)/*Rh*-TA [7] was inoculated into 10 mL of 2xTY medium (containing 50 μg.mL^{-1} kanamycin). Starter cultures were incubated overnight at 37 °C, with shaking at 250 rpm.

2. A 1 mL sample of overnight starter cultures was added to 100 mL of fresh 2xTY medium (containing 50 µg.mL^{-1} kanamycin). The expression cultures were incubated at 37 °C for 2 hr, whilst shaking at 250 rpm.
3. Expression was induced by the addition of 100 µL of 1 M IPTG. Cultures were incubated at 30 °C for a further 16 hr prior to harvesting, whilst shaking at 250 rpm.
4. Cells were harvested by centrifugation at 4 °C, 10 000 rpm for 10 min. Supernatant was removed and the cells were stored at −20 °C until lysis.
5. The cell pellet was thawed and suspended in lysis buffer (4 mL).
6. The resuspended protein was sonicated on ice using 10 cycles of 10 sec on, 10 sec off at 10 W.
7. The insoluble portion of the lysate was pelleted by centrifugation at 4 °C, 10 000 rpm for 15 min. The supernatant was removed and divided into 0.5 mL samples. The clarified lysate was stored at −20 °C. The total protein concentration was determined using the Bradford method and was typically 18–20 mg.mL^{-1}.

4.5.1.2.2 Amination of D-Xylose.

1. The reaction mixture was prepared by the addition of the sugar substrate D-xylose (112.5 mg, 0.75 mmol, 25 mM) IPA (7.5 mL, 1 M in 0.1 M phosphate buffer, pH 8.0 containing PLP 0.5 mM) and clarified cell lysate of *Rh*-TA (2 mL) to 50 mM phosphate buffer, pH 8.0 containing PLP 2 mM (20.5 mL). The reaction was incubated at 30 °C and 250 rpm.
2. After 4 hr, another aliquot of clarified cell lysate of *Rh*-TA (2 mL) was added, together with PLP (5 mg).
3. The reaction was incubated at 30 °C for 48 hr before being quenched with MeOH (20 mL) and centrifuged at 4000 rpm for 10 min to remove the protein precipitate.
4. The supernatant was basified to pH 12 with 1 M NaOH(aq) and concentrated by distillation under reduced pressure to remove excess IPA, leaving a yellow residue.
5. The yellow residue was dissolved in milliQ water (50 mL) and the pH was adjusted to 8.0 with 1 M NaOH(aq). The solution was loaded on to a prewashed Dowex 50WX8 ion-exchange resin (which had previously been rinsed thoroughly with milliQ water, until the washing water became colourless) and washed several times with water (100 mL) to remove excess salt. The amine product was then eluted with 28% aq NH_3 (60 mL), and the eluent was concentrated by distillation under reduced pressure. The off-white residue obtained was resuspended in 1 M HCl(aq) and concentrated by distillation under reduced pressure to dryness to give 1-amino-1-deoxy-D-xylitol hydrochloride **1** (117.5 mg, 79%; 95% purity by ^{1}H NMR – traces of NH_4Cl present). A batch of material was repurified (to remove any traces of NH_4Cl) with the Dowex resin to give the hydrochloric salt for further NMR analysis.

^{1}H NMR (700 MHz; D_2O) δ = 4.09 (dt, J = 9.3, 3.5, 1H, C*H*CH$_2$N), 3.88 (dd, J = 11.9, 2.9, 1H, C*H*CH$_2$O), 3.78 (dd, J = 11.9, 4.3, 1H, C*H*HOH), 3.73–3.69 (m, 2H, CH*H*OH and CHC*H*CH), 3.29 (dd, J = 13.1, 3.2, 1H, CH*H*N), 3.21 (dd, J = 13.1, 9.3, 1H, CH*H*N); ^{13}C NMR (176 MHz; D_2O with MeOH standard) δ = 71.6 (*C*HCH$_2$OH and CH*C*HCH), 67.9 (*C*HCH$_2$N), 62.5 (*C*H$_2$OH), 42.1 (*C*H$_2$N) ppm; Mp 110 ± 2 °C (with decomposition), HRMS (ESI+) calcd for $C_5H_{14}NO_4$ $[M+H]^+$ 152.0917, found 152.0914.

4.5.2 Procedure 2: Amination of D-Fructose

4.5.2.1 Materials and Equipment

- Glycerol stock of *E. coli* BL21(DE3)/*Mv*-TA [6]
- 2xTY broth (16 g.L^{-1} tryptone, 10 g.L^{-1} yeast extract, 5 g.L^{-1} NaCl)
- Kanamycin
- IPTG
- 0.1 M phosphate buffer
- PLP
- Lysis buffer (0.1 M potassium phosphate buffer, pH 8.0, containing PLP 0.5 mM)
- Glycerol
- D-fructose
- IPA
- MeOH
- 1 M Aqueous HCl
- 1 M Aqueous NaOH
- MilliQ water
- 28% aq NH_3
- Soniprep 150 sonicator (MSE Sanyo)
- Dowex 50WX8 ion-exchange resin

4.5.2.2 Procedure

4.5.2.2.1 Preparation of Mv-TA. See Procedure 1 for details.

1. The reaction mixture was prepared by the addition of the sugar substrate D-fructose (25 mM, 0.75 mmol), IPA (7.5 mL, 1 M in 0.1 M phosphate buffer, pH 8.0 containing PLP 0.5 mM) and clarified cell lysate of *Mv*-TA (3.6 mL) to 50 mM phosphate buffer, pH 8.0 containing PLP 2 mM (in 19 mL).
2. The reaction was incubated at 45 °C for 48 hr before being quenched with MeOH (20 mL) and centrifuged at 4000 rpm for 10 min to remove the protein precipitate.
3. The supernatant was basified to pH 12 with 1 M NaOH(aq) and concentrated under reduced pressure to remove excess IPA.
4. Product **2** was purified using a Dowex 50WX8 ion-exchange resin loaded on to an empty cartridge (5 cm high, ø 2 cm). The resin was initially rinsed thoroughly with milliQ water until the washing water became colourless.
5. The yellow residue remaining after removal of IPA under reduced pressure was redissolved in water (50 mL) and the pH was adjusted to 8.0. The solution was loaded on-to the prewashed Dowex 50WX8 ion-exchange resin and washed several times with water (100 mL) to remove excess salt. The amine product was then eluted with 28% aq NH_3 (60 mL) and the eluent was concentrated under reduced pressure. The off-white residue obtained was resuspended in 1 M HCl(aq) and concentrated to dryness to give the 2-amino-2-deoxy D-mannitol hydrochloride **2** as a single isomer (67 mg, 40%), with spectroscopic data consistent with that previously reported [7].

Table 4.3 *Further substrates.*

Sugar substrate	TA	Aminated product	Yield	Procedure
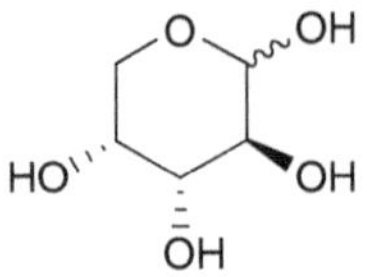	*Rh*-TA[5]	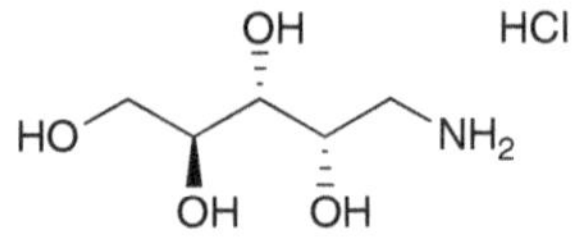 1-Amino-1-deoxy-L-arabinitol·HCl **3**	42 %	**1**
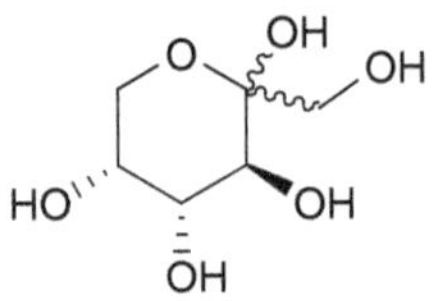	pQR2191[8]	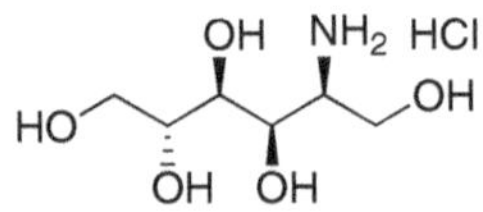 2-amino-2-deoxy-D-glucitol·HCl *epi*-**2**	21 %	**2**

^{1}H NMR (600 MHz; D_2O) δ = 4.21 (dd, *J* = 5.0, 1.8, 1H, C*H*CHN), 4.01 (dd, *J* = 12.3, 4.5, 1H, CH*H*CHN), 3.86 (dd, *J* = 12.3, 3.0, 1H, CH*H*CHN), 3.84 (dd, *J* = 12.0, 1.8, 1H, CH*H*CHOH), 3.79 (ddd, *J* = 8.5, 6.0, 3.0, 1H, C*H*CH_2OH), 3.69 (dd, *J* = 12.0, 6.0, 1H, CH*H*CHOH), 3.68 (dd, *J* = 8.5, 1.8, 1H, C*H*CHCH_2OH), 3.64–3.54 (m, 1H, C*H*N); ^{13}C NMR (151 MHz; D_2O with MeOH standard) δ = 71.0 (*C*HCH_2OH), 70.6 (*C*HCHCH_2OH), 66.9 (*C*HCHN), 63.0 (*C*H_2CHOH), 58.7 (*C*H_2CHN), 56.1 (*C*HN) ppm; HRMS (ES+) calcd for $C_6H_{16}NO_5$ $[M\text{-}Cl]^+$ 182.1028, found 182.1027.

4.5.3 Conclusion

A general methodology for the direct biocatalytic amination of carbohydrates has been developed using transaminases, which provides access to valuable acyclic aminopolyols in high yield in a single step. This methodology involves the use of enzyme-clarified lysates instead of purified proteins, which represents an advantage in terms of costs and potential industrialisation of the process. Moreover, purification with an ion-exchange resin also avoids tedious and time-consuming chromatographic methods. The methods shown here can be used for the facile preparation of different aminopolyols from many different carbohydrates; two further substrates are highlighted in Table 4.3 [9].

4.5.4 Acknowledgements

We gratefully acknowledge the UK Engineering and Physical Sciences Research Council (EPSRC) for providing financial support for this work to F.S. (EP/K014897/1) as part of its Sustainable Chemical Feedstocks programme. Input and advice from the project Industrial Advisory Board is also acknowledged. Furthermore, we gratefully acknowledge the UCL Mass Spectrometry and NMR Facilities in the Department of Chemistry UCL.

References

1. (a) Cárdenas-Fernández, M., Bawn, M., Hamley-Bennett, C. *et al.* (2017) *Faraday Discussions*, **202**, 415–431; (b) Werpy, T., Petersen, G., Aden, A. *et al.* (2014) *Top Value Added Chemicals from Biomass*, Vol. **I**, National Renewable Energy Lab.
2. (a) Ávalos, M., Babiano, R., Cintas, P. *et al.* (2008) *Chemistry: A European Journal*, **14**, 5656–5669; (b) Gómez, R.V. and Varela, O. (2009) *Macromolecules*, **42**, 8112–8117; (c) Ishikawa, T., Kudo, T., Shigemori, K. and Saito, S. (2000) *Journal of the American Chemical Society*, **122**, 7633–7637.
3. Sanchez-Vazquez, S.A., Hailes, H.C. and Evans, J.R.G. (2013) *Polymer Reviews*, **53**, 627–694.
4. Dangerfield, E.M., Plunkett, C.H., Win-Mason, A.L. *et al.* (2010) *Journal of Organic Chemistry*, **75**, 5470–5477.
5. Villegas-Torres, M.F., Martinez-Torres, R.J., Cazeres-Korner, A. *et al.* (2015) *Enzyme and Microbial Technology*, **81**, 23–30.
6. Höhne, M., Schätzle, S., Jochens, H. *et al.* (2010) *Nature Chemical Biology*, **6**, 807–813.
7. Wu, Y., Arciola, J. and Horenstein, N. (2014) *Protein & Peptide Letters*, **21**, 10–14.
8. Leipold, L., Dobrijevic, D., Jeffries, J.W.E. *et al.* (2019) *Green Chemistry*, **21**, 75–86.
9. Subrizi, F., Benhamou, L., Ward, J.M. *et al.* (2019) *Angewandte Chemie International Edition*, **58**, 3854–3858.

4.6 Converting Aldoses into Valuable ω-Amino Alcohols Using Amine Transaminases

Ryan Cairns,[1] Andrew Gomm,[1] James Ryan,[1] Thomas Clarke,[1] Evelina Kulcinskaja,[1] Kevin Butler[1] and Elaine O'Reilly[1,2]
[1]*School of Chemistry, University of Nottingham, Nottingham, Nottinghamshire, UK*
[2]*School of Chemistry, University College Dublin, Dublin, Ireland*

Amine transaminases (ATAs) are capable of converting aldehydes and prochiral ketones to the corresponding (chiral) amines in the presence of a pyridoxal 5'-phosphate (PLP) co-enzyme and sacrificial amine donor, and their use has been well established [1]. ATAs are one of a growing collection of enzymes that can now be considered alongside more traditional catalysts when designing synthetic routes to target molecules [2]. Current methodologies used to synthesise (agro)chemicals and pharmaceuticals that contain chiral amine functionality often rely on fossil fuel-derived chemicals and typically require the use of organic solvents, metal catalysts and high temperatures/pressures. There is an increasing need to develop low-cost and sustainable methods for the production of chiral building blocks [3]. The possibility of exploiting carbohydrates as an abundant feedstock for the production of valuable chiral amino alcohols/aminopolyols via direct amination of aldehyde and ketone functionality is an attractive prospect. Monosaccharides have largely been overlooked as substrates for ATAs, but recent publications have demonstrated their potential for the biocatalytic synthesis of chiral aminopolyols [4].

This methodology relies on the direct biocatalytic amination of aldoses, which exist at equilibrium as a mixture of their cyclic and open-chain forms (typically less than 1%), and allows access to a range of chiral ω-amino alcohols on a preparative scale.

4.6.1 Procedure: Preparative-Scale Conversion of D-Deoxyribose to (2*R*, 3*S*)-5-Aminopentane-1,2,3-Triol Using ATA-256

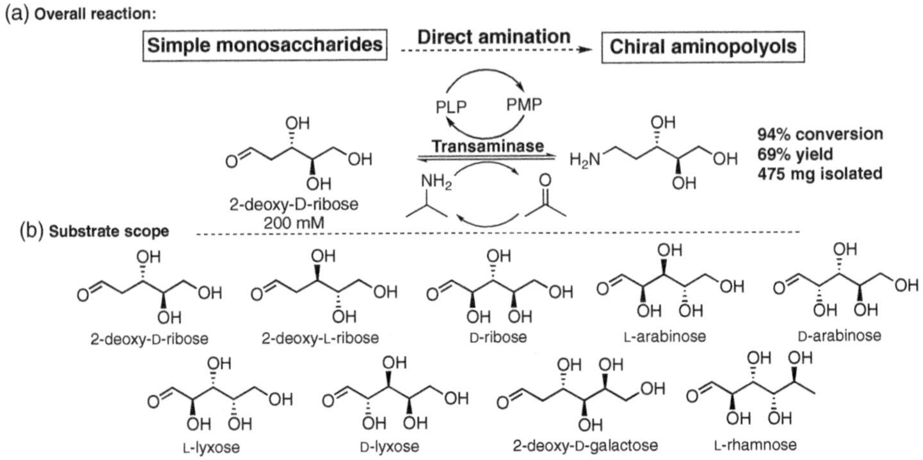

Scheme 4.9 (a) Overall transformation and conversion of D-deoxyribose to (2R,3S)-5-aminopentane-1,2,3-triol. (b) Panel of sugars that can be converted into the corresponding amino alcohols using ATAs.

4.6.1.1 Materials and Equipment

- (*S*)-Selective ATA-256 [5] from Codexis® (50 mg)
- Ammonium bicarbonate buffer (20 mL, 100 mM, pH 9.0)
- PLP hydrate
- Isopropylamine (IPA)
- 1M hydrochloric acid (aq)
- pH meter
- Conical centrifuge tube (Falcon, 50 mL)
- D-Deoxyribose
- Shaking incubator (Thermo Scientific MaxQ 8000)
- Benchtop centrifuge
- Maleic acid (standard for quantitative nuclear magnetic resonance, NMR)
- NMR tube (Wilmad 528-PP 500 MHz)
- NMR machine (500 MHz with cryoprobe)
- Deuterium oxide (D_2O)
- Round-bottom flask (250 mL)
- Rotary evaporator (Büchi)
- Methanol
- Sintered glass Büchner funnel
- Büchner flask
- Dowex 50WX8 hydrogen form
- Distilled water (dH_2O)
- Empty 25 g plastic flash column chromatography cartridge

- Glass pipette
- Chromatography pump (Asynt)
- Aqueous ammonium hydroxide (30%)
- TLC silica gel 60 F_{254} aluminium plates (Merck)
- Dichloromethane (DCM)
- $KMnO_4$ stain (1.0 g $KMnO_4$, 20.0 g K_2CO_3, 10.0 mL NaOH (5%), 150 mL H_2O)
- Ninhydrin stain (0.3 g ninhydrin, 100 mL *n*-butanol, 3 mL AcOH)

4.6.1.2 Procedure

1. Commercially available transaminase ATA-256 (50 mg) was rehydrated in ammonium bicarbonate buffer (20 mL, 100 mM, pH 9.0) containing PLP (1.0 mM) and IPA (800 mM) and the pH of the mixture was adjusted to 9.0 using 1 M aqueous HCl. This solution was prepared in a 50 mL conical centrifuge tube.
2. D-Deoxyribose (536 mg, 3.99 mmol) was added in a single portion to the rehydrated enzyme solution and the reaction was incubated on a shaking platform (200 rpm, 50 °C) for 48 hr, during which time the mixture generally turned from yellow to orange.
3. The solution was centrifuged at 15 000 rpm for 10 min and the conversion was analysed by NMR (see Scheme 4.9).
4. The contents of the reaction tube were transferred to a 250 mL round-bottom flask and concentrated by distillation under reduced pressure (20 mbar, 50 °C) to leave a dry residue.
5. Hot methanol (~50 °C, 3 × 20 mL) was added to the dry residue, in order to extract the starting material and product from it. After three extractions, the methanol was combined and filtered through a sintered glass Büchner funnel.
6. The methanol was removed *in vacuo* (340 mbar, 40 °C) to give a crude mixture of product and starting material, which was dissolved in dH_2O (10 mL).
7. Dowex 50WX8 hydrogen-form resin (20 g) was packed into an empty 25 g plastic flash column chromatography cartridge.
8. The resin was washed with dH_2O (100 mL) and the aqueous fractions were discarded.
9. The aqueous solution of product and starting material was added to the Dowex column using a glass pipette.
10. The Dowex column was washed with dH_2O (250 mL), using a chromatography pump to elute any starting material. The Dowex column was subsequently washed with 30% ammonium hydroxide solution (50 mL) and the chromatography pump was used to elute the desired product. TLC (DCM : methanol 7 : 3) followed by staining with KMnO4 or ninhydrin can be used to determine when the product has been completely eluted from the column, by spotting the eluent on silica and staining. Employing both staining solutions is useful because the starting sugar materials are active with $KMnO_4$ reagent but not with the ninhydrin stain, whereas the amino alcohol products are active with both.
11. The fractions containing product were combined, added to a round-bottom flask and evaporated (20 mbar, 50 °C).
12. HCl (1 M, 1 mL) was added to the dry residue in the round-bottom flask and the mixture was concentrated by distillation under reduced pressure (20 mbar, 50 °C) to give the (2*R*, 3*S*)-5-aminopentane-1,2,3-triol hydrochloride salt as a brown solid (475 mg, 69%

yield). ^{1}H NMR (400 MHz, D_2O) δ 3.83–3.69 (m, 2H), 3.67–3.50 (m, 2H), 3.28–3.06 (m, 2H), 2.07–1.95 (m, 1H), 1.87–1.73 (m, 1H); ^{13}C NMR (100 MHz, D2O) δ 74.3, 69.6, 62.3, 37.2, 29.3; LCMS (EI) m/z: calculated $C_5H_{14}NO_3^+$ $[M+H]^+$: 136.0968; found: 136.0971.

4.6.2 Analytical Method

Conversion was determined using NMR: a 500 μL aliquot of the crude reaction mixture was added to 50 μL of maleic acid (110 mM in D_2O) and analysed by NMR. (Water-supressed ^{1}H NMR spectra were recorded using a zgcppr pulse sequence on a Bruker AV(III)500 instrument fitted with a 5 mm autotunable dual ^{1}H/^{13}C (DCH) cryoprobe.) Data was collected with 64 k points with a sweep width of 20 ppm. Experiments were run with 64 scans using a relaxation delay of 10 sec and an acquisition time of 3.2 sec. Experiments were carried out at 298 K.

Post-NMR, manual phase correction was performed around the maleic acid peak, when required (automated baseline correction, Bernstein polynomial fit 'order 3'). The identifiable peak area of the aminoalcohol was compared to the maleic acid peak area and a conversion factor (×1.1) was applied to determine the concentration of aminoalcohol. The reactions were performed in triplicate and the average conversion and standard deviations were calculated.

4.6.3 Conclusion

This work demonstrates the first example of the biocatalytic conversion of cyclic aldoses to the corresponding ω-aminopolyols using ATAs. A selection of amino alcohols were synthesised on a preparative scale using commercially available ATA-256, allowing facile access to up to 475 mg of amino alcohol product (Table 4.4). This clearly demonstrates

Table 4.4 *Conversion of monosaccharides (100 or 200 mM) [4] to the corresponding amino alcohol products using ATA-256 and IPA (4 equiv.). Conversions were calculated by NMR after 48 hr at 50 °C and isolated yields were recorded after purification on Dowex® resin.*

Substrate	Conc. mM	Conv. %	Yield (%)	Isolated mg
2-deoxy-D-ribose	200	94	69	475
2-deoxy-L-ribose	200	94	69	474
D-ribose	100	62	47	176
L-arabinose	100	35	28	104
D-arabinose	100	75	38	144
2-deoxy-D-galactose	100	30	22	88
L-lyxose	200	24	16[a]	121
D-lyxose	200	21	11[a]	86
L-rhamnose	100	48	33	131

[a] After purification, compounds were still contaminated with IPA.HCl salt.

the scalability of the approach and its potential to become a widely used route for the synthesis of such amino alcohols starting from simple monosaccharides. The strategy has been successfully employed for the preparation of a range of ω-aminopolyols (Table 4.4).

References

1. (a) Gomm, A. and O'Reilly, E. (2018) *Current Opinion in Chemical Biology*, **43**, 106–112; (b) Guo, F. and Berglund, P. (2017) *Green Chemistry*, **19**, 333–360; (c) Slabu, I., Galman, J.L., Lloyd, R.C. and Turner, N.J. (2017) *ACS Catalysis*, **7**, 8263–8284.
2. (a) Turner, N.J. and O'Reilly, E. (2013) *Nature Chemical Biology*, **9**, 285–288; (b) de Souza, R.O.M.A., Miranda, L.S.M. and Bornscheuer, U.T. (2017) *Chemistry: A European Journal*, **23**, 12 040–12 063; (c) Hönig, M., Sondermann, P., Turner, N.J. and Carreira, E.M. (2017) *Angewandte Chemie International Edition*, **56**, 8942–8973; (d) Green, A.P. and Turner, N.J. (2016) *Perspectives on Science*, **9**, 42–48.
3. (a) Cherubini, F. (2010) *Energy Conversion and Management*, **51**, 1412–1421; (b) Chao-Jun, L. and Trost, B. (2008) *PNAS*, **105** (36), 13 197–13 202; (c) Lipschutz, B. and Ghorai, S. (2014) *Green Chemistry*, **16** (8), 3660–3679.
4. (a) Cairns, R., Gomm, A., Ryan, J. *et al.* (2019) *ACS Catalysis*, **9**, 1220–1223; (b) Subrizi, F., Benhamou, L., Ward, J. *et al.* (2019) *Angewandte Chemie International Edition*, **58**, 3854–3858.
5. Crowe, M., Foulkes, M., Francese, G. *et al.* (2013) Patent WO2013139987.

5

Other Carbon–Nitrogen Bond-Forming Biotransformations

5.1 Biocatalytic *N*-Acylation of Anilines in Aqueous Media

Anna Żądło-Dobrowolska[1] and Wolfgang Kroutil[2,3]
[1]*Institute of Organic Chemistry, Polish Academy of Sciences, Warsaw, Poland*
[2]*Institute of Chemistry, University of Graz, NAWI Graz, BioTechMed Graz, Graz, Austria*
[3]*Austrian Centre of Industrial Biotechnology, acib GmbH, Graz, Austria*

The development of efficient and environmentally friendly methods for the synthesis of amides is of significance because they constitute the key functional group in a wide variety of biological and synthetic structures, including peptides, pharmaceuticals, polymers and pesticides [1]. Generally, the chemical synthesis of amides is performed in organic solvents and requires acyl halides and harsh conditions. To fulfil the increasing demand for green chemistry procedures (minimising the use of organic solvents), enzymes have been investigated for C-N bond formation activity in water [2]. The application of enzymes offers the advantages of high selectivity, mild reaction conditions and catalyst biodegradability. For instance, C-acyltransferase from *Pseudomonas protegens* has been identified not only as catalysing C-C bond formation in a Friedel–Crafts-type acylation [3] but also as exhibiting promiscuous *N*-acetylation activity, transforming aniline derivatives into the corresponding acetanilides (Scheme 5.1) [4].

Semipreparative transformations were performed starting with 0.25 mmol of substrate. Anilines were converted into the corresponding acetanilides using phenyl acetate as acetyl donor within 24 hr and the products were isolated by column chromatography with high yields. Whilst aniline derivatives bearing an OH group in the *para*- or *ortho*-position were not converted, *meta*-substituted hydroxyaniline led to 96% product yield **2a**. Aniline, including anilines with various substituents in the *para*-position (Cl, *i*-Pr, Et), was well accepted, reaching yields up to 99%. Furthermore, anilines with an ethyl group in either the *meta* or the *ortho*-position were accepted, with the *para*-substituted substrate reacting fastest (**2e**; 99%), followed by the *meta*-substituted one (**2f**; 98%) [4].

Applied Biocatalysis: The Chemist's Enzyme Toolbox, First Edition. Edited by John Whittall and Peter W. Sutton.
© 2021 John Wiley & Sons Ltd. Published 2021 by John Wiley & Sons Ltd.

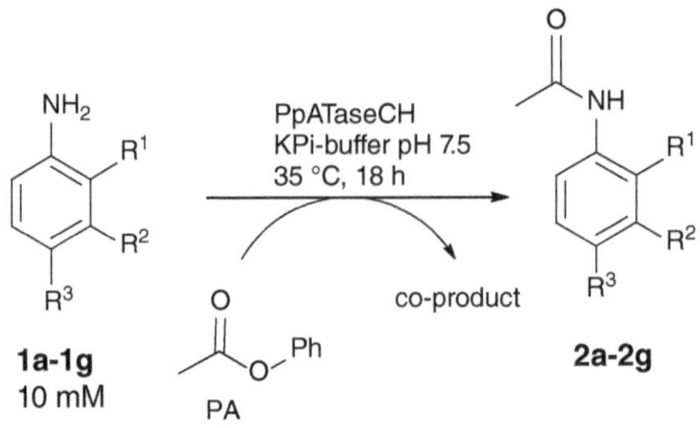

Product	R_1	R_2	R_3	Yield [%]
2a	H	OH	H	96
2b	H	H	H	70
2c	H	H	Cl	91
2d	H	H	*i*-Pr	83
2e	H	H	Et	99
2f	H	Et	H	98
2g	Et	H	H	13

Scheme 5.1 *Conversion of aniline derivatives to the corresponding acetanilides.*

5.1.1 Procedure 1: Biocatalytic Conversion of Aniline to Acetanilide

5.1.1.1 Materials and Equipment

- 3-Aminophenol (21.8 mg, 0.2 mmol, Sigma-Aldrich #100242)
- Potassium phosphate buffer (KPi-buffer; 100 mM, pH 7.5)
- Cell-free extract containing the *Pp*ATaseCH (see Section 6.4)
- Phenyl acetate (54.9 µL, Sigma-Aldrich #108723)
- Ethyl acetate (EtOAc)
- Brine (NaCl-saturated solution)
- Anhydrous Na_2SO_4
- *c*-Hexane
- Silica gel
- TLC silica gel 60 F_{254} aluminium plates (Merck)
- 50 mL Falcon tube
- 100 mL separating funnel
- Orbital shaker (InforsHT Multitron 2 Standard)
- Cooling centrifuge
- Rotary evaporator (Büchi)
- Potassium permanganate staining solution (1.5 g of $KMnO_4$, 10 g of K_2CO_3 and 1.25 mL of 10% NaOH in 200 mL of water)
- Cerium sulfate staining solution (10% w/v solution of cerium (IV) sulfate in 15% sulfuric acid)

5.1.1.2 Procedure

1. 3-Aminophenol (21.8 mg, 0.2 mmol, 10 mM final concentration) was dissolved in 100 mM KPi-buffer, pH 7.5 (20 mL) in a 50 mL Falcon tube.

2. Cell-free extract containing the *Pp*ATaseCH (2.5 mL, 1.65 U) was added to the reaction mixture and the bioacetylation was started by adding phenyl acetate (54.9 μL, 15 mM final concentration).
3. The mixture was further diluted with KPi-buffer (100 mM, pH 7.5) to a total volume of 25 mL.
4. The bioacetylation was run at 35 °C and 140 rpm for 24 hr.
5. The resulting suspension was extracted with EtOAc (2 × 50 mL). The organic layers were pooled in a separation funnel, washed with brine (2 × 40 mL) and dried over anhydrous Na_2SO_4, and the solvent was removed under reduced pressure.
6. The crude product was purified by flash chromatography over silica gel using *c*-hexane/EtOAc as eluent (9 : 1 v/v). Fractions were collected and checked for the desired substances via TLC. *c*-Hexane/EtOAc (9 : 1 v/v) was used as TLC solvent and staining was performed with cerium or permanganate staining solution. Fractions containing the desired product were combined and evaporated. *N*-(3-hydroxyphenyl)acetamide was obtained as a white solid (29.1 mg) with 96% yield.

^{1}H-NMR (300 MHz, acetone-*d6*): δ (ppm) = 2.02–2.09 (m, 3H), 6.53 (ddd, J_1 = 1.0 Hz, J_2 = 2.4 Hz, J_3 = 8.0 Hz, 1H), 6.92–7.02 (m, 1H), 7.08 (t, J = 8.0 Hz, 2H, Ar), 7.39 (t, J = 2.0 Hz, 1H), 8.38 (s, 1H, OH), 9.1 (s, 1H, NH); ^{13}C-NMR (75 MHz, acetone-*d6*): δC [ppm] = 23.4, 106.3, 110.2, 129.3, 140.7, 157.8, 168.0.

5.1.2 Conclusion

The described procedure enables *N*-acylation of aniline derivatives in aqueous reaction medium. Enzyme transformed various substituted anilines possessing substituents at the C2, C3 or C4 position.

References

1. (a) Pitman, I.H. (1981) *Medicinal Research Reviews*, **1**, 189–214; (b) Yu, C. and Mosbach, K. (1997) *Journal of Organic Chemistry*, **62**, 4057–4064; (c) Carey, J.S., Laffan, D., Thomson, C. and Williams, M.T. (2006) *Organic and Biomolecular Chemistry*, **4**, 2337; (d) Pitzer, J. and Steiner, K. (2016) *Journal of Biotechnology*, **235**, 32–46; (e) Żądło-Dobrowolska, A., Kłossowski, S., Koszelewski, D. *et al.* (2016) *Chemistry: A European Journal*, **22**, 16 684–16 689.
2. (a) Guranda, D.T., van Langen, L.M., van Rantwijk, F. *et al.* (2001) *Tetrahedron: Asymmetry*, **12**, 1645–1650; (b) Ulijn, R.V., Baragaña, B., Halling, P.J. and Flitsch, S.L. (2002) *Journal of the American Chemical Society*, **124**, 10 988–10 989; (c) Kuo, C.H., Lin, J.A., Chien, C.M. *et al.* (2016) *Journal of Molecular Catalysis B: Enzymatic*, **129**, 15–20; (d) Toplak, A., Nuijens, T., Quaedflieg, P.J. *et al.* (2016) *Advanced Synthesis & Catalysis*, **358**, 2140–2147; (e) Land, H., Hendil-Forssell, P., Martinelle, M. and Berglund, P. (2016) *Catalysis Science & Technology*, **6**, 2897–2900; (f) Kazemi, M., Sheng, X., Kroutil, W. and Himo, F. (2018) *ACS Catalysis*, **8**, 10 698–10 706.
3. (a) Schmidt, N.G., Pavkov-Keller, T., Richter, N. *et al.* (2017) *Angewandte Chemie International Edition*, **56**, 7615–7619; (b) Schmidt, N.G. and Kroutil, W. (2017) *European Journal of Organic Chemistry*, **39**, 5865–5871; (c) Schmidt, N.G., Żądło-Dobrowolska, A., Ruppert, V. *et al.* (2018) *Applied Microbiology and Biotechnology*, **102**, 6057–6068.
4. Żądło-Dobrowolska, A., Schmidt, N.G. and Kroutil, W. (2018) *ChemComm*, **54**, 3387–3390.

5.2 Enantioselective Enzymatic Hydroaminations for the Production of Functionalised Aspartic Acids

Haigen Fu, Jielin Zhang and Gerrit J. Poelarends*
Department of Chemical and Pharmaceutical Biology, Groningen Research Institute of Pharmacy, University of Groningen, Groningen, The Netherlands

Optically pure functionalised aspartic acid derivatives are highly valuable as tools for biological research and as chiral building blocks for pharmaceuticals, nutraceuticals and agrochemicals [1, 4]. Despite their broad applications, the direct asymmetric synthesis of functionalised aspartic acids remains a challenge. Current chemical strategies for the synthesis of enantioenriched *N*-alkyl-substituted aspartic acids are mainly based on the aza-Michael addition of appropriate amines to maleic acid, fumaric acid or their derivatives (ester, amide or monoalkali salt) [5, 6], followed by laborious purification and resolution to achieve the desired single enantiomer, and therefore lead to unsatisfactory product yields (<50%).

Biocatalysis provides a valuable alternative method for the production of chiral unnatural amino acids [7, 8]. We recently reported a biocatalytic route for the asymmetric synthesis of various *N*-(hetero)cycloalkyl-substituted L-aspartic acids using ethylenediamine-*N,N*'-disuccinic acid (EDDS) lyase as biocatalyst [9]. EDDS lyase is a homotetrameric protein that belongs to the aspartase/fumarase superfamily and exploits a deamination mechanism involving general base-catalysed formation of a carbanion stabilised as its *aci*-carboxylate form [10]. The enzyme naturally catalyses a reversible sequential two-step deamination of (*S,S*)-EDDS **1**, converting **1** via the intermediate (*S*)-*N*-(2-aminoethyl)aspartic acid (AEAA) **3** into ethylenediamine **4** and two molecules of fumaric acid **2** (Scheme 5.2a) [10]. In this section, we describe the expression and purification of the enzyme EDDS lyase, as well as its synthetic application in the enantioselective addition of heterocycloalkylamines **5a,b** to fumaric acid **2** to produce *N*-heterocycloalkyl-substituted L-aspartic acid derivatives **6a,b** (Scheme 5.2b).

5.2.1 Procedure 1: Expression and purification of EDDS lyase

5.2.1.1 Materials and Equipment

- Lysogenic broth (LB) agar plate with colonies of *Escherichia coli* TOP10 cells containing the pBADN (EDDS lyase-His) expression vector [10]
- Ampicillin (100 mg.mL^{-1} stock solution in water, filter-sterilised)
- Arabinose (200 mg.mL^{-1} stock solution in water, filter-sterilised)
- LB medium (1 L, sterilised)
- Sodium dodecyl sulfate (SDS) gels containing polyacrylamide (4–12%)
- Lysis buffer (50 mM Tris-HCl, 300 mM NaCl, 20 mM imidazole, pH 8.0), 200 mL
- Buffer A (50 mM Tris-HCl, 300 mM NaCl, 30 mM imidazole, pH 8.0), 200 mL
- Buffer B (50 mM Tris-HCl, 300 mM NaCl, 500 mM imidazole, pH 8.0), 200 mL
- Buffer C (20 mM Na_2HPO_4, pH 8.5), 500 mL
- Centrifuge capable of reaching 15 000 rpm with temperature control
- 50 mL conical-bottom centrifuge tube (CELLSTAR)

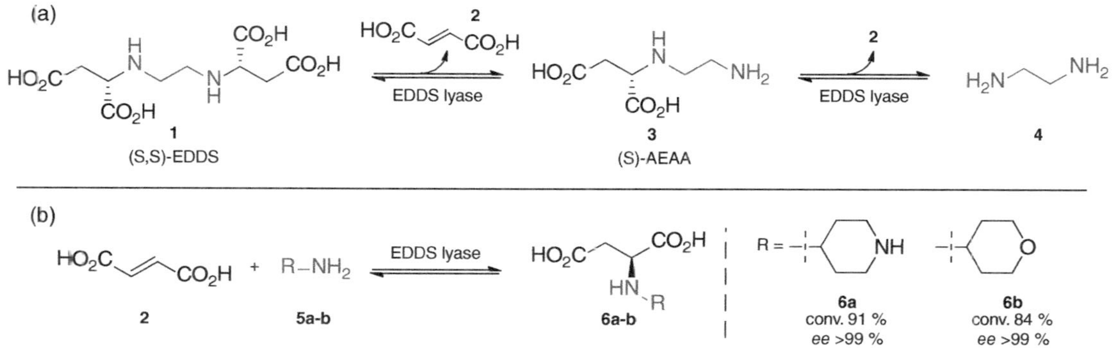

Scheme 5.2 *(a) Natural reaction catalysed by EDDS lyase. (b) EDDS lyase-catalysed addition of heterocycloalkyl-substituted amines* ***5a,b*** *to fumaric acid* ***2****, providing* N*-heterocycloalkyl-substituted L-aspartic acids* ***6a,b*** *as products.*

- Quartz cuvette with 10 mm light path (Hellma Analytics)
- Syringe filter, pore size 0.45 μm (Whatman)
- HiLoad 16/600 Superdex 200 pg column (GE Healthcare Bio-Sciences AB)
- Ni sepharose 6 fast-flow resin (GE Healthcare Bio-Sciences AB)
- AKTA explorer 10S protein purification system
- UV-VIS spectrophotometer
- Vortex mixer
- Rotary shaker with temperature control
- 15 mL filter cartridges (Screening Devices BV)
- 1.5 and 2 mL Eppendorf tubes
- Magnetic stirrer
- 3 L Erlenmeyer flask
- Sterile loop
- Sonicator (Branson Sonifier 450)

5.2.1.2 Procedure

1. *E. coli* TOP10 cells containing the pBADN (EDDS lyase-His) plasmid were collected from an LB plate and used to inoculate sterile LB medium (10 mL), containing ampicillin (100 μg.mL^{-1}), in a 50 mL conical-bottom centrifuge tube.
2. After overnight incubation at 37 °C and 200 rpm, the pre-culture was used to inoculate the rest of the LB medium (990 mL) containing ampicillin (100 μg.mL^{-1}) in a 3 L Erlenmeyer flask. The cells were grown at 37 °C and 200 rpm for about 4 hr until OD_{600} reached 0.8~1.0. Then, arabinose (0.05% w/v) was added to induce enzyme expression. Cultures were grown at 20 °C for 18 hr with shaking (200 rpm) on a rotary shaker.
3. The cells (4 g) were harvested by centrifugation (20 min at 6000× g and 4 °C). Cells harvested from 1 L of culture were resuspended in lysis buffer (15 mL, 50 mM Tris-HCl, 300 mM NaCl, 20 mM imidazole, pH 8.0).
4. Cells were disrupted by sonication for 4 × 40 sec (with a 5 min interval between each cycle) at a 60 W output. After centrifugation (45 min at 10 000× g and 4 °C), the soluble fraction (supernatant) containing EDDS lyase was separated from the insoluble part (pellet).
5. The supernatant was filtered through a pore filter (diameter 0.45 μm) and incubated with Ni sepharose 6 fast-flow resin (1 mL slurry in a filter cartridge used as column) at 4 °C for 18 hr, which had previously been equilibrated with lysis buffer.
6. The unbound proteins were eluted from the column by gravity flow. The column was washed first with lysis buffer (15 mL) and then with buffer A (30 mL, 50 mM Tris-HCl, 300 mM NaCl, 30 mM imidazole pH 8.0). Retained proteins were eluted with buffer B (5 mL, 50 mM Tris-HCl, 300 mM NaCl, 500 mM imidazole pH 8.0). Fractions were analysed by sodium dodecyl sulfate polyacrylamide gel electrophoresis (SDS-PAGE) on gels containing acrylamide (4–12%).
7. Fractions containing EDDS lyase were combined and loaded on to a HiLoad 16/600 Superdex 200 pg column, which had previously been equilibrated with buffer C (180 mL, 20 mM Na_2HPO_4, pH 8.5). The column was eluted with buffer C at 1 mL.min^{-1} for 1.2 column volumes. Fractions were collected and analysed by SDS-PAGE on gels containing acrylamide (4–12%).

8. The concentration of purified EDDS lyase in 20 mM Na_2HPO_4 buffer (pH 8.5) was determined with a spectrophotometer using the method of Waddell [11]. Around 60 mg (8.7 mg.mL^{-1}, 7 mL) of purified EDDS lyase was isolated from 1 L of cell culture following this protocol. The purified protein was divided into aliquots (200–500 μL), frozen by liquid nitrogen and stored at −20 °C for further use.

5.2.2 Procedure 2: Synthesis of Compound 6a

5.2.2.1 Materials and equipment

- 4-Aminopiperidine **5a** (200 mg, 2 mmol; Sigma-Aldrich)
- Fumaric acid **2** (0.2 mmol, 200 μL of 1 M stock solution in 20 mM Na_2HPO_4 buffer, pH 8.5)
- EDDS lyase in 20 mM Na_2HPO_4 buffer (8.7 mg.mL^{-1}, 2 mL, pH 8.5)
- Na_2HPO_4 buffer (20 mM, pH 8.5, 50 mL)
- NH_3 aqueous solution (2 M, 200 mL)
- HCl aqueous solution (1 M, 200 mL)
- HCl aqueous solution (2 M, 200 mL)
- NaOH solution (1 M, 200 mL)
- $CuSO_4$ aqueous solution (0.5 mM, 1 L)
- Ninhydrin solution in EtOH (1 g.L^{-1}, 100 mL)
- Distilled water (1 L)
- Balance
- Standard pipettes
- 50 mL measuring cylinder
- 50× 15 mL test tubes
- 50 mL round-bottom glass flask
- 100 mL round-bottom glass flask
- Dowex 50W X8 resin (hydrogen form, 100–200 mesh; Sigma-Aldrich)
- Dowex 1X8 resin (chloride form, 100–200 mesh; Sigma-Aldrich)
- TLC Silica Gel 60 F_{254} aluminium plates (Merck)
- Rotary evaporator connected to a vacuum pump
- Electrospray ionisation orbitrap high-resolution mass spectrometer
- pH meter
- Nuclear magnetic resonance (NMR) tube
- NMR spectrometer

5.2.2.2 Procedure

1. In a round-bottom glass flask (50 mL), 4-aminopiperidine **5a** (200 mg, 2 mmol) and fumaric acid **2** (0.2 mmol, 200 μL of 1 M stock solution) were added to 20 mM Na_2HPO_4 buffer, pH 8.5 (15 mL). The pH of the initial mixture was adjusted to 8.5 with HCl solution (1 M). To start the reaction, 1.9 mL of EDDS lyase (16.5 mg, 8.7 mg.mL^{-1}, 3 × 10^{-4} mmol, 0.15 mol% based on fumaric acid) was added, and the final volume of the reaction mixture was immediately adjusted to 20 mL with the same buffer.
2. The reaction mixture was incubated at room temperature for 7 days. After completion of the reaction, the enzyme was inactivated by heating to 70 °C for 10 min. The progress

of the enzymatic reaction was monitored using ^{1}H NMR spectroscopy by comparing signals corresponding to fumaric acid (6.5 ppm) and amino acid product.

3. An anion-exchange resin column (10 g of Dowex 1X8 resin, 100–200 mesh) was prepared by pretreatment with a solution of aqueous HCl (2 M, 4 column volumes), aqueous NaOH (1 M, 2 column volumes) and distilled water (4 column volumes). A cation-exchange resin column (10 g of Dowex 50W X8, 100–200 mesh) was prepared by pretreatment with aqueous NH_3 (2 M, 4 column volumes), aqueous HCl (1 M, 2 column volumes) and distilled water (4 column volumes).
4. The precipitated enzyme was removed from the reaction mixture by filtration (pore diameter 0.45 μm) and the filtrate was loaded slowly on to the activated anion-exchange column (10 g of Dowex 1X8 resin, 100–200 mesh). The column was washed with distilled water to remove unbound amine (2 column volumes) and the product was eluted with aqueous HCl (2 M, 2 column volumes).
5. The ninhydrin-positive fractions were collected and loaded on to the activated cation-exchange column (10 g of Dowex 50W X8, 100–200 mesh). The column was washed with water (3 column volumes) to remove the remaining fumaric acid and eluted with 2 M aqueous ammonia until the desired product was collected. The ninhydrin-positive fractions were collected, concentrated under vacuum and lyophilised to provide the desired product.
6. The amino acid product **6a** (91% conversion, 67% isolated yield, 29 mg, 1.3×10^{-4} mol, white solid) was identified by ^{1}H NMR, ^{13}C NMR and high-resolution mass spectrometry (HRMS).

^{1}H NMR (500 MHz, D_2O): δ 3.94 (dd, J = 10.1, 3.4 Hz, 1H), 3.58–3.49 (m, 3H), 3.13–3.07 (m, 2H), 2.80 (dd, J = 17.4, 3.5 Hz, 1H), 2.60 (dd, J = 17.4, 10.0 Hz, 1H), 2.43–2.33 (m, 2H), 1.96–1.85 (m, 2H). ^{13}C NMR (126 MHz, D_2O): δ 177.2, 173.5, 57.7, 51.9, 42.3 (2C), 36.2, 25.9, 25.4. HRMS (ESI$^+$): calcd for $C_9H_{17}O_4N_2$, 217.1183 $[M+H]^+$, found, 217.1181. The enantiomeric excess (>99% ee) was determined by chiral high-performance liquid chromatography (HPLC) analysis on a Nucleosil 5μ chiral-1 120A (250 × 4.6 mm) column, using isocratic 0.5 mM $CuSO_4$ as mobile phase at a flow rate of 1 mL.min^{-1}, 60 °C, UV detected at 240 nm, t_R = 6.4 min.

5.2.3 Procedure 3: Synthesis of Compound 6b

5.2.3.1 Materials and Equipment

- Tetrahydro-2*H*-pyran-4-amine **5b** (202 mg, 2 mmol; Sigma-Aldrich)
- Fumaric acid **2** (0.2 mmol, 200 μL of 1 M stock solution in 20 mM Na_2HPO_4 buffer, pH 8.5)
- EDDS lyase in 20 mM Na_2HPO_4 buffer (8.7 mg.mL^{-1}, 2 mL, pH 8.5)
- Na_2HPO_4 buffer (20 mM, pH 8.5, 50 mL)
- NH_3 aqueous solution (2 M, 200 mL)
- HCl aqueous solution (1 M, 200 mL)
- HCl aqueous solution (2 M, 200 mL)
- NaOH solution (1 M, 200 mL)
- $CuSO_4$ aqueous solution (0.5 mM, 1 L)
- Ninhydrin solution in EtOH (1 g.L^{-1}, 100 mL)

- Distilled water (1 L)
- Balance
- Standard pipettes
- 50 mL measuring cylinder
- 50× 15 mL test tubes
- 50 mL round-bottom glass flask
- 100 mL round-bottom glass flask
- Dowex 50W X8 resin (hydrogen form, 100–200 mesh; Sigma-Aldrich)
- Dowex 1X8 resin (chloride form, 100–200 mesh; Sigma-Aldrich)
- TLC Silica Gel 60 F_{254} aluminium plates (Merck)
- Rotary evaporator connected to a vacuum pump
- Electrospray ionisation orbitrap high-resolution mass spectrometry
- pH meter
- NMR tube
- NMR spectrometer

5.2.3.2 Procedure

1. In a 50 mL round-bottom glass flask, tetrahydro-2*H*-pyran-4-amine **5b** (228 mg, 2 mmol) and fumaric acid **2** (0.2 mmol, 200 μL of 1 M stock solution) were added to 20 mM Na_2HPO_4 buffer, pH 8.5 (15 mL). The pH of the initial mixture was adjusted to 8.5 with a 1 M aqueous HCl solution. To start the reaction, 1.9 mL of EDDS lyase (16.5 mg, 8.7 mg.mL^{-1}, 3×10^{-4} mmol, 0.15 mol% based on fumaric acid) was added, and the final volume of the reaction mixture was immediately adjusted to 20 mL with the same buffer.
2. The reaction mixture was incubated at room temperature for 7 days. After completion of the reaction, the enzyme was inactivated by heating to 70 °C for 10 min. The progress of the enzymatic reaction was monitored using ^{1}H NMR spectroscopy by comparing signals corresponding to fumaric acid (6.5 ppm) and amino acid product.
3. An anion-exchange resin column (10 g of Dowex 1X8 resin, 100–200 mesh) was prepared by pretreatment with a solution of aqueous HCl (2 M, 4 column volumes), aqueous NaOH (1 M, 2 column volumes) and distilled water (4 column volumes). A cation-exchange resin column (10 g of Dowex 50W X8, 100–200 mesh) was prepared by pretreatment with aqueous NH_3 (2 M, 4 column volumes), aqueous HCl (1 M, 2 column volumes) and distilled water (4 column volumes).
4. The precipitated enzyme was removed from the reaction mixture by filtration (pore diameter 0.45 μm) and the filtrate was loaded slowly on to the activated anion-exchange column (10 g of Dowex 1X8 resin, 100–200 mesh). The column was washed with distilled water to remove unbound amine (2 column volumes) and the product was eluted with aqueous HCl (2 M, 2 column volumes).
5. The ninhydrin-positive fractions were collected and loaded on to the activated cation-exchange column (10 g of Dowex 50W X8, 100–200 mesh). The column was washed with water (3 column volumes) to remove the remaining fumaric acid and eluted with 2 M aqueous ammonia until the desired product was collected. The ninhydrin-positive fractions were collected, concentrated under vacuum and lyophilised to provide the desired product.

6. The amino acid product **6b** (84% conversion, 41% isolated yield, 18 mg, 8.3×10^{-5} mol, white solid) was identified by ^{1}H NMR, ^{13}C NMR, and HRMS.

^{1}H NMR (500 MHz, D_2O): δ 4.07–4.03 (m, 2H), 3.98 (dd, $J = 9.0, 3.8$ Hz, 1H), 3.53–3.45 (m, 3H), 2.85 (dd, $J = 17.6, 3.8$ Hz, 1H), 2.71 (dd, $J = 17.7, 9.0$ Hz, 1H), 2.12–2.04 (m, 2H), 1.82–1.70 (m, 2H); ^{13}C NMR (126 MHz, D_2O): δ 177.3, 173.8, 65.7 (2C), 57.1, 53.4, 36.3, 29.5, 29.0. HRMS (ESI$^+$): calcd for $C_9H_{16}O_5N$, 218.1023 $[M+H]^+$, found, 218.1023. The enantiomeric excess (>99% ee) was determined by chiral HPLC analysis on a Nucleosil 5μ chiral-1 120A (250 × 4.6 mm) column, using isocratic 0.5 mM $CuSO_4$ as mobile phase at a flow rate of 1 mL.min^{-1}, 60 °C, UV detected at 240 nm, $t_R = 6.7$ min.

5.2.4 Analytical Method

The biocatalytic conversions of fumaric acid **2** into products **6a,b** were monitored using ^{1}H NMR spectroscopy. At regular time intervals (typically 24 hr), a sample (300 μL) was taken from the reaction mixture and heated at 70 °C for 10 min. The sample was then filtered through a pore filter (diameter 0.45 μm) and the filtrate was evaporated under vacuum. The resulting residue was dissolved in 0.5 mL of D_2O for ^{1}H NMR measurement. The conversion was estimated by comparing the signals of fumaric acid (6.5 ppm) and corresponding product in ^{1}H NMR spectra. The workup procedure was initiated when the fumaric acid (6.5 ppm) signal had almost vanished, indicating high conversion of the fumaric acid **2**.

5.2.5 Conclusion

The enzyme EDDS lyase can be readily produced and purified, with about 60 mg being obtained from 1 L of cell culture. After initial freezing in liquid nitrogen, the enzyme can be stored at −20 °C in a concentration of 8.7 mg.mL^{-1} (in 20 mM Na_2HPO_4 buffer, pH 8.5) for at least 6 months without significant loss of catalytic activity. We have showcased how EDDS lyase can accept the non-natural amine substrates **5a,b** in the enantioselective hydroamination of fumarate, yielding the corresponding *N*-heterocycloalkyl-substituted L-aspartic acids **6a,b** with high conversions (84–91%), good isolated yields (41–67%) and excellent stereoselectivity (>99% ee) (Table 5.1).

In addition, we previously reported that EDDS lyase can accept a panel of homo- and heterocycloalkyl amines (comprising four-, five- and six-membered rings) in the selective addition to fumarate, giving the corresponding *N*-(hetero)cycloalkyl-substituted L-aspartic acids with excellent optical purity (>99% ee in all cases) [9]. Except for the broad cycloalkylamines scope, we recently demonstrated that EDDS lyase can also accept a wide variety of structurally distinct amines for stereoselective addition to fumarate, providing (chemo)enzymatic access to various aminocarboxylic acids, including the natural product toxin A and aspergillomarasmine A [12], difficult-to-synthesise *N*-arylated aspartic acid derivatives [13] and substituted pyrazolidin-3-ones [13]. As such, EDDS lyase, with its remarkably broad amine scope, nicely complements the biocatalytic toolbox for the synthesis of noncanonical amino acids.

Table 5.1 *Enantioselective synthesis of N-heterocycloalkyl-substituted L-aspartic acids via the addition of amines **5a,b** to fumarate catalysed by EDDS lyase.*

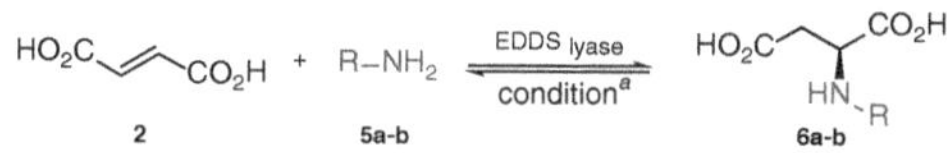

Entry	amines		Product		Conv.[b] (yield)[c] (%)	ee[d] (%)
1	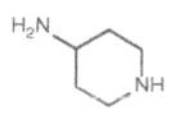	5a	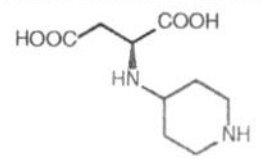	6a	91 (67)	>99
2	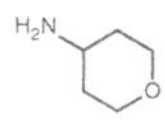	5b	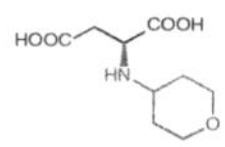	6b	84 (41)	>99

[a]Reaction conditions: the reaction mixture (20 mL) consisted of fumaric acid **2** (10 mM), amines **5a,b** (100 mM) and EDDS lyase (0.15 mol% based on fumaric acid) in buffer (20 mM Na_2HPO_4, pH 8.5), and was incubated at room temperature for 7 days.
[b]Conversions were determined using 1H NMR spectroscopy.
[c]Isolated yield after purification.
[d]The ee values were determined by chiral HPLC analysis using chemically synthesised reference compounds with known configuration.

References

1. King, A.M., Reid-Yu, S.A., Wang, W. *et al.* (2014) *Nature*, **510**, 503–506.
2. Maiti, M., Maiti, M., Rozenski, J. *et al.* (2015) *Organic and Biomolecular Chemistry*, **13**, 5158–5174.
3. McIlwain, B.C., Vandenberg, R.J. and Ryan, R.M. (2016) *Biochemistry*, **55**, 6801–6810.
4. Chattopadhyay, S., Raychaudhuri, U. and Chakraborty, R. (2014) *Journal of Food Science and Technology*, **51**, 611–621.
5. Piispanen, P.S. and Pihko, P.M. (2005) *Tetrahedron Letters*, **46**, 2751–2755.
6. Boros, M., Kökösi, J., Vámos, J. *et al.* (2007) *Amino Acids*, **33**, 709–717.
7. Parmeggiani, F., Weise, N.J., Ahmed, S.T. and Turner, N.J. (2018) *Chemical Reviews*, **118**, 73–118.
8. Xue, Y.-P., Cao, C.-H. and Zheng, Y.-G. (2018) *Chemical Society Reviews*, **47**, 1516–1561.
9. Zhang, J., Fu, H., Tepper, P.G. and Poelarends, G.J. (2019) *Advanced Synthesis & Catalysis*, **361**, 2433–2437.
10. Poddar, H., de Villiers, J., Zhang, J. *et al.* (2018) *Biochemistry*, **57**, 3752–3763.
11. Waddell, W.J. (1956) *Journal of Laboratory and Clinical Medicine*, **48**, 311–314
12. Fu, H., Zhang, J., Saifuddin, M. *et al.* (2018) *Nature Catalysis*, **1**, 186–191.
13. Fu, H., Prats Luján, A., Bothof, L. *et al.* (2019) *ACS Catalysis*, **9**, 7292–7299.

5.3 Biocatalytic Asymmetric Aza-Michael Addition Reactions and Synthesis of L-Argininosuccinate by Argininosuccinate Lyase ARG4-Catalysed Aza-Michael Addition of L-Arginine to Fumarate

Bernhard Schoenenberger,[1] Agata Wszolek,[2] Roland Meier,[1] Henrike Brundiek,[2] Markus Obkircher[1] and Roland Wohlgemuth[1]
[1]*Sigma-Aldrich/Merck KGaA, Buchs,Switzerland*
[2]*Enzymicals, Greifswald, Germany*
[3]*Institute of Technical Biochemistry, Technical University Lodz, Lodz,Poland*

Enzymes catalysing addition reactions of nucleophiles to conjugated acceptor systems are of great interest for generating complexity by enabling the formation of new bonds to carbon from simple building blocks. The formation of new carbon–carbon bonds by biocatalytic Michael addition reactions has been demonstrated in both natural metabolic pathways [1, 2] and the biocatalytic synthesis of more complex target compounds [3, 4]. Lipase B from *Pseudozyma antarctica* has been investigated for the catalysis of carbon–carbon [5] bond-forming Michael addition reactions, whereby a Ser 105 Ala mutant favoured the Michael addition reaction 100 times and at the same time suppressed the hydrolysis reaction more than 1000 times.

Biocatalysts are also of much interest for the construction of new carbon–nitrogen bonds by the addition of nitrogen nucleophiles to α,β-unsaturated carbonyl compounds in aza-Michael addition reactions, which can be considered hydroaminations of α,β-unsaturated carbonyl compounds. Aza-Michael addition reactions represent an important and well-studied class of reactions in synthetic organic chemistry, which offer great opportunities for new resource-efficient and sustainable chemistry, especially in cases where lengthy synthetic routes can be replaced by one step, protection-group free routes with high atom economy using highly enantioselective catalysts to avoid tedious purification of stereoisomer mixtures. Various lipases have been demonstrated to catalyse carbon–nitrogen bond-forming Michael addition reactions of various primary and secondary amines with a series of unsubstituted and substituted acrylic acid esters [6, 7], as illustrated by the generalised reaction at the top of Scheme 5.3. High chemoselectivity and good yields were achieved in lipase-catalysed aza-Michael addition reactions by careful choice of the lipase and the reaction conditions; however, in the case of the substituted acrylic acid esters, enantioselectivity was not reported [6]. As the catalytic asymmetric aza-Michael addition is an attractive reaction for the construction of new chiral centres and enzymes are inherently chiral catalysts by nature, it is worthwhile to search for enzymes capable of catalysing the direction of synthesis or degradation. Enzymes from EC class 4.3, the C-N-lyases, known to catalyse the cleavage of carbon–nitrogen bonds, can also be utilised to catalyse the formation of carbon–nitrogen bonds in the reverse direction. Such C-N Lyases are therefore excellent catalysts for enabling the highly regio- and stereoselective addition of nucleophiles containing terminal amino groups to conjugated carbon–carbon double bonds, as demonstrated recently by a number of examples from a great variety of aminocarboxylic acids [8, 12]. A selection of preparative applications of biocatalytic aza-Michael addition reactions using lipases and ethylenediamine-*N,N′*-disuccinic acid lyase (EDDS lyase) are illustrated in Scheme 5.3.

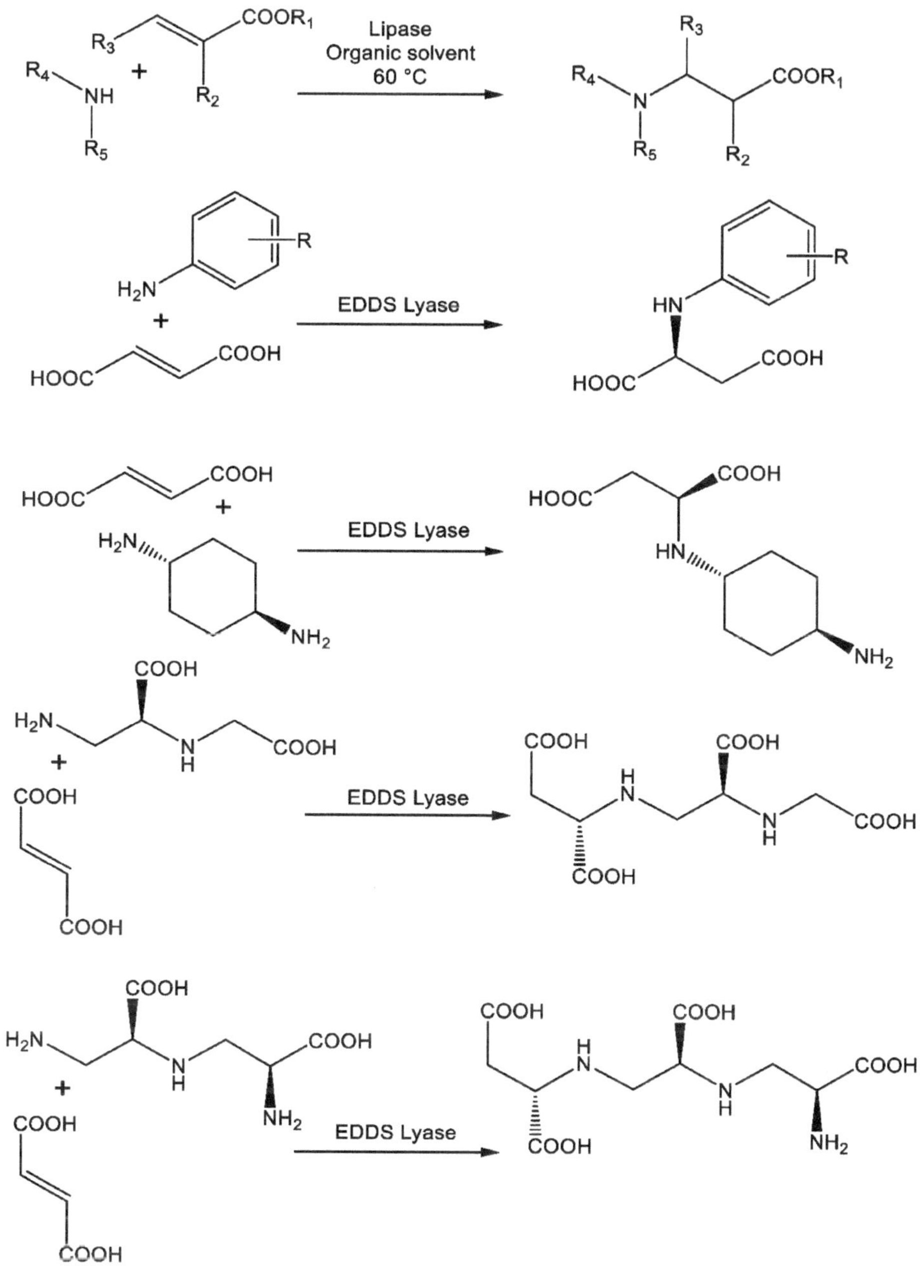

Scheme 5.3 *Selected enzymatic carbon nitrogen bond-forming Michael addition reactions.*

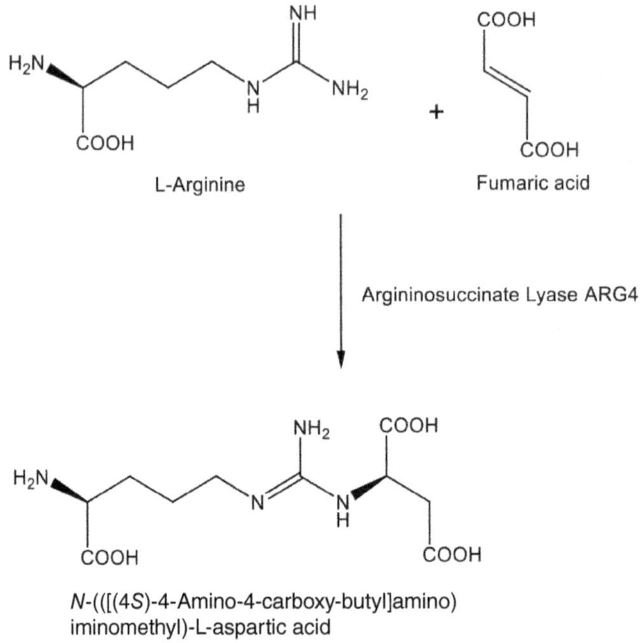

Scheme 5.4 *Argininosuccinate lyase ARG4-catalysed aza-Michael addition of L-arginine to fumarate.*

Argininosuccinate lyase EC 4.3.2.1, a C-N lyase highly conserved in bacteria and eukaryotes, is also able to catalyse the reverse reaction, the aza-Michael addition of L-arginine to fumarate to form the amino acid *N*-(([(4*S*)-4-amino-4-carboxybutyl]amino)iminomethyl)-L-aspartate; in short, L-argininosuccinate (Scheme 5.4), a nonproteinogenic natural amino acid and a key metabolite of the urea cycle and L-arginine metabolism. The synthesis of pure L-argininosuccinate has been achieved using recombinant argininosuccinate lyase ARG as catalyst for the synthesis of L-argininosuccinate in the lithium salt form at preparative scale.

The synthetic route in Scheme 5.4 is the reverse direction of the urea cycle in healthy humans, where the argininosuccinate lyase catalyses the cleavage reaction of the L-argininosuccinate, which is formed by the argininosuccinate synthase-catalysed condensation of aspartate and citrulline.

5.3.1 Procedure 1: Recombinant Expression of ARG4 in *E. coli* BL21(DE3)

5.3.1.1 Materials and Equipment

- *Escherichia coli* BL21(DE3) cells
- Lysogenic broth (LB) plates (15 g.L^{-1} agar, 10 g.L^{-1} tryptone, 10 g.L^{-1} NaCl, 5 g.L^{-1} yeast extract, pH 7.5) with 50 μg.mL^{-1} kanamycin

- LB medium (10 g.L^{-1} tryptone, 10 g.L^{-1} NaCl, 5 g.L^{-1} yeast extract, pH 7.5) with 50 µg.mL^{-1} kanamycin
- Terrific broth (TB) medium (12 g.L^{-1} casein, enzymatically digested, 24 g.L^{-1} yeast extract, 12.54 g.L^{-1} K_2HPO_4, 2.31 g.L^{-1} KH_2PO_4, pH 7.5) with 50 µg.mL^{-1} kanamycin
- Isopropyl-β-D-1-thiogalactopyranoside (IPTG)
- Potassium phosphate buffer pH 7.5 (50 mM)
- Sodium phosphate buffer at pH 7.5 (50 mM)
- Incubator
- Water bath
- 250 mL and 2 L Erlenmeyer flasks with baffles
- Orbital shaker
- Centrifuge
- Spectrophotometer
- Retsch Mixer Mill MM 200
- Glass beads (Ø 0.25–0.5 mm)
- Chromatography equipment
- Talon® resin
- Zeba spin columns

5.3.1.2 Procedure

1. The argininosuccinate lyase gene *ARG4* was derived from *Saccharomyces cerevisiae* (strain ATCC 204508/S288c) (UniProtKB: P04076). The synthetic gene was codon optimised for *E. coli* and equipped with an *N*-terminal 6x-his-tag and a TEV protease cleavage site for optional removal of the affinity tag. The gene was synthesised and subcloned into pET24a(+) via NdeI and NotI by Genscript, giving vector pET24-His(TV)-*ARG4-E. coli*(op).
2. Chemically competent BL21 (DE3) *E. coli* cells (100 µL; 100 µL aliquots were stored at −80 °C until usage) were transformed with 5 µL of the derived plasmid by using standard procedures [13] and plated on LB agar supplemented with 50 µg.mL^{-1} kanamycin for selection. Selected *E. coli* transformants were used for expression tests, in which different media, induction times and temperatures were optimised. The selected conditions were later used for the overexpression of *ARG4* in *E. coli* BL21 (DE3), as shown in the following steps.
3. A single *E. coli* colony harbouring *ARG4* gene was inoculated into 5 mL LB medium supplemented with kanamycin (50 µg.mL^{-1}) and incubated overnight at 30 °C and 180 rpm.
4. Overnight cultures of *E. coli* were used to inoculate 50 mL of TB medium supplemented with kanamycin (50 µg.mL^{-1}) in 250 mL baffled flasks. Cells were cultivated at 37 °C under vigorous shaking (150 rpm) until an OD_{600} of 0.5 was reached.
5. When the pre-culture reached the mid-log phase as follows, the main culture was started. Fresh TB medium (400 mL) in a 2 L Erlenmeyer flask was inoculated with the *E. coli* pre-culture (20 mL). Cells were cultivated at 37 °C under shaking (110 rpm) until an OD_{600} of 1.3 was reached.
6. After the culture reached OD_{600} 1.3, protein expression was induced with IPTG to a final concentration of 1mM and the incubation temperature was shifted to 25 °C

and the main culture shaken at 110 rpm for 20 hr. Cells were harvested by centrifugation at 5000 rpm and 4 °C for 30 min and washed once with 40 mL of ice-cold 50 mM sodium phosphate buffer at pH 7.5. The *E. coli* cell pellets containing the overexpressed enzyme (recombinant ARG4) were frozen at −20 °C.

7. For cell disruption, frozen cells were adjusted to OD_{600} 40 in 50 mM potassium phosphate buffer pH 7.5 and mixed with glass beads (Ø 0.25–0.5 mm) at an approximate ratio of 1 : 1. The cells were disintegrated mechanically by a cell disruptor (Retsch Mixer Mill MM 200: two cycles for 5 min at 600 rpm).
8. The samples were cooled on ice and insoluble cell debris was removed by centrifugation (5000 rpm, 4 °C, 30 min).
9. The supernatant containing the soluble protein fraction was purified by incubating the soluble protein fraction with Talon® resin on an orbital shaker (30 min, 4 °C).
10. A standard gravity flow protocol was applied to purify the ARG4 target protein. Therefore, the resin was washed at 4 °C with 10 column volumes of 50 mM potassium phosphate buffer pH 7.5 containing 10 mM imidazole to remove unspecifically bound protein and the target protein ARG4 was eluted with 1.5 column volumes of elution buffer (50 mM potassium phosphate buffer pH 7.5, 150 mM imidazole).
11. Purified ARG4 was obtained by removing imidazole on desalting columns using 50 mM potassium phosphate buffer pH 7.5 with Zeba Spin columns.

5.3.2 Procedure 2: ARG4-Catalysed Asymmetric Michael Addition Reaction of L-Arginine to Fumarate

5.3.2.1 Materials and Equipment

- L-Arginine (870 mg, 5 mmol)
- Fumaric acid (580 mg, 5 mmol)
- H_2O (100 mL)
- 1 M aqueous LiOH
- Soluble fraction of the purified cell extract of *E. coli* overexpressing recombinant argininosuccinate lyase (ARG4; 500 μL).
- Round-bottom flask
- n-Propanol
- Flash chromatography equipment
- Silica for flash chromatography
- Rotary evaporator with high vacuum
- Filter equipment
- Lyophiliser
- Silica thin-layer chromatography (TLC) plates
- UV cabinet consisting of UV lamp 254 nm and viewing box
- Ninhydrin
- Nuclear magnetic resonance (NMR) spectrometer ^1H 600 MHz

5.3.2.2 Procedure

1. A mixture of L-arginine (870 mg, 5 mmol) and fumaric acid (580 mg, 5 mmol) in H_2O (100 mL) was adjusted to pH 7.5 with 1 M aqueous LiOH. To this solution was added the soluble fraction of the purified cell extract of *E. coli* overexpressing recombinant argininosuccinate lyase (ARG4; 500 μL).
2. The solution was stirred at room temperature and the progress of the reaction was monitored periodically by ^{1}H-NMR and TLC.
3. After the conversion ceased (6 days), the mixture was concentrated by standard rotary evaporation under reduced pressure using membrane pumps and then dried at the even further reduced pressure of 0.1–0.01 mbar and room temperature, giving a white solid. This was dissolved in H_2O (5 mL), and then n-propanol (10 mL) was added.
4. The resulting emulsion was flash-chromatographed on silica with n-propanol/H_2O 2 : 1. The early fractions contained unreacted fumarate, followed by the L-argininosuccinate in later fractions.
5. Fractions containing the product, which eluted from the silica column as the lithium salt of L-argininosuccinate, were pooled, filtered, frozen and lyophilised to give 1.07 g (~70%) of the pure product *N*-(([(4S)-4-amino-4-carboxy-butyl]amino)-iminomethyl)-L-aspartic acid lithium salt (L-argininosuccinic acid lithium salt) in the form of a colourless solid.

N-(([(4*S*)-4-amino-4-carboxy-butyl]amino) imino-methyl)-L-aspartic acid lithium salt (L-argininosuccinic acid mono-lithium salt) NMR: δ_H (400 MHz; D_2O) 4.14 (1H, dd, *J*1 = 3.4, *J*2 = 9.9, succinyl-CH-NH), 3.65 (1H, dd, *J*1 = *J*2 = 6.2, arginyl-CαH), 3.17 (2H, t, *J* = 6.8, arginyl-CδH2), 2.71 (1H, dd, *J*1 = 3.4, *J*2 = 16.1, succinyl-CH), 2.41 (1H, dd, *J*1 = 9.9, *J*2 = 16.1, succinyl-CH′), 1.80 (2H, m, arginyl-CbH2), 1.58 (2H, m, arginyl-CgH2); dC (100.6 MHz; D2O) 178.9, 177.1, 174.4 (3 COOH), 155.8 (guanidino-C), 55.1, 54.3 (2 NH2-C-COOH), 40.6 (succinyl-CH2 and arginyl-Cd), 27.6 (arginyl-Cb), 23.9 (arginyl-Cg). H2O content (Karl Fischer titration): 14.1% elemental analysis, found: C, 33.3; H, 6.2; N, 15.5%. Calculated for $C_{10}H_{16}N_4O_6Li_2$. 14.1% H2O: C, 35.2; H, 6.2; N 16.4%; quantitative NMR (qNMR), content: 94.4% (calculated as di-lithium salt including 14.1% H_2O).

5.3.3 Analytical Method

The protein content of the purified protein was determined with the bicinchoninic acid assay (BCA) method using bovine serum albumin (BSA) as a standard. The analysis of ARG4 overexpression and purity was performed using 12.5% SDS-polyacrylamide gels. The gels were stained with Coomassie Blue.

Argininosuccinate lyase ARG4 activity was assayed at 25 °C spectrophotometrically by the absorbance change at 240 nm due to the formation of fumarate during the reaction ($\varepsilon = 2.44 \times 10^3\ M^{-1}.cm^{-1}$). The assay mixture contained 50 mM potassium phosphate buffer pH 7.5 and 1 mM of potassium L-argininosuccinate in a total volume of 1 mL and the reaction was initiated by the addition of the recombinant protein.

The TLC analysis was done with silica plates using n-propanol/H_2O 2 : 1 as solvent, detection by ninhydrin and exposition to short-wave UV light at 254 nm. Under these TLC conditions, the Rf of fumaric acid was 0.7 and that of arginine was 0.1. The final product, which was analysed with plate silica gel K60, mobile phase n-propanol/H_2O 2 : 1, detected with ninhydrin and scanned with CAMAG Linomat 5, showed a purity of 95.1% by this special TLC method.

5.3.4 Conclusion

A simple and straightforward biocatalytic asymmetric aza-Michael addition reaction has been established for the synthesis of the key metabolite *N*-(([(4*S*)-4-amino-4-carboxybutyl] amino) imino methyl)-L-aspartic acid, commonly referred to as L-argininosuccinate. This one-step addition reaction was developed by running part of the urea cycle in reverse. Argininosuccinate lyase from *S. cerevisiae* was selected as biocatalyst. The corresponding gene *ARG4* was expressed in *E. coli* for the production of the highly active and stable argininosuccinate lyase biocatalyst. The use of this argininosuccinate lyase and reaction monitoring by NMR enabled the development of a biocatalytic asymmetric aza-Michael addition reaction as a novel green chemistry route with high molecular economy for the synthesis of this important metabolite at gram scale.

5.3.5 Acknowledgement

Reproduced from Ref. [9] with permission from the Royal Society of Chemistry.

References

1. Kusebauch, B., Busch, B., Scherlach, K. *et al.* (2009) *Angewandte Chemie International Edition*, **48** (27), 5001–5004.
2. Miyanaga, A. (2019) *Natural Product Reports*, **36** (3), 531–547.
3. Garrabou, X., Macdonald, D.S., Wicky, B.I. and Hilvert, D. (2018) *Angewandte Chemie International Edition*, **130** (19), 5386–5389.
4. Biewenga, L., Saravanan, T., Kunzendorf, A. *et al.* (2019) *ACS Catalysis*, **9**, 1503–1513.
5. Svedendahl, M., Jovanović, B., Fransson, L. and Berglund, P. (2009) *ChemCatChem*, **1** (2), 252–258.
6. Steunenberg, P., Sijm, M., Zuilhof, H. *et al.* (2013) *Journal of Organic Chemistry*, **78** (8), 3802–3813.
7. Du, L.-H., Dong, Z., Long, R.-J. *et al.* (2019) *Organic and Biomolecular Chemistry*, **17**, 807–812.
8. Weiner, B., Poelarends, G.J., Janssen, D.B. and Feringa, B.L. (2008) *Chemistry: A European Journal*, **14** (32), 10 094–10 100.
9. Schoenenberger, B., Wszolek, A., Meier, R. *et al.* (2017) *RSC Advances*, **7**, 48 952–48 957.
10. Fu, H., Zhang, J., Saifuddin, M. *et al.* (2018) *Nature Catalysis*, **1** (3), 186–191.
11. Zhang, J., Fu, H., Tepper, P.G. and Poelarends, G.J. (2019) *Advanced Synthesis & Catalysis*, **361** (11), 2433–2437.
12. Fu, H., Prats Luján, A., Bothof, L., *et al.* (2019) *ACS Catalysis*, **9**, 7292–7299.
13. Hanahan, D. (1983) *Journal of Molecular Biology*, **166**, 557–580.

5.4 Convenient Approach to the Biosynthesis of C2,C6-Disubstituted Purine Nucleosides Using *E. coli* Purine Nucleoside Phosphorylase and Arsenolysis

Irina D. Konstantinova*, Alexey L. Kayushin, Alexandra O. Arnautova, Konstantin V. Antonov, Barbara Z. Yeletskaya, Mariya Ya. Berzina, Elena V. Dorofeeva, Olga I. Lutonina, Ilya V. Fateev, Roman S. Esipov and Anatoly I. Miroshnikov
Shemyakin and Ovchinnikov Institute of Bioorganic Chemistry, Russian Academy of Sciences, Moscow, Russian Federation

In recent years, the transglycosylation reaction catalysed by *Escherichia coli* nucleoside phosphorylases has been widely used in the synthesis of modified nucleosides [1].

During this reaction, a carbohydrate residue is transferred to a modified base (substituted purines [2], benzimidazoles [3] or 1,2,4-triazole-3-carboxamides [4]). The base can be used as is or as part of a riboside. Often, the use of this reaction is constrained by the need to shift the equilibrium towards the desired product. This is usually achieved by introducing into the reaction mixture an excess of one of the starting components or by selective removal of the poorly soluble target nucleosides (e.g. in case of Fludarabine) from the reaction medium [5]. However, even under those conditions, the maximum possible conversion of the starting nucleosides into target compounds is not higher than 50–60%. We found that the use of salts of *ortho*-arsenic acid during the transglycosylation reaction shifts the equilibrium towards the formation of the target arabinonucleosides [6]. The arsenolysis reaction can be expanded to obtain purine 3'-deoxyribosides. The shift of equilibrium towards the target product occurs due to the removal of ribose from the reaction: ribose arsenate obtained in the active centre of the enzyme is extremely unstable in solutions and decomposes, forming ribose and arsenate (Scheme 5.5). As a result, the equilibrium shifts towards the formation of the target nucleosides with conversion of 80–95%. Arsenate can be added both at the start of the reaction and after establishment of equilibrium.

5.4.1 Procedure 1: Biocatalytic Conversion of 3'-Deoxyinosine (3'-dI) and 6-*O*-Methylguanosine (6-*O*-Me-Gua-Rib) into 6-*O*-Me-Gua-3'-dR

5.4.1.1 Materials and Equipment

- 3'-Deoxyinosine (3'-dI) (0.1 g, synthesised as reported in [7])
- 6-*O*-Methylguanosine (0.05 g; CAS Number 7803-88-5, BOC Sciences)
- KH_2PO_4 (0.03 g)
- Distilled water (170 mL)
- Na_2HAsO_4 (Sigma, 0.003 g or 10 mL of 1 mM solution)
- MeOH
- EtOH (60 mL)
- Purine nucleoside phosphorylase *E. coli* (PNPase; 1230 units, prepared earlier [8])
- 0.25 L Schott bottle with screw cap
- pH meter

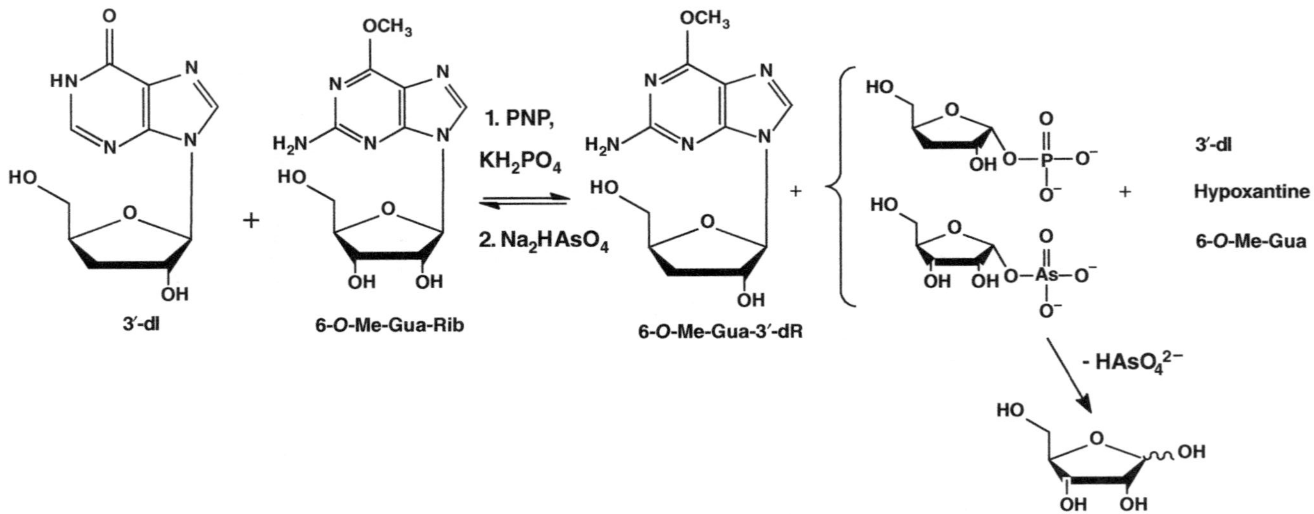

Scheme 5.5 *Synthesis of 3′-deoxyriboside 6-methoxyguanine (6-O-Me-Gua-3′-dR).*

- High-performance liquid chromatography (HPLC) system (Waters Breeze, column Nova-Pack C_{18}, 4.6 × 150 mm, 4 μm)
- Flash column (30 × 170 mm, Separon CGX C_{18}, 60 μm, Tessek Ltd)
- Rotary evaporator (Büchi)

5.4.1.2 Procedure

1. 3'-dI (100 mg, 0.38 mmol), 6-*O*-methylguanosine (50 mg, 0.17 mmol) and KH_2PO_4 (27.4 mg, 2 mmol) were dissolved in H_2O (170 mL) in a 0.25 L Schott bottle with screw cap. pH was adjusted to 7.0 with 3 M KOH (aqueous).
2. PNPase *E. coli* (1230 units) was added to the reaction mixture. The reaction was incubated at 50 °C. Progress was monitored by HPLC. Samples (5 μL) were analysed by HPLC System Waters Breeze with a Nova-Pack column (see Table 5.2 for the HPLC data).
3. (a) After achieving 76–78% conversion of 6-*O*-methylguanosine into 6-*O*-Me-Gua-3'-dR (7 days, determined by HPLC), Na_2HAsO_4 solution (1 mM, 10 mL) was added to the reaction mixture. In 1.5 hr, 98–99% conversion of starting 6-*O*-methylguanosine was achieved (HPLC data). (b) Arsenate can be added (1 mM, 2 mL) just after addition of PNPase *E. coli*. In this case, 98–99% conversion is achieved in 3 days. The reaction mixture was cooled to room temperature, EtOH (60 mL) was added and the solution was concentrated to ~10 mL by distillation under reduced pressure.
4. The target product was isolated by chromatography on Separon CGX C_{18} (elution by gradient of MeOH in H_2O, 0–70%, 500 mL, flow rate 6.8 $mL.min^{-1}$). Fractions containing the desired product (identified by HPLC) were combined, concentrated by distillation under reduced pressure and lyophilised. Yield was 35 mg (0.12 mmol, 71%). Purity was 99% (HPLC data).

UV: (H_2O, pH 7.0) λ_{max}, nm (ε, $M^{-1}.cm^{-1}$): 279 (8320); λ_{min}, nm: 226 (7853).

HRMS (ESI^+), *m/z* $[M+H]^+$: calculated for $C_{11}H_{17}N_5O_4$ 282.11856, found 282.11876; $[base+H]^+$: calculated for $C_6H_9N_5O$ 166.07293, found 166.07274; $[2M+Na]^+$: calculated for $C_{22}H_{34}N_{10}NaO_8$ 585.21078, found 585.21034.

Table 5.2 *Retention times for HPLC analysis.*

Substance	Retention time (min)	Percentage of components before addition of sodium arsenate	Percentage of components 1.5 h after addition of sodium arsenate
Hyp	5.54	22.89	32.09
Ino	6.06	7.75	1.08
3′ dI	6.64	45.34	42.84
6-*O*-Me-Gua	9.08	1.99 (8.27)[a]	3.80 (15.38)[a]
6-*O*-Me-Gua-Rib	9.46	3.24 (13.48)	0.46 (1.92)
6-*O*-Me-Gua-3′-dR	9.92	18.82 (78.25)	19.74 (82.70)

[a] Percentages in brackets reflect content of 6-*O*-Me-Gua and corresponding nucleosides only.

HPLC was performed on HPLC System Waters Breeze (Waters 1525, Waters 2487, Breeze 2). Eluent A: 20% MeOH/H_2O. Eluent B: MeOH. Detection at 260 nm. Elution: 0–100% B, 15 min, flow rate 0.5 $mL.min^{-1}$.

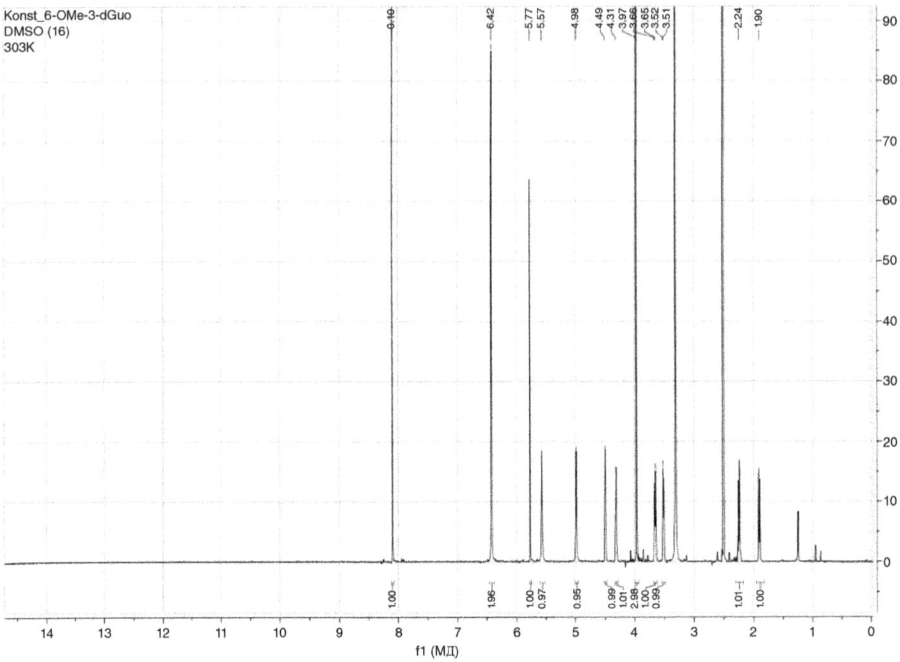

Figure 5.1 *1H-NMR spectrum.*

^{1}H NMR (700 MHz, DMSO-d6): δ = 8.10 (s, 1 H, H-2), 6.42 (br. sing., 2 H, NH), 5.77 (d, J = 2.23 Hz, 1 H, H-1'), 5.57 (br. d, J = 2.59 Hz, 1 H, OH-2'), 4.98 (br. t, J = 4.9 Hz, 1 H, OH-5'), 4.49 (m, 1 H, H-2'), 4.31 (m, 1 H, H-4'), 3.97 (s, 3 H, OCH_3), 3.66 (ddd, J = 11.8, 3.8, 3.9 Hz, 1 H, H-5a'), 3.51 (ddd, J = 11.9, 4.2, 4.3 Hz, 1 H, H-5b'), 2.24 (ddd, J = 14.6, 9.0, 5.8 Hz, 1 H, H-3a'), 1.90 (ddd, J = 13.2, 6.2, 2.9 Hz, 1 H, H-3b') (see Figure 5.1).

^{13}C NMR (176 MHz, DMSO-d6): δ = 160.58 (C6), 159.71 (C2), 153.60 (C4), 137.52 (C8), 113.84 (C5), 89.97 (C1'), 80.42 (C4'), 74.62 (C2'), 62.52 (C5'), 53.09 (OCH_3), 34.33(C3').

^{15}N NMR (71 MHz, DMSO-d6): δ = 240.90 (N7), 205.59 (N1), 188.44 (N3) 170.3 (N9), 81.32 (NH_2).

5.4.2 Conclusion

The described procedure allows the regio- and stereoselective production of the 3'-deoxyriboside-6-methoxyguanosine in high yield using standard chemical and microbiological laboratory equipment and methods. In principle, this procedure can be adapted for related substrates by the choice of suitable enzymes.

References

1. (a) Krenitsky, T.A., Koszalka, G.W. and Tuttle, J.V. (1981) *Biochemistry*, **20**, 3615–3621; (b) Mikhailopulo, I. and Miroshnikov, A. (2010) *Acta Naturae*, **2**, 36–58; (c) Fateev, I., Kharitonova, M., Antonov, K. *et al.* (2015) *Chemistry: A European Journal*, **21** (38), 13 401–13 419.

2. (a) Fateev, I.V., Antonov, K.V., Konstantinova, I.D. *et al.* (2014) *Beilstein Journal of Organic Chemistry*, **10**, 1657–1669; (b) Barai, V.N., Zinchenko, A.I., Eroshevskaya, L.A. *et al.* (2002) *Helvetica Chimica Acta*, **85**, 1893–1900; (c) Konstantinova, I.D., Antonov, K.V., Fateev, I.V. *et al.* (2011) *Synthesis*, **10**, 1555–1560.
3. (a) Konstantinova, I.D., Selezneva, O.M., Fateev, I.V. *et al.* (2013) *Synthesis*, **45** (2), 272–276; (b) Kharitonova, M.I., Fateev, I.V., Kaushin, A.L. *et al.* (2016) *Synthesis*, **48** (3), 394–406.
4. (a) Konstantinova, I.D., Chudinov, M.V., Fateev, I.V. *et al.* (2013) *Russian Journal of Bioorganic Chemistry*, **39**(1), 53–71; (b) Chudinov, M.V., Prutkov, A.N., Matveev, A.V. *et al.* (2016) *Bioorganic & Medicinal Chemistry Letters*, **26**, 3223–3225; (c) Chudinov, M.V., Matveev, A.V., Prutkov, A.N. *et al.* (2016) *Mendeleev Communications*, **26** (3), 214–216.
5. Bzowska, A., Kulikowska, E. and Shugar, D. (2000) *Pharmacology & Therapeutics*, **88**, 349–425.
6. Konstantinova, I.D., Fateev, I.V. and Miroshnikov, A.I. (2016) *Russian Journal of Bioorganic Chemistry*, **42** (4), 372–380.
7. Denisova, A.O., Tokunova, Y.A., Fateev, I.V. *et al.* (2017) *Synthesis*, **49**, 4853–4860.
8. Esipov, R.S., Gurevich, A.I., Chuvikovsky, D.V. *et al.* (2002) *Protein Expression and Purification*, **24** (1), 56–60.

5.5 Production of L- and D-Phenylalanine Analogues Using Tailored Phenylalanine Ammonia-Lyases

Emma Z.A. Nagy,[1] Souad D. Tork,[1] Alina Filip,[1] László Poppe,[1,2] Monica I. Toşa,[1] Csaba Paizs[1] and László C. Bencze*[1]

[1]*Biocatalysis and Biotransformation Research Center, Faculty of Chemistry and Chemical Engineering, Babeş-Bolyai University, Cluj-Napoca, Romania*
[2]*Department of Organic Chemistry and Technology, Budapest University of Technology and Economics, Budapest, Hungary*

The biocatalytic synthesis of optically pure L- and D-unnatural phenylalanines with high synthetic value is continuously providing new green synthetic procedures. Recently, by mapping the binding pocket of phenylalanine ammonia-lyase from *Petroselinum crispum* (*Pc*PAL), we developed tailored variants to use as biocatalysts for the synthesis of several phenylalanine analogues [1–3].

L-*p*-Br-phenylalanine L-**1b** is an important intermediate for the production of several valuable biarylalanines [4], whilst D-*p*-CH_3-phenylalanine D-**1a** has been successfully incorporated into Pin1 inhibitors [5] and into anti-inflammatory formyl-peptide receptor 1 antagonists [6]. The efficient *Pc*PAL-mediated synthesis of these products can be achieved by performing the corresponding ammonia elimination and ammonia addition reactions (Scheme 5.6) using *Escherichia coli* Rosetta cells, harbouring the plasmid carrying the gene of the tailored *Pc*PAL I460V variant, as whole-cell biocatalysts.

Within the kinetic resolution-type ammonia elimination reaction, the substrate is the racemic *p*-CH_3-phenylalanine *rac*-**1a**, from which the L-enantiomer L-**1a** is dehydroaminated to *p*-CH_3-cinnamic acid **2a**, yielding the unreacted D-*p*-CH_3-phenylalanine D-**1a** in a maximal theoretical yield of 50%. Meanwhile, the asymmetric synthetic route of the ammonia addition reaction, using *p*-Br-cinnamic acid **2b** as substrate, provides L-*p*-Br-phenylalanine L-**1b** in 100% theoretical yield.

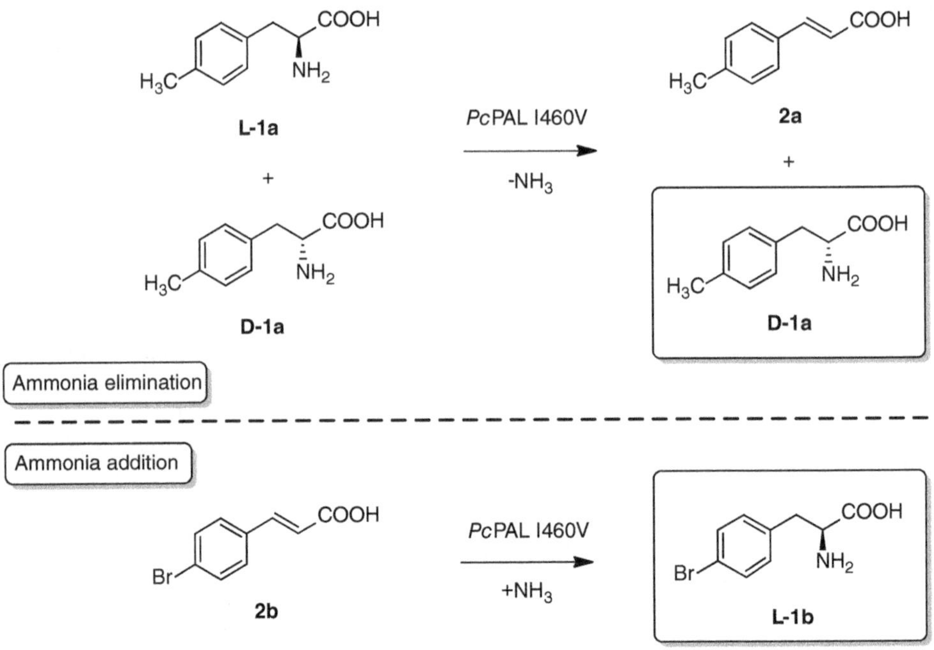

Scheme 5.6 *Ammonia elimination reaction of p-CH_3-phenylalanine rac-**1a** and ammonia addition to p-Br-cinnamic acid **2b** catalysed by E. coli whole cells harbouring the plasmid carrying the gene of PcPAL I460V.*

5.5.1 Procedure 1: Preparation of Whole-Cell Biocatalysts for Enzymatic Reactions

5.5.1.1 Materials and Equipment

- Lysogenic broth (LB)-agar plate with colonies of *E. coli* Rosetta (DE3) pLyS cells harbouring the expression vector pET19b carrying the *pcpal I460V* gene [2, 7]
- LB medium (premixed powder, 10 g for 500 mL dH_2O)
- Distilled water (dH_2O)
- Carbenicillin (stock solution of 50 mg.mL^{-1} in dH_2O, filter-sterilised)
- Chloramphenicol (stock solution of 30 mg.mL^{-1} in high-purity grade ethanol)
- Isopropyl-β-D-1-thiogalactopyranoside (IPTG; 0.5 M in dH_2O, filter-sterilised)
- Phosphate-buffered saline (PBS) buffer (20 mM NaH_2PO_4, 150 mM NaCl dissolved in dH_2O, pH adjusted to 8.0 with 10% NaOH)
- 50 mL polypropylene centrifuge tubes with polyethylene screw cap
- 100 mL Erlenmeyer flask (baffled) with cotton cap
- 2 L Erlenmeyer flask (baffled) with cotton cap
- Incubator (New Brunswick Innova 40 Benchtop Incubator Shaker)
- Autoclave (Raypa Steam Sterilizer)
- UV-Vis spectrophotometer (Varian Cary 50 UV-Vis spectrophotometer)
- Cooling centrifuge (Thermo Scientific SL 16R; 1751× g/4000 rpm)

5.5.1.2 Procedure

1. Premixed powder of LB medium was dissolved in dH_2O (10 g in 500 mL dH_2O and 1 g in 50 mL dH_2O) and autoclaved (25 min, 121 °C) in 2L and 100 mL Erlenmeyer flasks (baffled) with cotton caps to give sterile lysogenic broth.
2. The sterile LB medium (50 mL) was used to prepare the pre-culture by adding carbenicillin (50 μL) and chloramphenicol (56.5 μL) from their stock solutions to final concentrations of 50 and 34 $\mu g.mL^{-1}$, respectively. The solution was inoculated with a single colony of *E. coli* Rosetta (DE3) pLyS cells harbouring the expression vector pET19b carrying the *pcpal I460V* gene, then incubated overnight at 37 °C under continuous shaking (200 rpm).
3. The overnight culture (10 mL) was used to inoculate the 500 mL sterile LB medium and was incubated at 37 °C with continuous shaking (200 rpm) until the optical density OD_{600} reached a value of 0.6–0.8 (approx. 2 hr). The cell culture was cooled to 25 °C and protein expression was induced by adding 0.5 M IPTG solution (100 μL) in order to give a final IPTG concentration of 0.1 mM.
4. Cell growth was maintained at 25 °C and 200 rpm for another 16 hr, reaching a final OD_{600} value of 4–6 (stationary phase of growth). The culture volumes required for the biotransformations (Procedure 2) were harvested by centrifugation using 50 mL polypropylene centrifuge tubes at 4000 rpm (1751× g) and 4 °C for 25 min. The required volume of bacterial culture, providing the amount of whole-cell pellet needed for the preparative-scale reactions, can be calculated from the volume of the reactions, the optimal whole-cell biocatalyst concentration (defined in cell density OD_{600}) and the final OD_{600} value of the induced cells (see Procedure 2):

$$\textit{volume of bacterial culture (mL)} = \frac{\textit{reaction volume (mL)} \times OD_{600} \textit{ of whole cells in reaction}}{\textit{final } OD_{600} \textit{ of the cell culture}} \quad (5.1)$$

5. The whole-cell pellet was washed once with 40 mL PBS buffer (20 mM NaH_2PO_4, 150 mM NaCl, pH 8.0) then centrifuged (1751× g, 4 °C, 25 min). The supernatant was discarded. The cells were either handled further as described in Procedure 2 or stored at −20 °C for further use.

5.5.2 Procedure 2: Preparative-Scale Enzymatic Reactions

5.5.2.1 Materials and Equipment

5.5.2.1.1 Ammonia Elimination Reaction.

- Aqueous NH_4OH buffer (110 mL, 0.1 M with pH 9.5, adjusted with CO_2 in dH_2O; buffer solution for ammonia elimination reaction)
- *rac-p*-CH_3-phenylalanine (100 mg, 0.56 mmol)
- Whole-cell pellet (the amount resuspended in the reaction buffer provides a cell density of OD_{600} ~2.5; see Procedure 1 for the calculation of the volume of bacterial culture providing the required whole-cell pellet)
- 0.3 L Erlenmeyer flask (baffled) with cotton cap
- Incubator (New Brunswick Innova 40 Benchtop Incubator Shaker)
- Vortex mixer (Ika Vortex Genius 3)

5.5.2.1.2 ***Ammonia Addition Reaction.***

- Aqueous NH_4OH (220 mL, 6 M, pH 9.5, adjusted with CO_2 in dH_2O; buffer solution for ammonia addition reaction)
- *p*-Bromo-cinnamic acid (500 mg, 2.2 mmol)
- Whole-cell pellet (the amount resuspended in the reaction buffer provides a cell density of OD_{600} ~10; see Procedure 1 for the calculation of the volume of bacterial culture providing the required whole-cell pellet)
- 0.5 L Erlenmeyer flask (baffled) with cotton cap
- Incubator (New Brunswick Innova 40 Benchtop Incubator Shaker)
- Vortex mixer (Ika Vortex Genius 3)

5.5.2.2 *Procedure*

1. (a) The substrate solution was prepared by dissolving *rac-p*-CH_3-phenylalanine (100 mg) in 0.1 M NH_4OH buffer (112 mL, pH 9.5, adjusted with CO_2 in dH_2O) to give a final substrate concentration of 5 mM. (b) *p*-Br-Cinnamic acid (500 mg) was added to aqueous NH_4OH (220 mL 6 M, pH 9.5, adjusted with CO_2 in dH_2O) to give a final substrate concentration of 10 mM.
2. The *E. coli* cell pellets harbouring the plasmid with the *pcpal I460V* gene, prepared by Procedure 1, were gently resuspended in the correspondingly prepared substrate solution. (a) A 5 mM solution of *rac*-**1a** (112 mL) and cell pellet obtained from 47 mL bacterial culture of final OD_{600} ~6 is required to obtain a biocatalyst density of OD_{600} ~2.5 within the reaction volume. (b) A 10 mM solution of **2b** (220 mL) and cell pellet obtained from 366 mL bacterial culture of final OD_{600} ~6 is required to reach the optimal biocatalyst cell density of OD_{600} ~10.
3. The reactions were incubated at 200 rpm, 30 °C for 48 hr. Conversion values were monitored by reversed-phase high-performance liquid chromatography (HPLC) (Procedure 3).

5.5.3 Procedure 3: Monitoring the Biotransformations and Isolation of Products

5.5.3.1 *Materials and Equipment*

- Aqueous H_2SO_4 (10%)
- Ethyl acetate
- NH_4OH (2 M aqueous)
- Na_2SO_4, anhydrous
- Distilled water (dH_2O)
- Dowex® 50WX2 resin (15 mL in 10% H_2SO_4)
- pH paper
- Erlenmeyer flasks
- Rotatory evaporator (Heidolph Hei-VAP Ultimate)
- Cooling centrifuge (Thermo Scientific SL 16R)
- 50 mL centrifuge tubes PP with screw cap PE
- Separatory funnel (500 mL)

Table 5.3 *HPLC methods for the monitoring of enzymatic reactions (determination of conversion values) and for the assessment of the enantiomeric excess (ee) of the products.*

Compound	Eluent[a] (% B)	Retention time (min)		Response factor[b]	Wavelength of UV detection
		1	2	2 vs. *rac*-1	
1a, 2a	25–40/8 min	5.0	8.3	3.43	220
1b, 2b	35–50/6 min	5.7	8.5	1.59	220

[a]Mobile phase: A: NH_4OH buffer (0.1 M, pH 9.0); B: MeOH; flow rate: 1.0 mL.min^{-1}, measurements performed at 25 °C using Gemini NX-C18 column (150 × 4.5 mm; 5 μm).
[b]The relative response factor of the cinnamic acid derivatives **2a,b** compared to the corresponding amino acids **1a,b** was determined by injecting samples from their mixture solutions of controlled molar *ratio* of *rac*-**1a(b)** : **2a(b)**, ranging from 1 : 100 to 100 : 1.

5.5.3.2 *Procedure*

1. The conversions of the preparative-scale enzymatic reactions were determined using the HPLC methods presented in Table 5.3.
2. When the enzymatic reactions reached the stationary conversions (after ~30 hr at c = 49% in the case of *rac-p*-CH_3-phenylalanine, and after ~48 hr at c = 86% in the case of *p*-Br-cinnamic acid), the reaction mixture was acidified to pH 1.5 by dropwise addition of aqueous H_2SO_4 (10%, 50% w/v). The pH of the mixture was verified using pH paper.
3. The precipitate and cell debris were removed by centrifugation of the acidified reaction mixture in 50 mL polypropylene centrifuge tubes with polyethylene screw caps at 10 000 rpm (10 947× g), 4 °C for 20 min.
4. After discarding the precipitate, the produced or remaining cinnamic acid (*p*-CH_3-cinnamic acid **2a** or *p*-Br-cinnamic acid **2b**, correspondingly) was recovered by extracting the reaction mixture with ethyl acetate (3 × 40 mL) using a separatory funnel (500 mL). After drying of the organic phase with anhydrous Na_2SO_4 and concentration by distillation under reduced pressure, the cinnamic acid (**2a** or **2b**) was recovered.
5. The amino acid (D-**1a** or L-**1b**) found within the extracted aqueous phase was purified by ion-exchange chromatography using Dowex® 50WX2 resin (15 mL in 10% H_2SO_4). First, Dowex® 50WX2 resin (15 mL) was activated by washing the resin with 10% H_2SO_4 (100 mL), and then the extracted aqueous phase was loaded.
6. The column was washed with H_2SO_4 (10% – 3 × 10 mL), then with distilled water (dH_2O), until pH of 6–7 was reached (monitored by pH paper)
7. For the elution of the amino acid (D-**1a** or L-**1b**), 100 mL 2 M NH_4OH solution was used.
8. The column was further washed with distilled water (dH_2O) until a pH of 6–7 was again established.
9. The resin was washed with 10% H_2SO_4 (50 mL) and stored in 10% H_2SO_4 for further use.
10. The fraction eluted with the 2 M NH_4OH solution was concentrated by distillation under reduced pressure, affording the pure amino acid (D-**1a** or L-**1b**).

Table 5.4 *HPLC methods for the determination of the enantiomeric excess (ee) values of the D- and L-amino acid products.*

Compound	Eluent[a] (% B)	Flow (mL.min^{-1})	Retention time (min)	
			D-1	L-1
rac-**1a**	20	0.4	4.3	9.1
rac-**1b**	30	0.4	3.7	7.2

[a]Mobile phase: A: aqueous $HClO_4$ (pH 1.5); B: ACN, measurements performed at 25 °C using Crownpak CR-I (+) column.

Table 5.5 *Reaction time, conversion, isolation yield and enantiomeric excess (ee) values of the preparative-scale enzymatic reactions.*

Substrate	*Pc*PAL variant	Reaction time (hr)	c (%)	Isolation yield (%)	ee (%)
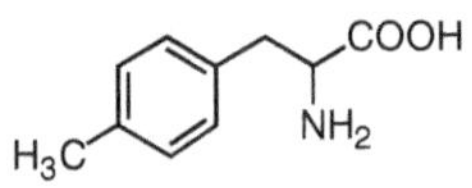	I460V	30	50	74	95
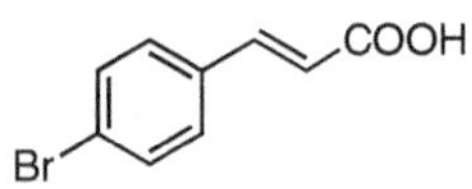	I460V	48	86	94	>99

11. The purity of the isolated products and their enantiomeric excess (ee) values (Table 5.5) were assessed by HPLC, using the methods from Tables 5.3 and 5.4.

5.5.4 Conclusion

We have described some highly efficient biocatalytic procedures for the synthesis of valuable enantiopure L- and D-phenylalanine analogues, such as L-*p*-Br-phenylalanine and D-*p*-CH_3-phenylalanine, using as biocatalyst recombinant *E. coli* whole cells harbouring the *Pc*PAL I460V variant. The procedure can be applied for the synthesis of several other optically pure phenylalanine derivatives using the corresponding tailored *Pc*PAL mutants [1, 2].

References

1. Nagy, E.Z.A., Tork, S.D., Lang, P.A. *et al.* (2019) *ACS Catalysis*, **9**, 8825–8834.
2. Filip, A., Nagy, E.Z.A., Tork, S.D. *et al.* (2018) *ChemCatChem*, **10**, 2627–2633.
3. Bencze, L.C., Filip, A., Bánóczi, G. *et al.* (2017) *Organic and Biomolecular Chemistry*, **15**, 3717–3727.
4. Ahmed, S.T., Parmeggiani, F., Weise, N.J. *et al.* (2015) *ACS Catalysis*, **5**, 5410–5413.

5. Dong, L., Marakovits, J., Hou, X. *et al.* (2010) *Bioorganic & Medicinal Chemistry Letters*, **20**, 2210–2214.
6. Hwang, T.L., Hung, C.H., Hsu, C.Y. *et al.* (2013) *Organic and Biomolecular Chemistry*, **11**, 3742–3755.
7. Dima, N.A., Filip, A., Bencze, L.C. *et al.* (2016) *Studia Universitatis Babeş-Bolyai Chemia*, **61** (2), 21–34.

5.6 Asymmetric Reductive Amination of Ketones Catalysed by Amine Dehydrogenases

Vasilis Tseliou,[1] Wesley Böhmer,[1] Maria L. Corrado,[1] Marcelo F. Masman,[1] Tanja Knaus[1] and Francesco G. Mutti*[1]

[1] *Van 't Hoff Institute for Molecular Sciences, HIMS-Biocat, University of Amsterdam, Amsterdam, The Netherlands*

Amine dehydrogenases (AmDHs) are NADH/NADPH-dependent enzymes that catalyse the reversible conversion of ketones and aldehydes into enantiomerically pure amines at the sole expense of ammonia and a hydride source. The latter is required for co-factor recycling and can usually be formate or glucose [1, 2]. A second enzyme such as a formate dehydrogenase (FDH) or a glucose dehydrogenase (GDH) catalyses the co-factor recycling. The AmDH-FDH (or GDH) system offers elevated atom economy, as the buffer of the reaction ($HCOONH_4$) directly provides the source of nitrogen and the reducing equivalents. Thus, only a catalytic amount of NAD(P)H is needed, whilst carbonate and water are the byproducts (Scheme 5.7).

There are currently 14 AmDH family members. They have either been created by enzyme engineering starting from L-amino acid dehydrogenases (L-AADHs) [2] or been discovered as native (nat-AmDHs) using metagenomic data [3]. The substrate scope of these AmDHs already covers a significant range of structurally diverse carbonyl compounds, giving access to approximately 80 α-chiral amines in high optical purity. This section describes the detailed procedure for the preparation and application of four engineered AmDHs: (i) Bb-PhAmDH, engineered from *Bacillus badius* phenylalanine dehydrogenase [4, 5]; (ii) cFL1-AmDH, a chimera obtained by domain shuffling from two first-generation AmDH variants [5, 6]; (iii) Rs-PhAmDH, engineered from *Rhodococcus* sp. *M4* phenylalanine dehydrogenase [5, 7]; and (iv) LE-AmDH-v1, engineered from *Geobacillus sterothemophilus* ε-deaminating L-lysine dehydrogenase [8]. The enzymes

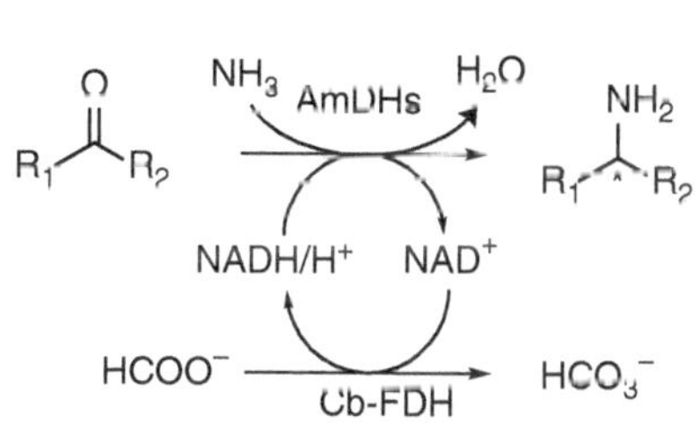

Scheme 5.7 *Biocatalytic amination using AmDHs with co-factor recycling.*

can be applied in tandem with FDH from *Candida boidinii* [9] for the reductive amination of carbonyl compounds. Overall, these AmDHs can give access to *R*-configured amines, as shown in Table 5.6.

5.6.1 Procedure 1: Expression of the Catalysts

5.6.1.1 *Materials and Equipment*

- Sterilised lysogenic broth (LB) medium (330 mL; 10 g.L^{-1} tryptone (Oxoid LP0042), 5 g.L^{-1} yeast (Oxoid LP0021), 5 g.L^{-1} NaCl)
- Kanamycin (stock solution, 50 mg.mL^{-1})
- *Escherichia coli* cells harbouring the plasmids coding for the AmDHs (Bb-PhAmDH, Rs-PhAmDH, cFL1-AmDH, LE-AmDH-v1; see Table 5.7 for enzyme sources and references for the plasmids)
- *E. coli* cells harbouring the plasmid coding for the Cb-FDH (see Table 5.7)
- Isopropyl-β-D-1-thiogalactopyranoside (IPTG stock 1 M)
- Lysis buffer (pH 8.0, 50 mM KH_2PO_4, 300 mM NaCl, 10 mM imidazole)
- Incubator for *E. coli* cell cultivation
- Centrifuge
- Eppifuge
- OD_{600} spectrophotometer
- Precast sodium dodecyl sulfate (SDS)-based protein gels
- Sample buffer, Laemmli 2× Concentrate (4% SDS, 20% glycerol, 10% 2-mercaptoethanol, 0.004% bromophenol blue and 0.125 M Tris HCl, pH approx. 6.8)
- Deionised water (dH_20)

5.6.1.2 *Procedure*

1. A sterilised LB medium (800 mL for cFL1-AmDH, 3× 800 mL for Rs-PhAmDH, LE-AmDH-v1 or Cb-FDH and 4× 800 mL for Bb-PhAmDH) supplemented with kanamycin (50 μg.mL^{-1}) was inoculated with 15 mL (for each flask) of an overnight culture (150 mL) and cells were grown at 37 °C and 170 rpm until the OD_{600} reached 0.7–1.0 (ca. 3 hr). The overnight culture was prepared by inoculating a single colony of transformed *E.coli* cells – harbouring the corresponding plasmid encoding for the AmDH or FDH – into 150 mL of LB media (sterilised) supplemented with kanamycin (50 μg.mL^{-1}). The overnight culture was grown in an incubator at 37 °C and shaken at 170 rpm.
2. Overexpression of the AmDHs or Cb-FDH was induced by addition of IPTG (0.5 mM). Flasks were shaken overnight at 25 °C and 170 rpm for cFL1-AmDH, Rs-PhAmDH, LE-AmDH-v1 and FDH and at 20 °C for Bb-PhAmDH.
3. The next day, cells were harvested by centrifugation at 4500 rpm and 4 °C for 20 min. The cell pellets were washed by resuspension in a lysis buffer (pH 8.0) and centrifuged at 4500 rpm and 4 °C. The obtained cells were weighed (approx. 5 g of cells per 800 mL culture) and stored at −20 °C.
4. It is recommended that the level of enzyme expression by SDS-gel page be estimated as follows: a sample (1 mL) of the main culture was collected prior to induction with IPTG and the OD_{600} was measured. From this sample, the volume (mL) equal to $1/(OD_{600} \times 2)$

Table 5.6 *Substrates accepted by at least one of the reported AmDHs.*

Compound	Structure	Conversion (%)[a]	ee (%)
1a	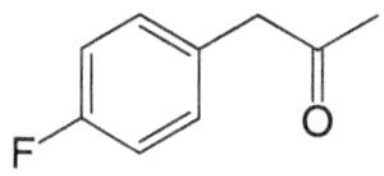	Bb-PhAmDH (93%) $_{50/50/48}$ cFL1-AmDH (93%) $_{50/30/24}$	>99 (*R*) >99 (*R*)
2a	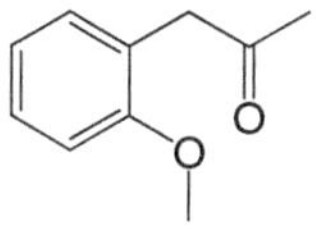	cFL1-AmDH (>99%) $_{50/130/48}$	>99 (*R*)
3a	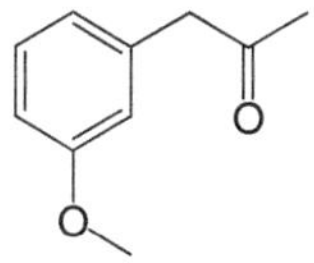	Rs-PhAmDH (98%) $_{50/50/24}$	>99 (*R*)
4a	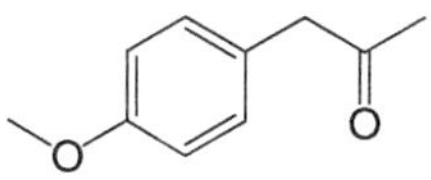	Rs-PhAmDH (>99%) $_{50/50/24}$	>99 (*R*)
5a	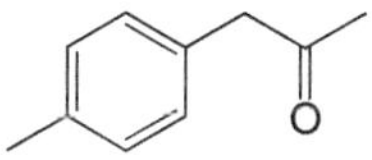	Rs-PhAmDH (98%) $_{50/130/48}$	>99 (*R*)
6a	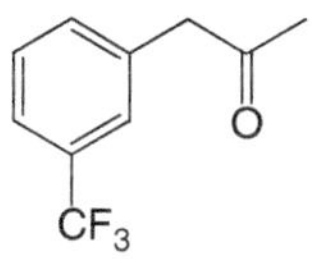	Rs-PhAmDH (98%) $_{50/130/48}$	>99 (*R*)
7a	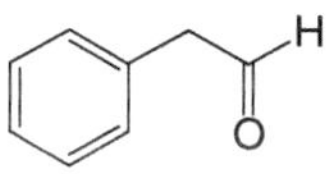	Bb-PhAmDH (34%) $_{50/50/48}$	N.A
8a	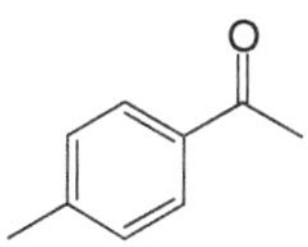	cFL1-AmDH (9%) $_{50/130/48}$	>99 (*R*)
9a	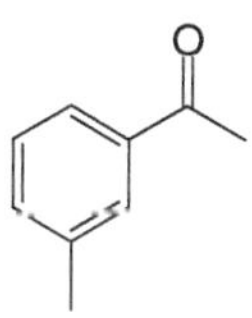	cFL1-AmDH (39%) $_{50/50/48}$ LE-AmDH-v1 (73%) $_{50/90/48}$	>99 (*R*) >99 (*R*)
10a	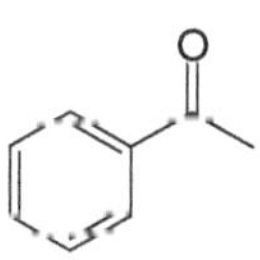	cFL1-AmDH (34%) $_{50/130/48}$ LE-AmDH-v1 (>99%) $_{50/90/48}$	>99 (*R*) >99 (*R*)

(*continued*)

Table 5.6 *(Continued)*

Compound	Structure	Conversion (%)[a]	ee (%)
11a	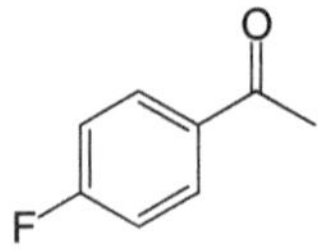	cFL1-AmDH (22%) $_{50/130/48}$ LE-AmDH-v1 (92%) $_{50/90/48}$	>99 (*R*) >99 (*R*)
12a	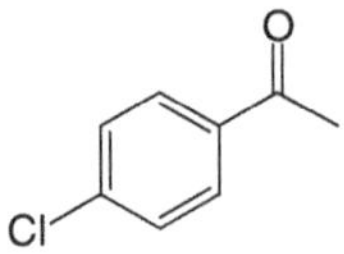	Rs-PhAmDH (33%) $_{50/100/48}$	>99 (*R*)
13a	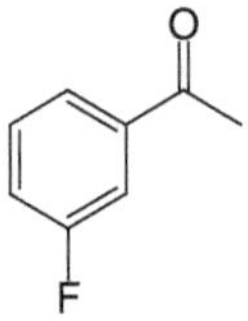	cFL1-AmDH (43%) $_{50/50/48}$ LE-AmDH-v1 (96%) $_{50/90/48}$	>99 (*R*) >99 (*R*)
14a	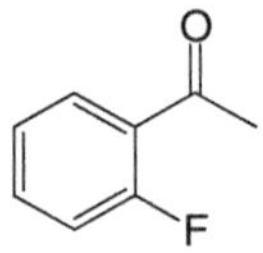	LE-AmDH-v1 (98%) $_{50/90/48}$	>99 (*R*)
15a	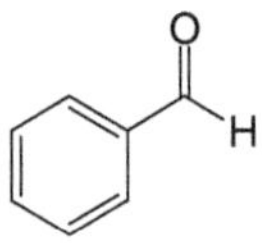	LE-AmDH-v1 (>99%) $_{50/90/48}$	N.A
16a	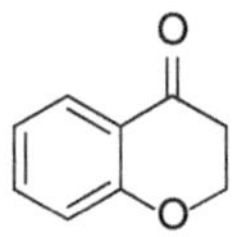	LE-AmDH-v1 (55%) $_{50/90/48}$	>99 (*R*)
17a	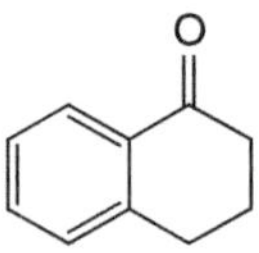	LE-AmDH-v1 (50%) $_{50/90/48}$	>99 (*R*)
18a	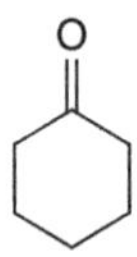	LE-AmDH-v1 (84%) $_{50/90/48}$	N.A
19a	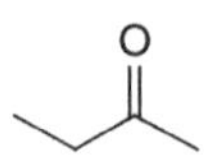	cFL1-AmDH (8%) $_{50/30/24}$ Rs-PhAmDH (4%) $_{50/50/24}$	N.D. N.D.

(continued)

Table 5.6 *(Continued)*

Compound	Structure	Conversion (%)[a]	ee (%)
20a	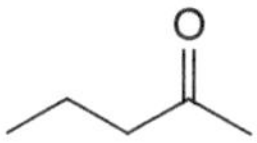	cFL1-AmDH (75%) $_{50/130/48}$ LE-AmDH-v1 (81%) $_{50/90/48}$	>99 (*R*) 89 (*R*)
21a	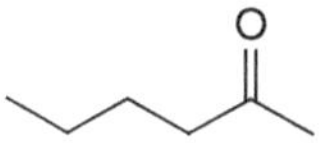	cFL1-AmDH (92%) $_{50/30/24}$ LE-AmDH-v1 (71%) $_{50/90/48}$	>99 (*R*) >99 (*R*)
22a	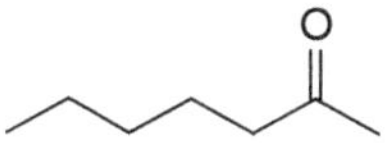	cFL1-AmDH (98%) $_{50/30/24}$ Rs-PhAmDH (99%) $_{50/50/24}$	>99 (*R*) >99 (*R*)
23a	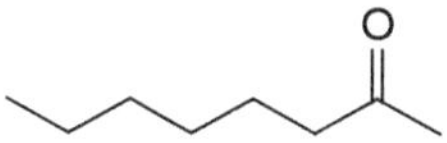	cFL1-AmDH (50%) $_{50/30/48}$ Rs-PhAmDH (93%) $_{50/50/48}$	>99 (*R*) >99 (*R*)
24a	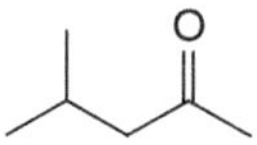	cFL1-AmDH (96%) $_{50/130/48}$ Rs-PhAmDH (91%) $_{50/130/48}$ LE-AmDH-v1 (53%) $_{50/90/48}$	>99 (*R*) >99 *R* 94 (*R*)
25a	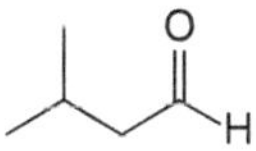	cFL1-AmDH (>99%) $_{50/30/24}$ Rs-PhAmDH (99%) $_{50/130/48}$	N.A. N.A.
26a	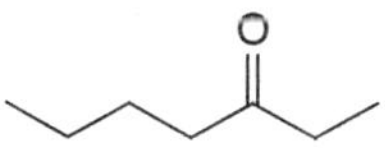	cFL1-AmDH (57%) $_{50/90/48}$ Rs-PhAmDH (32%) $_{50/100/48}$	>99 (*R*) >99 (*R*)
27a	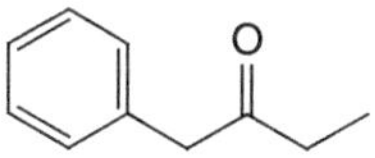	Rs-PhAmDH (99%) $_{50/130/48}$	>99 (*R*)
28a	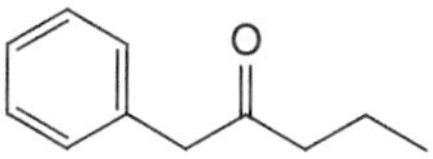	Rs-PhAmDH (71%) $_{50/100/48}$	>99 (*R*)
29a	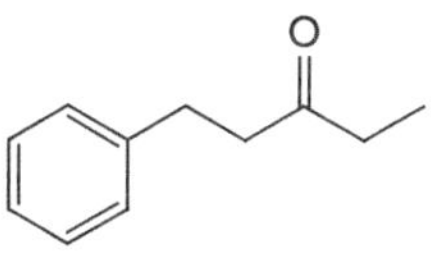	Rs-PhAmDH (87%) $_{50/100/48}$	>99 (*R*)
30a	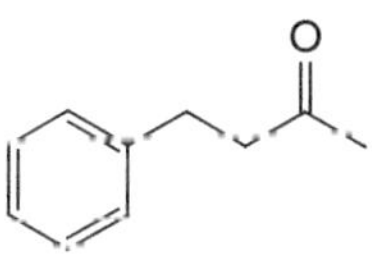	Rs-PhAmDH (>99%) $_{50/50/24}$	>99 (*R*)
31a	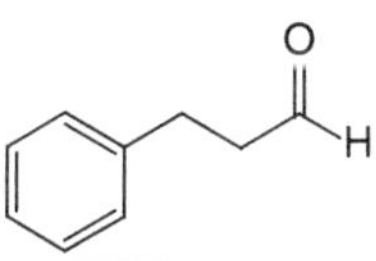	Rs-PhAmDH (96%) $_{50/100/48}$	N.A.

(continued)

Table 5.6 *(Continued)*

Compound	Structure	Conversion (%)[a]	ee (%)
32a	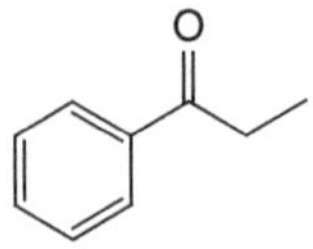	cFL1-AmDH (8%) $_{50/90/48}$ LE-AmDH-v1 (95%) $_{50/90/48}$	N.D >99.9 (*R*)
33a	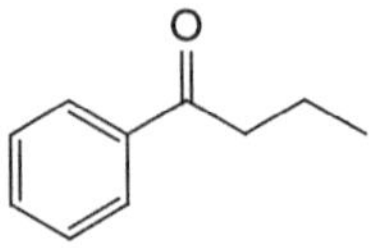	LE-AmDH-v1 (65%) $_{50/90/48}$	>99 (*R*)
34a	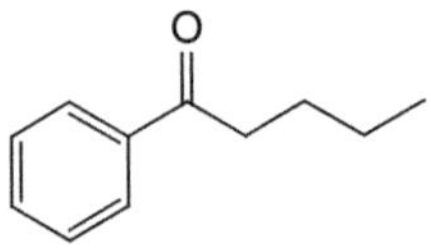	LE-AmDH-v1 (26%) $_{50/90/48}$	N.D.
35a	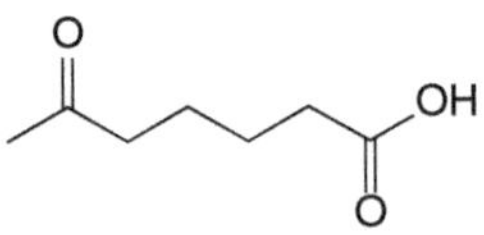	LE-AmDH-v1 (>99%) $_{50/90/48}$	>99 (*S*)
36a	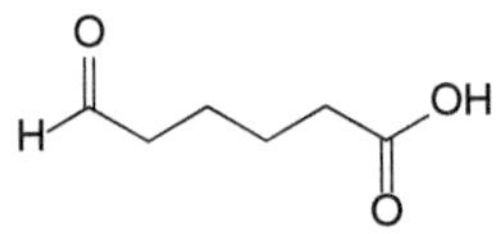	LE-AmDH-v1 (>99%) $_{50/90/48}$	N.A.
37a	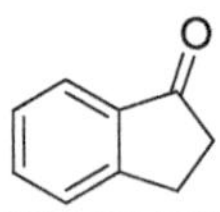	LE-AmDH-v1 (>99%) $_{20/90/48}$	>99 (*R*)

[a] The three subscripts after the percentage denote substrate concentration (mM), enzyme concentration (μM) and reaction time (hr), respectively. For Bb-PhAmDH, cFL1-AmDH and Rs-PhAmDH, the biocatalytic reactions were run at 30 °C, while for LE-AmDH-v1 at 50 °C.
N.A., not applicable; N.D., not determined.

Table 5.7 *Enzyme sources.*

Enzyme	Engineered from	Ref.
Bb-PhAmDH	*Bacillus badius* phenylalanine dehydrogenase	[4]
cFL1-AmDH	Chimeric enzyme	[6]
Rs-PhAmDH	*Rhodococcus* sp. *M4* phenylalanine dehydrogenase	[7]
LE-AmDH-v1	*Geobacillus sterothemophilus* ε-L-lysine dehydrogenase	[8]
Cb-FDH	*Candida boidinii* (wild-type)	[9]

was transferred to an Eppendorf tube (1.5 mL) and centrifuged (14 000 rpm for 1 min). The cells were resuspended in 25 μL of lysis buffer following addition of 25 μL of sample buffer (sample I). Another sample (200 μL culture + 800 μL dH_2O) was collected after the overnight induction and the OD_{600} was measured. The total volume (mL), equal to $1/(OD_{600} \times 2)$, was transferred to an Eppendorf tube and centrifuged (14 000 rpm for 1 min), then the cells were resuspended in 25 μL of lysis buffer with the addition of 25 μL of sample buffer (sample II). Equal amounts of sample I and sample II were run on an SDS-based protein gel to ensure comparable results. If the expression is accomplished successfully, it will be observed as a much thicker band for sample II compared with sample I (based on the enzyme's molecular weight). A higher enzyme expression level will result in a more intense SDS band for sample II.

5.6.2 Procedure 2: Purification of the Catalysts

5.6.2.1 Materials and Equipment

- Lysis buffer (pH 8.0, 50 mM KH_2PO_4, 300 mM NaCl, 10 mM imidazole, filtered and degassed)
- Wash buffer (pH 8.0, 50 mM KH_2PO_4, 300 mM NaCl, 25 mM imidazole, filtered and degassed)
- Elution buffer (pH 8.0, 50 mM KH_2PO_4, 300 mM NaCl, 300 mM imidazole, filtered and degassed)
- Phosphate buffer (pH 8.0, 50 mM)
- 0.45 μm Syringe Filter Whatman®
- Ultrasonicator
- Beckman J2-21 centrifuge (for lysate)
- Eppendorf 5430R centrifuge for Vivaspin®
- Centrifuge tubes Oakridge style
- NTA HisTrap FF columns (GE Healthcare Life Sciences)
- Peristaltic pump
- Vivaspin® centrifugal concentrator
- Dialysis tubes

5.6.2.2 Procedure

1. Cell pellets were defrosted and resuspended in a lysis buffer (5–10 mL of lysis buffer per gram of cell pellet).
2. The mixture containing *E. coli* cells was disrupted by sonication (amplitude = 45%; pulse on = 10 sec; pulse off = 10 sec; total sonication time = 10–20 min). Sonication was performed until the suspension became darker and less viscous. During and after sonication, the enzyme solutions were kept cooled in an ice bath in order to minimise any possible partial loss of enzymatic activity.
3. The disrupted cells were centrifuged at 16 000 rpm and 4 °C for 60 min (Beckman J2-21 centrifuge).
4. The supernatant was collected and filtered through a 0.45 μm filter. Small samples of the pellet (insoluble part) and the filtrate (soluble part) were taken for SDS-PAGE analysis (see Procedure 1, Step 4).

5. Protein purification was performed by column chromatography using prepacked Ni-NTA His trap TM HP columns (5 mL, 1 column volume, CV) according to the manufacturer's instructions. All solutions utilised during chromatography had first been filtered and degassed for 1 hr. The column was washed with dH_2O (5 CV, 25 mL) and conditioned with a lysis buffer (10 CV, 50 mL), then the filtered supernatant was loaded into the column. The column was washed with a lysis buffer (5 CV, 25 mL) and sufficient amounts of washing buffer (ca. 50 mL, 10 CV). The bound protein was recovered with an elution buffer (ca. 25 mL, aliquots of 5 mL each).
6. Note: All washing and elution fractions were analysed by SDS-PAGE. Each sample to be analysed by SDS should be diluted twice with a sample buffer (Laemmli 2× Concentrate)
7. Pure fractions containing sufficient purified protein were pooled and dialysed overnight against potassium phosphate buffer (50 mM, pH 8.0, 8 L) at 4 °C.
8. The pooled protein solution was concentrated using a Vivaspin® centrifugal concentrator (centrifuge: Eppendorf 5430R) to a volume of 5–10 mL.

5.6.3 Procedure 3: Determination of the Concentration of the Catalysts

5.6.3.1 Materials and Equipment

- Potassium phosphate buffer (pH 8.0, 50 mM)
- UV-Vis spectrophotometer
- Cuvettes for UV-Vis (1 cm path length)

5.6.3.2 Procedure

1. A 50 mM potassium phosphate buffer solution, pH 8.0 (1 mL), was transferred into a UV cuvette and a spectrum was measured from 200 to 800 nm in the UV-Vis spectrophotometer. This spectrum served as a 'blank'.
2. Pooled concentrated enzymatic solution (20 μL) was added into the phosphate buffer (980 μL, 50 mM, pH 8.0; final dilution 1 : 50). The mixture was transferred into a cuvette and the spectrum was recorded from 200 to 800 nm.
3. The enzyme concentration (M) was determined using the Lambert–Beer equation $A_{280} = \varepsilon\ (M^{-1}.cm^{-1}) \times c\ (M) \times l\ (cm)$ by considering the absorbance at 280 nm. At this point, the dilution of the enzyme (e.g. 50×) should be taken into consideration by multiplying the obtained value (M) by the dilution factor (e.g. 50). The $mg.mL^{-1}$ of the enzyme can be calculated by multiplying the calculated M by the molecular weight of the enzyme. The extinction coefficients and the molecular weight of the enzymes are reported in Table 5.8.
4. The second step was repeated using different concentrations of the enzyme to verify the obtained results.

5.6.4 Procedure 4: Representative Asymmetric Reductive Amination Catalysed by AmDH on Preparative Scale

5.6.4.1 Materials and Equipment

- Purified enzyme stock solution of Rs-PhAmDH (1.02 mL from 48.8 $mg.mL^{-1}$ stock)
- Purified enzyme stock solution of Cb-FDH (233 μL from 80.7 $mg.mL^{-1}$ stock)
- NAD^+ (20 mg, 0.03 mmol)

Table 5.8 Extinction coefficient (ε) at λ of 280 nm and molecular weight of the enzymes prepared using Procedure 3.

Enzyme	ε ($M^{-1}.cm^{-1}$)	Molecular weight (kDa)
Bb-PhAmDH	25.000	43.5
cFL1-AmDH	21.100	43.6
Rs-PhAmDH	26.600	38.9
LE-AmDH-v1	17.900	44.6
Cb-FDH	51.400	42.4

- Deionised water (dH_2O, 30 mL)
- Ammonium formate (1.90 g, 30 mmol)
- Concentrated ammonia solution (6.4 M in dH_2O, 0.274 mL)
- *p*-Fluorophenyl-2-propanone **1a** (50 mM, 195 μL, 1.27 mmol)
- Dimethyl sulfoxide (DMSO; 3 mL)
- Round-bottom flask (100 mL)
- HCl (1 M aqueous solution)
- KOH (10 M aqueous solution)
- Methyl *tert*-butyl ether (MTBE)
- $MgSO_4$ (anhydrous)
- Orbital shaker

5.6.4.2 *Procedure*

1. Ammonium formate (1.90 g, 30 mmol) was dissolved in dH_2O (25 mL), and ammonia solution (0.274 mL of a 6.4 M solution in dH_2O) was added. The pH of the ammonium formate buffer solution (ca. 1 M) was adjusted to 8.7 and dH_2O was added to a total volume of 30 mL.
2. Purified enzyme stocks of Rs-PhAmDH (49.8 mg, 43 μM) and Cb-FDH (18.8 mg, 15 μM) were added to the ammonium formate buffer (30 mL total volume, ca. 1 M, pH 8.7), which contained NAD^+ (1 mM). *p*-Fluorophenyl-2-propanone **1a** (195 μL, 1.27 mmol, 50 mM) in DMSO (3 mL, 10% v/v) was added to the reaction mixture. The reaction was incubated at 30 °C, 180 rpm for 48 hr.
3. Workup was performed by acidification of the reaction mixture to pH 2–4 via addition of 1 M HCl (aq). The water layer was washed with MTBE (15 mL) to remove any possible remaining ketone starting material. The pH of the reaction mixture was then increased (to 12–14) with 10 M KOH (aq), and the desired amine product was extracted with MTBE (2× 15 mL). The organic layers were combined and dried over $MgSO_4$. After filtration and evaporation of the solvent, the desired product was obtained in a pure form. Column chromatography was not required. The authenticity of the product was confirmed by ^{1}H-NMR.

5.6.5 Analytical Method

^{1}H NMR (400 MHz, $CDCl_3$): δ 7.12 (d, J = 8.6 Hz, 2H), 6.86 (d, J = 8.7 Hz, 2H), 3.81 (s, 3H), 3.13 (m, 1H), 2.67 (dd, J = 13.4, 5.3 Hz, 1H), 2.47 (dd, J = 13.4, 8.1 Hz, 1H), 1.12 (d, J = 6.3 Hz, 3H).

5.6.5.1 Analytics for Determination of Conversion

Column: Agilent J&W DB-1701 (30 m, 250 μm, 0.25 μm); carrier gas H_2; constant pressure 6.9 psi, split ratio 40 : 1, T injector 250 °C. Temperature program: T initial 60 °C, hold 6.5 min; gradient 20 °C.min^{-1} up to 100 °C, hold 1 min; gradient 20 °C.min^{-1} up to 280 °C, hold 1 min.

5.6.5.2 Analytics for Determination of Optical Purity

Column: Varian Chrompack Chirasel Dex-CB (25 m, 320 μm, 0.25 μm); carrier gas H_2; gas chromatography (GC) program parameters: constant flow 1.4 mL.min^{-1}, split ratio 40 : 1, T injector 200 °C. Temperature program: T initial 100 °C, hold 2 min; gradient 1°C.min^{-1} up to 130 °C, hold 5 min; gradient 10 °C.min^{-1} up to 170 °C, hold 10 min; gradient 10 °C.min^{-1} up to 180 °C; hold 1 min.

5.6.6 Conclusion

AmDHs enable the one-step conversion of ketones and aldehydes into amines, as shown in Table 5.6. The reductive amination of ketones resulted in enantiomerically pure α-chiral amines (>99% ee) possessing *R* configuration, the only exception being LE-AmDH-v1 with substrate **35a**, which afforded the corresponding *S*-configured amine (>99% ee). Substrates **20a** (89% ee (*R*)) and **24a** (94% ee (*R*)) were obtained with imperfect stereoselectivity.
LE-AmDH-v1, Rs-PhAmDH, Bb-PhAmDH and cFL1-AmDH show complementary substrate scopes. On the one hand, LE-AmDH-v1can be used for the preparation of acetophenone (**10a**), acetophenone derivatives (**9a**, **11a**, **13a**, **14a**) or bulky-bulky ketones such as **32–34a**, in which the carbonyl moiety is adjacent to the phenyl group. Similarly, LE-AmDH-v1 can give access to 4-aminochromane (from **16a**), 1-aminotetralin (from **17a**) and 1-aminoindan (from **37a**). On the other hand, Rs-PhAmDH, Bb-PhAmDH and cFL1-AmDH accept similar types of substrates, but Rs-PhAmDH is the preferable biocatalyst for the preparation of amphetamines (**1–6a**). Notably, Rs-PhAmDH is active towards ketone substrates in which the carbonyl moiety is located two or three carbons away from the phenyl group (e.g. **27–31a**). Finally, cFL1-AmDH is the preferable biocatalyst for the reductive amination of linear or branched aliphatic ketones possessing from five to eight carbons (e.g. **20–26a**).
Large-scale reductive amination resulted in amine products possessing excellent chemical and optical purities, simplifying the workup procedure. In fact, the unreacted ketone can first be extracted under acidic conditions; after the change in pH to basic, the enantiomerically pure amine is extracted.
The members and diversity of the AmDH family are expected to grow in the coming years, increasing their applicability towards the synthesis of more complex α-chiral amines. Further information about the enzyme sources can be found in Table 5.7.

References

1. Abrahamson, M.J., Vazquez-Figueroa, E., Woodall, N.B. *et al.* (2012) *Angewandte Chemie International Edition*, **51**, 3969–3972.
2. Patil, M.D., Grogan, G., Bommarius, A. and Yun, H. (2018) *ACS Catalysis*, **8**, 10 985–11 015.
3. Mayol, O., Bastard, K., Beloti, L. *et al.* (2019) *Nature Catalysis*, **2**, 324–333.
4. Abrahamson, M.J., Wong, J.W. and Bommarius, A.S. (2013) *Advanced Synthesis & Catalysis*, **355**, 1780–1786.

5. Knaus, T., Böhmer, W. and Mutti, F. G. (2017) *Green Chemistry*, **19**, 453–463.
6. Bommarius, B.R., Schurmann, M. and Bommarius, A.S. (2014) *ChemComm*, **50**, 14 953–14 955.
7. Ye, L.J., Toh, H.H., Yang, Y. *et al.* (2015) *ACS Catalysis*, **5**, 1119–1122.
8. Tseliou, V., Knaus, T., Masman, M.F. *et al.* (2019) *Nature Communications*, **10**, 3717.
9. Schutte, H., Flossdorf, J., Sahm, H. and Kula, M.-R. (1976) *European Journal of Biochemistry*, **62**, 151–160.

5.7 Utilisation of Adenylating Enzymes for the Formation of *N*-Acyl Amides

Shona M. Richardson, Piera M. Marchetti, Noor M. Kariem and Dominic J. Campopiano*
School of Chemistry, University of Edinburgh, Edinburgh, UK

Although an abundantly used chemical conversion, current methods for amide bond formation from carboxylic acids are inefficient. They suffer from poor atom economy with stoichiometric amounts of coupling agents, harsh conditions and the production of waste [1–3]. However, the high polarity, stability and conformational diversity make them desirable functional groups [4]. Enzymatic routes for amide bond formation bypass these restraints through the favoured reaction conditions of an aqueous environment with benign components. One such enzyme class is the amide bond synthetases (ABSs), a subgroup of which are members of the adenylating enzyme superfamily (ANLs). This diverse family of enzymes activate carboxylic acids for nucleophilic attack through a two-step catalytic reaction. First, a reactive acyl-adenylate intermediate is formed by condensing adenosine triphosphate (ATP) with the carboxylic acid. This intermediate is subsequently captured by the incoming amine nucleophile to form the desired amide product [5–10]. Figure 5.2 shows the differences between the chemical and biocatalytic methods of amide bond formation, emphasising the requirement for an additional coupling reagent in the case of the chemical reaction. Using a biocatalyst as an alternative synthetic strategy has its own drawbacks, however, with the rigid substrate specificity limiting the enzymatic applications. Recent investigations into our enzyme of interest, *Pseudoalteromonas tunicata* TamA, a member of the ANL superfamily, have shown significant promiscuity with respect to the fatty acid (FA) and amine substrates, making this enzyme an ideal new member of the biocatalyst toolbox [11, 12]. Here, we describe the experimental details that show how TamA can be used to screen a range of FA and amino acid substrates and to prepare a series of *N*-acyl amides such as the *N*-C_{12} L-histidine natural product (2*S*)-2-(dodecanoylamino)-3-(1*H*-imidazol-5-yl)propanoic acid, also known as *N*-dodecanoyl-L-histidine (*N*-C_{12}-L-histine) (Figure 5.2).

5.7.1 Procedure 1: Screening the TamA ANL Biocatalyst with Various Fatty Acid and Amino Acid Substrates

5.7.1.1 Materials and Equipment

- Purified recombinant TamA ANL domain from *P. tunicata* expressed in *E. coli* BL21 (DE3) using plasmid pET-HisTev_TamA ANL (expressed and purified as reported in [12])
- Carboxylic acid stock (typically 25 mM) in dimethyl sulfoxide (DMSO) (C_{10}–C_{14})
- Amino acid stock (1 M, pH 9.0) (histidine, methionine or proline)

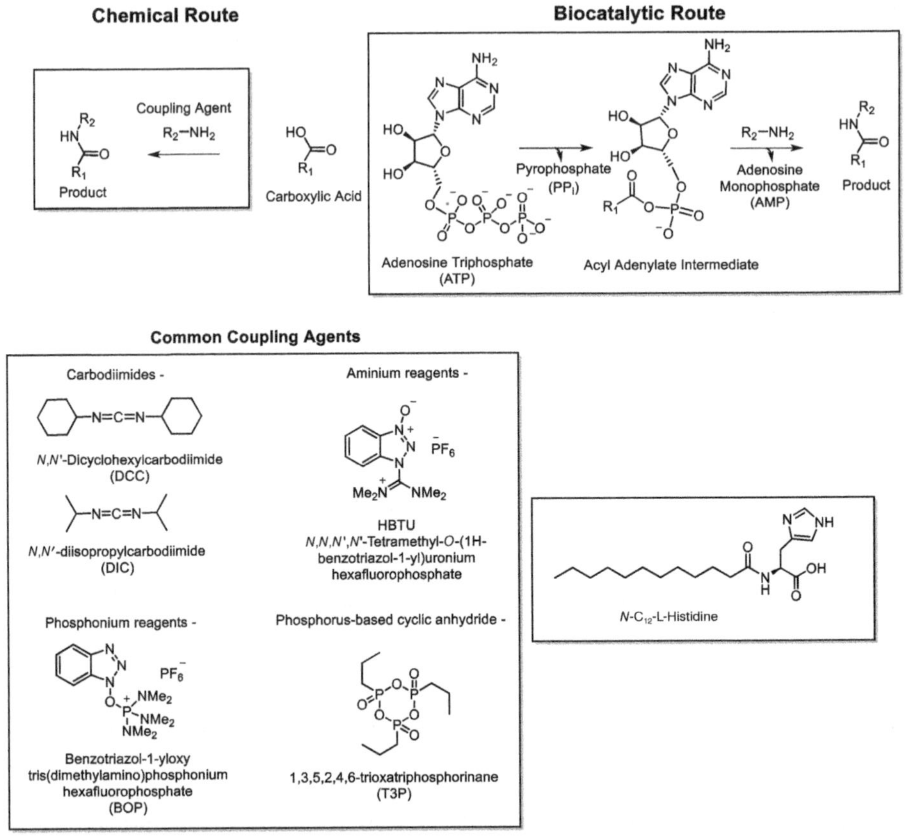

Figure 5.2 *Comparison of the chemical and biocatalytic routes to the synthesis of amides. Highlighted are the structures of the common coupling agents used in chemical synthesis and the* N-*C_{12}-L-histidine amide target synthesised using the TamA biocatalyst.*

- Sodium carbonate/bicarbonate buffer (1 : 8) (200 mM, pH 9.0)
- ATP stock (31.5 mM)
- $MgCl_2$ stock (100 mM)
- Acetonitrile (with 0.01/0.1% triflouroacetic acid, TFA)
- Thermoshaker
- 1.5 mL Eppendorf flask

5.7.1.2 Procedure

1. To a 1.5 mL Eppendorf, 500 µL of sodium carbonate/bicarbonate buffer (200 mM, pH 9.0) was added.
2. The rest of the reaction components were added to final concentrations of: fatty acid, 1 mM, 40 µL, 4% DMSO; ATP, 5 mM, 160 µL; $MgCl_2$, 10 mM, 100 µL; and amine, pH 9.0, 100 mM, 100 µL.
3. The reaction was initiated by the addition of the enzyme (10 µg.mL^{-1} final, 100 µL) to give a final reaction volume of 1 mL.

4. The reactants were shaken at 37 °C and 250 rpm for 24 hours.
5. The reaction was stopped by the addition of an equal volume of acetonitrile with TFA (1 mL) – use 0.1% TFA for high-performance liquid chromatography (HPLC) analysis or 0.01% TFA for liquid chromatography electrospray ionisation-mass spectrometry (LC ESI-MS) analysis.
6. The samples were centrifuged at 13 300 rpm for 10 min.
7. The samples were analysed by HPLC and LC ESI-MS (Table 5.9 shows some of the accepted carboxylic acids and amines).

5.7.2 Procedure 2: Preparative-Scale Synthesis of the *N*-C_{12}-L-Histidine Amide Using Recombinant TamA ANL

5.7.2.1 *Materials and Equipment*

- Purified recombinant TamA ANL domain from *P. tunicata* expressed in *E. coli* BL21 (DE3) using plasmid pET-HisTev_TamA ANL (expressed and purified as reported in [12])
- Dodecanoic acid stock (20 mg in 4 mL) in DMSO
- Sodium carbonate/bicarbonate buffer (200 mM, pH 9.0, 50 mL)
- ATP stock solution (132 mg in 16 mL water)
- $MgCl_2$ stock solution (190 mg in 20 mL water)
- Histidine stock solution (155 mg in 10 mL water, pH 9.0)
- Acetonitrile (0.01/0.1% TFA)
- Thermoshaker
- 250 mL flask
- Ethanol

5.7.2.2 *Procedure*

1. Dodecanoic acid (20 mg, 0.1 mmol) in a 4% w/v DMSO solution, ATP (132 mg, 0.24 mmol), $MgCl_2$ (190 mg, 2 mmol) and L-histidine (155 mg, 1 mmol) in a 100 mM solution of sodium carbonate/sodium bicarbonate buffer, pH 9.0 (50 mL) were added to a 250 mL flask to a final volume of 100 mL.
2. The reaction was initiated by the addition of the TamA ANL enzyme (1 mg, 1 mL).
3. Enzyme was continually added at 1 $mg.hr^{-1}$ until a total of 5 mg had been added.
4. The reaction was shaken at 37 °C and 250 rpm.
5. After 24 hr, a white precipitate had formed, which was isolated by filtration and washed with water and ethanol to afford *N*-C_{12}-L-histidine as a white precipitate (29 mg, 86%) (see Scheme 5.8). This was analysed on a 500 MHz Bruker NMR and by matrix assisted laser desorption ionisation, time of flight mass spectrometry (MALDI ToF) on a Bruker Ultraflex using 3,5-dimethoxy-4-hydroxy-cinnamic acid (sinapinic Acid) as the matrix.

δ H (500 MHz, DMSO-d^6) 7.90 (s,1H), 7.56 (d, J = 10, 1 H),6.74 (s, 1H), 4.35 (m, 1 H), 2.94 (m, 2H), 2.19 (t, J = 7.4, 2H), 2.06 (t, J = 7.5, 2H), 1.50 (t, J = 7.5 , 2H), 1.25 (s, 16 H), 0.86 (t, J = 2.5, 3 H). *m/z* (MALDI) for $[C_{18}H_{31}N_3O_3]^+$ predicted: 338.24, observed: $[M+H]^+$ 338.239, 339.242 and 340.244.

Table 5.9 *Panel of fatty acid and amine substrates used to screen the TamA-catalysed formation of* N*-acyl amides.*

Fatty acid substrate	Amino acid substrate	Conversion (%)
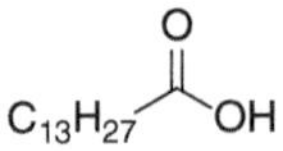	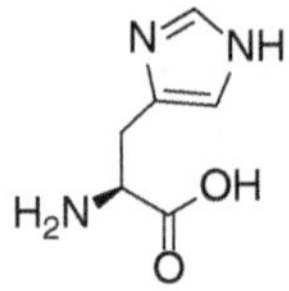	1
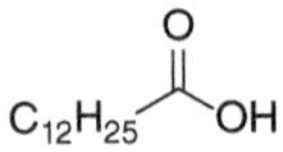	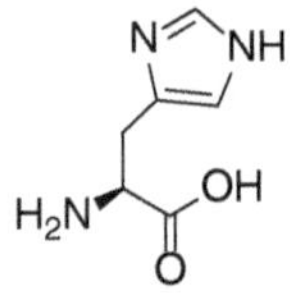	5
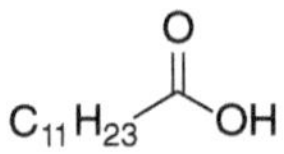	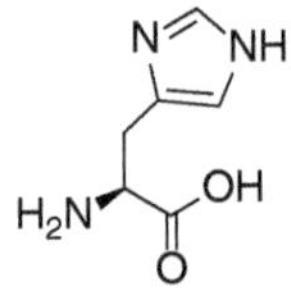	12
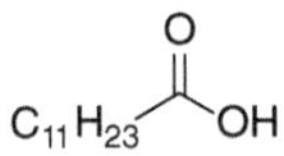	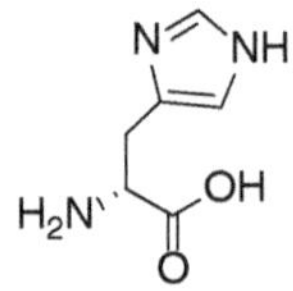	19
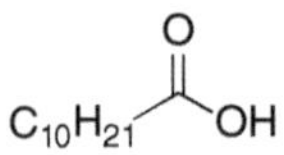	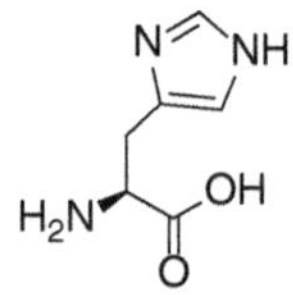	15
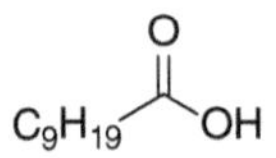	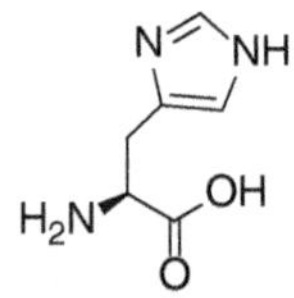	34
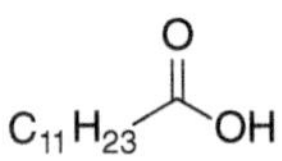	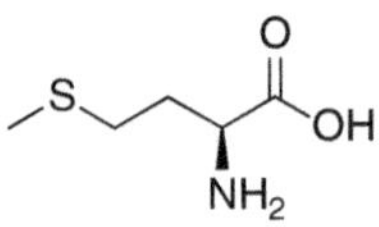	6
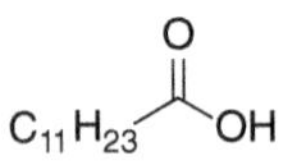	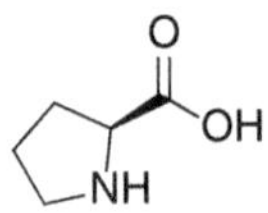	34

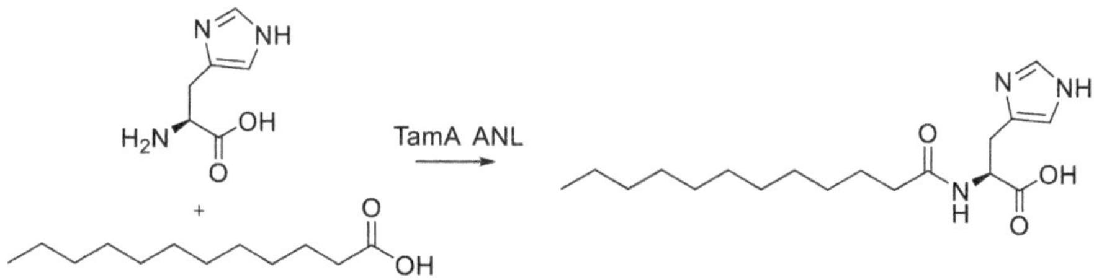

Scheme 5.8 Preparative-scale synthesis of the N-C_{12}-L-histidine amide using recombinant TamA ANL.

5.7.3 Analytical Method

5.7.3.1 *Product Analysis of* N-*acyl Amides by LC ESI-MS*

- Aliquots of the reaction (50 μL) were periodically taken, quenched with an equal volume of acetonitrile (0.01% TFA), centrifuged at 13 300 rpm for 10 min, loaded on to a Synapt G2-Si Q-TOF (Waters) instrument and analysed via LC ESI-MS with a Phenomenex Jupiter C18 5 μm 300 Å LC column coupled to an ESI source.
- Products were eluted using a gradient from 5% acetonitrile and 95% water with 0.1% formic acid to 95% acetonitrile over 30 min. The MS source was set at 120 °C, back pressure 2 mbar and sampling cone voltage 54 V. Extracted ion chromatograms (EICs) and masses were determined on MassLynx V4.1 software.

5.7.3.2 *HPLC to Determine Conversion with TamA ANL for the Formation of* N-C_{12}-*L-Histidine*

- Authentic *N*-acyl-amino acid samples were chemically synthesised using the coupling agent T3P [13], pyridine and DMF following published methods [14]. Samples for HPLC analysis were prepared in a 1 : 1 volume of acetonitrile (0.1% TFA)/water.
- Both authentic standards and reaction samples were centrifuged at 13 300 rpm for 10 min before 10 μL of the supernatant was injected on to a Phenomenex Luna 5 μm C18 100 Å HPLC column.
- Samples were eluted with water (0.1% TFA) for 5 min, followed by a 30 min gradient to 95% acetonitrile (0.1% TFA). This concentration was maintained for 5 min and then returned to 100% water (0.1% TFA) for 5 min.
- The eluent was monitored at 210 nm.

5.7.4 Conclusion

Having tested both the promiscuity and the synthetic utility of the TamA ANL domain, this ATP-dependent enzyme has been shown to be a useful tool in the synthesis of a series of long-chain fatty amides (*N*-acyl amides). The TamA ANL biocatalyst displays a broad FA substrate scope for chain lengths C_{10}–C_{14}, presumably via the corresponding acyl-adenylate intermediate. The experimental data also confirms the ability of the enzyme to accept a range of amino acid substrates, including L- and D-histidine, which provides evidence to support the hypothesis that the second step of the mechanism (reaction of amine with acyl-adenylate) is not enzyme-catalysed. Future work could see the use of *P. tunicata* TamA ANL for the preparation of clinically useful amide targets [7] and expansion of the substrate scope of the biocatalyst by structure-guided rationale engineering [15] combined with high-throughput screening.

References

1. Pattabiraman, V.R. and Bode, J.W. (2011) *Nature*, **480**, 471–479.
2. Pitzer, J. and Steiner, K. (2016) *Journal of Biotechnology*, **235**, 32–46.
3. Sabatini, M.T., Boulton, L.T., Sneddon. H.F. and Sheppard, T.D. (2019) *Nature Catalysis*, **2**, 10–17.
4. Sasaki, K. and Crich, D. (2011) *Organic Letters*, **13**, 2256–2259.
5. Gulick, A.M. (2009) *ACS Chemical Biology*, **4**, 811–827.
6. Schmelz, S. and Naismith, J.H. (2009) *Current Opininion in Structural Biology*, **19**, 666–671.
7. Adams, J.P., Brown, M.J.B., Diaz-Rodriguez, A. *et al.* (2019) *Advanced Synthesis and Catalysis*, **361**, 2421–2432.
8. Devine, P.N., Howard, R.M., Kumar, R. *et al.* (2018) *Nature Reviews Chemistry*, **2**, 409–421.
9. Hughes, D.L. (2018) *Organic Process Research and Development,* **22**, 574–584.
10. Petchey, M.R. and Grogan, G. (2019) *Advanced Synthesis and Catalysis*, **361**, 3895–3914.
11. Marchetti, P.M., Kelly, V., Simpson, J.P. *et al.* (2018) *Organic and Biomolecular Chemistry*, **16**, 2735–2740.
12. Marchetti, P.M., Richardson, S.M., Kariem, N.M. and Campopiano, D.J. (2019) *MedChemComm*, **10**, 1192–1196.
13. Hosamani, B. (2013) *Synthesis*, **45**, 1569–1601.
14. Petchey, M., Cuetos, A., Rowlinson, B. *et al.* (2018) *Angewandte Chemie- International Edition*, **57**, 11 584–11 588.
15. Weng, J.-Y., Bu, X.-L., He, B.-B. *et al.* (2019) *Chemical Communications*, **55**, 14 840–14 843.

6

Carbon–Carbon Bond Formation or Cleavage

6.1 An Improved Enzymatic Method for the Synthesis of (*R*)-Phenylacetyl Carbinol

Nan Liu, Xian Hui, Heng Li and Wen-Yun Gao*
College of Life Sciences, Northwest University, Xi'an, People's Republic of China

In both synthetic organic chemistry and medicinal chemistry, chiral α-hydroxyketones are versatile building blocks. (*R*)-Phenylacetyl carbinol (*R*-PAC) is one of the key chiral α-hydroxyketones that has been utilised as a synthon in the synthesis of a number of pharmaceuticals having α- and β-adrenergic properties, such as phenylpropanolamine, pseudoephedrine, L-ephedrine and norephedrine [1]. Currently, *R*-PAC is mainly manufactured by a fermentation process in which pyruvate decarboxylase (PDC) accomplishes the condensation of pyruvate with externally supplied benzaldehyde (BA) (Scheme 6.1) [2]. Such biotransformation suffers from a low reaction rate and inefficient conversion of substrates into *R*-PAC [2, 3]. To solve these problems, PDCs of different microbial origins have been tested, but no good result has been obtained [1,4]. Although the site-directed mutagenesis of *Zymomonas mobilis* PDC (ZmPDC) has led to some better PDC mutants, their catalytic efficiency is still very low [5].

Acetohydroxyacid synthase (AHAS) is a thiamine diphosphate (ThDP)-dependent enzyme that catalyses the condensation reaction of two molecules of pyruvate to form (*S*)-acetolactate [6] and can also mediate the reaction between pyruvate and BA to generate the chiral-hydroxyketone *R*-PAC (Scheme 6.1) [7]. To date, AHAS I of *Escherichia coli* has proved to be the best target for catalysing the ligation between pyruvate and BA and its analogues [7a,8], but this protein has its own drawbacks: it consists of two homodimers, one of which comprises two catalytic subunits (CSUs) that contain the full catalytic machinery needed to perform the chemical reaction, whilst the other contains

Applied Biocatalysis: The Chemist's Enzyme Toolbox, First Edition. Edited by John Whittall and Peter W. Sutton.
© 2021 John Wiley & Sons Ltd. Published 2021 by John Wiley & Sons Ltd.

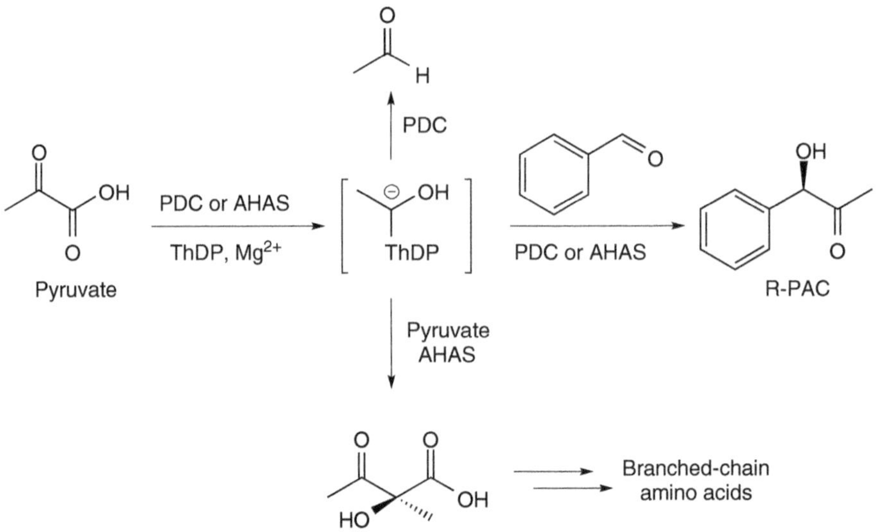

Scheme 6.1 *Enzymatic reactions catalysed by PDC and AHAS enzymes (Reproduced by permission from RSC Adv., 2017,7, 32664–32668).*

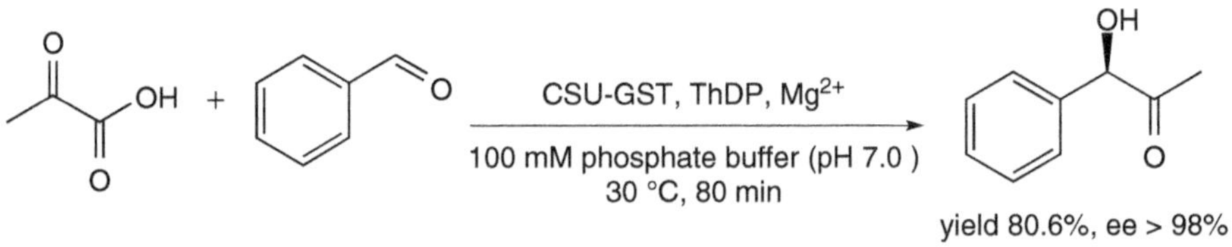

Scheme 6.2 *Enzymatic preparation of R-PAC using CSU-GST of E. coli AHAS I as a catalyst.*

two regulatory subunits (RSUs), which are necessary for stabilisation and regulation of the CSU [6]. The architecture of the protein makes it difficult to be prepared. Although it has been reported that the CSU of *E. coli* AHAS I is active, its activity is very weak in the absence of RSU. In addition, the CSU alone is even more unstable than AHAS I [9].

In one of our projects, we found the CSU of *E. coli* AHAS I with a glutathione S-transferase (GST) tag (CSU-GST) showed not only excellent stability, but also high activity [10]. Moreover, CSU-GST was able to mediate the condensation between pyruvate and BA to generate *R*-PAC as well. We thus extended the application of the protein to the synthesis of the chiral α-hydroxyketone *R*-PAC (Scheme 6.2).

6.1.1 Procedure 1: Preparation of CSU-GST of AHAS in *E. coli* BL21(DE3)

6.1.1.1 Materials and Equipment

- Autoclaved lysogenic broth (LB) medium
- Ampicillin (50 mg.mL^{-1} in ddH_2O, filter-sterilised)
- Isopropyl-β-D-1-thiogalactopyranoside (IPTG); 1 M in ddH_2O, filter-sterilised)

- LB agar plate with colonies of the strain *E. coli* BL21(DE3)-pPROEXTM HTb-*ilvB* (the *E. coli* BL21(DE3) strain harbouring the expression vector pPROEXTM HTb bearing the *E. coli ilvB* gene encoding the CSU of AHAS I) [10]
- Constant-temperature incubator shaker
- Cooling centrifuge
- Glutathione Sepharose 4B
- All other chemicals of analytical grade unless stated otherwise
- Lysis buffer (50 mM KH_2PO_4, 300 mM NaCl, 5 mM KCl, 5 mM BME, pH 7.5)
- Elution buffer (50 mM Tris-HCl containing 15 mM reduced glutathione, pH 8.0)
- Dialysis buffer (50 mM KH_2PO_4, 300 mM NaCl, 5 mM KCl, 5 mM BME, pH 7.5)

6.1.1.2 Procedure

1. To autoclaved LB medium (5 mL) containing ampicillin (final concentration 0.1 mg.mL^{-1}) in a 25 mL Erlenmeyer flask, a single colony of the *E. coli* BL21(DE3)-pPROEXTM HTb-ilvB strain was inoculated. The solution was shaken overnight at 37 °C and 200 rpm.
2. 1 mL of the solution was transferred to 100 mL autoclaved LB medium containing ampicillin (0.1 mg.mL^{-1}) and the mixture was shaken at 200 rpm and 37 °C for 12 hr.
3. 10 mL of the solution obtained in Step 2 was transferred to 1 L autoclaved LB medium containing ampicillin (0.1 mg.mL^{-1}) and the mixture was shaken at 200 rpm and 37 °C for 2~4 hr until the OD_{600} had reached 0.2~0.3.
4. IPTG (final concentration 0.5 mM) was added to the solution and the induction was performed at 30 °C and 180 rpm until the OD_{600} had reached 0.5~0.6 (ca. 6 hr).
5. Bacteria from the culture were harvested by centrifugation at 8000 rpm for 15 min (about 4 g).
6. The bacterial precipitation was resuspended in 8 mL lysis buffer (50 mM KH_2PO_4, 300 mM NaCl, 5 mM KCl, 5 mM BME, 1 mM PMSF and 5 mg.mL^{-1} lysozyme, pH 7.5) and the mixture was put on ice for 30 min. The lysed bacteria were subsequently centrifuged at 11 000 rpm and 4 °C for 30 min.
7. The supernatant was incubated under gentle shaking with 2.5 mL glutathione sepharose 4B resin at room temperature for 0.5 hr before being slowly poured out.
8. The residue resin was washed with lysis buffer (2 × 3 mL) and then eluted with elution buffer (3 mL).
9. The eluate was dialysed overnight at 4 °C against dialysis buffer (1 L). The LSU-GST protein thus obtained was analysed by SDS-PAGE and aliquoted into 0.5 mL Eppendorf tubes, flash-frozen and stored at −80 °C for later use.

6.1.2 Procedure 2: Preparation of *R*-PAC Mediated by CSU-GST

6.1.2.1 Materials and Equipment

- Sodium pyruvate
- Benzaldehyde (BA)
- Thiamine diphosphate (ThDP)
- Silica gel (100~200 mesh)

- Propanol, acetonitrile and *n*-hexane of high-performance liquid chromatography (HPLC) grade
- Chiralcel OD-H column (250 × 4.6 mm, 5 μm)
- Agilent RP-18 column (250 × 4.6 mm, 5 μm)
- Agilent 1200 HPLC system with an autosampler and a diode array spectrometer
- Varian Inova-400 MHz nuclear magnetic resonance (NMR) spectrometer
- Thermo Fisher LTQ XL mass spectrometer
- Polarimeter
- All other chemicals of analytical grade unless stated otherwise

6.1.2.2 Procedure

1. To 0.5 mL reaction mixture containing 100 mM potassium phosphate buffer (pH 7.0), 5.0 mM $MgCl_2$, 1.0 mM ThDP, 0.5 mM dithiothreitol, 6% DMSO, 40 mM BA and 100 mM pyruvate, 0.15 mg CSU-GST protein was added. The incubation was then carried out at 30 °C for 80 min.
2. The mixture was treated in a boiling water bath for 2 min, cooled and centrifuged at 10 000 rpm for 3 min.
3. The supernatant was extracted three times with $CHCl_3$ (1.0 mL × 3) and the extracts were pooled, back-washed twice with brine and dried over about 0.5 g anhydrous $MgSO_4$.
4. After 2 hr, the desiccant was removed through filtration and washed twice with $CHCl_3$ (1.0 mL × 2). The $CHCl_3$ was pooled and removed under reduced pressure.
5. The residue obtained was subsequently isolated on a silica gel column (1.0 × 10 cm, 10 g silica gel) developed by petroleum ether–ethyl acetate (3 : 1). The fractions containing the target molecule were combined and solvent was removed under reduced pressure. A white powder about 9.67 mg (yield 80.6%) was thus generated.
6. The purity of the product (>98%) was evaluated by HPLC on an Agilent RP-18 column (250 × 4.6 mm, 5 μm) using 30% aqueous acetonitrile as a mobile phase at a flow rate of 0.7 $mL.min^{-1}$, a detection wavelength of 280 nm and ambient temperature.
7. The enantiomeric excess value (>98% ee) of the enzymatic reaction was assessed by a chiral HPLC method on a chiral column Chiralcel OD-H (250 × 4.6 mm, 5 μm) at 25 °C and with a flow rate of 0.7 $mL.min^{-1}$. The mobile phase used was n-hexane/2-propanol 90 : 10. The UV detection wavelength was at 280 nm.
8. The product was further elucidated and confirmed by optical rotation, HRESI-MS, ^{1}H- and ^{13}C-NMR.

White powder, $[\alpha]_D{}^{20} = -177°$ (ca. 1.6, MeOH). ^{1}H-NMR (600 MHz, $CDCl_3$) δ 2.09 (s, 3H, Me), 4.29 (d, $J = 3$ Hz, 1H, OH), 5.09 (d, $J = 3$ Hz, 1H, CH), 7.32–7.39 (m, 5H, aromatic); ^{13}C-NMR (100 MHz, $CDCl_3$) δ 25.391 (CH_3), 80.224 (CH), 127.455 (aromatic CH × 2), 128.765 (aromatic CH), 129.164 (aromatic CH × 2), 137.979 (aromatic CH), 207.208 (C=O); HRMS (ESI) (negative mode) *m/z*: $[M-H]^-$ found 149.0605, $C_9H_9O_2$, requires 149.0603. All the data acquired were identical with those reported previously for *R*-PAC [11].

6.1.3 Conclusion

In this study, a practical enzymatic procedure for the preparation of the chiral α-hydroxyketone *R*-PAC using the GST-tag-fused CSU of *E. coli* AHAS I as a catalyst was established. Under the optimised reaction conditions, *R*-PAC could be efficiently produced in a good yield and with excellent stereoselectivity. We anticipate that this work could not only facilitate the preparation of *R*-PAC, but also promote the synthetic applications of *E. coli* AHAS I.

References

1. (a) Iding, H., Siegert, P., Mesch, K. and Pohl, M. (1998) *Biochimica et Biophysica Acta*, **1385**, 307–322; (b) Shukla, V.B. and Kulkarni, P.R. (2000) *World Journal of Microbiology & Biotechnology*, **16**, 499–506.
2. Rogers, P.L., Shin, H.S. and Wang, B. (1997) *Advances in Biochemical Engineering/ Biotechnology*, **56**, 33–59.
3. Long, A., James, P. and Ward, O.P. (1989) *Biotechnology and Bioengineering*, **33**, 657–660.
4. (a) Kren, V., Crout, D.H.G., Dalton, H. *et al.* (1993) *Journal of the Chemical Society, Chemical Communications*, **1993**, 341–342. (b) Ward, O.P. and Singh, A. (2000) *Current Opinion in Biotechnology*, **11**, 520–526.
5. (a) Bruhn, H., Pohl, M., Grotzinger, J. *et al.* (1995) *European Journal of Biochemistry*, **234**, 650–655. (b) Meyer, D., Walter, L., Kolter, G. *et al.* (2011) *Journal of the American Chemical Society*, **133**, 3609–3616.
6. Bunik, V.I., Tylicki, A. and Lukashev, N.V. (2013) *FEBS Journal*, **280**, 6412–6442.
7. (a) Engel, S., Vyazmensky, M., Geresh, S. *et al.* (2003) *Biotechnology and Bioengineering*, **83**, 833–840; (b) Sehl, T., Hailes, H.C., Ward, J.M. *et al.* (2013) *Angewandte Chemie International Edition*, **52**, 6772–6775.
8. Engel, S., Vyazmensky, M., Berkovich, D. *et al. Biotechnology and Bioengineering*, **88**, 825–831.
9. Vinogradov, V., Vyazmensky, M., Engel, S. *et al.* (2006) *Biochimica et Biophysica Acta*, **1760**, 356–363.
10. Li, H., Liu, N., Wang, W.T. *et al.* (2016) *Journal of Bioscience and Bioengineering*, **121**, 21–26.
11. (a) Sehl, T., Simon, R.C., Hailes, H.C. *et al.* (2012) *Journal of Biotechnology*, **159**, 188–194; (b) Brusee, J., Roos, E.C. and Der Gen, A.V. (1988) *Tetrahedron Letters*, **29**, 4485.

6.2 Tertiary Alcohol Formation Catalysed by a Rhamnulose-1-Phosphate Aldolase : Dendroketose-1-Phosphate Synthesis

Victor Laurent,[1] Virgil Hélaine,[1] Marielle Lemaire,[1] Marcel Salanoubat[2], Véronique de Berardinis[2] and Christine Guérard-Hélaine[1]
[1] *Université Clermont Auvergne, CNRS, SIGMA Clermont, Institut de Chimie de Clermont-Ferrand, Clermont-Ferrand, France*
[2] *Génomique Métabolique, Genoscope, Institut François Jacob, CEA CNRS, University of Evry, Evry, France*

Taking advantage of the promiscuous substrate activity of aldolases, we recently discovered that many diverse rhamnulose-1-phosphate aldolases (RhuAs) [1] were able to catalyse the addition of their natural nucleophile, dihydroxyacetone phosphate (DHAP) **1**, on to several activated ketones as the electrophile. This addition is exemplified here with

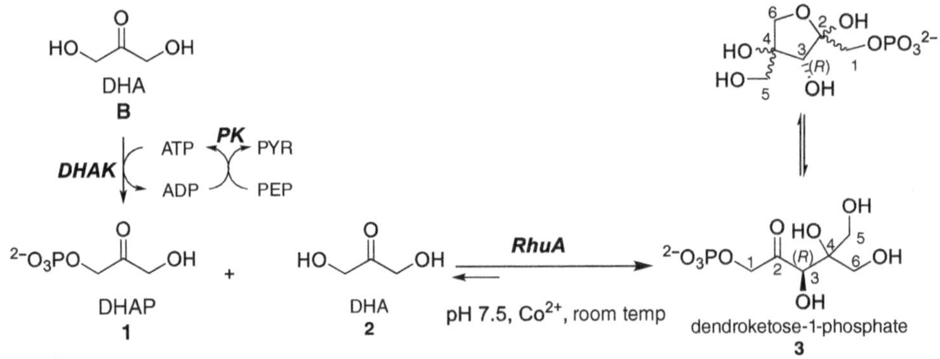

Scheme 6.3 *Preparation of dendroketose-1-phosphate* ***3*** *catalysed by rhamnulose-1-phosphate aldolase from Bacteroides thetaiotaomicron* ($RhuA_{Bthet}$).

dihydroxyacetone (DHA) **2**, allowing the preparation of the corresponding tertiary alcohol 3*R*-dendroketose-1-phosphate **3** (Scheme 6.3). Amongst these RhuAs isolated from biodiversity, the one from *Bacteroides thetaiotaomicron* ($RhuA_{Bthet}$; Uniprot Id: Q8A1A0) was the most efficient in catalysing this aldol reaction. Dendroketose-1-phosphate is a branched chain-sugar, a precursor of dendroketose. This latter compound was previously obtained through chemical self-aldolisation of dihydroxyacetone (DHA) [2]. Importantly, it could be dehydrated to 4-hydroxymethylfurfural [2b, 3]. Furfural derivatives, such as 4- or 5-hydroxymethylfurfural, are interesting building blocks for a large variety of compounds and are often cited in the literature as molecular platforms for the production of polymers and fine chemicals [4]. They can also be used to prepare biofuels [2b,5].

6.2.1 Procedure 1: $RhuA_{Bthet}$ Production and Purification

6.2.1.1 Materials and Equipment

- Stock of expression strain BL21-CodonPlus (DE3)-RIPL from Agilent Technologies containing the plasmid pET22b(+)-RhuABthet (Uniprot Id Q8A1A0), prepared as described in [6].
- α-Glycerophosphate dehydrogenase–triosephosphate isomerase (GPDH/TPI) mixture from rabbit muscle (type III, ammonium sulfate suspension, TPI 750–2000 $U.mg^{-1}$, GPDH 75–200 $U.mg^{-1}$, from Sigma-Aldrich).
- Distilled water (dH_2O)
- Lysogenic broth (LB) base, Miller's modified
- Equilibration sodium phosphate buffer (50 mM, pH 7.5, 300 mM NaCl)
- Elution sodium phosphate buffer (50 mM, pH 7.5 with 250 mM imidazole, 300 mM NaCl)
- Tris-HCl buffer solution in (50 mM, pH 7.5)
- Ampicillin solution (100 $mg.mL^{-1}$)
- Isopropyl-β-D-1-thiogalactopyranoside (IPTG) solution (0.5 M)
- Ni-NTA resin
- Bradford reagent
- β-Nicotinamine adenine dinucleotide (NADH) reduced disodium salt hydrate solution (20 mM)

- $MgSO_4$ solution (50 mM)
- Rotary shaker
- Spectrophotometer UV-Vis
- Centrifuge
- Magnetic stirrer
- Sonicator
- Freeze-drier
- Dialysis tubing cellulose membrane (average flat width: 33 mm)

6.2.1.2 Procedure

1. *Escherichia coli* BL21 (DE3)-RIPL colonies containing the plasmid pET22b(+)-$RhuA_{Bthet}$ (2 mL frozen stock) were pre-incubated in 100 mL LB containing a solution of ampicillin (100 µ, final concentration 100 µg.mL^{-1}) at 37 °C with shaking at 250 rpm.
2. An aliquot of pre-inoculum (10 mL) was transferred to 1 L LB containing ampicillin (1 mL, final concentration 100 µg.mL^{-1}). The resulting mixture was shaken at 250 rpm and 37 °C.
3. When the culture reached OD_{600} 0.5, the temperature was decreased to 30 °C and protein expression induced with IPTG (1 mL, final concentration 0.5 mM). The culture was incubated overnight at 30 °C with shaking at 250 rpm.
4. Cells were harvested by centrifugation at 10 000× g for 10 min at 4 °C, washed twice with Tris-HCl buffer (50 mL) and resuspended in Tris-HCl buffer (50 mL).
5. The cell suspension was kept cold in ice and was disrupted by sonication (40 min cycles of 10 sec of sonication followed by 12 sec of rest). Cell debris were discarded by centrifugation at 8000 rpm for 20 min.
6. The recombinant protein containing an *N*-terminal 6xHis tag was purified by immobilised metal-affinity chromatography (IMAC) on a Ni-NTA column (h = 5 cm/Ø = 3 cm) equilibrated with sodium phosphate buffer. $RhuA_{Bthet}$ was eluted with the elution buffer. The presence of enzyme in each fraction was monitored using the Bradford assay (aliquots of 10 µL in 200 µL of Bradford reagent).
7. In order to remove imidazole, a dialysis step was carried out. Fractions containing $RhuA_{Bthet}$ were pooled together in a dialysis membrane and incubated overnight at 4 °C in 5 L of mili-Q water, under gentle stirring. The dialysed protein was then freeze-dried to furnish 348 mg of protein as a white powder.

6.2.2 Procedure 2: DHAP Synthesis by Phosphorylation of DHA

6.2.2.1 Materials and Equipment

- dH_2O
- Dihydroxyacetone (DHA) dimer
- Phosphoenolpyruvic acid (PEP) monosodium salt hydrate solution (0.4 M)
- Adenosine triphosphate (ATP) disodium salt hydrate
- $MgCl_2$ $6H_2O$
- Tris-HCl buffer solution in dH_2O (50 mM, pH 7.5)
- NADH reduced disodium salt hydrate solution (20 mM)

- 1M NaOH (aq) solution
- 1M HCl (aq) solution
- $BaCl_2$,$2H_2O$
- Ethanol 95%
- Acetone 99%
- Pyruvate kinase (PK) from rabbit muscle (type II, ammonium sulfate suspension, 4750 U.mL^{-1}; Sigma-Aldrich)
- GPDH/TPI mixture from rabbit muscle (type III, ammonium sulfate suspension, TPI 750–2000 U.mg^{-1}, GPDH 75–200 U.mg^{-1}; Sigma-Aldrich)
- Dihydroxyacetone kinase (DHAK 5.8 units per mg), prepared as described in [7]
- Dowex HCR-W2 ion exchange resin, acidic form, 16–40 mesh
- Rotary shaker (100–200 rpm)
- Centrifuge (10 000× g)

6.2.2.2 Procedure

1. DHA (0.9 g, 10 mmol), $MgCl_2$ (0.66 g, 3.3 mmol), ATP (0.24 g, 0.43 mmol) and PEP solution (20 mL, 8 mmol) were added to dH_2O (220 mL). The pH was then adjusted to 7.5 with the 1M NaOH (aq) solution.
2. DHAK (200 U, 35 mg) and PK (2000 U, 0.42 mL) were added to the reaction solution and the volume was adjusted to 250 mL with dH_2O. The mixture was then allowed to gently shake for 4 hr.
3. The pH was adjusted to 6.5 with 1M HCl (aq) solution and the resulting precipitate discarded by centrifugation. Then $BaCl_2$ (3.9 g, 2 eq.) was added and the inorganic salts precipitate was removed by centrifugation. Ethanol (750 mL) was added and the flask was cooled to 4 °C and allowed to stand in the refrigerator overnight. The suspension obtained was then centrifuged and the solid washed with 1 : 3 water : ethanol (3 × 250 mL). The solid was dissolved in acetone and concentrated by distillation under reduced pressure to remove residual volatiles.
4. Then dH_2O (115 mL) was poured on to the solid and acidic ion-exchange resin (45 mL) was added until a clear solution was obtained. The resin was then removed by filtration over sintered glass to give a 49 mM DHAP solution.
5. The amount of DHAP formed was measured by enzymatic assay (reduction of DHAP catalysed by GPDH with concomitant oxidation of NADH to NAD^+). The NADH absorbance decrease at 340 nm was proportional to the DHAP concentration (ε_{NADH} = 6220 cm.M^{-1}). This spectrophotometric assay was run at room temperature for 10 min, in a total volume of 1 mL, containing Tris-HCl buffer (975 µL), NADH solution (20 µL), an aliquot of the reaction mixture (2.5–5µL) and a GPDH/TPI mixture (2 µL) [7].

6.2.3 Procedure 3: Rhamnulose Aldolase RhuA$_{Bthet}$ Biocatalysed Preparation of Dendroketose

6.2.3.1 Materials and Equipment

- dH_2O
- 49 mM DHAP solution prepared as described in Procedure 2

- $CoCl_2$ (aq) solution (10 mM)
- DHA
- Solutions of NH_4HCO_3 (aq) (0.33 and 1 M)
- Aldolase (RhuA$_{Bthet}$; Uniprot Id Q8A1A0) as described in Procedure 1
- Dowex 1 × 8 chloride form, 100–200 mesh
- 1M NaOH (aq) solution
- Rotary evaporator
- Rotary shaker
- Argon stream for degassing solution
- Thin-layer chromatography (TLC) silica gel 60 F_{254} aluminium plates
- Solution of 4 : 5 aqueous NH_4OH (32%) : EtOH (95%) as TLC eluent
- Vanillin solution (0.5 g of vanillin in 100 mL of 4/1 sulfuric acid (95%)/ethanol (95%)) as TLC revealing agent

6.2.3.2 *Procedure*

1. DHAP solution (8.1 mL, 0.4 mmol), $CoCl_2$ solution (400 µL, 4 µmol) and DHA (180 mg, 2 mmol, 5 equivs) were mixed in a 50 mL Erlenmeyer flask and diluted with dH_2O to a final volume of 12 mL. The pH was then adjusted to 7.5 with a 1M NaOH (aq) solution and the solution was degassed with an argon stream.
2. 30 mg of aldolase was added and the reaction mixture was allowed to gently shake (100–200 rpm) on a rotary shaker at room temperature under an argon atmosphere.
3. When DHAP has disappeared, as assessed by the spectrophotometric assay described for Procedure 2 (after around 5 hr reaction time), the solution was directly loaded on to a chromatography column packed with 3 mL of Dowex 1 × 8 resin preliminarily prepared (3 mL of Dowex 1 × 8 resin were washed with 10 volumes of an aqueous NaOH solution, then with 10 volumes of dH_2O, 10 volumes of an ammonium bicarbonate solution (1 M) and finally with 10 volumes of dH_2O).
4. The resin was washed with 4 volumes of dH_2O and the aldol adduct eluted with 4 volumes of an aqueous ammonium bicarbonate solution (0.33 M), the positive fractions being visualised by TLC using the appropriate eluent and revealing agent. These fractions were mixed and concentrated by distillation under reduced pressure. Then dH_2O (approx. 5 mL) was added to the residue and the solution was concentrated by distillation under reduced pressure. This procedure was repeated twice to ensure a total removal of ammonium bicarbonate to afford 99.4 mg of compound **3** as a white solid (90% yield).

6.2.3.3 *Analytical Data*

Data for **3**: mixture of diastereomers, only the major one being described.

^{1}H NMR (400 MHz, D_2O) δ 3.95 (s, 1H, H3), 3.88 (m, 2H, H6), 3,75 (m, 2H, H1), 3.55 (s, 2H, H5). ^{13}C NMR (101 MHz, D_2O) δ 102.44 (d, J = 9 Hz, C2), 78.01 (C4), 72.86 (C6), 70.96 (C3), 65.68 (d, J = 4 Hz, C1), 63.88 (C5). HRMS ESI-: m/z calcd for $[C_6H_{12}O_9P]^-$ = 259.0219; found 259.0278 (see Figure 6.1).

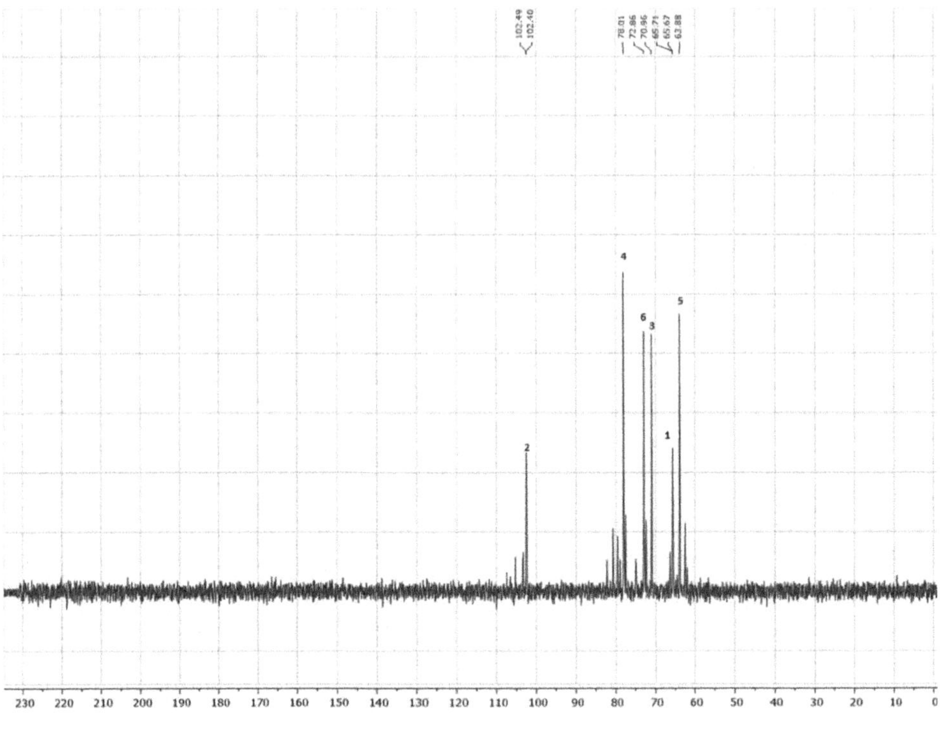

Figure 6.1 *NMR spectrum for dendroketose-1-phosphate **3**.*

6.2.4 Conclusion

This straightforward eco-friendly preparation of dendroketose-1-phosphate exemplifies the highly challenging access to the tertiary alcohol moiety, opening the scope of branched-chain sugars and analogues syntheses. It also demonstrates that some rhamnulose-1-phosphate aldolases display an unexpected but remarkable substrate promiscuity for activated ketones as electrophiles, as illustrated in Table 6.1 [1].

References

1. Laurent, V., Darii, E., Aujon, A. *et al.* (2018) *Angewandte Chemie International Edition*, **57**, 5467–5471.
2. (a) Utkin, L.M. (1950) *Proceedings of the USSR Academy of Sciences*, **67**, 301–304; (b) Deng, J., Pan, T., Xu, Q. *et al.* (2013) *Scientific Reports*, **3**, 1244–1250.
3. Wang, T., Nolte, M.W. and Shanks, B.H. (2014) *Green Chemistry*, **16**, 548–572.
4. van Putten, R.-J., van der Waal, J.C., de Jong, E. *et al.* (2013) *Chemical Reviews*, **113**, 1499–1597.
5. Rout, P.K., Nannawari, A.D., Prakash, O. *et al.* (2016) *Chemical Engineering Science*, **142**, 318–346.
6. Guérard-Hélaine, C., de Berardinis, V., Besnard-Gonnet, M. *et al.* (2015) *ChemCatChem*, **7**, 1871–1879.
7. Dürrenberger, M., Köhler, V., Wilson, Y.M. *et al.* (2016) Cascade reactions, in *Practical Methods for Biocatalysis and Biotransformations*, 3rd edn. (eds J. Whittall, P.W. Sutton and W. Kroutil), John Wiley & Sons, pp. 232–239.

Table 6.1 Activated ketones used as RhuA electrophiles.

Ketone	Aldol[a]	Yield (%)	Configuration	Diastereomeric excess (%)[b]
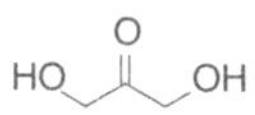	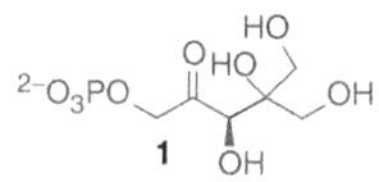 1	90	3R	–
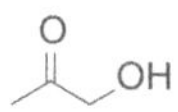	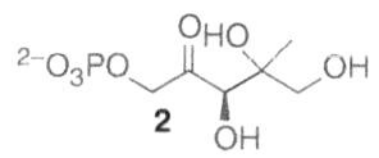 2	85	3*R*,4*R* (65%) 3*R*,4*S* (35%)	30
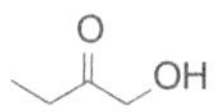	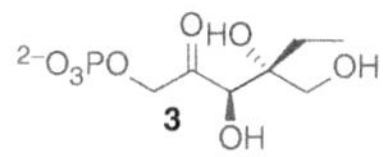 3	76	3*R*,4*R*	>95
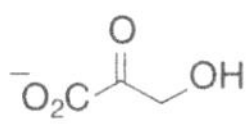	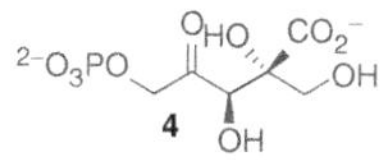 4	95	3*R*,4*R*	>95
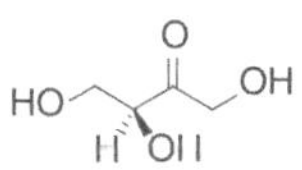	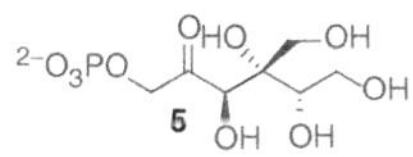 5	92	3*R*,4*R*	>95

[a] Aldols drawn in their linear forms for simplicity.
[b] Determined by ^{1}H NMR.
Reaction conditions: 30 mg RhuA$_{Bthet}$, 0.4 mmol DHAP, 400 μL $CoCl_2$ (10 mM), 5 eq. (dihydroxyacetone, hydroxypyruvate or L-erythrulose) or 15 eq. (hydroxyacetone or 1-hydroxybutanone) electrophile in 12 mL water under Ar at pH 7.5 and rt.

6.3 Easy and Robust Synthesis of Substituted L-Tryptophans with Tryptophan Synthase from *Salmonella enterica*

Kirsten Schroer, Mathieu Ligibel, Pierre-Yves Thouvenot and Radka Snajdrova
Novartis Institutes of BioMedical Research, Global Discovery Chemistry, Basel, Switzerland

The synthesis of halo- and methoxy-substituted tryptophan analogues is of high interest because of their importance as building blocks for the synthesis of synthetic drug substances and natural products. Whilst chemical synthesis of enantiopure substituted tryptophans is extensive and typically requires the separation of diastereoisomers [1], the direct enzymatic alkylation of indole with serine, catalysed by tryptophan synthase, constitutes a direct and efficient approach (Scheme 6.4) [2].

The commercially available recombinant *Escherichia coli* strain ATCC 37845 contains plasmid pSTB7, which expresses tryptophan synthase from *Salmonella enterica* (GenBank No. J01810.1). With this recombinant strain, the tryptophan synthase can be rapidly obtained by cultivation and subsequent cell lysis. The relaxed substrate scope of this tryptophan synthase allows for the functionalisation of numerous indole analogues, providing a biocatalytic synthesis of a wide range of substituted tryptophans. This route is

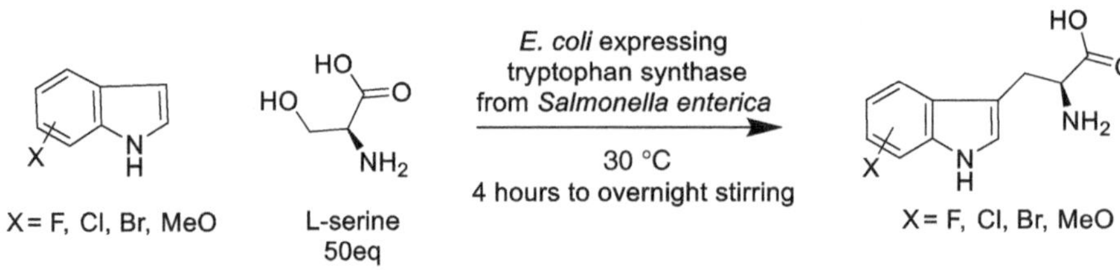

Scheme 6.4 *Preparation of substituted L-tryptophans.*

Table 6.2 *Typical conversion to substituted tryptophans catalysed by tryptophan synthase after 24 hr; analysed by HPLC-UV.*

Substrate	Conversion	Substrate	Conversion
4-F-Indole	>95%	5-Br-Indole	39%
5-F-Indole	>95%	6-Br-Indole	>95%
6-F-Indole	>95%	7-Br-Indole	>95%
7-F-Indole	>95%	4-MeO-Indole	>95%
4-Cl-Indole	>95%	5-MeO-Indole	>95%
5-Cl-Indole	>95%	6-MeO-Indole	>95%
6-Cl-Indole	>95%	7-MeO-Indole	>95%
7-Cl-Indole	>95%	5-F-Indazole	12%
4-Br-Indole	20%	6-Cl-5-F-Indazole	>95%

not limited to halo-tryptophans but is also appropriate for access to methyl-, methoxy- and amino-tryptophans [3]. Table 6.2 shows the typical conversions for differently substituted indoles and indazoles on an analytical scale (0.5 mL reaction volume with 20 mmol.L^{-1} substrate concentration).

6.3.1 Procedure 1: Preparation of 5-Fluoro-Tryptophan

6.3.1.1 Materials and Equipment

- *E. coli* strain (ATCC 37845)
- Lysogenic broth (LB), Miller's modified (Difco, Cat. No. 244620, preparation according to manufacturer's instruction)
- Turbo broth™ medium (AthenaES, Cat. No. 0115, preparation according to manufacturer's instruction)
- Reaction buffer, pH 7.8 (see Table 6.3)

Table 6.3 *Reaction buffer.*

Compound	Molecular weight (g.mol^{-1})	Concentration (mol.L^{-1})	Concentration (g.L^{-1})
KH_2PO_4	136.1	0.1	13.61
EDTA (ethylenediamine tetraacetic acid)	374.24 (disodium salt dihydrate)	0.005	1.86
PLP (pyridoxal phosphate)	247.1	0.0001	0.0024

6.3.1.2 Procedure

6.3.1.2.1 Cultivation and Protein Expression.

1. LB medium supplemented with 100 μg.mL^{-1} ampicillin (50 mL) was added into a 200 mL Erlenmeyer flask and inoculated with either a small portion of frozen cryoculture or one colony from a freshly grown culture of *E. coli* ATCC 37845 on LB agar. The flask was incubated overnight at 37 °C and 180 rpm as a pre-culture.
2. Turbo broth™ with 100 μg.mL^{-1} ampicillin (200 mL) was placed into a 1 L Erlenmeyer flask and inoculated with pre-culture of *E. coli* ATCC 37845 (5 mL). The flask was incubated at 37 °C and 180 rpm for 21 hr.
3. The cell culture was centrifuged at 7000× g for 10 min and the pellet was resuspended in 50 mL of reaction buffer. When several flasks were incubated in parallel, the resulting cell pellets were combined at this point. Cells were again centrifuged at 7000× g for 10 min and the cell pellet was resuspended in 40 mL reaction buffer.
4. Cells were lysed by sonication with 40% amplitude (2 sec pulses on/off) for a total of 10 min. After cooling on ice, they were centrifuged at 15 000× g for 10 min to remove cell debris. The crude supernatant was lyophilised overnight. From 1 L of cell culture, typically 1.5 g of crude lyophilised enzyme was obtained following this procedure.

6.3.1.2.2 Enzymatic Transformation and Product Purification.

1. Into a 1L Erlenmeyer flask was added reaction buffer (320 mL) and 5-fluoro-indole (1 g, 7.4 mmol) dissolved in methanol (10 mL).
2. L-Serine (38.9 g, 370.2 mmol, 50 equiv.) was dissolved in water (70 mL) and the enzyme preparation (600 mg) was added.
3. The reaction vessel was gently shaken at 30 °C.
4. The reaction was monitored by liquid chromatography–mass spectrometry (LCMS) analytics. After incubation overnight, complete conversion was observed.
5. The biotransformation mixture was washed with ethyl acetate (200 mL) and the organic layer was discarded. The aqueous layer was then acidified with 25% HCl to pH 1–2 and extracted with tetrahydrofuran (THF, 3 × 150 mL).
6. The organic layers were combined, dried over $MgSO_4$ and concentrated to dryness.
7. The solid was triturated in ethyl acetate to afford 1.3 g of the pure product as the HCl salt in the form of a white solid. Typical isolated yield is 50–70%.

In the case when the biotransformation does not reach full conversion, the same purification protocol is applied, followed by an additional chromatographic purification step. After the THF layers are combined, dried over $MgSO_4$ and concentrated to dryness, a yellow oil is typically obtained, which is subjected to reversed-phase chromatography under the conditions listed in Table 6.4

Fractions containing product were combined and concentrated to dryness, which resulted in the formation of product as formic acid salt (white solid). ^{1}H NMR (400 MHz, DMSO-d_6) δ 14.33–13.27 (br s, 1H), 11.18 (s, 1H), 8.63-7-95 (br s, 3H), 7.44–7.22 (m, 3H), 6.93 (td, J = 9.2, 2.6 Hz, 1H), 4.12 (t, J = 6.0 Hz, 1H), 3.25 (d, J = 6.0 Hz, 2H). X-ray analysis revealed the formation of only the (*S*)-enantiomer.

Table 6.4 Conditions for reversed-phase chromatography.

Mobile phase A	Water + 0.1% HCOOH
Mobile phase B	Acetonitrile + 0.1% HCOOH
Flow rate	40 mL.min^{-1}
Column	Nucleodur C18
Gradient	0% B during 5 min to 35% B during 30 min 35% B for 5 min
Detection	DAD 200–600 nm

6.3.2 Conclusion

The application of tryptophan synthase from *Salmonella enterica* has been demonstrated on a gram scale, enabling the synthesis of a broad range of substituted tryptophans. Due to high efficiency of this tryptophan synthase, despite the poor expression in the *E. coli* host strain, the target enzyme can be used in the form of a lyophilised crude lysate, which drives the cost for synthesis of the products down significantly.

The simplicity of cell cultivation, biotransformation and product purification facilitates an easy synthesis of tryptophan analogues on gram scale.

References

1. Perry, C.W., Brossi, A., Deitcher, K.H. *et al.* (1977) *Synthesis*, **7**, 492–494.
2. Goss, R.J.M. and Newill, P.L.A. *et al.* (2006) *ChemComm*, **47**, 4924–4925.
3. Winn, M., Roy, A.D., Grüschow, S. *et al.* (2008) *Bioorganic & Medicinal Chemistry Letters*, **18**, 4508–4510.

6.4 Biocatalytic Friedel–Crafts-Type *C*-Acylation

Anna Żądło-Dobrowolska[1] and Wolfgang Kroutil[2,3]
[1] *Institute of Organic Chemistry, Polish Academy of Sciences, Warszaw, Poland*
[2] *Institute of Chemistry, University of Graz, NAWI Graz, BioTechMed Graz, Graz, Austria*
[3] *Austrian Centre of Industrial Biotechnology, acib GmbH, Graz, Austria*

The design and development of catalytic and regioselective strategies for the *C*-acylation of aromatic compounds is of significant importance for organic synthesis. Since the discovery by Friedel and Crafts that aluminium chloride catalyses the condensation of acyl halides with various aromatic compounds, this reaction has become the first method of choice [1]. Although the classical approach is very broadly applied, it suffers from low regioselectivity. (Poly)phenolic compounds in particular remain challenging substrates, which can mostly be attributed to their electron-rich nature causing multiple *C*-acylations as well as *O*-acylations [2]. As an alternative to using chemical methods, an efficient, regioselective and environmentally friendly protocol was recently published [3]. It was found that multicomponent acyltransferase isolated from the bacterium *Pseudomonas*

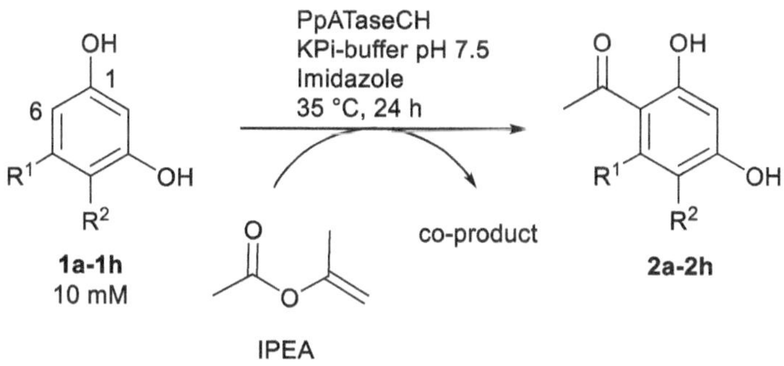

Product	R_1	R_2	Yield [%]
2a	H	H	68
2b	H	Cl	95
2c	H	CH_3	16
2d	H	C_2H_5	82
2e	H	*i*-Pr	81
2f	H	nC_4H_9	80
2g	H	nC_6H_{13}	52
2h	OMe	H	10

Scheme 6.5 *Conversion of resorcinol derivatives **1a–h** to the corresponding acetophenones **2a–h**.*

protegens (*Pp*ATaseCH) catalyses the *C*-acylation of resorcinol derivatives exclusively at the C6-position. *Pp*ATaseCH exhibits promiscuous activity, accepting not only the natural donor substrate 2,4-diacetylphloroglucinol (DAPG), but also non-natural, activated esters like isopropenyl acetate (IPEA), phenyl acetate and *N*-acetyl imidazole for the biocatalytic Friedel–Crafts acylation. Thus the enzyme breaks a C-O or C-N bond in the acetyl donor and enables subsequent C-C bond formation. Semipreparative-scale experiments were performed using commercially available IPEA as the donor substrate in the presence of imidazole, achieving isolated product yields (**2a–h**) of up to 95% after 24 hr (Scheme 6.5).

6.4.1 Procedure 1: Transformation of Competent *E. coli* Cells with Plasmid pEG332

6.4.1.1 Materials and Equipment

- Agar bacteriological (OxoidTM, #LP0011)
- Tryptone from casein (OxoidTM, #LP0042)
- Yeast extract (Oxoid, #LP0021)
- NaCl (Roth)
- Shaking incubator (Eppendorf)
- Ice bucket filled with ice

- Microcentrifuge tubes
- Two 1 L Schott bottles
- Sterile spreading device
- 1.5 mL Eppendorf tubes
- Ampicillin (100 mg.mL^{-1} in dH_2O, filter-sterilised)
- Competent *E.coli* BL21(DE3) cells
- Plasmid pEG332 DNA [3a]
- Autoclave
- Drying oven

6.4.1.2 Procedure

1. To prepare lysogenic broth (LB) medium: tryptone (5 g), yeast extract (2.5 g) and NaCl (5 g) were dissolved in dH_2O (500 mL) and autoclaved (20 min, 121 °C) in a 1 L Schott bottle with a screw cap to give sterile LB medium.
2. To prepare agar plates: tryptone (5 g), yeast extract (2.5 g), NaCl (5 g) and agar (7.5 g) were dissolved in dH_2O (500 mL) in a 1 L Schott bottle with a screw cap. The solution was sterilised by autoclaving (20 min, 121 °C) and cooled to 50 °C in a temperature-controlled water bath. 0.5 mL of 100 mg.mL^{-1} ampicillin was added and poured on to the plates.
3. *E. coli* cells were taken out of a −80 °C freezer and thawed on ice (approximately 15–20 min).
4. A 1.5 mL Eppendorf tube was pre-cooled on ice.
5. *E. coli* BL21(DE3) cells (100 μL) were mixed with an aqueous plasmid solution (22 ng in 5 μL).
6. The cell/DNA mixture was incubated on ice for 30 min, whilst gently mixing by flicking the bottom of the tube with the fingers every 5 min.
7. The cells were heat-shocked at 42 °C for 10 sec.
8. The tube was placed back on ice for 2 min.
9. LB medium (250 μL, not containing antibiotic) was added to the tube and incubated at 37 °C for 1 hr with shaking (350 rpm).
10. The transformation was plated on to an LB agar plate containing ampicillin.
11. The plates were incubated upside down at 30 °C overnight in the drying oven.

6.4.2 Procedure 2: Recombinant Expression of the *Pp*ATaseCH in *E. coli* BL21(DE3)

6.4.2.1 Materials and Equipment

- 1M solution of K_2HPO_4 (87.09 g.L^{-1})
- 1 M solution of KH_2PO_4 (68.045g.L^{-1})
- Distilled water (dH_2O)
- 10M HCl aqueous solution
- 10M KOH aqueous solution
- Ampicillin (100 mg.mL^{-1} in dH_2O, filter-sterilised)
- Anhydrotetracycline (AHTC; 2 mg.mL^{-1} in dH_2O, filter-sterilised)

- LB/ampicillin agar plate with colonies of *E. coli* BL21(DE3) harbouring the expression vector pEG332 bearing the genes encoding ATase inserted; see Procedure 1
- Three 0.5 L Schott bottles with screw caps
- 50 mL Falcon tube
- 5 L Erlenmeyer nonbaffled flask with a cap
- pH meter (Hanna Instruments)
- Orbital shaker (InforsHT Multitron 2 Standard)
- Autoclave
- Cooling centrifuge (min. 4500× g)

6.4.2.2 *Procedure*

1. To prepare LB medium: tryptone (10 g), yeast extract (5 g) and NaCl (10 g) were dissolved in dH_2O (1 L) and autoclaved (20 min, 121 °C) in a 5 L Erlenmeyer nonbaffled flask to give sterile broth (LB) medium.
2. To prepare 100 mM potassium phosphate buffer (KPi-buffer), pH 7.5: a 1M K_2HPO_4 aqueous solution (83.4 mL) was mixed with a 1M KH_2PO_4 aqueous solution (16.6 mL) and then diluted with dH_2O to 1L. The pH was measured by pH meter and adjusted to 7.5 using aqueous 10M HCl or KOH.
3. To prepare the overnight culture: sterile LB medium (10 mL) was placed into a sterile 50 mL Falcon tube and ampicillin stock solution (10 μL) was added to reach a final concentration of 100 $\mu g.mL^{-1}$.
4. The solution was inoculated with a single colony of *E. coli* BL21(DE3) harbouring pEG332 and shaken overnight at 30 °C and 140 rpm.
5. The next day, the LB medium (1 L) supplemented with ampicillin stock solution (1 mL; final conc. 100 $\mu g.mL^{-1}$) was inoculated with the LB overnight culture (10 mL).
6. The cells were grown at 37 °C and 140 rpm until the OD_{600} reached 0.7. Then the cultivation was cooled to 30 °C and expression of the recombinant enzyme was induced by the addition of 100 μL of AHTC stock solution (200 $\mu g.L^{-1}$ final conc.). The expression was performed for 21 hr at 30 °C.
7. The cells were harvested by centrifugation at 8000 rpm (12 028× g) and 4 °C for 15 min and washed once with 200 mL of KPi-buffer (50 mM, pH 7.5) before being recentrifuged (15 min, 8000 rpm, 12 028× g, 4 °C).

6.4.3 Procedure 3: Preparation of Cell-Free Extract

6.4.3.1 *Materials and Equipment*

- KPi-buffer (100 mM, pH 7.5); see Procedure 2
- Liquid nitrogen
- Ultrasonificator
- Ice bucket filled with ice
- Centrifuge (Hitachi)

6.4.3.2 *Procedure*

1. The cells from the enzyme expression (2 g) were gently resuspended in KPi buffer (14 mL) before being disrupted by ultrasonification (40% amplitude, 8 min, pulse 1 sec, pause 4 sec) in an ice bath.
2. After centrifugation (30 min, 14 000 rpm, 23 519× g), the cell-free extract was analysed by SDS-PAGE, aliquoted and shock-frozen in liquid nitrogen. The aliquots were stored at −20 °C for direct use in biotransformations.

6.4.4 Procedure 4: Biocatalytic Conversion of Resorcinol

6.4.4.1 *Materials and Equipment*

- Resorcinol (110 mg, 1.0 mmol; Sigma-Aldrich #8.22303)
- Isopropenyl acetate (1086 µL, IPEA; Sigma-Aldrich #117781)
- Imidazole (Sigma-Aldrich, #792527)
- KPi buffer (50 mM, pH 7.5); see Procedure 2
- Cell-free extract containing *Pp*ATaseCH; see procedure 3
- Ethyl acetate (EtOAc)
- Brine (NaCl-saturated solution)
- 6 N HCl
- CH_2Cl_2
- Anhydrous Na_2SO_4
- Silica gel (Merck)
- Thin-layer chromatography (TLC) silica gel 60 F_{254} aluminium plates (Merck)
- 200 mL baffled shaking flask
- 100 mL separating funnel
- Orbital shaker (InforsHT Multitron 2 Standard)
- Centrifuge
- Rotary evaporator (Büchi)
- Cinnamaldehyde/HCl staining solution (abs. EtOH (72.2 vol%), conc. HCl (3.6 vol%), trans-cinnamaldehyde (3.6 vol%)

6.4.4.2 *Procedure*

1. Resorcinol (110 mg, 1.0 mmol, 10 mM final conc.) was dissolved in KPi buffer (50 mM, pH 7.5, 70.5 mL) in a baffled shaking flask.
2. Imidazole (10 mL, 100 mM final conc.) from a 1 M stock solution prepared in KPi buffer was added. During imidazole addition, the pH changed from 7.5 to 8.3.
3. The mixture was preheated to 35 °C for 10 min.
4. Meanwhile, cell-free extract containing *Pp*ATaseCH (165 U, ~18.5 mL) was thawed at 21 °C, then briefly preheated period (35 °C, 10 min).
5. The enzyme solution was added to the reaction mixture and the bioacetylation was started by adding IPEA (1086 µL, 100 mM final concentration).
6. The bioacetylation (100 mL total volume, final pH 8.30) was run at 35 °C and 140 rpm for 24 hr.

7. The reaction was stopped by acidification (pH ~ 1.0) with aqueous HCl (6 M). Shaking was continued for a further 10 min to precipitate protein.
8. The resulting suspension was extracted with CH_2Cl_2 (3 × 50 mL). The organic layers, which still contained imidazolium salt or protein precipitate, were centrifuged (15 min, 4000 rpm) in order to remove the remaining solids. The cleared organic layers were pooled in a separation funnel, washed with brine (2 × 40 mL) and dried over anhydrous Na_2SO_4.
9. Solvent was removed under reduced pressure and the crude product was purified by flash chromatography over silica gel using CH_2Cl_2/EtOAc (85 : 15 v/v) as eluent. Fractions were checked for the desired substances via TLC using CH_2Cl_2/EtOAc (85 : 15 v/v) as TLC eluent and staining was performed with cinnamaldehyde/HCl. Fractions containing the desired product were combined and evaporated. 1-(2,4-Dihydroxyphenyl)ethan-1-one was obtained as a colourless solid (103 mg, 0.68 mmol, 68%), m.p. 120–125 °C.

^{1}H-NMR (300 MHz, acetone-*d6*): δ [ppm] = 2.56 (s, 3H), 6.33 (ds, J = 2.3 Hz, 1H), 6.45 (dd, J = 8.7 Hz, 2.3 Hz, 1H), 7.78 (d, J = 8.7 Hz, 1H), 9.43 (s, 1H, Ar-OH), 12.75 (s, 1H, Ar-OH). ^{13}C-NMR (75 MHz, acetone-*d6*): δ [ppm] = 25.4, 102.6, 107.9, 113.4, 133.4, 164.6, 165.2 (6×), 202.8. GC-MS (EI$^+$, 70 eV): m/z (%) = 152 [M$^+$] (39), 137 [$C_7H_5O_3^+$] (100), 109 [$C_6H_5O_2^+$] (3).

6.4.5 Conclusion

The described procedure demonstrates the synthetic utility of an acyltransferase for the regioselective *C*-acylation of unprotected C4- or C5-substituted resorcinols into the corresponding mono-*C*-acylated 2,4-dihydroxyphenyl ketones in a Friedel–Crafts-type acylation reaction. Compared to the chemical method, this highly efficient, catalytic and regioselective methodology enables reactions to be performed in aqueous media. In addition to the reported IPEA, native donor DAPG and non-native *N*-acylimidazole can be used as donor substrates with slightly lower conversion [3a].

References

1. (a) Friedel, C. and Crafts, J.M. (1877) *Comptes Rendus Chimie*, **84**, 1392; (b) Effenberger, F. and Epple, G. (1972) *Angewandte Chemie International Edition*, **11**, 300–301; (c) Sartori, G. and Maggi, R. (2009) *Advances in Friedel–Crafts Acylation Reactions: Catalytic and Green Processes*, CRC Press.
2. (a) Kroutil, W., Hagmann, L., Schuez, T.C. *et al.* (2005) *Journal of Molecular Catalysis B: Enzymatic*, **32**, 247–252; (b) Murashige, R., Hayashi, Y., Ohmori, S. *et al.* (2011) *Tetrahedron*, **67**, 641–649.
3. (a) Schmidt, N.G., Pavkov-Keller, T., Richter, N. *et al.* (2017) *Angewandte Chemie International Edition*, **56**, 7615–7619; (b) Schmidt, N.G. and Kroutil, W. (2017) *European Journal of Organic Chemistry*, **39**, 5865–5871; (c) Schmidt, N.G., Żądło-Dobrowolska, A., Ruppert, V. *et al.* (2018) *Applied Microbiology and Biotechnology*, **102**, 6057–6068; (d) Pavkov-Keller, T., Schmidt, N.G., Żądło-Dobrowolska, A. *et al.* (2019) *ChemBioChem*, **20**, 88–95; (e) Żądło Dobrowolska, A., Schmidt, N.G. and Kroutil, W. (2019) *ChemCatChem*, **11**, 1064–1068.

6.5 MenD-Catalysed Synthesis of 6-Cyano-4-Oxohexanoic Acid

Martina Sudar,[1] Alexandra Walter,[2] Michael Müller,[2] and Zvjezdana Findrik Blažević*[1]

[1]*University of Zagreb, Faculty of Chemical Engineering and Technology, Department of Reaction Engineering and Catalysis, Zagreb, Croatia*

[2]*Institute of Pharmaceutical Sciences, Albert-Ludwigs-Universität Freiburg, Freiburg, Germany*

MenD (2-succinyl-5-enolpyruvyl-6-hydroxy-3-cyclohexene-1-carboxylate synthase) provides biocatalytic access to new types of products not related to those currently accessible by thiamine diphosphate (ThDP)-dependent enzyme catalysis [1]. It catalyses the formation of substituted 1,4-diketones and 4-oxo acids via enzymatic Stetter reaction. The Stetter reaction is well known for cyanide-catalysed transformations of mostly aromatic aldehydes. Enzymatic Stetter reactions, however, have been largely unexplored, especially with respect to preparative transformations [2]. We have established a MenD-catalysed 1,4-addition of α-ketoglutaric acid to acrylonitrile in order to synthesise 6-cyano-4-oxohexanoic acid on gram scale [3]. The relevance of this reaction is in the potential application of 6-cyano-4-oxohexanoic acid as a building block, since this molecule contains a keto group, an acid function and a nitrile group that can be used to synthesise other functional groups in a straightforward manner [4].

The synthesis of 6-cyano-4-oxohexanoic acid **3** (Scheme 6.6) is a conjugate addition (Stetter reaction) of α-ketoglutaric acid **1** with decarboxylation to acrylonitrile **2**. It is catalysed by ThDP-dependent enzyme MenD (EC 2.2.1.9) from *Escherichia coli* with the addition of co-factor Mg^{2+} to the reaction mixture, providing 6-cyano-4-oxohexanoic acid **3** on preparative scale in 60% isolated yield.

6.5.1 Procedure 1: Synthesis of 6-Cyano-4-Oxohexanoic Acid

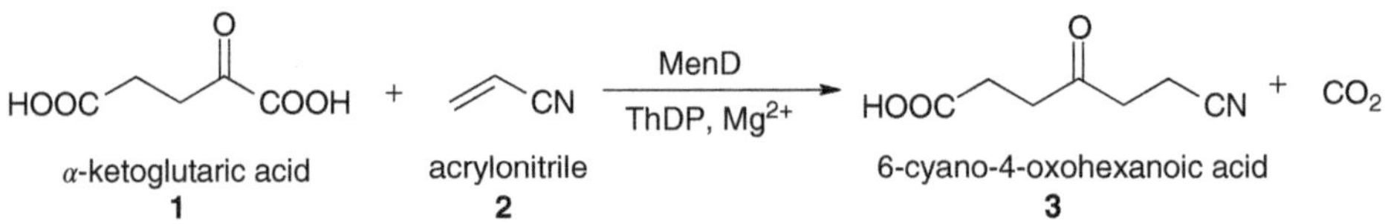

Scheme 6.6 *MenD-catalysed synthesis of 6-cyano-4-oxohexanoic acid* ***3****.*

6.5.1.1 Materials and Equipment

- Potassium phosphate buffer (400 mM, pH 8.0)
- Acrylonitrile (stabilised with 2% of *p*-methoxyphenol, 0.08 g, 1.5 mmol; Sigma-Aldrich)
- α-Ketoglutaric acid disodium salt dihydrate (α-ketoglutarate, 0.3 g, 1.5 mmol, 95%; Sigma-Aldrich)
- Thiamine diphosphate (ThDP; Sigma-Aldrich)
- Magnesium chloride ($MgCl_2$; Acros Organics)
- Pure protein suspension of MenD in 3.2 M $(NH_4)_2SO_4$ solution ($\gamma_{MenD} = 2.3$ mg.mL^{-1}) [5]

- Solution for analytical derivatisation (130 mM of *O*-benzylhydroxylamine hydrochloride ($BnONH_2 \cdot HCl$) in pyridine/methanol/water, 33 : 15 : 2)
- Methanol
- Acetonitrile
- Trifluoroacetic acid (TFA)
- Ultrapure water
- 5 M HCl solution in water
- Ethyl acetate
- Magnetic stirrer
- pH meter
- High-performance liquid chromatography (HPLC) system with UV detection
- HPLC columns (HP Hypersil APS C18 column (5 µm, 4.6 × 200 mm) and Phenomenex LiChrospher C18 column (5 µm, 4 × 250 mm))
- Vacuum evaporator

6.5.1.2 *Procedure*

1. Stock solutions of α-ketoglutarate (0.5 mL, final concentration 100 mM), ThDP (0.5 mL, final concentration 1 mM) and Mg^{2+} (0.5 mL, final concentration 2 mM) were added to phosphate buffer (0.967 mL of 400 mM, pH 8.0) in a reaction vessel of 10 mL (glass bottle with a cap to avoid evaporation, V_r = 5 mL) and stirred on a magnetic stirrer (150 rpm) in a thermostatic water bath at 30 °C. Undiluted acrylonitrile solution (15.175 M) was then added to the reaction vessel (33 µL, final concentration 100 mM).
2. The reaction was started with the addition of pure protein suspension of MenD (2.5 mL in 3.2 M $(NH_4)_2SO_4$ solution) to achieve a final enzyme concentration of approximately 1.15 mg.mL^{-1}.
3. Additional substrates, α-ketoglutarate (0.1 g) and acrylonitrile (33 µL), were added to the reactor after 24 hr, and again after 48 hr. α-Ketoglutarate was added to the reactor in the form of a powder to avoid changing the reaction mixture volume.
4. The reaction was monitored by HPLC. Samples for the analysis of α-ketoglutarate and acrylonitrile were diluted with water, filtered through a 0.2 µm filter and analysed on an HP Hypersil APS C18 column (see Analytical Method for details).
5. Samples (5 µL) for the analysis of 6-cyano-4-oxohexanoic acid were derivatised with the solution for derivatisation (50 µL of *O*-benzylhydroxylamine hydrochloride solution) at 25 °C and 1000 rpm for 20 min [6]. Methanol (450 µL) was added and the mixture was centrifuged at 14000 rpm for 2 min. In this manner, the enzyme precipitated, and the upper phase was used for analysis on a Phenomenex LiChrospher C18 column (see Analytical Method for details).
6. After 72 hr, 259 mM of 6-cyano-4-oxohexanoic acid was obtained.
7. The reaction mixture was filtered through a 0.2 µm filter to remove the enzyme.
8. The pH of the mixture was lowered below 2 with 5 M HCl solution.
9. Ethyl acetate was added to the mixture in the ratio 1 : 1 (three times) to extract the product from the aqueous phase. The sample was mixed on a homogeniser and then centrifuged at 4500 rpm for 5 min to separate the phases.

10. Three phases containing ethyl acetate were pooled and ethyl acetate was evaporated (40 °C and 200 mbar).
11. The product 6-cyano-4-ketohexanoic acid was obtained as a transparent oily liquid (116 mg, 60% isolated yield).

6.5.2 Analytical Method

6.5.2.1 Analysis of α-Ketoglutarate and Acrylonitrile

α-Ketoglutarate and acrylonitrile were analysed on an HP Hypersil APS C18 column (5 μm, 4.6 × 200 mm) with UV detection at 195 nm and 30 °C. Ultrapure water plus TFA (0.1% v/v) was used as mobile phase. The method used was isocratic, with a flow rate of the mobile phase of 1 mL.min^{-1}. Retention times for acrylonitrile and α-ketoglutarate were 3.95 and 6.70 min, respectively.

6.5.2.2 Analysis of 6-Cyano-4-Oxohexanoic Acid

6-Cyano-4-oxohexanoic acid was analysed using a Phenomenex LiChrospher C18 column (5 μm, 4 × 250 mm) with UV detection at 215 nm and 30 °C. Two mobile phases were used: phase A was water/TFA (0.1% v/v), whilst phase B acetonitrile/water/TFA 80 : 20 : 0.095 v/v. The method used was a gradient (from 10 to 70% B within 25 min) with a flow rate of 1.2 mL.min^{-1}. The retention time of the derivatised 6-cyano-4-oxohexanoic acid was 19.5 min.

6.5.3 Conclusion

The procedure described here enables the synthesis and isolation of 6-cyano-4-oxohexanoic acid in good yield. In principle, it could be adapted for the related substrates acrylate, methacrylate and methyl vinyl ketone [1].

References

1. (a) Fang, M., Macova, A., Hanson, K.L. *et al.* (2011) *Biochemistry*, **50**, 8712–8721; (b) Beigi, M., Gauchenova, E., Walter, L. *et al.* (2016) *Chemistry: A European Journal*, **22**, 13 999–14 005; (c) Beigi, M., Waltzer, S., Zarei, M. and Müller, M. (2014) *Journal of Biotechnology*, **191**, 64–68.
2. (a) Dresen, C., Richter, M., Pohl, M. *et al.* (2010) *Angewandte Chemie International Edition*, **49**, 6600–6603; (b) Kasparyan, E., Richter, M., Dresen, C. *et al.* (2014) *Applied Microbiology and Biotechnology*, **98**, 9681–9690; (c) Dresen, C., Kasparyan, E., Walter, L.S. *et al.* (2016) Stetter reactions catalysed by thiamine diphosphate-dependent enzymes, in *Practical Methods for Biocatalysis and Biotransformations*, Volume 3 (eds J. Whittall, P.W. Sutton and W. Kroutil), John Wiley & Sons, pp. 81–84.
3. (a) Sudar, M., Vasić-Rački, Đ., Müller, M. *et al.* (2018) *Journal of Biotechnology*, **268**, 71–80; (b) Sudar, M., Dejanović, I., Müller, M. *et al.* (2019) *Chemical and Biochemical Engineering Quarterly*, **32**, 501–510.
4. Zou, T., Yu, X., Feng, X. and Bao, M. (2015) *Chemical Communications*, **51**, 10 714–10 717.
5. Kurutsch, A., Richter, M., Brecht, V. *et al.* (2009) *Journal of Molecular Catalysis B: Enzymatic*, **61**, 56–66.
6. Garrabou, X., Castillo, J.A., Guérard-Hélaine, C. *et al.* (2009) *Angewandte Chemie International Edition*, **48**, 5521–5525.

6.6 Production of (*R*)-2-(3,5-Dimethoxyphenyl)propanoic Acid Using an Aryl Malonate Decarboxylase from *Bordetella bronchiseptica*

Scott P. France*, Caroline A. Blakemore*, Roger M. Howard and Aaron C. Smith
Pfizer Worldwide Research and Development, Groton, CT, USA

(*R*)-2-(3,5-Dimethoxyphenyl)propanoic acid is a valuable chiral building block for use in medicinal chemistry. A scalable asymmetric synthesis of this compound was developed utilising an aryl malonate decarboxylase (AMDase) to perform a desymmetrising decarboxylation of the corresponding malonic acid. Evaluation of the wild-type AMDase from *Bordetella bronchiseptica* [1] showed promise as a biocatalyst for this application, and rapid process optimisation and intensification led to a procedure that enabled high conversion and excellent enantioselectivity to be achieved at good substrate loading (Scheme 6.7). The process is operationally simple as the AMDase is not co-factor-dependent and therefore the only reaction components are the starting material and a lyophilised biocatalyst cell-free extract in buffer. Additionally, the reaction profile is very clean, with CO_2 the only byproduct. This process was successfully demonstrated on a 5.0 g scale and the desired product was isolated in 99% ee and 97% yield.

6.6.1 Procedure 1: Biocatalytic Synthesis of (*R*)-2-(3,5-Dimethoxyphenyl)propanoic Acid

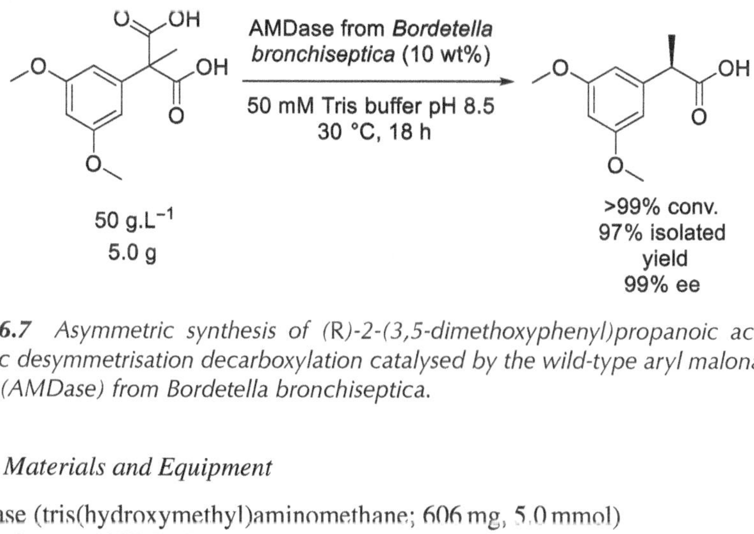

Scheme 6.7 Asymmetric synthesis of (R)-2-(3,5-dimethoxyphenyl)propanoic acid via an enzymatic desymmetrisation decarboxylation catalysed by the wild-type aryl malonate decarboxylase (AMDase) from Bordetella bronchiseptica.

6.6.1.1 Materials and Equipment

- Tris base (tris(hydroxymethyl)aminomethane; 606 mg, 5.0 mmol)
- Distilled water (107.5 mL)
- 35–38% HCl solution
- 2-(3,5-Dimethoxyphenyl)-2-methylmalonic acid (5.0 g, 19.7 mmol; Pfizer)
- 10 M NaOH aqueous solution
- AMDase from *B. bronchiseptica* (accession no. Q05115) lyophilised cell-free extract (0.50 g, 10 wt%; Prozomix)
- Methyl *tert* butyl ether (MTBE; 120 mL)
- Celite (5.0 g)

- Anhydrous $MgSO_4$
- Easymax 250 mL reactor equipped with overhead stirrer, pH probe, internal temperature probe and condenser
- Agilent 1200 Series Infinity II SFC
- Chiral SFC column Chiral Tech IB (250 mm × 4.6 mm × 5 μm)
- 4 : 1 acetonitrile : water (for sample preparation)
- Filter vials
- Rotary evaporator
- Nuclear magnetic resonance (NMR; 1H: 400 MHz, ^{13}C: 101 MHz) spectrometer

6.6.1.2 Procedure

1. Tris base (606 mg, 5.0 mmol) was charged to an Easymax 250 mL reactor equipped with overhead stirrer, pH probe, temperature probe and condenser.
2. Distilled water (90 mL) was added, the jacket temperature set to 25 °C and the stir speed set to 300 rpm.
3. The pH of the solution was adjusted to 8.5 by the addition of 35–38% HCl aqueous solution (50 μL).
4. 2-(3,5-Dimethoxyphenyl)-2-methylmalonic acid (5.0 g, 19.7 mmol) was charged to the reactor and rinsed in with distilled water (2 × 1 mL). The pH of the solution fell to ~2, with the undissolved starting material forming a slurry.
5. The pH of the mixture was carefully readjusted to 8.5 by the addition of 10 M aqueous NaOH (3.36 mL). As the pH was readjusted, the starting material dissolved completely.
6. *B. bronchiseptica* AMDase lyophilised cell-free extract (0.50 g, 10 wt%) was dissolved in distilled water (5.0 mL) and charged to the reactor with a rinse of distilled water (0.5 mL). The total final reaction volume was ~100 mL.
7. The internal reaction temperature was set to maintain 30 °C and the reaction stirred at 300 rpm.
8. The reaction was monitored by chiral SFC (see Analytical Method for details). Conversion was 92% after 2 hr and was complete after 18 hr.
9. Once full conversion was achieved, the internal reaction temperature was set to 20 °C.
10. Celite (2.5 g, 0.5 $g.g^{-1}$ starting material) was added and the reaction mixture was stirred at 400 rpm for 1 hr.
11. The reaction mixture was filtered through a pad of celite (2.5 g, 0.5 $g.g^{-1}$ starting material) over a 7 cm-diameter Buchner funnel and the filter cake was washed with distilled water (2 × 5 mL). The filtration was rapid (5 min).
12. The pH of the filtrate was adjusted to 2 by the addition of 35–38% aqueous HCl solution. As the pH was adjusted, some foaming occurred at pH ~6 as dissolved CO_2 was released. The reaction was stirred at the intermediate pH for 15 min to control the foaming and allow off-gassing to occur before acidification resumed. As the pH was lowered, the product precipitated, resulting in a fine suspension.
13. The product was extracted into MTBE (2 × 50 mL, 10 $mL.g^{-1}$). The phase separation was rapid, with a clean phase split.
14. The organic layers were combined, dried over anhydrous $MgSO_4$ and filtered, and the filter cake was washed with MTBE (2 × 10 mL).

Table 6.5 Chiral SFC method.

Time (min)	% A	% B
0	95	5
1.5	95	5
9	40	60
9.5	40	60
10	95	5

15. The solvent was removed under vacuum at 30 °C and the product was dried under high vacuum at 20 °C.
16. The final compound was isolated as a white solid (4.0 g, 19.0 mmol, 97% yield, 99% ee) with 99% NMR purity.

^1H NMR (400 MHz, DMSO-d_6) δ 12.24 (s, 1H), 6.45–6.40 (m, 2H), 6.40–6.36 (m, 1H), 3.72 (s, 6H), 3.59 (q, J = 7.0 Hz, 1H), 1.32 (d, J = 7.0 Hz, 3H). ^{13}C NMR (101 MHz, DMSO-d_6) δ 175.1, 160.4, 143.5, 105.6, 98.4, 55.1, 44.8, 18.4.

6.6.2 Analytical Method

6.6.2.1 Sampling Procedure

A 5 µL sample of the reaction mixture was diluted with 395 µL of 4 : 1 acetonitrile/water then passed through a filter vial. The liquors were analysed by chiral SFC.

6.6.2.2 Chiral SFC Method

- Column: Chiral Tech IB 250 mm × 4.6 mm × 5 µm.
- Gradient: A (CO_2) and B (methanol + 0.2% v/v 7 M ammonia in methanol; Table 6.5).
- Flow rate: 3 mL.min^{-1}.
- Detection: 210 nm.
- Retention times: (*R*)-2-(3,5-dimethoxyphenyl)propanoic acid (3.3 min); (*S*)-2-(3,5-dimethoxyphenyl)propanoic acid (3.6 min); 2-(3,5-dimethoxyphenyl)-2-methylmalonic acid (6.0 min).

6.6.3 Conclusion

The asymmetric synthesis of (*R*)-2-(3,5-dimethoxyphenyl)propanoic acid was achieved in high yield, enantioselectivity and purity using a wild-type AMDase from *B. bronchiseptica*. The operational simplicity, substrate concentration (50 g.L^{-1}) and clean reaction profile make this process highly amenable to large-scale application.

References

1. (a) Miyamoto, K. and Ohta, H. (1990) *Journal of the American Chemical Society*, **112**, 4077–4078; (b) Miyamoto, K. and Ohta, H. (1992) *European Journal of Biochemistry*, **210**, 475–481; (c) Kato, J., Kuroda, A., Hirota, R. *et al.* (2007) *Applied and Environmental Microbiology*, **73**, 5676–5678; (d) Miyamoto, K. and Kourist, R. (2016) *Applied Microbiology and Biotechnology*, **100**, 8621–8631.

7

Reductive Methods

7.1 Synthesis of Vibegron Enabled by a Ketoreductase Rationally Designed for High-pH Dynamic Kinetic Reduction

Birgit Kosjek

Process Research and Development, Merck & Co. Inc., Rahway, NJ, USA

A dynamic kinetic ketone reduction featuring pH > 10, temperature 45 °C and 50 vol % solvent was employed to establish the challenging (*R,R*) (C1',C2) stereochemistry in a commercially viable route to Vibegron [1]. Typical conditions reported for ketoreductase (KRED)-mediated reactions are pH <8 and <35 °C, limiting the substrate scope to easily epimerisable substrates such as β-keto esters and activated ketones [2]. We were unable to tune the functional groups in ketone intermediate **1** to lower its pK_a whilst maintaining our goal of a concise synthesis (Scheme 7.1). Thus, we employed computationally and structurally guided directed evolution to design a biocatalyst fit for the required challenging conditions. The key was to change the enzyme's affinity to bind the pH-sensitive NADP(H) co-factor more tightly, limiting its release from the active site and thus protecting it from the high-pH environment. Subsequent rounds of evolution further improved the reactivity and solvent tolerance of the KRED, enabling high substrate loading and increasing the overall economics of the process.

7.1.1 Procedure 1: Dynamic Kinetic Ketone Reduction of *tert*-Butyl (1-Oxo-1-Phenylhex-5-yn-2-yl)carbamate 1

7.1.1.1 Materials and Equipment

- Sodium tetraborate decahydrate (18.7 g)
- Deionised water (1.0 L)
- 5 M NaOH solution, for pH adjustment
- KRED-p301 (2.0 g, Codexis Inc.)

Applied Biocatalysis: The Chemist's Enzyme Toolbox, First Edition. Edited by John Whittall and Peter W. Sutton.
© 2021 John Wiley & Sons Ltd. Published 2021 by John Wiley & Sons Ltd.

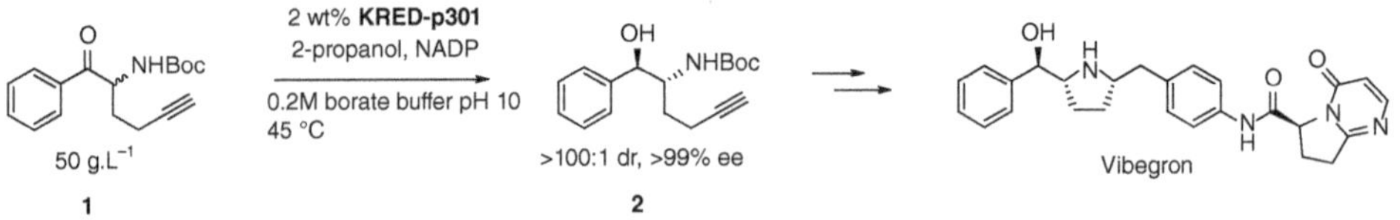

Scheme 7.1 Ketoreductase (KRED) designed for activity at high pH and elevated temperature enabling the asymmetric synthesis of tert-butyl ((1R,2R)-1-hydroxy-1-phenylhex-5-yn-2-yl)carbamate **2**, setting two stereocentres in dynamic fashion.

- β-Nicotinamide adenine dinucleotide phosphate sodium (β-NADP-Na, 0.2 g, 0.26 mmol)
- *tert*-Butyl (1-oxo-1-phenylhex-5-yn-2-yl)carbamate **1** (100 g, 0.35 mol)
- 2-Propanol (1.0 L)
- Methyl *tert*-butyl ether (MTBE)

7.1.1.2 Procedure

1. *tert*-Butyl (1-oxo-1-phenylhex-5-yn-2-yl)carbamate **1** was chemically synthesised.
2. A 0.2 M borate buffer was prepared by dissolving sodium tetraborate decahydrate (18.7 g) in deionised water (1.0 L). The pH was adjusted to 10.0 at room temperature using 5 M NaOH.
3. KRED-p301 (2.0 g) and NADP (0.2 g, 0.26 mmol) were added and slowly dissolved at room temperature.
4. A solution of **3** (100 g, 0.35 mol) in 2-propanol (1.0 L) was added at ambient temperature with good mixing.
5. The reaction was heated to 45 °C and aged for >24 hr.
6. At >95% conversion as determined by high-performance liquid chromatography (HPLC) analysis, the reaction was cooled to room temperature and MTBE (1.0 L) was added.
7. The phases were separated and the aqueous layer was back-extracted with MTBE/2-propanol (1 : 1, 1.0 L). The combined organic layers were washed with deionised water (1.0 L) and concentrated under vacuum to a final volume of ~0.3 L.
8. The solution was analysed by HPLC, showing a 94 g assay yield for the desired *tert*-butyl ((1*R*,2*R*)-1-hydroxy-1-phenylhex-5-yn-2-yl)carbamate **2** (93% assay yield, >100 : 1 dr, >99% ee). The solution was used directly in the downstream chemistry.

For structure identification, alcohol **2** could be crystallised from toluene in heptane. ^{1}H NMR (400 MHz, $CDCl_3$): major rotomer: δ 7.34 (m, 4 H), 7.28 (m, 1 H), 4.75 (s, br, 1 H), 4.73 (m, 1 H), 3.82 (m, 1 H), 3.27 (s, 1 H), 2.27 (m, 2 H), 1.97 (t, J = 2.6 Hz, 1 H), 1.83 (m, 1 H), 1.72 (m, 1 H), 1.37 (3, 9 H); ^{13}C NMR (100 MHz, d_6-DMSO): major rotomer: δ 156.7, 141.9, 128.5, 127.9, 126.5, 83.9, 79.9, 76.0, 69.1, 56.2, 30.6, 28.5, 15.7.

7.1.2 Analytical Method

Conversion and diastereomeric ratio were determined by reversed-phase HPLC using a Chiralpak IC column (250 × 4.6 mm, 5 μm) under isocratic conditions of 1 : 1 acetonitrile/water (0.1% phosphoric acid) at 1 mL.min^{-1} and 25 °C for 11 min. Detection at 210 nm. Undesired diastereomer pair (not resolved): 5.55 min. Desired diastereomer pair: 6.10 (desired) and 7.81 min. Ketone enantiomers: 8.67 and 9.87 min.

7.1.3 Conclusion

The asymmetric synthesis of Vibegron offers succinct, atom-efficient chemistry with minimal functional group manipulations. The key enabling feature is a dynamic kinetic biocatalytic ketone reduction, setting the challenging C1',C2 stereocentres in excellent selectivity. Directed evolution includes structurally designed mutations that tailor the enzyme's properties to the conditions required for substrate epimerisation, overcoming typical pH and temperature limitations for KRED performance. The optimised route has been demonstrated on a manufacturing scale.

References

1. Xu, F., Kosjek, B., Cabriol, F.L. *et al.* (2018) *Angewandte Chemie International Edition*, **57**, 6863.
2. (a) Xu, F., Chung, J.Y.L., Moore, J.C. *et al.* (2013) *Organic Letters*, **15**, 1342; (b) Applegate, G.A. and Berkowitz, D.B. (2015) *Advanced Synthesis & Catalysis*, **357**, 1619.

7.2 Synthesis of a GPR40 Partial Agonist Through a Kinetically Controlled Dynamic Enzymatic Ketone Reduction

Birgit Kosjek
Process Research and Development, Merck & Co. Inc., Rahway, NJ, USA

A robust and efficient route was developed to simultaneously install the chiral centres of the tricyclic headpiece of a GPR40 agonist [1]. We established an intramolecular enolate alkylation from a *trans*-alcohol which could be accessed from the ketone precursor **1** via dynamic kinetic reduction (Scheme 7.2). Epimerisation of **1** at close to typical KRED conditions was enabled through a keto-enol pathway. In the initial reactions, KRED-208 afforded desired *trans*-**2** in excellent enantioselectivity but only at 2.4 : 1 diastereomeric ratio. We found that the enzyme's intrinsic diastereoselectivity initially formed a slight excess of the *cis*-alcohol, but as the co-factor recycling system continuously produced acetone from 2-propanol, the reaction became reversible and cycled the alcohol products to *trans*-**2** until a thermodynamically determined ratio of 2.4 : 1 was reached. We evolved the KRED to overcome the thermodynamic equilibrium by increasing the diastereoselectivity to favour *trans*-**2** at >30 : 1 and implemented a nitrogen sweep to remove acetone and prevent reversibility. Additional improvements in enzyme activity and stability resulted in variant KRED-264 with excellent performance at elevated pH and temperature to economically prepare *trans*-alcohol **2** in high yield and selectivity.

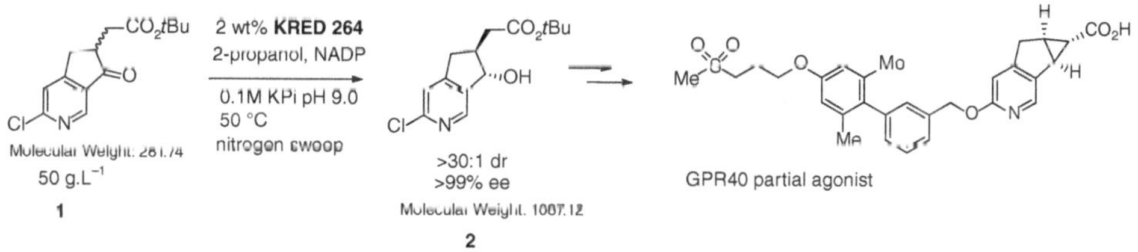

Scheme 7.2 Scalable synthesis of a GPR40 partial agonist enabled by a dynamic kinetic reduction using an evolved KRED.

7.2.1 Procedure1: Preparation of *tert*-Butyl 2-((6*S*,7*S*)-3-Chloro-7-Hydroxy-6, 7-Dihydro-5H-Cyclopenta[c]pyridin-6-yl)acetate 2 via Dynamic Kinetic Ketone Reduction

7.2.1.1 Materials and Equipment

- Potassium phosphate dibasic (0.35 g)
- Deionised water (20 mL)
- Ketoreductase KRED-264 (40 mg, Codexis Inc.)
- β-Nicotinamide adenine dinucleotide phosphate sodium (β-NADP-Na, 5 mg, 0.007 mmol)
- *tert*-Butyl 2-(3-chloro-7-oxo-6,7-dihydro-5H-cyclopenta[c]pyridin-6-yl)acetate **1** (2.0 g, 7.1 mmol)
- 2-Propanol (20 mL)
- Methyl *tert*-butyl ether (MTBE)

7.2.1.2 Procedure

1. *tert*-Butyl 2-(3-chloro-7-oxo-6,7-dihydro-5H-cyclopenta[c]pyridin-6-yl)acetate **1** [1] (2.0 g, 7.1 mmol) and 2-propanol (20 mL) were added to a cylindrical vessel with overhead mixing and heated to 60 °C until ketone **1** was completely dissolved.
2. The solution was cooled to ~35 °C and 0.1 M phosphate buffer pH 9.0 (15 mL) was added. The buffer was prepared by dissolving potassium phosphate dibasic (K_2HPO_4, 0.35 g) in deionised water (20 mL) and adjusting to pH 9.0 with 2 N NaOH at room temperature.
3. A solution of KRED-264 (40 mg) and NADP (5 mg, 0.007 mmol) in 0.1 M phosphate buffer pH 9.0 (5 mL) was added and the reaction was stirred at 50 °C.
4. After 2.5 hr reaction time, a nitrogen sweep was started to facilitate acetone removal. Solvent loss was monitored and replenished with a mixture of buffer/2-propanol (1 : 1).
5. At 22 hr, the reaction was cooled to 35 °C and extracted with a mixture of MTBE/2-propanol (5 : 1, 25 mL). The assay yield of *tert*-butyl 2-((6S,7S)-3-chloro-7-hydroxy-6, 7-dihydro-5H-cyclopenta[c]pyridin-6-yl)acetate **2** in the organic layer was determined by reversed-phase high-performance liquid chromatography (HPLC) as 1.93 g (97.2%, >30 : 1 dr, >99% ee).
6. The organic layer was partially concentrated and solvent switched to MTBE with azeotropic drying to reach a water content of 567 ppm and 10 volumes. This solution was used directly in the downstream chemistry [1].

Evaporation to dryness of a solution aliquot provided isolated material for structure confirmation. ^{1}H NMR ($CDCl_3$, 500 MHz) δ: 8.37 (s, 1H), 7.15 (s, 1H), 5.01 (br s, 1H), 3.99 (d, J = 3.4 Hz, 1H), 3.13 (dd, J = 16.1, 6.7 Hz, 1H), 2.65–2.59 (m, 1H), 2.63 (d, J = 1.0 Hz, 2H), 2.56–2.51 (m, 1H), 1.48 (s, 9H). ^{13}C NMR ($CDCl_3$, 125 MHz) δ: 173.12, 153.88, 150.67, 145.87, 139.68, 120.42, 81.93, 78.88, 46.67, 39.19, 35.94, 28.27.

7.2.2 Analytical Method

Conversion and diastereomeric ratio were determined by reversed-phase HPLC using a Chiralcel AS-3R column (100 × 4.6 mm, 3 μm) under isocratic conditions of 35 : 65

acetonitrile/water (0.1% phosphoric acid) at 1 mL.min^{-1} and 25 °C for 10 min. Detection at 210 nm. *Cis* diastereomers: 5.1 and 5.8 min. *Trans* diastereomers: 4.6 and 7.0 min (desired). Keto ester enantiomers: 8.8 and 9.1 min.

7.2.3 Conclusion

This economical dynamic kinetic reduction of an unactivated ketone presents a powerful example of the adaptability of biocatalysts to conditions outside their typical performance window, enabling a viable reaction sequence for large-scale production of a key intermediate. Detailed investigation of the kinetic and thermodynamic mechanisms was essential to inform the directed evolution strategy and develop the final enzymatic process.

Reference

1. Hyde, A.M., Liu, Z., Kosjek, B. *et al.* (2016) *Organic Letters*, **18**, 5888.

7.3 Lab-Scale Synthesis of Eslicarbazepine

Naga Modukuru, Joly Sukumaran, Steven J. Collier, Ann Shu Chan, Anupam Gohel, Gjalt W. Huisman, Raquel Keledjian, Karthik Narayanaswamy, Scott Novick, S.M. Palanivel, Derek Smith, Zhang Wei, Brian Wong, Wan Lin Yeo and David A. Entwistle*

Codexis Inc., Redwood City, CA, USA

The generic carbazepine series of active pharmaceutical ingredients (APIs), which range from carbamazepine **1** to the single *S*-enantiomer acetyl carbazepine **4**, has been developed over time to have improved product profiles in treating epileptic seizures (Scheme 7.3). Eslicarbazepine **3** has been synthesised by asymmetric variations of methods such as reduction/hydroboration of **2** [1] and transfer hydrogenation of **2** [2]. Racemic **3** has also been resolved by enantioselective enzymatic hydrolysis of a methoxyacetyl ester [3] or resolution of diastereomeric esters [4].

An asymmetric ketoreductase (KRED)-mediated method of synthesis was developed that is more efficient than any of the resolution procedures. This method does not suffer from the air sensitivity of catalysts and reagents used in the previously described asymmetric syntheses [5].

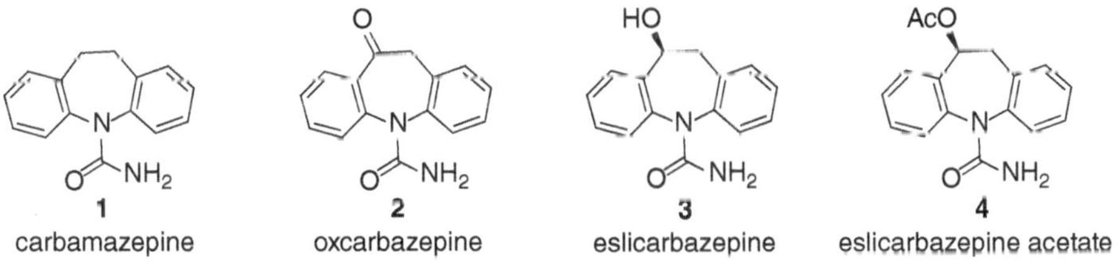

Scheme 7.3 Series of carbazepine APIs.

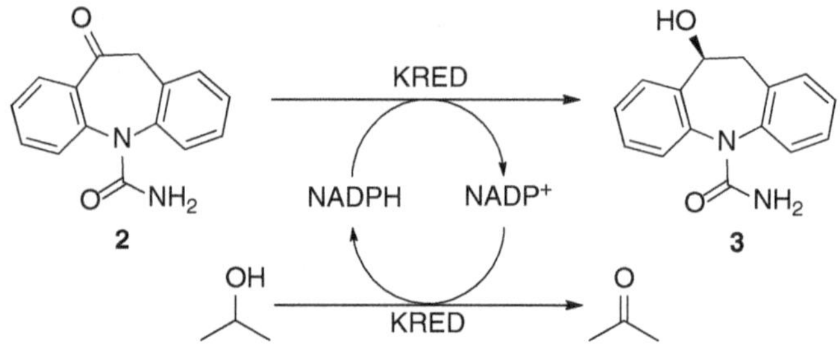

Scheme 7.4 Ketoreductase (KRED)-mediated synthesis of **3**.

The KRED used, CDX-021, was engineered at Codexis from a variant of a KRED from *Lactobacillus kefir* over four rounds of evolution [5]. A 1% w/w loading of the isolated CDX-021 successfully catalysed a reaction of 100 g.L^{-1} (10 L.kg^{-1}) of oxcarbazepine **2**, yielding eslicarbazepine **3** in >99% ee (Scheme 7.4).

A high conversion of >99% was required as **2** was difficult to remove by crystallisation and the product needed to meet a high purity specification.

7.3.1 Procedure 1: Preparation of Crude Eslicarbazepine 3

7.3.1.1 Materials and Equipment

- Triethanolamine (TEoA; 13.3 g)
- Water (1.2 L)
- $MgSO_4.7H_2O$ (0.25 g)
- Isopropanol (IPA; 400 mL)
- Oxcarbazepine (50 g)
- CDX-021 (0.5 g)
- $NADP^+$ (50 mg)
- Heptane (100 mL)
- Crude Eslicarbazepine (10 g)
- Methanol (120 mL)
- Celite (2.0 g)
- Water (70 mL)

7.3.1.2 Procedure

1. A 0.1 M pH 10 TEoA solution containing 1 mM magnesium was prepared by dissolving TEoA (13.3 g) and $MgSO_4.7H_2O$ (0.25 g) in deionised water (1 L) at room temperature. This solution was used as is, with no further pH adjustment.
2. IPA (300 mL), the buffer solution from Step 1 (190 mL) and oxcarbazepine (50 g) were sequentially added into a 1 L jacketed vessel. The agitation was set to 200 rpm and the contents were heated to 55 °C. The pH of the resultant mixture was 8.6.
3. In a separate vessel, a solution of CDX-021 (0.5 g) and NADP+ (50 mg) was prepared in the buffer from Step 1 (10 mL) and added to the main reactor contents.
4. Stirring in the main reactor was continued at 55 °C and a nitrogen sweep was applied across the surface of the reactor contents (0.8 L.min^{-1} for the vessel used).

5. In a separate vessel, a solution containing 60% v/v IPA and 40% v/v of the buffer from Step 1 was prepared (110 mL).
6. The volume of the main reactor was kept approximately constant by the occasional portion-wise addition of the IPA/buffer mixture from Step 5.
7. The reaction progress was monitored by taking samples and analysing by HPLC Method 1 for conversion and GC Method 3 for acetone content.
8. After 24 hr, once the conversion was >99%, the reactor content was drained into a flask and the IPA partially removed by rotary evaporation at 50 °C and 75 torr.
9. Water (100 mL) was added to the flask and the contents were further distilled to remove the final vestiges of IPA
10. The crude product was collected by filtration on a Buchner funnel, washed with water (100 mL) and heptane (200 mL) and dried for 24 hr in a vacuum oven (2 mbar) at 30 °C. Crude Eslicarbazepine **3** (48.0 g, 96% yield) was obtained as an off-white solid with a chemical purity of 98.7% (HPLC Method 4) with >99.9% ee (HPLC Method 2).
11. A suspension of crude **3** (10.0 g) in methanol (100 mL) was heated to 40 °C.
12. Celite (2.0 g) was added and the mixture was stirred at 40 °C for approximately 20 min.
13. The celite was removed by filtration and the residue was washed with preheated methanol (20 mL, 40 °C).
14. The filtrate was distilled under reduced pressure to approximately 30 mL volume, whereupon product precipitation occurred. The thick mixture was cooled to 5 °C.
15. Cold water (50 mL, 5 °C) was added over 30 min to the white mixture to precipitate the remaining product and the resulting slurry was stirred at 5 °C for a further 30 min.
16. The precipitated product was filtered, rinsed and washed with 20 mL water and dried in a vacuum oven for 16 hr (30 °C, 2 mbar).
17. Purified product (9.0 g, 90% recovery) was isolated as a white solid with 99.6% chemical purity (HPLC Method 4).

7.3.2 Analytical Method

7.3.2.1 HPLC Method 1

Column: Agilent Eclipse XDB C18 4.6 × 150 mm, 5 μm; MeCN/0.1% AcOH in water (40/60, isocratic); 25 °C at 1.4 mL.min^{-1}; product retention time 1.38 min; substrate retention time 1.91 min; wavelength 210 nm.

7.3.2.2 HPLC Method 2

Column: ChiraDex®, LiChroCART® 250-4, 5 μm; 95% Na_2HPO_4 (100 mM, pH 7.0) and 5% MeOH (isocratic); 15 °C at 1.0 mL.min^{-1}; *R*-enantiomer 15.4 min; *S*-enantiomer 18.3 min; wavelength 254 nm.

7.3.2.3 GC Method 3

Column: Roticap WAX Capillary, 50 m × 250 μm (ID) × 0.25 μm (FT); split 60 : 1; helium flow rate 1.1 mL.min^{-1} (constant-pressure mode); inlet 190 °C; detector 200°C; acetone retention time 3.9 min; IPA retention time 4.4 min.

7.3.2.4 HPLC Method 4

Method to compare area to authentic reference standard. Column: Agilent Eclipse XDB C18 4.6×150 mm, 5 µm; MeCN/0.1% AcOH in water (25/75, isocratic); 25 °C at 1 mL.min^{-1}; 230 nm.

7.3.3 Conclusion

An engineered KRED, CDX-021, mediated a high-yield process to give the desired product in very high enantiomeric excess (>99% ee). Whilst the reaction is indeed efficient, the methods for isolating and purifying the crude product would benefit from further optimisation.

References

1. (a) Katkam, S., Sagyam, R.R.; Buchikonda, R. *et al.* (2011) Patent WO2011091131; (b) Satyanarayana, R.M., Eswaraia, S., Kondal, R.B. and Venkatesh, M. (2011) Patent WO2011138795; (c) Biswas, S., Dubey, S.K., Bansal, V. *et al.* (2012) Patent WO2012120356.
2. (a) Mathes, C., Sedelmeier, G., Blatter, F. *et al.* (2004) Patent WO2004031155; (b) Learmonth, D.A., Grasa, G.A. and Zanotti-Gerosa, A. (2007) Patent WO2007012793; (c) Wisdom, R., Jung J. and Meudt, A. (2011) Patent EP2383261; (d) Daqing, C. and Guoliang Z. (2011) Patent CN102250005.
3. Husain, M. and Datta, D. (2010) Patent WO2011045648.
4. (a) Learmonth, D.A. (2002) Patent WO02092572; (b) Gharpure, M.M., Rane, D., Zope, S.S. *et al.* (2012) Patent WO2012156987; (c) Crasta, S.R.F., Joshi, A.V. and Bhanu, M.N. (2012) Patent WO2012121701; (d) Desai, S.J., Pandya, A.K., Sawant, S.P. and Mehariya K.R. (2011) WO2011117885.
5. Modukuru, N.K., Sukumaran, J., Collier, S.J. *et al.* (2014) *Organic Process Research & Development*, **18** (6), 810–815.

7.4 Direct Access to Aldehydes Using Commercially Available Carboxylic Acid Reductases

Margit Winkler,*[1] Anna Schwarz,[1] Gernot A. Strohmeier,[1] Jeffrey Kohrt[2] and Roger M. Howard[2]
[1]*acib GmbH, Graz, Austria*
[2]*Pfizer Worldwide Research and Development, Applied Synthesis Technologies – Biocatalysis, Groton, CT, USA*

Aldehydes are useful intermediates in medicinal chemistry. Carboxylic acid reductases (CARs) are a class of enzymes that selectively reduce widely available carboxylic acids to aldehydes [1] – a highly challenging chemical transformation. Protocols to use these enzymes in cell-free systems for medicinal chemistry applications would prove valuable. The proof of concept for the accomplishment of co-factor recycling at the expense of glucose and polyphosphate has recently been shown with purified enzyme preparations; the reaction system allowed quantitative reduction of various carboxylic acids with full recycling of all co-factors (Scheme 7.5) [2]. In this section, we show the scope and limitations

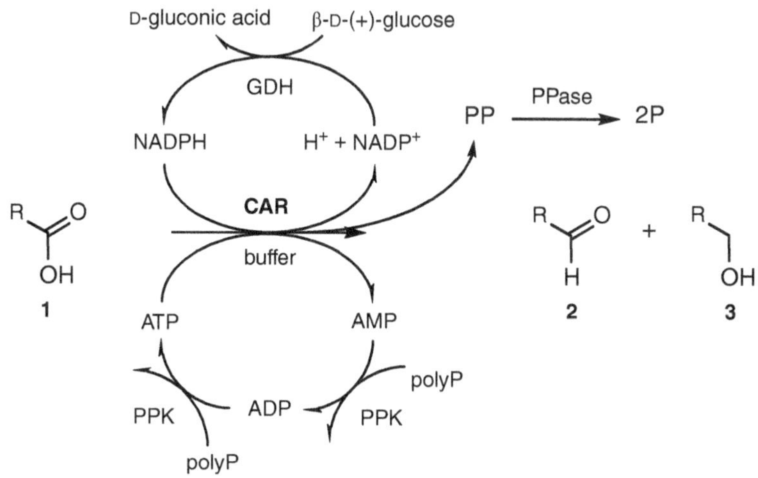

Scheme 7.5 Reduction of carboxylic acid **1** to aldehyde **2** with cell-free recycling of all co-factors. CAR, crude carboxylate reductase preparation (Prozomix); PPK, crude polyphosphate kinase preparations (Prozomix); GDH, commercial glucose dehydrogenase 105 preparation (Codexis); PPase, crude pyrophosphatase preparation (Prozomix); PolyP, polyphosphate; PP, pyrophosphate; P, ortho-phosphate.

of employing equivalent commercially available lyophilised crude CARs, which offer significant advantages in terms of cost and convenience.

7.4.1 Procedure 1: Screening-Scale Reduction of Carboxylates to Aldehydes

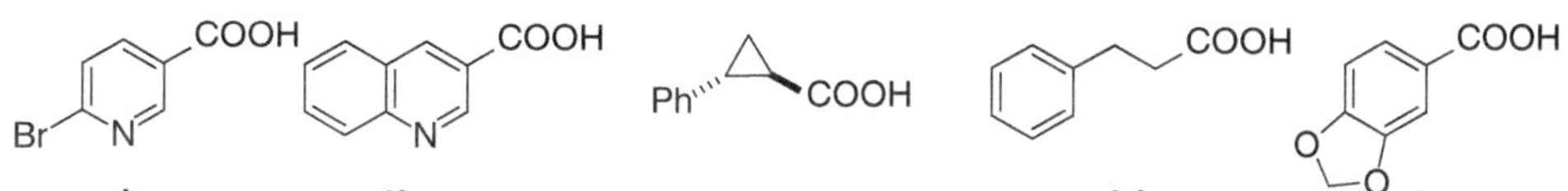

Scheme 7.6 Selected carboxylic acids.

7.4.1.1 Materials and Equipment

- 6-Bromo-pyridine-3-carboxylic acid **1a** (Aldrich 646989); quinoline-2-carboxylic acid **1b** (Aldrich 177148); *trans*-2-phenyl-cyclopropane carboxylic acid (rac)-**1c** (Aldrich P22354); 3-phenylpropionic acid **1d** (Aldrich 56670); piperonylic acid **1e** (Aldrich P49805) (Scheme 7.6)
- 3-(*N*-Morpholino)propanesulfonic acid (MOPS) buffer pH 7.50, 400 mM
- $NADP^+$ (C. Roth AE13.3, disodium salt)
- Adenosine triphosphate (ATP; Aldrich A26209, disodium salt hydrate)
- $MgCl_2$ (C. Roth 2189.1, hexahydrate)
- β-D-(+)-Glucose (C. Roth 6780, monohydrate)
- ddH_2O (Fresenius)

- Sodium polyphosphate (polyP) (Merck 1.065.291.000, lot K46879329603; medium chain length, n = 25)
- GDH-105 (40 U.mg^{-1}; Codexis)
- CAR005; CAR005-PPT (A); CAR005-PPT (C); CAR005-PPT (D); CAR001; CAR001-PPT (A); CAR001-PPT (B); CAR001-PPT (D) (Prozomix)
- *Ec*PPase WP_000055075.1 containing CFE powder (Prozomix) or purified *Ec*PPase [2a]
- *Sm*PPK WP_010968631.1 containing CFE powder (Prozomix) or purified *Sm*PPK [2a]
- *Mr*PPK WP_015586734.1 containing CFE powder (Prozomix) or purified *Mr*PPK [2a]
- Bromothymol blue (Sigma-Aldrich 114413)
- 1 mL Eppendorf tubes
- Thermomixer Comfort (Eppendorf)
- MeOH (Chemlab CL00.1371.2500)
- Formic acid (Fluka 06440)
- Eppendorf tabletop centrifuge 5415R
- V-shaped polypropylene 96-well plates with sealing foil or high-performance liquid chromatography (HPLC) vials
- HPLC system with UV detector or mass spectrometry (MS) detector
- HPLC column: EC 150/3 Nucleodur C18 Gravity 3.0 μm (SN: N0060018, Batch: 37408033, Macherey-Nagel)
- HPLC-grade acetonitrile (J.T. Baker)
- HPLC-grade ammonium acetate (Fluka)

7.4.1.2 Procedure

1. To prepare a 2× buffer mix consisting of 50% of the final volume of MOPS buffer (pH 7.50, 400 mM), $MgCl_2$ and β-D-(+)-glucose were added as solids, to give final concentrations of 140 and 200 mM, respectively.
2. 250 mM substrate stock solutions were prepared in 250 mM KOH.
3. A 1 M polyP stock solution was prepared (the molarity is based on a calculation using the molecular mass of an *ortho*-phosphate unit, $NaPO_3H$, 102.96 g.mol^{-1}). As an example, 2.74 mL of H_2O was added to 309 mg of sodium polyphosphate in a U-shaped 10 mL glass tube. 260 μL of KOH was added. Vortexing facilitated dissolution.
4. $NADP^+$ and ATP stock solutions (50 mM in ddH_2O) were prepared and kept on ice for later use.
5. Lyophilised crude CFE containing CAR, *Mr*PPK, *Sm*PPK or *Ec*PPase was taken out of the −20 °C freezer. When the containers reached room temperature, powders were gently dissolved in MOPS buffer (100 mM, pH 7.5) and stored on ice. Typically enzyme stock solutions are 5–30 mg.mL^{-1}.
6. An enzyme master mix consisting of *Mr*PPK, *Sm*PPK, *Ec*PPase and GDH-105 was prepared.
7. Per substrate, the following master mix was prepared: 32 μL of H_2O in an Eppendorf tube was mixed with 125 μL of 2× buffer mix, 50 μL polyP stock (1 M), 10 μL of substrate stock solution, 2.5 μL $NADP^+$ stock solution and 5 μL of ATP stock solution.
8. The enzyme master mix was added to 225 μL of the master mix from Step 7, to give final enzyme concentrations of 200 μg.mL^{-1} of *Mr*PPK preparation or 100 μg.mL^{-1} of purified *Mr*PPK, 80 μg.mL^{-1} of *Sm*PPK preparation or 40 μg.mL^{-1} of purified

*Sm*PPK, 50 μg.mL^{-1} of *Ec*PPase preparation or 25 μg.mL^{-1} of purified *Ec*PPase and 0.2 U.mL^{-1} GDH-105 preparation.

9. The reaction was started by the addition of typically 1 mg.mL^{-1} of CAR preparation to give a total reaction volume of 250 μL. Acid reduction proceeded at 28°C in a Thermomixer set at 30 °C and 700 rpm for 3–16 hr.
10. Optionally, bromothymol blue solution (5 μL, 8 mM in 8 mM KOH) can be added to monitor the pH. Acidification of the reaction – as indicated by a colour change from blue to yellow – indicates reaction progress and may optionally be counteracted by the addition of aqueous KOH.
11. Reactions were terminated and analysed.

7.4.1.3 *Analytical Method*

Typically, 100 μL of reaction mixture was quenched with 400 μL sample diluent (5% formic acid in MeOH), followed by centrifugation for 5 min at 13 200 rpm. 100 μL of supernatant was transferred into V-shaped 96-well polypropylene microtitre plates or glass vials with inserts for analysis.

All analytes described in this section can be detected with an HPLC system at 210 nm using a mobile phase consisting of acetonitrile and aqueous ammonium acetate (5 mM) and a reversed-phase C18 column. The results shown in Figure 7.1 were obtained using a Nucleodur C18 Gravity column at a flow rate of 1 mL.min^{-1} and 35 °C column temperature. Acid, alcohol and aldehyde can be separated with the following stepwise gradient: 0–0.7 min 5% acetonitrile; 0.71–3.0 min to 85% acetonitrile; 3.01–3.8 min 85%; 3.81–5.5 min re-equilibration to 5% acetonitrile. Injection volume: 4 μL. Results of the biotransformations on screening scale (10 mM substrate) are given in Figure 7.1.

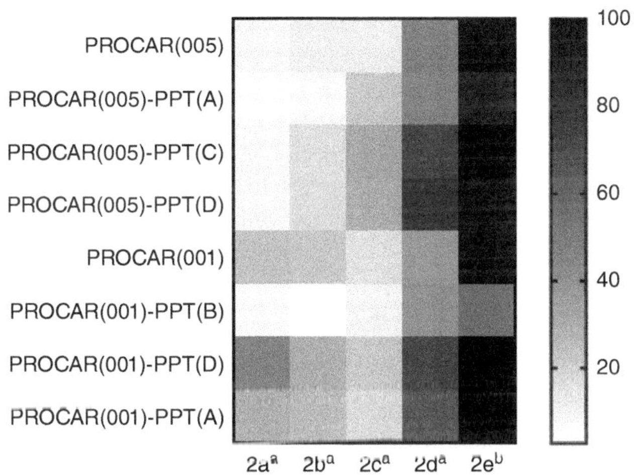

Figure 7.1 *Screening results of carboxylic acid synthons as substrates for aldehyde formation with crude carboxylic acid reductase (CAR) preparations. Substrate concentration: 10 mM. Purified PPK and PPase were applied. [a]Amount of lyophilised CAR preparation: 125 μg.mL^{-1}. [b]Amount of lyophilised CAR preparation: 250 μg.mL^{-1}.*

7.4.2 Procedure 2: Preparative-Scale Reduction of *trans*-2-Phenyl-Cyclopropane Carboxylic Acid 1c to *trans*-2-Phenyl-Cyclopropane Carboxaldehyde with *In Situ* Product Removal (ISPR)

7.4.2.1 Materials and Equipment

- Racemic *trans*-2-phenyl-cyclopropane carboxylic acid (Aldrich P22354)
- MOPS buffer pH 7.50, 400 mM
- $NADP^+$ (C. Roth AE13.3, disodium salt)
- ATP (Aldrich A26209, disodium salt hydrate)
- $MgCl_2$ (C. Roth 2189.1, hexahydrate)
- β-D-(+)-Glucose (C. Roth 6780, hydrate)
- ddH_2O (Fresenius)
- Sodium polyphosphate (Merck order no. 1.065.291.000, lot K46879329603; medium chain length n = 25, 2 M based on *ortho*-phosphate units in H_2O, pH 7.50)
- GDH-105 (40 $U.mg^{-1}$; Codexis)
- CAR005-PPT (A) (Prozomix)
- *Ec*PPase WP_000055075.1 containing CFE powder (Prozomix)
- *Sm*PPK WP_010968631.1 containing CFE powder (Prozomix)
- *Mr*PPK WP_015586734.1 containing CFE powder (Prozomix)
- Conical 50 mL polypropylene tube (CAPP)
- Cyclohexane (C. Roth 6886)
- Titramax 1000 (Heidolph)
- Temperature-controlled incubator IPP110 (Memmert)
- Hydrochloric acid (aq), 3 M
- Eppendorf tabletop centrifuge 5810R
- Vortex Genie 2 (Scientific Industries)
- MeOH (Chemlab CL00.1371.2500)
- Formic acid (Fluka 06440)
- MeOH/formic acid 19 : 1
- Glass vials with crimp caps (Machery-Nagel)
- HPLC system with UV detector or MS detector (Agilent)
- Column: Kinetex 2.6 μm Biphenyl 100A 00F4622-AN (Phenomenex 37408033 SN: H18-295961, Batch: H18-295961), equipped with a SecurityGuard™ Ultra Cartridge (Phenomenex AJ0-8775)
- HPLC-grade acetonitrile (J.T. Baker)
- HPLC-grade ammonium acetate (Fluka)

7.4.2.2 Procedure

1. A 2× buffer mix consisting of 50% of the final volume of MOPS buffer (pH 7.50, 400 mM) was prepared. $MgCl_2$ and β-D-(+)-glucose were added as solids, to give final concentrations of 140 and 200 mM, respectively.
2. 250 mM **1c** substrate stock (245 mg in 5.8 mL aqueous 280 mM KOH) was prepared.
3. A 2 M polyP stock solution was prepared. As an example, 8 mL of H_2O was added to 2.057 g of sodium polyphosphate in a U-shaped 10 mL glass tube. KOH was added until a pH of 7.50 was reached (1.07 mL), and the solution was filled up to 10 mL with H_2O.

4. $NADP^+$ and ATP stock solutions (50 mM in ddH_2O) were prepared and kept on ice for later use.
5. Lyophilised crude CFEs containing CAR005-PPT (A), *Mr*PPK, *Sm*PPK or *Ec*PPase were taken out of the −20 °C freezer and left to reach room temperature. Powders were gently dissolved to 20 mg.mL^{-1} (CAR), 10 mg.mL^{-1} (*Mr*PPK, *Sm*PPK and *Ec*PPase, respectively) in MOPS buffer (100 mM, pH 7.5) and stored on ice.
6. An enzyme master mix was prepared. For 40 mL final reaction volume, this master mix consisted of 0.8 mL *Mr*PPK stock, 0.32 mL *Sm*PPK stock, 0.2 mL *Ec*PPase stock, 2 mL CAR stock solution and 0.16 mL GDH-105.
7. H_2O, buffer mix, polyP stock solution and substrate stock solution were prewarmed to 30 °C.
8. The following components were added to 9.4 mL of H_2O in a 50 mL polypropylene tube: 4 mL polyP (2 M), 20 mL of 2× buffer mix, 1.92 mL of substrate stock solution, 0.4 mL $NADP^+$ stock solution and 0.8 mL of ATP stock solution. The reaction was started by the addition of 3.48 mL of enzyme master mix. To promote ISPR, 10 mL of cyclohexane was added and the tube was placed on a shaking platform and agitated at 28 °C and 750 rpm for 20 hr.
9. Aqueous HCl (3 M, 1 mL) was added to terminate the reaction.
10. Products and unreacted substrate were extracted by agitation of the tube on a titramax at 1000 rpm for 15 min, followed by centrifugation at 4000 rpm on a tabletop centrifuge. The organic layer was removed.
11. Fresh cyclohexane (10 mL) was added to the aqueous phase, which contained a pellet and a protein layer. The mix was homogenised by vortexing followed by agitation on a titramax at 1000 rpm for 15 min then centrifugation at 4000 rpm. The organic layer was removed and pooled with the first organic layer. Extraction was repeated two more times.
12. Samples of the organic phase were diluted with 4 volumes of MeOH/formic acid 19 : 1 and analysed by HPLC.
13. The assay yield of aldehyde, alcohol and unreacted carboxylic acid was quantified by linear interpolation versus calibration curves with authentic standards.
14. Conversion of **1c** was 99%. The combined organic layers consisted of 1% **1c**, 90% aldehyde **2c** and 9% alcohol **3c** (relative amounts). 0.4 mM of **3c** remained in the aqueous phase. The total assay yields for **2c** and **3c** are given in Scheme 7.7.

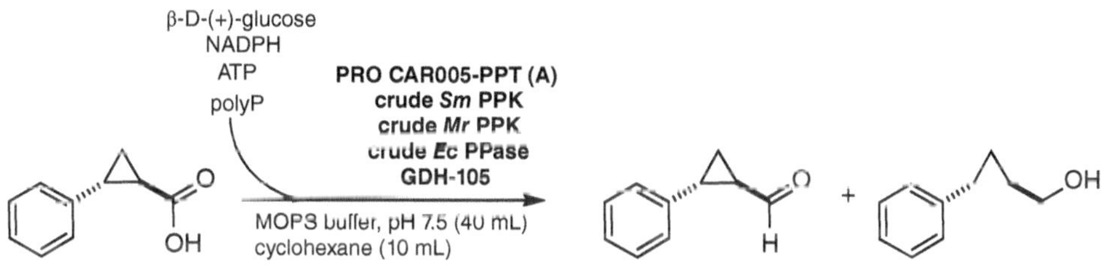

(rac)-**1c** (80 mg) **2c** (87%) **3c** (12%)

Scheme 7.7 *Preparative-scale reduction of racemic **1c** to aldehyde **2c** using crude enzyme preparations in combination with in situ product removal by application of a water-immiscible phase.*

7.4.2.3 Analytical Method

1–3c were separated with the following gradient: 0 min 30% acetonitrile; 0–2.0 min to 40% acetonitrile; 2.0–6.0 min to 60% acetonitrile; 6.0–9.0 min to 90% acetonitrile, until 13.0 min 90% acetonitrile; 13.0–19 min re-equilibration to 30% acetonitrile. Column: Kinetex Biphenyl. Injection volume: 5 µL. Compounds were detected at 210 nm. The enantiomeric excess values of the products were not determined.

7.4.3 Conclusion

The described procedure outlines how to screen for suitable commercial CAR preparations for a given substrate. The co-factor demand can be met by addition of 1.1 eq each of NADPH and ATP or by the cell-free recycling system shown herein. Either crude preparations of recycling enzymes or purified enzymes may be used. As an example on the multimilligram preparative scale, we have demonstrated the synthesis of *trans*-2-phenyl-cyclopropane carboxaldehyde using a combination of crude enzyme preparations for the reduction of the carboxylate to the aldehyde and the required recycling of NADPH and ATP. Using ISPR on preparative scale with cyclohexane as extracting agent, conversion was increased from 65 to 99%, whilst aldehyde overreduction slightly decreased from 15 to 12%, as compared to a single-phase reaction. This protocol could be further optimised to decrease the undesired overreduction of aldehydes to the corresponding alcohols, which typically increases with the load of introduced host-background proteins.

Acknowledgements

This work has been supported by the Federal Ministry of Science, Research and Economy (BMWFW), the Federal Ministry of Traffic, Innovation and Technology (bmvit), the Styrian Business Promotion Agency SFG, the Standortagentur Tirol, the Government of Lower Austria and Business Agency Vienna through the COMET-Funding Program managed by the Austrian Research Promotion Agency FFG. A.S. received funding from the Austrian Research Promotion Agency FFG (FEM-Tech stipend, grant no. 870897). The authors also thank Prozomix, UK (Prof. S.J. Charnock, Mr R.A.M. Duncan and Dr D. Cook) for the production of the crude lyophilised enzymes employed.

References

1. Winkler, M. (2018) *Current Opinion in Chemical Biology*, **43**, 23–29.
2. (a) Strohmeier, G.A., Eiteljörg, I.C., Schwarz, A. and Winkler, M. (2019) *Chemistry: A European Journal*, **25**, 6119–6123; (b) Strohmeier, G.A., Schwarz, A., Andexer, J.N. and Winkler, M. (2019) *Journal of Biotechnology*, **307**, 202–207.

7.5 Preparation of Methyl (*S*)-3-Oxocyclohexanecarboxylate Using an Enoate Reductase

Alba Diaz-Rodriguez,*[1] Timin Hadi,*[2] Radka Snajdrova,[1,4] Douglas Fuerst[2] and Gheorghe Doru Roiban[3]

[1]*Chemical Development, Medicinal Science & Technology, GlaxoSmithKline, Medicines Research Centre, Stevenage, Hertfordshire, UK*

[2]*Synthetic Biochemistry, Medicinal Science & Technology, GlaxoSmithKline, King of Prussia, PA, USA*

[3]*Synthetic Biochemistry, Medicinal Science & Technology, GlaxoSmithKline, Stevenage, Hertfordshire, UK*

[4]*Global Discovery Chemistry, Novartis Institute for Biomedical Research, Novartis Pharma AG, Lichtstrasse, Basel CH-4056 Basel, Switzerland*

Cyclohexanes with different substitutions and stereochemistries can be found in a number of pharmaceutically relevant compounds. During recent active pharmaceutical ingredient (API) route-scouting exercises at GSK, optically pure methyl (*S*)-3-oxocyclohexanecarboxylate **2** was identified as a key chiral intermediate. Several biocatalytic approaches have been used on multigram scale to access chiral 3-substituted cyclohexanones under mild reaction conditions, one of the most efficient options being enoate reductase (ERED) reduction of the methyl 3-oxocyclohex-1-enecarboxylate **1** [1]. ERED-mediated reduction represents an excellent alternative to traditional chemical methods and produces the desired compound with high isolated yields and enantiomeric excesses [2]. In the procedure discussed in this section, whole cells and crude-cell lysates were used to quickly generate material of sufficient (*S*)-**2** quality and quantity for downstream chemistry applications (Scheme 7.8).

7.5.1 Procedure 1: Biotransformation to Methyl (*S*)-3-Oxocyclohexanecarboxylates 2

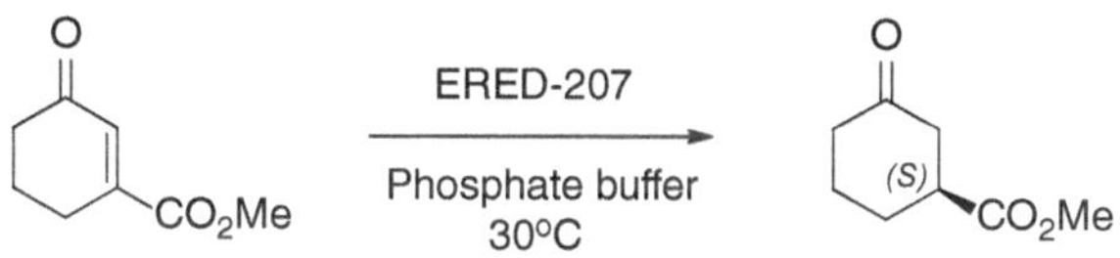

1 (*S*)-**2**

Scheme 7.8 *ERED-207-mediated reduction of methyl 3-oxocyclohex-1-ene-1-carboxylate **1** to access methyl (S)-3-oxocyclohexanecarboxylate **2** [1].*

7.5.1.1 *Materials and Equipment*

- Methyl 3-oxocyclohex-1-enecarboxylate **1** (55.0 g, 0.36 mol)
- Frozen cell paste of ERED-207 (37.5 g; available from Codexis Inc. as a lyophilised powder)

- NADP+ from Bontac Bio-Engineering (2.5 g)
- Glucose dehydrogenase GDH-CDX-901 from Codexis Inc. (3.0 g)
- D-Glucose (73.0 g, 0.41 mol)
- Potassium phosphate buffer (KH_2PO_4/K_2HPO_4) 200 mM, pH 7.0 (4.0 L)
- NaOH (5 M)
- Schott bottle (5 L)
- Ethyl acetate (EtOAc, 4 L)
- Anhydrous $MgSO_4$ (2.0 g)
- Cotton wool
- Controlled laboratory reactor (CLR, 5 L)
- Fisher Scientific sonicator FB705
- Agilent 7890B gas chromatography (GC) system with flame ionisation detector (FID)
- Achiral GC column (HP5 30 m × 320 μm ID × 0.25 μm film)
- Chiral GC column (Hydrodex β-TBDAc 25 m × 250 μm ID × 0.25 μm film)
- Rotary evaporator
- Bruker Advance 400 (^{1}H: 400 MHz, ^{13}C: 101 MHz) spectrometer using tetramethylsilane (TMS) as internal standard

7.5.1.2 Procedure

1. The phosphate buffer was prepared in a Schott bottle by adding K_2HPO_4 (492 mL, 1 M) and 308 mL KH_2PO_4 (308 mL, 1 M) and diluting with distilled water to a final volume of 4 L. At this point, pH indicated 7.0.
2. CLR jacket temperature was set to 30°C and overhead stirrer to 215 rpm.
3. D-Glucose (73.0 g, 1.25 eq) in 1 L potassium phosphate buffer 200 mM, pH 7.0 was charged into the CLR.
4. GDH-CDX-901 was charged into the CLR (3.0 g, 2.5% w/w).
5. NADP+ was added to the CLR (2.5 g, 0.01 eq).
6. Phosphate buffer (600 mL) was added to the CLR.
7. Frozen cell paste of ERED-207 (37.5 g) was resuspended in potassium phosphate buffer (200 mL) and disrupted using a sonicator (40% amplitude, 10 sec on, 30 sec off for 10 min) (37.5 g whole cells corresponds to approximately 2.7 g of commercial lyophilised ERED-207 enzyme powder).
8. The pH of the mixture was adjusted to 7.0 using 5 M aqueous NaOH.
9. The reaction was initiated by the addition of methyl 3-oxocyclohex-1-enecarboxylate **1** (55.0 g, 0.36 mol).
10. Stirring was carefully increased to 320 rpm and the CLR contents were maintained at 30 ± 2 °C.
11. Titration to pH 7.0 was carried out during the course of the reaction using 5M aqueous NaOH.
12. The reaction was sampled at different time points (30 min, 1 hr, 2 hr and 3 hr) and analysed by gas chromatography (GC) for conversion and enantiomeric excess determination. The reaction typically takes 3 hr to reach completion.
13. Quenching procedure: ethyl acetate (2 L) was charged to CLR, stirring speed adjusted to 150 rpm.
14. Contents were stirred for 15 min.
15. Aqueous (2.5 L) and organic (1.8 L) phases were recovered after first separation.

16. Aqueous phase and additional ethyl acetate (2 L) were charged to CLR and stirred for 10 min.
17. Ethyl acetate was collected (1.8 L) and organic phases were combined.
18. Ethyl acetate phase was filtered through cotton wool and solvent was removed by distillation under reduced pressure to afford a neat mixture of the desired methyl (*S*)-3-oxocyclohexanecarboxylate (50.1 g, 90% yield, 96.2% GC purity, 97% ee).

^{1}H-NMR (400 MHz, $CDCl_3$) δ ppm 3.7 (s, 3 H) 2.8 (m, 1 H) 2.6 (m, 2 H) 2.4–2.3 (m, 2 H) 2.2–2.0 (m, 2 H) 1.9–1.7 (m, 2 H). ^{13}C-NMR (101 MHz, $CDCl_3$) δ ppm 209.1 (1 C) 174.0 (1 C) 51.9 (1 C) 43.0 (1 C) 42.9 (1 C) 40.8 (1 C) 27.6 (1 C) 24.3 (1 C).

7.5.2 Analytical Method

7.5.2.1 Sampling Protocol

500 μL of reaction mixture was withdrawn from the stirred reaction mixture using a plastic pipette and transferred to a test tube. 100 μL of this mixture was quenched with EtOAc (1 mL). The mixture was then vortexed and spun down, and 600 μL of the organic layer was obtained and analysed by GC.

7.5.2.2 Achiral GC Method

An Agilent 7890B GC system with FID detection, equipped with an achiral HP5 column (30 m × 320 μm × 0.25 μm), was used to analyse samples. Method details: initial temperature 85 °C; 85 to 180 °C using 20 °C.min^{-1} ramp; 1.5 bar H_2, flow 12 mL.min^{-1}. Methyl 3-oxocyclohexanecarboxylate elutes at 1.8 min [2a].

7.5.2.3 Chiral GC Method

An Agilent 7890B GC system with FID detection, equipped with a chiral Hydrodex β-TBDAc column (25 m × 250 μm ID × 0.25 μm). Method details: initial temperature 70 °C; 70 to 155 °C using 10 °C.min^{-1} ramp; 155 to 220 °C using 25 °C.min^{-1} ramp; 1.9 bar H_2, flow 7.5 mL.min^{-1}. (*S*)-Methyl 3-oxocyclohexanecarboxylate elutes at 7.9 min (undesired (*R*)-methyl 3-oxocyclohexanecarboxylate at 7.8 min).

7.5.3 Conclusion

The bespoke procedure described in this section allows the stereoselective preparation of methyl (*S*)-3-oxocyclohexanecarboxylate using ERED-207 with very high yields and enantiomeric excess. This protocol could be further optimised to provide increased yields and product recovery by using different solvent systems. The opposite enantiomer can be obtained using commercially available ERED enzymes [1].

References

1. Hadi, T., Díaz-Rodríguez, A., Khan, D. *et al.* (2018) *Organic Process Research & Development*, **22**, 871–879.
2. (a) Turrini, N.G., Cioc, R.C., van der Niet, D.J.H. et al. (2017) *Green Chemistry*, **19**, 511–518; (b) Agudo, R. and Reetz, M.T. (2013) *ChemComm*, **49**, 10914–10916; (c) Bougioukou, D.J., Kille, S., Taglieber, A. and Reetz, M.T. (2009) *Advanced Synthesis & Catalysis*, **351**, 3287–3305.

8

Oxidative Methods

8.1 Macrocyclic Baeyer–Villiger Monooxygenase Oxidation of Cyclopentadecanone on 1 L Scale

Jan Brummund, Catharina Kleist and Martin Schürmann*
InnoSyn BV, Geleen, The Netherlands

The enzyme-catalysed Baeyer–Villiger oxidation of (cyclic) ketones into the corresponding esters and lactones is of interest because of the usually very high enantio- and regioselectivity of Baeyer–Villiger monooxygenases (BVMOs) [1]. Even for unsubstituted cyclic ketones, oxidation to the corresponding lactones by BVMOs is attractive because the use of harmful chemical oxidants as in chemical Baeyer–Villiger oxidation can be avoided and no undesired regioisomers as in the hydroxylation of fatty acids with cytochrome P450s are formed.

In this section, we describe the Baeyer–Villiger oxidation of cyclopentadecanone (CPD) to pentadecanolide (PDL) catalysed by cyclododecanone monooxygenase from *Rhodococcus ruber* SC1 (RrCDMO) [2] on 1 L scale. The reaction proceeds with pure oxygen as oxidant and glucose dehydrogenase as auxiliary enzyme for regeneration of NADPH using glucose as sacrificial substrate (Scheme 8.1).

Key to the reaction optimisation towards synthetically relevant substrate and product concentrations and productivities was the application of methanol as co-solvent of the almost water-insoluble CPD (5.3 $mg.L^{-1}$ or 0.024 $mmol.L^{-1}$ at 20 °C in 50 mM potassium phosphate buffer, pH 8.0). Up to 25% v/v of methanol could be applied in the target reaction to solubilise the substrate (24.9 $mg.L^{-1}$ or 0.111 $mmol.L^{-1}$), because the RrCDMO employed was highly stable in a 25% methanol/75% aqueous potassium phosphate buffer and retained 95% residual activity when stored for 7 days at 20 °C. Both enzymes, RrCDMO and GDH-01, were produced in fed-batch fermentations of recombinant *Escherichia coli* strains (data not shown, commercially available from InnoSyn) and applied as liquid enzyme formulations.

Applied Biocatalysis: The Chemist's Enzyme Toolbox, First Edition. Edited by John Whittall and Peter W. Sutton.
© 2021 John Wiley & Sons Ltd. Published 2021 by John Wiley & Sons Ltd.

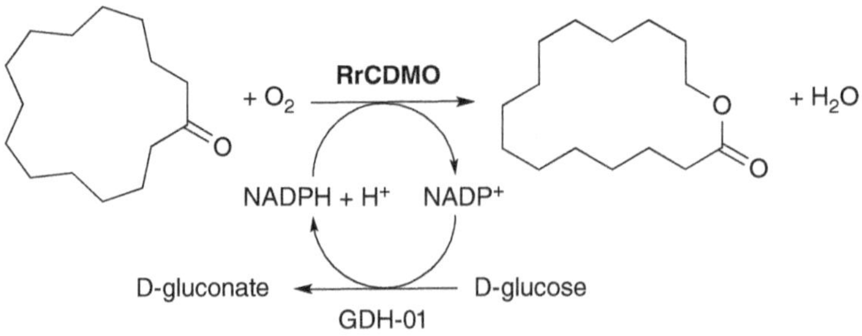

Scheme 8.1 *Baeyer–Villiger monooxygenase (BVMO)-catalysed oxidation of cyclopentadecanone (CPD) to pentadecanolide (PDL) using pure oxygen as oxidant and NADPH as reduction equivalent, producing H_2O as byproduct. NADPH was recycled using glucose dehydrogenase (GDH) and D-glucose as sacrificial substrate.*

For the process development from 1 L scale onwards, a controlled oxygen supply (eventually as compressed pure O_2 from a gas bottle) and the determination of oxygen dissolved in the reaction mix and in the off-gas were essential. The reactor (Figure 8.1) contained a frit to introduce the pressurised oxygen. The oxygen concentration was continuously monitored both in the liquid phase and in the gas outlet, and nitrogen (100 mL.min^{-1}) was blown into the headspace to keep the oxygen concentration below 8% v/v. The oxygen measurements enabled certain oxygen control from the operator, ensuring that oxygen was always available in the reactor. Autotitration with 5 M NaOH was applied to compensate for the gluconate formed by the NADPH regeneration system and to keep the pH stable at 7.5.

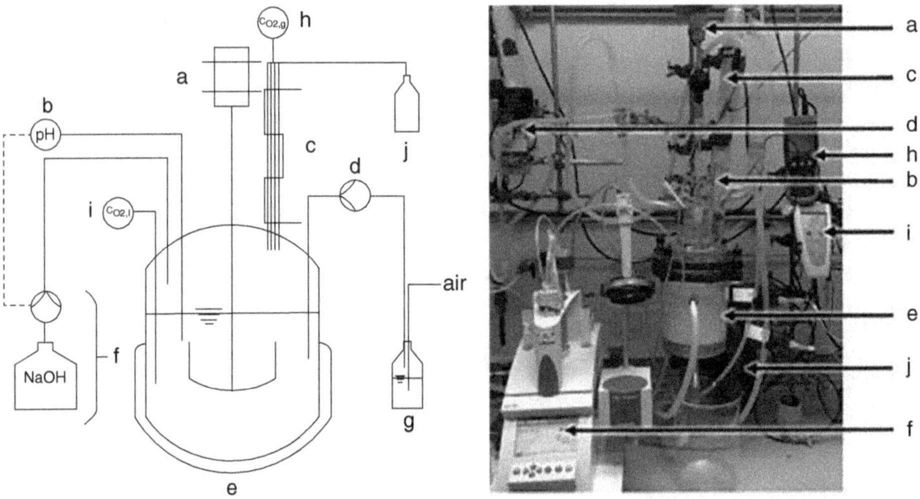

Figure 8.1 *1 L-scale reactor setup used for the Baeyer–Villiger oxidation of cyclopentadecanone (CPD) to pentadecanolide (PDL) by RrCDMO. Left, scheme; right, photo of the setup: (a) stirrer, (b) pH electrode, (c) reflux-condenser, (d) controlled gas supply, (e) reactor, (f) automatic titration device connected to pH electrode, (g) gas washing bottle filled with water, (h) oxygen sensor in gas outlet, (i) oxygen sensor in reaction mixture, (j) cooling trap. Not shown: compressed O_2 bottle and gas flow controller.*

8.1.1 Procedure 1: Biocatalytic Conversion of Cyclopentadecanone to Pentadecanolide

8.1.1.1 Materials and Equipment

- CPD (a generous gift from Givaudan Schweiz AG)
- RrCDMO; liquid enzyme formulation (commercially available from Innosyn)
- GDH-01; liquid enzyme formulation (commercially available from Innosyn)
- Potassium phosphate buffer (pH 7.5, 100 mM)
- Nicotinamide dinucleotide phosphate disodium salt ($NADP^+$)
- Methanol
- 5 M aqueous sodium hydroxide
- 1 L reactor setup
- Top stirrer
- pH electrode
- Reflux condenser
- Oxygen sensor in reaction mixture
- Oxygen sensor in gas outlet
- Mass flow meters for controlled oxygen, air and nitrogen supply
- Gas inlet (frit)
- Oxygen
- Nitrogen
- Automatic titration device connected to pH electrode
- Thermostat for heating the reaction
- Cryostat for cooling the condenser
- Dicalite 4208
- Glass filter (P3)
- Separation funnel (1 L)
- Agilent 7693 GC System with a PTV (Programmed Temperature Vaporisation) and FID detection
- Gas chromatography (GC) column: Rtx 5 Sil MS (30 m × 0.25 mm, 0.50 μm)

8.1.1.2 Procedure

1. The reactor jacket temperature was set to 30 °C.
2. Potassium phosphate buffer (pH 7.5, 100 mM; 100 mL) was charged to the reactor.
3. Water was added (666 mL).
4. $NADP^+$ disodium salt was added (400 mg).
5. The stirrer speed was set at 200 rpm.
6. RrCDMO was added (180 mL liquid enzyme formulation).
7. GDH-01 was added (2 mL liquid enzyme formulation).
8. The stirrer speed was set at 400 rpm.
9. The automatic titration device was set at pH 7.5 (titration via 5 M NaOH).
10. The oxygen flow rate was set at 25 $mL.min^{-1}$ (and decreased to 10 $mL.min^{-1}$ over the course of the reaction).
11. Cyclopenatadecanone (52 g, 231 mmol) dissolved in methanol was added in nine portions (initially 10 and ~5 g portions over time) at regular intervals over 5.5 hr.

12. Samples were taken and analysed by GC to follow the reaction progress.
13. After 8 hr, the CPD conversion was >95%.
14. The stirrer speed was set at 200 rpm.
15. *n*-Heptane (500 mL) was added.
16. The reactor was heated to 75 °C for 1 hr.
17. 25 g Dicalite 4208 and 25 g sodium sulfate were added as filter aid.
18. After 30 min stirring, the mixture was filtered over a glass filter (P3) containing a pre-coat of Dicalite 4208.
19. The filter cake was washed two times with *n*-heptane (250 mL) at 75 °C.
20. The filtrate was separated into the organic product layer and a water layer.
21. From the organic product layer, *n*-heptane was evaporated under vacuum at ~60 °C.
22. The overall PDL yield was of 47 g or 90.7% in the product oil of 98% chemical purity.
23. The determined product losses in the extraction step (1.2%), the filtration step (1.8%) and the solvent evaporation step (<0.1%) were low.

8.1.2 Analytical Method

Samples were taken from the reactor whilst stirring. The sample amount was weighed (~200 mg), diluted 10 times with acetonitrile containing naphthalene as internal standard (1.0–1.5 $mg.mL^{-1}$) and weighed again. The mixture was shaken and centrifuged. The clear supernatant was analysed. The retention times of CPD and PDL were 13.2 and 13.4 min, respectively (naphthalene 5.4 min).

8.1.2.1 GC Parameters

8.1.2.1.1 Temperature Program.

- Initial temperature: 150 °C
- Hold 1: 1 min
- Rate 1: 10 $°C.min^{-1}$
- Temperature: 290 °C
- Hold 2: 2 min
- Injector temperature: 280 °C
- Detector temperature: 280 °C
- Analysis time: 17 min

8.1.3 Conclusion

Dosing the CPD substrate dissolved in methanol in portions over time and keeping the maximum methanol concentration below 25% v/v enabled an almost complete substrate conversion of 230 mM CPD within 8 hr reaction time (Figure 8.2). The mass balance was ≥90% throughout the whole reaction despite a challenging sampling of the reaction mixture. At first, the base consumption (titrating the gluconic acid from the NADPH cofactor regeneration) was well in line with the substrate dosing and the product formation, demonstrating a high coupling efficiency of the BVMO reaction, whilst after about 6–7 hr when the BVMO reaction stopped, titration continued at a lower rate, indicating NADPH-consuming or other acid-producing background reactions (Figure 8.2).

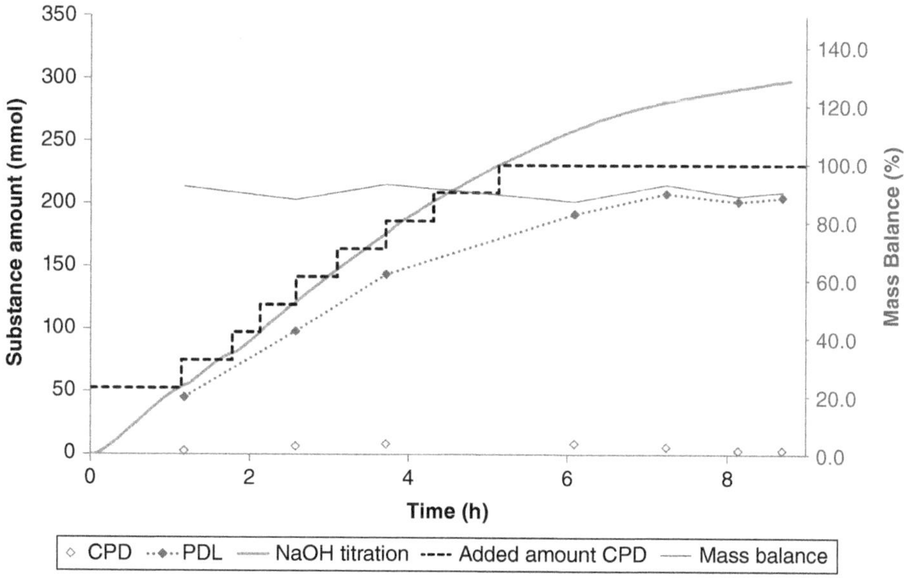

Figure 8.2 *Progress curve of cyclopentadecaone (CPD) oxidation to pentadecalactone (PDL) with RrCDMO and GDH-01 on 1 L scale. Reaction conditions: V = 1 L, T = 30 °C, pH 7.5, 10–25 mL.min^{-1} O_2, stirring rate = 400 rpm; addition of CPD portions in methanol, 1.5 eq. D-glucose, 0.5 mM $NADP^+$, finally 20% v/v methanol, 4 mM Tween-80, 10 mM potassium phosphate pH 7.5, 0.2% v/v GDH-01 liquid enzyme formulation, 18% v/v RrCDMO liquid enzyme formulation. Open diamonds, CPD substrate; filled diamonds, PDL product; bold line, NaOH titration; broken line, amount of CPD added; thin line, total mass balance of CPD and PDL based on GC analysis.*

With the implementation of the described process conditions, >95% CPD conversion was achieved within a short reaction time. The downstream procedure in combination with the applied reaction protocol resulted in an overall PDL yield of 90.7% in the product oil of >98% chemical purity, with unreacted CPD being the main impurity. The determined product losses in the extraction step (1.2%), the filtration step (1.8%) and the solvent evaporation step (<0.1%) were comparably low.

Finally, the developed process protocol was scaled up 100-fold for demonstration in our 200 L pilot plant reactor. In a 100 L reaction, more than 4 kg pentadecalactone was produced and isolated according to the scaled-up protocol.

Acknowledgements

This work has received funding from the European Union (EU) project ROBOX (grant agreement no. 635734) under the EU's Horizon 2020 Program Research and Innovation actions H2020-LEIT BIO-2014-1. The views and opinions expressed herein are those of the authors alone and do not necessarily reflect those of the European Union Research Agency. The European Union is not liable for any use that may be made of the information contained herein.

Cyclopentadecanone was generously supplied by Givaudan Schweiz AG.

References

1. Kamerbeek, N.M., Janssen, D.B., van Berkel, W.J.H. and Fraaije, M.W. (2005) *Advanced Synthesis & Catalysis*, **345** (67), 667–678.
2. Kostichka, K., Thomas, S.M., Gibson, K.J. *et al.* (2001) *Journal of Bacteriology*, **183** (21), 6478–6486.

8.2 Regioselective Lactol Oxidation with O_2 as Oxidant on 1 L Scale Using Alcohol Dehydrogenase and NAD(P)H Oxidase

Harrie Straatman, Jan Brummund and Martin Schürmann*
InnoSyn B.V., Geleen, The Netherlands

Chiral δ-lactones are naturally occurring flavour and fragrance (F&F) molecules mainly found in tropical trees and used in food and personal care applications. Currently, such δ-lactones – typically the (*R*)-enantiomers – are either extracted from plant material in enantiopure form or synthesised chemically as racemates. To save natural resources and still efficiently produce (*R*)-δ-lactones in enantiopure form, we developed a synthesis route based on aldolase technology as enabling step (Scheme 8.2). This route consists of an enantioselective aldolase reaction of an aliphatic aldehyde **1** and two molecules of acetaldehyde **2** catalysed by 2-deoxyribose-5-phosphate aldolase (DERA), followed by an oxidation of the lactol intermediate to a hydroxy-lactone, water elimination to an unsaturated lactone of the Massoia lactone type and optional double bond reduction to a δ-lactone [1].

As part of our effort to develop a synthesis route to such δ-lactones as natural flavours from natural aldehyde raw materials via (mainly) enzymatic transformations, we developed a process step consisting in the enzymatic oxidation of lactol **9** derived from the reaction of DERA with pentanal and acetaldehyde to the corresponding hydroxy-lactone **10** (Scheme 8.3) as an alternative to chemical oxidation. This was achieved by applying the

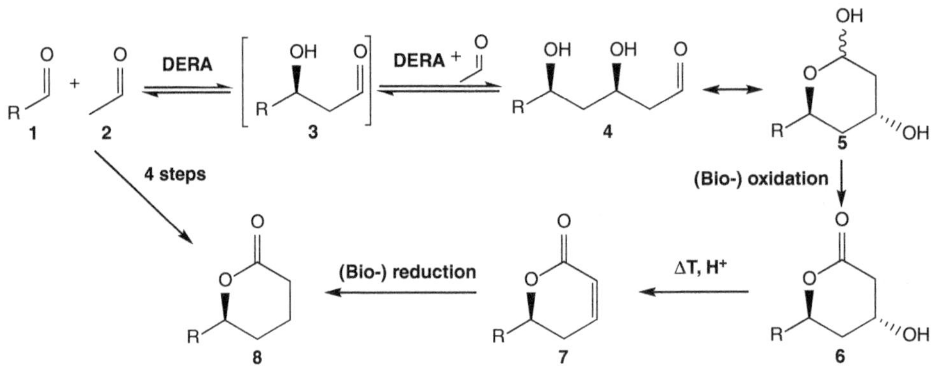

__Scheme 8.2__ General synthesis route to δ-lactones __8__ via: 2-deoxyribose-5-phosphate aldolase (DERA) reaction of an aliphatic aldehyde __1__ with two equivalents of acetaldehyde __2__ (via mono-aldol intermediate __3__ and sponateous ring-closure of hydroxy-aldehyde __4__); oxidation of the lactol intermediate __5__ to a hydroxy-lactone __6__; water elimination to α,β-unsaturated lactone __7__; and double bond reduction to δ-lactone __8__.

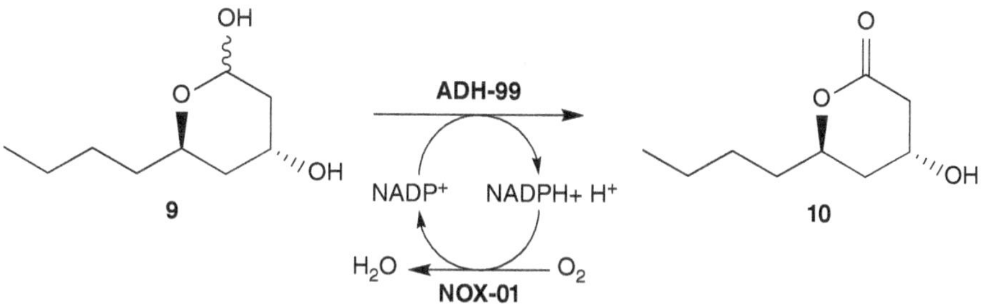

Scheme 8.3 Oxygen and NADP+ oxidation of lactol **9** derived from DERA reaction with pentanal and acetaldehyde to the corresponding hydroxy-lactone **10** using commercial ADH-99 (c-LEcta) and water, forming NOX-01 (InnoSyn) for NADPH regeneration to NADP+.

ADH-99 (c-LEcta) in the oxidation direction, consuming $NADP^+$ and producing NADPH and H^+, which was regenerated to $NADP^+$, with the water forming NAD(P)H-oxidase NOX-01 (InnoSyn) as auxiliary enzyme; this is a variant of the *Streptococcus mutans* NOX [2]. Both enzymes, ADH-99 and NOX-01, have been produced in fed-batch fermentations (data not shown), and they are commercially available from c-LEcta and InnoSyn, respectively.

For the process development from 1 L scale upwards, a controlled oxygen supply (eventually as compressed pure oxygen from a gas bottle) and determination of dissolved oxygen in the reaction mix and in the off-gas were essential. Furthermore, to enable an efficient oxidation of lactol **9** to hydroxy-lactone **10**, it was necessary to select an organic solvent that would dissolve the substrate without inactivating the NOX enzyme, because the solubility limit of lactol **9** is only around 35 $g.L^{-1}$ and the application of higher concentrations (required for a high productivity) resulted in a second (substrate) phase and inactivated the NOX. Co-solvents resulting in monophasic reaction systems also resulted in NOX inactivation in combination with the high substrate concentrations. Finally, 2-ethyl-1-hexanol was selected as second-phase solvent for process development and scale-up.

8.2.1 Procedure 1: Biocatalytic Conversion of Lactol 9 into Hydroxy-Lactone 10

8.2.1.1 *Materials and Equipment*

- Lactol **9** (78% w/w chemical purity as obtained from the DERA reaction)
- ADH-99 (c-LEcta); lyophilised powder
- NOX-01; liquid enzyme formulation
- Potassium phosphate buffer (pH 7.0, 100 mM)
- Nicotinamide dinucleotide phosphate disodium salt ($NADP^+$)
- 2-Ethyl-1-hexanol
- 1 L reactor setup (Figure 8.3)
- Top stirrer
- pH electrode
- Reflux condenser
- Oxygen sensor in reaction mixture
- Oxygen sensor in gas outlet

- Mass flowmeters for controlled oxygen, air and nitrogen supply
- Gas inlet
- Oxygen
- Automatic titration device connected to pH electrode
- Thermostat for heating the reaction
- Cryostat for cooling the condenser
- Dicalite 4208
- Glass filter (P3)
- Separation funnel (1 L)
- Agilent 7693 GC System with a PTV (Programmed Temperature Vaporisation) and FID detection
- Gas chromatography (GC) column: Rtx 5 Sil MS (30 m × 0.25 mm, 0.50 μm)

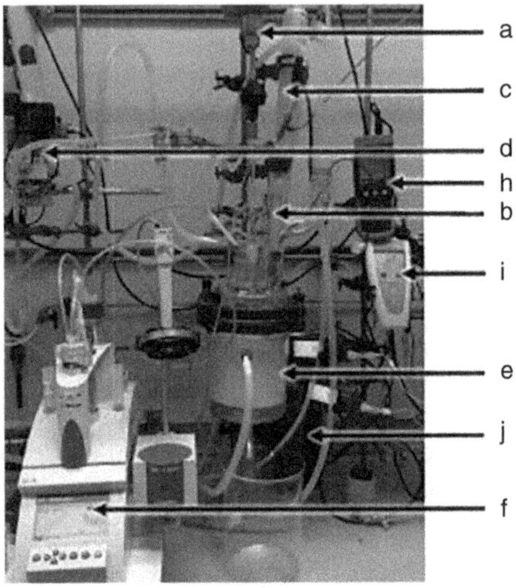

Figure 8.3 *1 L-scale reactor setup used for the oxidation of lactol* ***9*** *to hydroxy-lactone* ***10*** *by ADH-99 and NOX-01: (a) stirrer, (b) pH electrode, (c) reflux-condenser, (d) controlled air supply, (e) reactor, (f) automatic titration device connected to pH electrode, (g) gas washing bottle filled with water, (h) oxygen sensor in gas outlet, (i) oxygen sensor in reaction mixture, (j) cooling trap. Not shown: compressed* O_2 *bottle and gas flow controller.*

8.2.1.2 Procedure

1. The reactor jacket temperature was set to 28 °C.
2. Potassium phosphate buffer (pH 7.0, 100 mM; 125 mL) was charged to the reactor.
3. Water was added (100 mL).
4. $NADP^+$ disodium salt was added (250 mg).
5. The stirrer speed was set at 200 rpm.
6. ADH-99 was added (500 mg, 2950 units).
7. Lactol **9** (78% w/w, 30 g, 0.134 mol) was added.
8. 2-Ethyl-1-hexanol was added (220 mL).
9. The stirrer speed was set at 350 rpm, and the mixture was stirred for 15 min.
10. NOX-01 was added (25 mL liquid enzyme formulation, 18 280 units).

11. The oxygen flow rate was set at 20 mL.min^{-1}.
12. Samples were taken and analysed by GC to follow the reaction progress.
13. After 4 hr, the lactol conversion was >99%.
14. The stirrer speed was set at 200 rpm.
15. Dicalite 4208 was added as filter aid (25 g).
16. After 30 min stirring, the mixture was filtered over a glass filter (P3) containing a pre-coat of Dicalite 4208.
17. The filtrate was separated into the organic product layer and a water layer.
18. The filter cake was washed three times with 125 mL 2-ethyl-1-hexanol.
19. The water layer was extracted three times with the 2-ethyl-1-hexanol washes from the filter.
20. All organic layers were combined, resulting in a final product solution of hydroxy-lactone **10** in 2-ethyl-1-hexanol. Yield: 88.7%.

8.2.2 Analytical Method

Sampling was performed whilst stirring. The sample amount was weighed (~200 mg), diluted 10 times with acetonitrile containing naphthalene (1.0–1.5 mg.mL^{-1}, internal standard) and weighed again. The mixture was shaken and centrifuged. The clear supernatant was analysed by GC. The retention times were 8.14 min for lactol **9** and 8.86 min for lactone **10**.

8.2.2.1 GC Parameters

8.2.2.1.1 Injector Program.

- Initial temperature: 50 °C
- Hold: 0 min
- Rate 1: 100 °C.min^{-1}
- Temperature: 280 °C

8.2.2.1.2 Temperature Program.

- Initial temperature: 50 °C
- Hold: 2 min
- Rate 1: 30 °C.min^{-1}
- Temperature: 260 °C
- Hold: 1 min
- Injector temperature: 280 °C
- Detector temperature: 280 °C
- Analysis time: 10 min

8.2.3 Conclusion

The reactor (Figure 8.3) had a frit to introduce the pressurised oxygen. The pH was measured but not corrected. The oxygen concentration was continuously monitored both in the liquid phase and in the gas outlet, and nitrogen (100 mL.min^{-1}) was blown into the headspace to keep the oxygen concentration below 20% v/v. The oxygen measurements

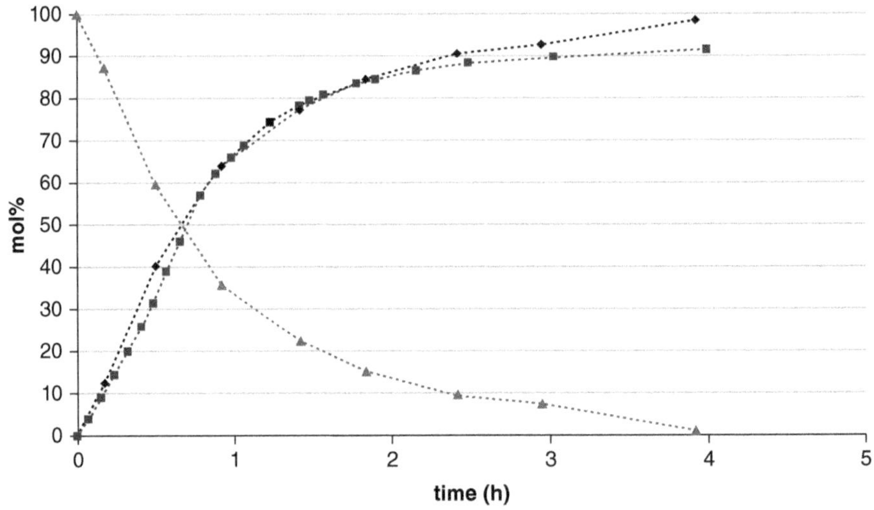

Figure 8.4 *Progress curve of lactol **9** oxidation with ADH-99 and NOX-01 on 500 mL scale. Reaction conditions: 500 mg ADH-99 (1.0 mg.mL^{-1}), 5% v/v NOX-01 liquid enzyme formulation, 4.9% w/w (280 mM) crude lactol **9** (78% chemical purity). T = 28 °C, 100 mM KPi buffer pH 7.0, 20 mL oxygen.min^{-1}, 250 mL 2-ethyl-1-hexanol including substrate plus 250 mL buffer including NOX. Triangles, lactol **9** conversion; diamonds, hydroxy-lactone **10** yield based on GC analysis; squares, hydroxy-lactone **10** yield based on oxygen consumption.*

enabled certain oxygen control from the operator, ensuring that oxygen was always available in the reactor. Furthermore, the oxygen mass balance (oxygen consumption) became a good estimation of the reaction progress (Figure 8.4). The reaction in this 1 L reactor was performed on 500 mL scale – the same filling level as anticipated for the pilot plant demonstration batch (100 L reaction volume in 200 L reactor).

The substrate (78% w/w chemical purity, as obtained from the DERA reaction) concentration was close to 5% w/w and the ADH-99 amount used at 1 mg.mL^{-1}, NOX-01 at 5% v/v liquid enzyme formulation. The reaction proceeded smoothly; after 4 hr, nearly 100% lactol **9** conversion was reached, with >95% hydroxy-lactone **10** yield (Figure 8.4). The yield based on oxygen consumption was very comparable to the lactone **10** yield based on GC analysis.

After completion of the reaction, the water layer and the organic layer (containing the product) were separated by filtration and the residual hydroxy-lactone **10** in the water layer was extracted into the organic phase. The final product remained in the 2-ethyl-1-hexanol phase, from which hydroxy-lactone **10** could be distilled or further processed (not investigated here). The hydroxy-lactone **10** was isolated in a good yield of 88.7%, as a solution in 2-ethyl-1-hexanol, with acceptable mass balance of 95%.

This protocol was linearly scaled up to 75 L reaction volume in our 200 L pilot plant reactor. After 7.5 hr, the lactol **9** conversion was >95% and the hydroxy-lactone **10** yield 95%.

Acknowledgements

This work has received funding from the European Union (EU) project ROBOX (grant agreement no. 635734) under EU's Horizon 2020 Program Research and Innovation actions

H2020-LEIT BIO-2014-1. The views and opinions expressed herein are those of the authors alone and do not necessarily reflect those of the European Union Research Agency. The European Union is not liable for any use that may be made of the information contained herein.

References

1. Mink, D., Wolberg, M., Schürmann, M. and Hilker, I. (2007) Patent WO2007068498A1.
2. Petschacher, B., Staunig, N., Müller, M. *et al.* (2014) *Computational and Structural Biotechnology Journal*, **9**, e201402005.

8.3 Synthesis of (3*R*)-4-[2-Chloro-6-[[(*R*)-Methylsulfinyl]methyl]-Pyrimidin-4-yl]-3-Methyl-Morpholine Using BVMO-P1-D08

Keith R. Mulholland,* Bradley Adams, Helen Benson, Julie Demeritt, William R.F. Goundry Steven McKown, Amy Robertson, Paul Siedlecki, Paula Tomlin and Kevin Vare

Departments of Pharmaceutical Sciences and Pharmaceutical Technology and Development, AstraZeneca, Macclesfield, Cheshire, UK

(3*R*)-4-[2-Chloro-6-[[(*R*)-methylsulfinyl]methyl]-pyrimidin-4-yl]-3-methyl-morpholine **2** is a chiral intermediate required for the synthesis of AZD6738 (a compound undergoing development for the treatment of colon and haematological cancers). AZD6738 contains a chiral sulfoxamine, which is derived from the chiral sulfoxide **2**. The synthesis of **2** has been performed on a pilot plant scale using the Codexis Baeyer–Villiger monooxygenase (BVMO) enzyme BVMO-P1-D08 following a screening programme of Codexis BVMO enzymes (Scheme 8.4) [1].

8.3.1 Procedure 1: Plant-Scale Preparation of Chiral Sulfoxide 2 Using BVMO P1-D08

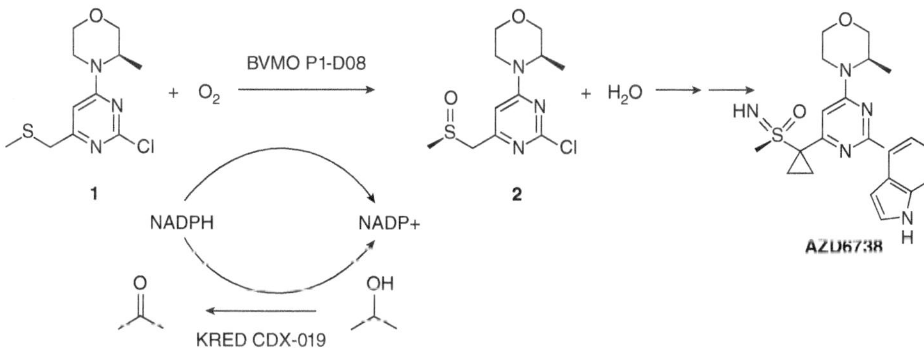

***Scheme 8.4** Synthesis of the chiral sulfoxide **2** using Codexis BVMO P1-D08.*

8.3.1.1 Materials and Equipment

- 5000 L stainless steel reactor (Vessel 1)
- 5000 L glass-lined reactor (Vessel 2)

- 1000 L stainless steel reactor (Vessel 3)
- (3*R*)-4-[2-Chloro-6-[[(*R*)-methylsulfanyl]methyl]-pyrimidin-4-yl]-3-methyl-morpholine (72.2 kg, 264 mol)
- 2-Propanol (174 L, 2.4 rel. vol.)
- Water (1395 L, 19.3 rel. vol.)
- Triethanolamine (22.6 kg, 151.5 mol)
- 1M HCl (15.0 L)
- 2-Propanol line wash (19.7 L, 0.27 rel. vol.)
- NADP (0.7 kg, 0.94 mol)
- Codexis KRED CDX-019 (1.4 kg)
- Codexis BVMO-P1-D08 (2.8 kg)
- Water (340 L) – Reactor 2
- 2-Propanol (38 L) – Reactor 2
- Water (100 L) – Reactor 2, second charge
- 2-Butanone (309 L, 4.3 rel. vol.)
- 2-Butanone (1020 L, 14.1 rel. vol.)
- Dichloromethane (1316 L, 18.2 rel. vol.)
- Dichloromethane (309 L, 4.3 rel. vol.)
- 2-Butanone (204 L, 2.8 rel. vol.)
- 2-Butanone (635 L, 8.8 rel. vol.)
- 2-Butanone (300 L, 4.2 rel. vol.)
- 2-Butanone (103 L, 1.4 rel. vol.)
- High-performance liquid chromatography (HPLC) system with photodiode array (PDA) detection and autosampler
- HPLC column (Symmetry Shield RP18, 150 × 3.9 mm, 5 µm)
- Chiral HPLC Column (Chiralpak AS-H, 250 × 4.6 mm, 5 µm)
- Bruker Advance A400 (^{1}H: 400 MHz, ^{13}C: 101 MHz) spectrometer using tetramethylsilane (TMS) as internal standard (d = 0)

8.3.1.2 *Procedure*

1. Vessel 1 was charged with water (1395 L), triethanolamine (22.6 kg, 151.5 mol) and 1M HCl (15.0 L).
2. The pH of the reaction mixture was checked (and adjusted, if necessary, to 8.8–9.2).
3. Vessel 1 was charged with a solution of (3*R*)-4-[2-chloro-6-[[(*R*)-methylsulfanyl]-methyl]-pyrimidin-4-yl]-3-methyl-morpholine (72.2 kg, 264 mol) in 2-propanol (174 L).
4. The charging lines were rinsed with 2-propanol (19.7 L).
5. The reaction mixture was stirred and warmed to 24–27 °C.
6. NADP (0.7 kg), KRED CDX-019 (1.4 kg) and BVMO P1-D08 (2.8 kg) were added to Vessel 1.
7. 2-Propanol (38 L) and water (392 L) were charged to the Vessel 1 receiver.
8. Compressed air was bubbled through the Vessel 1 receiver into Vessel 1 at a rate of 2–4 $m^3.h^{-1}$
9. The reaction mixture was stirred at 24–27 °C for 60 hr.
10. Water (100 L) was added to the Vessel 1 receiver across the 60 hr period.

11. Following reaction completion at 60 hr, 2-butanone (309 L) was charged to Vessel 1.
12. The contents of Vessel 1 were distilled under reduced pressure, keeping the temperature below 45 °C, until the volume remaining was ~1200 L.
13. 2-Butanone (1020 L) was added to Vessel 1.
14. The temperature in Vessel 1 was adjusted to 48–50 °C and the contents were stirred for 30 min.
15. The reaction mixture was allowed to settle.
16. The aqueous layer was separated and retained.
17. The organic layer was transferred to Vessel 2 and distilled under reduced pressure at a temperature of <45 °C until the residual volume was ~100 L.
18. The distilled organic layer was discharged from Vessel 2 and retained
19. The 2-butanone distillates from Vessel 2 were transferred back to Vessel 2 along with the aqueous layer retained in Step 16.
20. The aqueous layer in Vessel 2 was extracted with the 2-butanone distillates.
21. The aqueous layer was discharged from Vessel 2 and retained.
22. The 2-butanone extract was distilled under reduced pressure at <45 °C until the residual volume was ~100L.
23. Steps 18–21 were repeated three times to ensure complete extraction of **2** from the aqueous layer.
24. Dichloromethane (1316 L) was charged to the concentrated 2-butanone extracts (~100 L volume) in Vessel 2.
25. The mixture was stirred for 30 min.
26. The organic layer was separated and retained.
27. The aqueous layer was extracted with dichloromethane (309 L).
28. The aqueous layer was discarded.
29. The combined organic layers were retained in Vessel 2
30. 2-Butanone (204 L) was charged to Vessel 3.
31. The contents of Vessel 3 were distilled under reduced pressure at <45 °C until the residual volume was ~124 L. Additionally, the combined organic layers from Vessel 2 were slowly added.
32. A further charge of 2-butanone (635 L) was added to Vessel 3.
33. The contents of Vessel 3 were distilled under reduced pressure at <45 °C until the residual volume was ~250 L.
34. 2-Butanone (300 L) was added to Vessel 3 and the reaction mixture was heated to 60 °C.
35. The reaction mixture in Vessel 3 was held at 60 °C for 1 hr.
36. The reaction mixture in Vessel 3 was cooled to −10 °C over 10.5 hr.
37. The reaction mixture in Vessel 3 was held at −10 °C for 5 hr.
38. The crystalline product in Vessel 3 was isolated using a centrifuge.
39. The isolated solid in the centrifuge was washed with pre-cooled (−10 °C) 2-butanone (103 L).
40. The isolated solid in the centrifuge was dried for 36 hr at 45 °C under reduced pressure.
41. The isolated solid was discharged to give (3*R*)-4-[2-chloro-6-[[(*R*)-methylsulfinyl]-methyl]-pyrimidin-4-yl]-3-methyl-morpholine **2** as an off-white solid (53.9 kg, 68.3% yield).

^1H NMR (400 MHz, $CDCl_3$) δ 1.20 (d, $J = 6.8$ Hz, 3H), 2.52 (m, 1H), 2.63 (s, 3H), 3.21 (m, 1H), 3.44 (m, 1H), 3.58 (dd, $J = 11.6$ Hz, 3.1 Hz, 1H), 3.72 (d, $J = 11.5$ Hz), 1H), 3.92 (m, 3H), 4.07 (d, $J = 12.4$ Hz, 1H), 6.80 (s, 1H); assay (HPLC) 99%; assay (qualitative nuclear magnetic resonance, QNMR) 100%; chiral purity (HPLC) *(R,R)*-diastereoisomer 99.6%, *(R,S)*-diastereoisomer 0.4%.

8.3.2 Analytical Method

8.3.2.1 Sampling Procedure

Sampling was performed on the stirred reaction mixture (20 µL was diluted with 10 mL 8 : 2 MeCN : H_2O) and analysed by HPLC.

8.3.2.2 HPLC Method

- Non-chiral for in-process control, related substances (area%) and assay (% w/w).
- Instrument: HPLC equipped with PDA detector and autosampler
- Column: Symmetry Shield RP18 (150 × 3.9 mm, 5 µm)
- Wavelength: 254 nm
- Oven temperature: 20 °C
- Flow rate: 1.0 mL.min^{-1}
- Injection volume: 5 µL
- Mobile phase A: H_2O
- Mobile phase B: MeCN

8.3.2.3 Gradient Programme

See Table 8.1.

- Run time: 24 min
- Diluent: MeCN : H_2O 8 : 2

8.3.2.4 Chiral HPLC Method (for Analysis of Isolated 2)

- Instrument: HPLC equipped with PDA detector and autosampler
- Column: Chiralpak AS-H (250 × 4.6 mm, 5 µm)
- Wavelength: 254 nm

Table 8.1 *Gradient programme.*

Time (min)	A%	B%
0	95	5
10	25	75
15	5	95
18	5	95
18.01	95	5
24	95	5

- Column temperature: 35 °C
- Column flow: 0.8 mL.min^{-1}
- Injection volume: 5 µL
- Acquisition time: 30 min
- Run time: 30 min
- Diluent: Ethanol
- Mobile phase: n-hexane : ethanol 11 : 9 (v/v)

8.3.3 Conclusion

The described procedure enabled the stereoselective synthesis of (3*R*)-4-[2-chloro-6-[[(*R*)-methylsulfinyl]methyl]-pyrimidin-4-yl]-3-methyl-morpholine **2** using the Codexis BVMO enzyme P1-D08 along with the Codexis ketoreductase enzyme CDX-019. The successful transfer of the reaction to pilot plant scale required efficient air sparging and reaction agitation to ensure efficient mass transfer of oxygen into the reaction mixture [1].

Reference

1. Goundry, W.R.F., Adams, B.A., Benson, H. *et al.* (2017) *Organic Process Research & Development*, **21**, 107–113

8.4 Oxidation of Vanillyl Alcohol to Vanillin with Molecular Oxygen Catalysed by Eugenol Oxidase on 1 L Scale

Harrie Straatman, Jan Brummund and Martin Schürmann*
InnoSyn B.V., Geleen, The Netherlands

In order to demonstrate the scalability of alcohol oxidase reactions, which use molecular oxygen to oxidise alcohols to carbonyls and produce hydrogen peroxide as a byproduct, we optimised the reaction conditions and scaled up the oxidation of vanillyl alcohol to vanillin catalysed by eugenol oxidase (EUGO). The EUGO from *Rhodococcus jostii* RHA1 is described as catalysing this reaction with a k_{cat} value of 12 sec^{-1} and a K_m value of 40 µM, whilst the physiological substrate eugenol is oxidised to coniferyl alcohol (Scheme 8.5) four times more slowly but with 40-fold higher affinity [1].

The challenges of vanillyl alcohol solubility and the formation of the byproduct hydrogen peroxide were initially addressed by the addition of acetone as co-solvent, which is generally compatible with the requirements of flavour and fragrance production and the addition of catalase to break down H_2O_2 into H_2O and O_2 [2]. During hazard and safety studies, however, which were part of the scale-up of the reaction to 100 L pilot plant scale, we considered that the combination of the low affinity catalases typically have for H_2O_2 and the resulting presence of H_2O_2 together with acetone could lead to significant explosion risks. Hydrogen peroxide and acetone are known to form tri-acetone-triperoxide (TATP), which is an explosive compound. Risk assessment further identified as potential problems that catalase could be quickly inactivated or its addition could be forgotten.

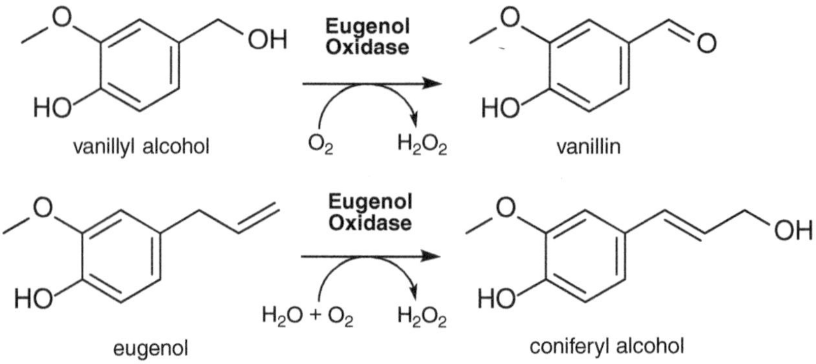

Scheme 8.5 Reactions catalysed by eugenol oxidase (EUGO): target reaction of vanillyl alcohol with O_2 into vanillin and H_2O_2; and physiological oxidation of eugenol with oxygen and water into coniferyl alcohol and hydrogen peroxide.

Initial reactions with vanillyl alcohol, EUGO, catalase and aeration with air (Scheme 8.6), omitting acetone as co-solvent, revealed incomplete mass balances and a dark colour formation, likely due to a base-catalysed Dakin oxidation of the vanillin product by hydrogen peroxide. Reactions at increasing pH resulted in higher EUGO activity that was indeed accompanied by increased colour formation, indicating that catalase was not sufficient to quench all the hydrogen peroxide produced in the EUGO reaction. Therefore, we investigated chemical H_2O_2 quenching by sodium sulfite (generating sodium sulfate as byproduct; Scheme 8.6) and optimised the reaction on 30–500 mL scale in reactors equipped for efficient aeration (Figure 8.5). Applying sodium sulfite and pure oxygen from a pressurised gas bottle, we developed a scalable recipe, which was scaled up to pilot plant scale to produce more than 4 kg of vanillin from vanillyl alcohol.

EUGO was produced in our labs by high cell-density 10 L-scale fed-batch fermentation of a recombinant *Escherichia coli* strain harbouring a synthetic codon-optimised gene for EUGO from *R. jostii* RHA1 (data not shown). As biocatalyst-formulation fermentation broth, in which the *E. coli* cells containing EUGO were disrupted by sonication or homogenisation, performed comparably well as cell-free extract, whilst nontreated broth performed slightly less well (data not shown). Therefore, such a homogenised fermentation broth was used as our biocatalyst formulation.

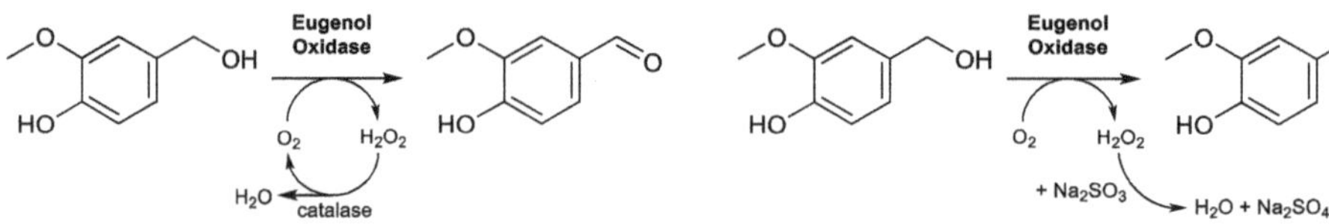

Scheme 8.6 Oxidation of vanillyl alcohol to vanillin by molecular oxygen catalysed by eugenol oxidase (EUGO) and quenching of the hydrogen peroxide byproduct with catalase (regenerating O_2, left) or sodium sulfite (producing sodium sulfate, right).

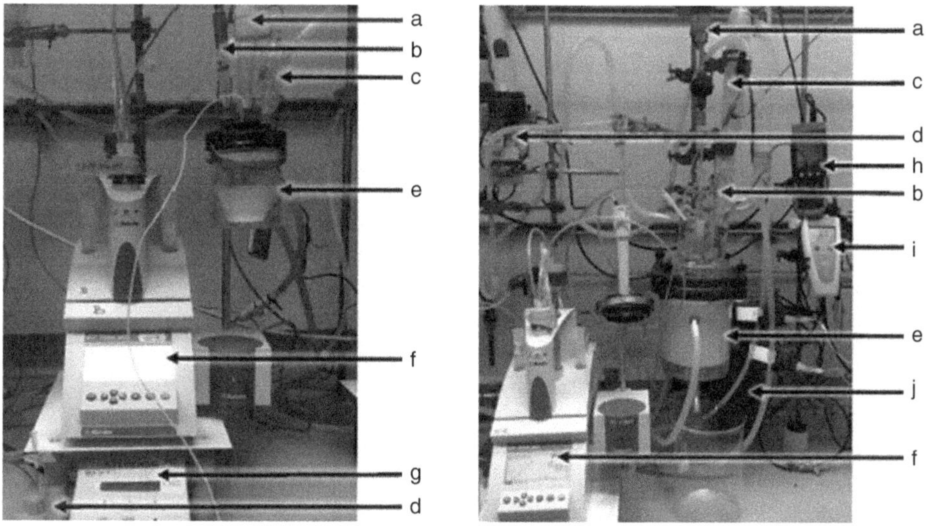

Figure 8.5 Left: 30 mL- and Right: 1 L-scale reactor set-up used for the oxidation of vanillyl alcohol to vanillin by eugenol oxidase (EUGO): (a) stirrer, (b) pH electrode, (c) reflux-condenser, (d) controlled air supply, (e) reactor, (f) automatic titration device connected to pH electrode, (g) gas washing bottle filled with water, (h) oxygen sensor in gas outlet, (i) oxygen sensor in reaction mixture, (j) cooling trap. Not shown: compressed O_2 bottle and gas flow controller.

8.4.1 Procedure 1: Biocatalytic Conversion of Vanillyl Alcohol to Vanillin

8.4.1.1 Materials and Equipment

- Vanillyl alcohol
- EUGO (homogenised *E. coli* fermentation broth)
- Sodium sulfite
- Oxygen
- Sodium hydroxide, 5 M
- Sulfuric acid, 20% in water
- Isopropyl acetate
- 1 L reactor setup
- Top stirrer
- pH electrode
- Reflux condenser
- Oxygen sensor in reaction mixture
- Oxygen sensor in gas outlet
- Mass flowmeters for controlled oxygen, air and nitrogen supply
- Gas inlet
- Automatic titration device connected to pH electrode
- Thermostat for heating the reaction
- Cryostat for cooling the condenser
- Dicalite 4208
- Glass filter (P3)

- Separation funnel (1 L)
- High-performance liquid chromatography (HPLC) system

8.4.1.2 Procedure

1. A 1 L reactor was used for 500 mL scale.
2. The thermostat temperature was set at 25 °C.
3. The nitrogen flow was set at 100 mL.min^{-1}.
4. Water (425 mL) and sodium sulfite (25 g, 400 mM) were added and stirred until dissolved.
5. Vanillyl alcohol was added (25 g, 330 mM).
6. The pH was adjusted to 9.5 with 5 M aqueous NaOH.
7. EUGO sonicated broth was added (25 g).
8. The oxygen flow was set at 10 mL.min^{-1}.
9. The pH was kept constant at 9.5 with titration of 5 M aqueous NaOH.
10. After completion of the reaction, the pH was adjusted to 2.9 with 20% Sulfuric acid.
11. Isopropyl acetate was added (250 mL).
12. The mixture was heated to 35–40 °C and stirred for 1 hr.
13. Dicalite 4208 was added (20 g).
14. After 15 min stirring, the mixture was filtered over a precoated glass filter (P3).
15. The filter cake was washed three times with 150 mL isopropyl acetate.
16. The filtrate was separated into a clear water layer and a reddish isopropyl acetate layer.
17. The water layer was extracted three times more with the filter cake washes.
18. The isopropyl acetate layers were combined, resulting is a 4.5% solution of vanillin.

8.4.2 Analytical Method

Samples were withdrawn from the reactor whilst stirring. The sample amount was weighed (~250 mg) into a 25 mL volumetric flask, two drops of 1 M phosphoric acid were added and the flask was filled up to the mark with acetonitrile. The mixture was shaken and centrifuged. The clear supernatant was analysed by HPLC (Figure 8.6).

8.4.2.1 HPLC Parameters

- Column: halo C18 150 × 4.6 mm, 2.7 μm particles
- Eluent: isocratic
  - Water + 0.1% formic acid = 75%
  - Acetonitrile + 0.1% formic = 25%
- Flow: 0.8 mL.min^{-1}
- Temperature: 35 °C
- UV detector: 231 nm
- Retention time compounds:
  - Vanillyl alcohol: 2.2 min
  - Vanillic acid: 2.6 min
  - Vanillin: 3.6 min

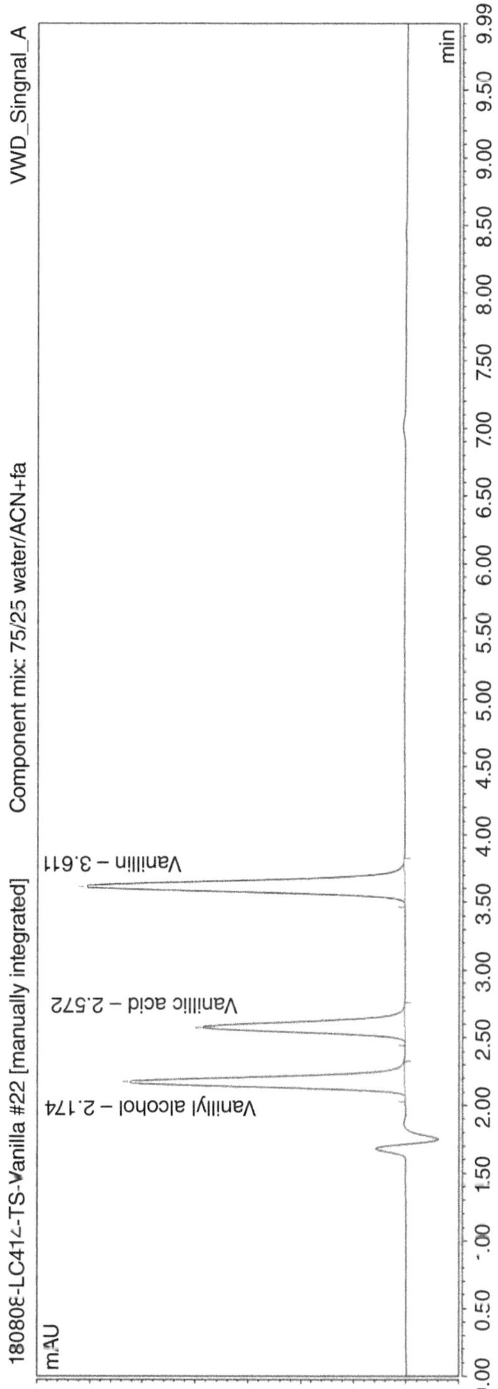

Figure 8.6 *Example chromatogram of a reference solution.*

8.4.3 Conclusion

The reaction progress could be followed by oxygen consumption or by the base titration of the acidic phenolic hydroxyl group of the vanillin produced to keep the reaction pH at 9.5. A somewhat more than 10% excess of sodium sulfite (370 mM) was used to quench the hydrogen peroxide equimolarly formed from the EUGO oxidation of the substrate vanillyl alcohol (330 mM). After ~3 hr reaction time and a substrate conversion of 80%, the sodium sulfite seemed to be consumed, as indicated by the increase of base titration (likely due to the formation of Dakin oxidation side products) and a drop in the mass balance from more than 95% to 90% (Figure 8.7). The addition of ~20% more sodium sulfite let the reaction continue to 98% substrate conversion and a product assay yield of 91% after 5 hr reaction time. At this stage, the aeration and the reaction were stopped. No vanillic acid formation could be detected at any stage of the reaction, underlining the excellent chemoselectivity of the EUGO enzyme.

After completion of the reaction, the product vanillin was present as a sodium salt. In principle, vanillin can be precipitated from water by acidification. However, the protein and cell debris in the reaction prevented its direct precipitation. Therefore, vanillin was extracted from the reaction mixture into an organic solvent after neutralisation and filtration of the proteins and the cell debris and was isolated in a good yield of 90.5% as a solution in isopropyl acetate. The HPLC analyses showed that the isopropyl acetate layer was very clean, containing 4.5% w/w vanillin and only a negligible 0.01% w/w vanillyl alcohol. In the water layer, some vanillin was detected (0.1% w/w), along with a small amount of an impurity that co-eluted with vanillyl alcohol. Vanillin can be crystallised from isopropyl

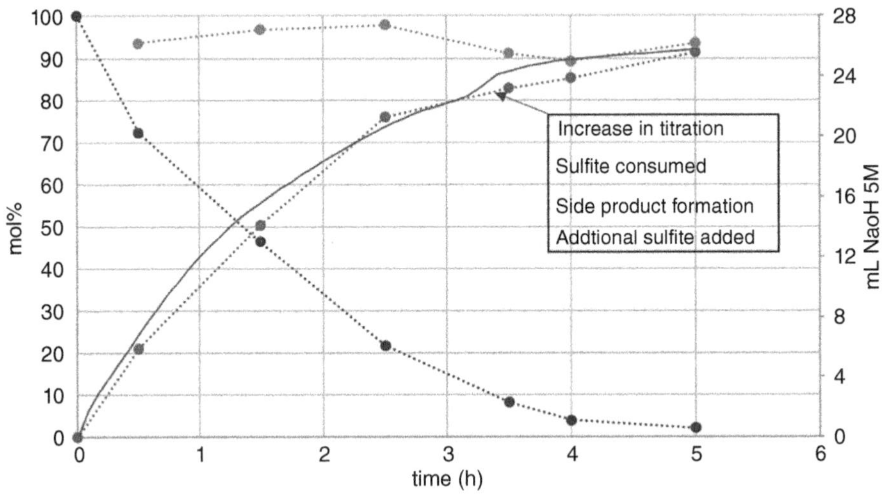

Figure 8.7 *Progress curve of vanillyl alcohol oxidation by eugenol oxidase (EUGO). Reaction conditions: 5% w/w vanillyl alcohol (330 mM), 4.5% w/w sodium sulfite (370 mM), 5% w/w EUGO (homogenised E. coli fermentation broth), T = 25 °C, 25 mL oxygen min^{-1}, pH set to 9.5 at the beginning of the reaction with 5 M NaOH and titrated over the reaction course with 5 M NaOH. Dark grey dots, vanillyl alcohol; grey dots, vanillin formed; light grey dots, mass balance; solid line, NaOH titration.*

acetate solution after concentration or addition of an antisolvent, but this was not done in this work.

This procedure was scaled up in our 200 L pilot plant reactor at 100 L reaction volume to produce, from 5 kg vanillyl alcohol, more than 4 kg vanillin (extracted into isopropyl acetate).

Acknowledgements

This work has received funding from the European Union (EU) project ROBOX (grant agreement no. 635734) under EU's Horizon 2020 Program Research and Innovation actions H2020-LEIT BIO-2014-1. The views and opinions expressed herein are those of the authors alone and do not necessarily reflect those of the European Union Research Agency. The European Union is not liable for any use that may be made of the information contained herein.

References

1. Jin, J., Mazon, H., van den Heuvel, R.H. *et al.* (2007) *FEBS Journal*, **274**, 2311–2321.
2. García-Bofill, M., Sutton, P.W., Guillén, M. and Álvaro, G. (2019) *Applied Catalysis A: General*, **582**, 117117.

8.5 Synthesis of Syringaresinol from 2,6-Dimethoxy-4-Allylphenol Using an Oxidase/Peroxidase Enzyme System

Mohamed H. Habib,[1,2] Milos Trajkovic[1] and Marco W. Fraaije[1]
[1]Molecular Enzymology, Groningen Biomolecular Sciences and Biotechnology Institute, University of Groningen, Groningen, The Netherlands
[2]Department of Microbiology and Immunology, Faculty of Pharmacy Cairo University, Cairo, Egypt

The importance of one-pot biocatalytic cascades has increased considerably in recent years. Biocatalytic cascades are attractive because they are efficient in time and costs when compared with conventional single-step one-enzyme reactions. An example of such an enzymatic cascade, which we recently developed, is the synthesis of a lignan, syringaresinol **3**, starting from a simple phenolic compound, 2,6-dimethoxy-4-allylphenol **1**.

Syringaresinol is a lignan formed when two sinapyl alcohol **2** units are linked via a β-β linkage. The compound is of significant importance in both the medical and the industrial fields. In the medical field, syringaresinol is used to inhibit the motility of *Helicobacter pylori* in the stomach, and thus it plays a role in protecting against gastric cancer [1]. In its glucoside form, it is used in the healing of wounds [2,3]. It increases the production of nitric oxide from nitric oxide synthase and thus plays a role in the vasorelaxation of blood vessels [4]. Syringaresinol is characterised by being very rigid, owing to the presence of a *cis*-fused bis-furan moiety in its structure. It may develop as an alternative to replace bisphenol A in industrial resins [5,6].

For the synthesis of syringaresinol **3** from 2,6-dimethoxy-4-allylphenol **1**, two enzymes are added to the reaction at the same time. In the first step, eugenol oxidase

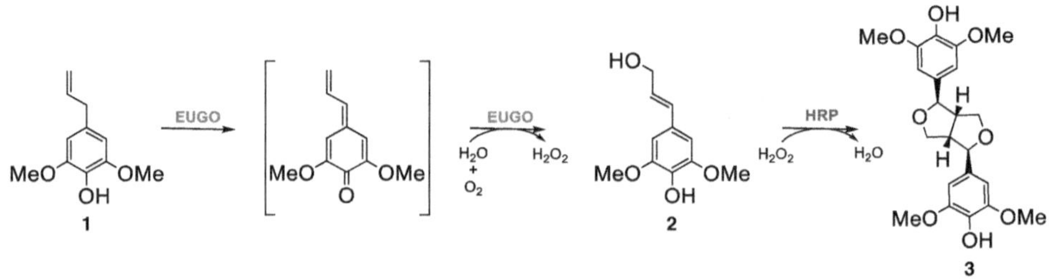

Scheme 8.7 Conversion of 2,6-dimethoxy-4-allylphenol **1** to syringaresinol **3** using EUGO and HRP. Sinapyl alcohol **2** is formed as an intermediate product.

(EUGO) catalyses the conversion of 2,6-dimethoxy-4-allylphenol **1** into sinapyl alcohol **2**, generating hydrogen peroxide as a byproduct. Horseradish peroxidase (HRP) then converts the generated sinapyl alcohol **2** into syringaresinol **3** using the hydrogen peroxide formed. Hence, the system is fully coupled. For efficient hydroxylation of 2,6-dimethoxy-4-allylphenol **1**, a mutant of EUGO has been engineered (EUGO I427A). For the coupling of sinapyl alcohol **2**, the natural plant enzyme HRP can be used.

8.5.1 Procedure 1: EUGO I427A Expression, Purification and Storage

8.5.1.1 Materials and Equipment

- 2 bacto tryptone (1 g)
- 2 yeast extract (0.5 g)
- 2 sodium chloride (1 g)
- Agar (1.5 g)
- 4 bacto tryptone (6 g)
- 4 yeast extract (12 g)
- 4 glycerol (2 mL)
- 4 KH_2PO_4 (1.15 g)
- 4 K_2HPO_4 (8.2 g)
- 1 L KH_2PO_4 (1 M)
- 1 L K_2HPO_4 (1 M)
- Glycerol (250 mL)
- NaCl (96.4 g)
- pBAD-EUGO-I427A-His plasmid (GECCO-Biotech)
- Ampicillin (50 mg.mL^{-1})
- Arabinose (20% w/v)
- Demi water
- Infors Incubator Shaker
- Beckman Coulter, Avanti JE centrifuge
- pH meter
- Sonics Vibra-Cell VCX130 probe sonicator
- ÄKTA purifier
- HiPrep 26/10 desalting column

8.5.1.2 Procedure

1. Fresh lysogenic broth (LB) agar plates were prepared containing 50 μg.mL^{-1} ampicillin. Competent *Escherichia coli* NEB 10-β cells or similar *E. coli* cells were transformed with pBAD-EUGO-I427A-His. Colonies were grown overnight on LB agar at 37 °C.
2. LB medium was prepared: bacto tryptone (1 g), yeast extract (0.5 g) and sodium chloride (1 g) were dissolved in 100 mL demi water and divided into 5 mL fractions in 20 mL test tubes. They were then sterilised by autoclaving at 121 °C for 20 min.
3. A single colony was used to inoculate a single test tube containing 5 mL LB medium. The tube was placed in an incubator operating at 37 °C/135 rpm overnight. This was the pre-culture.
4. Terrific broth (TB) medium: bacto tryptone (6 g), yeast extract (12 g) and glycerol (2 mL) were added to a 2 L flask and completed with water to a 450 mL volume. KH_2PO_4 (1.15 g) and K_2HPO_4 (8.2 g) were added to a separate 100 mL Schott bottle and completed to 50 mL with demi water. Both vessels were autoclaved at 121 °C for 20 min. After autoclaving, the potassium salts were added to the rest of the TB medium in the flask to make 500 mL medium.
5. The 5 mL pre-culture was used to inoculate the TB medium. Ampicillin was added to a final concentration of 50 μg.mL^{-1}.
6. The culture was placed in a 37 °C/135 rpm Infors shaker and the optical density (OD) routinely checked. When the OD reached a value of 0.6, arabinose was added to the flask to give a final concentration of 0.02% v/v for induction of protein expression.
7. The cultures were transferred to 30 °C for 18 hr.
8. Cells were harvested by centrifugation at 6000× g in a Beckman Coulter, Avanti JE centrifuge using a JLA 14 rotor at 4 °C for 20 min. The supernatant was decanted and the pellet (200 mg) resuspended in 15 mL buffer A (50 mM potassium phosphate (KPi) buffer, 0.5 M NaCl, 5% v/v glycerol, pH 8).

   Buffer A: 1 M aqueous K_2HPO_4 (47 mL), 1 M aqueous KH_2PO_4 (3 mL), NaCl (29.2 g) and glycerol (50 mL) were added to a 1 L Schott bottle and completed with demi water to 1 L volume. The pH of the buffer was adjusted to a value of 8.0 using 1 M KOH and 1 M HCl solutions and a pH meter.
9. The cells were disrupted by sonication using a Sonics Vibra-Cell VCX130 probe sonicator operating for 10 sec ON, 10 sec OFF at 70% amplitude for 10 min.
10. The cell-free extract was obtained by centrifugation at 12 000× g and 4 °C for 45 min.
11. The extract was filtered using a Whatman FP 30/0.45 CA-S membrane syringe to remove the remaining cell debris.
12. The EUGO variant was purified using an ÄKTA purifier fitted with a 5 mL HisTrap HP column. The column was first equilibrated using 5 column volumes Buffer A then loaded with the cell-free extract.

    Buffer B (50 mM KPi, 5 mM imidazole, 0.5 M NaCl, 5% v/v glycerol, pH 8) was used to wash off any nonspecific proteins from the column.

    Buffer C: 1 M aqueous K_2HPO_4 (47 mL), 1 M aqueous KH_2PO_4 (3 mL), NaCl (29.2 g), imidazole (0.34 g) and glycerol (50 mL) were added to a 1 L Schott bottle and completed with demi water to 1 L volume. The pH of the buffer adjusted to a value of 8.0 using 1 M KOH and 1 M HCl solutions and a pH meter.

13. The enzyme (which should appear as a yellow colour) was eluted from the column using a gradient of Buffer C (50 mM KPi, 500 mM imidazole, 0.5 M NaCl, 5% v/v glycerol, pH 8) from 0 to 500 mM imidazole within 30 min.
14. After purification, the protein was desalted using a HiPrep 26/10 desalting column. The column was first equilibrated using Buffer D (50 mM KPi, 0.15 M NaCl, 10% v/v glycerol, pH 8).

    Buffer D: 1 M aqueous K_2HPO_4 (40 mL), 1 M aqueous KH_2PO_4 (10 mL), NaCl (8.8 g) and glycerol (100 mL) were added to a 1 L Schott bottle and completed with demi water to 1 L volume. The pH of the buffer adjusted to a value of 8.0 using 1 M KOH and 1 M HCl solutions and a pH meter.
15. The enzyme was frozen using liquid nitrogen and stored at −20 °C for further use. The total volume of the enzyme was 4 mL and its concentration was 50 μM.

8.5.2 Procedure 2: Determination of the Steady-State Kinetic Parameters of EUGO Using 2,6-Dimethoxy-4-Allylphenol 1 as Substrate

8.5.2.1 Materials and Equipment

- Quartz cuvette
- JASCO V-650 spectrophotometer
- 50 mM KPi buffer pH 7.5
- Serial dilutions of 2,6-dimethoxy-4-allylphenol **1**

8.5.2.2 Procedure

1. The concentration of EUGO was determined based on the absorbance of the covalently bound flavin co-factor of EUGO (extinction coefficient at 441 nm = 14.2 $mM^{-1}.cm^{-1}$).
2. The steady-state kinetic parameters of the mutant EUGO using 2,6-dimethoxy-4-allylphenol **1** as a substrate were determined by following the increase in absorbance at 270 nm due to sinapyl alcohol formation (extinction coefficient at 270 nm = 14.1 $mM^{-1}.cm^{-1}$) in 50 mM KPi pH 7.5 at 25 °C).
3. A series of different dilutions of 2,6-dimethoxy-4-allylphenol **1** in dimethyl sulfoxide (DMSO) were prepared.
4. To a quartz cuvette, 400 μL of KPi pH 7.5 buffer was added, followed by 50 μL of the 2,6-dimethoxy-4-allylphenol **1** substrate dilution and finally 50 μL of 500 nM EUGO. The increase in absorbance at 270 nm due to product formation was followed. The k_{obs} for each substrate concentration was determined. Using the k_{obs} and substrate concentrations, a graph was plotted. Finally, the k_{cat} and K_M were determined.

8.5.3 Procedure 3: Biocatalytic Synthesis of Sinapyl Alcohol 2 and Analysis of its Formation via HPLC

8.5.3.1 Materials and Equipment

- 15 mL screw-cap Pyrex tubes
- 20 mM KPi buffer pH 7.0
- DMSO
- 500 mM 2,6-dimethoxy-4-allylphenol **1** in DMSO
- 50 μM EUGO I427A

- HPLC column – Grace Alltima HP C18 column (3 µm, 2.1 × 100 mm with a pre-column of the same material)
- Millex FH – PTFE filter (0.45 µm pore size)
- Buffer A (milliQ water with 0.1% v/v formic acid) and buffer B (acetonitrile with 0.08 % formic acid)
- 4 mM stock solutions of 2,6-dimethoxy-4-allylphenol **1** and sinapyl alcohol **2**

8.5.3.2 Procedure

1. To a 15 mL screw-cap Pyrex tube, an amount equivalent to 20 mM KPi pH 7, 5 µM EUGO I427A mutant, 5% v/v DMSO and 3.9 mg or 10 mM 2,6-dimethoxy-4-allylphenol **1** was added, for a final reaction volume of 2 mL.
2. The tubes were placed in an INNOVA 40 New Brunswick Incubator Shaker operating at 30 °C and 50 rpm.
3. At various time intervals, 200 µL samples were withdrawn from the reaction mixture.
4. The samples were heated at 95 °C for 10 min to inactivate the enzyme.
5. The samples were centrifuged at 14 000 rpm for 10 min.
6. The supernatant was filtered using a Millex-FH, PTFE filter (0.45 µM pore size).
7. A sample (10 µL) was injected into the JASCO high-performance liquid chromatography (HPLC) system using a Grace Alltima HP C18 column (3 µm, 2.1 × 100 mm, with a pre-column of the same material).
8. Sinapyl alcohol **2** and 2,6-dimethoxy-4-allylphenol **1** 4 mM stock solutions were also injected.
9. The solvents used for the system were (a) water with 0.1% v/v formic acid and (b) acetonitrile with 0.08% formic acid.
10. HPLC method: 2 min 10% B, 18 min increase gradient 10–70% B, 3 min 70% B, decrease gradient 70–10% B and re-equilibration for 7 min. Detection at a wavelength of 280 nm.
11. The retention time for 2,6-dimethoxy-4-allylphenol **1** was 15.3 min and the R_T for sinapyl alcohol **2** was 4.3 min. The 2,6-dimethoxy-4-allylphenol **1** was converted to sinapyl alcohol **2** within 22 hr (>90% conversion).

8.5.4 Procedure 4: Biocatalytic Synthesis of Syringaresinol 3 from 2,6-Dimethoxy-4-Allylphenol 1 Using EUGO I427A Mutant and HRP (For Test Amounts)

8.5.4.1 Materials and Equipment

- 15 mL screw-cap Pyrex tubes
- 20 mM KPi buffer pH 7
- DMSO
- 500 mM 2,6 dimethoxy-4-allylphenol **1** in DMSO
- 50 µM EUGO I427A
- 65 µM HRP (Sigma-Aldrich)
- Ethyl acetate
- Brine (saturated aqueous solution of sodium chloride)
- Anhydrous magnesium sulfate
- Silica gel for flash chromatography
- Dichloromethane
- Methanol

8.5.4.2 Procedure

1. To a 15 mL screw-cap Pyrex tube, an amount equivalent to 20 mM KPi pH 7, 5 μM EUGO I427A mutant, 5% v/v DMSO, 0.65 μM HRP and 3.9 mg or 10 mM 2,6-dimethoxy-4-allylphenol **1** was added, for a final reaction volume of 2 mL.
2. The tubes were placed in an INNOVA 40 New Brunswick Incubator Shaker operating at 30 °C and 50 rpm.
3. At various time intervals, 200 μL samples were withdrawn from the reaction mixture.
4. The samples were heated for 10 min at 95 °C to inactivate the enzyme.
5. The samples were centrifuged at 14 000 rpm for 10 min.
6. The supernatant was filtered using a Millex-FH, PTFE filter (0.45 μM pore size).
7. A sample (10 μL) was injected into the JASCO HPLC system.
8. The conditions for the HPLC system and HPLC method were as mentioned in Procedure 3.
9. The retention time for 2,6-dimethoxy-4-allylphenol **1** was 15.3 min and the R_T for sinapyl alcohol **2** was 4.3 min. Syringaresinol **3** had a retention time of 14 min. After 3.5 hr, sinapyl alcohol **2** was formed as a major product with traces of syringaresinol **3**. At 25 hr, however, all of the 2,6-dimethoxy-4-allylphenol **1** and the sinapyl alcohol **2** were converted to syringaresinol **3**.

8.5.5 Procedure 5: 1 g Conversion of 2,6-Dimethoxy-4-Allylphenol 1 Using EUGO I427A Mutant and HRP

8.5.5.1 Materials and equipment

- 2 L Erlenmeyer flask
- 20 mM KPi buffer pH 7
- DMSO
- 2,6-Dimethoxy-4-allylphenol **1** (1 g, 1.28 mM)
- 50 μM EUGO I427A mutant
- 65 μM HRP (Sigma-Aldrich)
- Ethyl acetate
  - Brine (saturated aqueous solution of sodium chloride)
  - Anhydrous magnesium sulfate
  - Silica gel for flash chromatography
  - Column for flash chromatography
  - Dichloromethane
  - Methanol

8.5.5.2 Procedure

1. To a 2 L Erlenmeyer flask, an amount equivalent to 20 mM KPi, pH 7, 3 μM EUGO I427A mutant, 5% v/v DMSO, 0.65 μM HRP and 2,6-dimethoxy-4-allylphenol **1** (1g, 1.28 mM) was added, for a final reaction volume of 250 mL.
2. The flask was placed in an INNOVA 40 New Brunswick Incubator Shaker operating at 30 °C and 50 rpm.
3. At various time intervals, 200 μL samples were withdrawn from the reaction mixture for analysis for substrate depletion.

4. The samples were heated at 95 °C to inactivate the enzyme.
5. The samples were centrifuged at 14 000 rpm for 10 min.
6. The supernatant was filtered using a Millex-FX, PTFE filter (0.45 μM pore size).
7. A sample was injected into the JASCO HPLC system. The size of the sample for injection was 10 μL.
8. The conditions for the HPLC system and HPLC method were as mentioned in Procedure 3.
9. After the HPLC results showed that there was complete depletion of 2,6-dimethoxy-4-allylphenol **1** (22 hr), the reaction was stopped by filtration through a pre-weighed 47 mm glass-fibre filter from PALL® under vacuum.
10. The filtrate was extracted with an equal volume of ethyl acetate. This step was repeated three times with fresh ethyl acetate each time.
11. The combined organic extracts were collected and combined together, then washed with brine.
12. The sample was dried over anhydrous $MgSO_4$, filtered and concentrated by distillation under reduced pressure.
13. The residue was purified by flash chromatography (SiO_2, dichloromethane/methanol 99 : 1).
14. The yield of syringaresinol **3** produced was 870 mg (81%) from the 1 g starting material. It appeared as pale yellow crystals. In addition to the syringaresinol **3**, 80 mg of an insoluble product was formed.

8.5.6 Conclusion

Syringaresinol **3** is synthesised in a one-pot two-enzyme conversion starting from the relatively cheap 2,6-dimethoxy-4-allylphenol **1**. The two enzymes used are the EUGO I427A mutant and the commercially available HRP. EUGO was engineered to improve substrate acceptance in the active site through formation of the I427A mutant. A 1 g-scale conversion of 2,6-dimethoxy-4-allylphenol **1** using both the aforementioned enzymes produced syringaresinol **3** in 81% yield. This method for producing syringaresinol **3** has several advantages over conventional methods: the yield is relatively high, it is much safer (chemical synthesis methods involve the use of toxic materials) and it is relatively inexpensive. The method described in this section is a relatively green approach to the production of syringaresinol **3**.

References

1. Fujikawa, T., Yamaguchi, A., Morita, I. *et al.* (1996) *Biological and Pharmaceutical Bulletin*, **19**, 1227–1230.
2. Jaufurally, A.S., Teixeira, A.R.S., Hollande, L. *et al.* (2016) *ChemistrySelect*, **1**, 5165–5171.
3. Semenov, A.A., Enikeev, A.G., Khobrakova, V.B. *et al.* (2013) *International Journal of Biomedicine*, **3**, 287–290.
4. Chung, B.H., Kim, S., Kim, J.D. *et al.* (2012) *Experimental & Molecular Medicine*, **44**, 191–201
5. Janvier, M., Hollande, L., Jaufurally, A.S. *et al.* (2017) *ChemSusChem*, **10**, 738–746.
6. Janvier, M., Ducrot, P.H. and Allais, F. (2017) *ACS Sustainable Chemistry & Engineering*, **5**, 8648–8656.

8.6 Biocatalytic Preparation of Vanillin Catalysed by Eugenol Oxidase

Miquel Garcia-Bofill, Marina Guillén, Peter W. Sutton# and Gregorio Álvaro
Bioprocess Engineering and Applied Biocatalysis Group, Department of Chemical Biological and Environmental Engineering, Autonomous University of Barcelona, Bellaterra, Spain
 # Current address: Glycoscience SL, Bizkaia, Spain

Biosynthesis of flavours is of increasing interest because most flavours can be labelled as 'natural' when they are produced by natural means [1]. Naturalness of products is important for consumers and can thus greatly increase their market price [2]. Different methods of producing vanillin, with different biocatalysts and substrates, are under study [3], but no biosynthetic pathways are currently economically viable. One way of optimising an enzyme reaction is by reusing the biocatalyst. This can be achieved by immobilising the enzyme.

We produced vanillin through the oxidation of vanillyl alcohol catalysed by eugenol oxidase (EUGO) covalently immobilised on to an epoxy-agarose-UAB support. The oxidation produces hydrogen peroxide, which can deactivate the enzyme [4], so we added catalase to the reaction mixture to eliminate it (Scheme 8.8). The immobilised derivative can be reused to perform five reaction cycles on preparative scale.

8.6.1 Procedure 1: Preparation of Epoxy-Agarose-UAB Support

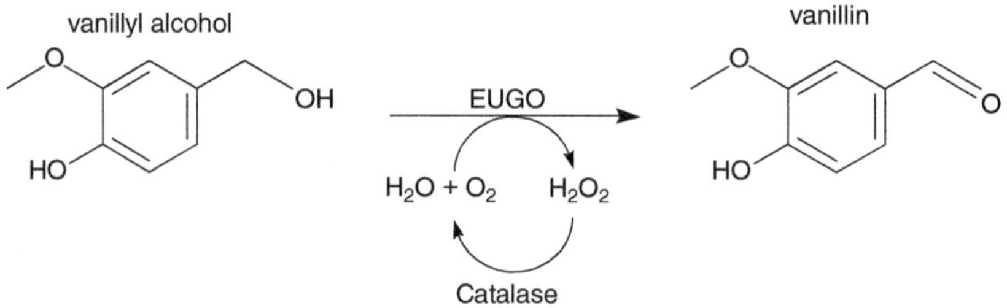

Scheme 8.8 *Oxidation of vanillyl alcohol (4-(hydroxymethyl)-2-methoxyphenol) to vanillin (4-hydroxy-3-methoxybenzaldehyde), catalysed by eugenol oxidase (EUGO) and catalase.*

8.6.1.1 Materials and Equipment

- Nonfunctionalised 6% BCL agarose bead standard from Agarose Beads Technology® (ABT®) brands (10 g)
- 0.6 M NaOH (aq) (10 mL)
- $NaBH_4$ (58.7 mg)
- 1,4 Butanediol diglycidyl ether (10 mL)
- Immobilisation buffer: 1 M potassium phosphate, pH 8 (100 mL)
- Rotary evaporator (Rotavap)
- Distilled water (dH_2O)
- Isopropanol (50 mL)
- Glass funnel filter
- 30 mL bottle with screw cap

8.6.1.2 Procedure

1. 6% BCL agarose (10 g) was washed with dH_2O (100 mL) and filtered through a glass filter funnel.
2. Washed 6% BCL agarose (10 g), 0.6M NaOH (10 mL), $NaBH_4$ (58.7 mg) and 1,4 butanediol diglycidyl ether (10 mL) were incubated under low agitation conditions at 25 °C in a rotary evaporator for 8 hr.
3. The resultant support was filtered and washed with dH_2O (100 mL) through a glass filter funnel diluted with isopropanol (50 mL) and stored at 4 °C for further use.
4. Before use, the support was washed with dH_2O (100 mL) and then with immobilisation buffer (100 mL) and filtered [5].

8.6.2 Procedure 2: Immobilisation of EUGO on an Epoxy-Agarose-UAB Support

8.6.2.1 Materials and Equipment

- EUGO solution (approx.. 218 $U.mL^{-1}$; InnoSyn)
- Epoxy-agarose-UAB (1.6 g)
- dH_2O
- Immobilisation buffer (1 M potassium phosphate, pH 8, 110 mL)
- Reaction Buffer 1 (50 mM potassium phosphate buffer, pH 7.5)
- β-Mercaptoethanol (210 μl)
- Glass filter funnel
- 30 mL bottle with screw cap
- Spectrophotometer: Cary 50 Bio US-Vis spectrophotometer

8.6.2.2 Procedure

1. Epoxy-agarose-UAB (2 g) was washed with dH_2O (50 mL) and then with immobilisation buffer (100 mL) and filtered, leaving moist beads.
2. The washed epoxy-agarose-UAB (1.6 g, or approximately 1.5 mL) was suspended in immobilisation buffer (10mL) and EUGO solution (3.5 mL) was added.
3. The mixture was incubated under mild agitation conditions at 25 °C for 4 hr.
4. A 0.2 M aqueous solution of β-mercaptoethanol (210 μL) was added to the mixture and incubated at 4 °C for a further 4 hr in order to open any residual epoxy groups.
5. Finally, the mixture was washed with 30% v/v acetone in Reaction Buffer 1 (100 mL) and used immediately.
6. The solution can be stored for more than 14 days by washing with 100 mM potassium phosphate and pH 6 buffer and submerging it in the same buffer, where it preserves more than 70% of its initial activity. Before use, the immobilised derivative must be washed with reaction buffer and filtered.
7. The obtained immobilised biocatalyst will have an approximate activity of 179 $U.mL^{-1}$.
8. Activity of the enzyme was measured as described in [4]. One unit of EUGO activity (1 U) is defined as the amount of enzyme required to produce 1 μmol of vanillin per minute. One unit of catalase activity (1 U) is defined as the enzyme required to convert 1 μmol of H_2O_2 per minute.

8.6.3 Procedure 3: Biocatalytic Conversion of Vanillyl Alcohol into Vanillin by EUGO with Catalase to Eliminate Hydrogen Peroxide

8.6.3.1 Materials and Equipment

- EUGO, either soluble (885 μL CFE) or immobilised on to epoxy-agarose-UAB (1.07 g) as prepared in Procedure 2 (both will have the same approximate activity in reaction: 18 U.mL^{-1})
- Catalase from bovine liver (90 mg, 4000 U.mg^{-1}; Sigma-Aldrich)
- Vanillyl alcohol (900 mg, 5.7 mmol for each reaction)
- Reaction Buffer 2 (145 mM potassium phosphate, pH 7.5 buffer, 5 mL for each reaction)
- Washing solution: 30% v/v acetone in 50 mM potassium phosphate pH 7.5 buffer (1 L)
- Acetone (4.3 mL for each reaction)
- Antifoam Buseti GLANAPON 2000 diluted in dH_2O (1 : 100 v/v, 1 mL)
- dH_2O
- Air (hydrated with 30% acetone solution water, 1 vvm)
- Magnetic stirrer with temperature control
- 30% v/v acetone in dH_2O
- Beaker (50 mL)
- Volumetric flask (10 mL)
- Syringe reactor (20 mL open syringe barrel with polyethylene a frit (V200PE100, Multisyntech GmbH), with a hose connected to the tip so that air can be passed through the frit into the reaction mixture contained in the barrel)
- Ethyl acetate (100 mL)
- Methyl benzoate as internal standard
- Gas chromatography (GC) system with flame ionisation detector (FID; 7890A Agilent GC)
- Innowax GC column (19095N-123, 30 m × 530 μm ID, 1 μm film)
- Centrifuge (min. 13 600 rpm)
- Syringe (1 mL)
- Filters suitable for organic solutions (0.45 μm)
- Vacuum filter

8.6.3.2 Procedure

1. The substrate solution was prepared by mixing vanillyl alcohol (899 mg, 5.7 mmol) with acetone (4.286 mL) and Reaction Buffer 2 (5 mL) in a 10 mL volumetric flask, topped up to 10 mL with dH_2O. Gentle warming to 35 °C and shaking were required to dissolve the substrate.
2. The substrate solution was added to the syringe reactor and immobilised with EUGO (1.07 g, 218 U). Catalase (9 mg.mL^{-1}, 360 kU) was added to the reactor with a magnetic stirrer.
3. The syringe was inserted into a water bath to control the temperature at 25 °C and stirring was started.
4. Air (1 vvm), previously saturated with a 30% acetone solution in water, was passed through the mixture.
5. Samples of 10 μL of antifoam were added to the reaction to prevent foaming.

Table 8.2 GC method.

Temperature program (r = °C.min^{-1})	Duration
100 °C 5 min – 10 r → 240 °C 2 min	21 min

Table 8.3 Retention times for GC analysis.

Substance	Retention (min)
Methyl benzoate (standard)	6.9
Vanillin	17.7
Vanillyl alcohol	19.6

6. Reaction samples (50 μL) were taken periodically and extracted with ethyl acetate supplemented with 5 mM methyl benzoate as reference standard (950 μL). The organic phase was centrifuged (1 min, 13 600 rpm), filtered (0.45 μm filters) and analysed by GC with an Innowax 19095N-123 column (see Table 8.2 for the GC method and Table 8.3 for retention times).

On reaction completion with EUGO on epoxy-agarose, 90% conversion and 50% solution yield were obtained, with 300 mg of vanillin produced. When the immobilised derivative was reused five times, 1340 mg of vanillin was obtained, increasing the biocatalyst yield from 1.6 to 3.3 mg vanillin.U^{-1}.

8.6.4 Analytical Method

8.6.4.1 GC Method

Samples were extracted with ethyl acetate (1 : 20 v/v) containing methylbenzoate (5 mM) as internal standard. The organic phase was analysed using a 7890A Agilent GC equipped with an Innowax 19095N-123 (30 m × 530 μm × 1 μm) column. The injector was kept at 225 °C; for the FID, the temperature was 250 °C. Helium was used as carrier gas at a constant pressure of 10 psi. The temperature programme and retention times are shown in Tables 8.2 and 8.3, respectively.

8.6.5 Conclusion

This procedure describes a method to produce vanillin from vanillyl alcohol in high yield using standard chemical, enzymes and laboratory chemical equipment. The immobilised derivative can be reused more than five times, optimising the biocatalyst yield.

References

1. El Parlamento Europeo y el Consejo de la Unión Europea (2008) Reglamento(CE) No. 1334/2008 Del Parlamento Europeo y del Consejo de 16 de diciembre de 2008 sobre los aromas y determinados ingredientes alimentarios con propiedades aromatizantes utilizados en los alimentos y por el que se modifican el Reglamento (CEE) no. 34–50.

2. Román, S., Sánchez-Siles, L.M. and Siegrist, M. (2017) *Trends in Food Science and Technology*, **67**, 44–57.
3. (a) Furuya, T., Kuroiwa, M. and Kino, K. (2016) *Journal of Biotechnology*, **243**, 25–28; (b) Yoon, S.H., Li, C., Lee, Y.-M. *et al.* (2005) *Biotechnology and Bioprocess Engineering*, **10,** 378–384; (c) Suresh, B. and Ravishankar, G.A. (2005) *Plant Physiology and Biochemistry*, **43**, 125–131.
4. García-Bofill, M., Sutton, P.W., Guillén, M. and Álvaro, G. (2019) *Applied Catalysis A: General*, **582**, 117117.
5. Sunberg, L. and Porath, J. (1974) *Journal of Chromatography A*, **90**, 87–98.

8.7 Vanillyl Alcohol Oxidase-Catalysed Production of (*R*)-1-(4′-Hydroxyphenyl)ethanol

Tom A. Ewing[1] and Willem J.H. van Berkel[2]

[1] *Wageningen Food & Biobased Research, Wageningen University & Research, Wageningen, The Netherlands*

[2] *Laboratory of Food Chemistry, Wageningen University & Research, Wageningen, The Netherlands*

The introduction of a chiral centre into a nonchiral starting material is hard to achieve using traditional methods of chemical synthesis. Nevertheless, it is very important to obtain efficient methods by which to do so, as pure enantiomers of chiral compounds are of great importance to sectors such as the pharmaceutical industry. One way in which chirality may be created is through the application of enzymes that are capable of enantioselectively functionalising a target molecule. For example, the flavin-dependent oxidoreductase vanillyl alcohol oxidase from *Penicillium simplicissimum* (VAO) can catalyse the enantioselective hydroxylation of 4-alkylphenols at the Cα position, leading to the formation of the (*R*)-enantiomer of the corresponding alcohol in high enantiomeric excess [1]. To demonstrate that this enzyme can be used to produce chiral secondary alcohols on a synthetically relevant scale, we employed it in the enantioselective hydroxylation of the nonchiral aromatic compound 4-ethylphenol to yield (*R*)-1-(4′-hydroxyphenyl)ethanol (Scheme 8.9) [2]. The procedure described here provides a detailed practical description of how this synthesis can be performed at a scale of 10 g of starting material. Readers interested in the production of the (*S*)-enantiomer are referred to recent work describing its enzymatic preparation [3].

8.7.1 Procedure 1: Production of *Escherichia coli* Cells Expressing the His-VAO Biocatalyst

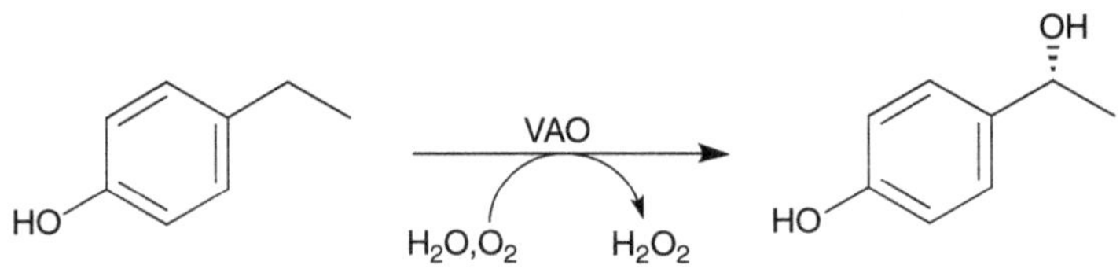

Scheme 8.9 *VAO-catalysed conversion of 4-ethylphenol to* (R)*-1-(4′-hydroxyphenyl)ethanol.*

8.7.1.1 Materials and Equipment

- pJ404-His-VAO plasmid [4] (available from authors upon request)
- Chemically competent BL21 *E. coli* cells (commercially available)

- Lysogenic broth (LB) medium mix (commercially available)
- LB agar mix (commercially available from Duchefa Biochemie Haarlem NL)
- Ampicillin
- Isopropyl-β-D-1-thiogalactopyranoside (IPTG)
- Deionised water
- 30 mL sterile syringe
- 0.45 μm pore size sterile syringe filter
- 50 mL sterile conical tube
- 2 × 100 mL Schott bottles
- 12 × 2 L Erlenmeyer flasks (no baffles)
- Centrifuge buckets
- Plastic cuvettes
- Petri dishes
- Cell spreader
- Parafilm
- Aluminium foil
- Plate incubator
- Autoclave
- Orbital shaker
- Spectrophotometer
- Tabletop Eppendorf centrifuge
- Centrifuge

8.7.1.2 Procedure

1. Ampicillin (2 g) was dissolved in deionised water (20 mL, final concentration 100 mg.mL^{-1}). The solution was transferred to a 30 mL syringe and sterilised by passing it through a 0.45 μm sterile syringe filter. Aliquots of this stock solution can be stored at −20 °C for up to 1 year.
2. IPTG (3.8 g) was dissolved in deionised water (20 mL, final concentration 190 mg.mL^{-1}, 0.8 M) and sterilised as described in Step 1. The solution was stored at −20 °C.
3. 100 mL LB agar was prepared according to manufacturer's instructions and transferred to 100 mL Schott bottle. The bottle was sterilised by autoclaving for 20 min at 120 °C, then removed from the autoclave and allowed to cool. Once cooled sufficiently to be comfortable to the touch, but before the LB agar had set, 100 μL of the 100 mg.mL^{-1} ampicillin stock solution was added. The solution was mixed by gentle swirling, being careful not to introduce air bubbles into the liquid. Once mixed, it was poured into petri dishes (approx. 15 mL per dish) and allowed to set. Plates are best prepared as fresh as possible, but can be stored at 4 °C for up to 2 weeks if necessary.
4. The pJ404 His VAO plasmid was transformed into competent BL21 *E. coli* cells according to the supplier's instructions. Using the cell spreader, the transformed cells were plated out on the prepared LB agar plates for selection and incubated in the plate incubator at 37 °C for 16 hr. Isolated colonies should be visible on the plates after this time. The edges were wrapped with parafilm and the colonies stored at 4 °C until it was time to pick one (preferably as soon as possible, at most for 1 week).
5. 6 L LB medium was prepared according to the supplier's instruction and divided equally over the 12 2 L Erlenmeyer flasks. The necks of the flasks were covered with a piece of aluminium foil, folded over once. Separately, 100 mL LB medium was prepared and

transferred to a 100 mL Schott bottle. The medium was sterilised by autoclaving for 20 min at 120 °C.

6. 10 mL LB was transferred from the Schott bottle into a 50 mL conical tube and 10 μL of the 100 mg.mL^{-1} ampicillin stock solution was added. The medium was inoculated with a colony of BL21 *E. coli* containing the pJ404-His-VAO plasmid by picking a colony using a sterile pipette tip and transferring the cells to the medium. The tube was transferred to an orbital shaker and incubated at 37 °C for 16 hr, shaking at 200 rpm.
7. After 16 hr, the tube contained a dense cell suspension. It was centrifuged for 15 min at 4000× g, 21 °C, to yield a clear supernatant and a cell pellet. The supernatant was removed and the pellet quickly resuspended in 10 mL fresh LB medium. 0.5 mL of 100 mg.mL^{-1} ampicillin solution was added to each of the 2 L Erlenmeyer flasks, followed by 0.5 mL of the cell suspension. The flasks were then transferred to an orbital shaker and incubated at 37 °C, shaking at 200 rpm. Periodically, a 1 mL sample of the culture was taken from one of the flasks and transferred to a cuvette, and the optical density at 600 nm was measured using the spectrophotometer.
8. Once the optical density of the cultures was 0.6, all Erlenmeyer flasks were removed from the shaker and the shaker temperature was set to 25 °C. 0.5 mL of the 0.8 M IPTG stock solution was added to each of the flasks, and they were returned to the shaker and incubated at 25 °C for 16 hr, shaking at 200 rpm.
9. After 16 hr, the cultures were transferred to centrifuge buckets and centrifuged for 15 min at 4000× g, 4 °C to yield a clear supernatant and a cell pellet (35–40 g). The supernatant was discarded. From here, one can proceed immediately to Procedure 2 with the cell pellet, or else the pellet can be stored at −20 °C and Procedure 2 can be performed at a later point.

8.7.2 Procedure 2: Extraction of the His-VAO Biocatalyst

Note: To prevent inactivation of the His-VAO biocatalyst, keep biocatalyst-containing solutions on ice whenever possible.

8.7.2.1 Materials and Equipment

- Potassium phosphate monobasic (KH_2PO_4)
- Potassium phosphate dibasic (K_2HPO_4)
- Dithiothreitol (DTT)
- cOmplete™ protease inhibitor cocktail pills (Roche)
- DnaseI (Roche)
- Magnesium sulfate heptahydrate ($MgSO_4 \cdot 7\ H_2O$)
- Ultrapure water
- 1.5 mL Eppendorf tubes
- 50 mL conical tubes
- 250 mL graduated cylinder
- 2 × 250 mL Schott bottle
- Magnetic stirring bar
- Ice bucket
- French pressure cell press
- Centrifuge

- Centrifuge tubes suitable for use at 39 000× g
- pH meter

8.7.2.2 Procedure

1. DTT (770 mg) was dissolved in ultrapure water (10 mL, final concentration 77 mg.mL^{-1}, 0.5 M). Aliquots of the solution can be stored at −20 °C for later use.
2. $MgSO_4{\cdot}7H_2O$ (2.46 g) was dissolved in ultrapure water (10 mL, final concentration 246 mg.mL^{-1}, 1 M) and stored at 4 °C.
3. A stock solution of 100 mM potassium phosphate buffer, pH 7.0 was prepared. KH_2PO_4 (2.72 g) was dissolved in ultrapure water (200 mL, final concentration 13.6 g.L^{-1}, 100 mM). Likewise, K_2HPO_4 (3.48 g) was dissolved in ultrapure water (200 mL, final concentration 17.4 g.L^{-1}, 100 mM). The KH_2PO_4 solution (80 mL) was then mixed with the K_2HPO_4 solution (120 mL) in the 250 mL Schott bottle, and the pH adjusted to 7.0 by adding KH_2PO_4 solution to make it more acidic or K_2HPO_4 solution to make it more basic. The 100 mM potassium phosphate buffer, pH 7.0, stock solution was stored at room temperature.
4. To prepare 200 mL lysis buffer, 100 mL 100 mM potassium phosphate buffer, pH 7.0, stock solution, 80 mL ultrapure water, 200 μL DTT solution and 100 μL $MgSO_4$ solution were mixed in a 250 mL graduated cylinder. Four cOmplete™ protease-inhibitor cocktail pills and DNaseI (4 mg) were added and the solution was stirred using a magnetic stirring bar until they had dissolved. The stirring bar was then removed and the volume adjusted to 200 mL by adding ultrapure water. The buffer was transferred to a 250 mL Schott bottle and cooled in a bucket of ice.
5. Cold lysis buffer was added to the cell pellets from Step 1. The volume of buffer added was minimised in order to obtain a biocatalyst with a high activity per volume: a volume of 70–100 mL (2–2.5 mL.g^{-1} cells) should be sufficient to obtain a suitable suspension. The cells were resuspended in the lysis buffer by pipetting up and down, and kept on ice.
6. The cell suspension was passed through a pre-cooled French pressure cell press three times, cooling well between cycles by putting the samples on ice in order to prevent inactivation of the biocatalyst. The suspension obtained after lysis was collected in 50 mL conical tubes.
7. The suspension was transferred to centrifuge tubes and centrifuged for 45 min at 39 000× g and 4 °C to obtain a clear yellow/brownish supernatant and a beige pellet . The supernatant was removed and transferred to 50 mL conical tubes. Typically, the volume of the obtained enzyme solution is 50–70 mL. A sample (0.5 mL) was taken and transferred to a 1.5 mL Eppendorf tube, in order to analyse the activity of the extract. The remainder of the supernatant was frozen using liquid nitrogen and stored at −20 °C. This was the His-VAO biocatalyst.

8.7.3 Procedure 3: Determination of the Activity of the His-VAO Biocatalyst

8.7.3.1 Materials and Equipment

- Vanillyl alcohol (3-methoxy-4-hydroxybenzyl alcohol, 308 mg, 2 mmol)
- Potassium phosphate monobasic (KH_2PO_4)

- Potassium phosphate dibasic (K_2HPO_4)
- Ethanol
- Ultrapure water
- His-VAO biocatalyst (prepared as described in Procedure 2)
- pH meter
- Parafilm
- Spectrophotometer with a thermostatted cell holder
- Hamilton syringe
- Quartz cuvettes, pathlength 1 cm, volume 600 µL

8.7.3.2 Procedure

1. 50 mM potassium phosphate buffer, pH 7.5, was prepared. KH_2PO_4 (0.68 g) was dissolved in ultrapure water (100 mL, 6.8 g.L^{-1}, 50 mM). Likewise, K_2HPO_4 (0.87 g) was dissolved in ultrapure water (100 mL, 8.7 g.L^{-1}, 50 mM). 40 mL of the K_2HPO_4 solution was then mixed with 10 mL of the KH_2PO_4 solution and the pH adjusted to 7.5 by adding KH_2PO_4 solution to make it more acidic or K_2HPO_4 solution to make it more basic.
2. Vanillyl alcohol (308 mg, 2 mmol) was dissolved in ethanol (10 mL, final concentration 30.8 mg.mL^{-1}, 200 mM). 100 µL of the vanillyl alcohol stock solution was added to 9.9 mL of the 50 mM potassium phosphate buffer, pH 7.5, to give a 2 mM vanillyl alcohol stock solution.
3. 590 µL of the 2 mM vanillyl alcohol stock solution was transferred to a quartz cuvette. The cuvette was covered with parafilm and incubated in the thermostatted cell holder of the spectrophotometer, set at 25 °C, for 10 min.
4. Using a Hamilton syringe, 10 µL of the His-VAO biocatalyst sample was added to the cuvette, puncturing the parafilm with the syringe. The solution was mixed by inverting the cuvette three times and transferred to the cell holder of the spectrophotometer, thermostatted at 25 °C. The conversion of vanillyl alcohol to vanillin was monitored by measuring the absorption of the solution at 340 nm (ε = 14 000 M^{-1}.cm^{-1}). The rate of the reaction should be such that an initial reaction rate could be determined by fitting a linear curve to the trace of the absorption. If the reaction rate was too high, the His-VAO biocatalyst sample was diluted in 50 mM potassium phosphate buffer, pH 7.5, and the measurement repeated. Once an adequate dilution was performed the measurement was repeated twice for certainty.
5. To calculate the activity of the His-VAO biocatalyst solution, the average initial rate of change in absorption at 340 nm from the reaction traces was determined by fitting linear curves to the data. Subsequently, the activity of the biocatalyst was calculated in U.mL^{-1}, where 1 U is the amount of enzyme required to convert 1 µmol of vanillyl alcohol to vanillin in 1 min. The enzyme solution thus prepared typically has an activity of 5–8 U.mL^{-1}.

8.7.4 Procedure 4: Enzymatic Synthesis of (*R*)-1-(4′-Hydroxyphenyl)ethanol

Note: This procedure involves working under a pure oxygen atmosphere. This represents a fire and explosion hazard. Please ensure appropriate safety measures are taken.

8.7.4.1 Materials and Equipment

- 4-Ethylphenol (10 g, 82 mmol)
- His-VAO biocatalyst (prepared as described in Procedure 2)
- Potassium phosphate monobasic (KH_2PO_4)
- Potassium phosphate dibasic (K_2HPO_4)
- Ultrapure water
- Acetone
- Ethyl acetate
- Hexane
- Anhydrous sodium sulfate (Na_2SO_4)
- Brine (saturated aqueous NaCl solution)
- Oxygen cylinder
- 4 L three-neck round-bottom flask
- Thermometer
- Three-way balloon adapter
- Vacuum pump
- Balloon
- Overhead stirrer
- Büchner funnel
- Erlenmeyer flask
- Filter papers
- Separation funnel
- Water bath
- Rotary evaporator
- Oil bath
- Vacuum oven
- pH meter

8.7.4.2 Procedure

1. 100 mM potassium phosphate buffer, pH 7.5, was prepared. KH_2PO_4 (13.6 g) was dissolved in ultrapure water (1 L, 13.6 g.L^{-1}, 100 mM). Likewise, K_2HPO_4 (69.6 g) was dissolved in ultrapure water (4 L, 17.4 g.L^{-1}, 100 mM). 3.2 L of the K_2HPO_4 solution was mixed with 800 mL of the KH_2PO_4 solution and the pH adjusted to 7.5 by adding KH_2PO_4 solution to make it more acidic or K_2HPO_4 solution to make it more basic.
2. 4-Ethylphenol (10 g, 82 mmol) was dissolved in acetone (345 mL) and transferred into a 4 L three-neck round-bottom flask, thermostatted at 30 °C using a water bath. The temperature was checked using a thermometer inserted through one of the side necks of the flask. 100 mM potassium phosphate buffer, pH 7.5 (3075 mL), was added and the solution incubated until the temperature was stable at 30 °C. Then, 202 U His-VAO biocatalyst was added (the volume depending on the activity of the extract, typically 25–40 mL). The overhead stirrer was inserted through the top neck of the flask.
3. The vacuum pump and a balloon filled with oxygen from the oxygen cylinder were attached to the three-way balloon adapter. The adapter was attached to the remaining side neck. The reaction vessel was evacuated and filled with oxygen from the balloon. This process was repeated twice.

4. The reaction mixture was incubated at 30 °C under stirring at 230 rpm for 9 hr, after which the oxygen atmosphere was removed. The reaction was then incubated under atmospheric pressure for a further 15 hr.
5. Using a rotary evaporator, the solvent was removed under reduced pressure whilst heating to 50 °C using a water bath to remove acetone and reduce the volume of solvent. Some white solid may be formed during the concentration process.
6. Once the volume had been reduced sufficiently (to approximately 1 L), the solution was filtered through a Büchner funnel and transferred to a separation funnel and 1 L of ethyl acetate was added. The solution was mixed, allowing separation and collecting the aqueous and organic phases separately. The aqueous phase was then returned to the separation funnel and the extraction process repeated three times with a further 500 mL ethyl acetate each time.
7. The organic phases were pooled and transferred to the separation funnel. 1 L of brine was added and the solution was mixed, allowing separation and then collecting the organic phase. To the organic phase, anhydrous Na_2SO_4 was added until dry; this was then removed by filtration.
8. The solvent was removed by distillation under reduced pressure to give a solid crude product (green beige solid). The crude product was dissolved in ethyl acetate (40 mL) then heated to 90 °C using an oil bath. Subsequently, the solution was cooled to room temperature, allowing a white solid to crystallise.
9. The suspension was filtered through a Büchner funnel. The remaining solid was washed with a 1 : 1 ethyl acetate/hexane mixture, collected and dried in a vacuum oven until no further decrease in weight was observed (4.10 g, 36%, 97% ee).

8.7.5 Analytical Method

8.7.5.1 GC and ^{1}H-NMR Analysis

8.7.5.1.1 Materials and Equipment.

- Gas chromatograph (GC) with flame ionisation detector (FID)
- Hydrodex β-6TBDM chiral GC column (25 m × 2.45 mm, Machery-Nagel)
- 400 MHz nuclear magnetic resonance (NMR) spectrometer

8.7.5.1.2 Procedure.

- The product was analysed by GC and ^{1}H-NMR to confirm its identity and purity.

8.7.5.1.3 GC Method. GC experiments were performed on a GC-2010 plus (Shimadzu) with an FID detector (260 °C) and a Hydrodex β-6TBDM column (25 m × 2.45 mm, Machery-Nagel). Hydrogen was used as a carrier gas at a flow rate of 1.5 $mL.min^{-1}$ and the injector temperature was 80 °C. Runs were performed using a 20 min gradient from 80 to 220 °C followed by 10 min at 220 °C. (*R*)-1-(4′-hydroxyphenyl)ethanol had a retention time of 22.3 min.

8.7.5.1.4 ^{1}H-NMR. (400 MHz, d6-DMSO) δ = 9.10 (s, 1H, Ar-OH), 7.12 (dt, J = 8.02 Hz, 2H, Ar-H), 6.68 (dt, J = 8.69 Hz, 2H, Ar-H), 4.85 (d, J = 4.29 Hz, 1H, aliph-OH), 4.63–4.57 (m, 1H, CH), 1.27 (d, J = 6.44 Hz, 3H, CH_3).

8.7.6 Conclusion

This procedure enables the enzymatic synthesis of the chiral secondary alcohol (*R*)-1-(4′-hydroxyphenyl)ethanol from the achiral starting material 4-ethylphenol. A 94% conversion of 10 g 4-ethylphenol was achieved after 24 hr and the product was obtained in 36% isolated yield and 97% ee. The procedure could potentially also be used for the production of chiral secondary alcohols from other short-chain 4-alkylphenols [1,4].

References

1. Drijfhout, F.P., Fraaije, M.W., Jongejan, H. *et al.* (1998) *Biotechnology and Bioengineering*, **59**, 171–177.
2. Ewing, T.A., Kühn, J., Segarra, S. *et al.* (2018) *Advanced Synthesis and Catalysis*, **360**, 2370–2376.
3. (a) Wuensch, C., Gross, J., Steinkellner, G. *et al.* (2013) *Angewandte Chemie International Edition*, **52**, 2293–2297; (b) Payer, S.E., Pollak, H., Glueck, S.M. and Faber, K. (2018) *ACS Catalysis*, **8**, 2438–2442.
4. Ewing, T.A., van Noord, A., Paul, C. E. and van Berkel, W.J.H. (2018) *Molecules*, **23**, 164.

8.8 Enzymatic Synthesis of Pinene-Derived Lactones

Antonino Biundo,[1,2] Arne Stamm,[1,2] Uwe T. Bornscheuer[3] and Per-Olof Syrén*[1,2,4]
[1]*School of Engineering Sciences in Chemistry, Biotechnology and Health, Department of Fibre and Polymer Technology KTH Royal Institute of Technology, Stockholm, Sweden*
[2]*Science for Life Laboratory, Division of Protein Technology, KTH Royal Institute of Technology, Solna, Sweden*
[3]*Institute of Biochemistry, Dept. of Biotechnology & Enzyme Catalysis, Greifswald University, Greifswald, Germany*
[4]*Wallenberg Wood Science Center, KTH Royal Institute of Technology, Stockholm, Sweden*

Plastics are versatile materials that find fundamental applications in our everyday lives. Annually, more than 300 million tons of plastics are produced globally [1]. Currently, only 1% of these synthetic materials derive from renewable resources, leading to so-called biobased plastics. The market for bioplastics is growing in alignment with the global effort to replace petroleum-dependent chemistries with sustainable options [2]. In particular, polyesters constitute one class of polymers with high industrial relevance in applications ranging from packaging and coatings to engineering and medicine [3]. They can be produced by polycondensation, transesterification and ring-opening polymerisation (ROP). ROP is a form of chain-growth polymerisation that uses lactones as monomeric building blocks, which in turn can be produced by Baeyer–Villiger oxidation of the corresponding cyclic ketones [4]. Cyclic ketones can be derived from renewable resources, such as sugar-based biomass, terpenes and terpenoids [5]. Enzymes belonging to the Baeyer–Villiger monooxygenase (BVMO) superfamily [6] can perform Baeyer–Villiger oxidation in the presence of oxygen and co-factors. Turpentine, a byproduct from the paper and pulp industry, contains high amounts of terpenes, especially (−)-α-pinene **1** and

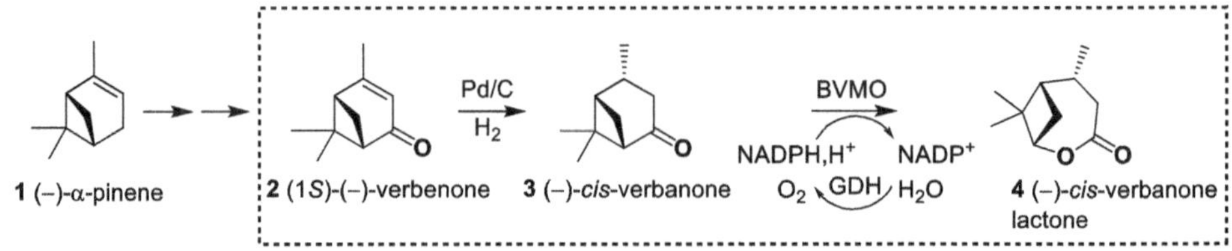

Scheme 8.10 *Conversion of (−)-cis-verbenone **2** to the normal pinene-derived lactone **4** by chemoenzymatic catalysis. BVMO, Baeyer–Villiger monooxygenase; GDH, glucose dehydrogenase as co-factor regeneration system.*

(−)-β-pinene. Compound **1** has been thoroughly explored as a starting material for the production of various compounds, including renewable chemicals, solvents and polymers [7]. However, the inert backbone of **1** prevents its efficient and controlled polymerisation. Through oxidation of **1**, the compound (1*S*)-(−)-verbenone **2** can be produced [8]. Our approach describes the chemoenzymatic transformation of the available **2** into the reduced verbanone **3** and further into pinene-derived lactone **4** for the production of terpene-derived polyesters (Scheme 8.10) [9]. This strategy includes screening of BVMO variants towards challenging substrates, such as **3**, in microtitre plate assays. The key enzymatic oxidation step for monomer activation reduces the utilisation of harsh chemicals such as *meta*-chloroperbenzoic acid, other peracids and hydrogen peroxide. Furthermore, the enzymatic process with BVMOs generates only water as a byproduct, avoiding the stoichiometric amounts of carboxylic acid produced as waste in chemical-catalysed Baeyer–Villiger oxidation.

8.8.1 Procedure 1: Synthesis of (−)-*cis*-Verbanone 3

8.8.1.1 Materials and Equipment

- Round-bottom flask equipped with magnetic stir bar and sealed with rubber septum
- Lab balloon/football
- Magnetic stirrer
- (−)-*cis*-Verbenone **2** (10 g, 67 mmol)
- 10% w/w palladium on carbon (Pd/C, 0.5 g)
- Dichloromethane (DCM, 50 mL)
- H_2 gas (~3 bar)

8.8.1.2 Procedure

1. (−)-*cis*-Verbenone **2** (10 g, 67 mmol) was added into a round-bottom flask.
2. 10% w/w of Pd/C (0.5 g) was added together with DCM (50 mL) into the flask.
3. The flask was sealed with a septum and stirred under H_2 pressure (~3 bar) for 4 hr.
4. The solution was filtered through a glass filter (1.6 μm pore size) and concentrated by distillation under reduced pressure. Verbanone **3** was obtained quantitatively (10.2 g, 99.6% conversion). The product was analysed by ^{1}H-nuclear magnetic resonance (NMR) and confirmed to purely contain verbanone according to reference spectra [9].

8.8.2 Procedure 2: Transformation of Plasmids into *E. coli* BL21(DE3) Competent Cells

8.8.2.1 Materials and Equipment

- Tryptone (8 g)
- Yeast extract (5 g)
- NaCl (4 g)
- Distilled water (dH_2O)
- Nutrient agar for microbiology (18.4 g)
- Kanamycin (50 mg.mL^{-1} in dH_2O, filter-sterilised)
- Nutrient agar plates containing kanamycin (40 μg.mL^{-1})
- Heat-shock competent cells (100 μL)
- Plasmid preparation (5 μL, 78 ng.μL^{-1} of pET28b(+) harbouring the mutated gene of the thermostable variant of cyclohexanone monooxygenase from *Acinetobacter calcoaceticus* (CHMO_QM) [6c])
- Autoclave (Systec VX-150)
- Two 1 L Schott glass bottles with screw caps
- Tabletop thermomixer

8.8.2.2 Procedure

1. Tryptone (8 g), yeast extract (4 g) and NaCl (4 g) were dissolved in dH_2O (800 mL) and autoclaved for 20 min at 121 °C in a 1 L Schott glass bottle to produce sterile lysogenic broth (LB) medium.
2. Nutrient agar for microbiology (18.4 g) was added to distilled H_2O (800 mL) and autoclaved for 20 min at 121 °C in a 1 L Schott glass bottle to produce sterile nutrient agar medium. The solution was cooled to approximately 55 °C and kanamycin (640 μL) stock solution was added to reach a final concentration of 40 μg.mL^{-1}.
3. Heat-shock competent cells (100 μL) were thawed on ice for 5 min. An aliquot (5 μL) of plasmid preparation (78 ng.μL^{-1} of pET28b(+) harbouring the mutated gene of the thermostable variant of cyclohexanone monooxygenase from *Acinetobacter calcoaceticus* (CHMO_QM) [6c]) was mixed with the competent cells and incubated on ice for 20 min. Afterwards, the cells were placed at 42 °C for 45 sec and again on ice for 2 min.
4. Preheated (42 °C) LB medium (900 μL) was added to the cell suspension. The cells were incubated at 37 °C and 300 rpm for 1 hr.
5. An aliquot (100 μL) of suspension was plated on to nutrient agar–kanamycin plates and incubated overnight at 37 °C.

8.8.3 Procedure 3: Recombinant Expression of a Cyclohexanone Monooxygenase Library in *E. coli* BL21(DE3)

8.8.3.1 Materials and Equipment

- Tryptone, bacteriological (12.8 g)
- Yeast extract, bacteriological (8 g)
- NaCl (4 g)
- dH_2O

- Kanamycin (50 mg mL^{-1} in dH_2O, filter-sterilised)
- Isopropyl-β-D-1-thiogalactopyranoside (IPTG; 0.1 M in dH_2O, filter-sterilised)
- Nutrient agar plates supplemented with kanamycin (40 μg.mL^{-1}), containing colonies of *E. coli* BL21(DE3) transformed with expression vector pET28b(+) harbouring mutants of the gene encoding CHMO_QM from Procedure 2
- Sterile 96 deep-well plate
- Sterile 24 deep-well plates covered with aluminium foil
- Orbital shaker (InforsHT Ecotron)
- Plate-reader spectrophotometer (SpectraMax i3x)
- Cooling centrifuge

8.8.3.2 Procedure

1. Tryptone (12.8 g), yeast extract (8 g) and NaCl (4 g) were dissolved in 1 L distilled H_2O and autoclaved for 20 min at 121 °C in a 1 L Schott glass bottle to produce sterile yeast extract tryptone (2xYT) medium.
2. To prepare the pre-culture, a single colony of *E. coli* BL21(DE3), each harbouring a mutated BVMO gene residing in vector pET28b(+), was picked from a freshly streaked plate and inoculated in 1 mL 2xYT–kanamycin in a 96 deep-well plate at 37 °C and 200 rpm overnight.
3. The OD_{600} of each 1 mL culture was measured using an aliquot (100 μL) transferred to a 96-well plate. Values were corrected by measurement of a blank. Analysis was performed in a plate-reader spectrophotometer using a factor of 0.1496 for comparison to a 1 cm cuvette.
4. The culture was diluted in 2xYT medium (3 mL) supplemented with kanamycin (2.4 μL) stock solution (40 μg.mL^{-1} final concentration) to an OD_{600} of 0.1 in 24 deep-well plates covered with aluminium foil.
5. The cells were grown at 37 °C and 200 rpm until the OD_{600} reached 0.6–0.8. The cultures were then cooled to 25 °C and protein expression was induced with 1.5 μL IPTG stock solution (0.05 mM final concentration) at 180 rpm for 20 hr.
6. The cells were harvested by centrifugation at 2276× g and 4 °C for 15 min. The supernatant was discarded and the cell pellet (0.16 g.$well^{-1}$) was stored at –20 °C for further analysis.

8.8.4 Procedure 4: Generation of Cell-Free Extracts of BVMO Variants

8.8.4.1 Materials and Equipment

- Cell pellets from Procedure 3
- Ethanol (EtOH)
- Cyclohexanone (1 M in EtOH)
- Nicotinamide adenine dinucleotide phosphate reduced (NADPH, 0.1 M in 50 mM Tris-HCl pH 8.5)
- B-PER™ Bacterial Protein Extraction reagent (5 mL.g^{-1} of cell pellet; Thermo Scientific)
- Tabletop plate shaker (750 rpm)
- Cooling centrifuge

- 96-well plates
- Plate-reader spectrophotometer (Spark TECAN)

8.8.4.2 Procedure

1. The cell pellet of a 3 mL culture was resuspended with 5 mL of B-PER™ Bacterial Protein Extraction Reagent per gram of cell pellet (0.8 mL for a 0.16 g pellet) and incubated at 750 rpm on a tabletop plate shaker at room temperature for 15 min.
2. A centrifugation at 2276× g and 4 °C was carried out for 15 min to remove the cell debris.
3. A reaction solution (20 mL) was prepared using cyclohexanone (12 μL, 0.6 mM final concentration) and NADPH (50 μL, 0.25 mM final concentration)
4. An aliquot of the cell-free extracts (20 μL) was placed in 96-well plates and the reaction solution (200 μL) was added to start the reaction. The consumption of NADPH was determined at 340 nm for 120 sec using a molar extinction coefficient of $\varepsilon_{NADPH} = 6.4\,mM^{-1}.cm^{-1}$. The volumetric activity of the cell-free extract of CHMO_QM was calculated at $1.01 \pm 0.07\,U.mL^{-1}$.

8.8.5 Procedure 5: Screening of Biocatalytic Oxidation of (−)-*cis*-Verbanone 3 with Cell-Free Extracts

8.8.5.1 Materials and Equipment

- (−)-*cis*-Verbanone (from Procedure 1, 1 M in DMF)
- Tris-HCl (50 mM, pH 8.5)
- NADPH (0.1 M in 50 mM Tris-HCl, pH 8.5)
- Flavin adenine dinucleotide (FAD, 62 mM in 50 mM Tris-HCl, pH 8.5)
- Glucose dehydrogenase (GDH from Codexis, $3\,mg.mL^{-1}$ in 50 mM Tris-HCl, pH 8.5)
- D-Glucose (1 M in 50 mM Tris-HCl, pH 8.5)
- Catalase from bovine liver ($11895\,U.mg^{-1}$)
- Ethyl acetate spiked with decane (2 mM)
- Cell-free extracts from Procedure 4
- pH meter
- 1 L Schott glass bottle with screw cap
- Orbital shaker (InforsHT Ecotron)
- 2 mL glass vials sealed with a breathable seal
- Orbital shaker (InforsHT Ecotron)
- Hamilton glass syringes (1 mL)
- Vortex
- Cooling centrifuge
- GCMS QP2010 system coupled with FID and mass spectrometry (MS) detector (Shimadzu)
- GC capillary column (Rxi-5ms: 30 m × 250 μm ID × 0.25 μm film)
- Sky Liner PTV 2010 liner with glass wool
- AOC-20i auto-injector (Shimadzu)

8.8.5.2 Procedure

1. (−)-*cis*-Verbanone **3** stock solution (1 μL) was added to a final concentration of 2 mM into a solution containing the BVMO (CHMO_QM) cell-free extract (40 μL, 0.04 U).

Stock solutions of NADPH (1 μL, 0.2 mM final concentration), FAD (0.5 μL, 62 μM final concentration), GDH (15 μL, 0.09 mg.mL^{-1} final concentration), D-glucose (100 μL, 0.2 M final concentration) and catalase (0.5 μL, 11.9 U.mg^{-1}) were added to the reaction mixture.

2. The reactions were performed at 30 °C and 200 rpm for 24 hr.
3. The reactions were quenched with ethyl acetate (2 × 500 μL) spiked with decane (2 mM) as reference standard. Samples were vortexed for 5 sec and centrifuged at room temperature and 9000× g for 10 min.
4. The upper phase was collected and analysed by GC with a Rxi-5ms capillary column (see Table 8.4 for the GC method and Table 8.5 for retention times).

8.8.6 Procedure 6: Scale-Up of the Biocatalytic Transformation

8.8.6.1 Materials and Equipment

- 2xYT medium supplemented with kanamycin (40 μg.mL^{-1})
- Nutrient agar plates supplemented with kanamycin (40 μg.mL^{-1}), containing colonies of *E. coli* BL21(DE3) transformed with expression vector pET28b(+) harbouring mutants of the gene encoding CHMO_QM from Procedure 1
- IPTG (0.1 M in dH_2O, filter-sterilised)
- Tris-HCl buffer (50 mM, pH 8.5)
- 1 L Schott glass bottles with screw caps
- Orbital shakers (InforsHT Ecotron)
- 0.1 L Erlenmeyer flasks (unbaffled)
- 0.5 L Erlenmeyer flasks (unbaffled)
- Plate-reader spectrophotometer (SpectraMax i3x)
- Cooling centrifuge
- Misonix sonifier cell disruptor S-4000 probe
- 2 L Erlenmeyer flasks (baffled)
- Hamilton glass syringes (1 mL)
- Vortex

Table 8.4 *GC method.*

Temperature program (r = °C min^{-1})	Duration
70 °C – 20 r → 300 °C – 3 r → 340 °C 10 min	35 min

Table 8.5 *Retention times for GC analysis.*

Substance	Retention time (min)
Decane	3.79
(−)-*cis*-Verbanone	5.42
Verbanone lactone	7.02

8.8.6.2 Procedure

1. A pre-culture (30 mL) was prepared in 2xYT medium supplemented with kanamycin (40 μg mL^{-1}) and grown at 37 °C and 200 rpm overnight.
2. A culture (200 mL) was inoculated with the pre-culture to an OD_{600} of 0.1. When the OD_{600} reached 0.6–0.8, the culture was cooled to 25 °C at 200 rpm and protein expression was induced with IPTG stock solution (100 μL) to 0.05 mM final concentration. The expression was carried out for 20 hr.
3. The cells (2.4 g) were harvested at 2276× g and 4 °C for 15 min and the supernatant was discarded. The cells were stored at −20 °C for further analysis.
4. To produce the cell-free extracts, the cells were resuspended in 50 mM Tris-HCl buffer pH 8.5 (25 mL) and sonicated for 1 min with 1 sec pulse and 2 sec pause, 60% duty cycle under ice cooling (Misonix sonifier cell disruptor ultrasonic S-4000 probe).
5. Cellular debris was removed by centrifugation at 40 000× g and 4 °C for 20 min.
6. The cell-free extract (16 mL, 1.46 $U.mL^{-1}$) was added to a 2 L baffled Erlenmeyer flask containing the reaction mixture (200 mL) as described in Procedure 5 and sealed with a breathable seal. The reaction was performed at 30 °C and 200 rpm for 24 hr.
7. An aliquot (500 μL) of the reaction was tested via GC, showing 39% conversion.
8. The product was extracted by washing the aqueous phase with ethyl acetate (3 × 200 mL). The collected organic phases were dried with magnesium sulfate, filtered and concentrated by distillation under reduced pressure. The slightly yellow oil thus obtained was analysed by ^{1}H-NMR and shown to have a relative abundance of 30%, which relates to around 25 mg of lactone [9].
9. The organic phase was purified using medium-pressure liquid chromatography (MPLC; 0% EtOAc in heptane over 2 column volumes (CV), then up to 30% EtOAc in heptane over 8 CV) and concentrated to afford the normal, most substituted lactone as a colourless oil. The optimal yield was approximatively 90% and the purified product was confirmed to be pinene-derived lactone **4** by ^{1}H-NMR analysis [9].

8.8.7 Conclusion

The described procedures highlight the great potential of utilising enzyme engineering on BMVOs to enlarge their active sites in order to increase their substrate promiscuity. Beyond identifying and applying variants capable of accepting bulky compounds such as the (−)-α-pinene-derived (−)-*cis*-verbanone, the concepts discussed in this section should be widely applicable to other challenging substrates. Moreover, the increase of oxygen content in the shake flask reaction enabled an increased yield of the product.

Acknowledgements

This work was generously funded by a FORMAS young-research leader grant (#942-2016-66). We acknowledge support from the Swedish Research Council (VR, #2016-06160).

References

1. Zhu, Y., Romain, C. and Williams, C.K. (2016) *Nature*, **540**, 354–362.
2. Ellen MacArthur Foundation (2017) The New Plastics Economy: Rethinking the Future of Plastics & Catalysing Action.
3. Zhang, X., Fevre, M., Jones, G.O. and Waymouth, R.M. (2018) *Chemical Reviews*, **118**, 839–885.
4. Baeyer, A. and Villiger, V. (1899) *Berichte der deutschen chemischen Gesellschaft*, **32**, 3625–3633.
5. Quilter, H.C., Hutchby, M., Davidson, M.G. and Jones, M.D. (2017) *Polymer Chemistry*, **8**, 833–837.
6. (a) Balke, K., Bäumgen, M. and Bornscheuer, U.T. (2017) *ChemBioChem*, **18**, 1627–1638; (b) Balke, K., Schmidt, S., Genz, M. and Bornscheuer, U.T. (2016) *ACS Chemical Biology*, **11**, 38–43; (c) Schmidt, S., Scherkus, C., Muschiol, J., *et al.* (2015) *Angewandte Chemie International Edition*, **54**, 2784–2787.
7. (a) Brown, H.C. and Ramachandran, P.V.A. (1992) *Accounts of Chemical Research*, **25**, 16–24; (b) Eryılmaz, S., Gül, M., İnkaya, E. & Taş, M. (2016) *Journal of Molecular Structure*, **1108**, 209–222; (c) Wróblewska, A., Miądlicki,P., Tołpa, J., *et al.* (2018) *Microporous and Mesoporous Materials*, **258**, 72–82.
8. Horn, E.J., Rosen, B.R., Chen, Y., *et al.* (2016) *Nature*, **533**, 77–81.
9. Stamm, A., Biundo, A., Schmidt, B., *et al.* (2019) *ChemBioChem*, **20**, 1664–1671.

8.9 Enzymatic Preparation of Halogenated Hydroxyquinolines

Fuchao Xu and Jixun Zhan*
Department of Biological Engineering, Utah State University, Logan, UT, USA

Quinolines, hydroxyquinolines and chloroquinolines are commonly present in bioactive compounds such as Camptothecin, Chloroquine and Montelukast. Preparation of halogenated compounds has mainly relied on chemical methods. Chemical halogenation typically requires harsh reaction conditions, including high temperatures/pressures and toxic chemical reagents. Furthermore, chemical halogenation lacks selectivity, which often results in unspecific halogenation and undesired byproducts, making product purification challenging [1]. Enzymatic halogenation represents a great alternative that allows specific and efficient halogenation of target substrates.

Rdc2 is a flavin-dependent halogenase from *Pochonia chlamydosporia* that has shown relaxed substrate specificity [2]. Through the introduction of a His_6-tag to both the *N*- and *C*-termini, the isolation yield of Rdc2 from *Escherichia coli* using Ni-NTA affinity chromatography was increased by threefold. *In vitro* reaction of Rdc2 and a flavin reductase (Fre) with different hydroxyquinolines revealed that 3-hydroxyquinoline, 5-hydroxyquinoline, 6-hydroxyquinoline and 7-hydroxyquinoline can be specifically halogenated (Scheme 8.11). These products were prepared by incubating the corresponding substrates with *E. coli* BL21(DE3)/Rdc2. After purification, the halogenated products were characterised as 4-chloro-3-hydroxyquinoline, 6-chloro-5-hydroxyquinoline, 5-chloro-6-hydroxyquinoline and 8-chloro-7-hydroxy by nuclear magnetic resonance (NMR) and mass spectrometry (MS) analyses [1]. This work represents the first enzymatic preparation of chlorohydroxyquinolines and provides a green method for the synthesis of this group of medicinally important compounds.

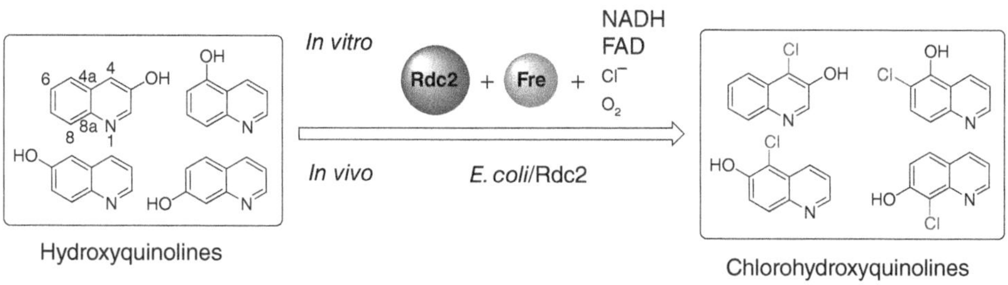

Scheme 8.11 *In vitro and in vivo enzymatic chlorination of hydroxyquinolines by Rdc2.*

8.9.1 Procedure 1: Construction of the Expression Plasmid for Rdc2 with Both *N*-Terminal and *C*-Terminal His_6-Tags

8.9.1.1 Materials and Equipment

- pZJ54 (pET28a-*rdc2,* previously constructed) or cDNA of *P. chlamydosporia* [2]
- CloneJET polymerase chain reaction (PCR) cloning kit (cloning vector, Thermo Fisher Scientific)
- Phusion high-fidelity DNA polymerase (New England Biolabs)
- *E. coli* XL-1 Blue
- Lysogeny broth (LB) medium
- Agar
- Restriction enzymes (*Nde*I and *Hind*III, New England Biolabs)
- T4 DNA ligase (New England Biolabs)
- Gel purification kit
- Plasmid miniprep kit
- Ampicillin
- Kanamycin
- PCR
- DNA electrophoresis system
- Gel imaging system
- Incubator
- Shaker

8.9.1.2 Procedure

1. The intron-free *rdc2* gene was amplified from the previously constructed pZJ54 by PCR with Phusion High Fidelity DNA Polymerase using a pair of specific primers, 5'-aaCATATGTCGGTACCCAAGTCTTG-3' (the *Nde*I site is underlined) and 5'-aaAAGCTTAACTTTGTTGAGGCCAA-3' (the *Hind*III site is underlined). The PCR components and conditions are described in Tables 8.6 and 8.7, respectively.
2. The PCR product was gel-purified and ligated into the cloning vector pJET1.2 with T4 DNA ligase by following the instructions of the CloneJET PCR cloning kit. The ligation mixture was transferred into *E. coli* XL-1 Blue competent cells by heat shock [3] and correct clones were selected on LB agar with 50 μg.mL^{-1} ampicillin. Colonies were picked for plasmid extraction. The correct plasmid was confirmed by digestion check with *Nde*I and *Hind*III and named as pFC55.

Table 8.6 *Reaction components for PCR amplification of rdc2.*

Component	20 μL reaction	Final concentration
10× Standard Phusion reaction buffer	2.0 μL	1×
10 mM dNTPs	0.4 μL	200 μM
Dimethyl sulfoxide (DMSO)	0.4 μL	
Forward primer	0.2 μL	0.2 μM
Reverse primer	0.2 μL	0.2 μM
pZJ54	0.2 μL	<1000 ng
Phusion high-fidelity DNA polymerase	0.2 μL	
Nuclease-free water	16.4 μL	

Table 8.7 *PCR conditions for amplification of rdc2.*

Step	Temp	Time
Initial denaturation	95 °C	5 min
30 cycles	95 °C	40 sec
	68 °C	40 sec
	72 °C	1.5 min
Final extension	72 °C	10 min
Hold	4 °C	

3. The *rdc2* gene was then excised from pFC55 with *Nde*I and *Hind*III and ligated into pET28a between the same sites using T4 DNA ligase. The ligation product was introduced into *E. coli* XL-1 Blue. Correct clones were selected on LB agar with 50 μg.mL^{-1} kanamycin and the plasmids were checked by digestion with *Nde*I and *Hind*III. The correct plasmid was named as pFC56.

8.9.2 Procedure 2: Expression and Purification of Rdc2 and Fre from *E. coli*

8.9.2.1 Materials and Equipment

- pFC56 (pET28a-*rdc2*)
- pZJ62 (pET28a-*fre*, previously constructed) [2]
- *E. coli* BL21(DE3)
- *E. coli* BL21-CodonPlus (DE3)-RIL
- LB medium
- Kanamycin
- Chloramphenicol
- Isopropyl-β-D-1-thiogalactopyranoside (IPTG)
- Ni-NTA resin
- Buffer A (100 mM Tris-HCl, 2.5 mM EDTA, 0.5 M NaCl, pH 7.9)
- MilliporeSigma™ Amicon™ Ultra Centrifugal Filter Units
- Water bath
- Incubator shaker
- Fridge
- Ultrasonic cell disruptor
- Refrigerated centrifuge

8.9.2.2 Procedure

1. The plasmid pFC56 was introduced into *E. coli* BL21(DE3) through a heat-shock transformation method for Rdc2 expression [3]. Similarly, pZJ62 was introduced into *E. coli* BL21-CodonPlus (DE3)-RIL for the expression of the flavin reductase Fre.
2. *E. coli* BL21(DE3)/pFC56 was grown overnight at 37 °C in 5 mL of LB medium supplemented with 50 μg.mL^{-1} kanamycin as the seed. Similarly, *E. coli* BL21-CodonPlus (DE3)-RIL/pZJ62 was grown overnight at 37 °C in 5 mL of LB medium supplemented with 25 μg.mL^{-1} chloramphenicol and 50 μg.mL^{-1} kanamycin.
3. 1 mL of each seed culture was inoculated into a 2 L flask containing 500 mL of LB medium with appropriate antibiotics. The flask was then placed in a shaker incubator at 37 °C and 250 rpm for 3–4 hr until an OD_{600} of 0.4 was reached. The cultures were cooled to 25 °C and protein expression was induced by adding 200 μM IPTG. The induced cultures were maintained at 25 °C and 250 rpm for an additional 18 hr.
4. The cells were harvested by centrifugation at 4500× g and 4 °C for 10 min, and then resuspended in 20 mL of lysis buffer by vortexing. The cells were disrupted by sonication on ice at 20 W for 5 min.
5. The cell lysates were centrifuged at 18 000 rpm for 30 min and the supernatants were incubated with 1.0 g of Ni-NTA resin for 3 hr. The purification was conducted by washing the resin with increasing concentrations of imidazole in Buffer A. The target proteins were eluted with Buffer A containing 250 mM imidazole.
6. Purified proteins were concentrated and the buffer was exchanged with MilliporeSigma™ Amicon™ Ultra Centrifugal Filter Units by centrifugation. The yields of Rdc2 and Fre were 18.5 and 10.3 mg.L^{-1}, respectively. Both enzymes were stored in 50% glycerol at −20 °C.

8.9.3 Procedure 3: *In Vitro* Chlorination of Hydroxyquinolines

8.9.3.1 Materials and Equipment

- Purified Rdc2 in 50% glycerol
- Purified Fre in 50% glycerol
- NADH
- FAD
- NaCl
- 3-Hydroxyquinoline, 5-hydroxyquinoline, 6-hydroxyquinoline and 7-hydroxyquinoline (15 mg of each)
- Phosphate buffer (pH 7.0)
- Agilent 1200 high-performance liquid chromatography (HPLC)
- Agilent 6130 liquid chromatography–mass spectrometry (LCMS)
- Water (HPLC-grade)
- Acetonitrile (HPLC-grade)
- Formic acid
- Eclipse Plus C18 reversed-phase analytical column (5 μm, 4.6×150 mm)
- Centrifuge

8.9.3.2 Procedure

1. The halogenation assay mixture (100 μL) consisted of 100 μM FAD, 10 mM NADH, 10 mM NaCl, 0.1 mM substrate, 20 μM Fre and 20 μM Rdc2 (enzymes from

Procedure 2) in 100 mM phosphate buffer, pH 7.0. Negative control had the same components except that Rdc2 was boiled.
2. The reaction mixtures were incubated at 30 °C for 2 hr and then quenched with 50 µL of methanol.
3. The mixtures were centrifuged at 15 000 rpm for 5 min to remove the precipitated proteins.
4. All samples were analysed on an Agilent 1200 HPLC coupled with an Agilent 6130 Single Quad LCMS using an Eclipse Plus C_{18} reversed-phase analytical column (5 µm, 4.6×150 mm) at 250 nm, with a flow rate of 1 mL.min^{-1}. A gradient of acetonitrile–water (5–50%) containing 0.1% formic acid was programmed over 30 min.

8.9.4 Procedure 4: *In Vivo* Chlorination of Hydroxyquinolines

8.9.4.1 Materials and Equipment

- *E. coli* BL21(DE3)/pFC56
- Kanamycin
- LB medium
- 3-Hydroxyquinoline (15 mg), 5-hydroxyquinoline (15 mg), 6-hydroxyquinoline (15 mg) and 7-hydroxyquinoline (15 mg)
- Centrifugation equipment
- Methanol
- Ethyl acetate
- Shaker incubator
- Separatory funnel
- Rotary evaporator
- Bruker AvanceIII HD Ascend-500 NMR instrument
- Methanol-d_4

8.9.4.2 Procedure

1. *E. coli* BL21(DE3)/pFC56 was grown in LB medium (500 mL) with 50 µg.mL^{-1} kanamycin and induced with 200 µM IPTG as described in Procedure 2.
2. After induction for 3 hr, substrate (15 mg, 0.103 mmol) dissolved in methanol (200 µL) was added. The cultures with different substrates were maintained at 28 °C and 250 rpm for an additional 48 hr.
3. Each biotransformation broth was centrifuged at 4500× g for 5 min to harvest the supernatant and pellet. The resulting pellet was extracted with methanol (100 mL) and the supernatant with ethyl acetate (3 × 500 mL).
4. The extracts were combined and the solvents were evaporated under reduced pressure.
5. The residue was dissolved in methanol and purified on an Agilent 1200 HPLC using the same HPLC conditions described in Procedure 3.
6. After purification, 4.0 mg of 4-chloro-3-hydroxyquinoline, 3.2 mg of 6-chloro-5-hydroxyquinoline, 4.5 mg of 5-chloro-6-hydroxyquinoline and 4.3 mg of 8-chloro-7-hydroxy were isolated in pure form.

7. The purified products were dissolved in methanol-d_4 and the NMR spectra were acquired on a Bruker AvanceIII HD Ascend-500 NMR instrument. The chlorination positions for each product were determined based on the 1D and 2D NMR spectra.

4-Chloro-3-hydroxyquinoline: ^{1}H NMR (CD_3OD, 500 MHz): δ 8.62 (1H, s, H2), 8.17 (1H, d, J = 7.9 Hz, H5), 7.99 (1H, d, J = 7.9 Hz, H8), 7.63–7.70 (2H, m, H6 and H7); ^{13}C NMR (CD_3OD, 125 MHz): 148.2 (C3), δ 142.5 (CH2), 128.2 (CH8), 127.9 (CH6), 126.9 (CH7), 122.5 (CH5), 107.3 (C4a), 102.9 (C4), 102.2 (C8a).

6-Chloro-5-hydroxyquinoline: ^{1}H NMR (CD_3OD, 500 MHz): δ 8.83 (1H, d, J = 3.0 Hz, H2), 8.72 (1H, d, J = 8.5 Hz, H4), 7.68 (1H, d, J = 9.0 Hz, H8), 7.56 (1H, dd, J = 8.5, 4.0 Hz, H3), 7.55 (1H, d, J = 9.0 Hz, H7); ^{13}C NMR (CD_3OD, 125 MHz): δ 156.4 (C5), 149.9 (CH2), 143.8 (C8a), 131.7 (CH4), 130.7 (CH8), 126.5 (C4a), 120.6 (CH3), 119.8 (CH7), 94.8 (C6).

5-Chloro-6-hydroxyquinoline: ^{1}H NMR (CD_3OD, 500 MHz): δ 8.80 (1H, d, J = 4.4 Hz, H2), 8.49 (1H, d, J = 8.4 Hz, H4), 7.94 (1H, d, J = 9.0 Hz, H8), 7.59 (1H, dd, J = 8.4, 4.1 Hz, H3), 7.58 (1H, d, J = 9.0 Hz, H7); ^{13}C NMR (CD_3OD, 125 MHz): δ 150.1 (C6), 147.9 (CH2), 144.0 (C8a), 130.4 (CH4), 129.6 (CH8), 126.9 (C4a), 122.1 (CH3), 121.3 (CH7), 112.0 (C5).

8-Chloro-7-hydroxyquinoline: ^{1}H NMR (CD_3OD, 500 MHz): δ 8.84 (1H, d, J = 4.2 Hz, H2), 8.31 (1H, d, J = 8.1 Hz, H4), 7.79 (1H, d, J = 8.9 Hz, H5), 7.44 (1H, dd, J = 8.1, 4.4 Hz, H3), 7.34 (1H, d, J = 8.9 Hz, H6); ^{13}C NMR (CD_3OD, 125 MHz): δ 155.3 (C7), 150.2 (CH2), 145.2 (C8a), 137.3 (CH4), 127.3 (CH5), 124.1 (C4a), 119.2 (CH3), 118.8 (CH6), 113.8 (C8).

8.9.5 Conclusion

Halogenases are becoming a useful biocatalytic tool for the preparation of halogenated compounds in synthetic chemistry. Rdc2 is a fungal flavin-dependent halogenase that has shown promise in introducing chlorine/bromine atoms into a variety of compounds. The purification of Rdc2 in *E. coli* was improved by the procedures described in this section. Selective halogenation of hydroxyquinolines was accomplished using Rdc2 both *in vitro* an *in vivo*. It is expected that this enzyme can be used to halogenate more hydroxyquinolines via a similar approach.

References

1. Xu, F., Merkley, A., Yu, D. and Zhan, J. (2016) *Tetrahedron Letters*, **57** (47), 5262–5265.
2. Zeng, J. and Zhan, J. (2010) *ChemBioChem*, **11** (15), 2119–2123.
3. Froger, A. and Hall, J. (2007) *Journal of Visualised Experiments*, **6**, e253.

9

Hydrolytic and Dehydratase Enzymes

9.1 Synthesis of (*S*)-3-(4-Chlorophenyl)-4-Cyanobutanoic Acid by a Mutant Nitrilase

Shanshan Yu, Peiyuan Yao, Qiaqing Wu and Dunming Zhu
National Engineering Laboratory for Industrial Enzymes, Tianjin Engineering Research Center of Biocatalytic Technology, Tianjin Institute of Industrial Biotechnology, Chinese Academy of Sciences, Tianjin, People's Republic of China

Desymmetrisation of prochiral 3-(4-chlorophenyl) glutaronitrile **1** to optically active (*S*)-3-(4-chlorophenyl)-4-cyanobutanoic acid **2** offers an attractive approach to the marketed drug (*R*)-Baclofen, which is used as a muscle relaxant and an antispastic agent [1]. We have screened a number of wild-type nitrilases, but their catalytic efficiency and stereoselectivity towards **1** did not meet industrial requirements [2]. For this reason, we established an *in vivo* enzymatic reaction performed by a triple mutant (*Ss*NIT-P194A/I201A/F202V) with high activity and stereoselectivity for the production of active **2**. This triple mutant comes from the *Synechocystis* sp. strain PCC 6803 (*Ss*NIT) and was engineered by directed evolution [3].

9.1.1 Procedure 1: Recombinant Expression of the Mutant *Ss*NIT-P194A/I201A/F202V in *Rosetta* 2(DE3)pLySs

9.1.1.1 Materials and Equipment

- Tryptone from casein (10 g)
- Yeast extract (5 g)
- NaCl (10 g)
- Distilled water (dH_2O)
- Ampicillin ($100\ mg.mL^{-1}$ in dH_2O, filter-sterilised)

Applied Biocatalysis: The Chemist's Enzyme Toolbox, First Edition. Edited by John Whittall and Peter W. Sutton.
© 2021 John Wiley & Sons Ltd. Published 2021 by John Wiley & Sons Ltd.

- *Rosetta* 2(DE3)pLySs harbouring pET15b(+)-*Ss*NIT-P194A/I201A/F202V [3]
- Isopropyl-β-D-1-thiogalactopyranoside (IPTG; 1 M in dH_2O, filter-sterilised)
- Lysogenic broth (LB) agar plate
- 0.1 L Erlenmeyer flask with seal film
- 2 L Erlenmeyer flasks with seal film
- Vertical autoclave (YXQ-LS-75S II)
- Cooling centrifuge (5804R)

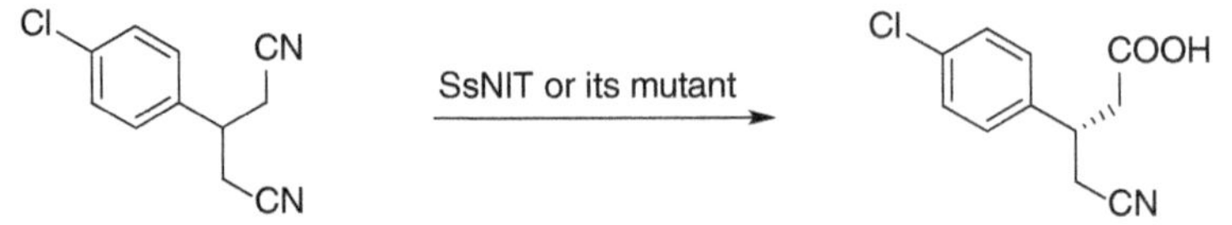

3-(4-chlorophenyl) glutaronitrile
1

(*S*)-3-(4-chlorophenyl)-4-cyanobutanoic acid
2

Scheme 9.1 *Conversion of 3-(4-chlorophenyl) glutaronitrile **1** into optically active (S)-3-(4-chlorophenyl)-4-cyanobutanoic acid **2**.*

9.1.1.2 Procedure

1. Peptone (0.2 g), yeast extract (0.1 g) and NaCl (0.2 g) were dissolved in dH_2O (20 mL) and autoclaved (20 min, 121 °C) in a 0.1 L Erlenmeyer flask to give sterile LB medium.
2. To prepare the pre-culture, ampicillin stock solution (20 μL) was added to a final concentration of 100 μg.mL^{-1}. The solution was inoculated with a single colony of *Rosetta* 2(DE3)pLySs harbouring pET15b(+)-*Ss*NIT-P194A/I201A/F202V and shaken at 37 °C and 200 rpm overnight.
3. The next day, LB medium (800 mL) supplemented with ampicillin (0.8 mL) stock solution to a final concentration of 100 μg.mL^{-1} (prepared as in Steps 1 and 2) contained in 2 L Erlenmeyer flasks with seal film was evenly inoculated from the LB pre-culture to give an OD_{600} of 0.1.
4. The cells were grown at 37 °C and 200 rpm until an OD_{600} of 0.8 was reached. Then the cultivation was cooled to 30 °C and expression of the mutant enzyme was induced by the addition of IPTG stock solution (80 μL, final conc. 0.1 mM). The expression was performed at 30 °C for 10 hr.
5. The final OD_{600} (3.5–4.5) was measured and the cells were harvested by centrifugation at 6000 rpm and 4 °C for 15 min. The supernatant was discarded and the cells were handled further as described in Procedure 2. 2.6 g cell paste was obtained.

9.1.2 Procedure 2: Preparation of Resting Cells

9.1.2.1 Materials and Equipment

- NaCl (0.9 g)
- $Na_2HPO_4 \cdot 12\ H_2O$ (2.75 g)
- $NaH_2PO_4 \cdot 2\ H_2O$ (0.54 g)
- pH meter

- dH_2O
- 0.25 L Erlenmeyer flasks with seal film

9.1.2.2 Procedure

1. NaCl (0.9 g) was dissolved in dH_2O and the volume was adjusted to 100 mL to give a saline solution.
2. $Na_2HPO_4 \cdot 12\ H_2O$ (2.75 g) and $NaH_2PO_4 \cdot 2\ H_2O$ (0.54g) were dissolved in dH_2O and the volume was adjusted to 100 mL to give sodium phosphate buffer.
3. The cells from the enzyme expression were gently resuspended in saline solution (100 mL) and harvested by centrifugation at 6000 rpm and 4 °C for 15 min to wash once. 2.6 g of cell paste was obtained.
4. 2.5 g of wet whole cells were gently resuspended in 47.5 mL sodium phosphate buffer.

9.1.3 Procedure 3: Biocatalytic Conversion of 3-(4-Chlorophenyl)glutaronitrile into (*S*)-3-(4-Chlorophenyl)-4-Cyanobutanoic Acid

9.1.3.1 Materials and Equipment

- 3-(4-Chlorophenyl)glutaronitrile (1.023 g in 2.5 mL dimethyl sulfoxide, DMSO)
- Ethyl acetate (EtOAc)
- Aqueous HCl (6 M)
- dH_2O
- Na_2SO_4, dry
- Methanol
- Dichloromethane
- Isopropanol
- Hexane
- Trifluoroacetic acid
- Resting cell suspension from Procedure 2
- 1 L separating funnel
- 7× 0.25 L Erlenmeyer flasks with seal film
- Orbital shaker (ZWY-211C)
- Silica gel chromatography
- Agilent 1200 series high-performance liquid chromatograph (HPLC)
- CHIRALPAK OD-H column (5 μm, 4.6 × 250 mm)
- UV detector at 210 nm
- Cooling centrifuge (5804 R)
- Rotary evaporator
- Thin-layer chromatography (TLC) Silica Gel 60 F254 aluminium plates
- UV detector with 254 nm capability
- Differential Scanning Calorimeter 250 (DSC 250)

9.1.3.2 Procedure

1. 3-(4-Chlorophenyl)glutaronitrile was chemically synthesised as recently reported [2].
2. The resting cell suspension obtained in Procedure 2 was placed in a 0.25 L Erlenmeyer flask with seal film.

3. 3-(4-Chlorophenyl)glutaronitrile (1.023 g in 2.5 mL DMSO) was added to the resting cells (final conc. 20.46 mg.mL^{-1}). The reaction was performed at 30 °C and 200 rpm in an orbital shaker.
4. For reaction monitoring by chiral HPLC, the cell suspension (100 μL) was quenched by 6 M HCl and extracted with ethyl acetate (EtOAc; 300 μL). After drying by anhydrous sodium sulfate, EtOAc was removed under reduced pressure and the sample was dissolved in the solvent mixture of the mobile phase. Samples were analysed by chiral HPLC with a CHIRALPAK OD-H column (see Analytical Method).
5. On 99% conversion as determined via HPLC, the cells were centrifuged at 7000 rpm and 4 °C for 10 min. The aqueous layer was set aside and the cells were diluted with 50 mL water and washed twice, and the aqueous layer was collected.
6. The combined aqueous layers were acidified with 6.0 M HCl to pH 1–2 and extracted with EtOAc (2 × 50 mL).
7. The combined organic layers were dried over Na_2SO_4 and filtered, and the solvent was removed under reduced pressure.
8. The crude product was purified by chromatography over silica using methanol/dichloromethane 2 : 98 as eluent. Fractions were collected and checked for the desired substances via TLC. Fractions containing the desired product were combined and evaporated by distillation under reduced pressure.
9. The desired (*S*)-3-(4-chlorophenyl)-4-cyanobutanoic acid **2** was obtained as a white solid in 1.017 g (91% yield, >99% ee; M.pt was 63.7 °C as detected by DSC 250).

9.1.4 Analytical Method

The methods and results of HPLC and TLC analysis are given in Tables 9.1–9.4.

Table 9.1 *HPLC method.*

Eluent	Isopropanol, hexane and trifluoroacetic acid (30 : 70 : 0.1, v/v/v)
Oven column temperature	30 °C
Flow rate	0.5 mL.min^{-1}
UV detector	210 nm

Table 9.2 *Retention times for HPLC analysis.*

Substance	Retention (min)
3-(4-Chlorophenyl)glutaronitrile	36.4
(*S*)-3-(4-Chlorophenyl)-4-cyanobutanoic acid	12.0

Table 9.3 *TLC method.*

Solvent	Methanol/dichloromethane 5 : 95
UV detector	254 nm

Table 9.4 *Retention factors for TLC analysis.*

Substance	Retention factor (Rf)
3-(4-Chlorophenyl)glutaronitrile	2/3
(*S*)-3-(4-Chlorophenyl)-4-cyanobutanoic acid	1/3

9.1.5 Conclusion

The described procedure enabled the regio- and stereoselective production of (*S*)-3-(4-chlorophenyl)-4-cyanobutanoic acid in high yield using standard chemical and microbiological laboratory equipment and methods. In principle, this procedure can be adapted for related 3-substituted glutaronitrile substrates by the choice of suitable enzymes.

References

1. (a) Patel, R. and Dickenson, A.H. (2016) *Pharmacology Research & Perspectives*, **4**, e00205; (b) Frampton, J.E. (2014) *CNS Drugs*, **28**, 835–854.
2. Duan, Y., Yao, P., Ren, J. *et al.* (2014) *Science China Chemistry*, **57** (8), 1164–1171.
3. Yu, S., Yao, P., Li, J. *et al.* (2019) *Catalysis Science & Technology*, **9** (6), 1504–1510.

9.2 Nitrilase-Mediated Synthesis of a Hydroxyphenylacetic Acid Substrate via a Cyanohydrin Intermediate

Gareth J. Brown[1] and Thomas S. Moody[1,2]
[1]*Almac Sciences, Craigavon, County Armagh, UK*
[2]*Arran Chemical Company, Unit 1 Monksland Industrial Estate, Athlone, Co. Roscommon, Ireland*

Almac selectAZyme™ nitrilase technology was used to access hydroxyphenylacetic acid derivative **2** starting from cyanohydrin intermediate **1** (Scheme 9.2). This approach has previous literature precedent [1].

Initial screening was carried out with pre-formed cyanohydrin **1** to identify nitrilases with the desired activity and selectivity.

9.2.1 Procedure 1: Screening

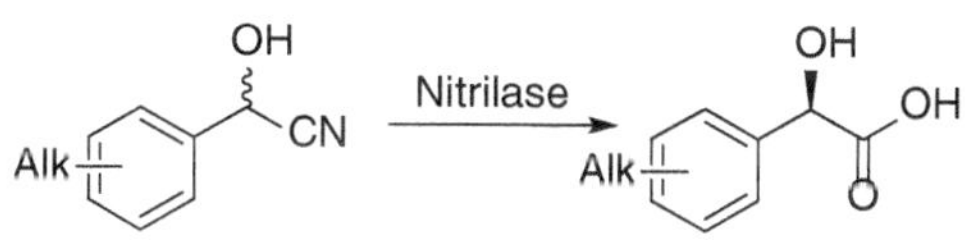

Scheme 9.2 *Nitrilase-mediated synthesis of hydroxyphenylacetic acid derivative **2**.*

9.2.1.1 Materials and Equipment

- Almac selectAZyme™ NITESK panel (lyophilised cell-free extracts; 10 mg)
- 96-well plate

- Na_2CO_3 buffer 0.1M, pH 9 (0.9 mL)
- Ethyl acetate (EtOAc; 0.1 mL per well)
- Pre-formed cyanohydrin **1** (10 mg) [2]
- 1 M aqueous HCl (1 mL)

9.2.1.2 Procedure

1. A 96-well plate containing Almac selectAZyme™ nitrilases (10 mg) was charged with 0.1 M Na_2CO_3 buffer, pH 9 (0.9 mL) in each well.
2. Cyanohydrin **1** (10 mg) in EtOAc (0.1 mL) was added.
3. The plate was sealed and shaken at 20 °C, 1000 rpm overnight.
4. The next day, the screening reactions were quenched with 1 M aqueous HCl (1 mL) and extracted into EtOAc (1 mL).
5. The layers were separated and the EtOAc phase was dried over $MgSO_4$.

9.2.1.3 Analytical Method

Determination of percentage conversion and enantioselectivity for screening reactions was achieved by HPLC analysis using the method detailed in Table 9.5.

Table 9.5 *HPLC conditions for assessing Nitrilase screening reactions.*

Column	CHIRALPAK IC-3, 4.6 x 250 mm, 3 μm
Flow rate	1 mL.min^{-1}
Eluent	Hexane with 0.1% TFA: 80%; ethyl acetate with 0.1% TFA: 20%
Injection volume	10.0 μL
Temperature	20 °C
Detector	PDA
Wavelength	275 nm
Run time	20 min

9.2.2 Procedure 2: Preparative Hydrolysis (Preparative-Scale Production of Hydroxyphenylacetic Acid Derivative)

9.2.2.1 Materials and Equipment

- Cyanohydrin **1** (4.8 g)
- Nitrilase crude lyophilised cell-free extract
- EtOAc
- Na_2CO_3 buffer 0.1 M, pH 8.8
- 100 mL round-bottom flask
- Stirrer bar
- pH probe
- 1 M HCl solution
- Toluene
- Separating funnel
- Rotary evaporator

9.2.2.2 Procedure

1. Buffer (0.1 M Na_2CO_3 buffer, pH 8.8, 36 mL) was charged to the 100 mL jacketed vessel containing a stirrer bar and equipped with pH probe (pH maintained by addition of 20% Na_2CO_3 solution as required).
2. Nitrilase crude lyophilised cell-free extract (0.12 g) was added to the solution.
3. Cyanohydrin (4.8 g) in EtOAc (4 mL) was added to the vigorously stirred solution.
4. The reaction was stirred at 20 °C for 24 hr until deemed complete by HPLC analysis.
5. The reaction mixture was extracted with EtOAc (2 × 20 mL) and the organic layer was set aside.
6. The aqueous layer was adjusted to pH 4 with 1 M HCl solution.
7. The aqueous layer was extracted with EtOAc (3 × 20 mL).
8. The combined organics were dried over $MgSO_4$, filtered and concentrated *in vacuo* to leave a solid.
9. The crude solid was recrystallised from toluene to give the product an overall 80% yield with 98% ee as per the method in Table 9.5.

9.2.3 Conclusion

Almac selectAZyme™ nitrilase technology has been applied for the enantioselective synthesis of a 2-hydroxyphenylacetic acid derivative **2** from the cyanohydrin intermediate **1** in high yield and enantioselectivity.

Almac has a wide range of nitrilases (~200) in stock that can be used for the hydrolysis of nitriles to carboxylic acids under mild conditions (neutral buffer, 20–30 °C). The advantages of the nitrilase approach are in the avoidance of the harsh acidic/basic conditions and high temperatures often required to perform such transformations chemically and the potential for resolution of racemic substrates.

References

1. (a) Liu, Z.Q., Zhang, X.H., Xue, Y.P. *et al.* (2014) *Journal of Agricultural and Food Chemistry*, **62**, 4685–4694; (b) Kamila, S., Zhu, D. and Biehl, E.R. (2006) *Organic Letters*, **8**, 4429–4431.
2. Ridge, D.N., Hanifin, J.W., Harten, L.A. *et al.* (1979) *Journal of Medicinal Chemistry*, **22**, 1385–1389.

9.3 Production of (*R*)-2-Butyl-2-Ethyloxirane Using an Epoxide Hydrolase from *Agromyces mediolanus*

Gheorghe-Doru Roiban,*[1] Peter W. Sutton,[2,6] Andrew Fosberry,[3] Christopher Morgan,[4] and Douglas Fuerst[5]

[1]*Synthetic Biochemistry, Medicinal Science & Technology, GlaxoSmithKline, Stevenage, Hertfordshire, UK*
[2]*Chemical Development, Medicinal Science & Technology, GlaxoSmithKline, Hertfordshire, UK*
[3]*Medicinal Science & Technology, GlaxoSmithKline, Stevenage, Hertfordshire, UK*
[4]*Chemical Development, Medicinal Science & Technology, GlaxoSmithKline, King of Prussia, PA, USA*
[5]*Synthetic Biochemistry, Medicinal Science and Technology, GlaxoSmithKline, King of Prussia, PA, USA*
[6]*Current address: Glycoscience SL, Bizkaia, Spain*

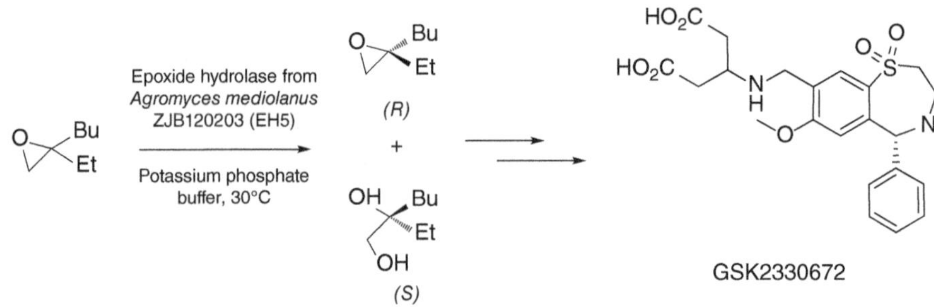

Scheme 9.3 *Epoxide hydrolase (EH) from* Agromyces mediolanus *catalysed enantioselective ring opening of racemic 2-butyl-2-ethyloxirane.*

(*R*)-2-Butyl-2-ethyloxirane is a valuable chiral synthon used for the preparation of GSK2330672, an ileal bile acid transport (iBAT) inhibitor indicated for diabetes type II and cholestatic pruritus [1]. The active pharmaceutical ingredient (API) contains two stereogenic centres and the previous medicinal chemistry route used classical resolution to access an enantiopure activated aminoalcohol or aziridine intermediate. To avoid the need for the wasteful classical resolution, we wanted to access the enantioenriched epoxide (*R*)-2-butyl-2-ethyloxirane starting from racemic 2-butyl-2-ethyloxirane. After an extensive search and reaction optimisation, epoxide hydrolase (EH5) from *Agromyces mediolanus* was identified as a suitable biocatalyst (Scheme 9.3) [2].

9.3.1 Procedure 1: Recombinant Expression of Epoxide Hydrolase EH5 in *E. coli* BL21(DE3)

9.3.1.1 Materials and Equipment

- Super optimal broth with catabolite repression (SOC) medium (150 μL)
- BL21(DE3) competent cells
- Lysogenic broth (LB) agar Petri plates + glucose (1%)
- Glycerol (80% w/v)
- Kanamycin (50 mg.mL^{-1})
- LB medium (tryptone (Bacto) 10.0 g.L^{-1}, yeast extract (Bacto) 5.0 g.L^{-1}, NaCl 5.0 g.L^{-1})
- Terrific broth (TB) (tryptone 12.0 g.L^{-1}, yeast extract 24.0 g.L^{-1}, KH_2PO_4 2.2 g.L^{-1}, K_2HPO_4 9.4 g.L^{-1})
- Antifoam (DC1520)
- Isopropyl-β-D-1-thiogalactopyranoside (IPTG; 0.5 M)
- Sterile spreader
- Incubator
- Water bath
- Thermomixer
- 500 mL Corning Erlenmeyer flask
- 50 L Pierre Guerin Biolaftte bioreactor
- BPS VirTris SP Scientific Advantage Pro lyophiliser
- Kuhner Shaker Climo ISF1-X

- Sorvall RC 12 BP
- Spectrophotometer Cecil CE2041

9.3.1.2 Procedure

1. EH5 plasmid (2 μL) [2a] was added into an aliquot of BL21(DE3) competent cells and left on ice for 15 min.
2. The cells were heat-shocked at 42 °C in a water bath for 45 sec then left on ice for 2 min.
3. SOC medium (150 μL) was added and the cells were shaken in a thermomixer at 37 °C, 1400 rpm for 1 hr.
4. The culture was spread on an LB plate containing kanamycin (50 μg.mL^{-1}) and glucose (1%) with a sterile spreader.
5. The plate was incubated at 37 °C overnight.
6. A single EH5 colony was inoculated into LB medium (10 mL) containing kanamycin (final conc. 50 μg.mL^{-1}) and stirred at 37 °C for 5 hr in a Kuhner shaker (220 rpm).
7. Inoculum (750 μL) was added over glycerol (250 μL, 80% w/v) and the mixture was frozen at −80 °C for long-term storage.
8. A sterile Corning Erlenmeyer flask (500 mL) containing LB medium (100 mL) and kanamycin (final conc. 50 μg.mL^{-1}) was inoculated with glycerol stock EH5 cells (10 μL) and incubated at 37 °C overnight with shaking in a Kuhner shaker (200 rpm).
9. A secondary seed stage was inoculated with 1% of the primary seed into eight sterile Erlenmyer flasks (500 mL) containing LB medium (100 mL), kanamycin (final conc. 50 μg.mL^{-1}) and glucose (final conc. 1%) and incubated at 30 °C overnight in a Kuhner shaker (200 rpm).
10. Production fermentations were carried out in a 50 L Pierre Guerin Biolaftte bioreactor using TB medium (35 L), glycerol (1%), antifoam (DC1520; 2 mL.L^{-1}), kanamycin (50 μg.mL^{-1}) and inoculum (2%) from the secondary seed stage. The fermenter was incubated at 37 °C, stirred at 200 rpm with air supplied at 30 L.min^{-1} under a pressure of 0.2 bar. At inoculation, the culture was at OD_{600} 0.2.
11. The culture was incubated at 37 °C for ~90 min, with continuous OD monitoring until it reached OD_{600} 2.0.
12. The temperature was reduced to 25 °C, the medium was induced by the addition of IPTG in one shot (final conc. 0.5 mM) and the culture was incubated for a further 23 hr.
13. The broth was harvested by centrifugation using a Sorvall RC 12BP centrifuge (6427 RCF) for 20 min and pellets (total wet cells 0.9 kg) were frozen and stored at −80 °C.
14. The frozen cell paste (8.0 g) was lyophilised using a BPS VirTris SP Scientific Advantage Pro lyophiliser to provide a lyophilised whole-cell powder of EH5 (2.0 g), which was stored in a −80 °C freezer for use in the reaction.

9.3.2 Procedure 2: Biocatalytic Conversion to Synthesise (*R*)-2-Butyl-2-Ethyloxirane

9.3.2.1 Materials and Equipment

- Racemic 2-butyl-2-ethyloxirane (22.6 g, 27.2 mL, 0.832 g.mL^{-1}, 177 mmol) [2a]
- EH5 lyophilised whole cells from Procedure 1

- Ethyl acetate (EtOAc)
- Potassium phosphate buffer pH 7.4, 100 mM
- Celite grade 545
- Brine (NaCl-saturated solution)
- 250 mL round-bottom flask
- 250 mL controlled laboratory reactor (CLR)
- Buchner filter with Whatman wet strengthened filter paper 114
- Cooling centrifuge
- Agilent 7890B GC System with flame ionisation detector (FID)
- Chiral gas chromatography (GC) column (Agilent 112-6632, CycloSil-B, 35–260 °C (280 °C): 30 m × 250 μm ID × 0.25 μm film)
- −80 °C freezer
- Rotary evaporator
- Bruker Advance 400 (^{1}H: 400 MHz, ^{13}C: 101 MHz) spectrometer using tetramethylsilane (TMS) as internal standard

9.3.2.2 *Procedure*

1. CLR jacket temperature was set to 30 °C.
2. Potassium phosphate buffer pH 7.4, 100 mM (40.8 mL) was charged to the CLR.
3. Lyophilised whole cells containing EH5 were taken out of the −80 °C freezer and allowed to reach room temperature.
4. EH5 lyophilised whole cells (2.0 g) were charged to the CLR.
5. Time 0 min: Stirring was cautiously started initially at 100 rpm and then increased gradually until the lyophilised enzyme became fully suspended in the buffer solution.
6. Time 2 min: When the lyophilised whole-cell lumps were small enough not to destroy the impeller, stirring was increased to 350 rpm.
7. Time 8 min: Almost all lyophilised enzyme had been resuspended.
8. Time 10 min: Stirring was adjusted to 300 rpm.
9. The reaction was started by the addition of racemic 2-butyl-2-ethyloxirane (22.6 g, 177 mmol) to the CLR.
10. The reaction contents were stirred in the CLR at 30 ± 2 °C for up to 1 hr.
11. Time 1 hr: The reaction was sampled and analysed, and enantiomeric excess and conversion were determined.
12. The reaction was left overnight, after which further reaction points were taken. The reaction was considered to be completed after 15 hr, when chiral GC indicated an enantiomeric excess of (*R*)-2-butyl-2-ethyloxirane ≥95%; typically, conversion was around ≥62 ± 2%.
13. Quenching procedure: EtOAc (53.4 mL) was charged to the CLR, the stirring speed was adjusted to 300 rpm and the contents were stirred for 5 min.
14. The CLR contents were transferred to a Buchner filter with a Whatman wet strengthened filter paper 114 filled with wet Celite grade 545 (34.5 g, diameter 9 cm, depth 1.5 cm). Filtration of the enzyme took approximately 4 min.
15. Additional EtOAc (30 mL) was charged to the CLR and stirred for 2 min in order to clean it.
16. EtOAc was collected and used to wash the Buchner funnel involved in filtration.
17. Organic phases were combined prior to loading back into the cleaned CLR.

18. The aqueous and organic layers were separated for 3 min. Between the layers, a small layer of emulsion was noticed.
19. The aqueous layer was discharged to waste and organic phase was retained in the vessel.
20. Saturated brine solution (53.4 mL) was charged. The mixture was stirred for 3 min and let to stand until complete layer separation was achieved (5 min). Additional stirring at 300 rpm helped the organic layer to clarify.
21. The aqueous layer was discharged to waste and EtOAc was collected into a round-bottom flask for rota-evaporation.
22. Solvent was removed by distillation under reduced pressure to afford a neat mixture of the desired epoxide (*R*)-2-butyl-2-ethyloxirane and diol (*S*)-2-ethylhexane-1,2-diol.
23. The neat mixture was distilled at 90 °C under 20 ± 5 mbar to give the desired epoxide (*R*)-2-butyl-2-ethyloxirane (3.00 g, 13.2% yield, 97% ee).

^{1}H NMR (400 MHz, $CDCl_3$) δ 2.61 (d, J = 4.9 Hz, 1H), 2.59 (d, J = 4.9 Hz, 1H), 1.72–1.46 (m, 4H), 1.42–1.26 (m, 4H), 0.99–0.87 (m, 6H). ^{13}C NMR (125 MHz, $CDCl_3$) δ 60.2, 52.2, 33.7, 27.0, 26.9, 22.9, 14.0, 8.9.

9.3.3 Analytical Method

9.3.3.1 Sampling Procedure

Sampling was performed whilst stirring, as follows:

1. The reaction mixture (100 μL) was quenched with EtOAc (1 mL), shaken vigorously and centrifuged at 12 000 rpm and 4 °C for 2 min to remove the cell debris.
2. The organic layer (100 μL) prepared in Step 1 was sampled and diluted with EtOAc (1 mL).
3. The sample prepared in Step 2 was injected into chiral GC.
4. Enantiomeric excess and conversion were calculated.

9.3.3.2 GC Method

Initial 60 °C, 60–80 °C 3 °C.min^{-1} ramp; from 80 to 210 °C, 40 °C ramp; 1.4 bar H_2, flow 4.2 mL.min^{-1}. (*S*)-2-butyl-2-ethyloxirane elutes at 5.91 min, (*R*)-2-butyl-2-ethyloxirane elutes at 6.08 min.

9.3.4 Conclusion

The described procedure enabled the stereoselective synthesis of (*R*)-2-butyl-2-ethyloxirane at high substrate loadings (~330 g.L^{-1}) using a wild-type enzyme from *A. mediolanus*. Enzyme was easily expressed using a laboratory fermenter and the reaction was carried out in a standard controlled laboratory reactor. This protocol could be further optimised to provide increased yields of (*R*)-2-butyl-2-ethyloxirane.

References

1. Guo, J.L., Martin, M.T., Mitchell, M.B. and Zhou, X. (2016) Patent WO2016020785A1.
2. (a) Roiban, G.-D., Sutton, P.W., Splain, R. *et al.* (2017) *Organic Process Research & Development* **21**, 1302–1310; (b) Barcan, G., Guo, J., Morgan, C. *et al.* (2018) Patent WO2018/002827A.

9.4 Preparation of (*S*)-1,2-Dodecanediol by Lipase-Catalysed Methanolysis of Racemic Bisbutyrate Followed by Selective Crystallisation

Jaan Parve,*[1,2] Ly Villo,[2] Imre Vallikivi,[3] Indrek Reile,[4] Tõnis Pehk,[4] Riina Aav,[2] Lauri Vares[1] and Omar Parve[2]
[1]*Institute of Technology, University of Tartu, Tartu, Estonia*
[2]*Department of Chemistry and Biotechnology, Tallinn University of Technology, Tallinn, Estonia*
[3]*TBD-Biodiscovery, Tartu, Estonia*
[4]*National Institute of Chemical Physics and Biophysics, Tallinn, Estonia*

Enantiomeric long-chain 1,2-alkanediols are valuable chiral building blocks for the synthesis of biologically active compounds such as insect sex pheromones and acetogenins [1]. Such diols can also be incorporated into the molecular framework of chiral derivatising agents useful for the stereochemical analysis of chiral carboxylic acids. Such applications require 1,2-alkanediols of very high enantiomeric purity.

Enantiomeric 1,2-alkanediols have been prepared by chemical hydrolytic kinetic resolution of terminal epoxides using Jacobsen's catalyst [2], as well as by asymmetric dihydroxylation of olefins [1]. In some cases, lipase-catalysed kinetic resolution of racemic mixtures, based on stereoselective acyl transfer reactions, has allowed separation of 1,2-alkanediol enantiomers [3]. However, prior to development of the current method, there were no easily scalable procedures for the separation of long-chain 1,2-alkanediol enantiomers. The method described herein provides product with high purity and stereochemistry without the use of preparative chromatography [4].

This procedure involves a simple chemical butyrylation of racemic 1,2-dodecanediol **1**, followed by stereoselective methanolysis of the crude bisbutyrate **2** catalysed by *Candida antarctica* lipase B (CALB) and highly chemo- and stereoselective crystallisation of the target (*S*)-1,2-dodecanediol (*S*)-**1** from the 'doubly crude' product of the enzymatic reaction (Scheme 9.4) [4, 5]. The procedure is notable for the very high reliability of all three steps, allowing considerable modifications of the process regimes. The engineering and duration of the enzymatic reaction, as well as the mixing regime and temperature of crystallisation,

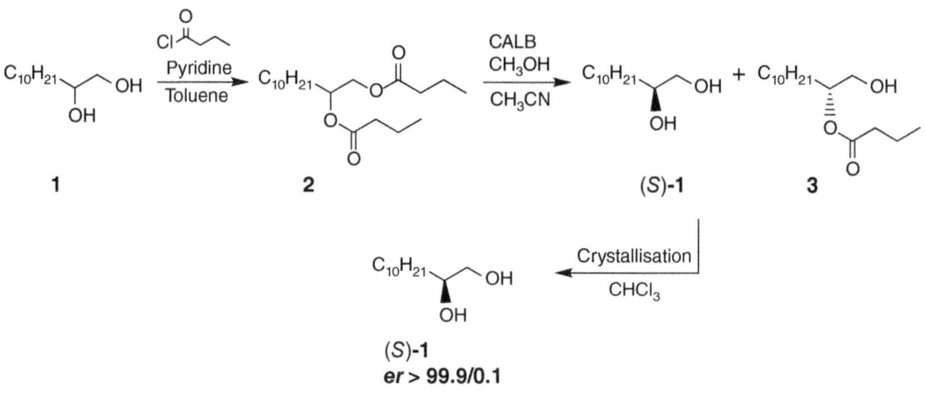

Scheme 9.4 *Procedure for the separation of enantiopure (S)-1,2-dodecanediol (S)-**1** from racemic mixture.*

can be extensively modified. For instance, biocatalyst may be used in a removable container [4], colloquially referred to as a 'teabag', and a reduced solvent volume may be employed to allow direct separation of the simultaneously crystallised target compound from the reaction mixture. In that case, however, the yield of product (*S*)-**1** will be lower compared with the current method. The present procedure has made high-quality (*S*)-1,2-dodecanediol (*S*)-**1** readily accessible for researchers and students.

9.4.1 Procedure 1: Synthesis of Racemic 1,2-Dodecanediol Bisbutyrate 2

9.4.1.1 Materials and Equipment

- Racemic 1,2-dodecanediol **1** (90% purity; 5.06 g, 25 mmol Sigma-Aldrich #213721, CAS No. 1119-87-5)
- Butyryl chloride (6.2 mL, 60 mmol)
- Pyridine (6.04 mL, 75 mmol)
- Toluene
- 2 × 250 mL round-bottomed flasks
- Magnetic stirrer
- Thin-layer chromatography (TLC) Silica Gel 60 F_{254} aluminium plates (Merck)
- Anisaldehyde dip reagent (150 mL EtOH, 7.0 mL conc H_2SO_4, 7.0 mL p-anisaldehyde) and heating device
- Methanol
- Petroleum ether (PE) 40–60 °C
- Ethyl acetate (EtOAc)
- TLC eluent (PE/EtOAc 10 : 3 v/v)
- Glass filter (Por. 3) with flask for vacuum filtration
- Filter aid, Hyflo®Super Cel® (diatomaceous earth, flux-calcined; Sigma-Aldrich #56678)
- Saturated $NaHCO_3/H_2O$ solution
- Brine
- 250 mL separating funnel
- 100 and 250 mL Erlenmeyer flasks
- Na_2SO_4, anhydrous
- Rotary evaporator

9.4.1.2 Procedure

1. In a 250 mL round-bottom flask, racemic 1,2-dodecanediol **1** (5.06 g, 25 mmol) was dissolved in a mixture of toluene (75 mL) and pyridine (6.04 mL, 75 mmol) on slight heating.
2. Butyryl chloride (6.2 mL, 60 mmol) was added dropwise with vigorous magnetic stirring.
3. After 20 min, when the reaction was complete by TLC, methanol (1.0 mL, 24.72 mmol) was added to the reaction mixture to react with any excess butyryl chloride. Stirring was continued for an additional 10 min. TLC: **2** Rf = 0.39 (eluent: PE/EtOAc 10 : 0.4).
4. The reaction mixture was filtered through a glass filter covered with filter aid and the reaction flask and filter cake were washed with toluene (25 mL).

5. The solution was transferred to a separating funnel and washed once with saturated aqueous $NaHCO_3$ solution (20 mL) and twice with brine (2 × 20 mL). The organic phase was dried over anhydrous Na_2SO_4 and filtered.
6. The solvent was evaporated under reduced pressure. Care had to be taken to remove as much residual pyridine as possible, as the presence of a base in the reaction mixture of Procedure 2 promotes intramolecular acyl migration in the monoester **3**, diminishing the 'apparent stereoselectivity' of the enzymatic reaction [5]. The oily crude product obtained (9.03 g, yield 105.5%) was used in the following enzymatic step without further purification.

Characteristics (measured for a purified sample) of bisbutyrate **2**: ^{1}H NMR (400 MHz, $CDCl_3$): δ 5.02 (1H, m, H2), 4.16 (1H, dd, J = 3.4 and 11.9 Hz, H1), 3.96 (1H, dd, J = 6.8 and 11.9 Hz, H1), 2.22 (4H, t, J = 7.4 Hz, Bu2), 1.58 (4H, m, Bu3), 1.49 (2H, m, H3), 1.20 (16H, m, H4–H11), 0.88 and 0.87 (2 × 3H, t, J = 7.4 Hz, Bu4), 0.81 (3H, t, J = 7.0 Hz, H12); ^{13}C NMR (101 MHz, $CDCl_3$): δ 173.4 and 173.2 (Bu1), 71.3 (C2), 64.9 (C1), 36.3 and 36.0 (Bu2), 31.9 (C10), 30.8 (C3), 29.6, 29.5, 29.4, 29.4, 29.3 (C5–C9), 25.1 (C4), 18.5 and 18.4 (Bu3), 14.1 (C12), 13.6 and 13.6 (Bu4); MS (m/z): 343, 255, 241, 211, 201, 184, 173, 166, 153, 144, 142, 138, 124, 110, 97, 82, 71; IR (neat, cm^{-1}): 1090, 1176, 1253,1381,1461, 1742, 2856, 2928.

9.4.2 Procedure 2: Lipase-Catalysed Methanolysis of *rac*-1,2-Dodecanediol Bisbutyrate 2 and Crystallisation of the Target (*S*)-1,2-Dodecanediol (*S*)-1

9.4.2.1 *Materials and Equipment*

- Crude racemic 1,2-dodecanediol bisbutyrate **2** (9.03 g, 25 mmol from Procedure 1)
- Acetonitrile (CH_3CN)
- Methanol
- *C. antarctica* lipase B immobilised on a polymer carrier, currently marketed as Novozym® 435 (CAS 9001-62-1; Novozymes A/S); enzyme used was from LOT 32760800, purchased from Strem Chemicals Inc. (06-3123, 25 g)
- Magnetic stirrer
- 250 and 100 mL flasks
- 2× flasks suitable for crystallisation
- TLC Silica Gel 60 F_{254} aluminium plates (Merck)
- Anisaldehyde dip reagent (150 mL EtOH, 7.0 mL conc H_2SO_4, 7.0 mL p-anisaldehyde) and heating device
- TLC eluent (PE/EtOAc 10 : 3 v/v)
- 4× glass filters (Por. 3) with flasks for vacuum filtration
- Filter aid, Hyflo®Super Cel® (diatomaceous earth, flux-calcined; Sigma-Aldrich #56678)
- Chloroform ($CHCl_3$)
- PE 40–60 °C

9.4.2.2 *Procedure*

1. Crude bisbutyrate **2** (9.03 g, 25 mmol) in a 250 mL round-bottom flask was dissolved in CH_3CN (96 mL). Methanol (4 mL) was added, followed by Novozym® 435 (1.7 g).

2. The reaction mixture was kept at 20 °C with slow magnetic stirring until only a trace of the starting material was detected by TLC (22 hr). TLC of the products: **1** Rf = 0.10; **3** Rf = 0.68 (eluent: PE/EtOAc 10 : 3).
3. The reaction mixture was filtered through a glass filter covered with filter aid and the solvent was evaporated under reduced pressure. Care had to be taken, because the diol could start to crystallise as the solvent volume decreased, generating multiple boiling centres and causing the solution to bump into the evaporator.
4. The crude product was dissolved in $CHCl_3$ (10 mL) on heating and the solution was kept at room temperature for 24 hr, allowing the target diol (*S*)-**1** to crystallise slowly, followed by 1 hr at +6 °C to increase the yield.
5. Resulting crystals were filtered off, washed sparingly with cold (−18 °C) PE (3 mL) and dried under reduced pressure yielding 1.2 g (yield: 23.7%) of enantiomerically pure diol (*S*)-**1** (*er* >99.9/0.1).
6. The mother liquor was evaporated and the residue was dissolved in a mixture of CH_3CN (48 mL) and methanol (2 mL) and allowed to incubate with Novozym® 435 (1.0 g) for an additional 22 hr.
7. After 22 hr, the enzyme was filtered off and the solvent was evaporated. The crystallisation procedure from $CHCl_3$ (5 mL) was repeated, yielding another 723 mg of (*S*)-**1** (*er* >99.9/0.1), bringing the overall yield to 38% (theoretical yield was ~45% – the declared chemical purity of the starting commercial racemic mixture was 90%).

Characteristics of the target (*S*)-**1**: ^{1}H NMR (800 MHz, $CDCl_3$, 30°C) δ 3.71 (1H, m, H2), 3.66 (1H, dd, *J* = 2.9, 11.2 Hz, H1), 3.44 (1H, dd, *J* = 7.8, 11.2 Hz, H1), 2.30 (2H, bs, OH), 1.43 (3H, m, 2 × H3, H4), 1.33-1.25 (15H, m, H4–H11), 0.88 (3H, t, *J* = 7.2 Hz, H12). ^{13}C NMR (201 MHz, $CDCl_3$, 30 °C) δ 72.33 (C2), 66.79 (C1), 33.10 (C3), 31.88 (C10), 29.64 (C5), 29.59 (C8), 29.58 (C7), 29.55 (C6), 29.32 (C9), 25.55 (C4), 22.67 (C11), 14.11 (C12). IR (KBr; cm^{-1}): 530, 582, 662, 720, 838, 871, 972, 992, 1006, 1032, 1053, 1073, 1105, 1135, 1311, 1470, 2850, 2919, 3240, 3323, 3478. Mp = 68–70 °C; $[\alpha]_D^{20}$ −14 (c 1.0, EtOH).

9.4.3 Procedure 3: Synthesis of (*S*)-1-Tosyl-1,2-Dodecanediol (*S*)-4 for HPLC Analysis

9.4.3.1 *Materials and Equipment*

- (*S*)-1,2-Dodecanediol (*S*)-**1** (20.2 mg, 0.1 mmol)
- Dichloromethane (CH_2Cl_2)
- Triethylamine (Et_3N)
- Tosyl chloride (21 mg, 0.11 mmol)
- Di-n-butyltin oxide (Bu_2SnO)
- 100 mL round-bottom flask
- TLC Silica Gel 60 F_{254} aluminium plates (Merck)
- Anisaldehyde dip reagent (150 mL EtOH, 7.0 mL conc H_2SO_4, 7.0 mL p-anisaldehyde) and heating device
- TLC eluent (PE/EtOAc 10 : 2 v/v)
- EtOAc
- PE 40–60 °C
- Saturated $NaHCO_3/H_2O$ solution

- Brine
- 250 mL separating funnel
- 100 and 250 mL Erlenmeyer flasks
- Na_2SO_4, anhydrous
- Glass filter (Por. 3) with flask (250 mL) for filtration
- Silica gel for column chromatography, 60 (Merck)
- Glass chromatography column and receptacles for fractions
- HPLC system
- HPLC column with chiral stationary phase: Daicel Chiralpak IA (0.46 × 25 cm)

9.4.3.2 Procedure

1. (*S*)-1,2-Dodecanediol (*S*)-**1** (20.2 mg, 0.1 mmol) was dissolved in dichloromethane (5 mL). Et_3N (17 μL, 0.122 mmol) and tosyl chloride (21 mg, 0.11 mmol) were added with stirring at room temperature, followed by a catalytic amount of Bu_2SnO (2.5 mg, 0.01 mmol).
2. After stirring for 24 hr at room temperature, EtOAc (50 mL) was added. The solution was transferred to a separating funnel, washed once with sat. $NaHCO_3$ (10 mL) and twice with brine (2 × 10 mL), then dried over anhydrous Na_2SO_4.
3. The solution was filtered and the solvent evaporated under reduced pressure. The resulting crude product was purified by flash chromatography over silica gel (eluent PE/EtOAc 10 : 1.6 v/v), affording (*S*)-1-tosyl-1,2-dodecanediol (*S*)-**4** (25.4 mg, 71.2% yield) (Scheme 9.5).

Characteristics measured for (*S*)-**4**: ^{1}H NMR (400 MHz, $CDCl_3$): δ 7.83 (2H, dm, $J = 8.3$ Hz, Tos-2,6), 7.38 (2H, dm, $J = 8.3$ Hz, Tos-3,5), 4.07 (1H, dd, $J = 2.7$ and 9.8 Hz, H1), 3.90 (1H, dd, $J = 7.1$ and 9.8 Hz, H1), 3.84 (1H, m, H2), 2.48 (3H, s, Tos-CH_3), 2.12 (1H, bs, 2-OH), 1.43 (3H, m, H3, H4), 1.28 (15H, m, H4, H5–H11), 0.90 (3H, t, $J = 2 \times 7.1$ Hz, H12). ^{13}C NMR (101 MHz, $CDCl_3$): δ 145.1 (Tos-4), 132.7 (Tos-1), 130.0 (Tos-3,5), 128.0 (Tos-2,6), 74.0 (C1), 69.5 (C2), 32.7 (C3), 31.9 (C10), 29.59, 29.55, 29.47, 29.45, 29.33 (C5–C9), 25.2 (C4), 22.7 (C11), 21.7 (Tos-CH_3), 14.1 (C12). MS (m/z): 357, 339, 326, 308, 263, 243, 215, 171, 157, 156, 140, 125, 111, 97, 92, 83, 71, 57, 55, 43, 41. IR (KBr, cm^{-1}): 444, 499, 530, 557, 592, 674, 721, 817, 845, 887, 902, 958, 1048, 1107, 1174, 1291, 1308, 1348, 1471, 1597, 2851, 2922, 3549. TLC: Rf = 0.34 (eluent: PE/EtOAc 5/1).

HPLC determination of the enantiomeric ratio of product (*S*)-**1** by analysis of the corresponding 1-tosylates (*S*)-**4** and (*R*)-**4** on a Daicel Chiralpak IA column; eluent 90/10 n-hexane/iPrOH; flow rate 1.0 mL.min^{-1}; detection UV 254 nm. Retention times of the 1,2-dodecanediol 1-tosylates: 8.69 min for (*R*)-**4**, 10.50 min for (*S*)-**4**.

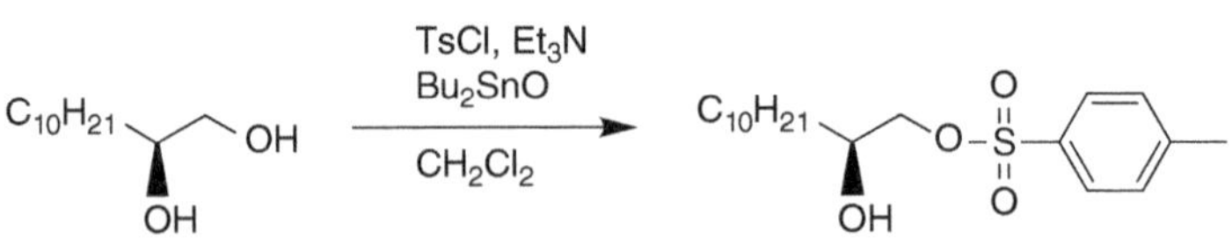

(*S*)-**1** (*S*)-**4**

Scheme 9.5 *Tosylation of product (S)-**1** for HPLC analysis [6].*

9.4.4 Conclusion

The described procedure allows the separation of crystalline (*S*)-1,2-dodecanediol (*S*)-**1** of very high chemical and stereochemical purity in high yield from a commercial racemic mixture of modest chemical purity. Lipase-catalysed kinetic resolution is used, followed by chemo- and stereoselective crystallisation of the target compound from the crude product dissolved in a limited amount of chloroform. The procedure is easy to scale up due to the technical simplicity and high reliability of all steps and the fact that no chromatographic separation is needed.

In addition to 1,2-dodecanediol, the described method has been successfully used for semipreparative separation of 1,2-octanediol enantiomers [5] and for the separation of all three stereoisomers of 1,2,7,8-octanetetrol (unpublished results). In principle, the method is suitable for the separation of stereoisomers of terminal vicinal alkanediol and alkanetetrol homologues that are amenable to crystallisation but whose corresponding perbutyrates remain soluble in acetonitrile.

References

1. Kolb, H.C., VanNieuwenhze, M.S. and Sharpless, K.B. (1994) *Chemical Reviews*, **94**, 2483–2547.
2. (a) Tokunaga, M., Larrow, J.F., Kakiuchi, F. and Jacobsen, E.N. (1997) *Science*, **277**, 936–938; (b) Chow, S. and Kitching, W. (2001) *ChemComm*, **2001**, 1040–1041.
3. (a) Poppe, L., Novák, L., Kajtár-Peredy, M. and Szántay, C. (1993) *Tetrahedron: Asymmetry*, **4**, 2211–2217; (b) Virsu, P., Liljeblad, A., Kanerva, A. and Kanerva, L. (2001) *Tetrahedron: Asymmetry*, **12**, 2447–2455.
4. Parve, O., Reile, I., Parve, J. *et al.* (2013) *Journal of Organic Chemistry*, **78**, 12 795–12 801.
5. Parve, J., Reile, I., Aid, T. *et al.* (2015) *Journal of Molecular Catalysis B: Enzymatic*, **116**, 60–69.
6. Martinelli, M.J., Nayyar, N.K., Moher, E.D. *et al.* (1999) *Organic Letters*, **1**, 447–450.

9.5 Biocatalytic Synthesis of *n*-Octanenitrile Using an Aldoxime Dehydratase from *Bacillus* sp. OxB-1

Alessa Hinzmann and Harald Gröger*
Faculty of Chemistry, Bielefeld University, Bielefeld, Germany

Nitriles represent a class of compounds in organic chemistry that have gained importance for a wide range of applications from different industrial segments such as pharmaceuticals and specialty and bulk chemicals [1]. For example, fatty acid-derived nitriles are used as intermediates in the production of fatty amines, which find use in many industrial applications, including as emulsifiers, corrosion inhibitors, dispersants and lubricating additives [1c, 2]. In general, a typical prerequisite for efficient and cost-attractive production on industrial scale is running processes with high substrate loading and simple downstream processing, thus enabling high space–time yields.

In continuation of our ongoing studies [3] on the chemoenzymatic synthesis of nitriles, we recently reported an efficient synthesis of various *n*-alkylnitriles that proceeds at very high substrate loading of up to 1.4 kg.L^{-1} of aqueous buffer as reaction medium [3e]. This process, which is based on an aldoxime dehydratase-catalysed dehydration reaction,

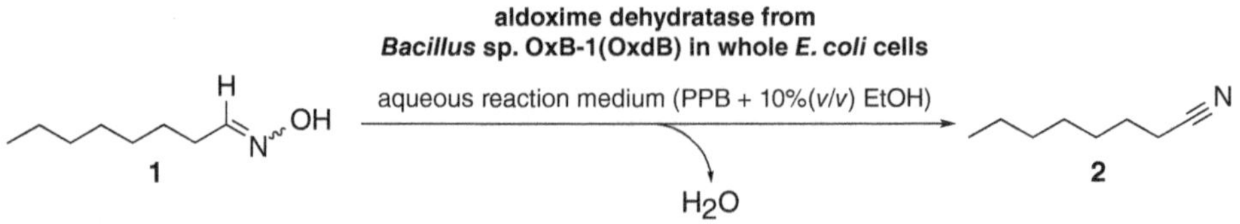

Scheme 9.6 *Conversion of* n*-octanaloxime* ***1*** *to* n*-octanenitrile* ***2*** *using the aldoxime dehydratase OxdB in whole cells as a biocatalyst.*

is exemplified for the synthesis of *n*-octanenitrile **2**, resulting in >99% conversion when operating at a substrate loading of 665 g.L^{-1} (Scheme 9.6) [3e]. Even at a further increased substrate loading of 1.4 kg.L^{-1} of aldoxime **1** with aqueous buffer, the reaction proceeds well, leading to 93% conversion.

9.5.1 Procedure 1: Recombinant Expression of Aldoxime Dehydratase from *Bacillus* sp. OxB-1 (OxdB) in *E. coli* BL21(DE3)CodonPlus-RIL

9.5.1.1 Materials and Equipment

- Lysogenic broth (LB) medium (sterile)
- Terrific broth (TB) medium (sterile)
- D-Lactose (20 g.L^{-1} solution in dH$_2$O, sterile)
- D-Glucose (50 g.L^{-1} solution in dH$_2$O, sterile)
- K_2HPO_4/KH_2PO_4-buffer (PPB; 50 mM, pH 7.0)
- Distilled water (dH$_2$O)
- Carbenicillin (100 mg.mL^{-1} in dH$_2$O, filter-sterilised)
- Chloramphenicol (25 mg.mL^{-1} in ethanol, filter-sterilised)
- LB agar plate with colonies of *E. coli* BL21(DE3)CodonPlus-RIL harbouring the vector pUC18 bearing the gene encoding OxdB inserted [3e]
- Schott bottles with screw caps
- 100 mL and 2 L Erlenmeyer flasks with alumina caps
- Orbital shaker (InforsHT Multitron 2 Standard)
- Autoclave (Tuttnauer 5075 EL)
- Cooling centrifuge (min. 4500× g)

9.5.1.2 Procedure

1. Premixed LB medium (25 g) was dissolved in dH$_2$O (1 L). Premixed TB medium (80.8 g) and glycerol (5 g) were dissolved in dH$_2$O (1 L). D-Lactose (20 g) was dissolved in dH$_2$O (1 L). D-Glucose (50 g) was dissolved in dH$_2$O (1 L). Solutions were autoclaved at 121 °C for 20 min to give sterile media (components).
2. To prepare the pre-culture, sterile LB medium (20 mL) was placed into a sterile 100 mL Erlenmeyer flask. Stock solutions of carbenicillin (20 μL) and chloramphenicol (20 μL) were added to reach final concentrations of 100 and 25 μg.mL^{-1}, respectively. The solution was inoculated with a single colony of *E. coli* BL21(DE3)CodonPlus-RIL harbouring pUC18_OxdB and shaken at 180 rpm and 37 °C overnight.

3. TB autoinduction medium (1.6 L total volume) was prepared from D-lactose solution (160 mL), D-glucose solution (16 mL) and TB medium (1.424 L) in a sterile 2 L Erlenmeyer flask. Stock solutions of carbenicillin (1.6 mL) and chloramphenicol (1.6 mL) were added prior to inoculation with the pre-culture (16 mL).
4. The main culture was shaken at 150 rpm and 37 °C for 2 hr. Afterwards, the temperature was decreased to 30 °C and the culture was shaken at 150 rpm for a further 72 hr (typical OD_{600} ~13).
5. The cells were harvested by centrifugation at 5000× g and 4 °C for 15 min. The supernatant was discarded and the cells were washed with PPB (2 × 15 mL). Finally, the cells (10 g) were resuspended in PPB (20 mL) and stored at 4 °C for later use. Cells produced this way can be used for approximately 6 weeks before a (significant) loss of activity is observed.

9.5.2 Procedure 2: Preparation of *n*-Octanaloxime 1

9.5.2.1 Materials and Equipment

- Hydroxylamine hydrochloride (334 g, 4.8 mol)
- Sodium carbonate (254 g, 2.4 mol)
- *n*-Octanal (500 mL, 3.2 mol)
- Ethanol (100 mL)
- dH_2O (1.9 L)
- *n*-Hexane (250 mL)
- 4 L round-bottom flask
- Magnetic stirrer (IKA RCT Classic)
- Magnetic stirring bar

9.5.2.2 Procedure

1. Hydroxylamine hydrochloride was dissolved in dH_2O (2 L) containing 5% v/v ethanol.
2. Sodium carbonate was added to the stirred aqueous solution.
3. *n*-Octanal was added dropwise whilst *n*-octanaloxime **1** started precipitating.
4. The reaction mixture was stirred for 12 hr at room temperature.
5. *n*-Octanaloxime **1** was filtered and washed with dH_2O (~160 mL).
6. *n*-Octanaloxime **1** (307 g, 2.1 mol, 67%) was obtained as colourless crystals after recrystallisation from *n*-hexane (250 mL).

9.5.3 Procedure 3: Biocatalytic Conversion of *n*-Octanaloxime 1 into *n*-Octanenitrile 2

9.5.3.1 Materials and Equipment

- *n*-Octanaloxime **1** from Procedure 2 (250.25 g, 1.75 mol)
- PPB, 50 mM, pH 7.0 (200 mL)
- Ethyl acetate (EtOAc; 100 mL)
- Resting cells from Procedure 1 (25 mL)
- 50 mL Falcon tubes
- 4 L Round-bottom flask

- Magnetic stirrer (IKA RCT Classic)
- Magnetic stirring bar
- 2× 50 mL syringe
- 2× sterile filter (Filtropur S 0.2, Sarstedt AG & Co. KG)
- Gas chromatography (GC) system with flame ionisation detector (FID; Shimadzu GC-2010 Plus)
- Chiral GC column (SGE Analytik B6B-174: 30 m × 0.25 mm ID, 0.25 µm film)
- Cooling centrifuge (min. 5000× g)
- Rotary evaporator (Büchi)

9.5.3.2 *Procedure*

1. *n*-Octanaloxime **1** (35.75 g, 0.25 mol) was placed into a 1 L round-bottom flask.
2. Ethanol (25 mL) was added and *n*-octanaloxime **1** was dissolved partly by stirring using a magnetic stirring bar.
3. PPB (200 mL) was added whilst stirring at 30 °C.
4. Resting cell suspension (25 mL of 333 $g.L^{-1}$ stock solution (Procedure 1), 33 $g.L^{-1}$ final wet whole-cell concentration) was added.
5. After reaction times of 0.5, 1, 1.5, 2, 2.5 and 4 hr, additional *n*-octanaloxime **1** portions (each portion 35.75 g, 0.25 mol; total amount after all additions 250.25 g, 1.75 mol) were added.
6. The reaction mixture was stirred at 30 °C for 24 hr.
7. The magnetic stirrer was stopped and the phases (aqueous PPB phase and organic *n*-octanenitrile **2** phase) were separated.
8. *n*-Octanenitrile **2** was filled into a 50 mL syringe and filter-sterilised through a 0.2 µm filter (Filtropur S 0.2, Sarstedt AG & Co KG).
9. The aqueous phase was extracted using EtOAc (2 × 50 mL). The EtOAc phases were combined and dried over $MgSO_4$ and the solvent was removed *in vacuo*, resulting in *n*-octanenitrile **2**, which was filled into a 50 mL syringe, filter-sterilised through a 0.2 µm filter (Filtropur S 0.2, Sarstedt AG & Co KG) and combined with the *n*-octanenitrile **2** from Step 8.
10. *n*-Octanenitrile **2** (188 g, 86%) was isolated as a yellowish oil.

9.5.4 Analytical Method

The method and results of GC analysis are given in Tables 9.6 and 9.7, respectively.

Table 9.6 *GC method.*

Temperature program (r = °C · min^{-1})	Duration
140 °C 1 min – 20 r → 190 °C 0.5 min – 50 r → 200 °C	4.2 min

Table 9.7 *Retention times for GC analysis.*

Substance	Retention (min)
n-Octanaloxime **1**	2.4
n-Octanenitrile **2**	2.7

Table 9.8 *Aliphatic nitrile synthesis using OxdB in 10 mL scale for different substrates.*

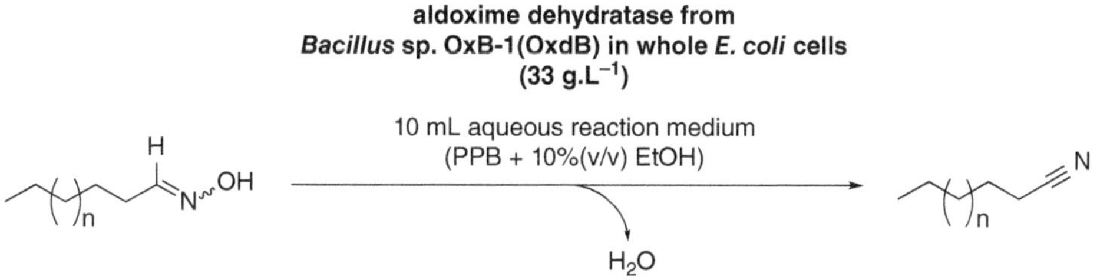

Entry	n	Final conc (g.L^{-1})	Conversion /%	Isolated yield /%
1	1	288	>99	81
2	3	665	>99	98
3	3	1430	93	
4	5	342	>99	83.5
5	5	428	93	

9.5.5 Conclusion

The described procedure enabled a practical and highly productive synthesis of *n*-octanenitrile **2** using aldoxime dehydratases as biocatalyst. In an aqueous reaction medium, *n*-octanaloxime **1** as a water-insoluble substrate was converted into the corresponding desired nitrile product **2** via enzymatic dehydration [3e]. Furthermore, this method enabled a simple product isolation by phase separation of the aqueous reaction medium, consisting of buffer and whole-cell catalyst and the organic nitrile phase, followed by extraction of the aqueous phase with an organic solvent. A successful extension of the substrate scope of this method towards related substrates with different alkyl chain lengths such as *n*-hexanaloxime or *n*-decanaloxime has been also demonstrated, albeit at somewhat lower substrate loadings to date [3e].

References

1. (a) Pollak, P., Romeder, G., Hagedorn, F. and Gelbke, H.-P. (2000) *Nitriles, in Ullmann's Encyclopedia of Industrial Chemistry*, Wiley-VCH; (b) Arpe, H.-J. (2010) *Industrial Organic Chemistry*, 5th edn, Wiley-VCH; (c) Reck, R.A. (1985) *Journal of the American Oil Chemists' Society*, **62**, 355–365; (d) Fleming, F.F., Yao, L., Ravikumar, P.C. *et al.* (2010) *Journal of Medicinal Chemistry*, **53**, 7902–7917; (e) Kleemann, A., Engels, J., Kutscher, B. and Reichert, D. (2001) *Pharmaceutical Substances: Syntheses, Patents, Applications*, 4th edn, Thieme Verlag.
2. (a) Visek, K. (2012) Fatty amines, in Kirk-Othmer Encyclopedia of Chemical Technology, Vol. 2, Wiley Interscience, pp. 518–532; (b) Breitbach, Z.S., Weatherly, C.A., Woods, R.M. *et al.* (2014) *Journal of Separation Science*, **37**, 558–565.
3. (a) Metzner, R., Okazaki, S., Asano, Y. and Gröger, H. (2006) *ChemCatChem*, **6**, 3105–3109; (b) Betke, T., Rommelmann, P., Oike, K. *et al.* (2017) *Angewandte Chemie International Edition*, 56, 12 361–12 366; (c) Betke, T., Maier, M., Gruber-Wölfler, H. and Gröger, H. (2018) *Nature Communications*, **9**, 5112; (d) Plass, C., Hinzmann, A., Terhorst, M. *et al.* (2019) *ACS Catalysis*, **9**, 5198–5203; (e) Hinzmann, A., Glinski, S., Worm, M. and Gröger, H. (2019) *Journal of Organic Chemistry*, **84**, 4867–4872.

9.6 Access to (*S*)-4-Bromobutan-2-ol through Selective Dehalogenation of *rac*-1,3-Dibromobutane by Haloalkane Dehalogenase

Zbynek Prokop,[1] Johannes Gross[2] and Mélanie Hall*[2]
[1]*Loschmidt Laboratories, Department of Experimental Biology and RECETOX, Masaryk University, Brno, Czech Republic*
[2]*Department of Chemistry, University of Graz, Graz, Austria*

The enzymatic dehalogenation of haloalkanes is catalysed by haloalkane dehalogenases (EC 3.8.1.5), which are hydrolytic enzymes showing activity on a broad variety of halogenated compounds. The reaction consists of the formal replacement of a halogen (Cl, Br, I) with a hydroxyl group. With secondary haloalkanes, an inversion of configuration is observed due to the mechanism of the reaction: a nucleophilic attack of the chiral centre by an aspartate residue via S_N2 generates an enzyme–ester intermediate, which gets hydrolysed by water in the subsequent step [1].

The dehalogenation of *rac*-1,3-dibromobutane by haloalkane dehalogenase LinB from *Sphingobium japonicum* UT26 [2] proceeds in a sequential manner and competition between the primary and the secondary positions is responsible for the formation of two different intermediate products, before further hydrolysis to the diol depletes the monohalogenated monoalcohols (Scheme 9.7) [3].

Owing to a measurable degree of regioselectivity [4], as well as enantioselectivity [5], in the early phase of the reaction, the intermediate product (*S*)-4-bromobutan-2-ol preferentially accumulates and can be accessed by simply stopping the reaction after a suitable amount of time. By letting the reaction proceed further, more of this product accumulates – but the complementary (*R*)-enantiomer gets formed too, yielding higher amounts of the compound but with an eroded enantiomeric excess. Extended reaction times should be avoided, otherwise further hydrolysis will lead to the predominant recovery of the diol as the final product (Scheme 9.7).

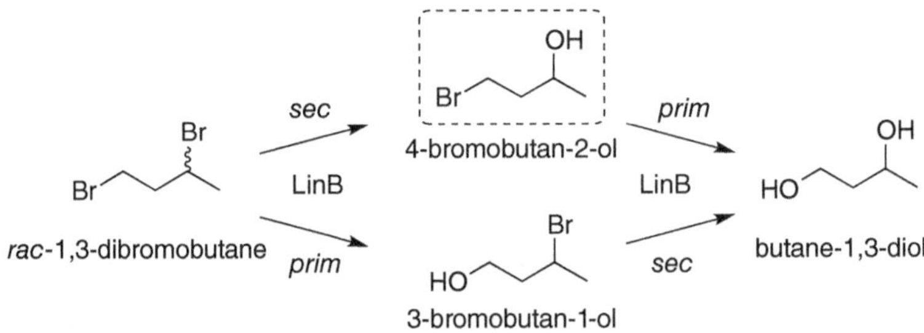

Scheme 9.7 *Biocatalytic sequential dehalogenation of* rac-*1,3-dibromobutane with haloalkane dehalogenase LinB from* Sphingobium japonicum *UT26 (prim and sec describe hydrolysis at the primary and secondary halogenated positions, respectively).*

9.6.1 Procedure 1: Preparation of Haloalkane Dehalogenase LinB

Haloalkane dehalogenase LinB from *S. japonicum* UT26 was discovered and its gene was cloned by Nagata and co-workers in 1994 [6]. Recombinant overexpression in *Escherichia coli* and protein purification were later reported [7].

9.6.1.1 Materials and Equipment

9.6.1.1.1 Chemicals.

- Bovine serum albumin (BSA), ≥96% (Merck)
- Ampicillin sodium (Sigma-Merck)
- Bradford reagent (Sigma-Merck)
- DNase-S (Sigma-Merck)
- Imidazole (Sigma-Merck)
- Isopropyl-β-D-1-thiogalactopyranoside (IPTG; Duchefa)
- Lysogenic broth (LB; Sigma-Merck)
- Potassium dihydrogen phosphate, p.a. (Sigma-Merck)
- Sodium chloride (Sigma-Merck)
- Potassium phosphate dibasic (Sigma-Merck)
- Super optimal broth with catabolite repression (SOC) outgrowth medium (New England Biolabs)
- Mercury thiocyanate (Sigma-Merck)
- Ferric ammonium sulfate (Sigma-Merck)
- Sodium bromide (Merck)
- Sodium chloride (Merck)

9.6.1.1.2 Biological Material.

- Synthetic codon-optimised His-tagged *linB* gene flanked with restriction sites *NdeI* and *HindIII* in pET21b (gene and protein sequence reported in Appendix)
- *E. coli* BL21(DE3) (New England Biolabs)

9.6.1.1.3 Instruments.

- Ultra-low-temperature freezer Innova U725 (New Brunswick Scientific)
- Shaking incubator NB-205 (N-Biotek)
- Orbital incubator Innova 44 (New Brunswick Scientific)
- Ultrasonic processor UP200S (Hielscher)
- Centrifuge 6K15 (Sigma)
- High-performance centrifuge Avanti J-30I (Beckman Coulter)
- Acta fast protein liquid chromatography (FPLC; GE Healthcare)
- NanoDrop™ (Thermo Scientific)
- Spectrophotometer Ultrospec 1000 (Pharmacia Biotech)
- Freeze dryer Scanvac CoolSafe (Labogene)
- Thermostatic water bath GLS400 (Grant)
- Microplate reader Sunrise (Tecan)

9.6.1.1.4 Solutions and Media.

- *LB medium:* LB broth (20 g) dissolved in 1 L of deionised water (dH_2O); – pH was adjusted to 7.2 with 1 M sodium hydroxide
- *Potassium phosphate buffer (50 mM):* potassium phosphate dibasic (7.14 g), potassium dihydrogen phosphate (1.23 g); dissolved in 1 L of dH_2O; pH was adjusted to 7.5 with 10 M phosphoric acid

- *Purification buffer A (equilibrating):* 2.85 g of potassium phosphate dibasic (2.85 g), potassium dihydrogen phosphate (0.49 g), sodium chloride (29.22 g), imidazole (0.68 g); dissolved in 1 L of dH_2O; pH was adjusted to 7.5 with 10 M phosphoric acid; filtration; sonication
- *Purification buffer B (eluting):* potassium phosphate dibasic (2.85 g), potassium dihydrogen phosphate (0.49 g), sodium chloride (29.22 g), imidazole (34.04 g); dissolved in 1 L of dH_2O; pH was adjusted to 7.5 with 10 M phosphoric acid; filtration; sonication
- *Iwasaki solution I:* mercury thiocyanate (0.45 g); dissolved in 150 mL of ethanol
- *Iwasaki solution II:* ferric ammonium sulfate (18.48 g); dissolved in 108 mL of concentrated HNO_3 and 192 mL of dH_2O
- *Glycine buffer (100 mM):* glycine (7.5 g); dissolved in 1 L of dH_2O; pH adjusted to 8.6 with 1 M sodium hydroxide

9.6.1.2 *Procedure*

9.6.1.2.1 Recombinant Expression of Haloalkane Dehalogenase.

1. The codon-optimised gene *linB* was commercially synthesised and cloned into recombinant plasmid pET21b using restriction enzymes NdeI and *Hind*III. The pET21b : *linBHis* plasmid was transformed into *E. coli* BL21(DE3) cells using a heat shock. Briefly, 1 μL of plasmid DNA (100 ng.μL^{-1}) was gently mixed with 10 μL of *E. coli* BL21(DE3) cell suspension. The mixture was incubated on ice for 20 min. The heat shock was induced by placing the mixture on a thermoblock at 42 °C for 1 min. The cells were then incubated on ice for 5 min. 1 mL of SOC outgrowth medium was added to the transformed cell suspension and the mixture was incubated at 37 °C for 1 hr. Regenerated cells were spread on LB agar plate (1.5% agar in LB medium containing 100 μg.mL^{-1} ampicillin) and cultivated overnight at 37 °C.
2. The transformed cell colonies were collected from the Petri dish and used for inoculation of 10 mL of precultivated culture. The pre-culture was cultivated in LB medium containing ampicillin (100 μg.mL^{-1}) at 37 °C for 15 hr, with shaking at 180 rpm (in a shaking incubator, NB-205).
3. For overexpression of LinB, 5 mL of precultivated culture was used to inoculate 1 L of LB media containing ampicillin (100 μg.mL^{-1}). The cultivation was carried out in an Erlenmeyer flask at 37 °C and 115 rpm (in an orbital incubator, Innova 44) for 3–4 hr until an OD_{600} ~0.5 was reached.
4. The culture was cooled to 20 °C. Induction of dehalogenase expression was initiated by the addition of IPTG to a final concentration of 200 μM. After induction, the culture was incubated at 20 °C and 115 rpm overnight.
5. The cells were harvested by centrifugation at 4 °C and 3000× g for 15 min centrifuge (6K15). The pellet (3–4 g.$L_{culture}^{-1}$) was resuspended in purification buffer A and the cell suspension was frozen at −70 °C using an ultra-low-temperature freezer (Innova U725) and stored at −70 °C for later use.

9.6.1.2.2 Purification of His-Tagged Haloalkane Dehalogenase.

1. The *E. coli* cell suspension containing LinB was disrupted by sonication with an ultrasonic processor (UP200S). Before sonication, 100 μg of DNase-S was added to 50 mL of cell suspension. The cell suspension was kept cool on ice during the entire sonication procedure.

2. The sonication was performed in three cycles, with 5 min breaks for cooling in between. Each sonication cycle was set as a repetition of 5 sec sonication and 5 sec pause lasting for 2 min in total.
3. After the sonication, a cell-free extract was isolated by centrifugation at 4 °C and 21 000× g for 1 hr using a high-performance centrifuge (Avanti J-30I).
4. The extract was filtered through a 0.45 μm-pore-size microfilter before loading on a chromatographic column. His-tagged LinB was purified using a HighTrap Chelating HP 5 mL column charged by Ni^{2+} ions attached to an Acta FPLC system. The enzyme was bound to the resin in purification buffer A. Unbound and weakly bound proteins were washed with a mixture of 7.5% purification buffer B in purification buffer A. His-tagged enzyme was then eluted with a mixture of 60% purification buffer B and 40% buffer A.
5. The enzyme was dialysed at 4 °C in 50 mM potassium phosphate buffer pH 7.5 overnight. The enzyme sample to buffer ratio was 1 : 1000 (5 mL of purified enzyme solution dialysed against 5 L of buffer).
6. The enzyme preparation could be used directly (stored as a solution at 4 °C) or lyophilised for long-term storage (see later).

9.6.1.2.3 Quantification of Haloalkane Dehalogenase Concentration.

1. For the assay, 30 μL of purified enzyme sample (in 50 mM potassium phosphate buffer at pH 7.5) obtained from the dialysis and 1 mL of Bradford reagent [8] were combined into a cuvette.
2. The mixture was incubated at laboratory temperature for 15 min.
3. The absorbance of the sample was measured at 595 nm on a spectrophotometer (Ultrospec 1000). Values were converted to concentrations according to the calibration curve, which was measured with albumin standard.
4. The concentration of the enzyme was measured via UV absorbance at 280 nm by NanoDrop™ without the need for cuvettes, where only 1 μL of enzyme solution was required for direct loading into the instrument.

9.6.1.2.4 Measurement of Dehalogenase Activity.

1. The enzymatic reaction was performed in 25 mL Erlenmeyer flasks closed by Mininert valves to prevent evaporation of the halogenated substrate. The substrate (10 μL of 1-chlorobutane or 1,2-dibromoethane) was injected using a glass Hamilton syringe into 10 mL of 100 mM glycine buffer (pH 8.6) in a closed Erlenmeyer flask and pre-incubated in a thermostatic water bath (GLS400) at 37 °C and 150 rpm for 30 min.
2. After pre-incubation, the enzymatic reaction was started by adding 0.2 mL of enzyme solution ($mg.mL^{-1\ of\ protein}$). During the reaction, 1 mL of the reaction mixture was taken at 0, 5, 10, 20 and 40 min and immediately mixed with 0.1 mL of 35% HNO_3. Blank was prepared by taking 1 mL of the reaction mixture before addition of the enzyme solution and mixing it with 0.1 mL of 35% HNO_3.
3. The concentration of halide released from the reaction with LinB was analysed using the Iwasaki method [9]. Each sample (100 μL) was pipetted into microtitre plate wells in eight replicates after the addition of 100 μL of Iwasaki solution I and 200 μL of Iwasaki solution II. The absorbance of the sample was measured at 460 nm using a microplate reader (Sunrise). Concentrations of product were calculated by subtracting the absorbance of the blank samples from the absorbance of the samples taken at 0, 5, 10, 20 and 40 min. Calculated values were converted to product concentrations

according to the calibration curve. Iwasaki solutions were calibrated by spiking the reaction buffer with solutions of sodium chloride or sodium bromide as a standard for detection of chloride and bromide ions. Enzyme activity was evaluated as a rate of product formation in time. The specific activity of LinB was 0.03 and 0.2 µmol.sec^{-1}.mg$_{enzyme}$$^{-1}$ for 1-chlorobutane and 1,2-dibromoethane, respectively.

9.6.1.2.5 Lyophilisation of Purified Haloalkane Dehalogenase.

1. 1 mL aliquots of the purified enzyme (concentration of 5 mg.mL^{-1} in 50 mM potassium phosphate buffer, pH 7.5) were filled into lyophilisation tubes covered by lyophilisation lids in a mode opened for sublimation.
2. The samples were initially frozen at −70 °C. The frozen enzyme samples were lyophilised overnight using a freeze dryer (Scanvac CoolSafe).
3. After the sublimation was completed, the lids were closed and the air was filled back to the lyophilisation chamber.
4. The dried enzyme powder was stored at 4 °C in tightly closed tubes under vacuum conditions for later use. Each tube contained 5 mg of pure enzyme and 8.37 mg of buffer salts. The overall yield was approximately 100 mg of pure LinB (approximately 0.27 g of the freeze-dried powder) obtained from 1 L of culture media.

9.6.2 Procedure 2: Biocatalytic Dehalogenation of *rac*-1,3-Dibromobutane Catalysed by LinB towards Formation of (*S*)-4-Bromobutan-2-ol

9.6.2.1 Materials and Equipment

- Tris-sulfate buffer (50 mM, pH 8.2, 2 mL)
- *rac*-1,3-Dibromobutane (0.6 µL)
- NaCl
- Ethyl acetate (EtOAc; 1 mL)
- Anhydrous Na_2SO_4
- Lyophilised LinB preparation (see Procedure 1)
- Eppendorf tubes (2 mL)
- Incubator shaker
- Biofuge centrifuge
- Glass GC vials

9.6.2.2 Procedure

1. The lyophilised preparation of LinB (1.5 mg) was diluted in 1 mL Tris-sulfate buffer (50 mM, pH 8.2) to achieve a concentration of 1.5 mg.mL^{-1}, and the stock solution was incubated at 21 °C and 120 rpm for 30 min in order to rehydrate the enzyme preparation.
2. Substrate *rac*-1,3-dibromobutane (0.6 µL, 5 µmol; final conc. 5 mM) was added to the 50 mM Tris-sulfate buffer solution, pH 8.2 (1 mL), followed by LinB (125 µL of the 1.5 mg.mL^{-1} stock solution prepared in Step 1, equivalent to a final enzyme concentration of 6 µM).
3. The reaction mixture was incubated at 21 °C and 120 rpm by mounting the samples horizontally on a Unitron AJ 260 shaker.

4. To reach the highest enantiomeric excess value for (*S*)-4-bromobutan-2-ol, the reaction was stopped after 10 min by saturating the reaction mixture with NaCl and extracting directly with EtOAc (2 × 500 μL). For higher concentrations of the product, the reaction was stopped after 60 min and worked up as described in Step 5.
5. Upon addition of EtOAc, the mixture was centrifuged at 13 000 rpm for 5 min to allow separation of the phases. The combined organic layers were dried over anhydrous Na_2SO_4 then centrifuged at 13 000 rpm for 5 min and the organic phase was transferred to GC vials for analysis.

9.6.3 Analytical Method

Conversion levels were calculated using a calibration curve obtained with authentic reference material [3]. Conversion and enantiomeric excess values were obtained by analysis via GC-FID using a chiral Hydrodex β-TBDAc column (50 m × 0.25 mm ID × 0.4 mm OD, Macherey-Nagel). The following temperature programme was employed: 130 °C for 7 min; 5 °C.min^{-1} to 135 °C; hold 5 min; 5 °C.min^{-1} to 140 °C; hold 5 min; 10 °C.min^{-1} to 170 °C. Retention times: *rac*-1,3-dibromobutane, 8.40 min (enantiomers could not be separated); (*S*)-4-bromobutan-2-ol, 12.72 min; (*R*)-4-bromobutan-2-ol, 13.30 min. Other products can also be analysed using the same method: (*S*)-3-bromo-butan-1-ol, 13.55 min; (*R*)-3-bromo-butan-1-ol, 13.90 min; (*S*)-butane-1,3-diol, 14.93 min; (*R*)-butane-1,3-diol, 15.55 min (Figure 9.1).

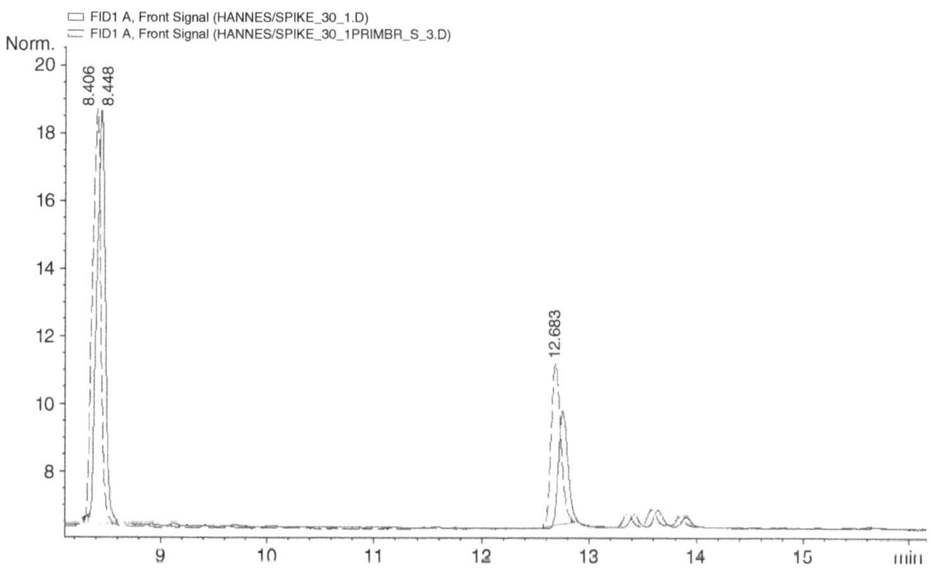

Figure 9.1 *GC-FID chromatogram trace obtained on a chiral column with separation of all compounds from the biotransformation of* rac-*1,3-dibromobutane with LinB. The dotted line was obtained from the sample resulting from the spiking of the biotransformation sample (full line) with authentic reference material for (*S*)-4-bromobutan-2-ol. Peak assignment in order of appearance:* rac-*1,3-dibromobutane, (*S*)-4-bromobutan-2-ol, (*R*)-4-bromobutan-2-ol, (*S*)-3-bromo-butan-1-ol, (*R*)-3-bromo-butan-1-ol.*

9.6.4 Remarks

In a typical experiment on 5 mM substrate (see Procedure 2), after 10 min (*S*)-4-bromobutan-2-ol is the only product monitored (0.68 mM in >99% ee). After 60 min reaction time, 2.55 mM of 4-bromobutan-2-ol is obtained as (*S*)-enantiomer in 73% ee, which was contaminated with 1.27 mM of 3-bromobutan-1-ol. After 2 hr, the trend is reversed, and 2.05 mM of (*S*)-3-bromobutan-1-ol can be obtained in 88% ee.

9.6.5 Conclusion

The regioselective and enantioselective formation of (*S*)-4-bromobutan-2-ol can be achieved by dehalogenation of *rac*-1,3-dibromobutane by haloalkane dehalogenase LinB from *S. japonicum* UT26. Due to competing hydrolysis reactions at the primary and secondary halogenated positions, the course of the reaction must be monitored, and the reaction should be stopped when it is judged to be ideal for the targeted application. For exquisite optical purity, shorter reaction times should be preferred, whilst higher yields can be obtained with slightly extended times. Longer incubations must be avoided to prevent the full consumption of the target product. In the future, it might be possible to engineer haloalkane dehalogenases to prevent hydrolysis at the primary halogenated positions [9].

LinB from *S. japonicum* UT26 is deposited in the Uniprot database with the accession number D4Z2G1 (see Appendix). The gene coding for this protein is labelled as D14594.2 (see Appendix). Crystal structures are available, and the data are deposited in the PDB (for atomic resolution, see 1MJ5 [10]). The protein material is available upon request from the spin-off company Enantis Ltd. (www.enantis.com).

Acknowledgements

The authors would like to express their thanks to the Czech Ministry of Education, Youth and Sports Infrastructure RECETOX (CZ.02.1.01/0.0/0.0/16_013/0001761 and LM2015051) and the European Union H2020 Project Rafts4Biotech (No. 720776).

Appendix

Nucleotide Sequence of LinB

GAAGGAGATATACATatgtcattaggcgcaaaaccattcggcgagaagaagtttattgagattaaaggtagaagaatg
gcctacatcgatgaaggtaccggtgacccgatcctgttccaacacggcaatccgacgagcagctacctgtggcgcaacatcatgc
cgcattgcgcaggcctgggtcgcctgattgcgtgtgatctgatcggtatgggtgacagcgataaactggacccgagcggtccgga
acgttatgcgtatgcggagcaccgtgactatctggatgccctgtgggaagcgctggatctgggcgatcgtgtggtcctggttgttca
cgactggggtagcgcattgggtttcgattgggctcgtcgtcatcgcgagcgtgttcagggtatcgcctacatggaagcgattgcga
tgccgatcgagtgggcggactttccggagcaggaccgcgatttgtttcaagcgttccgtagccaagcaggcgaagaactggtgtt
gcaggacaatgtctttgttgagcaggtcttgccaggcctgatcctgcgtccgctgagcgaggcggaaatggccgcgtaccgcgaa
ccgttcctggcagcgggtgaagcgcgtcgtccgactctgagctggccgcgtcagattccgattgccggcacgcctgctgatgtgg
tcgcgattgcgcgtgactacgcaggctggctgtctgagtccccgattccgaagctgttcatcaacgccgagccgggtgctctgacc
accggtcgcatgcgcgacttctgccgcacgtggccgaaccaaaccgagattaccgtggcgggtgctcactttatccaggaagatt
cgccggacgagattggtgcggccatcgcagcatttgttcgtcgtctgcgtcctgcacatcatcaccatcaccactaaAAGCTT

Amino Acid Sequence of LinB

MSLGAKPFGEKKFIEIKGRRMAYIDEGTGDPILFQHGNPTSSYLWRNIMPHCAGLG
RLIACDLIGMGDSDKLDPSGPERYAYAEHRDYLDALWEALDLGDRVVLVVHDWG
SALGFDWARRHRERVQGIAYMEAIAMPIEWADFPEQDRDLFQAFRSQAGEELVLQ
DNVFVEQVLPGLILRPLSEAEMAAYREPFLAAGEARRPTLSWPRQIPIAGTPADVV
AIARDYAGWLSESPIPKLFINAEPGALTTGRMRDFCRTWPNQTEITVAGAHFIQEDS
PDEIGAAIAAFVRRLRPAHHHHHH

References

1. (a) Janssen, D.B., Scheper, A., Dijkhuizen, L. and Witholt, B. (1985) *Applied and Environmental Microbiology*, **49**, 673–677; (b) Koudelakova, T., Bidmanova, S., Dvorak, P. *et al.* (2013) *Biotechnology Journal*, **8**, 32–45; (c) Nagata, Y., Ohtsubo, Y. and Tsuda, M. (2015) *Applied Microbiology and Biotechnology*, **99**, 9865–9881.
2. Nagata, Y., Miyauchi, K., Damborsky, J. *et al.* (1997) *Applied and Environmental Microbiology*, **63**, 3707–3710.
3. Gross, J., Prokop, Z., Janssen, D. *et al.* (2016) *ChemBioChem*, **17**, 1437–1441.
4. Kmunicek, J., Hynkova, K., Jedlicka, T. *et al.* (2005) *Biochemistry*, **44**, 3390–3401.
5. Prokop, Z., Sato, Y., Brezovsky, J. *et al.* (2010) *Angewandte Chemie International Edition*, **49**, 6111–6115.
6. Nagata, Y., Ohtomo, R., Miyauchi, K. *et al.* (1994) *Journal of Bacteriology*, **176**, 3117–3125.
7. Nagata, Y., Hynková, K., Damborský, J. and Takagi, M. (1999) *Protein Expression and Purification*, **17**, 299–304.
8. Bradford, M.M. (1976) *Analytical Biochemistry*, **72**, 248–254.
9. Iwasaki, I., Utsumi, S. and Ozawa, T. (1952) *Bulletin of the Chemical Society of Japan*, **25**, 226.
10. Oakley, A.J., Klvana, M., Otyepka, M. *et al.* (2004) *Biochemistry*, **43**, 870–878.

10

Glycosylation, Sulfation and Phosphorylation

10.1 Rutinosidase Synthesis of Glycosyl Esters of Aromatic Acids

Vladimír Křen
Institute of Microbiology of the Czech Academy of Sciences, Laboratory of Biotransformation, Prague, Czech Republic

Enzymatic glycosylation has now become a largely accepted method in the repertoires of organic chemists and biochemists. Various types of enzymes are used as catalysts, the most prominent being glycosyltransferases and glycosidases [1]. Glycosidases, unlike transferases, offer broader selectivity and flexibility, mainly towards glycosylation acceptors [2]. They also employ cheaper and more stable glycosyl donors – typically arylglycosides (e.g. *p*-nitrophenyl glycosides) or simple glycosides, such as lactose (for β-galactosidases). These enzymes, however, suffer from limitations, such as lower yields, lower selectivity towards polyolic acceptors and reluctance to glycosylate acidic hydroxyls (phenols). Glycosylation of other acceptor groups, such as carboxyl groups, has been occasionally described for glycosyltransferases, but not glycosidases.

We described a robust diglycosidase, which glycosylates a number of phenolic acceptors. This enzyme – rutinosidase from *Aspergillus niger* – is capable of transferring the rutinosyl residue (6-*O*-α-L-rhamnopyranosyl-β-D-glucopyranosyl) on to various alcohols, and more interestingly also to phenolic acceptors, which are generally difficult to glycosylate with glycosidases [3]. Crystal structure of rutinosidase from *A. niger* has been resolved and its high transglycosylation activity clarified [3b]. A major advantage of this transglycosylation reaction is the use of inexpensive and biocompatible rutin **1** as the glycosyl donor, whilst quercetin **2**, which is a byproduct of the reaction, precipitates and can be easily removed from the reaction mixture by filtration (precipitation – thermodynamic shift). This is one reason for the higher transglycosylation yield compared to other glycosidases. The

Applied Biocatalysis: The Chemist's Enzyme Toolbox, First Edition. Edited by John Whittall and Peter W. Sutton.
© 2021 John Wiley & Sons Ltd. Published 2021 by John Wiley & Sons Ltd.

rutinosides produced can be conveniently transformed *in situ* via a telescoping reaction with α-L-rhamnosidase to yield the relevant β-glucopyranosides [4].

Recently, we discovered that the rutinosidase from *A. niger* not only efficiently converts hydroxylated aromatic acids (e.g. coumaric and ferulic acids) into their respective phenolic rutinosides, but surprisingly also catalyses the formation of the respective glycosyl esters (Scheme 10.1) [5]. This glycosylation of the carboxylic group is limited only to derivatives of cinnamic acid bearing an OH group in the *p*- or *o*-position (e.g. *p*- and *o*-coumaric acid, ferulic acid and caffeic acid).

In this section, we describe the semipreparative glycosylation of *p*-coumaric acid as a representative example.

10.1.1 Procedure 1: Production and Purification of Recombinant Rutinosidase in *Pichia pastoris*

10.1.1.1 Materials and Equipment

- Yeast extract peptone dextrose (YPD) medium: yeast extract (Oxoid), 10 $g.L^{-1}$; bacteriological peptone (Oxoid), 20 $g.L^{-1}$; glucose, 20 $g.L^{-1}$
- Buffered glycerol-complex (BMGY) medium: yeast extract, 10 $g.L^{-1}$; peptone, 20 $g.L^{-1}$; 100 mM potassium phosphate, pH 6.0; yeast nitrogen base (YNB, Oxoid), 13.4 $g.L^{-1}$; biotin 0.0004 $g.L^{-1}$; glycerol, 10 $g.L^{-1}$
- Buffered methanol-complex (BMMY) medium: same composition as BMGY but with 0.5% v/v methanol in place of 1% w/v glycerol
- Buffered minimal glycerol (BMGH) medium: 100 mM potassium phosphate, pH 6.0; YNB, 13.4 $g.L^{-1}$; biotin, 0.0004 $g.L^{-1}$; glycerol, 10 $g.L^{-1}$
- Buffered minimal methanol (BMMH) medium: 100 mM potassium phosphate, pH 6.0; YNB, 13.4 $g.L^{-1}$; biotin, 0.0004 $g.L^{-1}$; methanol, 5 $g.L^{-1}$, used for the main culture
- Basal salt (BSM) medium: 85% H_3PO_4, 26.7 $mL.L^{-1}$; $CaSO_4.2H_2O$, 1.17 $g.L^{-1}$; K_2SO_4, 18.2 $g.L^{-1}$; $MgSO_4.7H_2O$, 14.9 $g.L^{-1}$; KOH, 4.13 $g.L^{-1}$; glycerol, 40 $g.L^{-1}$; supplemented with 4.35 $mL.L^{-1}$ PTM_1 (trace salts solution): $CuSO_4.5H_2O$, 6 $g.L^{-1}$; NaI, 0.08 $g.L^{-1}$; $MnSO_4.H_2O$, 3 $g.L^{-1}$; $Na_2MoO_4.2H_2O$, 0.2 $g.L^{-1}$; H_3BO_3, 0.02 $g.L^{-1}$; $CoCl_2$, 0.5 $g.L^{-1}$; $ZnCl_2$, 20 $g.L^{-1}$; $FeSO_4.7H_2O$, 65 $g.L^{-1}$; biotin, 0.2 $g.L^{-1}$; H_2SO_4 conc. 9.2 $g.L^{-1}$). Methanol added in fed-batch experiments was also supplemented with PTM_1 (1.2 $mL.L^{-1}$ pure methanol)
- 10 mM sodium acetate buffer, pH 3.6
- 10 mM citrate-phosphate buffer, pH 5.0 + 150 mM NaCl
- 0.1 M aqueous solution Na_2CO_3
- Zeocin (Sigma)
- *p*-Nitrophenol rutinoside (Sigma), 10 mM aqueous solution
- Fractogel EMD SO_3^- column (15 × 100 mm; Amersham)
- Superdex 200 10/300 GL column (10 × 300 mm; Amersham)
- 1 L Schott bottle with screw cap
- Distilled or Mili-Q water
- Cellulose membranes with a 10 kDa cut-off (Millipore)
- 50 mL Falcon® centrifuge tubes
- Orbital shaker incubator
- Centrifuge (min. 2500 rpm)
- 1 L separating funnel

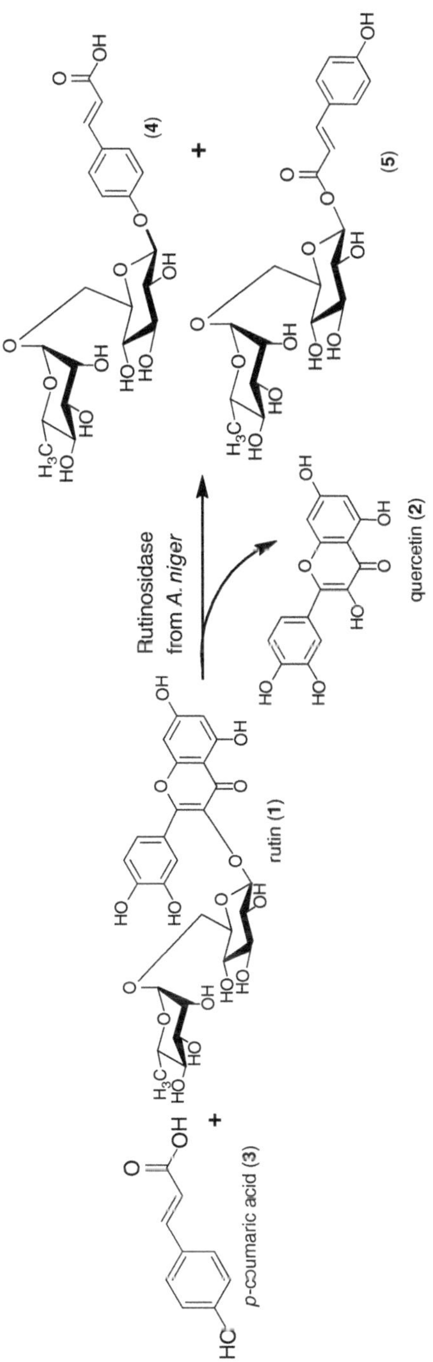

Scheme 10.1 *Glycosylation of p-coumaric acid with rutinosidase [5].*

- pH meter
- Aqueous NaOH (0.5 M)
- Aqueous HCl 2%
- Separating column (h: ~50 cm, d: ~5 cm)
- 50 mL round-bottom flask
- EasySelect *Pichia* Expression Kit (Invitrogen)
- Bradford reagent kit (Sigma)
- Bovine serum albumin (BSA; Sigma)

10.1.1.2 Procedure

1. The expression vector pPICZαA-RUT [3] was linearised with restriction endonuclease *Sac*I and the prepared competent *P. pastoris* cells KM71H were transformed with the linearised expression vector by electroporation according to the manufacturer's instructions (EasySelect *Pichia* Expression Kit) [3]. The electroporated cells were grown at various concentrations under the selection pressure of Zeocin (100 μg.mL^{-1}) on YPD agar plates at 28 °C for 2 days.
2. Recombinant rutinosidase production was carried out according to the manufacturer's instructions (EasySelect *Pichia* Expression Kit): the colonies were inoculated into 100 mL of BMGY medium, pH 6.0, and incubated overnight with shaking at 28 °C.
3. The cells were collected by centrifugation (5000× g, 10 min, 20 °C) and the pellet was resuspended in 30 mL of BMMY medium in a 300 mL baffled conical flask. The production of rutinosidase was induced by the addition of methanol (0.5% v/v) every 24 hr for 4 days. The flasks were incubated at 28 °C and 220 rpm.
4. Recombinant rutinosidase was purified from the culture medium of *Pichia pastoris* after 6 days of cultivation with methanol induction. The cells were harvested by centrifugation (5000× g, 10 min, 4 °C).
5. The supernatant was dialysed against 6 L of 10 mM sodium acetate buffer, pH 3.6, for 2 hr (dialysis tubing cellulose membrane, Sigma-Aldrich, cut-off 10 kDa). The pH of the solution was then adjusted to 3.6 with 10% acetic acid and filtered. This solution was loaded into a Fractogel EMD SO_3^- column (15 × 100 mm) in 10 mM sodium acetate buffer, pH 3.6. The protein was eluted using a linear gradient of 0–1 M NaCl (5 mL.min^{-1}). Fractions were collected and analysed for rutinosidase activity using *p*-nitrophenol rutinoside as substrate [1]. The fractions containing rutinosidase activity were concentrated by ultrafiltration using cellulose membranes with a 10 kDa cut-off (Millipore).
6. The concentrated protein was purified to homogeneity by gel filtration in a Superdex 200 10/300 GL column (10 × 300 mm, 10 mM citrate-phosphate buffer, pH 5.0, containing 150 mM NaCl).
7. Rutinosidase activity was determined spectrophotometrically using *p*-nitrophenol rutinoside as substrate. The reaction mixture contained *p*-nitrophenol rutinoside (10 mL of 10 mM aq solution), 10 mL of 50 mM citrate-phosphate buffer pH 5.0 and 30 mL of the enzyme solution. The reaction was incubated at 35 °C for 10 min and stopped by the addition of 1 mL of 0.1M Na_2CO_3. The released *p*-nitrophenol was determined spectrophotometrically at 420 nm. One unit of enzyme activity was defined as the amount of enzyme releasing 1.0 mmol of *p*-nitrophenol per minute under the assay conditions.

Protein concentrations were determined by Bradford assay calibrated for BSA. Typical yield of is ~21 U, for a total 5 mg of protein.

10.1.2 Procedure 2: Biotransformation Glycosylation of *p*-Coumaric Acid with Rutinosidase

10.1.2.1 Materials and Equipment

- 50 mM citrate phosphate (CP) buffer, pH 5
- Mobile phase for preparative high-performance liquid chromatography (HPLC): 10% acetonitrile in water + 0.1% formic acid
- Mobile phase for thin-layer chromatography (TLC) silica-gel EtOAc/MeOH/HCO_2H (4 : 1 : 0.05)
- Mobile phase for TLC RP-18 reverse-phase plates MeCN/H_2O (4 : 6)
- *p*-Coumaric acid as glycosyl acceptor (300 mg, 1.83 mmol; Sigma)
- Rutin (2234 mg, 3.66 mmol; Sigma)
- Dimethyl sulfoxide (DMSO; Sigma)
- Chromolith SemiPrep (100 × 10 mm; Merck, DE) column equipped with Chromolith RP-18e (5 × 4.6 mm) guard cartridge
- 1 L Schott bottle with screw cap
- Distilled or Mili-Q water
- Ethyl acetate (EtOAc)
- Methanol (MeOH)
- Ammonia (aqueous NH_3)
- TLC silica gel 60 F_{254} aluminium plates (Merck)
- RP-18 TLC plates (Merck)
- 50 mL Falcon® centrifuge tubes
- Orbital shaker incubator
- Centrifuge (min. 2500 rpm)
- 1 L separating funnel
- pH meter
- Aqueous NaOH (0.5 M)
- Aqueous HCl 2%
- Separating column (h: ~50 cm, d: ~5 cm)
- Amberlite XAD4® resin (Sigma)
- Rotary evaporator (rotavap)
- UV lamp
- 5% solution of concentrated sulfuric acid in EtOH
- 50 mL round-bottom flask

10.1.2.2 Procedure

1. *p*-Coumaric acid **3** (300 mg, 1.83 mmol, 1 eq) was dissolved in DMSO (3.5 mL). The resulting solution was added to the rutinosidase solution (0.3 U.mL^{-1}, 20 mL) and the pH was adjusted to 3.0 with 2% aqueous HCl. Rutin **1** (2234 mg, 3.66 mmol, 2 eq) was added in three portions every 90 min. The resulting heterogeneous mixture was shaken (180 rpm, Thermoshaker, Eppendorf, D) at 40 °C for a total of 6–7 hr. Reaction

was monitored by TLC: reaction aliquots were diluted five times with MeOH, analysed on silica plates using EtOAc/MeOH/HCO_2H (4 : 1 : 0.05) as a mobile phase or on reverse-phase plates (MeCN/H_2O; 4 : 6) and visualised under UV light and by charring with 5% H_2SO_4 in EtOH.

2. When the reaction was completed, the reaction mixture was diluted with 50 mM citrate-phosphate buffer, pH 4.5 (20 mL). Centrifuged (1500× g 10 min, 4 °C) supernatant was collected and the solids were resuspended in the same buffer volume and centrifuged again. Combined supernatants (pH 4.5) were extracted with EtOAc (2 × 20 mL) to remove the bulk of the unreacted acceptor, rutin and quercetin byproduct.
3. The EtOAc was removed by distillation under reduced pressure to leave a primarily aqueous solution. Non-ionic resin Amberlite XAD-4 (Sigma) was washed overnight with acetone, then washed extensively with water to remove all traces of organic solvents and filled into a column. The partially evaporated supernatant was loaded on to the column and washed with 4 column volumes of water. The resin was then washed with MeOH to elute the glycosylation products and the methanol solution was evaporated to yield a crude mixture of 170 mg (20%) of the glycosylated products, which were separated by preparative HPLC.
4. Preparative HPLC was performed with the same Shimadzu Prominence system, consisting of a DGU-20A_3 mobile-phase degasser, two LC-20AD solvent-delivery units, a SIL-20AC cooling autosampler, a CTO-10AS column oven, an SPD-M20A diode array detector and a fraction-collector FRC-10A (Shimadzu, Japan). The sample (10 mg) was dissolved in mobile phase (200 μL), centrifuged (1500× g, 5 min, 4 °C) and injected (25 μL) into the Chromolith SemiPrep (100 × 10 mm, Merck, DE) column, equipped with a Chromolith RP-18e (5× 4.6 mm) guard cartridge. The fractions (1.2 mL) were collected by eluting with a mobile phase of 10% acetonitrile in water, 0.1% formic acid at a flow rate of 1.2 mL.min^{-1}, 25 °C. Retention times of respective products: 2.6 (**4**), 3.9 (**5**) and 5.5 min (unreacted *p*-coumaric acid **3**). Fractions containing these respective products were combined and evaporated to dryness.
5. The preparative HPLC of the reaction mixture after XAD-4 workup (125.3 mg) yielded two fractions: 4-*O*-rutinosyl (*E*)-*p*-coumaric acid **4** (38.7 mg, 30.9%) white powder and 1-*O*-(*E*)-*p*-coumaroyl-β-rutinose **5** (3.9 mg, 3.1%) white powder. Full spectral data (MS, ^{1}H and ^{13}C NMR) are published in detail elsewhere [5].

10.1.3 Conclusion

This procedure describes rutinosylation of aromatic acids – derivatives of cinnamic acid. This is probably the first described glycosidase-catalysed glycosylation of a carboxylic group. Glycosylation of the phenolic moiety, which is rarely catalysed by glycosidases, proceeds in reasonable yields. The reaction employs rutin as glycosyl donor, which has a Generally Recognised as Safe (GRAS) status; therefore, it can be used in the production of biocompatible products. Generation of rather insoluble quercetin as a second product is an advantage in terms of higher yield and simple processing of the reaction mixture.

References

1. (a) Desmet, T., Soetaert, W., Bojarová, P. *et al.* (2012) *Chemistry: A European Journal*, **18**, 10 786–10 801; (b) Bojarová, P., Rosencrantz, R.R., Elling, L. and Křen, V. (2013) *Chemical Society Reviews*, **42**, 4774–4797.

2. (a) Bojarová, P. and Křen, V. (2009) *Trends in Biotechnology*, **27**, 199–209; (b) Bojarová, P., Kulik, N., Hovorková, M. *et al.* (2019) *Molecules*, **24**, 599.
3. (a) Šimčíková, D., Kotik, M., Weignerová, L. *et al.* (2015) *Advanced Synthesis & Catalysis*, **357**, 107–117; (b) Pachl, P., Kapešová, J., Brynda, J. *et al.* (2020) *FEBS Letters*, doi:10.1111/febs.15208.
4. Bassanini, I., Krejzová, J., Panzeri, W. *et al.* (2017) *ChemSusChem*, **10**, 2040–2045.
5. Bassanini, I., Kapešová, J., Petrásková, L. *et al.* (2019) *Advanced Synthesis & Catalysis*, **361**, 2627–2637.

10.2 Biocatalytic Synthesis of Kojibiose Using a Mutant Transglycosylase

Shari Dhaene, Jorick Franceus, Koen Beerens and Tom Desmet*
Centre for Synthetic Biology (CSB), Department of Biotechnology, Ghent University, Ghent, Belgium

Carbohydrates can offer a huge diversity in both structure and function. However, only a small number of saccharides are found in nature in sufficient amounts to allow their commercial exploration. In order to explore their potential, scalable processes are needed for rare sugars. In that context, kojibiose (2-*O*-α-D-glucopyranosyl-D-glucopyranose) is a promising compound. It has been shown that kojibiose is not metabolised by common oral bacteria and it has therefore attracted attention as a low-calorific sweetener for the prevention of tooth decay. Studies on its health-promoting properties are, however, hampered by its high price and limited availability [1].

Recently, a biocatalytic process was developed for the synthesis of kojibiose from sucrose and glucose. The enzyme sucrose phosphorylase originating from *Bifidobacterium adolescentis* (BaSP) can be used to produce the naturally occurring disaccharide kojibiose from the bulk sugars sucrose and glucose. However, the primary product is the α-(1,4)-bonded maltose and the target compound is only a side product that is difficult to isolate. By means of semirational mutagenesis and low-throughput screening, the selectivity of BaSP towards kojibiose could be increased to 95% by introducing two mutations, L341I and Q345S (Scheme 10.2) [2]. This double mutant has been utilised for process intensification, leading to the large-scale production of kojibiose and thus opening the door for its application testing. Thereby, the released fructose from sucrose was recycled to glucose by a glucose isomerase, leading to a higher atom efficiency (Scheme 10.3) [3]. In this section, we describe the expression of BaSP-L341I_Q345S in *Escherichia coli* and its use for the production of kojibiose from sucrose and glucose.

10.2.1 Procedure 1: Recombinant Expression of BaSP-L341I_Q345S in *E. coli*

10.2.1.1 Materials and Equipment

- Cryovial containing *E. coli* BL21(DE3) transformed with an expression plasmid pCXP34h bearing a gene encoding *B. adolescentis* sucrose phosphorylase (UniProt code A0ZZH6). The gene should harbour the mutations L341I and Q345S for improved synthesis of kojibiose, as well as the mutations Q331E, R393N, D445P, D446T, Q460E and E485H for improved thermostability [2, 4]
- Tryptone (300 g)
- Yeast extract (150 g)
- NaCl (150 g)

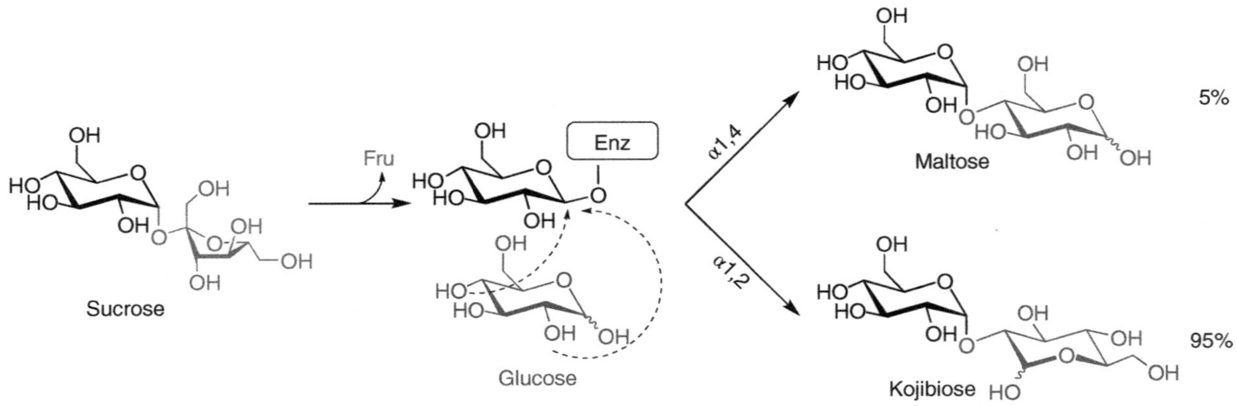

Scheme 10.2 *Transglucosylation of glucose by double-mutant sucrose phosphorylase with increased kojibiose selectivity.*

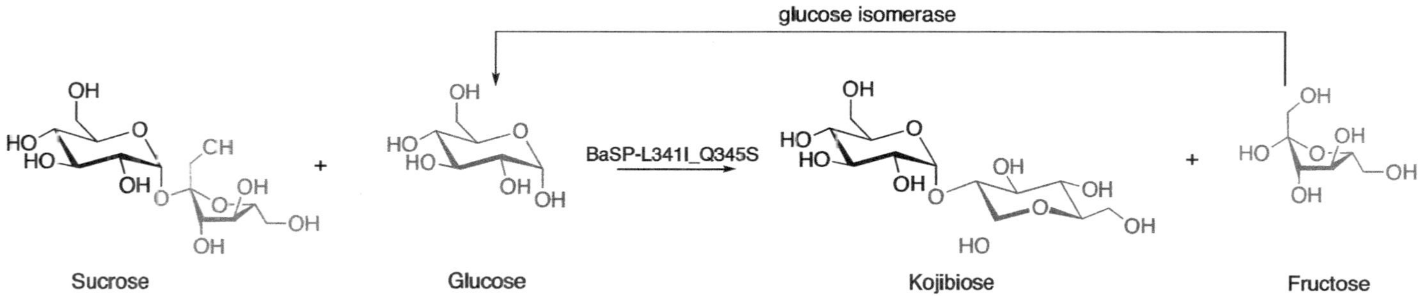

Scheme 10.3 *Conversion of sucrose and glucose into kojibiose.*

- Ampicillin (100 mg.mL^{-1}, 1.6 g in 16 mL dH_2O, filter-sterilised)
- Sterile glucose solution (450 g in 2 L dH_2O)
- Sterile NaOH (25% v/v, 1 L)
- Sterile 5 M H_2SO_4 (500 mL)
- Sterile antifoam (10% v/v silicone Snapsil RE 20; VWR BDH Prolabo, 500 mL)
- Distilled water (15 L)
- Glass test tube with stainless-steel cap (50 mL)
- Erlenmeyer flask with cotton plug (2 L)
- Four Schott bottles (2 L, 1 L, 500 mL, 500 mL) with screw caps
- Fermenter (30 L)
- Shaker (200 rpm)
- Spectrophotometer (600 nm)
- Cooled centrifuge (10 000× g)

10.2.1.2 Procedure

1. Double lysogenic broth (LB) was prepared by dissolving tryptone (20 g.L^{-1}), yeast extract (10 g.L^{-1}) and NaCl (5 g.L^{-1}) in 13 L distilled water. The double LB was added to a glass test tube with stainless-steel cap (5 mL), a 2 L Erlenmeyer flask (500 mL) and a fermenter (12.5 L). A glucose solution was prepared by adding glucose (450 g) to 2 L distilled water in a 2 L Schott bottle. Aqueous solutions of 25% v/v NaOH (1 L), 5 M H_2SO_4 (500 mL) and 10% v/v antifoam (500 mL) were prepared in Schott bottles of 1 L and 500 mL. The recipients were autoclaved (121 °C, 30 min).
2. Ampicillin solution (5 μL) was added to the sterile glass test tube prior to use to reach a final concentration of 100 μg.mL^{-1}, which was subsequently inoculated with the *E. coli* cells and incubated at 37 °C and 200 rpm for 8 hr. The culture was then added to the filled Erlenmeyer flask containing ampicillin (500 μL, final concentration 100 μg.mL^{-1}) and grown overnight under the same conditions (OD ~6).
3. The sterile fermenter was supplemented with ampicillin (15 mL, final concentration 100 μg.mL^{-1}) and glucose (1 L, final concentration of 30 g.L^{-1}). The culture broth was added to the fermenter for large-scale enzyme production. Temperature, pH, aeration and stirrer speed were set at 37 °C, 7.0, 1.2 vvm and 800 rpm, respectively. Drops of the NaOH and H_2SO_4 solutions were added to maintain the pH at 7. When necessary, antifoam was added manually.
4. The optical density of the culture was measured regularly using a spectrophotometer. When the OD_{600} stopped increasing (~8 hr, OD ~40), the cells were harvested by centrifugation at 4 °C (10 000× g, 20 min). The resulting pellets (~600 g) were frozen at −20 °C.

10.2.2 Procedure 2: Cell Lysis and Heat Purification of the Lysate

10.2.2.1 Materials and Equipment

- Lysozyme (500 mg)
- Phenylmethylsulfonyl fluoride (PMSF; 8.7 μg)
- 3-(*N*-Morpholino)propanesulfonic acid (MOPS; 209 mg)
- Distilled water (500 mL)

- Branson sonifier 450 (cell disruptor)
- Cooled centrifuge (10 000× g)
- Incubation bath (58 °C)
- Bicinchoninic acid assay (BCA) protein assay kit (Pierce)

10.2.2.2 Procedure

1. *E. coli* cell pellets obtained from Procedure 1 (100 g) were thawed and resuspended in 2 mM MOPS buffer, pH 7 (500 mL), to which lysozyme (1 $mg.mL^{-1}$) and PMSF (0.1 mM) were added. The suspension was incubated on ice for 30 min.
2. The suspension was placed in an ice bath and sonicated for 2 × 2 min (50% duty cycle, output level 3).
3. The suspension was partially purified by incubating at 58 °C for 15 min to denature any contaminating proteins. The partially purified lysate was then centrifuged again for 30 min (10 000× g, 4 °C).
4. The protein concentration of the mutant BaSP preparation was measured in triplicate with the BCA protein assay kit, using BSA as standard. The resulting heat-purified lysate (~20 $mg.mL^{-1}$) was stored at 4 °C.

10.2.3 Procedure 3: Enzymatic Synthesis of Kojibiose

10.2.3.1 Materials and Equipment

- Sucrose (6 kg)
- Glucose (350 g)
- Distilled water (10 L)
- Immobilised glucose isomerase (200 g, Sweetzyme® IT Extra, Novozymes™)
- Ultrapure water
- 25% v/v NaOH (aq) (100 mL)
- 7.5 mM NaOH (aq) (100 mL)
- 5 M H_2SO_4 (aq) (100 mL)
- 3× 100 mL Schott bottles
- Fermenter (30 L)
- Centrifuge (10 000× g)
- Fritted column with hot-water jacket (55 °C)

10.2.3.2 Procedure

1. Sucrose (6 kg, 1.8 M) and glucose (350 g, 0.2 M) were added to a reactor vessel in a total volume of 9.5 L distilled water. Aqueous solutions of 25% v/v NaOH (100 mL) and 5 M H_2SO_4 (100 mL) were prepared in 100 mL Schott bottles.
2. The mixture was supplemented with 500 mL of heat-purified BaSP mutant from Procedure 2 to a final concentration of 1 $mg.mL^{-1}$. The reaction was incubated at 55 °C and mixed gently at 35 rpm. The pH of the mixture was kept at 7.0 by adding drops of NaOH or H_2SO_4 solution when necessary.
3. After 4 hr immobilised glucose isomerase (20 $g.L^{-1}$) was added. Prior to its application, the immobilised Sweetzyme® IT Extra was allowed to equilibrate at room temperature

for 30 min in 20 volumes of ultrapure water and then washed five times with 20 volumes of ultrapure water at room temperature.
4. Samples were taken at regular time intervals and analysed by high-performance anion-exchange chromatography coupled with pulsed amperometric detection (HPAEC-PAD). Each sample (50 µL) was inactivated by adding 7.5 mM NaOH (aq) (450 µL) and leaving it on ice for 10 min, then centrifuging (10 000× g) for 5 min. The supernatant was adequately diluted in ultrapure water.
5. After 4 days, the reaction was stopped by removing the glucose isomerase via filtration over a fritted column with a hot-water jacket (55 °C) followed by heating to 90 °C for 10 min. The mixture was then cooled to 30 °C.

10.2.4 Procedure 4: Isolation of Kojibiose

10.2.4.1 Materials and Equipment

- Spray-dried baker's yeast, *Saccharomyces cerevisiae* (300 g, Algist Bruggeman)
- 2× Seitz® K Series Depth Filter
- Vacuum filtration system
- Ethanol (2 L)
- Refractometer
- Rotary evaporator (Rotavapor R-200; Büchi)

10.2.4.2 Procedure

1. The contaminating carbohydrates (glucose, fructose, sucrose and maltose) were removed by adding spray-dried baker's yeast 30 $g.L^{-1}$ followed by incubation at 37 °C (~12 hr until all carbohydrates were removed).
2. The yeast was separated from the mixture by vacuum filtration over a Seitz® K Series Depth Filter.
3. The residue was evaporated *in vacuo* at 50 °C to a Brix of 48, cooled overnight to room temperature and incubated at 4 °C and 20 rpm for 24 hr. (Brix is w/v % sugar content unit).
4. The obtained crystals were washed with 2 L of ethanol over a Seitz® K Series Depth Filter and dried to the air. The supernatant was crystallised a second time.
5. A white to off-white powder was obtained (3 kg of 99.8% purity after the first crystallisation and 1.2 kg of 99% purity after the second crystallisation, confirmed by HPAEC-PAD and nuclear magnetic resonance (NMR); see Figures 10.1 and 10.2).

10.2.5 Analytical Method

10.2.5.1 HPAEC-PAD Analysis

The reaction course and purity (%) were determined by HPAEC-PAD (Dionex ICS-3000) equipped with a CarboPac PA20 column. The samples were analysed at a constant flow rate of 0.5 $mL.min^{-1}$ at 30 °C for 30 min. After 13 min of isocratic elution of 30 mM NaOH, the concentration was gradually increased to 100 mM over 5 min, kept constant for 3 min and decreased again to 30 mM within 1 min, before undergoing an equilibration of 8 min (see Table 10.1 for the eluent profile and Table 10.2 for retention times).

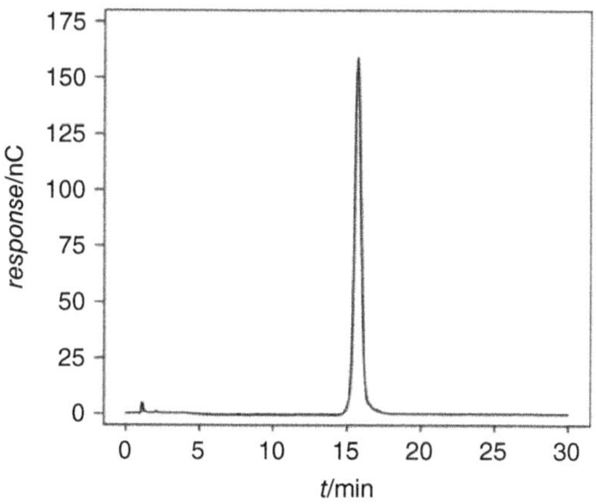

Figure 10.1 *HPAEC-PAD profile of purified kojibiose (note that the small peak at 1 min in the HPAEC profile is the injection peak).*

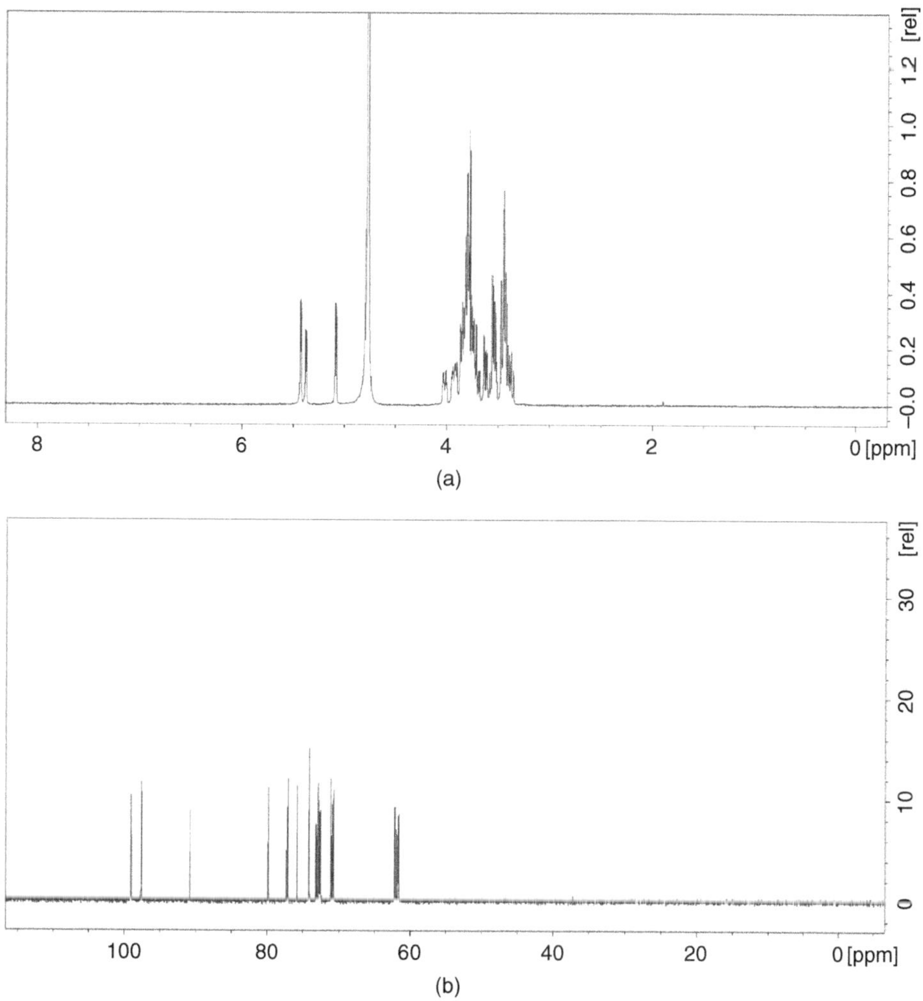

Figure 10.2 *(a) ^1H NMR and (b) ^{13}C NMR spectrum of purified kojibiose.*

Table 10.1 *Eluent profile for HPAEC-PAD analysis.*

Time (min)	Ultrapure water (%)	100 mM NaOH (Baker-analysed) in ultrapure water (%)
0	70	30
13	70	30
18	0	100
21	0	100
22	70	30
30	70	30

Table 10.2 *Retention times for HPAEC-PAD analysis.*

Substance	Retention time (min)
Trehalose	2.0
Glucose	6.1
Fructose	7.2
Sucrose	8.6
Isomaltose	12.1
Kojibiose	16.0
Maltose	26.4

10.2.5.2 NMR Analysis

Solutions for NMR were prepared in D_2O. ^1H NMR spectra were recorded at 400 MHz and ^{13}C NMR spectra at 100.6 MHz using a BRUKER AVANCE III-400.

10.2.6 Conclusion

The described procedure enabled the production of kojibiose from sucrose and glucose in a single biocatalytic step, with recycling of the byproduct fructose to glucose. The rare sugar can now be obtained in larger quantities, enabling a more thorough characterisation of its physicochemical, biological and sensory properties.

References

1. (a) Sanz, M.L., Gibson, G.R. and Rastall, R.A. (2005) *Journal of Agricultural and Food Chemistry*, **53**, 5192–5199: (b) Hodoniczky, J., Morris, C.A. and Rae, A.L. (2012) *Food Chemistry,* **132**, 1951–1958.
2. Verhaeghe, T., De Winter, K., Berland, M. *et al.* (2016) *ChemComm*, **52**, 3687–3689.
3. Beerens, K., De Winter, K., Van de Walle, D. *et al.* (2017) *Journal of Agricultural and Food Chemistry*, **65**, 6030–6041.
4. (a) Aerts, D., Verhaeghe, T., De Mey, M. *et al.* (2011) *Engineering in Life Sciences*, **11**, 10–19; (b) Cerdobbel, A., De Winter, K., Aerts, D. (2011) *Protein Engineering, Design and Selection*, **24**, 829–834.

10.3 Biocatalytic Synthesis of Nigerose Using a Mutant Transglycosylase

Jorick Franceus, Shari Dhaene, Koen Beerens and Tom Desmet*
Centre for Synthetic Biology (CSB), Department of Biotechnology, Ghent University, Ghent, Belgium

With their enormous diversity in structure and function, sugars hold great potential for practical applications. However, many sugars cannot be isolated from nature in significant amounts, hindering their characterisation and commercial exploitation. One of those rare sugars is nigerose, an α-(1,3)-bonded disaccharide of glucose that is present in small amounts in sake and other fermented products. It is also a building block of cell-wall polysaccharides such as nigeran. In spite of its low availability, a few interesting properties have already been reported [1].

Recently, a biocatalytic process was developed for the synthesis of nigerose from sucrose and glucose. When presented with those bulk sugars, the wild-type sucrose phosphorylase from *Bifidobacterium adolescentis* (BaSP) performs a transglucosylation reaction to form a mixture of the α-(1,2)-bonded glucobiose kojibiose and the α-(1,4)-bonded maltose [2]. The regioselectivity of BaSP can be switched by introducing a Q345F mutation, yielding nigerose and maltose instead [3]. The quadruple mutant R135Y/D342G/Y344Q/Q345F (BaSP-YGQF) further improves the enzyme's catalytic efficiency drastically, enabling a more straightforward and efficient synthesis in aqueous solution [4]. In this section, we describe the expression of BaSP-YGQF in *E. coli* and its use for the production of nigerose from sucrose and glucose (Scheme 10.4).

10.3.1 Procedure 1: Recombinant Expression of BaSP-YGQF in *E. coli*

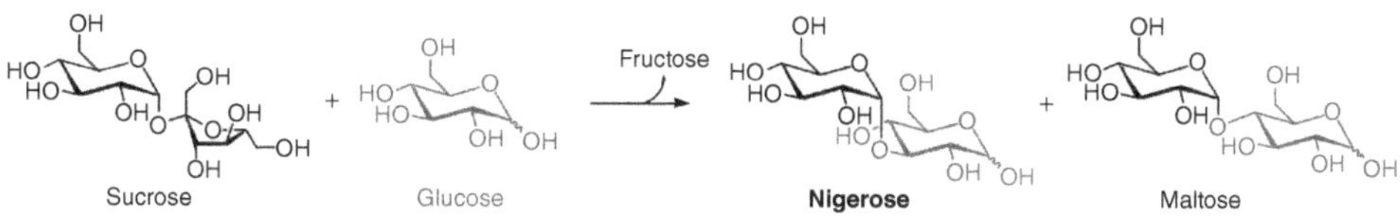

Scheme 10.4 *Transglucosylation of glucose catalysed by BaSP-YGQF. The disaccharides nigerose and maltose are obtained.*

10.3.1.1 Materials and Equipment

- Cryovial containing *E. coli* BL21(DE3) transformed with an expression plasmid pCXP34h bearing a gene encoding *B. adolescentis* sucrose phosphorylase (UniProt code A0ZZH6). The gene should harbour the mutations R135Y, D342G, Y344Q and Q345F for improved synthesis of nigerose, as well as the mutations Q331E, R393N, D445P, D446T, Q460E and E485H for improved thermostability [4, 5]
- Tryptone (300 g)
- Yeast extract (150 g)
- NaCl (150 g)
- Glucose (450 g)
- Ampicillin (1.5 g)
- 25% v/v aqueous NaOH

- Antifoam (10% v/v silicone Snapsil RE 20; VWR BDH Prolabo)
- Distilled water (15 L)
- Glass test tube with stainless-steel cap (50 mL)
- Erlenmeyer flask with cotton plug (2 L)
- 4× Schott bottles (1 L, 1 L, 500 mL, 100 mL) with screw caps
- Fermenter (30 L)
- Shaker (200 rpm)
- Spectrophotometer (600 nm)
- Cooled centrifuge (10 000× g)

10.3.1.2 Procedure

1. Double lysogenic broth (LB) was prepared by dissolving tryptone (20 g.L^{-1}), NaCl (10 g.L^{-1}) and yeast extract (10 g.L^{-1}) in 13 L distilled water. The double LB was added to a glass test tube with stainless-steel cap (5 mL), a 2 L Erlenmeyer flask (500 mL) and a fermenter (12.5 L). Glucose (450 g) was dissolved in 2 L distilled water and the solution was poured into two 1 L Schott bottles. Solutions of 500 mL of aqueous NaOH (25% v/v) and 100 mL of antifoam (10% v/v) were prepared in Schott bottles of 500 mL and 100 mL, respectively. All recipients were autoclaved. The medium was supplemented with ampicillin to a final concentration of 100 mg.L^{-1} prior to use.
2. The test tube was inoculated with the *E. coli* cells and incubated at 37 °C and 200 rpm for 8 hr. The culture was then added to the Erlenmeyer flask and grown overnight under the same conditions.
3. The culture broth and the glucose solution were added to the fermenter for large-scale enzyme production. Temperature, pH, aeration and stirrer speed were set at 37 °C, 7.0, 1.5 vvm and 800 rpm, respectively. Drops of 25% v/v aqueous NaOH were added to maintain the pH at 7. When necessary, antifoam was added manually.
4. The optical density of the culture was measured regularly using a spectrophotometer. When the OD_{600} stopped increasing (~8 hr, OD ~40), the cells were harvested by centrifugation at 10 000× g and 4 °C for 10 min. The resulting pellets (~600 g) were frozen at −20 °C.

10.3.2 Procedure 2: Cell Lysis and Heat Purification of the Lysate

10.3.2.1 Materials and Equipment

- Lysozyme (300 mg)
- Phenylmethylsulfonyl fluoride (PMSF; 5.2 mg)
- Distilled water (300 mL)
- Branson sonifier 450 (cell disruptor)
- Cooled centrifuge (10 000× g)
- Incubation bath (55 °C)
- NanoDrop ND-1000 spectrophotometer (Thermo Scientific)

10.3.2.2 Procedure

1. *E. coli* cell pellets obtained from Procedure 1 (60 g) were thawed and resuspended in distilled water (300 mL), to which lysozyme (300 mg) and PMSF (0.1 mM) were added. The suspension was incubated on ice for 30 min.

2. The suspension was placed in an ice bath and sonicated for 3 × 3 min (50% duty cycle).
3. To remove cell debris, the lysate was centrifuged at 10 000× g and 4 °C for 45 min. The debris was discarded.
4. The supernatant was incubated at 55 °C for 1 hr to denature any contaminating proteins. The partially purified lysate was then centrifuged again at 10 000× g and 4 °C for 45 min.
5. The protein concentration (~20 g.L^{-1}) of the BaSP-YGQF preparation was measured using a NanoDrop spectrophotometer. The extinction coefficient was 69 790 M^{-1}.cm^{-1} and the molecular weight 56.1 kDa.

10.3.3 Procedure 3: Enzymatic Synthesis of Nigerose

10.3.3.1 Materials and Equipment

- Sucrose (256 g)
- Glucose (135 g)
- Distilled water (250 mL)
- Aqueous NaOH (25% v/v)
- Aqueous H_2SO_4 (5 M)
- 2× 100 mL Schott bottles
- Reactor vessel (2 L)
- Centrifuge (10 000× g)
- Dionex ICS-3000 (high-performance anion-exchange chromatography, HPAEC) equipped with a CarboPac PA20 column and pulsed amperometric detection (PAD)
- Eluent: 100 mM NaOH ('Baker Analysed') and 10 mM NaOAc in ultrapure water

10.3.3.2 Procedure

1. Sucrose (256 g, 1.5 M) and glucose (135 g, 1.5 M) were added to a reactor vessel. Solutions of 25% v/v aqueous NaOH (100 mL) and 5 M H_2SO_4 (100 mL) were prepared in 100 mL Scott bottles.
2. Part of the heat-purified protein preparation (250 mL) was added to the reactor vessel and diluted with distilled water, so that the final protein concentration was 10 g.L^{-1}, with a total reaction volume of 500 mL. The reaction was incubated at 52 °C and mixed gently at 35 rpm. The pH of the mixture was kept between 6.5 and 7.0 by adding drops of aqueous NaOH or H_2SO_4 when necessary.
3. Samples were taken at regular time intervals and analysed by HPAEC-PAD. Each sample (10 μL) was analysed at a constant flow rate of 0.5 mL.min^{-1} at 30 °C for 9 min. The eluent composition was 100 mM NaOH and 10 mM NaOAc in ultrapure water (see Table 10.3 for retention times).

Table 10.3 *Retention times for HPAEC-PAD analysis.*

Substance	Retention time (min)
Glucose	2.0
Fructose	2.3
Sucrose	3.0
Nigerose	6.0
Maltose	6.7

4. When no additional synthesis of nigerose was observed (~52 hr), the reaction was stopped by heating the mixture to 95 °C for 15 min. The mixture was then cooled and centrifuged at 10 000× g for 30 min, after which the precipitate was discarded.
5. Nigerose was synthesised with a yield of 645 mM (221 $g.L^{-1}$) or 43% (mole/mole) relative to sucrose.

10.3.4 Procedure 4: Isolation of Nigerose

10.3.4.1 Materials and Equipment

- Glucoamylase (5000 U; Megazyme)
- Invertase (5000 U; Megazyme)
- Centrifuge (10 000× g)
- Sugar-purification system (Knauer): preparative liquid chromatography (LC) system equipped with a 75 °C heated Vertex Plus AX column (dimensions: 250 × 30 mm) loaded with Knauer Eurokat Na-resin (mesh 25–56 μm) and a refractive index detector
- Ultrapure water
- Refractometer
- Rotary evaporator (Rotavapor R-200, Büchi)

10.3.4.2 Procedure

1. The contaminating disaccharides (sucrose and maltose) were hydrolysed by adding glucoamylase and invertase (10 $U.mL^{-1}$), then incubated at 37 °C for 30 min, inactivated at 95 °C for 10 min and centrifuged at 10 000× g for 30 min. It is important to make sure that the added enzymes do not degrade the product. Minor activity on nigerose is possible once the other disaccharides have been hydrolysed. Fewer enzymatic units can be added to slow the reaction if desired, allowing tighter control over the process by taking samples and analysing them using HPAEC-PAD as described in Procedure 3.
2. Nigerose was isolated from the remaining monosaccharides and Maillard reaction products by preparative LC with a sugar-purification system (Knauer). The system was equipped with a 75 °C heated Vertex Plus AX column (250 × 30 mm) loaded with Knauer Eurokat Na-resin (mesh 25–56 μm) and a refractive index detector. The flow rate was 7.5 $mL.min^{-1}$. After flushing the system with ultrapure water for 1 min, 2 mL of the reaction mixture (Brix 19.5%) was injected on to the column (2 $mL.min^{-1}$). The eluent was ultrapure water. The nigerose fraction was collected, whilst the other fractions were discarded (see Table 10.4 for retention times). A new run was started automatically after 24 min.

Table 10.4 *Retention times for the sugar-purification system.*

Substance	Retention time (min)
Maillard products	7.5–11.0
Nigerose	11.2–13.0
Glucose and fructose	13.2–16.2

3. Water was removed from the nigerose fraction using a rotary evaporator under complete vacuum, with a 60 °C water bath.
4. A white to off-white powder was obtained (98 g; yield of 38% relative to sucrose).

10.3.5 Conclusion

The described procedure enabled the production of nigerose from sucrose and glucose in a single biocatalytic step. The rare sugar can now be obtained in larger quantities, enabling a more thorough characterisation of its physicochemical, biological and sensory properties.

References

1. (a) Murosaki, S., Muroyama, K., Yamamoto, Y. *et al.* (1999) *Bioscience, Biotechnology, and Biochemistry*, **63**, 373–378; (b) Sanz, M., Gibson, G. and Rastall, R. (2005) *Journal of Agricultural and Food Chemistry*, **53**, 5192–5199; (c) Al-Otaibi, N.A.S., Cassoli, J.S., Martins-de-Souza, D. *et al.* (2019) *Gigascience*, **8**, giy155.
2. Verhaeghe, T., De Winter, K., Berland, M. *et al.* (2016) *ChemComm*, **52**, 3687–3689.
3. Kraus, M., Gorl, J., Timm, M. and Seibel, J. (2016) *ChemComm*, **52**, 4625–4627.
4. Franceus, J., Dhaene, S., Decadt, H. *et al.* (2019) *ChemComm*, **55**, 4531–4533.
5. (a) Aerts, D., Verhaeghe, T., De Mey, M. *et al.* (2011) *Engineering in Life Sciences*, **11**, 10–19; (b) Cerdobbel, A., De Winter, K., Aerts, D. *et al.* (2011) *Protein Engineering, Design and Selection*, **24**, 829–834.

10.4 Easy Sulfation of Phenols by a Bacterial Arylsulfotransferase

Ron Wever* and Aloysius F. Hartog
Van 't Hoff Institute for Molecular Sciences, Faculty of Science, University of Amsterdam, Amsterdam, The Netherlands

The use of an enzymatic method to synthesise sulfated compounds is gaining more and more ground [1]. The popularity is mainly due to the availability of bacterial arylsulfotransferases (ASTs) and the convenient expression methods of the recombinant enzymes. Unlike the eukaryotic sulfotransferases, which use the very expensive 3'phosphoadenosine-5'phosphosulfate (PAPS) and are in general substrate-specific, these bacterial enzymes use cheap sulfate donors such as *p*-nitrophenyl sulfate (*p*-NPS) and *N*-hydroxysuccinimide sulfate [2]. Many different compounds, including nonphenolics, have been reported to be sulfated under mild conditions [3]. We have developed and used the recombinant AST from *Desulfitobacterium hafniense*, which is easily expressed in *Escherichia coli*. Here, we report as an example the enzymatic sulfation of *E*-resveratrol (Scheme 10.5). Through careful selection of the concentration of the donor and the incubation time, it was possible to obtain the four different sulfated resveratrol derivatives.

10.4.1 Procedure 1: AST Preparation

10.4.1.1 Materials and Equipment

- Lysogenic broth (LB) medium containing 50 mg.mL^{-1} kanamycin
- Isopropyl-β-D-1-thiogalactopyranoside (IPTG)

- 1 M Tris base
- Tris-HCl buffer (pH 8)
- Diethylaminoethyl (DEAE) sephacel
- Glass column (100 mL)
- 1 M aqueous NaCl
- POROS anion-exchange (Applied Biosystems)
- Amersham Pharmacia fast protein liquid chromatography (FPLC) system
- Tris-glycine buffer (pH 9.1)
- *p*-Nitrophenyl sulfate (PNS)
- Phenol

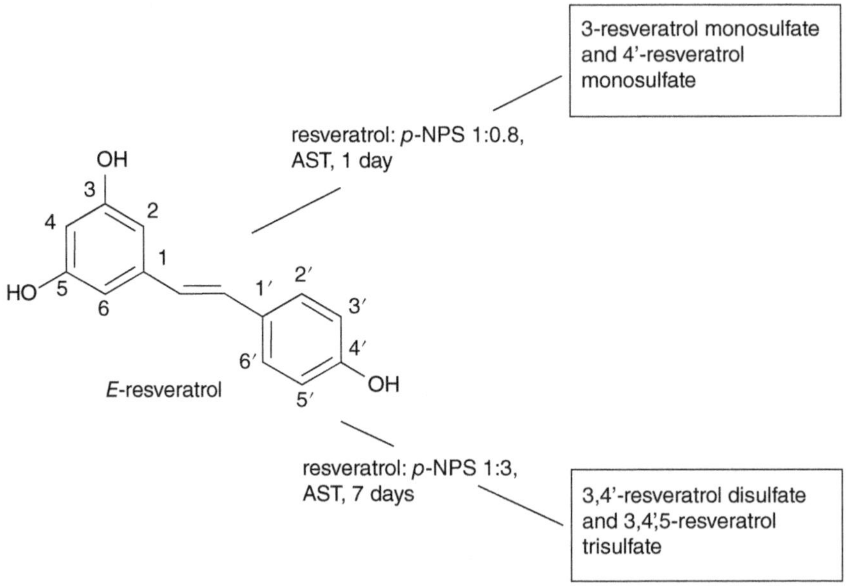

Scheme 10.5 *Enzymatic Sulfation of* E-*resveratrol by arylsulfotransferase (AST).*

10.4.1.2 Procedure

The sequence of AST from the dehalorespiring bacterium *D. hafniense* coding for a protein of 628 residues was found in the National Center for Biotechnology Information (NCBI) protein database. The gene coding for AST was amplified from *D. hafniense* genomic DNA by polymerase chain reaction (PCR) and cloned [1b] in vector pET26b, resulting in pJFT006. For details on this and on the sequences of the *N*- and *C*-terminal region of the enzyme, see [1b]. The AST was expressed and purified using a previously developed method [1b] with minor modifications. Briefly, *E. coli* BL21 (DE3) with pJFT006 (carrying the *D. hafniense* AST gene) was grown and 1 mM IPTG was used to induce expression at 25 °C overnight.

The cells were first frozen to −80 °C and then thawed. Freezing and thawing resulted in cell lysis and greatly increased the enzyme yield. After removal of the cell debris by centrifugation (20 000× g for 30 min), DEAE Sephacel was added, and the pH was increased to 8 with 1 M Tris base. The DEAE Sephacel was poured into a column, which was washed extensively with 100 mM NaCl in 50 mM Tris-HCl (pH 8). The AST was eluted by a stepwise NaCl gradient. Fractions containing AST activity were dialysed and applied on to a POROS column (Applied Biosystems) using an Amersham Pharmacia FPLC system and the enzyme was eluted using a linear gradient towards 1 M NaCl. The purified sample (10 U.mg^{-1}) was dialysed against Tris (50 mM, pH 8.0), concentrated in an Amicon Centricon to 5–10 U.mL^{-1} and stored at −20 or −80 °C. 1 U AST is defined as the amount of enzyme catalysing the formation of 1 μmol *p*-NP.min^{-1} in a system containing 5 mM of *p*-nitrophenyl sulfate as sulfate donor and 5 mM of phenol as acceptor in 50 mM Tris-glycine buffer (pH 9.1) at 30 °C. The extinction coefficient *p*-NP is 16.6 mM^{-1}.cm^{-1}.

10.4.2 Procedure 2: Biocatalytic Sulfation of Resveratrol to 3-Resveratrol, 4'-Resveratrol, 3,4'-Resveratrol Disulfate and 3,4',5-Resveratrol Trisulfate

10.4.2.1 Material and Equipment

- 100 mM Tris-glycine buffer (pH 9.1)
- Acetone
- *n*-Butanol
- Methanol
- 1 M acetic acid
- Ethyl acetate
- *E*-Resveratrol (Sigma R5010)
- *p*-Nitrophenyl sulfate K-salt (Sigma N3877)
- 20 × 20 cm TLC plate silica 60 F254 (Macherey-Nagel)
- Flash column filled with Silica 60 (Merck)
- $NH_4H_2PO_4$ aqueous buffer (25 mM, pH 6.3)
- 1 M tetrabutylammonium phosphate ion-pairing reagent (TBAP; Sigma)
- Acetonitrile

10.4.2.2 Procedure

10.4.2.2.1 E-Resveratrol-3-Sulfate Ester and E-Resveratrol-4'-Sulfate Ester.

1. AST (5 U) was added to a suspension containing *E*-resveratrol (30 mg, 0.14 mmol), 3 mL acetone, *p*-nitrophenyl sulfate K-salt (28 mg, 0.13 mmol) and 50 mM Tris-glycine buffer, pH 9.1 (17 mL).
2. After incubation at room temperature for 1 day, the solution became clear and yellow and high-performance liquid chromatography (HPLC) measurements showed that all *p*-NPS had been reacted.
3. Acetone was flushed off with nitrogen, and the *p*-nitrophenol formed was removed by extraction with ethyl acetate (2 × 20 mL).
4. A few drops of 1 M acetic acid were added to the reaction solution (pH 8.6) to bring the pH to 5, and the solution was again extracted with 1 × 20 mL ethyl acetate.
5. The sulfated compounds were extracted by 2 × 20 mL *n*-butanol. The *n*-butanol solution was removed by distillation under reduced pressure and the product was dissolved in 1–2 mL methanol.

6. To purify the sulfated compounds preparative TLC was used with the eluent n-butanol/methanol saturated with NH_3 19 : 1 (methanol was saturated with NH_3 and stored at 4 °C). The bands at Rf 0.7 (4'-sulfate), Rf 0.6 (3-sulfate) and Rf 0.0–0.1 (3,4'-disulfate) were scraped off and eluted with 3 mL methanol. After drying, 3 mg *E*-resveratrol-4'-sulfate ester (NH_4^+ salt), yield 7% and 13 mg *E*-resveratrol-3-sulfate ester (NH_4^+ salt), yield 31% were obtained.

10.4.2.2.2 E-Resveratrol-3,4'-Disulfate Ester and E-Resveratrol-3,4',5-Trisulfate Ester.

1. AST (75 U) was added to a suspension containing *E*-resveratrol (68 mg, 0.30 mmol), *p*-nitrophenyl sulfate K-salt (250 mg, 1.14 mmol) in 50 mM Tris-glycine buffer, pH 9.1 (30 mL, final volume).
2. After 1 day, all the resveratrol was solubilised and most was converted into the 3,4'-disulfated compound. The third OH group at position 5 was more slowly sulfated to about 40% over subsequent days.
3. The incubation was continued for 7 days. The pH was checked each day and corrected by addition of 1 M Tris, pH 9.1.
4. The reaction mixture ultimately consisted of 6 mM *E*-resveratrol-3,4'-disulfate ester and 4 mM *E*-resveratrol-3,4',5-trisulfate ester, as calculated from the HPLC data.
5. The solution was extracted with ethyl acetate (3 × 20 mL) and the pH was brought to 5 with 1 M acetic acid. Extraction was undertaken with ethyl acetate (2 × 20 mL) to remove *p*-NP.
6. Water was removed by distillation under reduced pressure and the residue was dissolved in acetonitrile and applied to a silica flash column. The sulfated compounds were eluted with acetonitrile containing 0.5% acetic acid/methanol, where the methanol concentration was increased in a stepwise fashion (5, 10, 15 and 20%). Fractions of 15 mL were collected, where early fractions contained pure 3,4'-disulfated resveratrol and later ones contained a mixture of di- and trisulfated resveratrol.
7. The mixture of di- and trisulfated resveratrol was further purified by preparative TLC on silica using 3 : 2 : 1 n-butanol/water/acetic acid as eluent. The band at Rf 0.6 was pure 3,4'-disulfate and that at Rf 0.3 was 3,4',5-trisulfate resveratrol. Both compounds were scraped off the plate and extracted from the silica with methanol. After evaporation of the solvent, 59 mg of solid 3,4'-disulfated *E*-resveratrol, yield 51% and 41 mg 3,4',5-trisulfated *E*-resveratrol, yield 29% were obtained.

General Remarks.

- The optimal pH of the enzyme is 9.2. When the pH of the incubation dropped below 8.8, it was raised to 9.1 using 1 M Tris.
- 15% acetone will solubilise resveratrol but rapidly decreases the enzyme activity to about 50%. This was used to obtain monosulfated resveratrol.
- All reactions and purification steps were monitored by HPLC.
- Note that under acidic conditions, these monosulfate esters are less stable.
- The 4'-monosulfate was converted into the 3,4'-disulfated compound more rapidly than was the 3-sulfate.

10.4.3 Analytical Method

10.4.3.1 HPLC and Columns

Samples from the reaction mixture were diluted 100 times with acetonitrile/water 1 : 1 and 25 µL was injected on to a Nucleosil 100-5 C18 HD column (plus guard column), which was equilibrated with solvent (A) $NH_4H_2PO_4$ buffer (25 mM, pH 6.3) containing 5 mM TBAP and 5% acetonitrile. An Agilent 1100 HPLC system was used with a 0–80% gradient (34 min run, 0.4 mL.min^{-1} flow rate) of solvent (B) 10% H_2O/90% acetonitrile and 5 mM TBAP. The UV absorbance of the compounds was monitored at 300 nm. To calculate the degree of sulfate transfer, peak areas of the detected substrates and products were used. Retention times: *E*-resveratrol, 20.7 min; 4'-resveratrol monosulfate, 22.8 min; 3-resveratrol monosulfate, 23.4 min; 3,4'-resveratrol disulfate, 24.7 min; 3,4',5-resveratrol trisulfate, 25.9 min.

10.4.3.2 NMR Data

E-Resveratrol. ^{1}H NMR (400 MHz, DMSO-D_6): δ 6.11 (1H, t), 6.47 (2H, d), 6.76 (2H, d), 6.80–6.96 (2H, q), 7.39 (2H, d), 9.19 (2 OH), 9.53 (1 OH).

^{1}H NMR (400 MHz, CD_3OD): δ 6.15–6.16 (1H, t), 6.45 (2H, d), 6.75–6.77 (2H, d), 6.78–6.98 (2H, q), 7.33–7.37 (2H, d).

^{13}C NMR (400MHz, CD_3OD): δ 159.67 (2xC), 158.39, 141.33, 130,43, 129.41, 128.82 (2xC), 127.03, 116.50 (2xC), 105.78 (2xC), 102.65.

E-Resveratrol 3-Sulfate. ^{1}H NMR (400 MHz, D_2O): δ 6.59–6.61 (1H, t), 6.85–6.88 (3H, m), 6.93 (1H, d), 6.98–7.24 (2H, q), 7.52–7.55 (2H, d).

^{1}H NMR (400 MHz, CD_3OD) as NH_4^+ salt (>90%): δ 6.69 (1H, t), 6.76–6.81 (3H, m), 7.00 (1H, t), 6.87–7.08 (2H, q), 7.39–7.41 (2H, dd).

^{13}C NMR (400 MHz, CD_3OD) as NH_4^+ salt: δ 130,67, 129.50 (2xC), 126.95, 127.03, 117.05 (2xC), 111.94, 111.18, 109.04.

E-Resveratrol 4'-Sulfate (3 mg 80% Pure, Contamination by 3-Sulfate). ^{1}H NMR (400 MHz, CD_3OD) as NH_4^+ salt: δ 6.20 (1H, t), 6.50–6.51 (2H, d), 6.94–7.07(2H, q), 7.29–7.32 (2H, d), 7.50–7.52 (2H, d).

^{13}C NMR (400MHz, CD_3OD): δ 160.62 (2xC), 154,35, 141.72, 136.38, 130.34, 129.63. 129.06 (2xC), 123.57 (2xC), 106.90 (2xC), 104.01.

E-Resveratrol 3,4'-Disulfate. ^{1}H NMR (500 MHz, CD_3OD): δ 6.72 (1H, t), 6.80 (1H, t), 7.01–7.05 (2H, m), 7.11–7.15 (1H, d), 7.30–7.32 (2H, d), 7.53–7.55.

^{13}C NMR (300MHz, CD_3OD): δ 160.26, 155.03, 154.44, 141.51, 136.25, 130.33, 129.83, 129.19 (2xC), 123.55 (2xC), 112.53, 111.76, 109.85.

E-Resveratrol 3,4',5-Trisulfate. ^{1}H NMR (400 MHz, CD_3OD): δ 7.18 (1H, t), 7.09–7.24 (2H, q), 7.32–7.34 (2H, d), 7.39 (2H, d), 7.56–7.58 (2H, d).

^{13}C NMR (400 MHz, CD_3OD): δ 130.83, 129.02 (2xC), 128.83, 123.24 (2xC), 117.23 (2xC), 115.52.

10.4.4 Conclusion

We have shown that it is possible to obtain the different sulfated derivatives of *E*-resveratrol by a simple enzymatic procedure and through careful selection of enzyme concentration and incubation times. The *p*-NP formed must be removed by repeated and time-consuming extraction using ethyl acetate. A new sulfate donor, *N*-hydroxysuccinimide sulfate, has been reported [2] that may replace *p*-NPS in the sulfation of various compounds. This donor hydrolyses easily and is much less toxic and safer in use compared to *p*-nitrophenol.

10.4.4.1 Other Substrates Sulfated by this Enzymatic System

The following substrates have been reported for the AST: estrone, 17-beta-estradiol, enkephalin, nonphenolic alcohols (2-phenylalcohol, glucose, cylcohexanol, glycerol), various phenolics including catechols, quercetin, silybin, bisphenol A, 4-hydroxycoumaric acid and lignin. Also, the carbohydrate derivatives *p*-NP-glycerol (*R*), *p*-NP-β-D-glucopyranoside, phenyl-β-D-glucopyranoside, *p*-NP-β-D-*N*- acetylglucosamine, *N*-acetyl glucosamine and cellobiose have been reported to be substrates for the AST [1, 3].

References

1. (a) Kobashi, K., Kim, D.H. and Morikawa, T. (1987) *Journal of Protein Chemistry*, **6**, 237–244; (b) Van der Horst, M.A., Van Lieshout, J.F.T., Bury, A. *et al.* (2012) *Advanced Synthesis & Catalysis*, **354**, 3501–3508; (c) Marhol, P., Hartog, A.F., Van der Horst, M.A. *et al.* (2013) *Journal of Molecular Catalysis B: Enzymatic*, **89,** 24–27; (d) Ayuso-Fernandez, I., Galmés M.A., Bastida, A. and Garcia-Junceda, E. (2014) *ChemCatChem*, **6**, 1059–1065.
2. Hartog, A.F. and Wever, R. (2016) *Journal of Molecular Catalysis B: Enzymatic*, **129**, 43–46.
3. (a) Van der Horst, M.A., Hartog, A.F., El Morabet, R. *et al.* (2015) *European Journal of Organic Chemistry*, **2015,** 534–541; (b) Hartog, A.F. and Wever, R. (2015) *Advanced Synthesis & Catalysis*, **357**, 2629–2632; (c) Purchartová, K., Valentová, K., Pelantová, H. *et al.* (2015) *ChemCatChem*, **7**, 3152–3162; (d) Prinsen, P., Narani, A., Hartog, A.F. *et al.* (2017) *ChemSusChem*, **610**, 2267–2273; (e) Islam, S., Laaf, D., Infanzon, B. *et al.* (2018) *Chemistry: A European Journal*, **24**, 17 117–17 124; (f) Islam, S., Mate, D.M., Martinez, R. *et al.* (2018) *Biotechnology and Bioengineering*, **115**, 1106–1115; (g) Islam, S., Mate, D.M., Martinez, R. *et al.* (2018) *Biotechnology and Bioengineering*, **115**, 1106–1115; (h) Ji, Y., Islam, S., Mertens, A.M. *et al.* (2019) *Applied Microbiology and Biotechnology*, **103**, 3761–3771.

10.5 Shikimate Kinase-Catalysed Phosphorylations and Synthesis of Shikimic Acid 3-Phosphate by AroL-Catalysed Phosphorylation of Shikimic Acid

Bernhard Schoenenberger,[1] Agata Wszolek,[2] Roland Meier,[1] Henrike Brundiek,[2] Markus Obkircher[1] and Roland Wohlgemuth[1,3]
[1] *Sigma-Aldrich/Merck KGaA, Buchs, Switzerland*
[2] *Enzymicals, Greifswald, Germany*
[3] *Institute of Technical Biochemistry, Technical University Lodz, Lodz,Poland*

Small-molecule kinases are attractive catalysts for chemo-, regio- and enantioselective *O*-phosphorylations of compounds with multiple hydroxy groups [1] in one reaction step without the use of protecting groups, such as the selective phosphorylation of cyclic trihydroxycarboxylic acids catalysed by shikimate kinase (Scheme 10.6). Shikimate kinase EC 2.7.1.71 is essential for microorganisms, certain parasites and plants. Depending on the microbial or plant species, different numbers of isozymes will be present. Its absence in humans and animals makes shikimate kinase an attractive target enzyme for the development of enzyme inhibitors aiming at treating infectious, parasitic or plant diseases.

As shown in Scheme 10.6, the high selectivity in catalysing the phosphorylation of only one out of three hydroxy groups is remarkable [2]. The structures, catalytic mechanisms and individual steps of the phosphorylation reactions catalysed by shikimate kinases of a number of microbial pathogens, such as *Mycobacterium tuberculosis* [3], *Helicobacter pylori* [4] and methicillin-resistant *Staphylococcus aureus* [5], are of major interest in the development of shikimate kinase inhibitors as new antimicrobials, for which the reaction product shikimate 3-phosphate is also needed [6]. Whilst bacterial and eukaryotic shikimate kinases belong to the NMP kinase superfamily, archaeal shikimate kinases have been found within the fold characteristic of the GHMP kinase superfamily of the small-molecule kinases [7]. In *Escherichia coli*, the phosphorylation of shikimic acid is catalysed by the two isoenzymes, shikimate kinases I and II, which differ in their K_m. The catalytically more active shikimate kinase II, corresponding to the *aroL*-encoded enzyme AroL, has a K_m of about 200 μM for shikimate, whereas shikimate kinase I, corresponding to the *aroK*-encoded enzyme AroK, has a low affinity with a K_m of more than 20 mM [8].

Shikimic acid 3-phosphate is a central microbial, parasite and plant metabolite as a product of the shikimate kinase-catalysed reaction and is also an enzyme substrate for the subsequent step catalysed by 5-enolpyruvoyl-shikimate 3-phosphate synthase, a target of anti-infectives and of herbicides. The various approaches to the chemical synthesis of shikimic acid 3-phosphate have been faced with several challenges due to the number of reaction steps, reaction times and hazardous/toxic chemicals. Therefore, an easily scalable one-step procedure has been developed for the phosphorylation of shikimic acid in the 3-position (see Scheme 10.7) without the use of protecting groups using recombinant *E. coli* K12 shikimate kinase AroL [9].

After completion of the AroL-catalysed phosphorylation, the shikimate 3-phosphate was purified by calcium salt precipitation, cation-exchange chromatography and final conversion into its lithium salt. This efficient biocatalytic method has overcome the bottlenecks and problems of the multistep chemical syntheses and is superior to any other method [9].

10.5.1 Procedure 1: Recombinant Expression of Shikimate Kinase AroL in *E. coli* BL21(DE3)

10.5.1.1 Materials and Equipment

- *E. coli* BL21(DE3) cells
- Lysogenic broth (LB) plates (agar, 15 g.L^{-1}; tryptone, 10 g.L^{-1}; NaCl, 10 g.L^{-1}; yeast extract, 5 g.L^{-1}; with kanamycin, 50 μg.mL^{-1}; pH 7.5)
- LB medium (tryptone, 10 g.L^{-1}; NaCl, 10 g.L^{-1}; yeast extract, 5 g.L^{-1}; with kanamycin, 50 μg.mL^{-1}; pH 7.5)
- Terrific broth (TB) medium (casein, enzymatically digested, 12 g.L^{-1}; yeast extract, 24 g.L^{-1}; K_2HPO_4, 12.54 g.L^{-1}; KH_2PO_4, 2.31 g.L^{-1}; with kanamycin, 50 μg.mL^{-1}; pH 7.5)

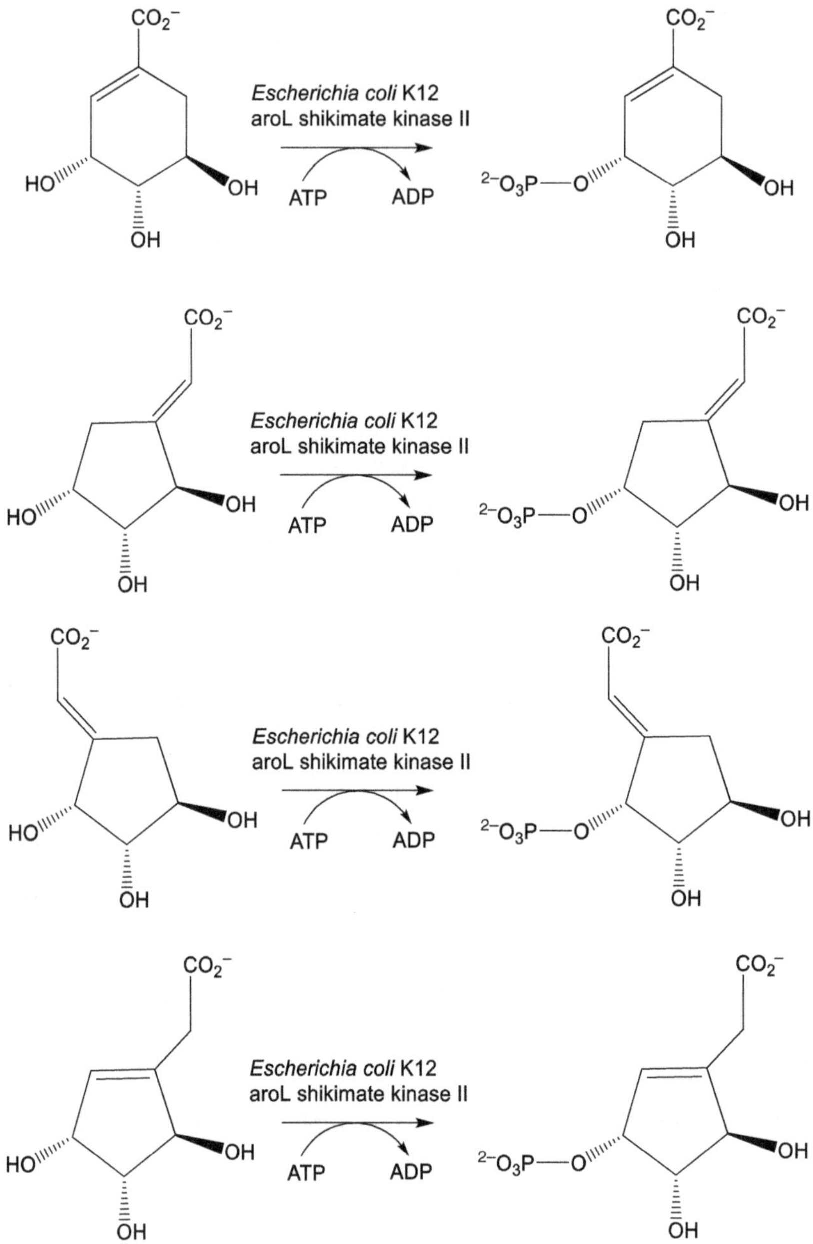

Scheme 10.6 *Shikimate kinase-catalysed phosphorylations of cyclic trihydroxycarboxylic acids.*

- Isopropyl-β-D-1-thiogalactopyranoside (IPTG)
- Sodium phosphate buffer pH 7.5 (50 mM)
- Tris-HCl buffer pH 7.5 (50 mM)
- Imidazole
- Incubator
- Water bath
- Erlenmeyer flasks with baffles (250 mL and 2 L)
- Orbital shaker
- Centrifuge
- Spectrophotometer
- Retsch Mixer Mill MM 200
- Glass beads (Ø 0.25–0.5 mm)
- Chromatography equipment
- Talon® resin
- Zeba spin columns

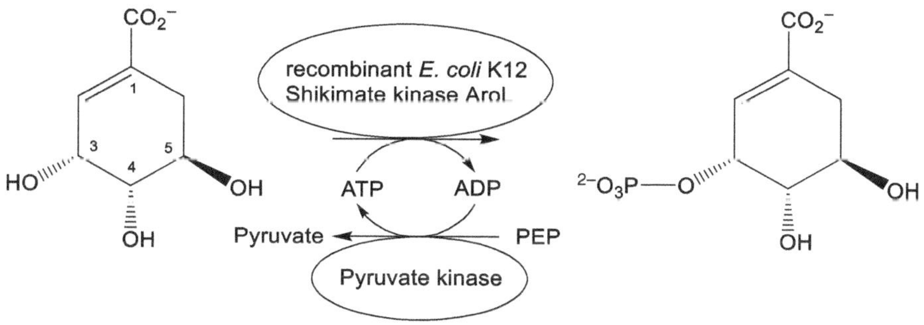

Scheme 10.7 *Recombinant* E. coli *K12 shikimate kinase AroL-catalysed phosphorylation of (3*R*,4*S*,5*R*)-3,4,5-trihydroxy-1-cyclohexene-1-carboxylic acid (Shikimic acid) to (3*R*,4*S*,5*R*)-3,4,5-trihydroxy-1-cyclohexene-1-carboxylic acid 3-phosphate (Shikimic acid 3-phosphate) using the phosphoenolpyruvate/pyruvate kinase-system for ATP regeneration.*

10.5.1.2 Procedure

1. The shikimate kinase gene *aroL* was derived from *E. coli* K12 (UniProtKB: P0A6E1) and a codon optimisation algorithm was employed for expression in *E. coli* [9]. The codon-optimised synthetic *aroL* gene was equipped with an *N*-terminal 6x-his-tag and a TEV protease cleavage site in order to have the option of removing the affinity tag. The gene was synthesised and subcloned into pET24a(+) via NdeI and NotI by Genscript, giving vector pET24-His(TV)-AroL-Ecoli(op).
2. Chemically competent BL21 (DE3) *E. coli* (100 μL) cells (100 μL aliquots of chemically competent *E. coli* BL21 (DE3) cells were stored at −80 °C until usage) were transformed with 5 μL of the derived plasmid by standard procedures [10] and plated on LB agar supplemented with 50 μg.mL^{-1} kanamycin for selection. Selected *E. coli* transformants were used for expression tests, in which different media, induction times and temperatures were optimised. The selected conditions were later used for the overexpression of *aroL* in *E. coli* BL21 (DE3), as shown in the following steps.

3. A single *E. coli* colony harbouring the *aroL* gene was inoculated into 5 mL LB medium with composition according to the standard recipe of Miller, consisting of tryptone ($10\,g.L^{-1}$), yeast extract ($5\,g.L^{-1}$), sodium chloride ($10\,g.L^{-1}$), pH 7.0 ± 0.2 in distilled water, autoclaved and supplemented with kanamycin ($50\,\mu g.mL^{-1}$). The solution was incubated at 30 °C and 180 rpm.
4. Overnight cultures of *E. coli* were used to inoculate 50 mL of TB medium supplemented with kanamycin ($50\,\mu g\,mL^{-1}$) in 250 mL baffled flasks. Cells were cultivated at 37 °C under vigorous shaking (150 rpm) until an OD_{600} of 0.5 was reached.
5. When the pre-culture reached the mid-log phase, the main culture was started. Fresh TB medium (400 mL) in a 2 L Erlenmeyer flask was inoculated with the *E. coli* pre-culture (20 mL). Cells were cultivated at 37 °C under shaking (110 rpm) until an OD_{600} of 1.3 was reached.
6. After the culture reached OD_{600} 1.3, protein expression was induced with IPTG to a final concentration of 1 mM and the incubation temperature was shifted to 25 °C and the main culture was shaken at 110 rpm for 20 hr. The *E. coli* cells were centrifuged at 5000 rpm and 4 °C for 30 min and washed once with 40 mL of ice-cold 50 mM sodium phosphate buffer at pH 7.5.
7. The *E. coli* cell pellets containing the recombinant shikimate kinase AroL were frozen at −20 °C.
8. The frozen *E. coli* cells were suspended in Tris-HCl buffer (50mM, pH 7.5), adjusted to OD_{600} 40 and mixed with glass beads (Ø 0.25–0.5 μm) at a ratio of approximately 1 : 1.
9. Disintegration of the cells was achieved mechanically using a cell disruptor (Retsch Mixer Mill MM 200), whereby two cycles of 5 min were performed at 600 rpm.
10. The samples were cooled on ice and insoluble cell debris was centrifuged at 5000 rpm and 4 °C for 30 min. The supernatant contained the soluble AroL.
11. The soluble AroL was incubated with Talon® resin on an orbital shaker at 4 °C for 30 min.
12. Following a standard gravity flow protocol, the resin was washed at 4 °C with 10 column volumes of Tris-HCl buffer (50mM, pH 7.5) containing 10 mM imidazole to remove unspecifically bound proteins.
13. The target enzyme AroL was eluted with 1.5 column volumes of elution buffer (50 mM Tris-HCl buffer, pH 7.5, 150 mM imidazole).
14. Purified AroL was obtained by removing imidazole on desalting columns using 50 mM Tris-HCl buffer pH 7.5 with Zeba Spin columns.

10.5.2 Procedure 2: AroL-Catalysed Phosphorylation of Shikimic Acid to Shikimic Acid 3-Phosphate Lithium Salt

10.5.2.1 Materials and Equipment

- Shikimic acid (4.6 g, 26.4 mmol)
- H_2O (400mL)
- KCl (3.4 g)
- $MgCl_2$ hydrate (1.9 g)
- Phosphoenolpyruvate (PEP) (4.9 g, 23.8 mmol, 0.9 eq)
- Adenosine 5'-phosphate (ATP) (0.4g, 1.34 mmol, 0.05 eq)

- 1 M NaOH
- Pyruvate kinase from rabbit muscle, type III, lyophilised powder (900 U), 350–600 U.mg^{-1} protein (Sigma-Aldrich/Merck P9136)
- Soluble protein fraction of recombinant shikimate kinase AroL (900 μL)
- 1 M HCl
- Calcium acetate (23 g)
- Acetone (400 mL)
- 1× 250 mL round-bottom flask
- 300 g cation exchanger Dowex 50W X8 Hydrogen Form, strongly acidic, 200–400 mesh (Sigma-Aldrich/Merck 44519)
- Ion-exchange chromatography (IEC) equipment
- 1 M aqueous LiOH
- Centrifuge
- Rotary evaporator
- Lyophiliser
- Nuclear magnetic resonance (NMR) spectrometer: ^{1}H-NMR 400 MHz, ^{31}P-NMR 162 MHz

10.5.2.2 Procedure

1. KCl (3.4 g), $MgCl_2$ hydrate (1.9 g), PEP (4.9 g, 23.8 mmol) and ATP (0.47 g, 1.34 mmol) were added to a solution of shikimic acid (4.6 g, 26.4 mmol) in water (90 mL) and the resulting solution was adjusted to pH 7 with 1 M aqueous NaOH.
2. Pyruvate kinase (900 units) and the soluble protein fraction of recombinant shikimate kinase AroL (900 μL) were added to the mxiture.
3. After 18 hr stirring at room temperature, NMR analysis of a sample of the reaction mixture showed complete consumption of PEP.
4. The mixture was adjusted to pH 4 with 1 M aqueous HCl and microfiltered after 1 hr.
5. A solution of calcium acetate (23 g) in water (100mL) was added to the filtrate with stirring. Acetone (400 mL) was then added slowly.
6. After 2 hr, the suspension was filtered and the residue was dried at room temperature under high vacuum and taken up in water (200 mL).
7. To this suspension, cation exchanger (Dowex) in its H^+-form (50 g) was added with stirring.
8. After 0.5 hr, the resulting thin suspension was passed through a column with 250 g of the same ion exchanger and eluted with 1.2 L further water. The eluate was partially concentrated on a rotary evaporator and transformed to the lithium salt by addition of an appropriate amount of 1 M LiOH to adjust to pH 7, which resulted in 3.83 g (53% yield) of a white solid after lyophilisation.

Analytical data for the isolated final product, shikimic acid 3-phosphate lithium salt: ^{1}H-NMR (D_2O, 400 MHz, δ): 6.36 (m, 1H, H(2)), ~4.7 (m, 1H, H(3), covered by HDO signal), 3.95 (m, 1H, H(5)), 3,69 (dd, 1H, H(4)), 2.60 (dd, 1H, H(6)), 2.08 (dd, 1H, H(6’)); ^{31}P-NMR (D_2O, 162 MHz, proton decoupled, δ): 4.32; 13C-NMR (D2O, 101 MHz, proton decoupled, 31P coupled, δ): 175.4(s), 135.3 (s), 130.6 (d), 71.4(d), 69.2 (d), 67.3 (s), 31.9(s). Purity by TLC (silica gel 60 on glass plates, n-propanol/H_2O/AcOH = 6 : 3 : 1), Rf (shikimic acid) 0.85, and Rf (shikimic acid 3-phosphate) 0.72. Purity by HPLC: >97%.

10.5.3 Analytical Method

The bicinchonic acid (BCA) method was used to measure the protein content of the purified protein, with bovine serum albumin (BSA) as standard [9]. AroL overexpression and purity were analysed with 12.5% sodium dodecyl sulfate–polyacrylamide gels (SDS–PAGE), which were stained with Coomassie brilliant blue G250.

The shikimate kinase AroL activity was assayed by adenosine diphosphate (ADP) release in an assay coupled with nicotinamide adenine dinucleotide (NADH) oxidation using pyruvate kinase and lactate dehydrogenase [9]. The activity assays were carried out at room temperature by monitoring the absorbance change of NADH upon oxidation at a wavelength of 340 nm ($\varepsilon = 6.22 \times 10^3$ $M^{-1}.cm^{-1}$). The assay mixture contained 100 mM Tris-HCl buffer pH 7.5, 1mM phosphoenolpyruvate, 5mM $MgCl_2$, 50 mM KCl, 0.15 mM NADH, 4 U of pyruvate kinase from rabbit muscle (Sigma-Aldrich/Merck), 8 U of L-lactic acid dehydrogenase from rabbit muscle (Sigma-Aldrich/Merck), 2.5 mM ATP and 1.6 mM of shikimic acid in a total volume of 1 mL. The reaction was started by the addition of the recombinant shikimate kinase AroL protein. One unit of enzyme activity is defined as the amount of enzyme that catalyses the phosphorylation of 1 µmol of shikimic acid per minute at pH 7.5 and 25 °C.

10.5.4 Conclusion

The production of the highly active and stable recombinant *E. coli* K12 shikimate kinase AroL has been achieved in high yield. AroL has been applied in the ATP-dependent phosphorylation of shikimic acid, which is coupled with the pyruvate kinase-catalysed ATP regeneration to avoid inhibition by ADP. In-process analysis by NMR has facilitated the development of an efficient biocatalytic one-step synthesis of shikimic acid 3-phosphate, which after a standard workup procedure can be prepared in >97% purity and with a 53%yield of the isolated pure lithium salt. This is an excellent starting point for process intensification, covering not only product recovery and purification but also the selection and optimisation of the shikimate kinase. Although the ATP-dependent shikimate kinase AroL was selected for this process due to its favourable kcat/Km and apparent turnover number, optimisation of the biocatalyst properties in order to achieve higher space–time yields at lower costs is of interest. The achieved shikimate kinase AroL-catalysed route has provided a number of new insights and a short preparative access to the key metabolite shikimate-3-phosphate, which previously had not been available globally. The new shikimate kinase AroL-catalysed phosphorylation of shikimic acid is a superior route, as other methods have the disadvantages of a tedious multistep synthesis, costly or commercially unavailable starting materials, use of toxic heavy-metal reagents and tedious workup procedures due to the presence of side products.

Acknowledgement

This section is derived in part from an article published in *Biotechnology Journal*, 16 April 2018, copyright 2018 Wiley-VCH Verlag GmbH & Co. KGaA, Weinheim, available online at https://doi.org/10.1002/biot.201700529 and is reproduced with permission from John Wiley and Sons and Copyright Clearance Center.

References

1. (a) Gauss, D., Schoenenberger, B., Molla, G.S. *et al.* (2016) Biocatalytic phosphorylation of metabolites, in *Applied Biocatalysis – From Fundamental Science to Industrial Applications* (eds A. Liese, L. Hilterhaus, U. Kettling and G. Antranikian), Wiley-VCH, pp. 147–177; (b) Wohlgemuth, R., Liese, A. and Streit, W. (2017) *Trends in Biotechnology*, **35**, 452–465.
2. An, M. and Bartlett, P.A. (2004) *Organic Letters*, **6** (22), 4065–4067.
3. Hartmann, M.D., Bourenkov, G.P., Oberschall, A. *et al.* (2006) *Journal of Molecular Biology*, **364**, 411–423.
4. Cheng, W.-C., Chen, Y.-F., Wang, H.-J. *et al.* (2012) *PLoS ONE*, **7** (3), e33481.
5. Favela-Candia, A., Téllez-Valencia, A., Campos-Almazán, M. *et al.* (2019) *Molecular Biotechnology*, **61**, 274–285.
6. Simithy, J., Gill, G., Wang, J. *et al.* (2015) *Analytical Chemistry*, **87**, 2129–2136.
7. Daugherty, M., Vonstein, V., Overbeek, R. and Osterman, A. (2001) *Journal of Bacteriology*, **183** (1), 292–300.
8. Whipp, M.J. and Pittar, A.J. (1995) *Journal of Bacteriology*, **177** (6), 1627–1629.
9. Schoenenberger, B., Wszolek, A., Meier, R. *et al.* (2018) *Biotechnology Journal*, **13**, 1700529.
10. Hanahan, D. (1983) *Journal of Molecular Biology*, **166**, 557–580.

10.6 Kinase-Catalysed Phosphorylations of Ketohexose Phosphates and LacC-Catalysed Synthesis of D-Tagatose-1,6-Diphosphate Lithium Salt

Bernhard Schoenenberger,[1] Stefanie Kind,[2] Roland Meier,[1] Thorsten Eggert,[2] Markus Obkircher[1] and Roland Wohlgemuth[1,3]

[1] *Sigma-Aldrich/Merck KGaA, Buchs, Switzerland*

[2] *Evoxx Technologies GmbH, Monheim am Rhein, Germany*

[3] *Institute of Technical Biochemistry, Technical University Lodz, Lodz,Poland*

Ketohexose 1-phosphates and ketohexose 6-phosphates and their selective enzymatic phosphorylation by kinases play important roles in the metabolism of ketohexoses such as D-fructose and D-tagatose, as the negative charge provided by the phosphate group keeps the ketohexose phosphates inside the cells by preventing them from crossing the cell membrane. Selective kinase-catalysed phosphorylation of either ketohexose 1-phosphates or ketohexose 6-phosphates adds another phosphate group to give the corresponding ketohexose 1,6-diphosphates as central metabolites, as enzymatic cleavage reactions are then facilitated by the repulsion of the two negatively charged phosphate groups. The two principally different phosphorylation routes towards these ketohexose 1,6-diphosphates both consist in kinase-catalysed phosphorylations but differ with respect to the enzymes and phosphorylation sites by catalysing the phosphorylation either of ketohexose 1-phosphates in the 6-position or of ketohexose 6-phosphates in the 1-position, as shown in Scheme 10.8 for D-fructose-1,6-diphosphate and D-tagatose-1,6-diphosphate.

The diversity of enzymes in nature that catalyse the phosphorylation of ketohexose 6-phosphates in the 1-position has been of much interest for the design of mild and resource-efficient synthetic phosphorylation methods [1] and for the phosphorylation of ketohexoses. Mild enzymatic phosphorylations have become well established as practical synthetic methods [2] and provide particular benefits when dealing with labile

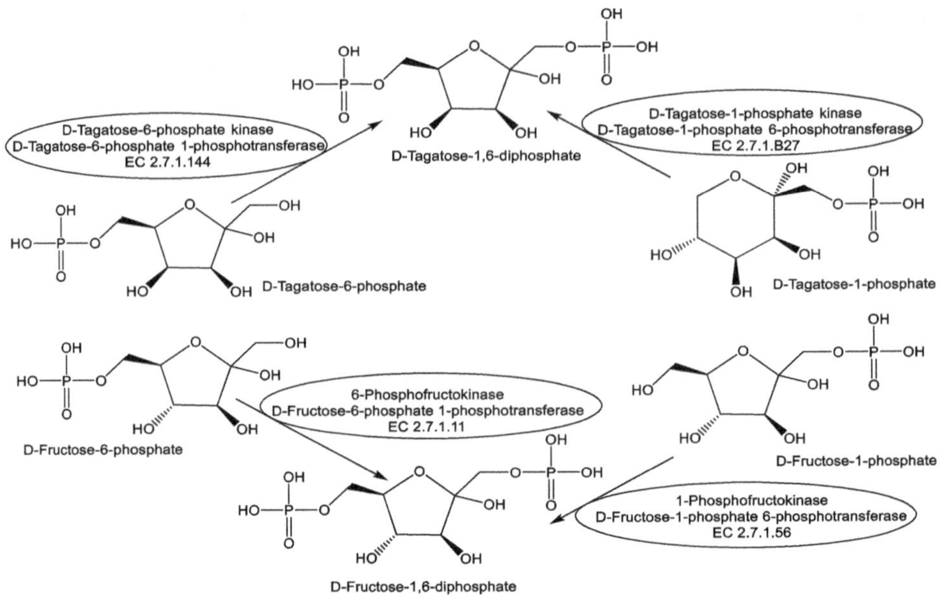

Scheme 10.8 *Selective kinase-catalysed phosphorylations of D-fructose phosphates and D-tagatose phosphates.*

substrates such as hexose-6-phosphates. In order to develop a viable D-tagatose-6-phosphate kinase-catalysed synthesis of D-tagatose-1,6-diphosphate, the *lacC* gene from *Lactococcus lactis* was synthesised for its recombinant expression in *E. coli* (Scheme 10.9) [3].

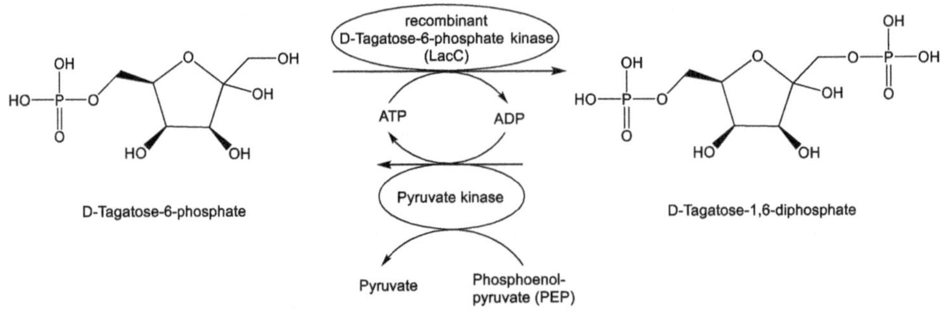

Scheme 10.9 *Enzymatic phosphorylation of D-tagatose-6-phosphate in the 1-position catalysed by recombinant D-tagatose-6-phosphate kinase (LacC) and the phosphoenolpyruvate/pyruvate kinase-system for ATP regeneration.*

10.6.1 Procedure 1: LacC-Catalysed Phosphorylation of D-Tagatose-6-Phosphate to D-Tagatose-1,6-Diphosphate Lithium Salt

10.6.1.1 Materials and Equipment

- D-Tagatose-6-phosphate lithium salt (876 mg, 3.22 mmol; Sigma-Aldrich/Merck product number 50661)

- Adenosine triphosphate (ATP; 89 mg, 0.05 eq)
- Phosphoenolpyruvate (PEP; 598 mg, 0.90 eq)
- $MgCl_2$ hexahydrate (280 mg)
- H_2O (400 mL)
- 1 M aqueous NaOH
- Pyruvate kinase from rabbit muscle, type III, lyophilised powder (280 U), 350–600 U.mg^{-1} protein (Sigma-Aldrich product number P9136)
- Soluble fraction of the crude cell extract of *E. coli* overexpressing recombinant D-tagatose-6-phosphate kinase (LacC; 70 μL); see [4] for all details required to produce this recombinant kinase.
- Aqueous 1 M HCl (6.5 mL)
- 1× Round-bottom flask
- 0.45 μm membrane filter and filtration device
- $BaCl_2$ hexahydrate (4.75 g in 15 mL H_2O)
- 120 mL acetone
- 8 mL 1 M LiOH aqueous solution
- Ion-exchange chromatography (IEC) equipment
- 70 g Strongly acidic cation exchange resin DOWEX 50W X8, 200–400 mesh (Sigma-Aldrich/Merck product number 44519)
- Lyophiliser
- Nuclear magnetic resonance (NMR) spectrometer ^1H 600 MHz, ^{13}C 151 MHz, ^{31}P 243 MHz

10.6.1.2 Procedure

1. D-Tagatose-6-phosphate lithium salt (876 mg, 3.22 mmol) was dissolved in H_2O (20 mL), then $MgCl_2$ hexahydrate (280 mg), PEP (598 mg, 0.90 eq) and ATP (89 mg, 0.05 eq) were added.
2. The pH was adjusted to 7.5 by careful addition of 1 M aqueous NaOH.
3. The reaction was started by adding 280 U pyruvate kinase (PK) and 4 U recombinant D-tagatose-6-phosphate kinase (LacC) (70 μL).
4. Reaction samples were prepared for NMR analysis as follows: a 100 μL reaction sample was diluted with 700 μL D_2O. Reaction monitoring was done by ^1H- and ^{31}P-NMR. After complete turnover, as shown by the absence of PEP peaks in ^1H- and ^{31}P-NMR (18 hr), the pH was adjusted to 3.5 by the addition of about 1 M HCl (6.5 mL).
5. The reaction mixture was filtered through a membrane filter (0.45 mm).
6. A solution of $BaCl_2$ hexahydrate (4.75 g in 15 mL H_2O) and then acetone (60 mL) were added to the filtrate with stirring.
7. The resulting suspension was centrifuged for 5 min at 4000 rpm and the solid residue was suspended in 60 mL acetone/H_2O (1 : 1) and centrifuged again.
8. The residue was vacuum-dried and then suspended in H_2O (35 mL), and 35 g of the strongly acidic cation-exchange resin DOWEX 50W X8 was added.
9. The mixture was stirred for 30 min, whereupon the residue had been dissolved completely.
10. The mixture was given on a column with an additional 35 g of the strongly acidic cation-exchange resin DOWEX 50W and then eluted with 280 mL water.

11. The eluent (pH 2.1) was carefully adjusted to pH 7.0 by slowly adding 1 M aqueous LiOH solution (about 8 mL) whilst stirring.
12. The solution was subsequently lyophilised to give 1.0 g of the pure D-tagatose-1, 6-diphosphate lithium salt, which was obtained as an almost white dry powder in 85% yield.
13. The purity of the powder was analysed by thin-layer chromatography (TLC) and NMR. Elemental analysis showed the D-tagatose-1,6-diphosphate to be present in the form of a lithium salt hydrate.

10.6.2 Analytical Method

Proteins were analysed for protein content by the Bradford method using bovine serum albumin (BSA) as standard, for protein purity (SDS-PAGE, Coomassie stain) and for enzyme activity (^{1}H-NMR, ^{31}P-NMR) [3].

Analytical data for the final product D-tagatose-1,6-diphosphate lithium salt hydrate: ^{1}H NMR (D_2O, 600 MHz, d): 4.48 (m, H–C(4)-a), 4.34 (m, H–C(4)-b), 4.27 (m, H–C(3)), 4.19 (m, H–C(5)), 4.10–4.03 (m, H0–C(6)), 3.97–3.89 (m, H–C(6)), 3.86–3.75 (m, H and H0–C(1)); ^{13}C-NMR (D_2O, 151 MHz, d): 104.3 (C(2)-a), 101.8 (C(2)-b), 79.2 (C(5)-b), 78.5 (C(5)-a, 76.5 (C(3)-a), 71.0 (C(4)-b and (C(4)-a), 70.6 (C(3)-b), 65.9 (C(1)-a), 65.6 (C(1)-b), 64.0 (C(6)-a and b); ^{31}P-NMR (D_2O, 243 MHz, d): 1.03 and 0.70 (2 phosphate groups), 0.00 (Pi). TLC: r_f 0.17 (Silica gel 60, n-BuOH : H2O : AcOH = 2 : 1 : 1, 2',7'-Dichlorofluorescein) LCMS: $(M–H)^- = 338.9$ (Ascentis Express C18 column, 2.7 μm, gradient elution with $(n\text{-}Bu)_3N$-buffer pH 4.95–MeOH); MS: ES^-, calculated molecular weight for tagatose-1,6-diphosphate (free acid form) 340.11.

10.6.3 Conclusion

The soluble and very highly active recombinant D-tagatose-6-phosphate kinase (LacC) and the phosphoenolpyruvate/pyruvate kinase-system for ATP regeneration have been used for the highly selective phosphorylation of D-tagatose-6-phosphate in the 1 position. This has enabled an efficient and straightforward biocatalytic synthesis of the important metabolite D-tagatose-1,6-diphosphate in its lithium salt form. This LacC-catalysed phosphorylation of D-tagatose-6-phosphate is easily scalable and has been successfully scaled up to the gram scale.

Acknowledgement

This section is derived in part from an article published in *Biocatalysis and Biotransformation*, 10 July 2019, copyright Taylor & Francis, available online at https://www.tandfonline.com/10.1080/10242422.2019.1634694.

References

1. Wen, L., Huang, K., Wei, M. *et al.* (2015) *Angewandte Chemie International Edition*, **54** (43), 12 654–12 658.

2. (a) Gauss, D., Schoenenberger, B., Molla, G.S. *et al.* (2016) Biocatalytic phosphorylation of metabolites, in *Applied Biocatalysis – From Fundamental Science to Industrial Applications* (eds A. Liese, L. Hilterhaus, U. Kettling and G. Antranikian), Wiley-VCH, pp. 147–177; (b) Wohlgemuth, R., Liese, A. and Streit, W. (2017) *Trends in Biotechnology*, **35**, 452–465.
3. Schoenenberger, B., Kind, S., Meier, R. *et al.* (2020) *Biocatalysis and Biotransformation*, **38** (1), 53–63.

10.7 Kinase-Catalysed Phosphorylations of Xylulose Substrates and Synthesis of Xylulose-5-Phosphate Enantiomers

Norman Hardt,[1] Stefanie Kind,[2] Bernhard Schoenenberger,[1] Thorsten Eggert,[2] Markus Obkircher,[1] and Roland Wohlgemuth[1,3]
[1]*Sigma-Aldrich/Merck KGaA, Buchs, Switzerland*
[2]*Evoxx Technologies GmbH, Monheim am Rhein, Germany*
[3]*Institute of Technical Biochemistry, Technical University Lodz, Lodz,Poland*

Enzymatic phosphorylations of monosaccharides using kinases in conjunction with regeneration of the phosphoryl donor have become very attractive in straightforward syntheses of phosphorylated monosaccharides [1], due to their excellent selectivity, enabling the use of nonprotected monosaccharide acceptors and phosphoryl donors, as well as the large number of enzymes known to catalyse the transfer of phosphoryl groups. Enantiocomplementary kinases from various sources have been discovered to be rather specific in catalysing the phosphorylation of xylulose in the 1 and 5 positions [2], as illustrated in Scheme 10.10. D-xylulokinase (EC 2.7.1.17) is specific for catalysing the 5-phosphorylation of D-xylulose and 1-deoxy-D-xylulose, whilst D-ribulose, xylitol and D-arabitol are phosphorylated much less efficiently. L-Xylulokinase (EC 2.7.1.53), on the other hand, is rather specific for catalysing the 5-phosphorylation of L-xylulose to L-xylulose-5-phosphate.

The phosphorylation of D-xylulose in the 1 position is catalysed by a fructokinase, also designated ketohexokinase (KHK), from *Homo sapiens*, whilst L-rhamnulokinase from *Thermotoga maritima* can be utilised for catalytic 1-phosphorylation of L-xylulose. Preparatively useful kinases for the 1- or 5-phosphorylation of D- and L-xylulose should have high activity and good operational and storage stability, whilst their enzyme activity and stability properties should ideally match with the stability range of the products.

Since their discovery in the last century, the D- and L-enantiomers of xylulose-5-phosphate have been found to be of vital importance in a number of different metabolic pathways and to play distinct roles according to their unique biological activity [2]. Chemical synthesis of D-xylulose-5-phosphate, after the preparation of the protected D-arabinose intermediate as starting material, involves the five reaction steps of acid hydrolysis, reduction with $NaBH_4$, introduction of the dibenzylphosphoryl group to the primary alcohol, Dess Martin periodinane oxidation in dichloromethane and removal of the protection group by hydrogenolysis [3]. Reducing the number of reaction and purification steps in lengthy synthetic routes to highly selective, resource-efficient and short preparative methodologies is therefore highly desirable, promising significant advantages if process design goals can be realised. As enzymatic approaches have thus been at the forefront of methods for the synthesis of D-xylulose-5-phosphate (Scheme 10.11), it has been of interest to focus attention on versatile and complementary synthetic methods to both the D- and

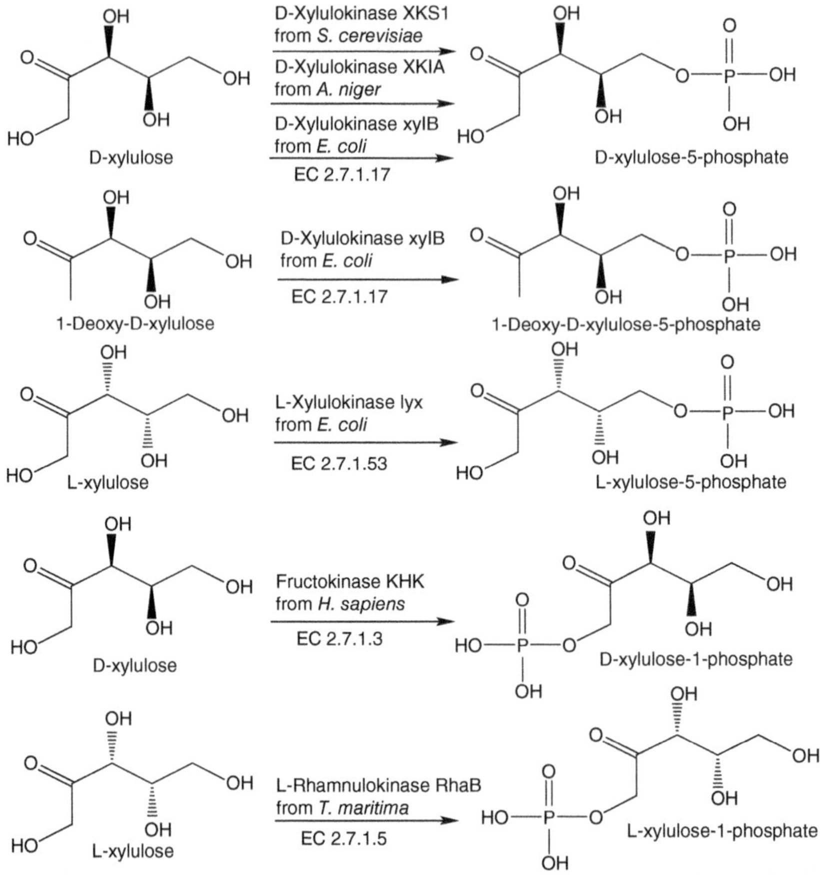

Scheme 10.10 *Selection of xylulokinase-catalysed phosphorylation reactions.*

L-enantiomers of xylulose-5-phosphate. Straightforward one-step syntheses of these xylulose-5-phosphate enantiomers with high yields and excellent purities by enzymatic phosphorylation of the corresponding monosaccharides require high selectivities and complete conversions in order to minimise subsequent purification work.

The *XKS1* gene from *Saccharomyces cerevisiae* and the *lyx* gene from *Escherichia coli* were synthesised and cloned into an appropriate expression vector to allow for the production of the xylulokinases in *E. coli* [2]. The recombinant D- and L-xylulokinases, adenosine triphosphate (ATP) regeneration and in-process analysis by ^{31}P- and ^{1}H-nuclear magnetic resonance (NMR) spectroscopy have been essential to the development of the enzymatic phosphorylation reactions of D- and L-xylulose in the 5 position.

10.7.1 Procedure 1: Enzymatic Synthesis of L-Xylulose-5-Phosphate Lithium Salt

10.7.1.1 Materials and Equipment

- L-Xylulose (200 mg, 1.3 mmol, 1.0 eq)
- ATP (36 mg, 66 μmol, 0.05 eq)

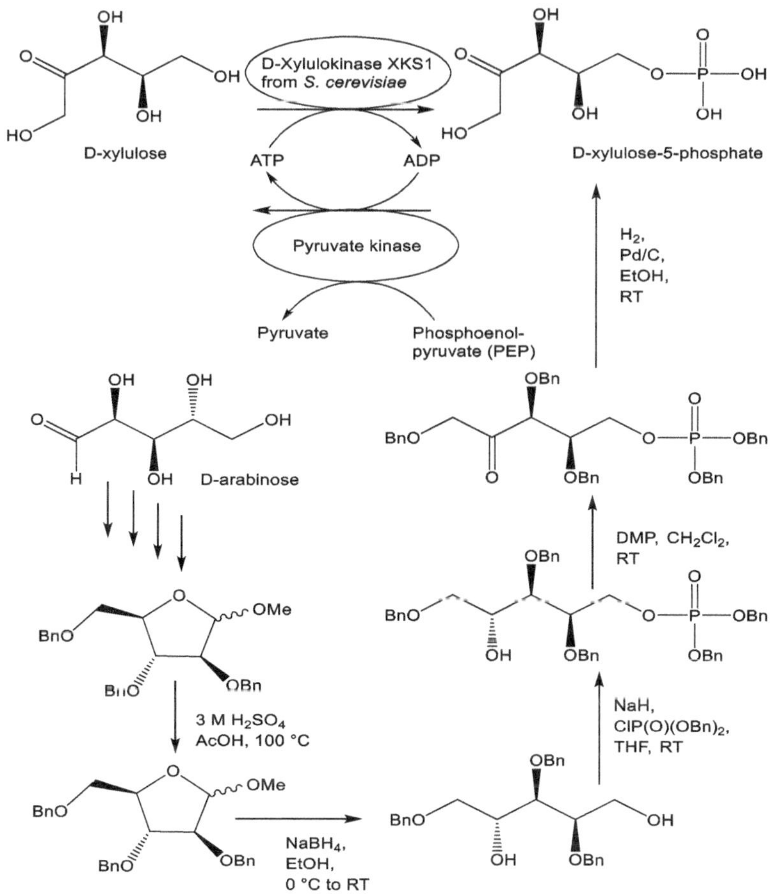

Scheme 10.11 *Comparison of the chemical route (bottom) to D-xylulose-5-phosphate starting from protected D-arabinose intermediate with the enzymatic route (top) of the ATP-dependent XKS1-catalysed phosphorylation of D-xylulose to D-xylulose-5-phosphate using ATP regeneration with pyruvate kinase and phosphoenolpyruvate.*

- Phosphoenolpyruvate (PEP) (240 mg, 1.2 mmol, 0.9 eq)
- $MgCl_2$ solution (40 mL, 10mM)
- 1 M aqueous NaOH
- Pyruvate kinase from rabbit muscle, type III, lyophilised powder (400 U), 350–600 U.mg^{-1} protein (Sigma-Aldrich product number P9136); a solution of 1 U.µL^{-1} in water was prepared – here, 400 U in 400 µL water
- Soluble fraction of the crude cell extract of *E. coli* overexpressing L-xylulose/3-keto-L-gulonate kinase (400 µL); see [3] for all details required to produce this recombinant kinase
- Calcium acetate monohydrate (0.690 g, 3.9 mmol, 3 eq)
- 1× round-bottom flask
- 10 000 MW Amicon Ultra-15 filter membrane

- Rotary evaporator
- Filtration device
- Ion-exchange chromatography (IEC) equipment
- Lithium-conditioned Dowex 50W X8 H+ resin
- NMR spectrometer (^{1}H 600 MHz, ^{13}C 151 MHz, ^{31}P 243 MHz)

10.7.1.2 Procedure

1. L-Xylulose (200 mg, 1.3 mmol, 1.0 eq), ATP (36 mg, 66 μmol, 0.05 eq) and PEP (240 mg, 1.2 mmol, 0.9 eq) were dissolved in 10 mM $MgCl_2$ solution (40 mL).
2. The pH was adjusted to 6 by adding 1 M NaOH (1.8 mL).
3. Subsequently, the enzymatic reaction was started by adding pyruvate kinase (400 U in 400 μL) and the soluble fraction of the crude cell extract of *E. coli* overexpressing L-xylulose/3-keto-L-gulonate kinase (about 5 U in 400 μL).
4. The solution was stirred gently at room temperature for 3.5 hr.
5. The conversion was followed by ^{1}H- and ^{31}P-NMR and the reaction samples were prepared for NMR analysis as follows: A reaction sample of 100 μL was diluted with 700 μL D_2O. After complete conversion, as monitored by the absence of PEP peaks in ^{1}H- and ^{31}P-NMR, the enzymes were removed by ultrafiltration using a 10 000 MWCO Amicon Ultra-15 filter membrane.
6. Calcium acetate monohydrate (690 mg 3.9 mmol, 3 eq) was added to the reaction solution to precipitate the L-xylulose-5-phosphate calcium salt as a white solid.
7. The resulting L-xylulose-5-phosphate calcium salt was filtered and dried.
8. The L-xylulose-5-phosphate calcium salt was converted into the corresponding lithium salt using a lithium-conditioned Dowex 50W X8 H+ resin.
9. The solution of the L-xylulose-5-phosphate lithium salt was evaporated to give L-xylulose-5-phosphate lithium salt 276 mg as a white solid in 91% yield.

^{1}H NMR (D_2O, 600 MHz): δ 4.58 (d, J = 19.4 Hz, 1H), 4.46 (d, J = 2.2 Hz, 1H), 4.46 (d, J = 19.4 Hz, 1H), 4.12 (td, J = 6.5, 2.2 Hz, 1H), 3.76 (td, J = 7.1, 6.6, 3.3 Hz, 2H). ^{13}C NMR (D_2O, 151 MHz): δ 212.99 (d, J = 6.6 Hz), 75.27 (d, J = 2.3 Hz), 71.31 (d, J = 7.0 Hz), 66.07, 63.75 (d, J = 4.2 Hz). ^{31}P NMR (D_2O, 243 MHz): δ 4.21 (1P). Mass spectrometry (MS): found, 229.01; calculated, 229.01 ($M–H^+$, $C_5H_{10}O_8P^-$).

10.7.2 Analytical Method

Proteins were analysed for protein content (Bradford method, bovine serum albumin (BSA) as protein standard) and for protein purity (SDS-PAGE, Coomassie stain) [2].

10.7.3 Conclusion

Selective 1- and 5-phosphorylations of nonprotected xylulose substrates can be achieved in one step by enzymatic phosphorylations using recombinant kinases and ATP regeneration with the phosphoenolpyruvate/pyruvate kinase system. *S. cerevisiae* xylulokinase and the *E. coli* L-xylulose/3-keto-L-gulonate kinase have been successfully overexpressed in *E. coli*. Soluble xylulokinase fractions, which have been obtained by straightforward downstream processing, have then been utilised for the enzymatic phosphorylations of D- and

L-xylulose substrates to D-xylulose-5-phosphate and L-xylulose-5-phosphate, which after subsequent isolation and purification were obtained in excellent purity and yield. These approaches thus represent straightforward and easily scalable access routes to the two important enantiomeric metabolites, D-xylulose-5-phosphate and L-xylulose-5-phosphate, which are very valuable for a variety of applications in the biomedical and life sciences.

Acknowledgement

This section is derived in part from an article published in *Biocatalysis and Biotransformation*, 22 June 2019, copyright Taylor & Francis, available online at https://www.tandfonline.com/10.1080/10242422.2019.1630385.

References

1. (a) Gauss, D., Schoenenberger, B., Molla, G.S. *et al.* (2016) Biocatalytic phosphorylation of metabolites, in *Applied Biocatalysis – From Fundamental Science to Industrial Applications* (eds A. Liese, L. Hilterhaus, U. Kettling and G. Antranikian), Wiley-VCH, pp. 147–177; (b) Wohlgemuth, R., Liese, A. and Streit, W. (2017) *Trends in Biotechnology*, **35**, 452–465.
2. Hardt, N., Kind, S., Schoenenberger, B. *et al.* (2020) *Biocatalysis and Biotransformation*, **38** (1), 35–45.
3. Wei, W.-C. and Chang, C.-C. (2017) *European Journal of Organic Chemistry*, **2017** (21), 3033–3040.

10.8 Phosphoramidates by Kinase-Catalysed Phosphorylation and Arginine Kinase-Catalysed Synthesis of N_{ω}-Phospho-L-Arginine

Bernhard Schoenenberger,[1] Agata Wszolek,[2] Thomas Milesi,[1] Henrike Brundiek,[2] Markus Obkircher[1] and Roland Wohlgemuth[1,3]
[1]*Sigma-Aldrich/Merck KGaA, Buchs, Switzerland*
[2]*Enzymicals, Greifswald, Germany*
[3]*Institute of Technical Biochemistry, Technical University Lodz, Lodz,Poland*

Phosphoramidates represent important structural features of metabolites, natural products and synthetic compounds such as catalysts, ligands and pharmaceuticals. In vertebrates and invertebrates, *N*-phosphorylated guanidino products, termed phosphagens, play important roles in the energy metabolism of living cells, where the rapid and timely availability of the cells' energy currency adenosine triphosphate (ATP) is essential. The development of resource-efficient and sustainable methodologies for the synthesis of phosphoramidates is therefore highly desirable. As biocatalytic phosphorylations have been successfully utilised in the synthesis of phosphorylated metabolites [1] and the synthesis of phosphoramidates in nature is catalysed by kinases, the identification, characterisation and production of nitrogen–phosphorus bond-forming kinases is attractive for the enzymatic synthesis of phosphoramidates (see Scheme 10.12) [2].

Short and sustainable synthetic routes to phosphoramidates can be achieved without using protecting groups on the other potential phosphorylation sites in the substrates by highly selective enzymatic phosphorylation on nitrogen.

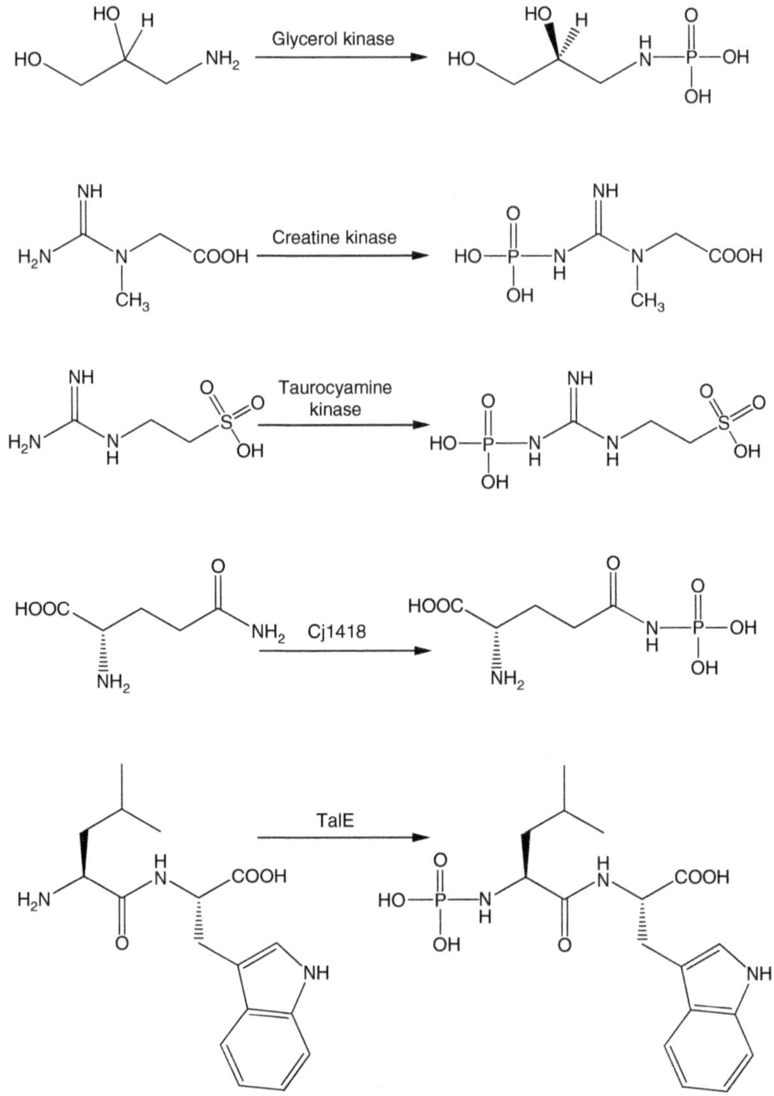

Scheme 10.12 *Selected N-phosphorylation reactions to phosphoramidates catalysed by nitrogen–phosphorus bond-forming kinases.*

The phosphagen N_ω-phospho-L-arginine, a main reserve of high-energy compounds that plays a key role in enzymatic phosphorylations in the invertebrate muscle, has been found in biosynthetic pathways of parasites that are totally different from those found in mammalian host tissues. This makes the metabolic pathway to N_ω-phospho-L-arginine an attractive therapeutic target for parasitic diseases. The lack of availability of N_ω-phospho-L-arginine is due to tedious or difficult procedures for the preparation of this important metabolite in pure form. The inherent instability of energy-rich N_ω-phospho-L-arginine, where the phosphate is attached to a nitrogen atom (*N*-phosphorylation), forming an acid-labile phosphoramidate bond, makes chemical multistep synthesis tedious and renders its isolation from biological sources difficult. Therefore, we aimed at implementing an improved and very efficient synthesis of N_{ω}-phospho-L-arginine in one reaction step by a simple enzymatic phosphorylation of L-arginine with a recombinant arginine kinase, because the isolation from natural biological sources would have been both inefficient and uneconomic compared with the recombinant expression of this enzyme.

The experimental evidence for arginine kinases, excluding protein arginine kinases, at the protein level and the characterisation of their function have been extended since their first discovery in crude extracts of crab muscle by Karl Lohmann in 1935 [3] to many species of invertebrates, protozoa and, most recently, bacteria, where genes coding for arginine kinase are cloned and expressed in *E. coli* and where arginine kinase activities have been confirmed [4]. The arginine kinase EC2.7.3.3 from the Atlantic horseshoe crab *Limulus polyphemus* [5], with its 357 amino acids and molecular mass of 40 239 Da, has been thoroughly investigated and found to be an ideal model system. Our fundamental understanding has increased tremendously towards a detailed characterisation of the structure, dynamics and thermodynamics of the Michaelis complex and the transition state of arginine kinase and along the whole enzymatic reaction cycle [6].

Therefore, the arginine kinase ArgK from *Limulus polyphemus* (ArgK-LP) has been overexpressed in *Escherichia coli* BL21(DE3) as a soluble protein in high yield [7] and around 30% has been localised in the soluble protein fraction of *E. coli*. As the recombinant protein carries an *N*-terminal His-tag, ArgK has been successfully purified by immobilised metal-affinity chromatography. The obtained yield of the purified ArgK was 24 mg.L^{-1} of *E. coli* culture. The specific activity was about 80 U.mg^{-1} of purified protein and 834 U.mL^{-1} in the crude extract. This highly active and stable arginine kinase, with an efficient regeneration system for ATP using the phosphoenolpyruvate/pyruvatekinase system as illustrated in Scheme 10.13 and in-process analysis with high information content, has enabled the development of a viable procedure for the biocatalytic synthesis of the important metabolite N_{ω}-phospho-L-arginine on a gram scale [7].

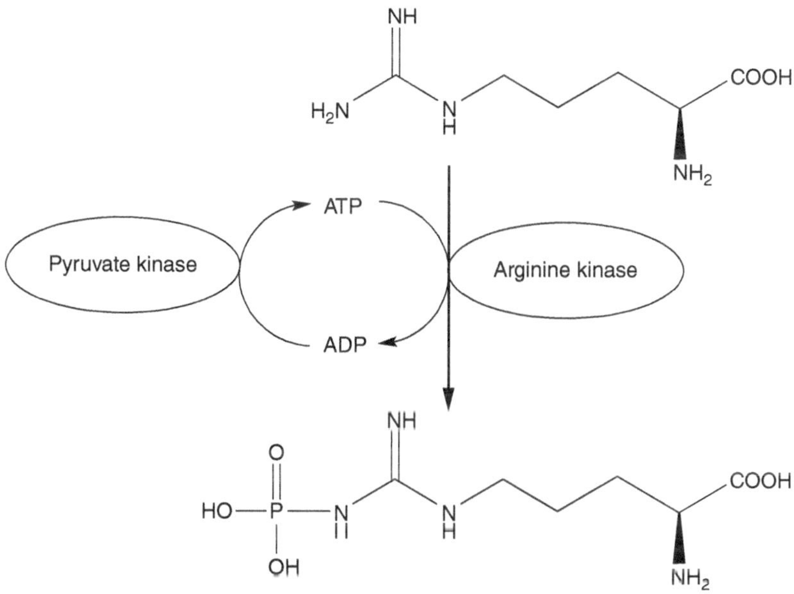

Scheme 10.13 *Arginine kinase-catalysed N_{ω}-phosphorylation of L-arginine to N_{ω}-phospho-L-arginine.*

10.8.1 Procedure1: Recombinant Expression of Arginine Kinase in *E. coli* BL21(DE3)

10.8.1.1 Materials and Equipment

- *E. coli* BL21(DE3) cells
- Lysogenic broth (LB) plates (agar, 15 g.L^{-1}; tryptone, 10 g.L^{-1}; NaCl, 10 g.L^{-1}; yeast extract, 5 g.L^{-1}; with kanamycin, 50 µg.mL^{-1}; pH 7.5)
- LB medium (tryptone, 10 g.L^{-1}; NaCl, 10 g.L^{-1}; yeast extract, 5 g.L^{-1}; with kanamycin, 50 µg.mL^{-1}; pH 7.5)
- Terrific broth (TB) medium (casein, enzymatically digested, 12 g.L^{-1}; yeast extract, 24 g.L^{-1}; K_2HPO_4, 12.54 g.L^{-1}; KH_2PO_4, 2.31 g.L^{-1}; with kanamycin, 50 µg.mL^{-1}; pH 7.5)
- Isopropyl-β-D-1-thiogalactopyranoside (IPTG)
- Sodium phosphate buffer pH 7.5 (50 mM)
- Tris-HCl buffer pH 7.5 (50 mM)
- Imidazole
- Incubator
- Water bath
- Erlenmeyer flasks with baffles (250 mL and 2 L)
- Orbital shaker
- Centrifuge
- Spectrophotometer
- Retsch Mixer Mill MM 200
- Glass beads (Ø 0.25–0.5 mm)
- Chromatography equipment
- Talon® resin
- Zeba spin columns

10.8.1.2 Procedure

1. The synthetic arginine kinase gene *ArgK* from *Limulus polyphemus* (ArgK-LP; UniProtKB: P51541), codon-optimised for *E. coli* and equipped with an *N*-terminal 6V-His-tag and a tobacco etch virus (HEV) protease cleavage site for optional removal of the affinity tag, was synthesised and subcloned into pET24a(+) via NdeI and NotI by Genscript, giving vector pET24-His(TV)-Karg-LimPo(op).
2. 100 µL of chemically competent BL21 (DE3) *E. coli* cells (100 µL aliquots of chemically competent *E. coli* BL21 (DE3) cells were stored at −80 °C until usage) were transformed with 5 µL of the derived plasmid by standard procedures [8] and plated on LB agar supplemented with 50 µg.mL^{-1} kanamycin for selection.
3. A single *E. coli* colony-harbouring *ArgK* gene was inoculated into 5 mL LB medium supplemented with kanamycin (50 µg.mL^{-1}) and incubated at 30 °C and 180 rpm overnight.
4. Overnight cultures of *E. coli* were used to inoculate 50 mL of TB medium supplemented with kanamycin (50 µg.mL^{-1}) in 250 mL baffled flasks. Cells were cultivated at 37 °C under vigorous shaking (150 rpm) until an OD_{600} of 0.5 was reached.
5. When the pre-culture reached the mid-log phase, the main culture was started. Fresh TB medium (400 mL) in a 2 L Erlenmeyer flask was inoculated with the *E. coli* pre-culture

(20 mL). Cells were cultivated at 37 °C under vigorous shaking (150 rpm) until an OD_{600} of 0.5 was reached.

6. After the culture reached OD_{600} 1.3, protein expression was induced with IPTG to a final concentration of 1 mM and the incubation temperature was shifted to 25 °C. The main culture was then shaken at 110 rpm for 20 hr.
7. Cells were harvested by centrifugation at 5000 rpm and 4 °C for 30 min and washed once with 40 mL of ice-cold 50 mM sodium phosphate buffer at pH 7.5. The *E. coli* cell pellets containing the overexpressed arginine kinase were frozen at −20 °C.
8. The frozen *E. coli* cells were suspended in Tris-HCl buffer (50 mM, pH 7.5), adjusted to OD_{600} 40 and mixed with glass beads (Ø 0.25–0.5 μm) at a ratio of approximately 1 : 1.
9. The cells were disintegrated mechanically by a cell disruptor (Retsch Mixer Mill MM 200: two cycles for 5 min at 600 rpm).
10. The samples were cooled on ice and insoluble cell debris was removed by centrifugation (5000 rpm, 30 min, 4 °C). The supernatant contained the soluble protein fraction.
11. The recombinant arginine kinase ArgK was purified by immobilised metal-affinity chromatography. The soluble protein fraction was incubated with Talon® resin (2 mL) in an orbital shaker (30 min, 4 °C). Then a standard gravity flow protocol was applied to purify the target protein. The resin was washed at 4 °C with 10 column volumes of Tris-HCl buffer (50 mM, pH 7.5) containing 10 mM imidazole to remove unspecifically bound protein.
12. The target protein arginine kinase ArgK was eluted with 1.5 column volumes of elution buffer (50 mM TrisHCl buffer, pH 7.5, 150 mM imidazole).
13. Purified arginine kinase was obtained by removing imidazole on desalting columns using 50 mM Tris-HCl buffer pH 7.5 with Zeba spin columns.

10.8.2 Procedure 2: Arginine Kinase-Catalysed Phosphorylation of L-Arginine to N_ω-phospho-L-Arginine Lithium Salt

10.8.2.1 Materials and Equipment

- L-Arginine (1000 mg, 5.75 mmol)
- ATP (158 mg, 0.05 eq)
- Phosphoenolpyruvate monopotassium salt (PEP) (1124 mg, 0.95 eq)
- Pyruvate kinase from rabbit muscle, type III, lyophilised powder (200 U), 350–600 U.mg^{-1} protein (Sigma-Aldrich P9136)
- Magnesium chloride (345 mg)
- H_2O (100 mL)
- Acetic acid (60 mL, 0.1 M aqueous)
- Soluble fraction of the purified cell extract of *E. coli* overexpressing recombinant arginine kinase (ArgK; 200 μL).
- 1× Round-bottom flask
- 1-Propanol
- 1 M aqueous NH_4OH
- Methanol
- Rotary evaporator with high vacuum
- Silica TLC plates

- UV cabinet consisting of a UV lamp 254 nm and viewing box
- Ninhydrin
- Nuclear magnetic resonance (NMR) spectrometer ^{1}H 400 MHz, ^{31}P 162 MHz

10.8.2.2 Procedure

1. Magnesium chloride (345 mg), ATP (158 mg, 0.05 eq) and PEP (1124 mg, 0.95 eq) were added to a solution of L-arginine (1000 mg, 5.75 mmol in 30 mL water, adjusted to pH 8.0 with 0.1 M acetic acid 55 mL) and the pH was adjusted to 8.0. Meanwhile, L-arginine (adjusted to pH 8.0), $MgCl_2$, ATP and PEP were added and the pH increased to 8.6; this had to be readjusted to 8.0 before adding the kinases.
2. Pyruvate kinase (200 U) and arginine kinase solution (200 mL) were added and the reaction mixture again adjusted to pH 8.0 by addition of a small amount of 0.1 M acetic acid with stirring.
3. The reaction progress was monitored by ^{31}P-NMR. The reaction mixture was concentrated after 2 days and dried.
4. The product was worked up by normal phase (NP) column chromatography on silica gel (methanol/water 1 : 1). The strong acidic ion exchanger Dowex 50 WX8 was loaded with LiOH then washed until neutral. The product was subsequently converted to the lithium salt, yielding 410 mg (28%) N_{ω}-phospho-L-arginine.

Analytical data: Single spot on TLC (silica gel, Rf = 0.67 with eluent H_2O/1-propanol/NH_4OH 25% = 11 : 6 : 3; Rf = 0.05 with eluent 1-propanol/NH_4OH 25%/ H_2O = 6 : 3 : 1); ^{1}H NMR (D_2O, 400 MHz): $\delta = 3.45$ (dd, 1H, $J1$, $J2 \approx 6$ Hz), 3.17 (t, 2H, $J = 6.8$ Hz), 1.82 (m, 2H), 1.60 ppm (m, 2H); ^{31}P NMR (D_2O, 162 MHz, CPD): $\delta = 3.26$ ppm.

10.8.3 Analytical Method

The protein content of the purified protein was determined according to the bicinchoninic acid assay (BCA) method with bovine serum albumin (BSA) as standard [7].

Analysis of ArgK overexpression and purity was performed using 12.5% SDS-polyacrylamide gels stained with Coomassie Blue.

Arginine kinase activity was measured spectrophotometrically at $\lambda = 340$ nm based on ADP release by a coupled assay with pyruvate kinase and lactate dehydrogenase [7]. Arginine kinase ArgK was assayed at room temperature by recording the oxidation of NADH ($\lambda = 340$ nm, $\varepsilon = 6.22 \times 10^3$ $M^{-1}.cm^{-1}$). The assay mixture contained 100 mM Tris-HCl buffer pH 7.5, 1 mM phosphoenolpyruvate, 5 mM $MgCl_2$, 50 mM KCl, 0.15 mM NADH, 4 U pyruvate kinase, 8 U lactate dehydrogenase, 2.5 mM ATP and 10 mM L-arginine in a total volume of 1 mL. The reaction was started by the addition of recombinant protein. One unit of enzyme activity is defined as the amount of enzyme catalysing the conversion of 1 μmol of substrate per minute at pH 7.5 and 25 °C.

10.8.4 Conclusion

A simple and scalable biocatalytic procedure has been developed for the selective N_{ω}-phosphorylation of L-arginine to N_{ω}-phospho-L-arginine at gram scale. This provides direct synthetic access to N_{ω}-phospho-L-arginine in pure form and overcomes the problems with previous tedious or difficult procedures for the preparation of this important metabolite. A highly active arginine kinase has been prepared by cloning and expressing the gene coding for ArgK from *Limulus polyphemus* in *E. coli.* ^{31}P-NMR spectroscopy enabled the rapid development of the arginine kinase-catalysed phosphorylation of L-arginine, which was combined with the phosphoenol-pyruvate/pyruvate kinase system for recycling adenosine diphosphate (ADP) back to the phosphorylating agent ATP. The biocatalytic procedure and the subsequent workup to the pure product N_{ω}-phospho-L-arginine have been successfully demonstrated and open up new opportunities for the selective biocatalytic *N*-phosphorylation of interesting low-molecular-weight compounds and metabolites. The continuing discovery of novel nitrogen–phosphorus bond-forming kinases looks promising for the further expansion of enzymatic methods for the synthesis of phosphoramidates.

Acknowledgement

This section is derived in part from an article published in *ChemCatChem*, 18 October 2016, copyright 2017 Wiley-VCH Verlag GmbH & Co. KGaA, Weinheim, available online at https://doi.org/10.1002/cctc.201601080 and is reproduced with permission from John Wiley and Sons and Copyright Clearance Center.

References

1. (a) Gauss, D., Schoenenberger, B., Molla, G.S. *et al.* (2016) Biocatalytic phosphorylation of metabolites, in *Applied Biocatalysis – From Fundamental Science to Industrial Applications* (eds A. Liese, L. Hilterhaus, U. Kettling and G. Antranikian), Wiley-VCH, pp. 147–177; (b) Wohlgemuth, R., Liese, A. and Streit, W. (2017) *Trends in Biotechnology*, **35**, 452–465.
2. (a) Crans, D.C. and Whitesides, G.M. (1985) *Journal of the American Chemical Society*, **107** (24), 7008–7018; (b) Shih, Y.S. and Whitesides, G.M. (1977) *Journal of Organic Chemistry*, **42** (25), 4165–4166; (c) Uda, K., Saishoji, N., Ichinari, S. *et al.* (2015) *Journal of Biological Chemistry*, **290** (20), 12 951–12 963; (d) Taylor, Z.W., Brown, H.A., Narindoshvili, T. *et al.* (2017) *Journal of the American Chemical Society*, **139** (28), 9463–9466; (e) Baulig, A., Helmle, I., Bader, M. *et al.* (2019) *Chemical Science*, **10**, 4486–4490.
3. Lohmann, K. (1935) *Biochemische Zeitschrift*, **282**, 109–119.
4. Fraga, D., Stock, K., Aryal, M. *et al.* (2019) *Comparative Biochemistry and Physiology Part B*, **233**, 60–71.
5. Strong, S.J. and Ellington, W.R. (1995) *Biochimica et Biophysica Acta* **1246** (2), 197–200.
6. Peng, Y., Hansen, A.L., Bruschweiler-Li, L. *et al.* (2017) *Journal of the American Chemical Society*, **139**, 4846–4853.
7. Schoenenberger, B., Wszolek, A., Milesi, T. *et al.* (2017) *ChemCatChem*, **9**, 121–126.
8. Hanahan, D. (1983) *Journal of Molecular Biology*, **166**, 557–580.

11

Enzymatic Cascades

11.1 Redox-Neutral Ketoreductase and Imine Reductase Enzymatic Cascade for the Preparation of a Key Intermediate of the Lysine-Specific Histone Demethylase 1 (LSD1) Inhibitor GSK2879552

Jonathan Latham,*[1] Markus Schober,[1] Joseph Hosford,[1] Mahesh J. Sanganee[2] and Gheorghe-Doru Roiban[1]

[1]*Synthetic Biochemistry, Medicinal Science & Technology, GlaxoSmithKline, Medicines Research Centre, Stevenage, Hertfordshire, UK*

[2]*Chemical Development, Medicinal Science & Technology, GlaxoSmithKline, Medicines Research Centre, Stevenage, Hertfordshire, UK*

Amine **4** is a key intermediate in the synthesis of GSK2879552 – an inhibitor of lysine-specific histone demethylase 1 (LSD1) investigated in clinical trials for the treatment of small-cell lung cancer and acute leukaemia [1]. The route to this compound was improved by use of an imine reductase (IRED) enzyme, which allowed preparation of **4** in excellent enantiopurity via racemic **3** without the use of borohydride – a reaction that was used for the kilogram-scale manufacture of active pharmaceutical ingredient (API) [2]. The route involving enzymatic reductive amination had room for improvement, since aldehyde **2** was prepared via Cu(I) oxidation of the corresponding alcohol **1** in a separate step [3]. An enzymatic cascade reaction was subsequently employed for direct conversion of **1** to **4** by combination of the IRED with a ketoreductase (KRED) for *in situ* oxidation of **1** to the corresponding aldehyde **2** (Scheme 11.1). Both the KRED and the IRED utilise the same redox co-factor ($NADP^+$/NADPH) but for opposite redox reactions. The KRED oxidation utilises $NADP^+$ as a hydride acceptor, generating NADPH. The IRED then uses this NADPH as a reductant, regenerating $NADP^+$ in the process. This internal co-factor recycling allows the cascade to be redox-neutral and to not require external co-factor recycling. This cascade obviates the need to isolate aldehyde **2** (an unstable and noncrystalline solid), thereby improving green metrics and process safety.

Applied Biocatalysis: The Chemist's Enzyme Toolbox, First Edition. Edited by John Whittall and Peter W. Sutton.
© 2021 John Wiley & Sons Ltd. Published 2021 by John Wiley & Sons Ltd.

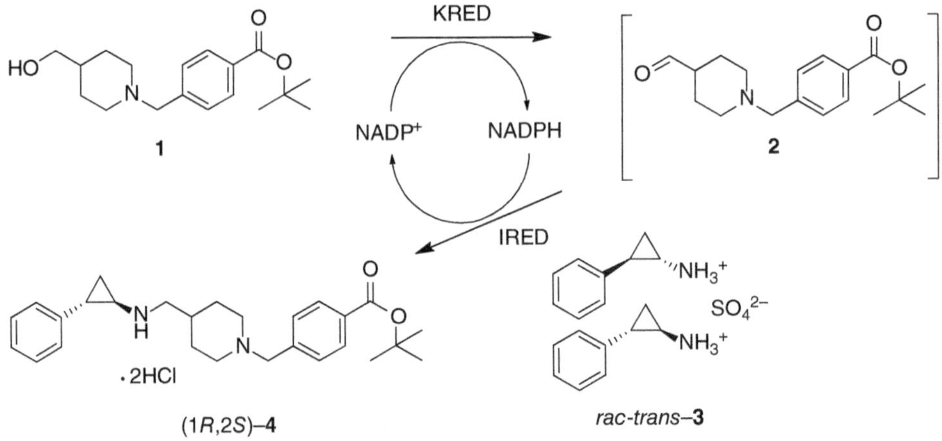

Scheme 11.1 *Enzymatic oxidation–reduction cascade for conversion of **1** into **4**.*

11.1.1 Procedure 1: Recombinant Expression of KRED from *Lactobacillus coleohominis*

11.1.1.1 Materials and Equipment

- Glycerol stock of BL21 (DE3) transformed with pET28(b) containing WP_006915881 gene insert [2b]
- Lysogenic broth (LB) medium (tryptone (Bacto, 10 $g.L^{-1}$), yeast extract (Bacto, 5 $g.L^{-1}$) and NaCl (5 $g.L^{-1}$))
- Glucose (50% v/v)
- Kanamycin (50 $mg.mL^{-1}$)
- Terrific broth (TB) medium (tryptone (Bacto, 12 $g.L^{-1}$), yeast extract (Bacto, 24 $g.L^{-1}$), K_2HPO_4 (9.4 $g.L^{-1}$) and KH_2PO_4 (2.2 $g.L^{-1}$))
- Isopropyl-β-D-1-thiogalactopyranoside (IPTG, 1.0 M)
- Potassium phosphate buffer (50 mM, pH 7.0)
- Corning disposable Erlenmeyer flask (250 mL)
- Thompson ultraYield shake flask (2.5 L)
- Thompson ultraYield flask seal
- Kuhner shaking climo ISF1-X incubator
- Jenway 6300 spectrophotometer
- ThermoScientific Heraeus megafuge 40 R
- Analytik Ltd M110Y microfluidiser
- BPS VirTris SP scientific advantage pro lyophiliser

11.1.1.2 Procedure

1. A sterile corning Erylenmeyer flask (250 mL) containing LB medium (50 mL), kanamycin (final conc. 50 $\mu g.mL^{-1}$) and glucose (final conc. 1.0% w/v) was inoculated with KRED glycerol stock (50 μL) and incubated at 30 °C overnight with shaking in a Kuhner shaker (220 rpm).

2. Four shake flasks (2.5 L) containing TB medium (1 L), kanamycin (final conc. 50 μg.mL^{-1}) and glycerol (final conc. 0.4% v/v) were inoculated with 10 mL of the starter culture from Step 1 and incubated at 30 °C with shaking in a Kuhner shaker (200 rpm).
3. OD_{600} was measured during incubation by taking 1 mL aliquots until it reached 0.6–0.8.
4. Cultures were induced by addition of IPTG (final conc. 1 mM), and incubation continued at 30 °C overnight with shaking in a Kuhner shaker (200 rpm).
5. After incubation for 24 hr total, broth was harvested by centrifugation (4000 rpm, 4 °C) for 30 min.
6. Cell paste (37.6 g) was resuspended in 50 mM potassium phosphate buffer, pH 7.0 (189 mL).
7. Resuspended cell paste was filtered through muslin cloth then disrupted by microfluidisation
8. Lysate was clarified by centrifugation (10 000 rpm, 4 °C) for 1 hr.
9. Clarified lysate was frozen at −80 °C then lyophilised.
10. Once lyophilisation was complete, the dried powder (2–3 g.L culture^{-1}) was stored at −20 °C for later use.

11.1.2 Procedure 2: Preparation of Compound 4 from Compound 1

11.1.2.1 Materials and Equipment

- *tert*-Butyl 4-((4-(hydroxymethyl)piperidin-1-yl)methyl)benzoate **1** (5.0 g)
- *trans*-2-Phenylcyclopropanamine **3** (6.26 g)
- KRED-lyophilised powder (5.0 g; see Procedure 1)
- IRED M3-lyophilised powder (1.0 g) [2b]
- Dimethyl sulfoxide (DMSO; 50 mL)
- Potassium phosphate buffer (100 mM, pH 7.0, 450 mL)
- Acetic acid 99.5% v/v (25 mL)
- Sodium chloride solution 25% w/v (35 mL)
- Nicotinamide adenine dinucleotide phosphate disodium salt ($NADP^+$; 128 mg)
- Water (50 mL)
- Acetone (20 mL)
- Controlled laboratory reactor (CLR) fitted with a polytetrafluoroethylene (PTFE) impeller (1 L)
- Whatman No. 113 filter paper
- Buchner funnel
- Buchner flask (2 L)
- Salvis lab vacuum oven
- High-performance liquid chromatograph (HPLC) equipped with an Agilent Bonus RP column (4.6 × 150 mm, 3.5 μm)
- HPLC equipped with a ChiralPak AD-RH (2.1 × 150 mm, 5 μm)

11.1.2.2 Procedure

1. CLR jacket temperature was set to 30 °C.
2. Compound **1** (5.0 g, 16.4 mmol, 1.0 eq) was charged to the CLR.

3. DMSO (50 mL) was added to the CLR.
4. 100 mM potassium phosphate buffer, pH 7.0 (200 mL) was added to the CLR.
5. Stirring was started at 300 rpm.
6. Compound **3** (6.26 g, 34.4 mmol, 2.1 eq) was added to the CLR.
7. KRED (5.0 g, 100% w/w), IRED (1.0 g, 20% w/w) and $NADP^+$ (128 mg, 0.16 mmol, 0.1 mol %) were added to a separate glass bottle.
8. 100 mM Potassium phosphate buffer, pH 7.0 (200 mL) was charged to the glass bottle.
9. The contents of the glass bottle were stirred and then charged to the CLR.
10. The bottle was washed with 100 mM potassium phosphate buffer, pH 7.0 (50 mL) and the washings were charged to the CLR.
11. The reaction was stirred at 300 rpm and 30 ± 2 °C for 24 hr.
12. Acetic acid 99.5% v/v (25 mL) was charged to the CLR, followed by 25% w/v sodium chloride solution (35 mL).
13. The CLR jacket temperature was set to 5 °C and the reaction left to stir at 300 rpm for 2 hr.
14. The reaction suspension was filtered through Whatman No. 113 filter paper in a Buchner funnel.
15. The filter cake was washed with water chilled to below 4 °C (4 × 12.5 mL), then by acetone chilled to below 4 °C (2 × 10 mL).
16. The collected solid was transferred to a glass dish and dried overnight in a vacuum oven (40 ± 3 °C, 20–100 mbar).
17. Compound **4** was isolated as a white solid (3.9 g, 7.9 mmol, 48.3% yield, 99.7% ee).

11.1.3 Analytical Method

11.1.3.1 Sampling Procedure

Sampling was performed whilst stirring:

1. Reaction mixture (50 μL) was added to acetonitrile (400 μL) and 1 M acetic acid (200 μL), followed by mixing with vortex and centrifugation (13 000 rpm, 4 °C) for 1 min.
2. Supernatant of the prepared sample was injected on chiral and achiral HPLC [2].
3. Enantiomeric excess and conversion were calculated.

11.1.3.2 Achiral HPLC and Chiral HPLC

The same methods were used as described in [2].

11.1.4 Conclusion

This protocol demonstrates a redox-neutral enzymatic cascade for the direct conversion of alcohols into secondary amines on 5 g scale. Product of excellent enantiomeric excess (99.7% ee) was isolated in acceptable yield. This work paves the way towards implementing similar cascade reactions on an industrially relevant scale.

References

1. Johnson, N.W. and Kasparec, J. (2014) Patent US 8853408 B2.
2. (a) Khan, D., Hosford, J., Rufell, K. *et al.* (unpublished) Imine reductase catalysed enantioselective reductive amination for the preparation of a key intermediate to lysine specific histone demethylase 1 (LSD1) inhibitor GSK2879552; (b) Schober, M., MacDermaid, C., Ollis, A.A. *et al.* (2019) *Nature Catalysis*, **2**, 909–915.
3. Ochen, A., Whitten, R., Aylott, H.E. *et al.* (2019) *Organometallics*, **38** (1), 176.

11.2 Asymmetric Synthesis of α-Amino Acids through Formal Enantioselective Biocatalytic Amination of Carboxylic Acids

Hui Liu,[1] Bernd Nidetzky[1,2] and Alexander Dennig*[1,2]

[1]*Institute of Biotechnology and Biochemical Engineering, Graz University of Technology, Graz, Austria*

[2]*Austrian Centre of Industrial Biotechnology (acib), Graz, Austria*

Enzymatic multistep one-pot reactions offer efficient, selective and sustainable routes to the production of α-amino acids (α-AAs) [1]. Current chemical routes towards enantiopure L-α-amino acids (L-α-AAs) often require cumbersome catalyst preparation or harsh reaction conditions and toxic HCN as reagent [2]. Here, we report a short bench protocol for a two-step, scalable and atom-efficient linear biocascade for the one-pot conversion of carboxylic acids into enantiopure L-α-AAs on 150 mL scale (Scheme 11.2) [1a,3]. In step 1, a cytochrome P450 peroxygenase ($P450_{CLA}$) [4] catalysed hydroxylation of a carboxylic acid with H_2O_2 as oxidant, affording the corresponding α-hydroxy acids (α-HAs). In step 2, a redox-neutral hydrogen borrowing cascade (NAD^+ as electron shuttle) allowed oxidation and reductive amination of α-HAs (**1b–5b**) into L-α-AAs (**1c–5c**) by using two stereocomplementary α-hydroxyisocaproate dehydrogenases (L- and D-HIC-DH) [5] and a L-phenylalanine dehydrogenase (L-Phe-DH) [6]. Enantiopure L-α-AAs could thus be obtained at $g.L^{-1}$ concentration and gram scale without intermediate purification, using widely available and renewable carboxylic acids as substrates, nontoxic reagents and easy-to-prepare catalysts.

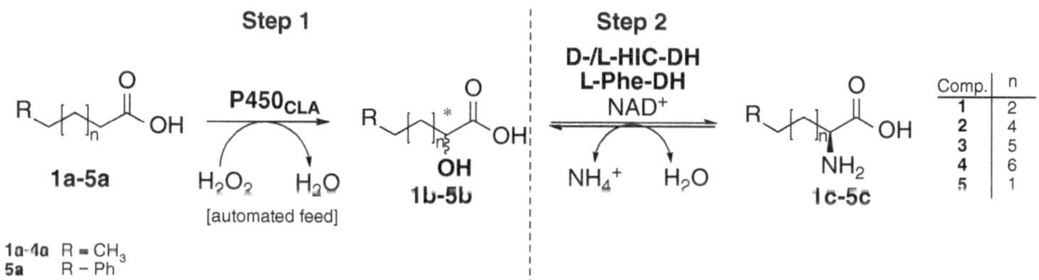

*Scheme 11.2 One-pot cascade for regio- and enantioselective transformation of selected carboxylic acids (**1a–5a**) into enantiopure L-α-AAs (**1c–5c**) [1a].*

11.2.1 Procedure 1: Catalyst Production/Recombinant Expression of Enzymes in *E. coli*

11.2.1.1 Materials and Equipment

- Tryptone (29 g)
- Yeast extract (50.5 g)
- NaCl (5 g)
- Glycerol (8 g)
- K_2HPO_4 (25 g)
- KH_2PO_4 (4.62 g)
- Distilled water (dH_2O)
- Ampicillin (100 mg.mL^{-1} stock solution in dH_2O, filter-sterilised)
- Kanamycin (50 mg.mL^{-1} stock solution in dH_2O, filter-sterilised)
- Isopropyl-β-D-1-thiogalactopyranoside (IPTG; 0.5 M in dH_2O, filter-sterilised)
- Lysogenic broth (LB) agar plate with colonies of *E. coli* BL21(DE3) harbouring the expression vector pET28a(+) bearing the genes encoding $P450_{CLA}$ [4] or the expression vector pET21a bearing the genes encoding either L- or D-HIC-DH [5]
- LB agar plate with colonies of *E. coli* BL21(DE3) harbouring the expression vector pEamTA bearing the gene encoding L-Phe-DH [6]
- 0.5 L and 1 L Schott bottles with screw caps
- 0.25 L and 1 L Erlenmeyer flasks (nonbaffled) with cotton caps
- Orbital shaker (Multitron Standar, Infors HT)
- Autoclave (Varioklav steam steriliser)

11.2.1.2 Procedure

1. Tryptone (5 g), yeast extract (2.5 g) and NaCl (5 g) were dissolved in dH_2O (500 mL). The mixture was autoclaved (20 min, 121 °C) to give sterile LB medium. Tryptone (24 g), yeast extract (48 g) and glycerol (8 g) were dissolved in dH_2O (1.8 L). K_2HPO_4 (25 g) and KH_2PO_4 (4.62 g) were dissolved in dH_2O (200 mL). Both solutions were autoclaved in separate vessels (20 min, 121 °C) and then mixed to give sterile terrific broth (TB) medium.
2. Sterile LB medium (50 mL) was filled into a sterile 0.25 L Erlenmeyer flask. Ampicillin or kanamycin stock solution (50 μL) was added: kanamycin to grow cells of *E. coli* BL21(DE3) harbouring pET28a(+)$P450_{CLA}$, ampicillin to grow cells of *E. coli* BL21(DE3) harbouring pET21a(+) or pEamTA(+). Cell material from agar plates was added with sterile tips to the culture media, which were then incubated at 37 °C and 120 rpm for 16 hr.
3. TB medium (200 mL) was filled into sterile 1 L Erlenmeyer flasks and supplemented with ampicillin (2 mL) or kanamycin (2 mL) stock solutions. Cells from the LB culture (2 mL) were then added to the TB medium.
4. The cells were grown at 37 °C and 200 rpm until an OD_{600} 0.6–1.0 was reached. Then cultures were cooled to 25 °C and expression of the respective enzyme was induced by adding IPTG (200 μL). The expression was performed at 25 °C and 120 rpm for 16–24 hr.
5. The cells were harvested by centrifugation at 4500× g and 4 °C for 20 min. Cell pellets were washed with KPi buffer (10 mM, pH 7.0, 4 °C; 20 mL) and then recentrifuged. Washed cell pellets (around 10 g_{paste}·$L_{culture\ media}{}^{-1}$) were stored on ice for later use.

11.2.2 Procedure 2: Catalyst Formulation

11.2.2.1 Materials and Equipment

- Cell pellets containing P450$_{CLA}$, L-HIC-DH, D-HIC-DH or L-Phe-DH
- Lysozyme from chicken egg white (40 000 U.mg^{-1}, 50 mg)
- Buffer A (0.1 M KPi buffer, pH 7.0, 0.3 M KCl and 50 mM imidazole; 1 L)
- Buffer B (0.1 M KPi buffer, pH 7.0, 0.3 M KCl and 400 mM imidazole; 1 L)
- Buffer C (KPi, 0.1 M, pH 7.0, 0.3 M KCl; 2 L)
- Sonicator (Sonic Dismembrator Model 505, Fisher Scientific)
- 5 mL His-tag column (HisTrap HP protein purification column)
- Äkta purification system (ÄKTA prime plus)
- Dialysis tube (cellulose) with 6–8 kDA cut-off (ZelluTrans from Roth)
- 1 L beaker
- Stirring plate and magnetic stirring bar
- 30 kDa cut-off centrifuge tube (Vivaspin Turbo from Satorius)
- Centrifuge (min. 4000× g)
- Centrifuge (min. 24 000× g)
- Nanodrop spectrophotometer (DeNovix DS-11 spectrophotometer)
- KPi buffer (10 mM, pH 7.0; 20 mL)
- KPi buffer (50 mM, pH 7.75; 15 mL)
- Liquid nitrogen (N_2)
- Freeze dryer (Christ ALPHA 1-4)

11.2.2.2 Procedure

11.2.2.2.1 Step 1: His-Tag Purification of P450$_{CLA}$ [7].

1. The frozen cell pellet (~20 g paste from 2 L TB culture) containing P450$_{CLA}$ was resuspended in buffer A (50 mL) containing lysozyme (50 mg). The resuspended cells were incubated at 37 °C for 2 hr.
2. Cells were disrupted by sonication (6 min total pulse time; programme: 2 sec pulse on, 4 sec pulse off; 60% amp).
3. Disrupted cells were centrifuged (24 000× g, 20 min, 4°C) and the clear supernatants were combined.
4. Clear supernatants were loaded (3 mL.min^{-1}) on to a His-tag column equilibrated in buffer A and connected to an Äkta purification system. After binding of proteins, the column was washed with buffer A (50 mL). P450$_{CLA}$ was eluted with buffer B (isocratic flow). The elution of P450$_{CLA}$ was monitored by UV signal.
5. Collected fractions (5 mL, strong red colouration) containing P450$_{CLA}$ were combined and filled into a dialysis bag, which was placed in a beaker (1 L) filled with buffer C (660 mL). The solution was stirred at 60 rpm and 4 °C for 12 hr. The buffer was replaced twice by buffer C (660 mL).
6. The dialysed protein was concentrated in a cut-off centrifuge tube (final volume 1–2 mL), then protein concentration was determined using a Nanodrop spectrophotometer.

11.2.2.2.2 Step 2: Formulation of HIC-DHs and Phe-DH.

1. Cell pellets containing HIC-DHs (~10 g paste) were resuspended in KPi buffer (10 mM, pH 7.0; 20 mL). The resuspended cells were frozen in liquid N_2 and freeze-dried. The obtained powders (~1 g) were stored at 4 °C for later use.
2. Cell pellets (~10 g paste) containing L-Phe-DH were resuspended in KPi buffer (50 mM, pH 7.75; 15 mL), then sonicated (4 min total pulse time; programme: 2 sec pulse on, 4 sec pulse off; 65% amp).
3. After centrifugation (24 000× g, 1 hr, 4 °C), supernatants were collected and frozen in N_2. The protein (typically 800–1000 U.mL^{-1}) was stored at −20 °C for later use.

11.2.3 Procedure 3: Hydroxylation and Amination of Octanoic Acid 2a

11.2.3.1 Materials and Equipment

- $P450_{CLA}$ in buffer C (purified enzyme, 6 μM or 43.2 mg)
- KPi buffer (pH 7.0, 20 mM; 139.5 mL)
- Round-bottom flasks (500 mL)
- Magnetic stirrer plate and stirrer
- pH titration unit (TitroLine alpha)
- Octanoic acid **2a** (3 mmol)
- EtOH (99%; 9 mL)
- 10 M KOH
- H_2O_2 solution (960 mM stock in dH_2O; 50 mL)
- Automatic/programmable pump (MCP-CPF Process IP65)
- 10 and 37% HCl
- Ethyl acetate (EtOAc; >99%; 0.1% v/v 1-octanol as internal standard)
- Na_2SO_4, anhydrous
- Methanol (MeOH, >99%)
- Trimethylsilyl diazomethane in diethyl ether ($TMSCHN_2$, 2 M)
- Gas chromatography/mass spectrometry (GC-MS) chromatograph (Agilent 7890A GC-system equipped with an Agilent 5975C mass-selective detector)
- GC column (Agilent HP-5MS column; 30 m × 320 μm, 0.25 μm film)
- Catalase from bovine liver (25 mg or 50 000 U)
- D- and L-HIC-DH (each 2 U.mL^{-1}, corresponding to ~30 mg cell powder)
- L-Phe-DH (8 U.mL^{-1}, corresponding to 0.5–1 mL cell lysate)
- NH_4Cl (1.6 g)
- NH_4CO_3 (0.29 g)
- Nicotinamide adenine dinucleotide (NAD^+, Na salt; 617 mg)
- Thin-layer chromatography (TLC) silica gel plates (Merck)
- TLC mobile phase: dH_2O/acetone/acetic acid/1-butanol in a ratio of 35 : 23 : 7 : 35
- Ninhydrin solution (4 g.L^{-1} in acetone)
- Centrifuge (min. 4000× g, 20 min, 4 °C)
- Liquid nitrogen (N_2)
- Freeze dryer (Christ ALPHA 1-4)
- Glass chromatography column
- Cation-exchange resin (Dowex® 50WX8; 50–100 mesh)

- Distilled water (dH_2O)
- 1 M HCl (1 L)
- 1 M NH_4OH (500 mL)
- Ethanol (EtOH, synthesis grade, >99%; 5 mL)
- Thionyl chloride ($SOCl_2$; 250 µL)
- Rotary evaporator (Laborota 4000)
- Dichlormethane (CH_2Cl_2, synthesis grade, >99%; 2.5 mL)
- Triethylamine (Et_3N; 125 µL)
- Benzoyl chloride (50 µL)
- High-performance liquid chromatograph (HPLC) with UV detector (Shimadzu LC-20AD HPLC system with SPD-M20A diode array detector)
- Chiral HPLC columns: Chiralcel OD-H column (25 × 0.46 cm, 5 µM), ChirobioticT column (25 × 1 cm, 5 µM)
- HPLC sample solvent (acetonitrile/dH_2O 1 : 1; 10 mL)
- HPLC-grade solvents for mobile phase: *n*-heptane, isopropanol, dH_2O, EtOH
- Cooling centrifuge (min. 4000× g)
- Magnetic stirrer and stirring bar

11.2.3.2 Procedure

11.2.3.2.1 **Step 1: Hydroxylation of 2a.**

1. Purified $P450_{CLA}$ (43.2 mg) in buffer C (3 mL) was pipetted into KPi buffer (pH 7.0, 20 mM; 139.5 mL) filled into a round-bottom flask (500 mL). The solution was continuously stirred at 200 rpm, 1 atm and rt.
2. **2a** (216 mg, 1.5 mmol) was dissolved in EtOH (7.5 mL) and added to the flask. The pH was readjusted to 7.0 with 10 M KOH (aq).
3. H_2O_2 (960 mM aq) was pumped into the reaction vessel at intervals of 30 min (total 19 pulses of 125 µL). The reaction proceeded at 22 °C for 10 hr.
4. Additional **2a** (216 mg, 1.5 mmol in 1.5 mL EtOH) was added to the reaction. The pH was set to 7.0 and the H_2O_2 supplementation was restarted as described in Step 3.
5. Substrate conversion was monitored by GC-MS. 500 µL of reaction sample was acidified with 10% HCl (50 µL) and extracted with EtOAc (300 µL). The organic phase was dried over Na_2SO_4. 120 µL of dried organic phase was mixed with 60 µL of MeOH, then 10 µL of 2 M $TMSCHN_2$ was added.
6. The obtained methyl esters were injected into a GC-MS. Separation was achieved using an Agilent HP-5MS column with He as carrier gas (see Tables 11.1 and 11.2). **2a** was converted by 99% and **2b** obtained in >99% yield (minor traces of the relevant β-hydroxy acid were formed).

Table 11.1 *GC-MS programmes used for analysis of α-hydroxy acids.*

Substrate	Detector	Column	Program
1a	GC-MS	HP-5MS	40 °C/hold 5 min; 20 °C.min^{-1} to 180 °C; hold 0 min
2a–5a	GC-MS	HP-5MS	100 °C/hold 5 min; 10 °C.min^{-1} to 300 °C; hold 0 min

Table 11.2 Retention times of substrates and products during GC-MS analysis.

Substrate	Retention time (min)	Product	Retention time (min)
1a	6.1	**1b**	9.1
2a	5.8	**2b**	7.3
3a	7.2	**3b**	8.5
4a	8.4	**4b**	9.4
5a	8.9	**5b**	9.8

11.2.3.2.2 Step 2: Amination of 2b.

1. Catalase (25 mg) was added to the reaction, which was then incubated at 200 rpm, 1 atm and rt for 30 min.
2. The following components were added: D- and L-HIC-DH (2 U.mL^{-1} of each; ~30 mg dry powder), L-Phe-DH (8 U.mL^{-1}, typically 0.5–1 mL of prepared lysate), NH_4Cl (1.6 g) and NH_4CO_3 (0.29 g). The pH was adjusted to 8.0 using 10 M KOH (aq).
3. NAD^+ (617 mg) was added to initiate the reaction.
4. Conversion of **2b** was monitored by following its depletion by GC-MS. Derivatisation and analysis of **2b** were carried out as already described.
5. Formation of α-AAs was monitored by TLC. 2 µL of the reaction mixture was spotted and separated on a silica plate. α-AAs were visualised by spraying ninhydrin solution on to the plate and then gently heating.
6. For product workup, the solution was acidified with 37% HCl (aq) to pH 1.0. Solids were removed by centrifugation (4000× g, 30 min, rt). The supernatant was frozen in liquid N_2 and lyophilised.
7. α-AAs were purified via cation-exchange chromatography. For this, the cation-exchange resin (100 g) was filled into a glass chromatography column, the column material was washed with dH_2O (500 mL) and 1 M HCl solution (500 mL), α-AAs were resuspended in 1 M HCl (10 mL) and loaded on-to the column, the column was washed with 1 M HCl solution (500 mL) and dH_2O (500 mL) and the α-AAs were eluted with 1 M NH_4OH solution (500 mL).
8. Fractions containing the target amino acid were collected and analysed by TLC. Positive fractions were combined, frozen in N_2 and freeze-dried.
9. The solid residue was washed with dH_2O (pH 7.0, 500 µL; pH was readjusted with a few µL of 0.1 M HCl or NaOH). Residual solvent was removed by freeze-drying. Structures and purity were determined by ^{1}H-/^{13}C-NMR. **2c** was obtained in 84% yield (257 mg; 59% isolated yield) and >99% ee.

11.2.4 Analytical Method

1. Purified amino acid **1c**-**4c** (0.1 mmol) was mixed with EtOH (1 mL) at 0 °C. $SOCl_2$ (neat, 51 µL) was added dropwise and the solution was warmed to rt before heating to reflux for 10 hr. The solution was cooled and solvents were removed under vacuum. The obtained solid was dissolved in CH_2Cl_2 (0.5 mL), then Et_3N (28 µL, 0.2 mmol) and

Table 11.3 Chiral UV-HPLC setup used for analysis of AAs.

Product	Detector	Column	Conditions/programme
1c–4c	HPLC-UV (254 nm)	Chiralcel OD-H	1.0 mL.min^{-1}, 30 °C, isocratic mobile phase: (a) 96 : 4 *n*-heptane : i-PrOH (**1c** and **2c**), (b) 92 : 8 *n*-heptane : i-PrOH (**3c**), (c) 90 : 10 *n*-heptane : i-PrOH (**4c**)
5c	HPLC-UV (212 nm)	ChirobioticT	0.7 mL.min^{-1}, isocratic mobile phase, 98 : 2 H_2O : EtOH, 25 °C

benzoyl chloride (11.6 μL, 0.1 mmol) were added. After stirring for 1 hr, the mixture was purified by preparative TLC. The derivatised α-AAs were analysed by chiral HPLC-UV (Table 11.3).

2. Amino acid **5c** (4 mg) was solubilised in HPLC sample solvent (2 mL). The sample was analysed by chiral HPLC (Tables 11.3 and 11.4).
3. **1c–5c** were obtained in optically pure form with >99% ee for all amino acids. Typically, isolated yields ranged between 88 and 248 mg (40–76%) of the respective α-AAs.

11.2.5 Conclusion

The described procedure enabled the production of enantiopure L-α-AAs under mild reaction conditions and using renewable materials and catalysts. L-α-AAs may thus be obtained at high yield in a three-step linear biocascade operated in one-pot fashion (Table 11.5). This

Table 11.4 Retention times of substrates and products for HPLC analysis.

Product	Retention time (min)
1c	D-**1c** = 7.79; L-**1c** = 13.43
2c	D-**2c** = 6.26; L-**2c** = 8.42
3c	D-**3c** = n.d.; L-**3c** = 7.0[a]
4c	D-**4c** = 4.93; L-**4c** = 6.11
5c	D-**5c** = 6.38; L-**5c** = 7.43

n.d., not determined.
[a]Optical rotation measurement confirmed the presence of a single enantiomer.

Table 11.5 Preparative synthesis of L-amino acids (**1c–5c**).

Substrate	Step 1 α-HAs (%)	Step 2 α-AA (%)	Iso yield (%)	ee (%)
1a	>99	78 (*S*)	40 (156 mg)	>99
2a	>99	84 (*S*)	59 (257 mg)	>99
3a	95	>99 (*S*)	72 (372 mg)	>99
4a	73	95	30 (171 mg)	>99
5a	94	69 (*S*)	48 (44 mg)[a]	>99

[a]Conversion done on 50 mL scale

synthetic concept can be extended to further substrates and offers an attractive and green alternative to the traditional cyanide-based asymmetric chemical synthesis of α-AAs.

References

1. (a) Dennig, A., Blaschke, F., Gandomkar, S. *et al.* (2019) *Advanced Synthesis & Catalysis*, **361**, 1348–1358; (b) Bossow, B. and Wandrey, C. (1987) *Annals of the New York Academy of Sciences*, **506**, 325–336.
2. Najera, C. and Sansano, J.M. (2007) *Chemical Reviews*, **107**, 4584–4671.
3. Dennig, A., Gandomkar, S., Cigan, E. *et al.* (2018) *Organic and Biomolecular Chemistry*, **16**, 8030–8033.
4. Girhard, M., Schuster, S., Dietrich, M. *et al.* (2007) *Biochemical and Biophysical Research Communications*, **362**, 114–119.
5. (a) Schutte, H., Hummel, W. and Kula, M.R. (1984) *Applied Microbiology and Biotechnology*, **19**, 167–176; (b) Hummel, W., Schutte, H. and Kula, M.R. (1985) *Applied Microbiology and Biotechnology*, **21**, 7–15.
6. Brunhuber, N.M.W., Banerjee, A., Jacobs, W.R. and Blanchard, J.S. (1994) *Journal of Biological Chemistry*, **269**, 16 203–16 211.
7. Dennig, A., Kuhn, M., Tassoti, S. *et al.* (2015) *Angewandte Chemie International Edition*, **54**, 8819–8822.

11.3 Enantioselective, Catalytic One-Pot Synthesis of γ-Butyrolactone-Based Fragrances

Ceyda Kumru,[1] Thomas Classen[2] and Jörg Pietruszka[1,2]

[1]*Institut für Bioorganische Chemie, Heinrich-Heine-Universität Düsseldorf im Forschungszentrum Jülich, Jülich, Germany*

[2]*IBG-1: Bioorganic Chemistry, Forschungszentrum Jülich, Jülich, Germany*

Enantiopure γ-butyrolactones are not only versatile building blocks for organic synthesis, they are also natural products of interest in their own right. A number of known flavour and fragrance compounds bear this structural motif. One common derivative is the *quercus lactone* **1** – also known as *whisky lactone* – a flavour component in alcoholic beverages (Scheme 11.3). Its taste and odour properties have been described as coconut-like with an earthy or celery-like note and depend on the configuration of the lactone [1]. Since the amount obtainable from natural sources is limited, alternative enantioselective approaches towards γ-butyrolactones are in demand [2]. The present gram-scale protocol for the consecutive chemoenzymatic cascade is based on the work of Kumru *et al.* [3]. It starts from olefin **2** using an ene reductase (ER) from *Zymomonas mobilis* (ER-Ncr_{ZM}) to afford intermediate **3**, which is reduced by an alcohol dehydrogenase (ADH) from *Ralstonia* sp. (ADH-R) to intermediate **4**, which undergoes lactonisation to compound **1** on acidic workup.

11.3.1 Procedure 1: Preparation of Enzyme Catalyst

11.3.1.1 Materials and Equipment

- *Escherichia coli* BL21(DE3) harbouring the pHT::ncrZM-vector (pET22b amp^R p_{T7}(6xhis:tev:ncr(ZM))) [3]

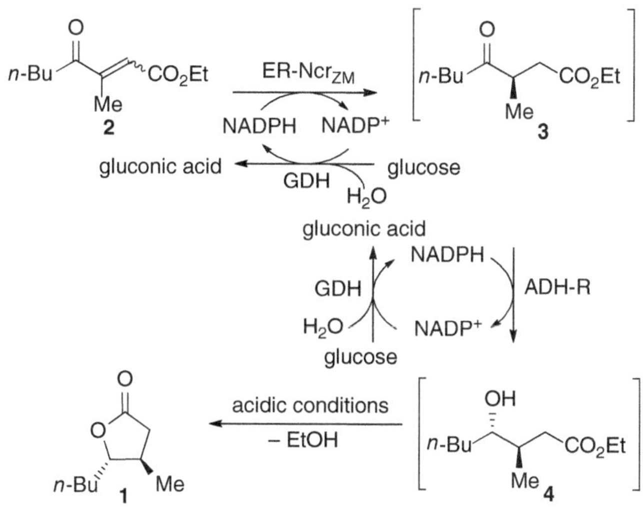

Scheme 11.3 *Chemoenzymatic synthesis of the quercus lactone* ***1*** *[3].*

- *E. coli* BL21(DE3) harbouring the pet22b::RADH-vector (pET22b amp^R p_{T7}) [4]
- Lysogenic broth containing ampicillin (LB_{amp}) medium (1.0% w/v tryptone, 0.5% w/v yeast extract, 1.0% w/v NaCl, 0.2% v/v ampicillin (c_{stock} = 100 mg.mL^{-1}); 5 mL)
- Terrific broth containing ampicillin (TB_{amp}) medium (1.2% w/v tryptone, 2.4% w/v yeast extract, 0.4% w/v glycerol, 0.231% w/v KH_2PO_4, 1.254% w/v K_2HPO_4, 0.1% v/v ampicillin (c_{stock} = 100 mg.mL^{-1}); 1 L)
- Isopropyl-β-D-1-thiogalactopyranoside (IPTG) 1 M in water (filter-sterilised)
- Buffer P_0: 20 mM potassium phosphate buffer pH 6.5
- Buffer P_{30}: P_0 with 30 mM imidazole pH 6.5 from 2 M stock solution
- Buffer P_{250}: P_0 with 250 mM imidazole pH 6.5 from 2 M stock solution
- Aqueous KP_i buffer (100 mM, pH 7)
- $MgCl_2$
- 3.5 L Fernbach flask (other flasks may work as well)
- Orbital shaker 25 °C, 120 rpm
- Cooled centrifuge for 50 mL reaction tubes 12 000 rpm
- Peristaltic pump (optional)
- 5 mL Ni-NTA column (e.g. NiNTA superflow cartridge from Qiagen)
- Ultrafiltration device (e.g. Vivaspin 20 MWCO 10 kDa from Satorius)
- PD10 desalting column (e.g. from GE Healthcare)
- Flavinmononucleotide (FMN, optional)
- 50 mL conical reaction tubes
- Lyophiliser (optional)

11.3.1.2 Procedure

11.3.1.2.1 **Expression.**

1. LB_{amp} medium (5 mL) was inoculated with several colonies of *E. coli* BL21(DE3) pHT::ncrZM or pet22b::RADH and incubated at 37 °C overnight under mild shaking.

2. For the main culture, TB_{amp} medium (1 L) in a baffled Fernbach flask was inoculated with the entire overnight culture (OD_{600} 0.6–0.8) and incubated at 25 °C and 120 rpm for 8–10 hr.
3. The main culture was induced with 1 M IPTG to a final concentration of 100 μM and incubated at 25 °C and 120 rpm (OD_{600} 6.0–8.0) for a further 14 hr.
4. The whole cells were harvested by centrifuging the suspension at 6430× g and 4 °C for 10 min.
5. The pellets (~10 g.L^{-1}) were either stored long-term frozen at −20 °C (up to 3 months) or used directly for cell disruption and purification, if needed.

11.3.1.2.2 Purification of Ncr_{ZM}.

1. Wet cells (5 g) were resuspended in buffer P_0 (25 mL).
2. The resulting 20% w/v suspension was subjected to cell disruption by ultrasonication (2 × 5 min).
3. The disrupted cells were pelleted by centrifugation at 12 000 rpm and 4 °C for 10 min.
4. The clarified supernatant was applied to a 5 mL NiNTA column equilibrated with buffer P_0 using a peristaltic pump, circulating the cell-free crude extract through the column five times. The loaded column was subjected to an elution programme using a protein medium-pressure liquid chromatography (MPLC) system: 6 column volumes (CV) with P_0 (this step is optional and can be skipped without loss of purity or yield), 6 CV with P_{30} and finally 6 CV of P_{250} to elute the protein. The elution chromatogram can be monitored at wavelengths of either 280 nm (all proteins) or 400 nm (only Ncr_{ZM}). If no automatic chromatographic system is available, the protocol is applicable to gravity-flow or syringe-driven chromatography as well.
5. The yellow eluate fractions, containing the desired protein, were combined and concentrated using ultrafiltration (10 kDa MWCO) up to 10 mg.mL^{-1}. The eluate can optionally be supplemented with 5 mM FMN during the concentration step to complement the fraction of apo-Ncr_{ZM} with FMN. This procedure is able to increase the specific activity by a factor of eight.
6. Using a (microvolume) photometer (blanked on ultrafiltrator flow-through) the degree of FMN saturation (*S*) and the protein concentration (β) can be estimated using the empirical formulae:

$$S = 3.39\frac{A_{458nm}}{A_{280nm}} = 4.53\frac{A_{432nm}}{A_{280nm}}$$

and:

$$\beta = \frac{A_{280nm}}{(1.015 \cdot S + 0.8)} \text{mg.mL}^{-1}$$

7. To remove the remaining imidazole (and the unbound FMN, if used), the buffer was exchanged to P_0 using a PD10 desalting column. The imidazole may decrease the stability of the protein in solution and cause impurities during the following workup. Remaining FMN will interfere with the photometric protein concentration but not with the reaction.
8. The protein (~60–90 mg) was either stored as lyophilisate or used directly.

11.3.1.2.3 Cell Disruption of *E. coli BL21(DE3) Harbouring ADH-R.*

1. Wet cells (5 g) were resuspended in KP_i buffer (100 mM, pH 7, 25 mL) with 1 mM $MgCl_2$.
2. The resulting 20% w/v suspension was subjected to cell disruption by ultrasonication (2 × 5 min).
3. The disrupted cells were pelleted by centrifugation at 12 000 rpm and 4 °C for 10 min.
4. The protein (~0.7–1.0 g) was either used as crude lysate or stored as lyophilisate.

11.3.2 Procedure 2: Enzymatic Synthesis of Lactone 1

11.3.2.1 Materials and Equipment

- Buffer P_0: 20 mM potassium phosphate buffer pH 6.5
- Aqueous KP_i buffer (100 mM, pH 7)
- $MgCl_2$
- Glucose dehydrogenase from *Bacillus subtilis* [5]
- Glucose
- Disodium $NADP^+$
- Ethyl 3-methyl-4-oxooct-2-enoate **2** [3]
- $MgSO_4$
- Cooled centrifuge for 50 mL reaction tubes 12 000 rpm
- 50 mL conical reaction tubes
- Magnetic stirrer
- pH-stat (e.g. Titrino plus 848 from Deutsche METROHM GmbH & Co. KG)
- Silica gel column
- Eluent *n*-pentane/acetone 98 : 2
- Diethyl ether
- Aqueous NaOH (1 M)
- Ncr from *Zymomonas mobilis*
- ADH from *Ralstonia* sp.

11.3.2.2 Procedure

Enoate **2** was synthesised by a Horner–Wadsworth–Emmons reaction as described in detail in [3].

1. Enoate **2** (1 g, 5.0 mmol) was mixed with $NADP^+$ (3.33 mL, 20 mM in buffer P_0) and glucose (61 mL, 0.1 M in buffer P_0); GDH (333 mL, ~150 U.mL^{-1}) and Ncr_{ZM} (2.5 U.mmol^{-1}; 1 U is defined as the amount of enzyme necessary to convert 1 μmol of cyclohex-2-enone in 1 min under the given conditions (30 °C, 20 mM KP_i pH 6.5, 0.15 mM NADPH, 1 mM cyclohex-2-enone, photometrically monitored via the NADPH consumption at 340 nm)) were added and the reaction was stirred overnight at 30 °C. The pH was kept constant at 6.5 by a pH-stat with aqueous NaOH (1 M).
2. After complete conversion, as determined by GC, ADH-R (139 U.mmol^{-1}) was added together with $NADP^+$ (3.33 mL, 20 mM in KP_i buffer; 100 mM, pH 7), GDH (333 mL,

~150 U.mL^{-1}), glucose (61 mL, 0.1 M in KP_i buffer; 100 mM, pH 7) and $MgCl_2$ (6.5 mL, 1 mM, 5 vol %). The reaction mixture was stirred at 30 °C until maximal conversion was reached, and the pH was kept constant at 6.5 by a pH-stat with NaOH (1 M).

3. After complete conversion, as determined by gas chromatography (GC), the reaction mixture was thoroughly extracted with diethyl ether (200 mL). Due to a poor phase separation, the interphases were centrifuged (5 min, max. RCF). The combined organic layers were dried over $MgSO_4$ and filtered, and the solvent was removed under reduced pressure.
4. The crude product was purified by chromatography over silica using *n*-pentane/acetone 98 : 2 as eluent to afford product **1** (578 mg, 3.70 mmol, 73%, 78% conv., >99% ee, sel. 0 : 98 : 2 : 0) as a colourless oil with coconut odour.

11.3.3 Analytical Method

The analytical data are in accordance with the literature [1]. Rf (*n*-pentane/acetone 95 : 5) = 0.12. ^{1}H-NMR (600 MHz, $CDCl_3$): δ (ppm) = 0.92 (t, *J* = 7.1 Hz, 3 H, 4′-H), 1.14 (d, *J* = 6.3 Hz, 3 H, 1′-H), 1.31–1.43 (m, 3 H, 3′-H, 2′-H), 1.47–1.54 (m, 1 H, 1″-H), 1.56–1.64 (m, 1 H, 1″-H), 1.64–1.71 (m, 1 H, 2″-H), 2.15–2.26 (m, 2 H, 2-H, 3-H), 2.63–2.71 (m, 1 H, 2-H), 4.01 (ddd, *J* = 8.2 Hz, *J* = 8.0 Hz, *J* = 3.9 Hz, 1 H, 4-H). ^{13}C-NMR (151 MHz, $CDCl_3$): δ (ppm) = 13.89 (C4′, CH_3), 17.50 (C1′, CH_3), 22.48 (C2′/C3′, CH_2), 27.83 (C2′/C3′, CH_2), 33.70 (C1″, CH_2), 36.07 (C3, CH), 37.13 (C2, CH_2), 87.46 (C4, CH), 176.59 (C1, C_{quat}). IR (film), $\tilde{\nu}$ (cm^{-1}): 2959, 2933, 2874, 1772 (lactone), 1459, 1424, 1383, 1331, 1284, 1255, 1210, 1170, 1124, 1078, 984, 926, 942, 855, 786, 733, 662. MS (EI, 70 eV): *m/z* (%) = 157 (4) [$(M+H)^+$], 138 (6) [$(M-H_2O)^+$], 99 (100) [$(M-C_4H_9)^+$], 84 (54) [$(C_4H_4O_2)^+$], 69 (15) [$(C_5H_9)^+$], 51 (47). GC/MS: t_R (min) = 7.79. GC: t_R (min) = 71.5 [(3*R*,4*S*), >99% ee, sel.: 0 : 98 : 2 : 0], 71.1 (3*S*,4*R*), 73.3 (3*S*,4*S*), 73.6 (3*R*,4*R*) (Hydrodex β-TBDAc, 90 °C – 60 min, 5 °C/min to 150 °C – 5 min). Optical rotation: $[\alpha]^{20}_D = -89.0$ (3*R*,4*S*) (c = 1, MeOH, >99% ee, >99% de) (literature: $[\alpha]^{20}_D = -95$ (3*R*,4*S*), +96 (3*S*,4*R*) (c = 1.0–1.6, MeOH)) [1].

11.3.4 Conclusion

The presented consecutive chemoenzymatic cascade utilising an ER and an ADH is not limited to the successful scalable synthesis of the title compound, but can be extended to a variety of γ-butyrolactones (transformation of ketones **5** into lactones **6**; Scheme 11.4). Factors influencing yield and selectivity are obviously the size and relative position of the substituents, but also the configuration of the substrate [3,6]. Whilst sterically more demanding substituents limit the flexibility when choosing an adequate enzyme, more options are provided when only methyl groups are present. Then, in some cases, all possible stereoisomers are accessible with the right combination of substrate and reductase, rendering the approach highly efficient not only for the synthesis of selected flavour compounds, but also for a variety of demanding chiral building blocks.

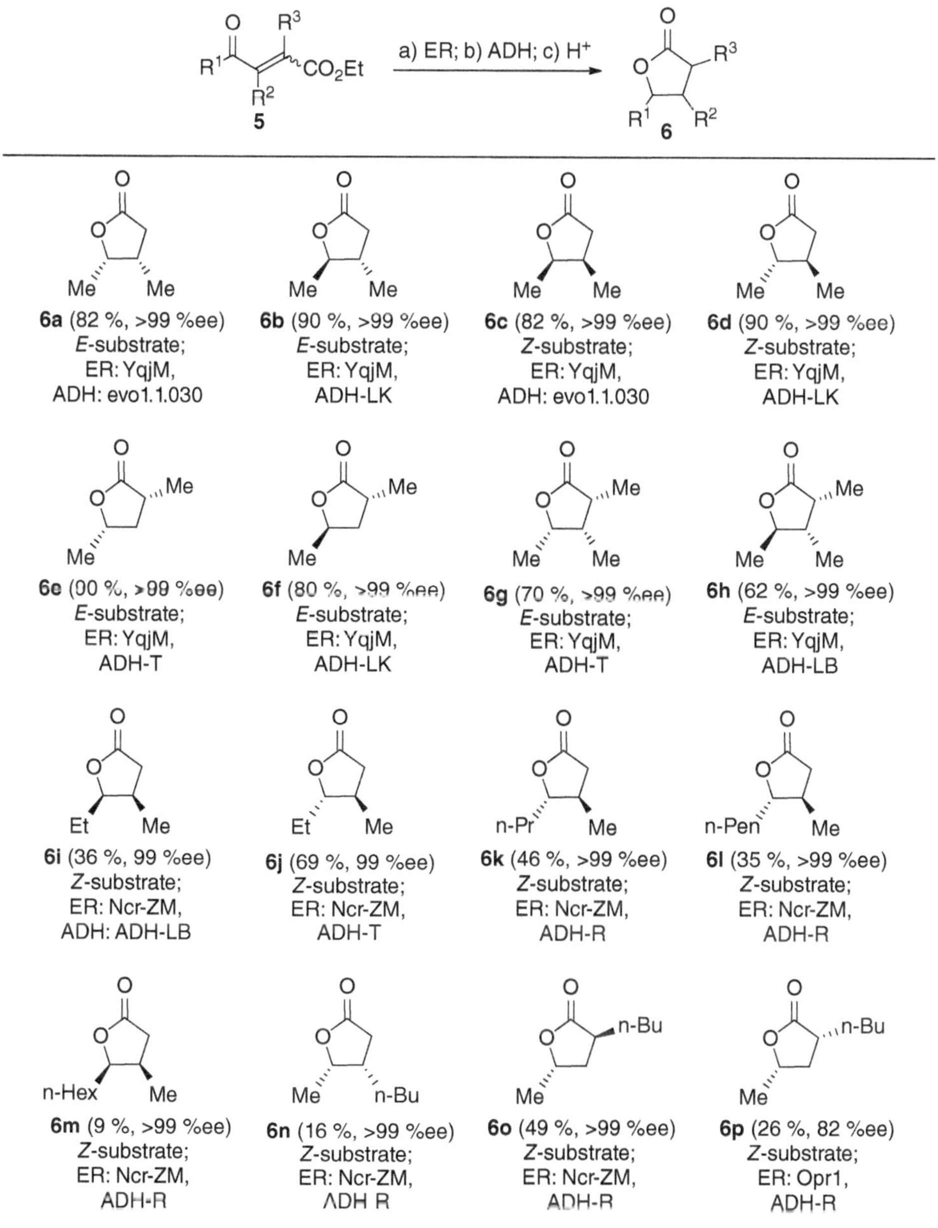

Scheme 11.4 Chemoenzymatic synthesis of lactone **6** [3,6]. Yields and selectivity for major isolated product given. Enzymes used: ene reductase (ER) from *Bacillus subtilis* (YqjM) and dienoate reductase from *Lycopersicon esculentum* (Opr1), alcohol dehydrogenases from *Lactobacillus brevis* (ADH-LB), *Lactobacillus kefir* (ADH-LK), *Thermoanaerobacter* sp. (ADH-T) and a commercial source (evo1.1.030).

References

1. Günther, C. and Mosandl, A. (1986) *Liebigs Annalen der Chemie*, 2112–2122.
2. Mao, B., Fañanás-Mastral, M. and Feringa, B.L. (2017) *Chemical Reviews*, **117**, 10 502–10 566.
3. Kumru, C., Classen, T. and Pietruszka, J. (2018) *ChemCatChem*, **10**, 4917–4926.
4. Holec, C., Sandkuhl, D., Rother D. *et al.* (2015) *ChemCatChem*, **7**, 3125–3130.
5. Weckbecker, A. and Hummel, W. (2005) Glucose dehydrogenase for the regeneration of NADPH and NADH, in *Microbial Enzymes and Biotransformations* (ed. J.L. Barredo), Humana Press, pp. 225–238.
6. Classen, T., Korpak, M., Schölzel, M. and Pietruszka, J. (2014) *ACS Catalysis*, **4**, 1321–1331.

11.4 Synthesis of Six out of Eight Carvo-Lactone Stereoisomers via a Novel Concurrent Redox Cascade Starting from (*R*)-and (*S*)-Carvones

Julia Jodlbauer,[1] Naseem Iqbal,[2] Jon D. Stewart,[3] Peter Macheroux,[4] Marko D. Mihovilovic[1] and Florian Rudroff*[1]
[1] *TU Wien, Institute of Applied Synthetic Chemistry, Vienna, Austria*
[2] *US–Pakistan Center for Advanced Studies in Energy (USPCAS-E), National University of Science and Technology (NUST), Islamabad,Pakistan*
[3] *University of Florida, Department of Chemistry and Biomedical Engineering, Gainesville, FL, USA*
[4] *TU Graz, Institute of Biochemistry, Graz, Austria*

Multistep one-pot reactions are an evolving field in asymmetric synthesis and have found broad application in the last decade [1]. Through the targeted combination of different, non-native biocatalysts, artificial enzymatic cascades can be constructed [1a]. Domino, tandem or cascade processes are particularly attractive, as the individual steps can be combined modularly. This allows the structural and functional complexity of a chemical entity to be significantly increased in a single step [2]. For this reason, we established a one-pot enzymatic redox cascade composed of five different enoate reductases (EREDs; SYE-4 [3], OPR1 [4], OPR3 [4], YqjM [5] and W116I [6]) from diverse bacterial origins and four Baeyer–Villiger monooxygenases (BVMOs; CHMO*Acineto* [7], CHMO*Brevi1* [8], CHMO*Brevi2* [8] and CHMO*Coma* [9]) with complementary regioselectivity.

Through the coupling and combining of different EREDs and BVMOs (Scheme 11.5), we achieved methods to produce six out of eight carvo-lactone stereoisomers (**3a**, **3a′**, **3b**, **3b′**, **3c** and **3c′**) starting from readily available natural carvones. Furthermore, we showed the applicability of the two-step cascade via representative preparative-scale experiments. One-pot enzymatic redox reactions of selected ERED–BVMO combinations led to good yields of 'normal' and 'abnormal' carvo-lactones **3a** (71%), **3c** (76%), **3b′** (71%) and **3b** (30%) [10].

11.4.1 Procedure 1: Preparation of Crude Cell Extracts from *E. coli* BL21 (DE3) with Respective EREDs and BVMOs

11.4.1.1 Materials and Equipment

- Bacto-peptone (5 g)
- Yeast extract (24 g)

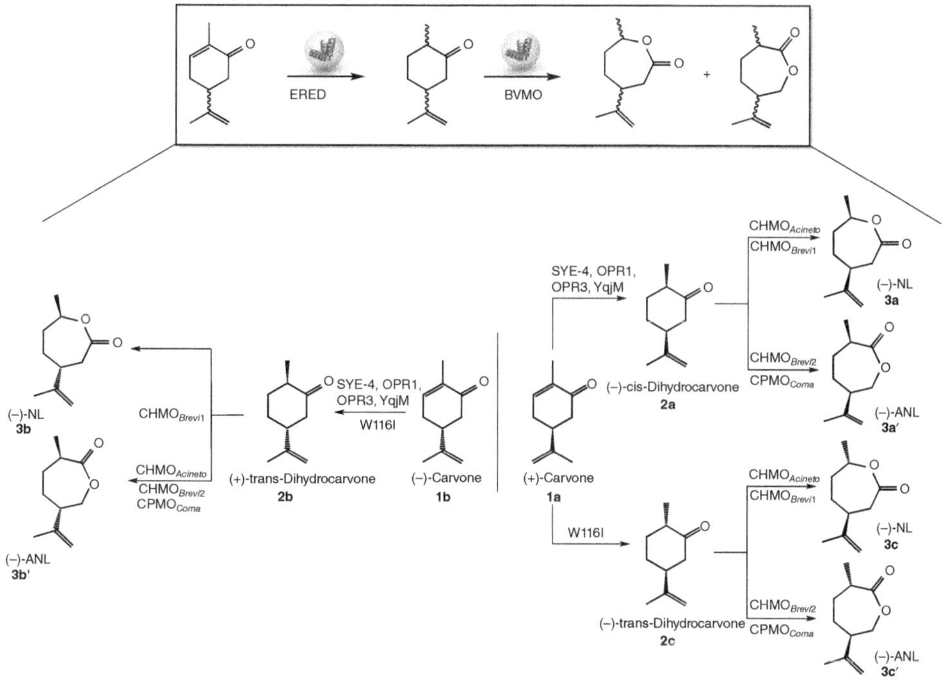

Scheme 11.5 *Catalysis of (+)- and (−)-carvone into six carvo-lactone stereoisomers* ***3a, 3a′, 3b, 3b′, 3c, 3c′*** *by different ERED and BVMO enzymes.*

- NaCl (13 g)
- Bacto-tryptone (24 g)
- Bacto-yeast (48 g)
- Glycerol (8 mL)
- $K_2HPO_4 \cdot 3\ H_2O$ (32.8 g)
- KH_2PO_4 (4.84 g)
- Distilled water (dH_2O)
- KCl (8 g)
- Na_2HPO_4 (1.44 g)
- HCl (1 M aq)
- Ampicillin (100 $mg.mL^{-1}$ in dH_2O, filter-sterilised)
- Kanamycin (50 $mg.mL^{-1}$ in H_2O, filter-sterilised)
- Lysogenic broth (LB) agar plates with colonies of *E. coli* BL21(DE3) strains (summarised in Table 11.6)
- Isopropyl-β-D-1-thiogalactopyranoside (IPTG; 0.5 M in dH_2O, filter-sterilised)
- Protease inhibitor (phenyl methane sulfonyl fluoride)
- 5× 0.5 L Schott bottles with screw caps
- 9× 0.1 L Erlenmeyer flasks (baffled) with cotton caps
- 9× 1 L Erlenmeyer flasks (baffled) with cotton caps
- Orbital shaker (Gerhard THO5)
- Table autoclave (Tuttnauer 2540EL)

- pH meter
- Cooling centrifuge (Sigma 3K30, rotor 372/C)
- Cooling centrifuge (Sigma 6K15, rotor 12155-H)
- Spectrophotometer (to measure OD_{590}/OD_{600})
- Ultrasonic wave generator (Bandelin Sonopuls HD 3200)
- Ultrasonic sonotrode (Bandelin KE 76)
- Bio-Rad Protein Assay

Table 11.6 *Enzyme expression strains and expression systems.*

Enzyme	Strain	Plasmid	Inducer	Selective marker	Reference
SYE-4	*E. coli* BL21 (DE3)	pGEX-4T-2	IPTG	Ampicillin	[3]
OPR1	*E. coli* BL21 (DE3)	pET21a(+)	IPTG	Ampicillin	[11]
OPR3	*E. coli* BL21 (DE3)	pET21a(+)	IPTG	Ampicillin	[11]
YqjM	*E. coli* BL21 (DE3)	pET21a(+)	IPTG	Ampicillin	[12]
W116I	*E. coli* BL21 (DE3)	pDJB5	IPTG	Kanamycin	[6]
$CHMO_{Acineto}$	*E. coli* BL21 (DE3)	pET22b(+)	IPTG	Ampicillin	[7]
$CHMO_{Brevi1}$	*E. coli* BL21 (DE3)	pQE-30	IPTG	Ampicillin	[8]
$CHMO_{Brevi2}$	*E. coli* BL21 (DE3)	pTrcHis-TOPO	IPTG	Ampicillin	[8]
$CHMO_{Coma}$	*E. coli* BL21 (DE3)	pCMP201	IPTG	Ampicillin	[13]

11.4.1.2 Procedure

1. Peptone (5 g), yeast extract (2.5 g) and NaCl (5 g) were weighed into a 0.5 L Schott bottle and dissolved in dH_2O (500 mL). Media were autoclaved (20 min, 121 °C) to give sterile LB medium.
2. Bacto-tryptone (24 g), bacto-yeast (48 g), glycerol (8 mL), $K_2HPO_4 \cdot 3\,H_2O$ (32.8 g) and KH_2PO_4 (4.6 g) were dissolved in dH_2O (2 L). The medium was distributed to four 0.5 L Schott bottles and autoclaved (20 min, 121 °C) to give sterile terrific broth (TB) medium.
3. NaCl (8 g), KCl (0.2 g), Na_2HPO_4 (1.44 g) and KH_2PO_4 (0.24 g) were dissolved in dH_2O (800 mL) and the pH was adjusted to 7.4 with HCl (1 M aq). The solution was filled to a final volume of 1 L with dH_2O and autoclaved (20 min, 121 °C) to give sterile phosphate-buffered saline (10 mM PBS).
4. The crude cell extracts were prepared for each ERED- and BVMO-expressing *E. coli* BL21(DE3) strain (summarised in Table 11.6). For each pre-culture, sterile LB medium (20 mL) was prepared in sterile baffled 0.1 L Erlenmeyer flask. Antibiotics ampicillin or kanamycin (see Table 11.6) were added to a final concentration of 100 or 50 μg. mL^{-1}, respectively. The medium was inoculated with a single colony of the relevant *E. coli* BL21(DE3) strain and incubated at 37 °C and 200 rpm overnight.
5. The next day, sterile 1 L Erlenmeyer flasks containing sterile TB medium (200 mL) were supplemented with appropriate antibiotic and inoculated with 2 vol % of overnight culture.
6. The cells were grown at 28 °C for 3 hr until an OD_{600} ~1 was reached. Then, the production of enzyme was induced by IPTG (0.5 mM) and the flask was shaken at 28 °C, 120 rpm for a further 24 hr.

7. After 24 hr, the cells were harvested by centrifugation at 6000 rpm and 4 °C for 15 min and the cell pellet was resuspended in PBS (5 mL). Protease inhibitor (phenyl methane sulfonyl fluoride) was added to make 1 μL.mL^{-1}.
8. Cells were ruptured by sonication (six times for 10 sec with a rest period of 1 min at an amplitude of 50%, 4 °C).
9. After sonication, the suspension was centrifuged at 10 000 rpm and 4 °C for 15 min. Cell lysate was taken and stored at −20 °C without any further manipulation. The additional glycerol within the TB medium and the protease inhibitor were shown to increase the final protein concentration.
10. Bradford assay was done to calculate the protein concentration in the crude cell extracts.

11.4.2 Procedure 2: Biotransformation of (−)/(+)-Carvones for Screening Experiments

11.4.2.1 Materials and Equipment

- Tris(hydroxymethyl)-aminomethane (3.0 g)
- HCl (1 M aq)
- NADP$^+$ (200 μM aqueous solution)
- Glucose-6-phosphate (4 mM)
- Glucose-6-phosphate dehydrogenase (1 U)
- (+)-Carvone (2 mM, stock solution in EtOH/H_2O 2 : 1)
- (−)-Carvone (2 mM, stock solution in EtOH/H_2O 2 : 1)
- High-purity ethanol
- Ethyl acetate
- Methyl benzoate
- Distilled water (dH_2O)
- Crude cell extract from Procedure 1
- Sterile multiwell plates (500 μL each well)
- Gas chromatography (GC) system with flame ionisation detector (FID; Thermo Finnigan Focus GC)
- Chiral GC column (BGB175: 30 m × 0.25 mm ID, 0.25 μm film)

11.4.2.2 Procedure

1. Tris(hydroxymethyl)-aminomethane (3 g) was dissolved in dH_2O (400 mL). The pH was adjusted to 8.0 with HCl (1 M aq). The solution was made up to 500 mL and then autoclaved (20 min, 121 °C) to give sterile Tris-HCl buffer (50 mM).
2. For the screening of the biotransformation of (−)/(+)-carvones, individual reactions were performed for all different ERED/BVMO combinations. The prepared crude cell extracts (see Procedure 1, EREDs and BVMOs 5 mg each, 100 μL) were placed in sterile multiwell plates.
3. Tris-HCl (90.5 μL), NADP$^+$ (0.4 μL), glucose-6-phosphate (8 μL), glucose-6-phosphate dehydrogenase (0.24 μL) and substrate ((−)/(+)-carvone, 0.8 μL) were added to the crude cell extract and incubated at 30 °C.
4. Samples (~200 μL) were collected after 6 hr (in case of all combinations with SYE-4, OPR1, OPR3 or YqjM) or 24 hr (for combinations with W116I).

5. Product was extracted with ethyl acetate with 1 mM methyl benzoate as reference standard (2 × 200 μL) and the samples were analysed by chiral GC (GC method described in Table 11.7).
6. The results of the one-pot redox reactions of (+)/(−)-carvone (**1a/1b**) with EREDs and BVMOs to the corresponding carvo-lactones (**3a**, **3a′**, **3b**, **3b′**, **3c**, **3c′**) are summarised in Table 11.8 and Table 11.9.

Table 11.7 *GC method.*

Temperature program (r = °C.min^{-1})	Duration
80 °C, 2 min → 80–160 °C, 5 °C.min^{-1}, 160 °C, 1 min → 160–220 °C, 10 °C.min^{-1}, 220 °C, 8 min	33 min

Table 11.8 *One-pot redox reaction of (+)-carvone **1a** into the corresponding carvo-lactones **3** by EREDs and BVMOs.*

EREDs/BVMOs	SYE-4[a]	OPR1	OPR3	YqjM	W116I[b]
CHMO$_{Acineto}$	100%[c] **3a** (99 : 1)[d]	100% **3a** (98 : 2)	99% **3a** (75 : 25)	100% **3a** (85 : 15)	95% **3c** (99 : 1)
CHMO$_{Brevi1}$	78% **3a** (100 : 0)	25% **3a** (100 : 0)	97% **3a** (79 : 21)	34% **3a** (100 : 0)	45% **3c** (99 : 1)
CHMO$_{Brevi2}$	19% **3a′** (0 : 100)	2% **3a′** (0 : 100)	7.5% **3a′** (0 : 100)	4% **3a′** (0 : 100)	Trace **3c′** (0 : 100)
CPMO$_{Coma}$	8% **3a′** (0 : 100)	3% **3a′** (0 : 100)	3% **3a′** (0 : 100)	Trace **3a′** (0 : 100)	Trace **3c′** (0 : 100)

[a]Samples taken after 6 hr for SYE-4, OPR1, OPR3 and YqjM.
[b]Samples taken after 24 hr.
[c]Conversion into the desired carvo-lactones **3** determined by GC.
[d]Normal to abnormal lactone ratio.

Table 11.9 *One-pot redox reaction of (−)-carvone **1b** into the corresponding carvo-lactones **3** by EREDs and BVMOs.*

EREDs/BVMOs	SYE-4[a]	OPR1	OPR3	YqjM	W116I[b]
CHMO$_{Acineto}$	97%[c] **3b′** (3 : 97)[d]	100% **3b′** (0 : 100)	95% **3b′** (1 : 99)	100% **3b′** (0 : 100)	82% **3b′** (3 : 97)
CHMO$_{Brevi1}$	65% **3b** (62 : 38)	33% **3b** (67 : 33)	79% **3b** (61 : 39)	55% **3b** (70 : 30)	35% **3b** (55 : 45)
CHMO$_{Brevi2}$	2.5% **3b′** (0 : 100)	Trace **3b′** (0 : 100)	2% **3b′** (0 : 100)	2% **3b′** (0 : 100)	Trace **3b′** (0 : 100)
CPMO$_{Coma}$	18% **3b′** (0 : 100)	Trace **3b′** (0 : 100)	3.5% **3b′** (0 : 100)	Trace **3b′** (0 : 100)	Trace **3b′** (0 : 100)

[a]Samples taken after 6 hr for SYE-4, OPR1, OPR3 and YqjM.
[b]Samples taken after 24 hr.
[c]Conversion into the desired carvo-lactones **3** determined by GC.
[d]Normal to abnormal lactones ratio.

11.4.3 Procedure 3: Selected Biotransformations of (+)/(−)-Carvones on a Preparative Scale

11.4.3.1 Materials and Equipment

- Tris-HCl buffer (pH 8, 50 mM; preparation in Procedure 2)
- $NADP^+$ (100 mM)
- Glucose-6-phosphate (100 mM)
- Glucose-6-phosphate dehydrogenase (1 U)
- (+)-Carvone (0.5 M solution (30–50 mg) in EtOH/H_2O 2 : 1)
- (−)-Carvone (0.5 M solution (30–50 mg) in EtOH/H_2O 2 : 1)
- High-purity ethanol
- Diethyl ether (Et_2O)
- Petrol ether (PE; 40–60 °C bp)
- Na_2SO_4, dry
- Distilled water (dH_2O)
- Crude cell extract from Procedure 1
- Sterile baffled Erlenmeyer flasks with cotton caps and aluminium foil
- Separating funnel (100mL)
- GC system with FID (Thermo Finnigan Focus GC)
- Chiral GC column (BGB175: 30 m × 0.25 mm ID, 0.25 μm film)
- Silica Gel 60 from Merck (40–63 μm)
- Bruker AC 200 (200 MHz) spectrometer
- Deuterated solvent with tetramethylsilane (TMS) as internal standard
- Anton Paar MCP 500 Polarimeter

11.4.3.2 Procedure

11.4.3.2.1 **Synthesis of (−)-(4*S*,7*R*)-7-Methyl-4-(prop-1-en-2-yl)oxepan-2-one 3a.**

1. Biotransformation of (+)-Carvone **1a** in preparative scale was performed in sterile baffled Erlenmeyer flasks with Tris-HCl buffer (16 mL), $NADP^+$ (28 μL), glucose-6-phosphate (1.12 mL), and glucose-6-phosphate dehydrogenase (41.3 μL).
2. Crude cell extracts (preparation in Procedure 1) of *E. coli* BL21(DE3) cells expressing SYE-4 (1.34 mL, 35.5 $mg.mL^{-1}$) and $CHMO_{Acineto}$ (2 mL, 17.5 $mg.mL^{-1}$) and (+)-Carvone **1a** (30 mg, 0.2 mmol, in 2:1 ethanol/water) were added to the Erlenmeyer flasks. The reaction was incubated at 30 °C for 24 hr.
3. The crude product **3a** was extracted with Et_2O (3 × 20 mL) in a separating funnel.
4. The combined organic layers were dried over Na_2SO_4 and filtered, and the solvent was removed under reduced pressure.
5. The crude product was purified over silica column chromatography by using silica gel with 3 : 1 PE/Et_2O as eluent.
6. The desired dihydrocarvo-lactone, (−)-(4*S*,7*R*)-7-methyl-4-(prop-1-en-2-yl)oxepan-2-one **3a**, was obtained as a colourless oil (24 mg, 71% yield).

Table 11.10 *Data and results from obtained product (−)-(4S,7R)-7-methyl-4-(prop-1-en-2-yl)oxepan-2-one* ***3a****.*

Chemical formula	$C_{10}H_{16}O_2$
Molecular weight	168.23
Optical rotation	$[\alpha]_D{}^{22}$ = −5.7 (c = 0.19, $CHCl_3$) (lit [14] $[\alpha]_D{}^{22}$ = −3.7 (c = 3.78, $CHCl_3$)
Appearance	Colourless oil
^{1}H NMR (200 MHz; $CDCl_3$; Me_4Si)	δ 1.30 (d, *J* = 6.4 Hz, 3H), 1.64–1.90 (m, 4H), 1.72 (s, 3H), 2.41–2.52 (m, 1H), 2.68–2.98 (m, 2H), 4.38–4.47 (m, 1H), 4.76–4.79 (m, 2H)
^{13}C NMR (50 MHz; $CDCl_3$; Me_4Si)	δ 21.3 (q, CH3), 21.8 (q), 29.7 (t), 33.2 (t), 38.4 (t), 38.6 (d), 75.4(d), 111.3 (t), 146.3 (s), 173.7 (s)

7. The optical rotation values were determined using an Anton Paar MCP 500 Polarimeter and calculated using the following equation:

$$[\alpha]_{\mathbf{D}}^{\mathbf{22}} = \mathbf{100} * \frac{\alpha}{c} * 1); c\left[\frac{g}{100\,mL}\right], 1[dm]$$

8. The optical rotation plus the chemical formula, molecular weight, appearance, nuclear magnetic resonance (^{1}H NMR (200 MHz; $CDCl_3$; Me_4Si); ^{13}C NMR (50 MHz; $CDCl_3$; Me_4Si)) and gas chromatography/mass spectrometry (GC-MS) data are summarised in Table 11.10.

11.4.3.2.2 Synthesis of (−)-(3*R*,6*S*)-3-Methyl-6-(prop-1-en-2-yl)oxepan-2-one 3b'.

1. The general protocol for redox biotransformation of carvones on preparative scale was used (described in section 'Synthesis of (−)-(4*S*,7*R*)-7-methyl-4-(prop-1-en-2-yl)oxepan-2-one **3a**').
2. (−)-Carvone **1b** (30 mg, 0.2 mmol in 2 : 1 ethanol/water) was converted into (−)-(3*R*,6*S*)-3-methyl-6-(prop-1-en-2-yl)oxepan-2-one **3b'** by crude cell extracts (preparation in Procedure 1) of *E. coli* BL21(DE3) cells expressing SYE-4 (1.34 mL, 35.5 mg.mL^{-1}) and $CHMO_{Acineto}$ (2 mL, 17.5 mg.mL^{-1}).
3. The desired dihydrocarvo-lactone, (−)-(3*R*,6*S*)-3-methyl-6-(prop-1-en-2-yl)oxepan-2-one **3b'**, was obtained as a colorless oil (24 mg, 71% yield). Further results are listed in Table 11.11.

Table 11.11 *Data and results from obtained product (−) -(3R,6S)-3-methyl-6-(prop-1-en-2-yl)oxepan-2-one* ***3b'****.*

Chemical formula	$C_{10}H_{16}O_2$
Molecular weight	168.23
Optical rotation	$[\alpha]_D{}^{22}$ = −31.4 (c = 0.70, $CHCl_3$) (lit [14] $[\alpha]_D{}^{22}$ = −34.6 (c = 0.9, $CHCl_3$)
Appearance	Colourless oil
^{1}H NMR (200 MHz; $CDCl_3$; Me_4Si)	δ 1.21 (d, *J* = 6.6 Hz, 3H), 1.55–1.83 (m, 4H), 1.76 (s, 3H), 2.27–2.35 (m, 1H), 2.72–2.80 (m, 2H), 4.14–4.20 (m, 1H), 4.81–4.83 (m, 2H)
^{13}C-NMR (50 MHz; $CDCl_3$; Me_4Si)	δ 18.4 (q), 21.8 (q), 31.8 (t), 34.2 (t), 37.2 (d), 46.4 (d), 71.6 (t), 111.1 (t), 145.6 (s), 177.8 (s)

11.4.3.2.3 Synthesis of (−) -(4*S*,7*S*)-7-Methyl-4-(prop-1-en-2-yl)oxepan-2-one 3c.

1. The general protocol for redox biotransformation of carvones on preparative scale was used (described in section 'Synthesis of (−)-(4*S*,7*R*)-7-methyl-4-(prop-1-en-2-yl)oxepan-2-one **3a**').
2. (−)-Carvone **1b** (30 mg, 0.2 mmol in 2 : 1 ethanol/water) was converted into (−)-(3*R*,6*S*)-3-methyl-6-(prop-1-en-2-yl)oxepan-2-one **3b'** by crude cell extracts (preparation in Procedure 1) of *E. coli* BL21(DE3) cells expressing SYE-4 (1.34 mL, 35.5 $mg.mL^{-1}$) and $CHMO_{Acineto}$ (2 mL, 17.5 $mg.mL^{-1}$).
3. The desired dihydrocarvo-lactone, (−)-(4*S*,7*S*)-7-methyl-4-(prop-1-en-2-yl)oxepan-2-one **3c**, was obtained as a colourless oil (25 mg, 76% yield). Further results are listed in Table 11.12.

11.4.3.2.4 Synthesis of (−)-(4*R*,7*R*)-7-Methyl-4-(prop-1-en-2-yl)oxepan-2-one 3b.

1. The general protocol for redox biotransformation of carvones on preparative scale was used (described in section 'Synthesis of (−)-(4*S*,7*R*)-7-methyl-4-(prop-1-en-2-yl)oxepan-2-one **3a**').
2. (−)-Carvone **1b** (30 mg, 0.2 mmol in 2 : 1 ethanol/water) was converted into (−)-(3*R*,6*S*)-3-methyl-6-(prop-1-en-2-yl)oxepan-2-one **3b'** by crude cell extracts (preparation in Procedure 1) of *E. coli* BL21(DE3) cells expressing SYE-4 (1.34 mL, 35.5 $mg.mL^{-1}$) and $CHMO_{Acineto}$ (2 mL, 17.5 $mg.mL^{-1}$).
3. The mixture of normal and abnormal lactone (**3b** and **3b′**) was purified with column chromatography using silica gel with 6 : 1 PE/Et_2O as eluent.
4. The desired dihydrocarvo-lactone, (−)-(4*R*,7*R*)-7-methyl-4-(prop-1-en-2-yl)oxepan-2-one **3b**, was obtained as a colourless oil (15 mg, 30% yield). Further results are listed in Table 11.13.

11.4.4 Conclusion

The described procedure enabled six out of eight carvo-lactone stereoisomers to be prepared by a combination of stereoselective bioreduction and regioselective biooxygenation, starting from the terpenone precursor carvone. Additional potential of this two-stage cascade was demonstrated by preparative-scale experiments, which showed high yields of

Table 11.12 *Data and results from obtained product (−)-(4S,7S)-7-methyl-4-(prop-1-en-2-yl)oxepan-2-one* ***3c****.*

Chemical formula	$C_{10}H_{16}O_2$
Molecular weight	168.23
Optical rotation	$[\alpha]_D^{22}$ = +42.3 (c = 1.13, $CHCl_3$) (lit [14] $[\alpha]_D^{22}$ = +45.5 (c = 1.48, $CHCl_3$)
Appearance	Colourless oil
^{1}H NMR (200 MHz; $CDCl_3$; Me_4Si)	δ 1.29 (d, *J* = 6.2 Hz, 3H), 1.63–1.93 (m, 4H), 1.75 (s, 3H), 2.41–2.52 (m, 1H), 2.68–2.96 (m, 2H), 4.38–4.49 (m, 1H), 4.77–4.79 (m, 2H)
^{13}C-NMR (50 MHz; $CDCl_3$; Me_4Si)	δ 21.3 (q), 21.8 (q), 29.7 (t), 33.2 (t), 38.4 (t), 38.6 (d), 75.4 (d), 111.3 (t), 146.3 (s), 173.7 (s)

Table 11.13 *Data and results from obtained product (−)-(4R,7R)-7-methyl-4-(prop-1-en-2-yl)oxepan-2-one* ***3b****.*

Chemical formula	$C_{10}H_{16}O_2$
Molecular weight	168.23
Optical rotation	$[\alpha]_D^{22}$ = −39.6 (c = 0.96, $CHCl_3$) (lit [14] $[\alpha]_D^{22}$ = −42.1 (c = 1.12, $CHCl_3$)
Appearance	Colourless oil
^1H NMR (200 MHz; $CDCl_3$; Me_4Si)	δ 1.38–1.42 (3 H, d, J 8.4 Hz), 1.62–1.97 (4 H, m), 1.73 (3 H, s), 2.24–2.35 (1 H, m), 2.59–2.78 (2 H, m), 4.39–4.53 (1 H, m), 4.73–4.76 (2 H, m)
^{13}C-NMR (50 MHz; $CDCl_3$; Me_4Si)	δ 20.1 (q), 22.6 (q), 34.3 (t), 35.8 (t), 40.1 (t), 41.7 (d), 76.3 (d), 110.1 (t), 148.4 (s), 174.6 (s)
GC-MS	*m/z* 168 (M+, 8%), 139 (15), 108 (63), 81 (39), 67 (100)

desired chiral carvolactones. In principle, this process can be adapted to related substrates and extended for further suitable enzymes.

References

1. (a) Bayer, T., Milker, S., Wiesinger, T. *et al.* (2015) *Advanced Synthesis & Catalysis*, **357**,1587–1618; (b) Muschiol, J., Peters, C., Oberleitner, N. *et al.* (2015) *ChemComm*, **51**, 5798–5811; (c) Wu, S. and Li, Z. (2018) *ChemCatChem*, **10**, 2164–2178; (d) Sperl, J.M. and Sieber, V. (2018) *ACS Catalysis*, **8**, 2385–2396; (e) Schmidt, S., Castiglione, K. and Kourist, R. (2018) *Chemistry: A European Journal*, **24**, 1755–1768.
2. Rudroff, F., Mihovilovic, M.D., Gröger, H. *et al.* (2018) *Nature Catalysis*, **1**, 12–22.
3. Brigé, A., Van Den Hemel, D., Carpentier, W. *et al.* (2006) *Biochemical Journal*, **394**, 335–344.
4. Hall, M., Stueckler, C., Kroutil, W. *et al.* (2007) *Angewandte Chemie International Edition*, **46**, 3934–3937.
5. Kitzing, K., Fitzpatrick, T.B., Wilken, C. *et al.* (2005) *Journal of Biological Chemistry*, **280**, 27 904–27 913.
6. Padhi, S.K., Bougioukou, D.J. and Stewart, J.D. (2009) *Journal of the American Chemical Society*, **131**, 3271–3280.
7. Donoghue, N.A., Norris, D.B. and Trudgill, P.W. (1976) *European Journal of Biochemistry*, **63**, 175–192.
8. Brzostowicz, P.C., Gibson, K.L., Thomas, S.M. *et al.* (2000) *Journal of Bacteriology*, **182**, 4241–4248.
9. Griffin, M. and Trudgill, P.W. (1976) *European Journal of Biochemistry*, **63**, 199–209.
10. Iqbal, N., Stewart, J.D., Macheroux, P. *et al.* (2018) *Tetrahedron*, **74**, 7389–7394.
11. Hall, M., Stueckler, C., Ehammer, H. *et al.* (2008) *Advanced Synthesis & Catalysis*, **350**, 411–418.
12. Fitzpatrick, T.B., Amrhein, N. and Macheroux, P. (2003) *Journal of Biological Chemistry*, **278**, 19 891–19 897.
13. Iwaki, H., Hasegawa, Y., Wang, S. *et al.* (2002) *Society*, **68,** 5671–5684.
14. Cernuchová, P. and Mihovilovic, M.D. (2007) *Organic and Biomolecular Chemistry*, **5**, 1715–1719.

11.5 One-Pot Biocatalytic Synthesis of D-Tryptophan Derivatives from Substituted Indoles and L-Serine

Fabio Parmeggiani,[1] Nicholas J. Turner[1] and Roberto A. Chica[2]
[1]Manchester Institute of Biotechnology, School of Chemistry, University of Manchester, Manchester, UK
[2]Department of Chemistry and Biomolecular Sciences, Centre for Catalysis Research and Innovation, University of Ottawa, Ottawa, ON, Canada

D-Tryptophan and its derivatives are important building blocks for the preparation of a broad range of natural products, pharmaceutical intermediates and bioactive molecules such as Tadalafil [1], Lanreotide [2], Mitragynine [3] and Necrostatins [4]. The development of enzymatic procedures to access these important synthons is therefore highly desirable in order to improve efficiency and selectivity, as well as to reduce the environmental footprint of the chemical process. Recently, we developed a one-pot, two-step procedure for the biocatalytic synthesis of D-tryptophan derivatives, which we tested and optimised on a preparative scale (Scheme 11.6) [5]. The first step, mediated by tryptophan synthase from *Salmonella enterica* serovar *typhimurium* (TrpS) [6], involved condensation of indoles **1** with L-serine to afford L-tryptophan derivatives L-**2** with perfect enantioselectivity. The second step was a stereoinversion cascade employing two *E. coli* whole-cell biocatalysts: one producing a wild-type L-amino acid deaminase from *Proteus myxofaciens* (LAAD) [7] that oxidised L-**2** into the corresponding imines, which spontaneously hydrolysed to ketoacids **3**, and the other producing an engineered variant of D-alanine aminotransferase from *Bacillus* sp. YM-1 (DAAT) [8] that converted ketoacids **3** into the target D-**2** with high enantioselectivity (91 to >99% ee). Whilst both TrpS and LAAD displayed excellent

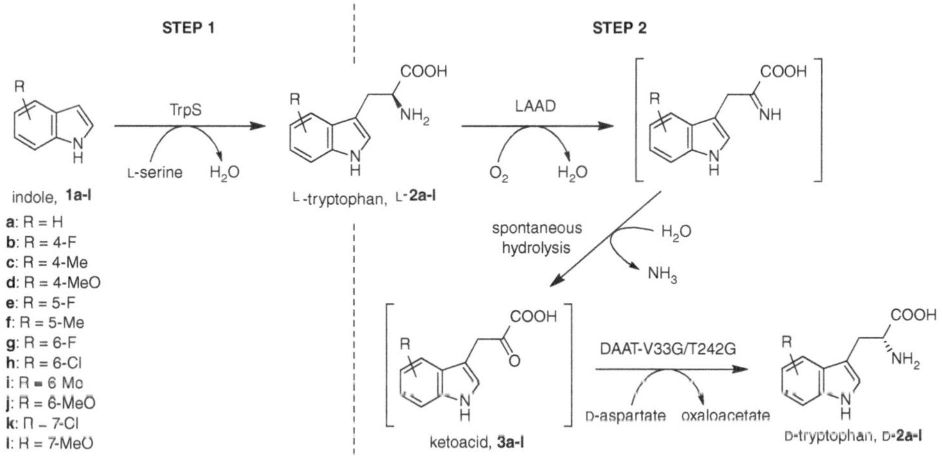

Scheme 11.6 *One-pot conversion of indoles **1** into D-tryptophans D-**2**.*

activity against their respective substrates, DAAT required the introduction of two mutations, V33G and T242G, to reach near-native catalytic efficiency against D-tryptophan [5].

Employing this system, preparative-scale syntheses were demonstrated for the one-pot conversion of indoles **1a–l** into D-**2a–l** using standard chemical and microbiological techniques. The three biocatalysts were employed in the form of crude lysates or whole cells, and the product was isolated directly from the biotransformation medium using preparative reverse-phase chromatography. The products were obtained in good yields (63–79%) and high to excellent enantioselectivity (91 to >99%). The detailed protocol is exemplified here with the synthesis of D-tryptophan D-**2a** from indole **1a**.

11.5.1 Procedure 1: Production of TrpS in *E. coli* BL21(DE3) and Preparation of Cell Lysate

11.5.1.1 Materials and Equipment

- Lysogenic broth (LB) agar plate containing colonies of *E. coli* BL21(DE3) harbouring the pSTB7 plasmid encoding TrpS (available from the American Type Culture Collection, product code ATCC 37845) [9]
- LB medium
- Distilled water (dH_2O)
- Ampicillin (100 mg.mL^{-1} stock in dH_2O, filter-sterilised)
- Potassium phosphate (KP_i) buffer (100 mM, pH 8.0)
- Hen egg white lysozyme (10 mg.mL^{-1} stock in dH_2O)
- DNAse I (1 mg.mL^{-1} stock in dH_2O)
- 15 mL sterile culture tubes
- Inoculating loops
- Autoclaved 2 L Erlenmeyer flask (baffled) with foam bung
- 300 mL centrifuge tubes
- 50 mL centrifuge tubes
- Dewar flask
- Liquid nitrogen
- Shaking incubator (Innova 44, New Brunswick Scientific)
- Refrigerated centrifuge (min. 25 000× g)
- Ultrasonic cell disruptor (Soniprep 150, Measuring and Scientific Equipment UK Ltd)

11.5.1.2 Procedure

1. A single colony of *E. coli* BL21(DE3) carrying the pSTB7 plasmid was used to inoculate LB medium (8 mL) supplemented with ampicillin (final conc. 100 μg.mL^{-1}).
2. The culture was grown overnight at 37 °C with shaking at 220 rpm and then used to inoculate LB medium (800 mL) containing ampicillin (final conc. 100 μg.mL^{-1}). Following incubation at 37 °C and 180 rpm for 7 hr, the temperature was lowered to 18 °C and the culture was grown overnight. Importantly, the pSTB7 plasmid bears a constitutive promoter, so no induction of protein expression was required.
3. Cells were harvested by centrifugation at 2000× g and 4 °C for 20 min and an aliquot (2.0 g wet weight) was resuspended in 20 mL KP_i buffer (100 mM, pH 8.0). Excess cell paste (typical wet weight 4.4 g per flask) could be stored at –20 °C for further use, if required.

4. Lysis was performed by adding 500 µL of lysozyme (10 mg.mL^{-1} stock) to the cell suspension and incubating at 37 °C and 220 rpm for 30 min.
5. The cell suspension was sonicated on ice (20 sec on, 20 sec off, 20 cycles) and treated with 100 µL DNAse I (1 mg.mL^{-1} stock) at 37 °C and 220 rpm for 30 min.
6. The mixture was centrifuged at 25 000× g and 4 °C for 30 min and the supernatant was filtered (0.2 µm syringe filter) to remove cellular debris. The raw lysate was then aliquoted in 2 mL portions, frozen rapidly in liquid nitrogen and stored at –20 °C for use in Procedure 4.

11.5.2 Procedure 2: Production of LAAD in *E. coli* BL21(DE3)

11.5.2.1 Materials and Equipment

- LB agar plate containing colonies of *E. coli* BL21(DE3) harbouring the pET28a-*Pmyx*LAAD plasmid [5]
- LB medium
- dH_2O
- Kanamycin (60 mg.mL^{-1} in dH_2O, filter-sterilised)
- Isopropyl-β-D-1-thiogalactopyranoside (IPTG; 1 M stock in dH_2O, filter-sterilised)
- KP_i buffer (100 mM, pH 8.0)
- 15 mL sterile culture tubes
- Inoculating loops
- Autoclaved 2 L Erlenmeyer flask (baffled) with foam bung
- 300 mL centrifuge tubes
- Shaking incubator (Innova 44, New Brunswick Scientific)
- Refrigerated centrifuge (min. 2000× g)

11.5.2.2 Procedure

1. A single colony of *E. coli* BL21(DE3) cells transformed with the pET28a-*Pmyx*LAAD plasmid was used to inoculate LB medium (8 mL) supplemented with kanamycin (final conc. 60 µg.mL^{-1}).
2. The culture was grown overnight at 37 °C with shaking at 220 rpm and then used to inoculate LB medium (800 mL) containing kanamycin (final conc. 60 µg.mL^{-1}). This culture was incubated at 37 °C and 160 rpm until an OD_{600} of 0.4–0.6 was reached, after which IPTG was added to a final concentration of 1 mM. The culture was then incubated at 30 °C and 160 rpm for 5 hr.
3. Cells were harvested by centrifugation at 2000× g and 4 °C for 20 min, washed with KP_i buffer (100 mM, pH 8.0) and harvested by centrifugation at 2000× g and 4 °C for 20 min. The cell paste (typical wet weight 2.8 g per flask) was aliquoted and stored at –20 °C for use in Procedure 4.

11.5.3 Procedure 3: Production of DAAT-V33G/T242G in *E. coli* BL21(DE3)

11.5.3.1 Materials and Equipment

- LB agar plate containing colonies of *E. coli* BL21(DE3) harbouring the pET11a-DAAT-V33G/T242G plasmid [5]

- LB medium
- dH_2O
- Ampicillin (100 mg.mL^{-1} stock in dH_2O, filter-sterilised)
- IPTG (1 M stock in dH_2O, filter-sterilised)
- KP_i buffer (100 mM, pH 8.0)
- 15 mL sterile culture tubes
- Inoculating loops
- Autoclaved 2 L Erlenmeyer flask (baffled) with foam bung
- 300 mL centrifuge tubes
- Shaking incubator (Innova 44, New Brunswick Scientific)
- Refrigerated centrifuge (min. 2000× g)

11.5.3.2 Procedure

1. A single colony of *E. coli* BL21(DE3) cells transformed with the pET11a-DAAT-V33G/T242G plasmid was used to inoculate LB medium (8 mL) supplemented with ampicillin (final conc. 100 µg.mL^{-1}).
2. The culture was grown overnight at 37 °C with shaking at 220 rpm and then used to inoculate LB medium (800 mL) supplemented with ampicillin (final conc. 100 µg.mL^{-1}). This culture was incubated at 37 °C and 160 rpm until an OD_{600} of 0.4–0.6 was reached, after which IPTG was added to a final concentration of 0.1 mM. The culture was then incubated at 15°C and 160 rpm for 5 hr.
3. Cells were harvested by centrifugation at 2000× g and 4 °C for 20 min, washed with KP_i buffer (100 mM, pH 8.0) and harvested again by centrifugation at 2000× g and 4 °C for 20 min. The cell paste (typical wet weight 3.9 g per flask) was aliquoted and stored at –20 °C for use in Procedure 4.

11.5.4 Procedure 4: Biocatalytic Conversion of Indole 1a to D-Tryptophan D-2a

11.5.4.1 Materials and Equipment

- dH_2O
- Indole **1a** (5 mmol, 585 mg)
- L-Serine (10 mmol, 1.05 g)
- D-Aspartic acid (45 mmol, 5.90 g)
- Pyridoxal 5′-phosphate (PLP, 500 mM stock in dH_2O)
- KP_i buffer (100 mM, pH 8.0)
- TrpS lysate (10 mL) prepared according to Procedure 1
- *Pmyx*LAAD cell paste (5 g wet weight) prepared according to Procedure 2
- DAAT-V33G/T242G cell paste (5 g wet weight) prepared according to Procedure 3
- Methanol
- Hydrochloric acid solution (0.5 M in dH_2O)
- Sodium hydroxide solution (5 M in dH_2O)
- Reverse-phase C_{18}-silica
- 1 L Schott bottle
- Glass column with sintered glass frit (internal diameter 2.5 cm, length 25 cm)
- 50 mL centrifuge tubes

- Collection tubes
- UV-compatible 96-well plate, flat bottom
- Shaking incubator (SI 500, Stuart)
- Refrigerated centrifuge (min. 2000× g)
- Centrifugal evaporator (Genevac EZ-2 Plus)
- High-performance liquid chromatography (HPLC) system equipped with UV detector (Agilent 1200 Series) and suitable eluents/columns (see Analytical Method)
- Peristaltic pump (Thermo Scientific FH15)
- Plate reader (Tecan Infinite M200 Pro)

11.5.4.2 Procedure

1. Indole **1a** (585 mg, 5 mmol) was dissolved in MeOH (12 mL) and added to a 1 L Schott bottle containing a solution of L-serine (1.05 g, 10 mmol) in KP_i buffer (178 mL, 100 mM, pH 8.0).
2. The reaction was initiated by addition of TrpS lysate (10 mL) and the bottle was incubated vertically in an orbital shaker at 37 °C and 140 rpm. Reaction progress was monitored by HPLC analysis on nonchiral phase, and complete conversion to L-**2a** was achieved in 3 hr.
3. D-Aspartic acid (5.9 g, 45 mmol) and PLP (2 mL of a 0.5 mM solution in water) were added to the reaction, and the pH was readjusted to 8.0 by addition of aqueous NaOH aliquots (5 M).
4. Cells producing *Pmyx*LAAD (5 g wet weight) and others producing DAAT-V33G/T242G (5 g wet weight) were added to the reaction mixture, and the bottle was incubated horizontally in an orbital shaker at 37 °C and 140 rpm. Reaction progress was monitored by HPLC analysis on chiral phase, and complete stereoinversion to D-**2a** was achieved in 2 hr.
5. The biotransformation mixture was centrifuged at 2000× g and 4 °C for 30 min and the supernatant was transferred to a centrifugal evaporator for concentration and drying.
6. Reverse-phase C_{18}-silica (~50 g) was suspended in MeOH (~100 mL) and the slurry was poured in portions into a glass column (internal diameter 2.5 cm, length 25 cm) until a compact layer of about 15 cm was obtained. The flow rate (10–15 mL.min^{-1}) was regulated by a peristaltic pump connected downstream.
7. The column was washed with MeOH (75 mL) and equilibrated with MeOH 0.5% v/v in dH_2O (30 mL). One-third of the solid obtained after evaporation of the biotransformation supernatant was resuspended in aqueous HCl (25 mL, 0.5 M) and centrifuged again to remove insoluble components. The supernatant was loaded on the column, which was then washed with MeOH 0.5% v/v in dH_2O (150 mL).
8. The product was eluted with MeOH 25% v/v in dH_2O (200 mL), and 15 mL fractions were collected in separate tubes.
9. Following elution, the column was washed again with MeOH (150 mL) and the purification procedure (Steps 7–8) was repeated twice, once with each of the remaining two-thirds of the dried biotransformation supernatant, using the same packed column.
10. The presence of D-tryptophan product was monitored spectrophotometrically at 280 nm in a 96-well plate (100 μL per well). All fractions containing the product were pooled and evaporated overnight in a centrifugal evaporator, yielding pure D-**2a** as a slightly yellow solid (673 mg, 66% yield).

11.5.5 Analytical Method

Conversion of indole **1a** into tryptophan **2a** was measured by reverse-phase HPLC on a nonchiral Zorbax Extend C18 column (50 mm × 4.6 mm × 3.5 μm, Agilent) according to the following protocol:

- Flow rate: 1.0 mL.min^{-1}
- Temperature: 40 °C
- Detection wavelength: 280 nm
- Mobile phase: aqueous NH_4OH 0.1 M, pH 10.0/MeOH
- Elution: 90 : 10 (0–1 min), 90 : 10–10 : 90 (1–18 min), 10 : 90 (18–21 min), 10 : 90–90 : 10 (21–23 min), 90 : 10 (23–30 min)
- Retention times: **1a**, 10.4 min; **2a**, 3.2 min

Enantiomeric excess values of the tryptophan product **2a** were measured by reverse-phase HPLC on a Crownpak CR(+) column (150 mm × 4 mm × 3.5 μm, Daicel) according to the following method:

- Flow rate: 1.0 mL.min^{-1}
- Temperature: 25 °C
- Detection wavelength: 280 nm
- Mobile phase: aqueous $HClO_4$ 1.14% w/v/MeOH
- Isocratic elution: 85 : 15
- Retention times: D-**2a**, 10.5 min; L-**2a**, 13.9 min

11.5.6 Conclusion

The procedure presented here enables the fully biocatalytic synthesis of enantiopure D-tryptophan from indole and L-serine, compensating for the current lack of a D-selective

*Table 11.14 Synthesis of D-tryptophan derivatives D-**2b–l** from substituted indoles **1b–l** [6].*

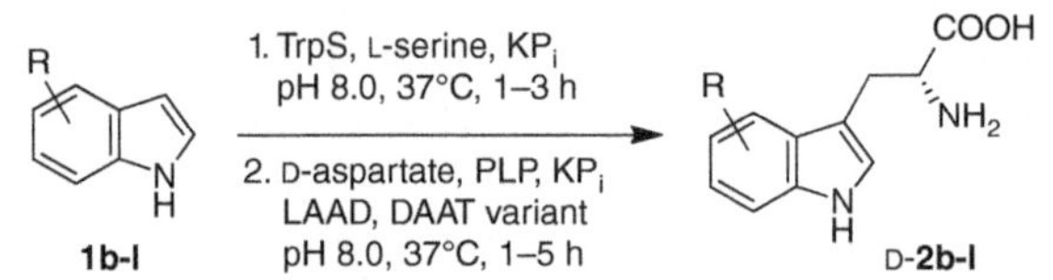

R	Substrate	Product	Conv. (%)	Isol. yield (%)	ee (%)
4-F	**1b**	D-**2b**	>99	68	97
4-Me	**1c**	D-**2c**	95	64	>99
4-MeO	**1d**	D-**2d**	>99	79	91
5-F	**1e**	D-**2e**	>99	78	>99
5-Me	**1f**	D-**2f**	81	63	>99
6-F	**1g**	D-**2g**	>99	76	>99
6-Cl	**1h**	D-**2h**	>99	68	>99
6-Me	**1i**	D-**2i**	>99	74	97
6-MeO	**1j**	D-**2j**	>99	71	94
7-Cl	**1k**	D-**2k**	84	66	>99
7-MeO	**1l**	D-**2l**	>99	76	97

tryptophan synthase. The approach is simple and efficient, requiring only crude lysates and whole-cell biocatalysts, eliminating the need for enzyme purification. The utility of our method has also been demonstrated for the synthesis of a broad range of substituted D-tryptophan derivatives D-**2b–l** from the corresponding commercially available indoles **1b–l** (Table 11.14), and we expect this procedure to be applicable with little modification to the synthesis of other valuable D-tryptophan analogues.

References

1. Shi, X.-X., Liu, S.-L., Xu, W. and Xu, Y.-L. (2008) *Tetrahedron: Asymmetry*, **19**, 435–442.
2. Gurjar, M.K., Tripathi, N.K., Pramanik, C. M. and Deshpande, A.P. (2017) Patent Application WO 2017/178950 A1.
3. Ma, J., Yin, W., Zhou, H. and Cook, J.M. (2007) *Organic Letters*, **9**, 3491–3494.
4. Teng, X., Degterev, A., Jagtap, P. *et al.* (2005) *Bioorganic & Medicinal Chemistry Letters*, **15**, 5039–5044.
5. Parmeggiani, F., Rué Casamajo, A., Walton, C.J.W. *et al.* (2019) *ACS Catalysis*, **9**, 3482–3486.
6. (a) Buller, A.R., Brinkmann-Chen, S., Romney, D.K. *et al.* (2015) *Proceedings of the National Academy of Sciences of the United States of America*, **112**, 14 599–14 604; (b) Francis, D., Winn, M., Latham, J. *et al.* (2017) *ChemBioChem*, **18**, 382–386.
7. (a) Molla, G., Melis, R. and Pollegioni, L. (2017) *Biotechnology Advances*, **35**, 657–668; (b) Parmeggiani, F., Ahmed, S.T., Thompson, M.P. *et al.* (2016) *Advanced Synthesis & Catalysis*, **358**, 3298–3306.
8. (a) Tanizawa, K., Masu, Y., Asano, S. *et al.* (1989) *Journal of Biological Chemistry*, **264**, 2445–2449; (b) Walton, C.J.W., Parmeggiani, F., Barber, J.E.B. *et al.* (2018) *ChemCatChem*, **10**, 470–474.
9. Kawasaki, H., Bauerle, R., Zon, G. *et al.* (1987) *Journal of Biological Chemistry*, **262**, 10 678–10 683.

11.6 *Escherichia coli* Lysate Multienzyme Biocatalyst for the Synthesis of Uridine-5′-Triphosphate from Orotic Acid and Ribose

Thomas Loan, Christopher Easton and Apostolos Alissandratos*
Research School of Chemistry, Australian National University, Canberra, ACT, Australia

Nucleoside triphosphates (NTPs) such as uridine-5′-triphosphate (UTP) are important synthetic targets with wide-ranging uses in molecular biology and diagnostics, but their chemical syntheses involve toxic reagents and multistep procedures, and usually require laborious purifications [1]. *In vitro* multienzymatic cascades offer an attractive alternative for efficient nucleotide synthesis in aqueous solution [2], but they are limited by the complexity of preparing the purified enzymes and the requirement for stoichiometric quantities of the expensive adenosine-5′-triphosphate (ATP) enzyme co-factor [3]. To overcome these limitations, we have developed a recombinant *E. coli* cell lysate, containing the necessary combination of endogenous and recombinant enzymes, for the one-pot transformation of simple starting materials (orotic acid **4** and ribose **1**) into UTP **8**, with >95% conversion over 2.5 hr (1.4 $g.L^{-1}$) [4]. The cell lysate is prepared with minimal processing from a single bacterial culture of *E. coli* BL21(DE3), transformed by expression plasmids encoding the requisite enzymes that are not otherwise present in native lysate. No ATP supplementation is necessary. Instead, adenine-related species, found in the lysate, which would otherwise be discarded as waste, are utilised and recycled with inexpensive acetyl phosphate and endogenous lysate enzymes. Although details on the synthesis of UTP **8** are presented here, the same lysate catalyses production of guanosine-5'-triphosphate (GTP) and ATP when orotic acid is substituted with guanine and adenine, respectively (see Procedure 3). The lysate-synthesised UTP **8** and other NTPs are suitable for *in situ* use in important biotechnological applications (e.g. cell-free protein synthesis, nucleic acid amplification) without any need for prior purification [3].

The UTP **8** synthesis cascade (Scheme 11.7) is initiated through the consecutive actions of recombinantly overexpressed ribokinase (Rbk) (EC: 2.7.1.15) and phosphoribosyl pyrophosphate (PRPP) synthetase (Pps) (EC: 2.7.6.1), in order to covert ribose **1** into ribose-5-phosphate **2** and then PRPP **3**. The sugar phosphate **3** is further transformed *in situ* by a four-step cascade catalysed by endogenous lysate enzymes of *de novo* pyrimidine biosynthesis [2a]. Initially, orotate phosphoribosyl transferase (Opt; EC: 2.4.2.10) catalyses the formation of an *N*-glycosidic bond between **3** and orotic acid **4** to produce the nucleotide

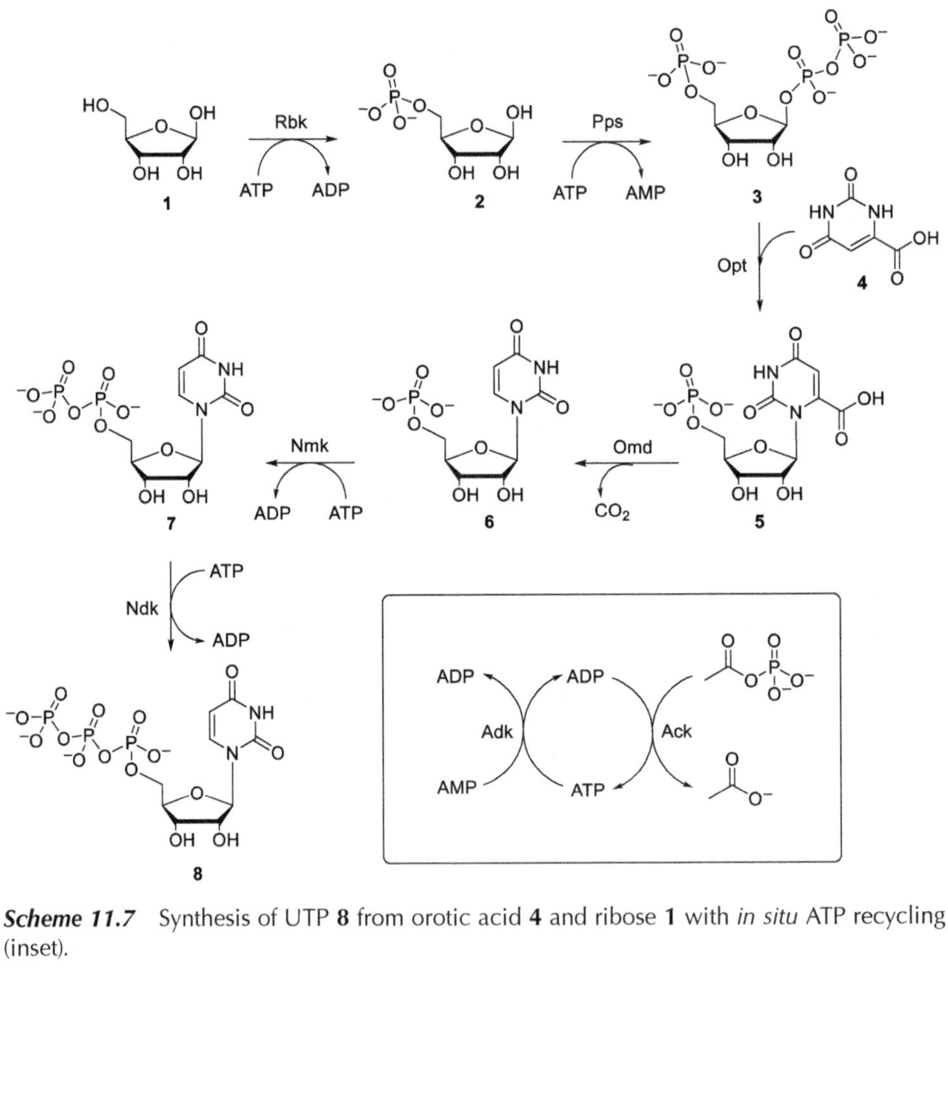

Scheme 11.7 Synthesis of UTP **8** from orotic acid **4** and ribose **1** with *in situ* ATP recycling (inset).

orotidine-5′-monophosphate (OMP) **5**, which is then decarboxylated by OMP decarboxylase (Omd; EC: 4.1.1.23) to uridine-5′-monophosphate (UMP) **6**. The monophosphate **6** is consecutively phosphorylated by nucleoside monophosphate and diphosphate kinases (Nmk and Ndk; EC: 2.7.4.4 and 2.7.4.6) to generate uridine-5′-diphosphate **7** and, finally, UTP **8**. ATP is consumed in four of these transformations, but enough is available from endogenous lysate ingredients (AMP, ADP and ATP), through *in situ* recycling with acetyl phosphate catalysed by endogenous acetate kinase (Ack; EC: 2.7.2.1) and adenylate kinase (Adk; EC: 2.7.4.3), that no exogenous ATP is required.

11.6.1 Procedure 1: Preparation of *E. coli* BL21(DE3) Harbouring the Recombinant Plasmids

11.6.1.1 Materials and Equipment

- Tryptone from casein (10 g)
- Yeast extract (5 g)
- NaCl (10 g)
- EtOH (abs, 1 mL)
- Ampicillin (50 mg)
- Chloramphenicol (25 mg)
- Isopropyl-β-D-1-thiogalactopyranoside (IPTG; 240 mg)
- Distilled water (dH_2O)
- Colonies of *E. coli* BL21(DE3) harbouring vector pACYC-RbK, encoding Rbk from *E. coli* BL21(DE3) (Uniprot: A0A140NF19), and vector pETMCSIII-Pps, encoding Pps from *E. coli* BL21(DE3) (Uniprot: A0A140NCS9) [4], grown on a lysogenic broth (LB) agar plate (25 mL, 1.5% w/v agar, 100 mm plate) supplemented with ampicillin (50 $\mu g.mL^{-1}$) and chloramphenicol (25 $\mu g.mL^{-1}$)
- 2 L Erlenmeyer flask (baffled)
- 50 mL sterile falcon tube
- Orbital incubator shaker (min. 37 °C, 160 rpm)
- High-speed centrifuge (min. 4000× g) and centrifuge tubes (2× min. 0.5 L)
- Autoclave

11.6.1.2 Procedure

1. Although genes encoding Rbk (*rbsK,* NCBI accession: WP_012767772) and Pps (*prsA*, NCBI accession: WP_001298109) are present in *E. coli*, the required enzymatic activities were not detected in native cell lysate, and therefore vectors for their recombinant overexpression were introduced. The recombinant cell strain harbouring both vectors is available upon request from the authors, or it may be prepared through basic molecular cloning procedures, using standard biochemical reagents and cell strains, as detailed elsewhere [3]. Briefly, expression vectors were constructed from empty parent vectors through standard restriction-based cloning of the polymerase chain reaction (PCR)-amplified target genes with restriction site-bearing overhangs from *E. coli* BL21(DE3) genomic DNA. Competent *E. coli* BL21(DE3) was transformed with both vectors through electrotransformation and maintained under antibiotic pressure (ampicillin and chloramphenicol) to avoid plasmid loss.

2. Antibiotic solutions were prepared by dissolving ampicillin (50 mg) in dH_2O (1 mL) and chloramphenicol (25 mg) in EtOH (1 mL). An IPTG stock solution (1 M) was prepared by dissolving IPTG (240 mg) in dH_2O (1 mL). These solutions were stored at −20 °C and thawed on ice before use.
3. Sterile LB medium was prepared by dissolving tryptone (10 g), yeast extract (5 g) and NaCl (10 g) in dH_2O (1 L) in a 2 L baffled Erlenmeyer flask and autoclaving (20 min, 121 °C). It was supplemented with 1 mL ampicillin solution (final conc. 50 $\mu g.mL^{-1}$) and 1 mL chloramphenicol solution (final conc. 25 $\mu g.mL^{-1}$) to produce LB-Amp-Chl.
4. To prepare a pre-culture, sterile LB-Amp-Chl (5 mL) was placed in a falcon tube, inoculated with a single colony of *E. coli* BL21(DE3) harbouring vectors pACYC-RbK and pETMCSIII-Pps and incubated at 37 °C and 160 rpm overnight.
5. Following incubation, the remaining sterile LB-Amp-Chl (~1 L) was inoculated with the entire pre-culture and incubated at 37 °C with shaking at 160 rpm.
6. Cells were grown until an OD_{600} of 0.6 was reached (~3 hr), at which point 1 mL of the IPTG stock solution was added (final conc. 1 mM) to induce protein expression from the recombinant vectors. Incubation was then continued for 4 hr at the lower temperature of 30 °C, with shaking at 160 rpm.
7. Following incubation, cells were harvested by centrifugation at 4000× g for 15 min, with cooling at 4 °C in pre-weighed centrifuge tubes. The supernatant was discarded, and the centrifuge tubes were weighed to determine the mass of the wet cell pellet (typically around 5 grams of wet cell pellet per litre of culture). The pellet was then frozen and stored at −80 °C.

11.6.2 Procedure 2: Preparation of Recombinant Cell-Lysate Biocatalyst

11.6.2.1 Materials and Equipment

- TrisOAc (3.6 g)
- KOAc (3.1 g)
- $Mg(OAc)_2 \cdot 4H_2O$ (6.0 g)
- Phenylmethylsulfonyl fluoride (PMSF; 17 mg)
- Dithiothreitol (15 mg)
- EtOH (abs 100 μL)
- NaOH (6 M aq, ~1 mL)
- dH_2O
- Ice in 1 L foam cooler
- Sonicator (Omni Sonic Ruptor 400 with ORT-375 tip)
- High speed centrifuge (min. 20 000× g) and centrifuge tubes (30 mL; Nalgene)

11.6.2.2 Procedure

1. A 10× stock of cell resuspension buffer was prepared in advance by dissolving 3.6 g TrisOAc (final conc. 100 mM), 3.1 g KOAc (final conc. 160 mM) and 6.0 g $MgOAc.4H_2O$ (final conc. 140 mM) in dH_2O (100 mL). The pH was adjusted to 8.25 with 6 M NaOH and the solution was sterilised by autoclaving (20 min, 121 °C) before storing at 4 °C.

2. A solution of PMSF (17 mg) in 1 mL EtOH (final conc. 100 mM) and a solution of dithiothreitol (15 mg) in 1 mL dH_2O (final conc. 100 mM) were prepared fresh before use. The 1× cell resuspension buffer was then prepared by diluting 1 mL of the concentrated stock (10×) with 8.8 mL dH_2O, 100 μL of the dithiothreitol solution (final conc. 1 mM) and 100 μL of the PMSF solution (final conc. 1 mM). The PMSF solution was added to the aqueous buffer, and not the other way around, in order to avoid PMSF phase separation.
3. Frozen cell pellet (Procedure 1) was thawed on ice, and 1 mL of the cell resuspension buffer mixture was added per gram of wet cell pellet. Cells were then gently resuspended by swirling over ice.
4. The suspended cells were lysed in an ice bath by sonication for 6 min at 50% power and 50% pulse length.
5. The resultant cell lysate was clarified by centrifugation at 20 000× g and 4 °C for 1 hr. The obtained supernatant was the cell lysate biocatalyst and could either be employed immediately or snap frozen in 500 μL aliquots for storage at −80 °C (stable over 12 months). The approximate total protein content of the cell lysate solution, determined by measuring absorbance at 280 nm, was 150 $mg.mL^{-1}$ (based on $\varepsilon_{280} = 1$ $[mg.mL.cm]^{-1}$).

11.6.3 Procedure 3: Enzymatic Synthesis of UTP 8

11.6.3.1 Materials and Equipment

- Acetic anhydride (19 mL, 0.2 mol)
- H_3PO_4 (85%, commercial; 6.5 mL, 0.1 mol)
- Ethyl acetate (1 L)
- $NaHCO_3$ (8.4 g)
- NaOH (10 M; 10 mL)
- Ice from dH_2O (25 g)
- dH_2O
- Orotic acid **4** (23.4 mg, 0.15 mmol)
- D-(−)-Ribose **1** (22.5 mg, 0.15 mmol)
- $NaH_2PO_4.H_2O$ (1.4 g)
- $MgCl_2.6H_2O$ (102 mg)
- NaOH (6 M; 1 mL)
- Ice in 1 L foam cooler
- Cooling bath thermostatted at 0 °C
- Water bath thermostatted at 37 °C
- Round-bottom flask (100 mL; acetyl phosphate synthesis)
- Separatory funnel (100 mL; acetyl phosphate synthesis)
- Eppendorf tube (2 mL), or a round bottom flask for larger reaction volumes (up to 50 mL) in UTP synthesis

11.6.3.2 Procedure

1. Although commercial acetyl phosphate may be used, the material used here was chemically synthesised as reported in [5]. Briefly, acetic anhydride (19 mL) was added dropwise to a mixture of phosphoric acid (6.5 mL) and ethyl acetate (60 mL) cooled in a 0 °C bath, and the solution was stirred for 6 hr. It was then added to a slurry of water

(50 mL), ice (25 g) and sodium bicarbonate (8.4 g) and stirred until carbon dioxide evolution ceased. The aqueous layer containing the product was washed with ethyl acetate (9 × 100 mL) (to remove acetic acid), neutralised with NaOH (10 M; ~5 mL) and stored in 1 mL aliquots at −20 °C for later use. The aqueous solution obtained contained ~1 M acetyl phosphate, determined enzymatically as described in [6]. Thawed aqueous solutions were employed directly for enzyme-catalysed synthesis.

2. Sodium phosphate buffer (0.5 M, pH 7.4) was prepared by dissolving $NaH_2PO_4 \cdot H_2O$ (1.4 g) in 10 mL dH_2O and adjusting the pH to 7.4 with 6 M NaOH (~100 μL).
3. Stock solutions (final conc. 100 mM) were prepared for orotic acid **4** (23.4 mg) in 1.5 mL dH_2O, for ribose (22.5 mg) in 1.5 mL dH_2O and for $MgCl_2$ (102 mg) in 5 mL dH_2O. These solutions were sufficient for the preparation of up to 50 mL of the synthesis reaction mixture, and any unused material was stored at −20 °C for later use.
4. A cell lysate aliquot (Procedure 2) was thawed on ice. This was enough for the preparation of up to 5 mL of the synthesis reaction mixture, and once thawed any unused material was stored at −20 °C for later use (within 2 months). The total cell lysate produced in Procedure 2 was sufficient to prepare up to 50 mL of the synthesis reaction mixture.
5. For every millilitre total UTP synthesis reaction volume, 290 μL dH_2O was mixed with 400 μL sodium phosphate buffer (final conc. 200 mM), 50 μL acetyl phosphate solution (final conc. 50 mM), 100 μL $MgCl_2$ solution (final conc. 10 mM), 30 μL orotic acid **4** (final conc. 3 mM) and 30 μL ribose solutions (final conc. 3 mM). Finally, the synthesis was initiated by addition of 100 μL cell lysate (Procedure 2).
6. The reaction mixture was incubated at 37 °C in a water bath for 2.5 hr, during which time the starting material was fully consumed.
7. The desired UTP **8** was obtained in >95% yield as a solution (1.4 $g.L^{-1}$ reaction) based on equimolar starting amounts of orotic acid and ribose (determined by high-performance liquid chromatography (HPLC) analysis; see Procedure 4).
8. The lysate prepared here was not limited to UTP **8** and also catalysed the synthesis of GTP and ATP from their respective nucleobases (Table 11.15).

Table 11.15 *Other substrates that can be synthesised via this procedure.*

Nucleobase substrate (replacing orotic acid)	Product	Unoptimised yield (nucleobase conversion after 2 hr)
Adenine (0.41 $g.L^{-1}$ reaction)	Adenosine-5′-triphosphate (ATP)	>50%
Guanine (0.45 $g.L^{-1}$ reaction)	Guanosine-5′-triphosphate (GTP)	>45%

11.6.4 Procedure 4: HPLC Analysis of the UTP Synthesis Mixture

11.6.4.1 Materials and Equipment

- Ammonium phosphate (6.9 g)
- Tetrabutylammonium phosphate monobasic solution (1 M in H_2O, commercial)
- NH_4OH (aqueous 28% w/w, commercial; ~100 μL)
- dH_2O (1 L)

- MeOH (1 L, HPLC-grade)
- Sodium dodecyl sulfate (SDS) solution (0.2% w/v, prepared in 100 mL dH_2O)
- UTP solution (100 mM, commercial standard in H_2O)
- ATP solution (100 mM, commercial standard in H_2O)
- Orotic acid (15.6 mg)
- AMP monohydrate (36.5 mg)
- UMP disodium salt (36.8 mg)
- ADP monopotassium salt dihydrate (50.1 mg)
- UDP disodium salt hydrate (44.8 mg)
- 5× 10 mL volumetric flasks
- Mobile phase filtration apparatus (250 mL solvent reservoir, 1 L flask) fitted with nylon filtration membrane (0.45 μm pores, 47 mm diameter) and flask side-arm connected to vacuum line
- Agilent 1100 Series HPLC system equipped with photodiode array (DAD)
- Alltima HP C18 column (5 μ 250 × 5.6 mm, equipped with a 190 Å 5 μ guard)
- Agilent HPLC vials (2 mL) and high-recovery glass inserts (400 μL flat bottom)

11.6.4.2 Procedure

1. Aqueous HPLC solvent (A) was prepared by dissolving ammonium phosphate (6.9 g) in 1 L dH_2O (final conc. 60 mM), adding 5 mL tetrabutylammonium phosphate solution (final conc. 5 mM) and adjusting to pH 5.0 with NH_4OH (~100 μL). The mixture was then filtered using the mobile phase filtration apparatus to remove any particulate matter (filtration is optional).
2. Organic HPLC solvent (B) was prepared by adding 5 mL of the tetrabutylammonium phosphate solution to 1 L HPLC grade MeOH (final conc. 5 mM) and filtering with the mobile phase filtration apparatus (filtration is optional).
3. UTP synthesis reaction progress and final yield were determined through analysis of the reaction mixture, using an HPLC method allowing separation and detection of UTP, orotic acid and nucleotide intermediates. Progress was monitored by withdrawing 50 μL samples from the reaction and quenching with an equal volume of the SDS solution (final conc. 0.1% w/v), in an HPLC vial equipped with a glass insert (necessary for reduced sample volume). Vials were loaded on to the Agilent 1100 HPLC autosampler tray for automated HPLC analysis.
4. An automated analysis method was carried out, involving: (a) injection of 5 μL sample, (b) separation of reaction components on the C18 column via the mobile phase gradient programme shown in Table 11.16 and (c) detection (DAD) of analyte absorbance at 259 nm (approx. λ_{max} for nucleobases tested).
5. Analytes were identified based on retention time (Table 11.17) through comparison with genuine standards, prepared as follows: UTP and ATP standard solutions (100 mM) were obtained commercially and 20 μL of each was diluted with 180 μL dH_2O (final conc. 10 mM). Orotic acid (15.6 mg), AMP (36.5 mg), UMP (36.8 mg), ADP (50.1 mg) and UDP (44.8 mg) were each dissolved in dH_2O (to final volumes of 10 mL in volumetric flasks; final conc. 10 mM) 40 μL of each 10 mM stock solution was then diluted with 160 μL dH_2O (final conc. 2 mM), and 100 μL of this dilute were used for the determination of HPLC retention time.

Table 11.16 *HPLC gradient method.*

Time (min)	Aqueous phase (A)	Organic phase (B)	Gradient program
0	83%	17%	Linear
18	83%	17%	Linear
19	70%	30%	Linear
23.2	70%	30%	Linear
24	83%	17%	Linear
35	End		Linear

Table 11.17 *Retention times for HPLC analysis.*

Substance	Retention (min)
Uridine-5′-monophosphate (UMP)	5.1
Orotic acid	5.8
Adenosine-5′-monophosphate (AMP)	7.6
Uridine-5′-diphosphate (UDP)	10.2
Adenosine-5′-diphosphate (ADP)	15.7
Uridine-5′-triphosphate (UTP)	22.0
Adenosine-5′-triphosphate (ATP)	24.3

6. For each species, $5 \times 100\,\mu L$ standards were prepared through twofold serial dilutions of the 2 mM standard and analysed by HPLC. Linear response curves were calculated through regression analysis of the obtained DAD responses (integrated peak area) and used for quantification of each species in unknown reaction samples.

11.6.5 Conclusion

The workflow described here enables the production of UTP **8** from nucleobase and ribose **1** in high yields and short timeframes (Figure 11.1). The only required biocatalyst is a single cell lysate, prepared through standard microbiological methodology and without laborious purification. Through the selection of appropriate starting materials and recombinant enzymes, this method may be extended to the synthesis of other natural and unnatural NTPs. Lysate-synthesised NTPs have been employed directly in cell-free protein synthesis and PCR without prior purification [3,7]. Nonetheless, preparative chromatographic methods reported elsewhere [2] are suitable for the isolation of UTP and other NTPs from enzymatic reaction mixtures. Finally, all ATP-dependent steps are supported by inexpensive acetyl phosphate and by endogenous lysate co-factors and enzymes. The same principle is applicable to any ATP-dependent enzyme that may be expressed recombinantly in *E. coli* [3,8].

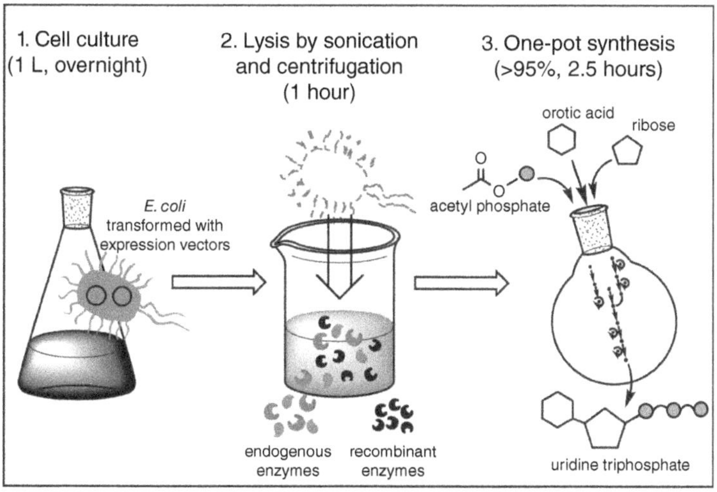

Figure 11.1 *Workflow for cell lysate biocatalyst preparation and application for UTP synthesis.*

References

1. (a) Burgess, K. and Cook, D. (2000) *Chemical Reviews*, **100**, 2047–2059; (b) Kaczynski, T.P. and Chmielewski, M.K. (2017) *Mini-Reviews in Organic Chemistry*, **14** (6), 448–452; (c) Roy, B., Depaix, A., Perigaud, C. and Peyrottes, S. (2016) *Chemical Reviews*, **116** (14), 7854–7897.
2. (a) Schultheisz, H.L., Szymczyna, B.R., Scott, L.G. and Williamson, J.R. (2008) *ACS Chemical Biology*, **3**, 499–511; (b) Schultheisz, H.L., Szymczyna, B.R., Scott, L.G. and Williamson, J.R. (2011) *Journal of the American Chemical Society*, **133** (2), 297–304.
3. Alissandratos, A., Caron, K., Loan, T.D. *et al.* (2016) *ACS Chemical Biology*, **11** (12), 3289–3293.
4. Loan, T.D., Easton, C.J. and Alissandratos, A. (2019) *New Biotechnology*, **49**, 104–111.
5. Crans, D.C. and Whitesides, G.M. (1983) *Journal of Organic Chemistry*, **48** (18), 3130–3132.
6. Whitesides, G.M., Siegel, M. and Garrett, P. (1975) *Journal of Organic Chemistry*, **40** (17), 2516–2519.
7. Loan, T.D., Easton, C.J. and Alissandratos, A. (2019) *Scientific Reports*, **9**, 15621.
8. Alissandratos, A., Hartley, C.J., French, N.G. *et al.* (2019) *ACS Sustainable Chemistry & Engineering*, **7** (9), 8522–8529.

11.7 Aerobic Synthesis of Aromatic Nitriles from Alcohols and Ammonia Using Galactose Oxidase

Jan Vilím, Tanja Knaus and Francesco G. Mutti*
Van 't Hoff Institute for Molecular Sciences, HIMS-Biocat, University of Amsterdam, Amsterdam, The Netherlands

The enzymatic synthesis of aromatic nitriles from alcohols and ammonia presents a valuable alternative to currently applied chemical methodologies. The synthesis of nitriles is traditionally accomplished through non-environmentally friendly approaches such as transition metal-catalysed cyanation, Sandmeyer reaction and Rosenmund–von Braun reaction [1]. Biocatalytic routes towards the synthesis of nitriles either utilise cyanide in combination with aldehydes (e.g. hydroxynitrile lyases, halohydrin dehalogenases) [2] or require prior *in situ* formation of the substrate (aldoxime dehydratases) [3]. Ammoxidation is a more sustainable alternative, as it consumes molecular oxygen and ammonia as reagents [4].

Galactose oxidase (GOx) is a copper-dependent oxidoreductase that naturally catalyses the aerobic oxidation of alcohol functions (as in galactose) to aldehydes. A variant of the GOx from *Fusarium* sp. (GOx_{M3-5}) was evolved in the laboratory to perform the oxidation of benzylic primary and secondary alcohols to the related carbonyl compounds [5]. We recently discovered that GOx_{M3-5} can also promiscuously catalyse the oxidation of aldehydes to nitriles in the presence of ammonia and air [6]. Furthermore, we developed a concurrent, two-step biocatalytic cascade that exploits both the natural and the promiscuous catalytic activities of GOx in order to perform one-pot oxidation of primary alcohols into their corresponding nitriles (Scheme 11.8). Finally, the biocatalytic process was applied for the preparative-scale synthesis of 2'-fluorobenzonitrile from 2'-fluorobenzyl alcohol (Scheme 11.9).

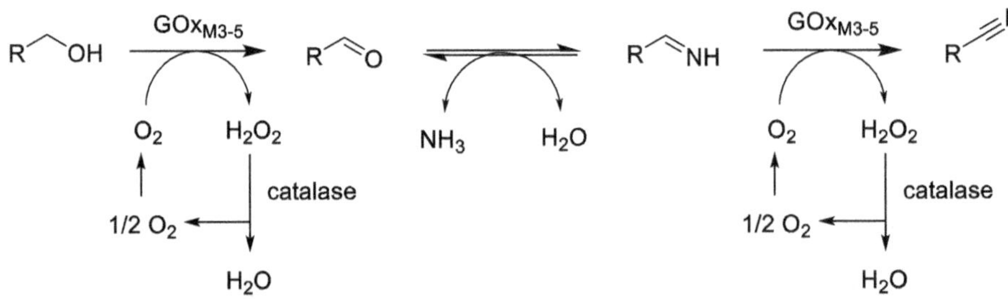

Scheme 11.8 *Biocatalytic cascade catalysed by* GOx_{M3-5} *for the synthesis of nitriles from alcohols, air and ammonia.*

11.7.1 Procedure 1: Recombinant Expression of GOx in *E. coli* BL21(DE3)

11.7.1.1 Materials and Equipment

- Lysogenic broth (LB; tryptone, 10 g.L^{-1}, yeast extract 5 g.L^{-1}, sodium chloride 10 g.L^{-1}; Carl Roth)
- Deionised water (dH_2O)
- Kanamycin (50 mg.mL^{-1} in dH_2O, filter-sterilised; Carl Roth)
- Isopropyl-β-D-1-thiogalactopyranoside (IPTG; 1 M in dH_2O, filter-sterilised; Carl Roth)
- $CuSO_4{\cdot}5H_2O$ (0.6 M in dH_2O, autoclaved; Sigma-Aldrich)
- 0.9% w/v NaCl
- 1× 0.25 L Erlenmeyer flask with foam cap
- 6× 2 L Erlenmeyer flasks with foam caps
- *C*-terminal Strep-tagged GOx_{M3-5}/pET42a(+) in *E. coli* BL21 (DE3) strain in the form of glycerol stock [6] (see Amino Acid Sequence)

- Autoclave
- Orbital shaker with temperature control (Infors HT, Multitron standard)
- Photometer (WPA CO 8000 cell-density meter)
- 1.5 mL disposable cuvettes (Brand)
- Centrifuge (4 °C)
- Freezer (−20 °C)

11.7.1.1.1 **Amino Acid Sequence of *C*-Terminal Strep-Tagged GOx$_{M3–5}$.** MASAPIGSAIPRNNWAVTCDSAQSGNECNKAIDGNKDTFWHTFYGANGDPKPPHTYTIDMKTTQNVNGLSVLPRQDGNQNGWIGRHEVYLSSDGTNWGSPVASGSWFADSTTKYSNFETRPARYVRLVAITEANGQPWTSIAEINVFQASSYTAPQPGLGRWGPTIDLPIVPAAAAIEPTSGRVLMWSSYRNDAFEGSPGGITLTSSWDPSTGIVSDRTVTVTKHDMFCPGISMDGNGQIVVTGGNDAKKTSLYDSSSDSWIPGPDMQVARGYQSSATMSDGRVFTIGGSFSGGVFEKNGEVYSPSSKTWTSLPNAKVNPMLTADKQGLYMSDNHAWLFGWKKGSVFQAGPSTAMNWYYTSGSGDVKSAGKRQSNRGVAPDAMCGNAVMYDAVKGKILTFGGSPDYTDSDATTNAHIITLGEPGTSPNTVFASNGLYFARTFHTSVVLPDGSTFITGGQRRGIPFEDSTPVFTPEIYVPEQDTFYKQNPNSIVRAYHSISLLLPDGRVFNGGGGLCGDCTTNHFDAQIFTPNYLYDSNGNLATRPKITRTSTQSVKVGGRITISTDSSISKASLIRYGTATHTVNTDQRRIPLTLTNNGGNSYSFQVPSDSGVALPGYWMLFVMNSAGVPSVASTIRVTQLELQNAPAHGWSHPQFEKRSAGSWSHPQFEKGAMTGWSHPQFEK

11.7.1.2 Procedure

1. LB (125 g) was dissolved in dH_2O (5 L). The obtained LB medium was divided into Erlenmeyer flasks (1 × 100 mL in 250 mL flask and 6 × 800 mL in 2 L flasks) and sterilised in an autoclave at 121 °C for 20 min.
2. A pre-culture was prepared in the sterilised LB medium (100 mL) contained in the 250 mL Erlenmeyer flask. Kanamycin (100 μL, final conc. 50 μg mL^{-1}) was added and the *E. coli* BL21 (DE3) strain harbouring the *C*-terminal Strep-tagged gene of GOx$_{M3–5}$ [5] in a pET42a(+) plasmid was inoculated from a glycerol stock [6]. The pre-culture was incubated in an orbital shaker at 37 °C and 170 rpm overnight.
3. The main cultures were prepared via the addition of kanamycin solution (800 μL) to the sterilised LB media (800 mL) contained in the 2 L Erlenmeyer flasks, followed by inoculation with 15 mL of the pre-culture from Step 2. Cultures were shaken in an orbital shaker at 37 °C and 170 rpm for 3–4 hr. Within this time, the OD_{600} was regularly measured using a cell-density meter until it reached a value between 0.6 and 0.9.
4. A stock solution of $CuSO_4$ (1.33 mL) was added to the main culture (final conc. 1 mM) prior to the induction of protein expression with IPTG (400 μL; final conc. 0.5 mM). The main cultures were incubated in an orbital shaker at 25 °C and 170 rpm overnight.
5. *E. coli* cells containing overexpressed GOx$_{M3–5}$ were harvested by centrifugation at 3400× g and 4 °C for 15 min. The resulting pellets were resuspended in 0.9% NaCl solution, centrifuged again following the same conditions and stored at −20 °C. Ultimately, an average of 5 g of wet cells were obtained per Erlenmeyer flask. Alternatively, cells could be used directly, as indicated in Procedure 2.

11.7.2 Procedure 2: Preparation of Cell-Free Extract

11.7.2.1 Materials and Equipment

- $HCOONH_4$ (5.96 g; Alfa-Aesar)
- NH_3 (8.1 M aq; 0.67 mL; Merck)
- dH_2O
- KOH (10 M in dH_2O; Carl Roth)
- 100 mL plastic beaker
- Spatula
- 50 mL falcon tubes
- *E. coli* cells (~5.1 g) containing overexpressed GOx_{M3-5}, as obtained in Procedure 1
- Ultrasonic processor
- Centrifuge (4 °C, >20 000× g)
- Ice

11.7.2.2 Procedure

1. An ammonium formate buffer (400 mM, pH 9.0) was prepared by dissolving ammonium formate (5.96 g) in dH_2O (230 mL) and adding concentrated aqueous ammonia (0.67 mL). The volume was brought to 245 mL and the pH adjusted by dropwise addition of 10 M KOH solution. Finally, the volume was adjusted again to 250 mL.
2. Fresh or frozen *E. coli* cells (5.1 g) harbouring the overexpressed GOx_{M3-5} were carefully resuspended in the ammonium formate buffer (30 mL, 400 mM, pH 9.0) using a spatula. Subsequently, the resuspended cells were transferred into a 100 mL plastic beaker and filled with the ammonium formate buffer up to 50 mL total volume (1 g of wet cells per 10 mL of buffer).
3. The resuspended cells were kept on ice and disrupted by ultrasonication (10 sec on, 10 sec off, 45% amp) for 10 min sonication time, then centrifuged at 35 000× g and 4 °C for 40 min to obtain a clear cell-free extract (the sonication time may vary depending on the instrument used; sonication should be carried out until the cell extract becomes a dark-brownish colour).
4. The cell-free extract was carefully decanted into two 50 mL falcon tubes and further used in the biocatalytic reaction. It is advised to keep the cell-free extract on ice prior to the addition to the reaction mixture. Moreover, depending on the speed of the centrifuge, it is necessary to shorten or prolong the centrifugation time to obtain an extract without any visible cell debris.

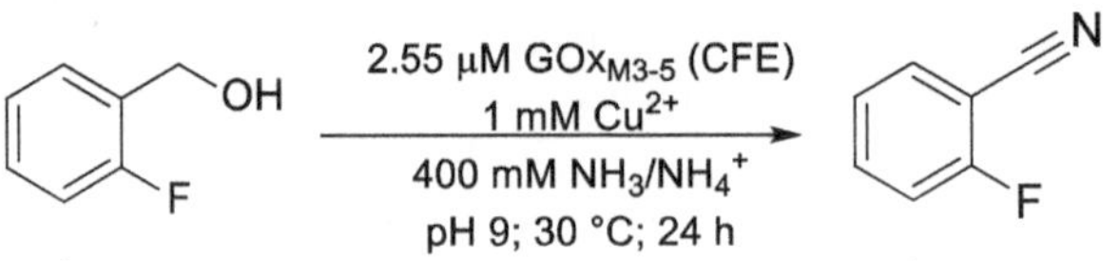

1 2

Scheme 11.9 *Biotransformation of 1.2 mmol (151 mg) 2′-fluorobenzyl alcohol* ***1*** *into 2′-fluorobenzonitrile* ***2****.*

11.7.3 Procedure 3: Biocatalytic Synthesis of 2′-Fluorobenzonitrile

11.7.3.1 Materials and Equipment

- Cell-free extract containing GOx_{M3-5}, as obtained in Procedure 2 (48 mL)
- Ammonium formate buffer 400 mM, pH 9.0 (59.4 mL)

- $CuSO_4$ (pentahydrate; Sigma-Aldrich; 10 mM concentration) in an ammonium formate buffer 400 mM, pH 9.0 (12 mL)
- Catalase (10 mg.mL^{-1} in dH_2O, 0.6 mL; Sigma-Aldrich)
- 2′-Fluorobenzyl alcohol **1** (1.2 mmol, 151 mg; Alfa-Aesar)
- 500 mL Erlenmeyer flask with foam cap
- Aluminium foil
- Ethyl acetate containing 10 mM toluene (both Merck)
- Glass vial
- Diethyl ether (Biosolv)
- Orbital shaker (with temperature control; Infors HT, Multitron standard)
- Centrifuge
- Nitrogen stream
- 50 mL falcon tubes
- Gas chromatograph with flame ionisation detector (GC-FID) equipped with a DB1701 column (30 m, 0.25 mm, 0.25 μM; Agilent)

11.7.3.2 Procedure

1. An ammonium formate buffer (400 mM. pH 9.0, 59.4 mL, as prepared in Procedure 2) was added into the 500 mL Erlenmeyer flask.
2. The stock solutions of $CuSO_4$ (12 mL; final conc. 1 mM) and catalase (0.6 mL; final conc. 0.05 mg.mL^{-1}) were added to the buffer.
3. The cell-free extract (48 mL), freshly prepared as described in Procedure 2, was added to the Erlenmeyer flask and the reaction mixture was gently mixed.
4. Compound **1** (129 μL, 1.2 mmol) was added (final conc. 10 mM). The Erlenmeyer flask was closed with the foam cap, the top was sealed with the aluminium foil and the reaction mixture was incubated in the orbital shaker at 30 °C and 170 rpm.
5. After 21 hr, a 0.5 mL aliquot was taken and extracted with EtOAc (650 μl) containing 10 mM toluene as an internal standard and dried over $MgSO_4$ to determine the conversion and analytical yield of 2'-fluorobenzonitrile **2**. The extracted sample was analysed by GC-FID equipped with a DB-1701 column (see Table 11.18 for the method specification, Table 11.19 for the retention time and Figure 11.2 for the chromatogram). The analytical yield for nitrile **2** formation was >99%.
6. To determine the isolated yield of the preparative-scale reaction of **2**, the reaction mixture was divided into six 50 mL falcon tubes (20 mL of solution in each), then extracted with Et_2O (three times, each 20 mL of Et_2O per tube per round of extraction).
7. The total volume of 360 mL Et_2O was combined and dried over $MgSO_4$, and the solvent was concentrated under reduced pressure to 1 mL. The compound was dried by evaporation of the remaining solvent using a stream of nitrogen in a 2 mL Eppendorf tube and subsequently transferred into a glass vial.
8. 108.8 mg of **2** was isolated as light-yellow liquid, corresponding to a 75% isolated yield.

Table 11.18 GC method.

Temperature program (r = °C.min^{-1})	Duration
$T_{injector}$ = 250 °C; constant pressure = 6.9 psi 80 °C, 6.5 min – 10 r → 160 °C, 5 min – 20 r → 200 °C, 2 min – 20 r → 280 °C, 1 min Column: DB-1701 30 m, 250 μm, 0.25 μm (Agilent J&W)	30 min

Table 11.19 Retention times for GC analysis.

Substance	Retention (min)
Toluene	3.0
2′-fluorobenzyl alcohol	11.7
2′-fluorobenzaldehyde	7.7
2′-fluorobenzonitrile	10.5

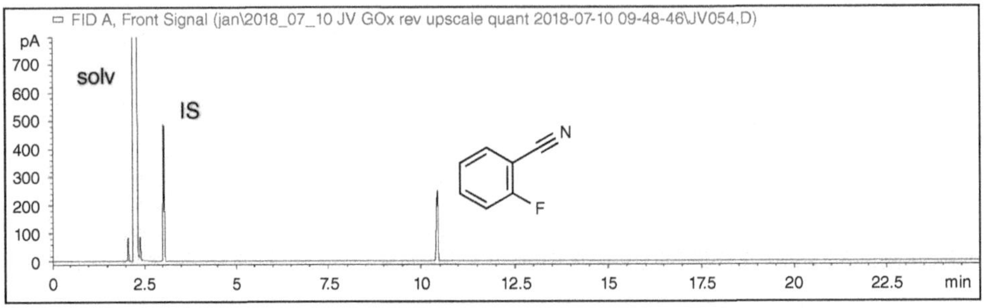

Figure 11.2 *Chromatogram of preparative-scale synthesis of* **2**. *solv, solvent; IS, internal standard.*

11.7.4 Additional Substrates

It is important to mention that this described method is not limited to the synthesis of 2'-fluorobenzonitrile. GOx_{M3-5} is also capable of converting various substituted benzyl alcohols, pyridylmethanols and cinnamyl alcohol and reaching TONs between 380 and 2820 $mol_{nitrile\ produced} \cdot mol_{biocatalyst}^{-1}$ (Scheme 11.10) [6]. Moreover, GOx_{M3-5} has been reported to also accept structural motifs that have not yet been investigated in relation to the formation of nitriles [7].

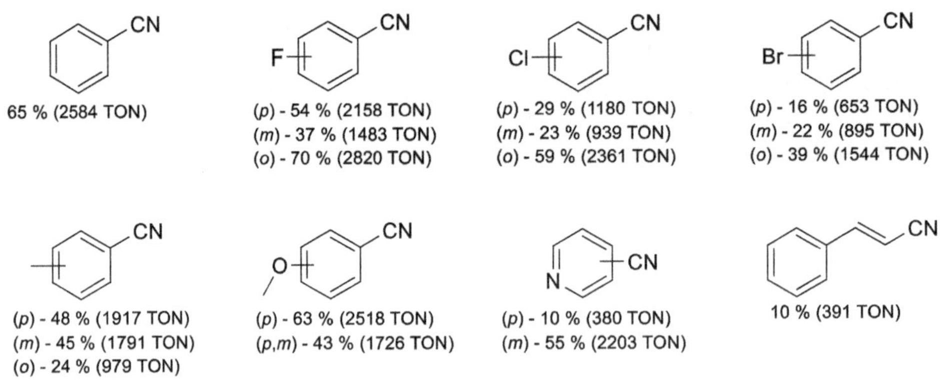

Scheme 11.10 *Elucidated list of substrates accepted by GOx_{M3-5} [6].*

11.7.5 Conclusion

The synthesis of nitriles starting from alcohols or aldehydes and employing ammonia is a synthetically valuable transformation, particularly if molecular oxygen in the form of air is used as an innocuous oxidant. Moreover, the use of the biocatalyst in the form of a cell-free extract enables the omission of the time-consuming enzyme-purification step. The procedure described here results in the synthesis of **2** through quantitative conversion and a high isolated yield (75%, 108.8 mg). To date, reports have described the conversion of 18 alcohol substrates into nitriles on analytical scale. Interestingly, the reaction has afforded a better yield of **2** on the preparative scale compared with the analytical. This difference can be attributed to the enhanced aeration of the reaction mixture in an Erlenmeyer flask. Further optimisation of the reaction conditions on preparative scale might result in even higher isolated yields. Notably, the TONs reported in Scheme 11.10 were calculated by considering the estimated concentration of GOx_{M3-5} according to the purification yield [6]. However, as purification yields generally fluctuate, it is advisable to express the enzyme loading in terms of units of catalytic activity per mass of wet cells. Overall, the presented procedure for the biocatalytic aerobic oxidation of alcohols to nitriles is robust and easy to use.

References

1. Kim, J., Kim, H.J. and Chang, S. (2012) *Angewandte Chemie International Edition*, **51**, 11 948–11 959.
2. (a) Bracco, P., Busch, H., von Langermann, J. and Hanefeld, U. (2016) *Organic and Biomolecular Chemistry*, **14**, 6375–6389; (b) Schallmey, A. and Schallmey, M. (2016) *Applied Microbiology and Biotechnology*, **100**, 7827–7839.
3. Asano, Y. and Okazaki, S. (2017) Enzymes in the aldoxime–nitrile pathway, in *Future Directions in Biocatalysis*, 2 edn (ed. T. Matsuda), Elsevier, pp. 173–187.
4. Martin, A. and Kalevaru, V.N. (2010) *ChemCatChem*, **2**, 1504–1522.
5. Escalettes, F. and Turner, N.J. (2008) *ChemBioChem*, **9**, 857–860.
6. Vilim, J., Knaus, T. and Mutti, F.G. (2018) *Angewandte Chemie International Edition*, **57**, 14 240–14 244.
7. Birmingham, W.R. and Turner, N.J. (2018) *ACS Catalysis*, **8**, 4025–4032.

11.8 Hydrogen-Borrowing Conversion of Alcohols into Optically Active Primary Amines by Combination of Alcohol Dehydrogenases and Amine Dehydrogenases

Maria L. Corrado, Vasilis Tseliou, Joseline A. Houwman, Wesley Böhmer, Jan Vilím, Marcelo F. Masman, Tanja Knaus and Francesco G. Mutti*
Van 't Hoff Institute for Molecular Sciences, HIMS-Biocat, University of Amsterdam, Amsterdam, The Netherlands

The direct amination of an alcohol possesses the inherent advantage that the reagent and product are in the same oxidation state; therefore, external redox equivalents are not required. However, the first challenge is to perform a tandem process in which an oxidative and a reductive step are running simultaneously. The second challenge is to perform a redox-neutral process in which the electrons liberated in the first oxidative step are

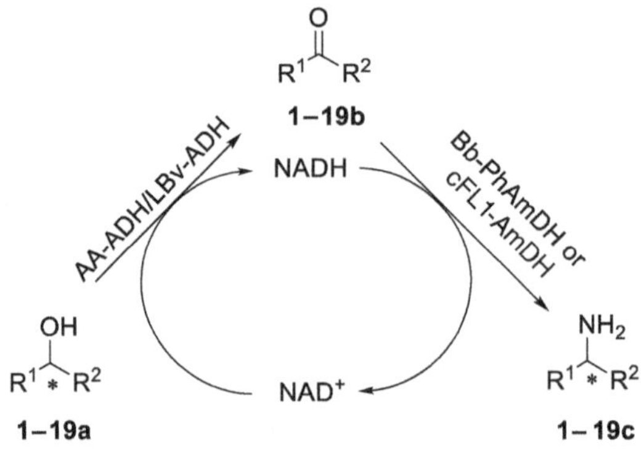

Scheme 11.11 One-pot hydrogen-borrowing amination of racemic alcohols into optically active α-chiral amines.

quantitatively recycled and consumed in the subsequent reductive step [1]. Such a process is defined as 'hydrogen-borrowing.' This section is particularly focused on the hydrogen-borrowing conversion of racemic alcohols into optically active α-chiral amines. In this process, the hydride released in the first step – the oxidation of the alcohol to the ketone catalysed by alcohol dehydrogenase (ADH) – is consumed in the second step, which is the reductive amination of the ketone catalysed by amine dehydrogenase (AmDH) (Scheme 11.11) [2]. The redox self-sufficient cycle uses ammonium ion/ammonia as a nitrogen source and generates only water as a byproduct. Moreover, the cascade requires only catalytic amounts of β-nicotinamide adenine dinucleotide (NAD^+) coenzyme to shuttle the hydride within the two steps.

These procedures report the use of two ADHs and two AmDHs, namely ADH from *Aromatoleum aromaticum* (AA-ADH) [3], LBv-ADH variant from *Lactobacillus brevis* [4], Bb-PhAmDH variant K78S-N277L from *Bacillus badius* [5] and cFL1-AmDH chimeric variant [5b,6]. The method has been successfully applied for the asymmetric amination of a panel of racemic secondary alcohols. Specifically, the procedures describe hydrogen borrowing amination of racemic alcohols by:

1. using isolated enzymes in solution (Procedures 1–3);
2. using co-immobilised enzymes (Procedure 4); and
3. using resting *E. coli* cells co-expressing the enzymes (Procedure 5).

11.8.1 Procedure 1: Preparation of Purified Amine Dehydrogenases Bb-PhAmDH and cFL1-AmDH

11.8.1.1 Materials and equipment

- Sterilised lysogenic broth (LB) medium ($10\,g.L^{-1}$ tryptone, Oxoid LP0042; $5\,g.L^{-1}$ yeast, Oxoid LP0021; $5\,g.L^{-1}$ NaCl)
- *E. coli* cells harbouring the plasmid coding for the Bb-PhAmDH [5a] and cFL1-AmDH [6]
- Deionised water (dH_2O)
- Agar plates supplemented with kanamycin ($50\,\mu g.mL^{-1}$)

- Kanamycin (50 mg.mL^{-1} stock in dH_2O, filter-sterilised)
- *E. coli* BL21 (DE3) competent cells
- Sterilised Erlenmeyer flasks
- Isopropyl-β-D-1-thiogalactopyranoside (IPTG 1 M stock in dH_2O, filter-sterilised)
- Incubator for *E. coli* cell cultivation
- Centrifuge for cell harvesting (3400× g, 4 °C)
- Aqueous solution of NaCl (5% w/w)
- Centrifuge (39 200× g, 4°C)
- Lysis buffer (10 mM imidazole buffer, pH 8.0, 50 mM KH_2PO_4, 300 mM NaCl)
- Lysozyme from chicken egg white (3.2 mg, Sigma L6876, lyophilised powder, protein 95%, >40 000 U.mg^{-1} protein)
- 5 mL HisTrap FF Ni^{2+} affinity columns (GE Healthcare)
- dH_2O (filtered and degassed)
- Ultrasonicator for cell disruption
- 0.45 μm syringe filter Whatman®
- 0.20 μm syringe filter Whatman®
- Washing buffer (imidazole 25 mM, pH 8.0, 50 mM KH_2PO_4, NaCl 300 mM)
- Elution buffer (imidazole 300 mM, pH 8.0, 50 mM KH_2PO_4, NaCl 300 mM)
- Potassium phosphate buffer (8 L, pH 8.0, 50 mM)
- Dialysis tubes
- Centripreps (Millipore)
- SDS-PAGE and electrophoresis setup
- UV-Vis spectrophotometer
- Quartz cuvettes
- Thrombin Cleavage Capture Kit

11.8.1.2 Procedure

1. The plasmids (pET28b) carrying either the Bb-PhAmDH variant or cFL1-AmDH were transformed in *E. coli* BL21 (DE3) according to the supplier's standard protocol (Novagen).
2. Colonies were cultivated overnight on a sterile agar plate supplemented with kanamycin (50 μg.mL^{-1}).
3. A single colony of *E. coli* BL21/His$_6$-tagged Bb-PhAmDH or cFL1-AmDH was taken from the agar plate for inoculation into an Erlenmeyer flask (100 mL) containing sterilised LB medium supplemented with kanamycin (50 μg.mL^{-1}). Culture was grown at 37 °C and 170 rpm for 16 hr.
4. Four large cultures (4 × 800 mL) containing sterilised LB medium supplemented with kanamycin (50 μg.mL^{-1}) were inoculated with the overnight culture (15 mL each). The OD_{600} was checked after 2 hr and was found to be around 0.7. Then, cultures were induced for protein expression with IPTG (final conc. 0.5 mM) and further shaken at 20 °C and 170 rpm for 24 hr.
5. Cells were harvested by centrifugation (3400× g, 4 °C) and the pellets were washed with an aqueous solution of NaCl (5% w/w). Pellets (~5.5 g.L^{-1}) were frozen and stored at −20 °C.
6. Two HisTrap FF columns charged with Ni^{2+} (5 mL each) were washed with dH_2O (50 mL) and conditioned with the lysis buffer (50 mL).

7. The wet cells containing the AmDHs (~9.2 g of cells containing Bb-PhAmDH variant or ~8 g of cells containing cFL1-AmDH variant) were suspended in the lysis buffer (40 mL).
8. Cells were initially disrupted with lysozyme (1 mg.mL^{-1}) by shaking at 20 °C and 150 rpm for 30–40 min. The mixture was diluted with the lysis buffer up to 80 mL. Then, the disruption was completed by sonication (5 min, amplitude 45%, pulse on 20 sec, pulse off 40 sec).
9. The suspension was centrifuged at 39 200 × g rpm and 4 °C for 1 hr and the supernatant was filtered first through a 0.45 μm filter and then through a 0.20 μm filter.
10. The solution was loaded into the previously conditioned HisTrap FF columns. Subsequently, the columns were washed with the washing buffer (75 mL).
11. The columns were eluted with the elution buffer (4 mL × 10 fractions).
12. Fractions were analysed by SDS-PAGE, showing the enzymes were obtained with high purity (>99%).
13. The fractions containing the highly purified enzymes were combined and dialysed overnight against the potassium phosphate buffer (8 L, pH 8.0, 50 mM).
14. The enzyme solution was concentrated (~5 mL) using centripreps and the concentration was measured spectrophotometrically (Bb-PhAmDH: ε = 25 000 M^{-1}.cm^{-1} at λ = 280 nm; MW = 43 474 kDa; enzyme yield = 477 mg; cFL1-AmDH: ε = 21 000 M^{-1}.cm^{-1} at λ= 280 nm; MW = 43 573 kDa; enzyme yield = 400 mg).
15. Due to the potential negative effect of the His-tag on the activity of ADHs [2], *N*-terminal His$_6$-tag was cleaved using biotinylated thrombin (Novagen) following the supplier-denoted procedure (Thrombin Cleavage Capture Kit), with a slight modification [7]. *N*-terminal His$_6$-tag of Bb-PhAmDH (126 mg in 2.25 mL of potassium phosphate buffer pH 8.0, 50 mM) and cFL1-AmDH (138 mg in 1.53 mL of potassium phosphate buffer pH 8.0, 50 mM) were generally cleaved using biotinylated thrombin (5 U) in a Tris-HCl buffer (50 mL, 20 mM, pH 8.4) supplemented with NaCl (150 mM) and $CaCl_2$ (2.5 mM) at 20 °C for 2 hr. The cleavage of the His$_6$-tag proceeded quantitatively with only a minor loss of catalytic activity. The residual biotinylated thrombin present in the solution was removed via the addition of streptavidin agarose beads. Upon biotin–streptavidin binding, the agarose beads were removed by centrifugation [7]. The cleaved His-tag was removed by flowing the final solution through a Ni^{2+} affinity column.
16. The final solutions of highly pure Bb-PhAmDH and cFL1-AmDH devoid of the His-tag were dialysed overnight in the phosphate buffer (pH 8.0, 50 mM), concentrated (1.1 mL for both AmDHs; typical activity was 600–700 U.mL^{-1}, measured with phenylacetone for Bb-PhAmDH and with 2-hexanone for cFL1-AmDH), flash-frozen in liquid nitrogen and stored at −80 °C. Prior to storage, the concentration of the Bb-PhAmDH and cFL1-AmDH (typical concentration 90 mg.mL^{-1} for both AmDHs) was determined spectrophotometrically using the extinction coefficient as described in Step 14.

11.8.2 Procedure 2: Preparation of Purified Alcohol Dehydrogenases AA-ADH and LBv-ADH

11.8.2.1 Materials and Equipment

- Crude cell extracts of AA-ADH (45 mL, purified enzyme content 800 mg) and LBv-ADH (22 mL, purified enzyme content 260 mg) from BASF, Ludwigshafen, Germany [3,4b]

- Centrifuge (39 200 × g rpm, 4 °C)
- HiPrep 26/10 desalting column (GE 8 Healthcare, bed volume 53 mL)
- Tris-HCl buffer (20 mM, pH 8.0)
- $MgCl_2$
- HiPrepQ HP 16/10 column (GE Healthcare)
- Start buffer (Tris-HCl buffer 20 mM, pH 8.0)
- Elution buffer (20 mM Tris-HCl, 1 M NaCl, pH 8.0)
- Superdex 200 16/10 (GE Healthcare)
- Tris-HCl (pH 8.0, 50 mM supplemented with 150 mM NaCl)
- Tris-HCl (pH 8.0, 50 mM)
- Centripreps (Millipore)
- SDS-PAGE and electrophoresis setup
- UV-Vis spectrophotometer
- Quartz cuvettes

Note: In the case of purification of LBv-ADH, all buffers were supplemented with $MgCl_2$ (1 mM).

11.8.2.2 Procedure

1. An aliquot of cell extract containing AA-ADH (45 mL) or LBv-ADH (22 mL) was thawed and centrifuged at 39 200× g and 40 min for 4 °C. The lysate was used for the enzyme purification, whilst cell debris was disposed of.
2. The lysate was first passed on to a HiPrep 26/10 desalting column and subsequently eluted with a Tris-HCl buffer (20 mM, pH 8.0).
3. The solutions AA-ADH and LBv-ADH were purified by anion-exchange chromatography using a HiPrepQ HP 16/10 column. The elution of the ADH was performed with a gradient between a start buffer and an elution buffer.
4. After SDS-PAGE, fractions containing the desired ADH in sufficient purity (>85%) were combined.
5. In the second step, AA-ADH and LBv-ADH were additionally purified via size-exclusion chromatography (Superdex 200 16/10) using a Tris-HCl buffer (pH 8.0, 50 mM supplemented with 150 mM NaCl).
6. The fractions containing the ADH in high purity (>95%) were combined and dialysed overnight in a Tris-HCl buffer (50 mM, pH 8.0).
7. The protein solutions were concentrated (AA-ADH: 18 mL; LBv-ADH: 5.1 mL) using Centripreps (Millipore) and the final protein concentration (AA-ADH: 45 $mg.mL^{-1}$; LBv-ADH: 51 $mg.mL^{-1}$) was determined at $\lambda = 280$ nm (LBv-ADH: $\varepsilon = 20\,000\ M^{-1}.cm^{-1}$, MW = 26.7 kDa; AA-ADH: $\varepsilon = 22\,500\ M^{-1}.cm^{-1}$, MW = 26.6 kDa).

11.8.3 Procedure 3: Hydrogen-Borrowing Amination of (*S*)-Phenyl-Propan-2-ol (*S*)-1a to Produce (*R*)-Phenyl-2-Propylamine (*R*)-3a on Preparative Scale

11.8.3.1 Materials and Equipment

- Round-bottom flask (100 mL)
- 2 M ammonium chloride buffer pH 8.7
- NAD^+ (final conc. 1 mM in reaction mixture)
- AA-ADH solution (2 mL from 44.5 $mg.mL^{-1}$ stock solution)

- Bb-PhAmDH solution (6.4 mL, 41 mg.mL^{-1})
- 20 mM *(S)*-**1a** (126 mg, 0.93 mmol)
- Orbital shaker
- 12 M aqueous HCl solution
- 10 M aqueous KOH solution
- CH_2Cl_2
- EtOAc
- $MgSO_4$
- GC-FID

11.8.3.2 Procedure

1. A 2 M ammonium chloride buffer solution, pH 8.7 (47 mL) containing catalytic NAD^+ (final conc. 1 mM) was added to a round-bottom flask.
2. AA-ADH stock solution (2 mL, 44.5 mg.mL^{-1}) and Ph-AmDH stock solution (6.4 mL, 41 mg.mL^{-1}) were added.
3. The substrate (*S*)-**1a** was added (129 μL, 0.925 mmol), and the reaction was shaken in an orbital shaker at 190 rpm and 30 °C for 48 hr.
4. A small aliquot of the reaction mixture was taken (500 μL), treated with a 10 M aqueous solution of KOH (200 μL) and extracted with CH_2Cl_2 (2 × 300 μL). The organic phase was dried with anhydrous $MgSO_4$ and analysed by gas chromatography (GC) to verify the high conversion level (typically >99%).
5. The reaction mixture was acidified with a 12 M aqueous solution of HCl (1 mL). The reaction was then extracted with EtOAc (2 × 10 mL) in order to remove the unreacted alcohol and the ketone intermediate.
6. The reaction mixture was treated with a 10 M aqueous solution of KOH until the pH reached >10. The basic reaction mixture containing the free amine product was extracted with EtOAc (3 × 10 mL). The combined organic phase was dried with anhydrous $MgSO_4$ and filtered, and the solvent was evaporated to afford (*R*)-**1c** as a yellow oil (107 mg, 0.79 mmol, 85% isolated yield, >99% ee). The product was analysed by GC with a flame ionisation detector (FID) for purity and enantiomeric excess.

11.8.4 Procedure 4: Co-immobilisation of AA-ADH and cFL1-AmDH on EziG™ Fe-Amber Beads and Hydrogen Amination of Racemic Alcohols on Analytical Scale

11.8.4.1 Materials and Equipment

- His_6-tagged cFL1-AmDH (23 nmol, 1 mg; purified according to Procedure 1)
- His_6-tagged AA-ADH (35 nmol, 1 mg; overexpressed in *E. coli* BL21 (DE3) using pET28b plasmid; cultivated and purified according to Procedure 1) [8]
- Tris-HCl buffer (50 mM, pH 8.0)
- EziG™ Fe-amber beads (40 mg; EnginZyme AB)
- C-Star Orbital Shaker No. 12846016 (Thermo Fisher Scientific)
- Eppendorf Thermomixer compact 5350 (Germany)
- Biorad protein-assay dye reagent concentrate (Carl Roth)
- UV-Vis spectrophotometer
- 7890A GC system (Agilent Technologies) equipped with FID (carrier gas: H_2)
- Ammonium chloride buffer (pH 8.7, 2 M)

- Deionised water (dH_2O)
- NAD^+ (Melford Biolaboratories)
- (*S*)-1-Phenyl-2-propanol **1a** (Sigma-Aldrich)
- 5 M KOH in dH_2O
- EtOAc
- $MgSO_4$

11.8.4.2 Procedure

1. Purified His_6-tagged cFL1-AmDH (29 µL from 35 mg mL^{-1} stock solution, 23 nmol) and AA-ADH (67 µL from 15 mg.mL^{-1} stock solution, 35 nmol) were combined in a 50 mM Tris-HCl buffer solution, pH 8.0 (1 mL) at 4 °C. EziG™ Fe-amber beads (40 mg; total enzyme loading on carrier 50 mg.$g_{beads}{}^{-1}$) were suspended in the enzyme solution and the suspension was incubated at 4 °C and 120 rpm for 2 hr.
2. The immobilisation process was monitored over time by taking a small aliquot of the previous solution and performing a Bradford assay (i.e. measuring absorbance at $\lambda = 595$ nm of a solution consisting of 980 µL ready-to-use Bradford solution plus 20 µL sample solution). The co-immobilised enzymes were collected by sedimentation and the remaining buffer solution was discarded. The final yield of immobilisation was determined using the following equation:

$$Yield\ of\ immobilisation\ [\%] = \frac{A_{595\ before\ immobil.} - A_{595\ after\ immobil.}}{A_{595\ before\ immobil.}} \times 100\%$$

   where $A_{595\ \text{before immobil.}}$ is the enzyme solution's absorbance before immobilisation and $A_{595\ \text{after immobil.}}$ is the residual supernatant solutions' absorbance after removal of the beads carrying the co-immobilised enzymes.
3. The co-immobilised enzymes were applied to the hydrogen-borrowing amination of (*S*)-**1a** as a substrate (1.37 µL, final conc. 9.4 µmol) in a NH_4Cl/NH_4OH buffer (0.5 mL, 2 M, pH 8.7) supplemented with NAD^+ (1 mM). The reactions were incubated in an orbital shaker at 30 °C for 48 hr.
4. At the end of the reaction, the co-immobilised enzymes were collected by sedimentation. The separated aqueous phase was basified with a 5 M aqueous solution of KOH (100 µL) and extracted with EtOAc (2 × 500 µL). The organic layer was dried over anhydrous $MgSO_4$ and analysed using GC-FID equipped with DB1701-60m (Method A). The biotransformation yielded (*R*)-**1c** with an average 78% yield over three cycles (1st cycle, 96%; 2nd cycle, 90%; 3rd cycle, 48%). (*R*)-**1c** was obtained with an optical purity of >99% ee.
5. After the previous reaction cycle, the co-immobilised enzymes were suspended in a fresh reaction buffer and reused for next cycle.

11.8.5 Procedure 5: Hydrogen-Borrowing Amination of Racemic 2-Hexanol Utilising *E. coli* Resting Cells Containing Co-expressed ADHs and AmDH on Preparative Scale

11.8.5.1 Materials and Equipment

- LB agar plate with colonies of *E. coli* BL21(DE3) harbouring two expression vectors: (a) pET28b(+) with the gene encoding for *N*-terminal GST-tagged LBv-ADH [4]; (b) pETDUET with two genes encoding for *N*-terminal His_6-tagged cFL1-AmDH

[5b,6] and AA-ADH [3]. The engineered strain is herein identified as *E. coli* (cFL1-AA-LBv) [9]

- Sterilised LB medium (10 g.L^{-1} tryptone, Oxoid LP0042; 5 g.L^{-1} yeast, Oxoid LP0021; 5 g.L^{-1} NaCl)
- dH_2O
- Ampicillin (100 mg.mL^{-1} in dH_2O, filter-sterilised; Carl Roth)
- Kanamycin (50 mg.mL^{-1} in dH_2O, filter-sterilised; Carl Roth)
- IPTG
- Lysis buffer (NH_4Cl/NH_3 1 M, pH 8.7)
- 2× Laemmli sample buffer
- Ammonium chloride buffer (NH_4Cl/NH_3 1 M, pH 8.7)
- Glucose in dH_2O (20 mM)
- *rac*-2-Hexanol **17a** (0.511 g, 5 mmol, 625 μL)
- KOH (10 M)
- EtOAc
- $MgSO_4$
- 4-Dimethylaminopyridine dissolved in acetic anhydride (50 mg.mL^{-1})
- Incubator for *E. coli* cell cultivation
- 2 mL Eppendorf tubes
- Orbital shaker
- GC-FID
- Agilent DB-1701 column
- Variant Chiracel DEX-CB column
- 500 mL sterile baffled flask
- Methyl *tert*-butyl ether (MTBE)

11.8.5.2 Procedure

1. Antibiotics (5 μL of each stock solution; final conc. 100 μg.L^{-1} ampicillin, 50 μL.L^{-1} kanamycin) were aseptically added to LB medium (4 × 5 mL) and the medium was subsequently inoculated with a single colony from the LB agar plate. The culture was incubated in the orbital incubator at 37 °C and 170 rpm overnight.
2. Sterile LB medium in Erlenmeyer flasks (4 × 800 mL) containing kanamycin and ampicillin (800 μL of each stock solution; final conc. 100 μg.L^{-1} ampicillin, 50 μL.L^{-1} kanamycin) was inoculated with a cell suspension from the overnight culture (4 × 2 mL of culture) and cells were grown at 37 °C and 170 rpm until an OD_{600} of 0.6–0.8 was reached (~6 hr). Overexpression of cFL1-AmDH, AA-ADH and LBv-ADH was induced by IPTG (0.5 mM) and the flasks were incubated at 25 °C and 170 rpm overnight.
3. Protein expression was verified by taking a sample of a 0.5 OD unit ($V_{culture}$ (mL) = $0.5/OD_{600}$) before and after overnight induction. Pellet samples and resuspend cells (25 μL of lysis buffer, 25 μL of sample buffer) were heated at 95 °C for 5 min. An aliquot of the prepared samples (10 μL) was used for SDS-PAGE analysis.
4. Cells were harvested by centrifugation at 3400× g for 20 min. Pellets were washed once with an ammonium chloride buffer (~100 mL, 1 M, pH 8.7). Cells were again pelleted by centrifugation at 3400× g for 20 min, weighed (wet cell weight ~16 g) and gently resuspended in the ammonium chloride buffer to reach a cell density of 67.4 $mg_{wet\ cells}$.mL^{-1}.

5. Biotransformations comprised a total of 250 mL cell suspension in the ammonium chloride buffer (i.e. 225 mL cell suspension, final conc. 60 $mg_{wet\ cells}$·mL^{-1}; 25 mL of 200 mM glucose, final conc. 20 mM; 625 µL (5 mmol) of *rac*-**17a**, final conc. 20 mM) in a 500 mL baffled flask. The baffled flask was used to maximise the aeration of the cells as well as to mix the cells during incubation. The flask was tightly sealed with aluminium foil, parafilm and a cap to prevent any possible evaporation of the substrate. Reactions were incubated at 30 °C and 170 rpm for 24 hr.
6. The reaction was monitored by GC, by working up 1 mL aliquot as follows. The reaction aliquot was quenched by adding 10 M aqueous KOH solution (200 µL) and vortexing for 2 sec, then extracting with EtOAc (2 × 500 µL). Due to the viscous nature of the reaction, the samples required vortexing for at least 10 sec after the addition of EtOAc in order to extract the maximum amount of reaction products. Subsequently, the samples were centrifuged at 18 000× g at 4 °C for 10 min. This resulted in a clear separation of the aqueous and organic layers; however, the viscosity of the aqueous layer resulted in a lower volume of accessible solvent, allowing the extraction of ~75% added volume of EtOAc (350–400 µL of the previously added 500 µL). The combined organic layers were dried over anhydrous $MgSO_4$ and conversion was determined by GC-FID with an Agilent DB-1701 column.
7. When the reaction reached completion, the preparative reaction mixture was acidified to pH 2–4 through the addition of concentrated HCl solution (12.5 mL) and stirred for 30 min. The reaction was divided over 10 × 50 mL Falcon tubes and extracted with MBTE (3 × 60 mL) to remove the unreacted alcohol and ketone intermediate (**17a,b**). The organic layers were combined and dried over anhydrous $MgSO_4$ and the MBTE was removed by distillation under reduced pressure. The residue was brownish in colour, possibly due to the extraction of some *E. coli* components.
8. The pH of the reaction was increased to basic through the addition of 10 M aqueous KOH (5 mL) to each Falcon tube and extraction was performed with MBTE (3 × 60 mL) to remove the amine. The organic layers were combined and dried with anhydrous $MgSO_4$ and the MBTE was removed by distillation under reduced pressure. The resulting fraction containing **17c** was yellowish in colour.
9. The purity of the amine and the ketone fractions was checked by nuclear magnetic resonance (^{1}H-NMR; 400 MHz, $CDCl_3$). Overall, the process yielded **17c** in 16% isolated yield (40% conversion) with high chemical (>99%) and optical (>99% ee) purity.

11.8.6 Analytical Method

11.8.6.1 Determination of Conversion

- Method A Column: Agilent J&W DB-1701 (60 m, 250 µm, 0.25 µm). GC programme parameters: injector 250 °C, constant pressure 32.07 psi; temperature programme: 80 °C/hold 6.5 min, 160 °C/rate 5 °C.min^{-1}/hold 2 min, 280 °C/rate 20 °C.min^{-1}/hold 1 min.
- Method B – Column: Agilent J&W DB-1701 (30 m, 250 µm, 0.25 µm). GC programme parameters: injector 250 °C, constant pressure 14.50 psi; temperature programme:

80 °C/hold 6.5 min, 160 °C/rate 10 °C.min^{-1}/hold 5 min, 200 °C/rate 20 °C.min^{-1}/hold 2 min, 280 °C/rate 20 °C.min^{-1}/hold 1 min.

- Method C – Column: Agilent J&W DB-1701 (30 m, 250 μm, 0.25 μm). GC programme parameters: injector 250 °C, constant pressure 14.60 psi; temperature programme: 60 °C/hold 6.5 min, 100 °C/rate 20 °C.min^{-1}/hold 1 min, 280 °C/rate 20 °C.min^{-1}/hold 1 min (see Table 11.20).

Table 11.20 *GC retention times and analysis methods for substrates, intermediates and products.*

Alcohol	Retention time (min)	Ketone	Retention time (min)	Amine	Retention time (min)	GC-FID method
1a	20.90	**1b**	21.03	**1c**	19.22	A
2a	22.02	**2b**	21.87	**2c**	20.03	A
3a	24.03	**3b**	24.38	**3c**	22.57	A
4a	27.46	**4b**	27.77	**4c**	26.50	A
5a	28.43	**5b**	28.43	**5c**	27.57	A
6a	23.98	**6b**	24.07	**6c**	22.60	A
7a	15.8	**7b**	15.09	**7c**	14.33	B
8a	15.24	**8b**	14.65	**8c**	14.83	B
9a	11.57	**9b**	11.22	**9c**	9.88	B
10a	11.62	**10b**	9.75	**10c**	9.90	B
11a	12.36	**11b**	10.88	**11c**	10.49	B
12a	12.20	**12b**	11.08	**12c**	10.37	B
13a	21.77	**13b**	21.99	**13c**	19.71	A
14a	13.48	**14b**	13.79	**14c**	12.19	B
15a	8.53	**15b**	8.28	**15c**	6.73	B
16a	5.57	**16b**	5.32	**16c**	4.26	B
17a	5.61	**17b**	5.16	**17c**	4.05	C
18a	3.48	**18b**	3.28	**18c**	2.82	C
19a	4.53	**19b**	4.00	**19c**	3.42	C

11.8.6.2 Determination of Enantiomeric Excess

- Method A – Column: Varian Chrompack Chiracel Dex-CB column (25 m, 320 μm, 0.25 μm). GC programme parameters: injector 200°C, constant flow 1.7 mL; temperature programme: 100 °C/hold 2 min, 130 °C/rate 1 °C.min^{-1}/hold 5 min, 170 °C/rate 10 °C.min^{-1}/hold 10 min, 180 °C/rate 10 °C.min^{-1}/hold 1 min.
- Method B – Column: Varian Chrompack Chiracel Dex-CB column (25 m, 320 μm, 0.25 μm). GC programme parameters: injector 200°C, constant flow 1.7 mL; temperature programme: 100 °C/hold 2 min, 118 °C/rate 1 °C.min^{-1}/hold 5 min, 170 °C/rate 10 °C.min^{-1}/hold 10 min, 180 °C/rate 10 °C.min^{-1}/hold 1 min.
- Method C – Column: Varian Chrompack Chiracel Dex-CB column (25 m, 320 μm, 0.25 μm). GC programme parameters: injector 200°C, constant flow 1.7 mL; temperature programme: 60 °C/hold 2 min, 100 °C/rate 5 °C.min^{-1}/hold 2 min, 180 °C/rate 10 °C.min^{-1}/hold 1 min (see Table 11.21).

Table 11.21 *GC retention times and analysis methods for derivatised amine enantiomers.*

(*S*)-amine	Retention time (min)	(*R*)-amine	Retention time (min)	GC-FID method
(*S*)-**1c**	39.99	(*R*)-**1c**	40.18	A
(*S*)-**2c**	48.40	(*R*)-**2c**	48.72	A
(*S*)-**3c**	42.04	(*R*)-**3c**	42.25	A
(*S*)-**4c**	44.24	(*R*)-**4c**	44.37	A
(*S*)-**5c**	46.60	(*R*)-**5c**	46.89	A
(*S*)-**6c**	41.17	(*R*)-**6c**	41.36	A
(*S*)-**7c**	43.11	(*R*)-**7c**	43.50	A
(*S*)-**8c**	42.90	(*R*)-**8c**	43.47	A
(*S*)-**9c**	34.13	(*R*)-**9c**	36.08	A
(*S*)-**10c**	28.45	(*R*)-**10c**	29.26	A
(*S*)-**11c**	35.70	(*R*)-**11c**	37.89	A
(*S*)-**12c**	37.44	(*R*)-**12c**	38.67	A
(*S*)-**13c**	39.40	(*R*)-**13c**	40.13	A
(*S*)-**14c**	40.40	(*R*)-**14c**	40.82	A
(*S*)-**15c**	25.60	(*R*)-**15c**	26.48	B
(*S*)-**16c**	17.96	(*R*)-**16c**	18.96	B
(*S*)-**17c**	15.79	(*R*)-**17c**	16.10	B
(*S*)-**18c**	14.11	(*R*)-**18c**	14.71	C
(*S*)-**19c**	15.01	(*R*)-**19c**	15.17	C

11.8.7 Conclusion

AmDH-ADH (Table 11.22) tandem hydrogen-borrowing amination of alcohols has proven an efficient system for the preparation of *R*-configured α-chiral primary amines starting from either enantiomerically pure alcohols or racemic mixtures. In the latter case, a combination of two enantiocomplementary ADHs is required to transform *rac*-**1–19a** into the corresponding amine (3rd column of Table 11.22). The applicability of this system has been demonstrated using co-immobilised dehydrogenases (50 $mg_{enzyme}.g_{beads}^{-1}$) on controlled-porosity glass (CPG)-Fe^{III} ion-affinity beads (EziG)™ by using 8.7 nmol of ADH and 23 nmol of AmDH with 20 mM of substrates **1a**, **2a**, **8a**, **16a** and **17a** (4th column). Finally, this system was implemented in resting *E.coli* cells for the amination of *rac*-**1a**, **15a–18a** (5th column). Overall, the biocatalytic asymmetric hydrogen-borrowing amination of alcohols is enabled by the exquisite stereoselectivity and inherent compatibility of ADHs and AmDHs. The reaction displays the highest possible atom efficiency, as it requires only catalytic amounts of NAD^+, consumes ammonia and generates water as the sole byproduct. This method was recently extended to the regio- and stereoselective conversion of vicinal diols into optically active 1,2-amino alcohols on preparative scale [10]. Enzyme sources are given in Table 11.23.

Table 11.22 *Biocatalytic synthesis of α-chiral amines from alcohols using the hydrogen borrowing cascade.*

Compound	Structure	Free enzymes [2] Conv (%)[a]/ee (%)	Immob. enzymes [8] Conv (%)[b]/ee (%)	Resting cells [9] Conv %[c]/ee (%)
1a	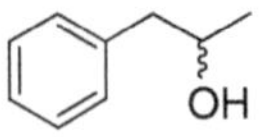	81, >99(*R*) AA-ADH+ LBv-ADH Bb-PhAmDH	96, >99(*R*) AA-ADH cFL1-AmDH	46 ± 14, >99(*R*) AA-ADH+ LBv-ADH cFL1-AmDH
2a	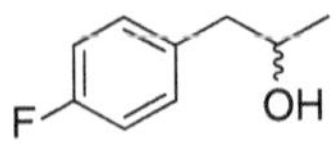	66, >99(*R*) AA-ADH+ LBv-ADH Bb-PhAmDH	95, >99(*R*) AA-ADH cFL1-AmDH	
3a	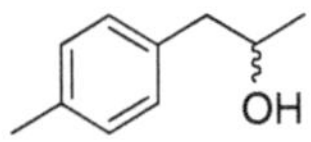	47, 97(*R*) AA-ADH+ LBv-ADH Bb-PhAmDH		
4a	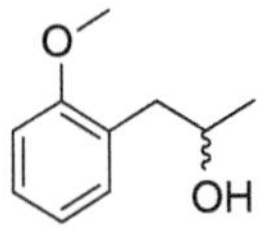	78, >99(*R*) AA-ADH+ LBv-ADH Bb-PhAmDH		
5a	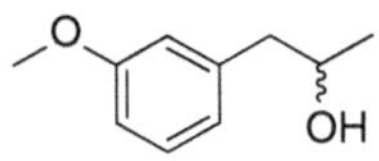	30, 82(*R*) AA-ADH+ LBv-ADH Bb-PhAmDH		
6a	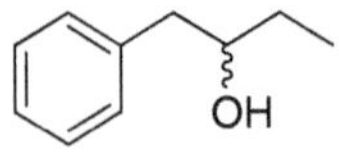	87, >99(*R*) AA-ADH+ LBv-ADH cFL1-AmDH		
7a	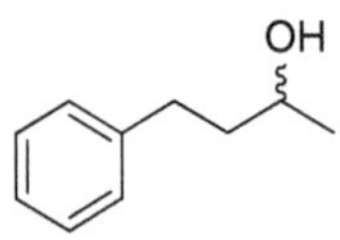	92, 83(*R*) AA-ADH+ LBv-ADH cFL1-AmDH		
8a	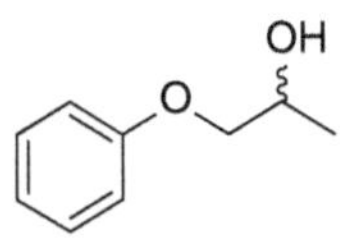	84, >99(*R*) AA-ADH+ LBv-ADH cFL1-AmDH	28, >99(*R*) AA-ADH cFL1-AmDH	
9a	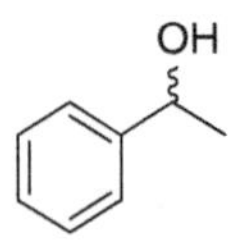	16, >99(*R*) AA-ADH+ LBv-ADH cFL1-AmDH		
10a	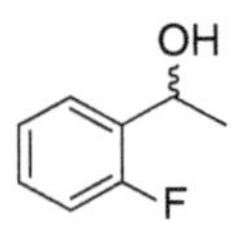	16, >99(*R*) AA-ADH+ LBv-ADH cFL1-AmDH		
11a	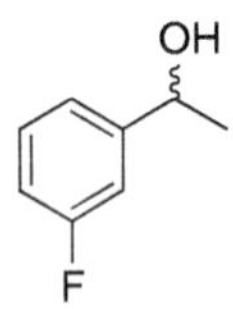	20, >99(*R*) AA-ADH+ LBv-ADH cFL1-AmDH		

(continued overleaf)

Table 11.22 *(continued)*

Compound	Structure	Free enzymes [2] Conv (%)[a]/ee (%)	Immob. enzymes [8] Conv (%)[b]/ee (%)	Resting cells [9] Conv %[c]/ee (%)
12a	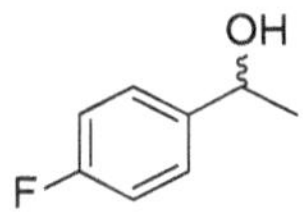	12, >99(*R*) AA-ADH+ LBv-ADH cFL1-AmDH		
13a	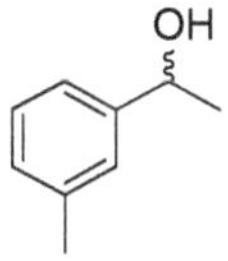	19, >99(*R*) AA-ADH+ LBv-ADH cFL1-AmDH		
14a	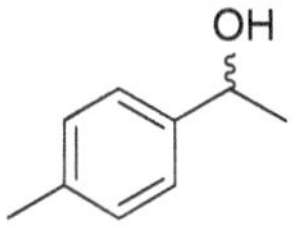	9, >99(*R*) AA-ADH+ LBv-ADH cFL1-AmDH		
15a	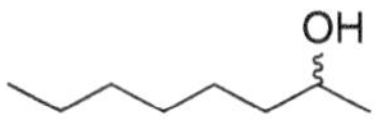	93, 99(*R*) AA-ADH+ LBv-ADH cFL1-AmDH		33 ± 5, >99(*R*) AA-ADH+ LBv-ADH cFL1-AmDH
16a	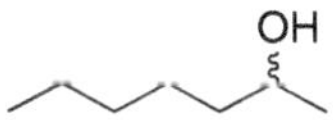	96, 99(*R*) AA-ADH+ LBv-ADH cFL1-AmDH	95, >99(*R*) AA-ADH cFL1-AmDH	81 ± 4, >99(*R*) AA-ADH+ LBv-ADH cFL1-AmDH
17a	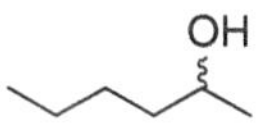	95, >99(*R*) AA-ADH+ LBv-ADH cFL1-AmDH	82, >99(*R*) AA-ADH cFL1-AmDH	82 ± 2, >99(*R*) AA-ADH+ LBv-ADH cFL1-AmDH
18a	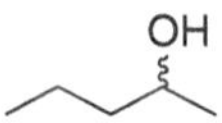	66, >99(*R*) AA-ADH+ LBv-ADH cFL1-AmDH		12 ± 4, >99(*R*) AA-ADH+ LBv-ADH cFL1-AmDH
19a	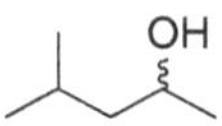	80, >99(*R*) AA-ADH+ LBv-ADH cFL1-AmDH		

[a]Conversions for the amination of *rac*-**1–19a** (20 mM) using AA-ADH (1.0–2.3 mg.mL^{-1}), LBv-ADH (1.0–2.6 mg.mL^{-1}) and AmDH (20 U) in ammonium formate buffer (pH 8.5, 2 M) supplemented with NAD$^+$ (1 mM) at 30 °C, 190 rpm for 48 hr; reaction volume 500 μL.

[b]Conversions for the amination of (*S*)-**1,2,8,16,17a** (20 mM) using AA-ADH (8.7 nmol) and AmDH (23 nmol) in ammonium chloride buffer (pH 8.7, 2 M) supplemented with NAD$^+$ (1 mM) at 30 °C, 700 rpm for 48 hr (Eppendorf thermomixer), reaction volume 500 μL.

[c]Conversions for the amination of *rac*-1,15-18a (5–20 mM) using resting *E. coli*[AA-ADH/cFL1-AmDH + LBv-ADH] (60 mg_{cww}.mL^{-1}) in ammonium chloride buffer (pH 8.7, 2 M) supplemented with glucose (1 eq. to the substrate) at 30 °C, 170 rpm for 24 hr; reaction volume 1 mL

Table 11.23 Enzyme sources.

Entry	Enzyme	Recombinant originated from	Ref.
1	AA-ADH	*Aromatoleum aromaticum* (wild-type)	[3]
2	LBv-ADH	*Lactobacillus brevis* variant	[4]
3	Bb-PhAmDH	*Bacillus badius* (K78S-N277L variant)	[5a]
4	cFL1-AmDH	Chimeric enzymes	[6]

References

1. (a) Knaus, T., Mutti, F.G., Humphreys, L.D. *et al.* (2015) *Organic and Biomolecular Chemistry*, **13**, 223–233; (b) Schrittwieser, J.H., Sattler, J., Resch, V. *et al.* (2011) *Current Opinion in Chemical Biology*, **15**, 249–256.
2. Mutti, F.G., Knaus, T., Scrutton, N.S. *et al.* (2015) *Science*, **349**, 1525–1529.
3. Hoffken, H.W., Duong, M., Friedrich, T. *et al.* (2006) *Biochemistry*, **45**, 82–93.
4. (a) Schlieben, N.H., Niefind, K., Muller, J. *et al.* (2005) *Journal of Molecular Biology*, **349**, 801–813; (b) Hummel, W. and Riebel, B. (1999) Patent Application WO 9947684.
5. (a) Abrahamson, M.J., Wong, J.W. and Bommarius, A.S. (2013) *Advanced Synthesis & Catalysis*, **355**, 1780–1786; (b) Knaus, T., Böhmer, W. and Mutti, F.G. (2017) *Green Chemistry*, **19**, 453–463.
6. Bommarius, B.R., Schurmann, M. and Bommarius, A.S. (2014) *ChemComm*, **50**, 14953–14955.
7. Novagen® (2011) User Protocol TB188 Rev. C 0311JN.
8. Böhmer, W., Knaus, T. and Mutti, F.G. (2018) *ChemCatChem*, **10**, 731–735.
9. Houwman, J.A., Knaus, T., Costa, M. and Mutti, F.G. (2019) *Green Chemistry*, **21**, 3846–3857.
10. Corrado, M.L., Knaus, T. and Mutti, F.G. (2019) *Green Chemistry*, **21**, 6246–6251.

11.9 Ene-Reductase-Mediated Reduction of C=C Double Bonds in the Presence of Conjugated C≡C Triple Bonds: Synthesis of (*S*)-2-Methyl-5-Phenylpent-4-yn-1-ol

Danilo Colombo,[1] Elisabetta Brenna,*[1,2] Francesco G. Gatti,[1] Maria Chiara Ghezzi,[1] Daniela Monti,[2] Fabio Parmeggiani[1] and Francesca Tentori[1]

[1]*Department of Chemistry, Materials and Chemical Engineering 'Giulio Natta', Politecnico di Milano, Milan, Italy*

[2]*Istituto di Chimica del Riconoscimento Molecolare, CNR, Milan, Italy*

Classical chemical methods are rarely chemoselective towards the hydrogenation of alkenes in the presence of alkynes. As for conjugated enynes, only a few examples are reported in the literature [1]. By contrast, the enzymatic version of the reduction, catalysed by ene-reductases (ERs), occurs with complete chemoselectivity thanks to its specific mechanism. Typical substrates of this reaction are functionalised alkenes bearing on the C=C bond at least one electron-withdrawing group (EWG) capable of establishing hydrogen bonds with specific amino acid residues in the active site of the enzyme. To date, only one example of ER-mediated reduction involving a similarly activated C≡C triple bond has been reported: the conversion of 4-phenyl-3-butyn-2-one into (*E*)-benzalacetone and then to 3-phenylbutan-2-one [2]. In ER-catalysed reactions, stereospecific *anti*-hydrogen

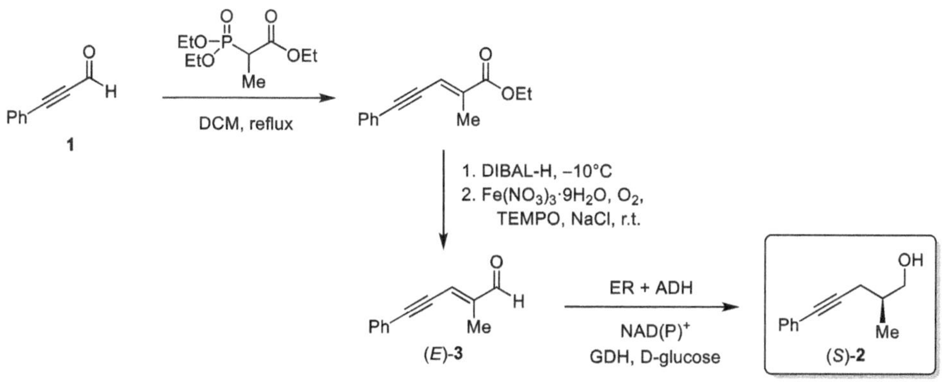

Scheme 11.12 Homologation of phenylpropiolaldehyde **1** into phenylpentynol (*S*)-**2**.

addition occurs exclusively at the multiple bond linked directly to the activating EWG, avoiding the reduction of other multiple bonds within the molecule. Thus, biocatalysed hydrogenations offer synthetic chemists the possibility to develop an effective generally applicable route for the reduction of alkenes in the presence of alkynes, enriching the toolbox of synthetic strategies that can be adopted and the variety of molecular skeletons that can be created.

With this aim, we described [3] the chemo-enzymatic homologation of phenylpropiolaldehyde **1** to afford phenylpentynol (*S*)-**2**, through phenylpentenynal (*E*)-**3** as the key intermediate (Scheme 11.12). A standard protocol was employed for the synthesis of **3** from **1**, consisting in a Horner–Wadsworth–Emmons (HWE) reaction followed by DIBAL-H reduction and final allylic oxidation. The key step of the whole procedure was an enzymatic sequential cascade obtained by combining the enantioselective reduction of the C=C double bond of derivative **3**, mediated by Old Yellow Enzyme 3 (OYE3), with the reduction of the carbonyl group into primary alcohol, catalysed by a commercial alcohol dehydrogenase EVO200. The two reactions required NADPH and NADH, respectively, as co-factors. Both were effectively regenerated by glucose dehydrogenase (GDH) from *Bacillus megaterium*, at the expense of glucose as sacrificial substrate, employing only catalytic amounts of $NAD(P)^+$. The final compound, (*S*)-2-methyl-5-phenylpent-4-yn-1-ol (*S*)-**2**, was obtained in 44% isolated yield (from **1**) and 95% ee.

11.9.1 Procedure 1: Production of OYE3 in *E. coli* BL21(DE3) and Purification of the Enzyme

11.9.1.1 Materials and Equipment

- Lysogenic broth (LB) agar plate with colonies of *E. coli* BL21(DE3) harbouring the plasmid pET30a-OYE3 (*Saccharomyces cerevisiae* OYE3 UniProtKB accession number P41816) [4]
- LB medium
- Distilled water (dH_2O)
- Kanamycin ($50\,mg.mL^{-1}$ stock in dH_2O, filter-sterilised)
- Potassium phosphate (KP_i) buffer (100 mM, pH 8.0)

- 500 mL Erlenmeyer flask (baffled)
- Inoculating loops
- 3 L Erlenmeyer flask (baffled) with foam bung
- Centrifuge tubes
- Shaking incubator (Innova 44, New Brunswick Scientific)
- Isopropyl-β-D-1-thiogalactopyranoside (IPTG; 1 M stock in dH_2O, filter-sterilised)
- Cooling centrifuge (min. 25 000× g)
- Ultrasonic homogeniser (Omni Ruptor 250, Omni International)
- Lysis buffer (20 mM potassium phosphate buffer pH 7.0, 300 mM NaCl, 10 mM imidazole)
- Elution buffer (20 mM potassium phosphate buffer pH 7.0, 300 mM NaCl, 500 mM imidazole)
- Glass column with sintered glass frit (internal diameter 1 cm, length 15 cm)
- Resin for IMAC (Ni-Sepharose Fast Flow, GE Healthcare)
- Peristaltic pump (Minipuls 3, Gilson)

11.9.1.2 Procedure

1. A single colony of *E. coli* BL21(DE3) carrying the pET-30a-OYE3 plasmid was used to inoculate LB medium (100 mL) supplemented with kanamycin (final conc. 100 μg.mL^{-1}).
2. The culture was grown at 37 °C and 220 rpm overnight, then used to inoculate LB medium (750 mL) containing kanamycin (final conc. 100 μg.mL^{-1}). This culture was incubated at 37 °C and 220 rpm until an OD_{600} of 0.5–0.6 was reached, then IPTG was added (final conc. 0.1 mM) and it was incubated at 30 °C and 200 rpm for 4 hr.
3. Cells were harvested by centrifugation at 5000× g and 4 °C for 30 min and the pellet (~8.5 g wet weight) was resuspended in 50 mL lysis buffer.
4. The cell suspension was sonicated on ice (20 sec on, 40 sec off, 10 cycles).
5. The mixture was centrifuged at 20 000× g and 4 °C for 30 min and the supernatant was collected.
6. The raw lysate was incubated with Ni-Sepharose resin (~5 mL) in ice with shaking for 2 hr and then transferred to a glass column. The protein was eluted with a mobile phase composed of 20 mM potassium phosphate buffer pH 7.0, 300 mM NaCl and 10–300 mM imidazole gradient, obtained by the combination of lysis buffer and elution buffers with different ratios.
7. All fractions containing OYE3 were collected and the protein concentration was estimated via Bradford assay [5], with an average concentration of 3 mg.mL^{-1} (total volume 20 mL). The enzyme solution was quickly stored at –80 °C. The activity of the purified enzyme was estimated at 0.5 U.mL^{-1} by spectrophotometric assay using 2-cyclohexen-1-one as standard substrate.

11.9.2 Procedure 2: Production of GDH from *Bacillus megaterium* DSM509 in *E. coli* BL21(DE3) and Purification of the Enzyme

11.9.2.1 Materials and Equipment

- LB agar plate with colonies of *E. coli* BL21(DE3) harbouring the plasmid pKTS-GDH (*Bacillus megaterium* GDH UniProtKB accession number P40288) [4,6]

- LB medium
- dH_2O
- Ampicillin (50 mg.mL^{-1} stock in dH_2O, filter-sterilised)
- KP_i buffer (100 mM, pH 8.0)
- 500 mL Erlenmeyer flask (baffled)
- Inoculating loops
- 3 L Erlenmeyer flask (baffled) with foam bung
- Centrifuge tubes
- Shaking incubator (Innova 44, New Brunswick Scientific)
- IPTG (1 M stock in dH_2O, filter-sterilised)
- Anhydrotetracycline (200 µg.mL^{-1} stock in dH_2O, filter-sterilised)
- Cooling centrifuge (min. 25 000× g)
- Ultrasonic homogeniser (Omni Ruptor 250, Omni International)
- Lysis buffer (20 mM potassium phosphate buffer pH 7.0, 300 mM NaCl, 10 mM imidazole)
- Elution buffer (20 mM potassium phosphate buffer pH 7.0, 300 mM NaCl, 500 mM imidazole)
- Glass column with sintered glass frit (internal diameter 1 cm, length 15 cm)
- Resin for IMAC (Ni-Sepharose Fast Flow, GE Healthcare)
- Peristaltic pump (Minipuls 3, Gilson)

11.9.2.2 Procedure

1. A single colony of *E. coli* BL21(DE3) carrying the pKTS-GDH plasmid was used to inoculate LB medium (100 mL) supplemented with ampicillin (final conc. 100 µg.mL^{-1}).
2. The culture was grown at 37 °C and 220 rpm overnight, then used to inoculate LB medium (750 mL) containing ampicillin (final conc. 100 µg.mL^{-1}). This culture was incubated at 37 °C and 220 rpm until an OD_{600} of 0.5–0.6 was reached, at which point IPTG (final conc. 1 mM) and anhydrotetracycline (final conc. 50 ng.mL^{-1}) were added and it was incubated at 30 °C and 200 rpm for 5 hr.
3. Cells were harvested by centrifugation at 2000× g and 4 °C for 20 min and the pellet (~8.0 g wet weight) was resuspended in 50 mL Lysis buffer.
4. The cell suspension was sonicated on ice (20 sec on, 40 sec off, 10 cycles).
5. The mixture was centrifuged at 20 000× g and 4 °C for 30 min and the supernatant was collected.
6. The raw lysate was incubated with Ni-Sepharose resin (~5 mL) in ice with shaking for 2 hr and then transferred to a glass column. The protein was eluted with a mobile phase composed of 20 mM potassium phosphate buffer pH 7.0, 300 mM NaCl and 10–300 mM imidazole gradient, obtained by the combination of lysis buffer and elution buffers with different ratios
7. All fractions containing GDH were collected and the protein concentration was estimated via Bradford assay [5], with an average concentration of 2 mg.mL^{-1} (total volume 15 mL). The enzyme solution was aliquoted and stored at –80 °C. Activity of the purified enzyme was estimated at 90 U.mL^{-1} by spectrophotometric assay using glucose as standard substrate.

11.9.3 Procedure 3: Biocatalytic Conversion of (*E*)-3 into (*S*)-2 with OYE3 and EVO200

11.9.3.1 Materials and Equipment

- Compound (*E*)-**3** (50 mg, 0.29 mmol)
- KP_i buffer (50 mM, pH 7.0)
- OYE3 prepared according to Procedure 1
- GDH prepared according to Procedure 2
- Isopropanol
- Ethyl acetate
- ADH EVO200 (from Evoxx)
- D-glucose
- Nicotinamide adenine dinucleotide phosphate ($NADP^+$)
- Nicotinamide adenine dinucleotide (NAD^+)
- Sodium sulfate
- 20 mL glass vial
- Shaking incubator (SI500, Stuart)
- Rotary evaporator (Büchi R200)
- Gas chromatography/mass spectrometry (GC-MS) system (Agilent) and suitable column (see Analytical Method)
- High-performance liquid chromatography (HPLC) system equipped with UV detector (Agilent 1200 Series) and suitable eluents/columns (see Analytical Method)

11.9.3.2 Procedure

1. Compound (*E*)-**3** (50 mg, 0.29 mmol) was dissolved in *i*PrOH (200 µL) and added to a 20 mL glass vial containing D-glucose (219 mg, 4 eq), $NADP^+$ (10 µmol, 7.4 mg), NAD^+ (6.6 mg, 10 µmol) and ADH EVO200 (4 mg) in 50 mM potassium phosphate buffer, pH 7.0 (3 mL).
2. 1700 µL of the OYE3 solution from Procedure 1 (containing ~5 mg of enzyme) and 500 µL of the GDH solution from Procedure 2 (containing ~1 mg of enzyme) were added to the mixture.
3. The solution was incubated vertically in an orbital shaker at 30 °C and 150 rpm. The reaction was followed by GC-MS analysis, reaching complete conversion in about 24 hr.
4. The mixture was extracted with EtOAc (3 × 3 mL). The organic phases were then combined, dried over anhydrous Na_2SO_4 and concentrated by distillation under reduced pressure using a rotary evaporator.
5. The desired compound, (*S*)-2-methyl-5-phenylpent-4-yn-1-ol ((*S*)-**2**), was obtained in 34.8 mg (69% yield, 95% ee).

11.9.4 Analytical Method

Conversion of (*E*)-**3** into (*S*)-**2** was monitored by GC-MS, using an HP-5MS column (30 m × 0.25 mm × 0.25 µm), according to the following method:

- Temperature programme: 60 °C (1 min)/6 °C.min^{-1}/150 °C (1 min)/12 °C.min^{-1}/280 °C (5 min)

- Total analysis time: 33 min
- Retention times: (*E*)-**3**, 19.44 min; (*S*)-**2**, 19.76 min

The enantiomeric excess values of compound (*S*)-**2** was determined by HPLC analysis on a Chiralcel OD column (4.6 mm × 250 mm, Daicel), using the following conditions:

- Flow rate: 0.6 mL.min^{-1}
- Detection wavelength: 215 nm
- Mobile phase: *n*-hexane/*i*-PrOH 95 : 5
- Retention times: (*R*)-**2**, 16.9 min; (*S*)-**2**, 20.7 min

11.9.5 Conclusion

The same synthetic procedure [3] was employed to convert 2-octynal **4** into (*S*)-2-methyldec-4-yn-1-ol (*S*)-**5** in 38% yield (from **4**) and 90% ee and to prepare achiral alcohols **6** and **7** from aldehydes **1** and **4**, respectively, in 34 and 35% isolated yield. Furthermore, the single-step OYE-mediated reduction applied to alkenynal intermediates (*E*)-**8** and (*E*)-**9** afforded 4-alkynals **10** and **11**, easily available from **1** and **4** in 37 and 35% isolated yield (Scheme 11.13).

This work shows the application of a novel chemo-enzymatic sequence involving a classical C=C bond formation reaction followed by an enzyme-mediated chemoselective hydrogenation in order to achieve the addition of a CH_2CH_2OH/CH_2CHO moiety to an alkynal. The target molecules are key intermediates of natural products [7], pheromones [8], cosmetics [9] and pharmaceuticals [10]. This approach avoids some of the troublesome and undesirable steps involved in the known routes to these derivatives: the use of metal-based reagents and catalysts, the hazards involved with the formation and use of alkynyllithium salts and the need for protection/deprotection steps.

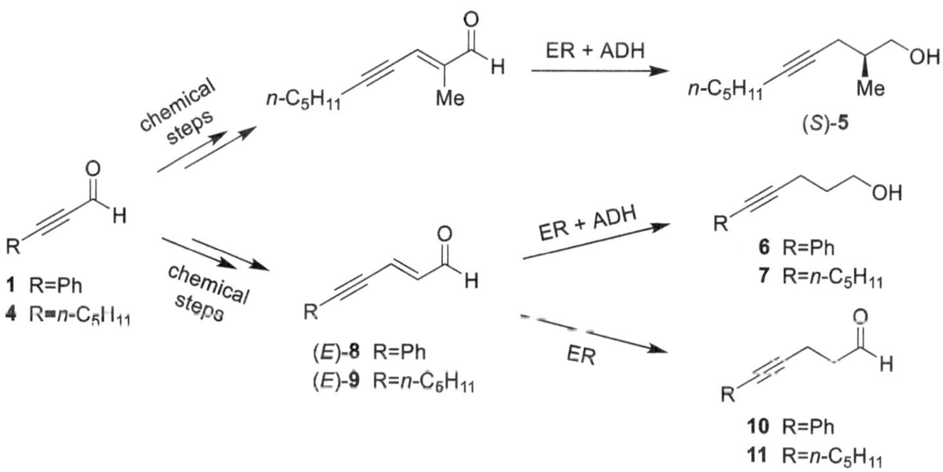

Scheme 11.13 Homologation of aldehydes **1** and **4** into alcohol derivatives (*S*)-**5**, **6** and **7** and into aldehydes **10** and **11**. Chemical steps: (i) HWE reaction; (ii) DIBAL-H reduction; (iii) allylic oxidation.

References

1. (a) Liu, T.-L., Wang, C.-J. and Zhang, X. (2013) *Angewandte Chemie International Edition*, **52**, 8416–8419; (b) Warren, S.A., Fox, M.E., Meek, G.A. *et al.* (2017) Patent WO2017/064627; (c) Trost, B.M., Taft, B.R., Masters, J.T. and Lumb, J.-P. (2011) *Journal of the American Chemical Society*, **133**, 8502–8505; (d) Trost, B.M., Masters, J.T., Taft, B.R. and Lumb, J.-P. (2016) *Chemical Science*, **7**, 6217–6231.
2. Müller, A., Stürmer, R., Hauer, B. and Rosche, B. (2007) *Angewandte Chemie International Edition*, **46**, 3316–3318.
3. Colombo, D., Brenna, E., Gatti, F.G. *et al.* (2019) *Advanced Synthesis & Catalysis*, **361**, 2638–2648.
4. Bechtold, M., Brenna, E., Femmer, C. *et al.* (2012) *Organic Process Research & Development*, **16**, 269–276.
5. Bradford, M.M. (1976) *Analytical Biochemistry*, **72**, 248–254.
6. Neuenschwander, M., Butz, M., Heintz, C. *et al.* (2007) *Nature Biotechnology*, **10**, 1145–1147.
7. (a) Li, X. and Ovaska, T.V. (2017) *Organic Letters*, **9**, 3837–3840; (b) Adrian, J. and Stark, C.B.W. (2016) *Journal of Organic Chemistry*, **81**, 8175–8186; (c) Lerchen, A., Knecht, T., Koy, M. *et al.* (2017) *Chemistry: A European Journal*, **23**, 12 149–12 152; (d) Vyvyan, J.R., Engles, C.A., Bray, S.L. *et al.* (2017) *Beilstein Journal of Organic Chemistry*, **13**, 2122–2127; (e) Akkachairin, B., Tummatorn, J., Supantanapong, N. *et al.* (2017) *Journal of Organic Chemistry*, **82**, 3727–3740; (f) Kikuchi, H., Ito, I., Takahashi, K. *et al.* (2017) *Journal of Natural Products*, **80**, 2716–2722; (g) Clark, R.C., Sang, Y.L. and Boger, D.L. (2008) *Journal of the American Chemical Society*, **130**, 12 355–12 369.
8. (a) Mori, C. and Yang, Y. (2016) *Tetrahedron*, **72**, 4593–4607; (b) Gries, G., Clearwater, J., Gries, R. *et al.* (1999) *Journal of Chemical Ecology*, **25**, 1091–1104; (c) Colby, E.A., O'Brien, K.C. and Jamison, T.F. (2005) *Journal of the American Chemical Society*, **127**, 4297–4307.
9. Dalko, M., Marat, X., Hitce, J. and Li, C.-J. (2018) US Patent US2018/57440.
10. (a) Carballeira, N.M., Montano, N., Reguera, R.M. and Balana-Fouce, R. (2010) *Tetrahedron Letters*, **51**, 6153–6155; (b) Takahashi, A. and Shibasaki, M. (1998) *Journal of Organic Chemistry*, **53**, 1227–1231; (c) Huffman, J.W., Liddle, J., Duncan, S.G. Jr. *et al.* (1998) *Bioorganic & Medicinal Chemistry*, **6**, 2383–2396; (d) Nielsen, M., Jacobsen, C.B., Paixão, M.W. *et al.* (2009) *Journal of the American Chemical Society*, **131**, 10 581–10 586.

12

Chemo-Enzymatic Cascades

12.1 Synergistic Nitroreductase/Vanadium Catalysis for Chemoselective Nitroreductions

Iustina Slabu,[1] Serena Bisagni,[1] Simon J. Hedley,[2] Alan H. Cherney,[2] Ahir Pushpanath,[1] Amin Bornadel,[1] Jacques LePaih,[1] Steven M. Mennen[2] and Beatriz Dominguez[1]

[1]*Johnson Matthey, Cambridge, UK*

[2]*Drug Substance Technologies, Amgen Inc., Cambridge, MA, USA*

Anilines are present in a number of active pharmaceutical ingredients (APIs) and agrochemicals [1] and are generally produced by reduction of the corresponding nitroarenes. Popular reduction procedures include catalytic methods involving metals such as palladium or platinum [1, 2] and more traditional methods using stoichiometric reducing agents [1]. A recently developed approach to this transformation exploits nitroreductases (NRs), an enzyme class able to reduce a broad range of nitroaromatic compounds [3], which is used in nature for detoxification [4], oxidative stress regulation [4] and bioluminescent processes [5] in living organisms. NRs have previously received limited attention in organic synthesis because the reduction of nitroaromatics in the presence of NRs is limited to the generation of hydroxylamines [1, 6]; further progress to the desired aniline is rare and results from spontaneous but slow disproportionation [7]. To promote complete reduction, inspiration was taken from chemocatalytic approaches that rely on metal salts to catalyse the disproportionation of the hydroxylamine intermediate to amine and nitroso compounds [8]. Gratifyingly, combining NRs with a first-row transition metal salt delivers full, chemoselective conversion of nitro compounds into the corresponding amines under mild conditions [9].

Herein, we showcase the effect of metal additives on the biocatalysed nitroreduction, the scope of this transformation and the course of reaction intensification.

Applied Biocatalysis: The Chemist's Enzyme Toolbox, First Edition. Edited by John Whittall and Peter W. Sutton.
© 2021 John Wiley & Sons Ltd. Published 2021 by John Wiley & Sons Ltd.

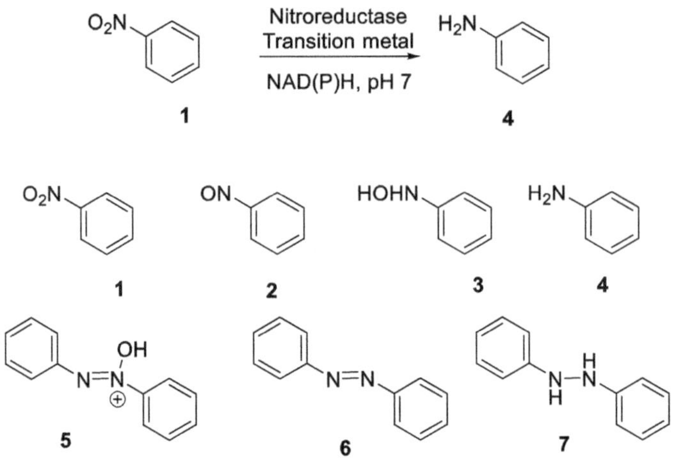

Scheme 12.1 *General reaction scheme for the reduction of nitroaromatics **1** to anilines **4**, showing intermediates (**2** and **3**) and byproducts (**5**, **6** and **7**).*

12.1.1 Procedure 1: Screening of Different Metal Additives with Nitroreductase NR-14 for the Reduction of 2-Methyl-5-Nitropyridine

12.1.1.1 Materials and Equipment

- Potassium phosphate buffer 250 mM pH 7.0 (for 500 mL, 11.7 g K_2HPO_4 and 7.9 g KH_2PO_4 dissolved in milliQ water)
- Nitroreductase NR-14 (2.5 mg; Johnson Matthey)
- Glucose dehydrogenase GDH-101 (1.0 mg; Johnson Matthey)
- $NADP^+$ anhydrous (7.4 mg, 0.01 mmol)
- D-Glucose (144 mg, 0.8 mmol)
- 250 mM Potassium phosphate buffer, pH 7.0 (9.5 mL)
- 2-Methyl-5-nitropyridine **1a** (55.3 mg, 0.4 mmol)
- Toluene (1 mL)
- V_2O_5 (3.6 mg, 0.02 mmol)
- $CuCl_2$ (2.7 mg, 0.02 mmol)
- $FeCl_2$ (2.5 mg, 0.02 mmol)
- $NaMoO_4$ (4.1 mg, 0.02 mmol)
- $CoCl_2$ (2.6 mg, 0.02 mmol)
- Toluene (1 mL)
- Water, high-performance liquid chromatography (HPLC)-grade
- Acetonitrile, HPLC-grade
- Trifluoroacetic acid, HPLC-grade
- Balance
- Thermoshaker
- Vortex
- Tabletop centrifuge
- 2 mL Eppendorf tubes

- Reversed-phase HPLC equipped with Zorbax SB-Phenyl column 5 μm, 4.6 × 250 mm (Agilent)

12.1.1.2 Procedure

1. Solution 1 was prepared by dissolving NR-14 (25 mg), GDH-101 (10 mg), D-glucose (144 mg, 0.8 mmol) and $NADP^+$ (7.4 mg, 0.01 mmol) in 250 mM potassium phosphate buffer, pH 7 (5 mL).
2. Solutions 2 were prepared by individually dissolving either V_2O_5 (3.6 mg, 0.02 mmol), $CuCl_2$ (2.7 mg, 0.02 mmol), $FeCl_2$ (2.5 mg, 0.02 mmol), $NaMoO_4$ (4.1 mg, 0.02 mmol) or $CoCl_2$ (2.6 mg, 0.02 mmol) in 250 mM potassium phosphate buffer, pH 7 (4.5 mL). Where there was no metal additive, Solution 2 was 250 mM potassium phosphate buffer, pH 7.
3. Solution 3 was prepared as a 0.4 M stock solution of 2-methyl-5-nitropyridine **1a** (55.3 mg, 0.4 mmol) in toluene (1 mL).
4. Solution 1 (500 μL) was added to an Eppendorf tube, followed by Solution 2 (450 μL) and then Solution 3 (50 μL), to give a final reaction volume of 1 mL. The reactions were incubated in a thermoshaker at 35 °C with shaking for 18 hr.
5. For analysis, the reaction samples (1 mL) were treated with 1 mL acetonitrile, vortexed and centrifuged at 13 000 rpm for 2 min in a tabletop centrifuge. Aliquots (900 μL) were transferred to HPLC glass vials for analysis (Scheme 12.2, Table 12.1).

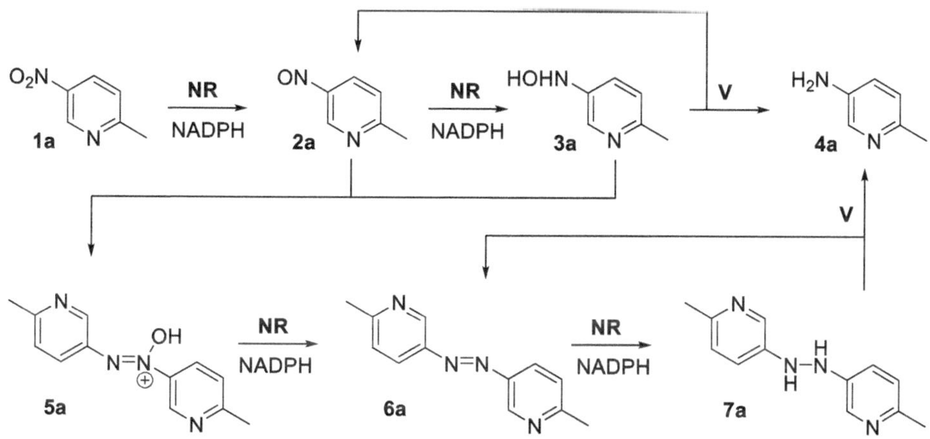

Scheme 12.2 *Reduction of 2-methyl-5-nitropyridine.*

12.1.2 Procedure 2: Substrate Scope of Four Nitroreductases in the Presence of Vanadium (V) Oxide

12.1.2.1 Materials and Equipment

- 250 mM potassium phosphate buffer, pH 7.0 (for 500 mL, 11.7 g K_2HPO_4 and 7.9 g KH_2PO_4 dissolved in milliQ water)
- Nitroreductases NR-04, NR-14, NR-17 and NR-24 (2.5 mg; Johnson Matthey)

Table 12.1 *Additive screening for reduction of NR-14.*

Entry	Substrate	Enzyme	Additive	**1** (%)	**3** (%)	**4** (%)	**5** (%)
1		NR-14	None	0	67	8	25
2	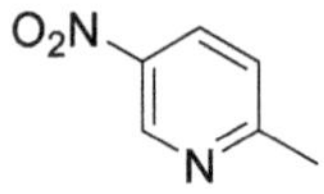	NR-14	V_2O_5	1	0	96	3
3		NR-14	$CuCl_2$	0	2	78	19
4		NR-14	$FeCl_2$	0	0	97	3
5	**1a**	NR-14	$NaMoO_4$	0	33	45	22
6		NR-14	$CoCl_2$	6	3	7	84

- Glucose dehydrogenase GDH-101 (1 mg; Johnson Matthey)
- $NADP^+$ anhydrous (7.4 mg, 0.01 mmol)
- D-Glucose (144 mg, 0.8 mmol)
- Substrate: 2-methyl-5-nitropyridine **1a**, nitrobenzene **1b**, 3-nitrostyrene **1c**, 1-chloro-2-nitrobenzene **1d**, 4-nitrobenzonitrile **1e** or 1-(benzoxy)-3-nitrobenzene **1f** (0.4 mmol)
- V_2O_5 (3.6 mg, 0.02 mmol)
- Toluene
- Water, HPLC-grade
- Acetonitrile, HPLC-grade
- Trifluoroacetic acid, HPLC-grade
- Balance
- Thermoshaker
- Vortex
- Tabletop centrifuge
- 2 mL Eppendorf tubes
- Reversed-phase HPLC equipped with Zorbax SB-Phenyl column 5 μm, 4.6 × 250 mm (Agilent)

12.1.2.2 Procedure

1. Solution 1 was prepared by dissolving NR (25 mg), GDH-101 (10 mg), D-glucose (144 mg, 0.8 mmol) and $NADP^+$ (7.4 mg, 0.01 mmol) in 250 mM potassium phosphate buffer, pH 7 (5 mL).
2. Solution 2 was prepared by dissolving V_2O_5 (3.6 mg, 0.02 mmol) in 250 mM potassium phosphate buffer, pH 7 (4.5 mL). Where there was no metal additive, Solution 2 was 250 mM potassium phosphate buffer, pH 7.
3. Solution 3 was prepared as a 0.4 M stock solution of nitroaromatic substrate **1a** (55.3 mg, 0.4 mmol), **1b** (49.2 mg, 0.4 mmol), **1c** (59.7 mg, 0.4 mmol), **1d** (63.0 mg, 0.4 mmol), **1e** (59.2 mg, 0.4 mmol) or **1f** (91.7 mg, 0.4 mmol) in toluene (1 mL).
4. Solution 1 (500 μL) was added to an Eppendorf tube, followed by Solution 2 (450 μL) and then Solution 3 (50 μL), to give a final reaction volume of 1 mL. The reactions were incubated in a thermoshaker at 35 °C with shaking for 18 hr.
5. For analysis, the reaction samples (1 mL) were treated with 1 mL acetonitrile, vortexed and centrifuged at 13 000 rpm for 2 min in a tabletop centrifuge. Aliquots (900 μL) were transferred to HPLC glass vials for analysis (Table 12.2).

Table 12.2 *Substrate scope of NR-04, NR-14, NR-17 and NR-24.*

Entry	Substrate	Enzyme	Additive	Conversion (%)	**4** (%)
1	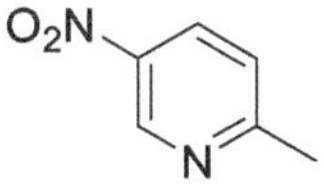	NR-04	None	>99	13
2		NR-04	V_2O_5	>99	97
3		NR-14	V_2O_5	99	96
4		NR-17	V_2O_5	99	95
5	**1a**	NR-24	V_2O_5	>99	96
6	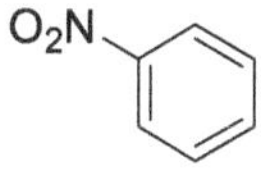	NR-04	None	97	17
7		NR-04	V_2O_5	>99	72
8		NR-14	V_2O_5	>99	64
9		NR-17	V_2O_5	>99	64
10	**1b**	NR-24	V_2O_5	>99	72
11	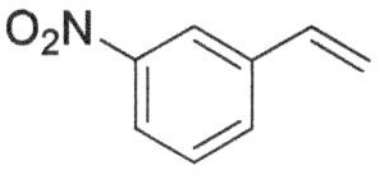	NR-04	None	>99	46
12		NR-04	V_2O_5	98	96
13		NR-14	V_2O_5	93	92
14		NR-17	V_2O_5	96	96
15	**1c**	NR-24	V_2O_5	97	97
16	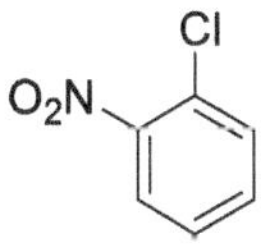	NR-04	None	>99	<1
17		NR-04	V_2O_5	99	99
18		NR-14	V_2O_5	>99	99
19		NR-17	V_2O_5	>99	99
20	**1d**	NR-24	V_2O_5	99	98
21	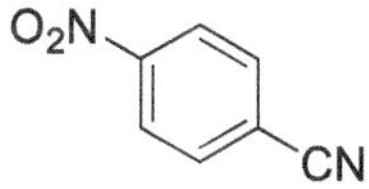	NR-04	None	>99	5
22		NR-04	V_2O_5	>99	89
23		NR-14	V_2O_5	97	94
24		NR-17	V_2O_5	97	94
25	**1e**	NR-24	V_2O_5	>99	86
26	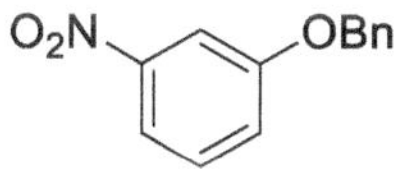	NR-04	None	98	6
27		NR-04	V_2O_5	96	90
28		NR-14	V_2O_5	71	67
29		NR-17	V_2O_5	97	93
30	**1f**	NR-24	V_2O_5	41	35

12.1.3 Procedure 3: Process Intensification of 2-Methyl-5-Nitropyridine Reduction

12.1.3.1 Materials and Equipment

- 250 mM potassium phosphate buffer, pH 7.0 (for 500 mL, 11.7 g K_2HPO_4 and 7.9 g KH_2PO_4 dissolved in milliQ water)
- Nitroreductase NR-14 (2.5 mg; Johnson Matthey)
- GDH-101 (1 mg; Johnson Matthey)
- $NADP^+$ anhydrous (18.6 mg, 0.025 mmol)
- D-Glucose (3.6 g, 20 mmol)
- 2-Methyl-5-nitropyridine **1a** (4 × 172.5 mg, 5 mmol)

- Metal additive: V_2O_5 (90.9 mg, 0.5 mmol)
- Toluene
- Water, HPLC-grade
- Acetonitrile, HPLC-grade
- Trifluoroacetic acid, HPLC-grade
- NaOH 10 M in water
- Balance
- Potentiometric titrator
- Three-neck 50 mL round-bottom flask
- Hotplate with thermocouple and oil bath
- Vortex
- Tabletop centrifuge
- Reversed-phase HPLC equipped with Zorbax SB-Phenyl column 5 μm, 4.6 × 250 mm (Agilent)

12.1.3.2 Procedure

1. D-Glucose (3.6 g, 20 mmol), $NADP^+$ (18.6 mg, 0.025 mmol), V_2O_5 (90.9 mg, 0.5 mmol), NR-14 (62.5 mg) and GDH-101 (25 mg) were added to a three-neck round-bottom flask with 250 mM potassium phosphate buffer, pH 7 to a final volume of 23.75 mL.
2. The pH probe of the titrator was inserted into the central neck of the round-bottom flask. A 10 M NaOH(aq) solution feed line was inserted into the lateral neck, in order to titrate pH changes during the reaction.
3. The reaction was heated at 35 °C with magnetic stirring and the titrator was activated to maintain the pH at 7 throughout the reaction.
4. 2-Methyl-5-nitropyridine **1a** (4 × 172.5 mg, 5 mmol) was weighed into four separate glass vials (690 mg total).
5. The reaction was started by adding the first aliquot of substrate **1a**. The remaining aliquots were added at intervals of 1 hr, completing the addition by the third hour. The reaction was incubated for a total of 6 hr.
6. 100 μL aliquots of the reaction were taken at intervals of 30 min and treated with 1 mL acetonitrile, vortexed and centrifuged at 13 000 rpm for 2 min in a tabletop centrifuge. Aliquots (900 μL) were transferred to HPLC glass vials for analysis (Figure 12.1).

12.1.4 Analytical Method

Conversion was measured by HPLC, determining the uncorrected relative peak area percentage. Separation was obtained using Zorbax SB-Phenyl 5 μm, 4.6 × 250 mm column at 25 °C, with detection at 230 nm and 1 $mL.min^{-1}$ flow. Mobile phase A (water with 0.1% TFA) and mobile phase B (acetonitrile with 0.1% TFA) were mixed in a gradient from 95 : 5 (A : B) to 5 : 95 (A : B) over 13 min with 2 min hold at the start and 5 min at the end.

12.1.5 Conclusion

Vanadium was demonstrated to be the most efficient metal co-catalyst for biocatalysed nitroreduction. Slower or partial formation of amine **4a** was obtained with Fe, Cu and

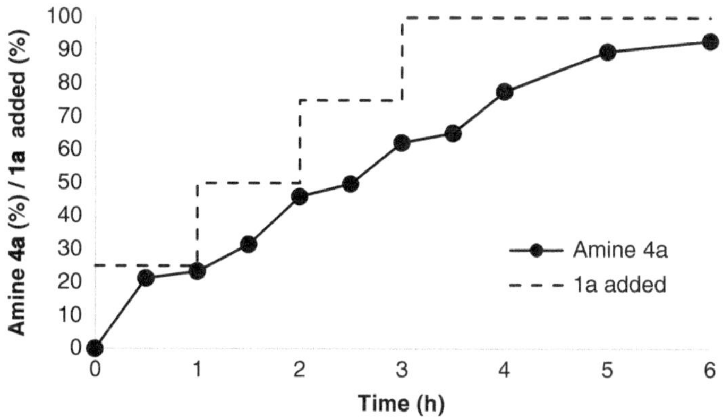

Figure 12.1 *Result of fed-batch nitroreduction of 2-methyl-5-nitropyridine.*

Mo, whilst full conversion to dimerisation product **5a** was obtained with Co additives (see Table 12.1).

NRs and V were tested for the nitroreduction of a variety of substrates with different substituents that are prone to overreduction or elimination under chemocatalysed reduction conditions. NRs showed an exquisite selectivity for the reduction of the nitro group (see Table 12.2), leaving the substituents intact, which makes the biocatalysed nitroreduction an interesting alternative to traditional nitroreduction methods when chemoselectivity is an issue.

Finally, the reaction was demonstrated to be scalable (Figure 12.1), and although further efforts are ongoing to optimise the process, biocatalysed nitroreduction has the potential to become a viable alternative to traditional reduction methods.

References

1. (a) Song, J., Huang, Z., Pan, L. *et al.* (2018) *Applied Catalysis B: Environmental*, **227**, 386–408; (b) Orlandi, M., Brenna, D., Harms, R. *et al.* (2018) *Organic Process Research & Development*, **22**, 430–445; (c) Kadam, H.K. and Tilve, S.G. (2015) *RSC Advances*, **5**, 83 391–83 407; (d) Hoogenraad, M., van der Linden, J., Smith, A. *et al.* (2004) *Organic Process Research & Development*, **8**, 469–476.
2. (a) Gallagher, W., Marlatt, M., Livingston, R. *et al.* (2012) *Organic Process Research & Development*, **16**, 1665–1668; (b) Corma, A. and Serna, P. (2006) *Science*, **313**, 332–334. (c) Kasparian, A., Savarin, C., Allgeier, A. and Walker, S. (2011) *Journal of Organic Chemistry*, **76**, 9841–9844; (d) Boymans, E.H., Witte, P.T. and Vogt, D. (2015) *Catalysis Science & Technology*, **5**, 176–183.
3. (a) Toogood, H. and Scrutton, N. (2018) *ACS Catalysis*, **8**, 3532–3549; (b) Williams, R. and Bruce, N. (2002) *Microbiology*, **148**, 1607–1614.
4. (a) Roldán, M., Pérez-Reinado, E., Castillo, F. and Moreno-Vivián, C. (2008) *FEMS Microbiology Reviews*, **32**, 474–500; (b) Marques de Oliveira, I., Bonatto, D. and Pegas Henriques, J. (2010) Nitroreductases: enzymes with environmental, biotechnological and clinical importance, in *Current Research, Technology and Education Topics in Applied Microbiology and Microbial Biotechnology* (ed. A. Méndez-Vilas), Formatex Research Center, pp. 1008–1019; (c) Xiao, Y.,

Wu, J., Liu, H. *et al.* (2006) *Applied Microbiology and Biotechnology*, **73**, 166–171; (d) Kwak, J., Lee, H. and Kim, D.H. (2003) *Applied and Environmental Microbiology*, **69**, 4390–4395.
5. Zenno, S., Koike, H., Kumar, A. *et al.* (1996) *Journal of Bacteriology*, **178**, 4508–4514.
6. (a) Yanto, Y., Hall, M. and Bommarius, A.S. (2010) *Organic and Biomolecular Chemistry*, **8**, 1826–1832; (b) Nadeau, L.J., He, Z. and Spain, J.C. (2000) *Journal of Industrial Microbiology and Biotechnology*, **24**, 301–305; (c) Pitsawong, W., Hoben, J. and Miller, A. (2014) *Journal of Biological Chemistry*, **289**, 15 203–15 214; (d) Miller, A., Park, J., Ferguson, K. *et al.* (2018) *Molecules*, **23**, 211–233.
7. (a) Haber, F. (1898) *Zeitschrift für Elektrochemie und Angewandte Physikalische Chemie*, **22**, 506–514; (b) Studer, M., Neto, S. and Blaser, H. (2000) *Topics in Catalysis*, **13**, 205–212; (c) Studer, M. and Baumeister, P. (1995) Patent WO1996036597A1; (d) Makaryan, J. and Savchenko, V. (1993) *Studies in Surface Science and Catalysis*, **1993**, 2439–2442.
8. (a) Baumeister, P., Blaser, H.U. and Studer, M. (1997) *Catalysis Letters*, **43**, 219–222; (b) Blaser, H.U. and Studer, M. (1999) *Applied Catalysis A: General*, **189**, 191–204; (c) Duan, Z., Ma, G. and Zhang, W. (2012) *Bulletin of the Korean Chemical Society*, **33**, 4003–4006; (d) Kadam, H.K. and Tilve, S.G. (2012) *RSC Advances*, **2**, 6057–6060.
9. Patent filed: P100446US01.

12.2 Chemo-Enzymatic Synthesis of (*S*)-1,2,3,4-Tetrahydroisoquinoline Carboxylic Acids Using D-Amino Acid Oxidase

Shuyun Ju,[1] Mingxin Qian,[2] Gang Xu,[1] Lirong Yang[1] and Jianping Wu[1]
[1] *Institute of Bioengineering, College of Chemical and Biological Engineering, Zhejiang University, Zhejiang, People's Republic of China*
[2] *Tongli Biomedical Co. Ltd, Zhangjiagang, Jiangsu, People's Republic of China*

Optically pure 1,2,3,4-tetrahydroisoquinolines (THIQs), especially 1-substituted THIQs, are a class of privileged structural moieties that have been found in a variety of natural products and synthetic pharmaceuticals [1]. Amongst the various synthetic methods, chemo-enzymatic deracemisation represents an attractive approach to the chiral THIQs, because of its high yield and excellent enantioselectivity. However, current applications mainly focus on the non-carboxyl-substituted THIQs, such as 1-alkyl-, 1-benzyl- and 1-aryl-substituted versions [2]. For chiral carboxyl-substituted THIQs, a comparable chemo-enzymatic approach is still challenging due to the lack of applicable oxidoreductases. In this context, an (*R*)-selective D-amino acid oxidase from *Fusarium solani* M-0718 (*Fs*DAAO) [3] with broad substrate scope and excellent enantioselectivity was found through genome mining. This enzyme was applied for the kinetic resolution of a range of racemic 1- and 3-carboxyl substituted THIQs with different substitutions on the aryl ring **1a–6a**. The oxidative dehydrogenation products of the (*R*)-enantiomers were dihydroisoquinoline carboxylic acids **1b–6b**, which could be reduced to the racemic substrates by a nonselective chemical reducing agent, ammonia-borane ($NH_3{\cdot}BH_3$). Consequently, by combination of *Fs*DAAO and $NH_3{\cdot}BH_3$ in one pot, these racemic carboxyl-substituted THIQs were deracemised to the corresponding (*S*)-enantiomers (Scheme 12.3) [4]. It should be noted that a large excess of $NH_3{\cdot}BH_3$ was essential to ensure the immediate reduction of the intermediate 1,4-dihydroisoquinoline-3-carboxylic acid **5b**, which proved to be unstable. Preparative-scale deracemisation of racemic 1,2,3,4-tetrahydroisoquinoline-1-carboxylic acid *rac*-**1a** and 1,2,3,4-tetrahydroisoquinoline-3-carboxylic acid *rac*-**5a** was further demonstrated.

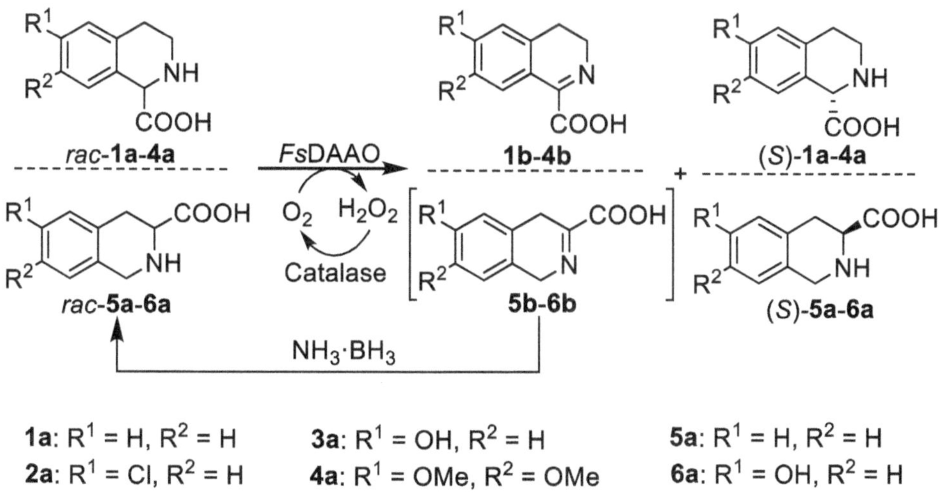

Scheme 12.3 *Chemo-enzymatic derecemisation of racemic carboxyl-substituted THIQs using FsDAAO.*

12.2.1 Procedure 1: Recombinant Expression of *Fs*DAAO in *E. coli* BL21(DE3)

12.2.1.1 Materials and Equipment

- Lysogenic broth (LB) broth medium (tryptone 10 g.L^{-1}, yeast extract 5 g.L^{-1}, NaCl 10 g.L^{-1})
- LB agar medium (tryptone 10 g.L^{-1}, yeast extract 5 g.L^{-1}, NaCl 10 g.L^{-1}, agar 15 g.L^{-1})
- Distilled water (dH_2O)
- Kanamycin (50 mg.mL^{-1} stock solution in dH_2O, filter-sterilised)
- Isopropyl-β-D-1-thiogalactopyranoside (IPTG, 0.5 M stock solution in dH_2O, filter-sterilised)
- *E. coli* BL21(DE3) competent cells (Takara)
- Autoclave
- Syringes
- 0.22 μm syringe filters (sterile)
- 15 mL Falcon tubes
- Silica gel plugs
- Petri dish
- Sterile loop
- 1 L baffled Erlenmeyer flasks with cotton caps
- UV-vis spectrophotometer
- Cuvette
- Pipettes and pipette tips
- 1.5 mL Eppendorf tubes
- Orbital shaker
- 100 mL centrifuge tubes
- Cooling centrifuge
- Ultra-low-temperature freezer

The DNA sequence encoding *Fs*DAAO (GenBank accession number: BAA00692.1) was codon-optimised, chemically synthesised and inserted into the expression vector pET-28a (+) with the cleavage sites of *Nde*I and *Hin*dIII by GENEWIZ (Genewiz Biotech Co. Ltd, China) (5 µg). The corresponding codon-optimised gene sequence is as follows:
ATGAGTAACACCATTGTTGTGGTTGGCGCCGGCGTTATTGGCCTGACCAGTGCC
CTGCTGCTGAGCAAAAATAAAGGCAATAAAATTACCGTGGTGGCCAAACACAT
GCCGGGTGATTATGATGTGGAATATGCAAGTCCGTTTGCAGGCGCAAATCATAG
CCCGATGGCCACCGAAGAAAGTAGCGAATGGGAACGCCGCACCTGGTATGAAT
TTAAACGCCTGGTGGAAGAAGTTCCGGAAGCAGGTGTTCATTTTCAGAAAAGC
CGTATTCAGCGTCGCAATGTTGATACCGAAAAAGCACAGCGCAGTGGCTTTCCG
GATGCACTGTTTAGCAAAGAACCGTGGTTTAAAAATATGTTTGAGGATTTTCGC
GAGCAGCATCCGAGCGAAGTGATTCCGGGTTATGATAGCGGTTGTGAATTTACC
AGCGTGTGTATTAATACCGCCATTTATCTGCCGTGGCTGCTGGGTCAGTGCATTA
AAAATGGTGTTATTGTTAAGCGTGCAATTCTGAATGATATTAGCGAAGCCAAAA
AACTGAGTCATGCAGGTAAAACCCCGAATATTATTGTGAATGCAACCGGTCTGG
GCAGCTATAAACTGGGCGGCGTGGAAGATAAAACCATGGCACCGGCCCGCGGT
CAGATTGTGGTGGTGCGCAATGAAAGTAGTCCGATGCTGCTGACCAGCGGTGT
GGAAGATGGCGGTGCAGATGTGATGTATCTGATGCAGCGTGCCGCAGGCGGCG
GTACCATTCTGGGCGGTACCTATGATGTTGGCAATTGGGAAAGCCAGCCGGATC
CGAATATTGCAAATCGCATTATGCAGCGTATTGTTGAAGTTCGTCCGGAAATTGC
CAATGGCAAAGGTGTTAAAGGTCTGAGTGTTATTCGTCATGCAGTTGGTATGCG
CCCGTGGCGCAAAGATGGTGTTCGCATTGAAGAAGAAAAACTGGATGATGAAA
CCTGGATTGTGCATAATTATGGCCATAGTGGTTGGGGTTATCAGGGTAGTTATGG
TTGTGCCGAAAATGTGGTTCAGCTGGTGGATAAAGTTGGTAAAGCAGCAAAAA
GCAAACTGTAA.

12.2.1.2 Procedure

1. The recombinant expression vector pET-28a (+) bearing a codon-optimised gene for *Fs*DAAO (pET-28a-*Fs*DAAO; 5 µg) was dissolved in 100 µL dH_2O.
2. 10 µL of pET-28a-*Fs*DAAO-containing solution was taken into 100 µL *E. coli* BL21(DE3) competent cells in an Eppendorf tube. The tube was thawed on ice for 30 min.
3. The tube was placed in a 42 °C water bath for 90 sec and quickly transferred into ice for 2–3 min.
4. 890 µL of LB culture solution (no resistance) was added. The tube was incubated at 37 °C with shaking at 200 rpm for 1 hr.
5. The cells were streaked on to LB agar plates containing kanamycin (50 $\mu g.mL^{-1}$) for selection at 37 °C overnight.
6. A single colony was picked and inoculated in sterile LB medium (5 mL) containing kanamycin (final conc. 50 $\mu g.mL^{-1}$) in a Falcon tube. The tube was incubated at 37 °C with shaking at 200 rpm for 7–8 hr.
7. The prepared culture (2 mL) was added into a 1 L Erlenmeyer flask containing sterile LB medium (200 mL) and kanamycin (final conc. 50 $\mu g.mL^{-1}$). The flask was incubated at 37 °C with shaking at 200 rpm until an OD_{600} of 0.6–0.8 was reached.
8. The flask was cooled to 18 °C and supplemented with an aqueous IPTG solution (40 µL) to a 0.1 mM final concentration. The flask was then incubated at 18 °C for a further 15 hr.

9. In 100 mL centrifuge tubes, the cells were harvested by centrifugation at 4000 rpm and 4 °C for 15 min. The tubes containing cell pellet (approximately 750 mg) were stored at −80 °C for later use.

12.2.2 Procedure 2: One-Pot Chemo-Enzymatic Deracemisation of *rac*-1a

12.2.2.1 Materials and Equipment

- Racemic 1,2,3,4-tetrahydroisoquinoline-1-carboxylic acid *rac*-**1a** hydrochloride salt, provided by Tongli Biomedical Co. Ltd
- dH_2O
- Ammonium hydroxide
- Ammonia-borane ($NH_3{\cdot}BH_3$)
- Catalase from bovine liver (Sigma-Aldrich)
- Ethyl alcohol absolute
- Three-necked flask (100 mL)
- Constant-temperature magnetic stirrer
- Millipore Amicon Ultra 15 mL centrifugal filter (3 kDa)
- Ultrasonic cell disruptor
- pH meter
- High-performance liquid chromatography (HPLC) system equipped with UV detector
- CHIRALPAK ZWIX® (−) (150 × 4 mm) column
- Methanol for HPLC, gradient grade, ≥99.9%
- Formic acid (FA)
- Diethyl amine (DEA)

12.2.2.2 Procedure

1. In a 100 mL three-necked flask, *rac*-**1a** hydrochloride salt (500 mg, 2.34 mmol) was dissolved in dH_2O (25 mL).
2. The pH was carefully adjusted to 8.0 by addition of aqueous ammonium hydroxide (10 M).
3. $NH_3{\cdot}BH_3$ (289 mg, 9.36 mmol) was added to the solution.
4. In a 100 mL centrifuge tube, the lyophilised *E.coli* cells containing *Fs*DAAO from 200 mL of fermentation broth were resuspended in dH_2O (25 mL). The tube was kept in ice-water mixture and the cell suspension was disrupted by sonication (400 W, cycles of 3 sec sonication followed by 7 sec rest for 35 min). The supernatant was obtained by centrifugation at 12 000 rpm and 4 °C for 20 min.
5. The supernatant (25 mL) and catalase (100 mg) were added into the substrate solution.
6. The flask was placed in a thermostatic water bath at 30 °C and the solution was magnetically stirred.
7. During the reaction, 50 μL solution was taken into 950 μL MeOH (containing 50 mM FA + 25 mM DEA) every 30 min. The mixture was then centrifuged at 12 000 rpm for 2 min and filtered.
8. The filtrate was analysed by chiral HPLC.
9. Upon completion of the reaction, the proteins were removed with an Amicon Ultra 15 mL centrifugal filter (3 kDa).

10. The solution was freeze-dried, taken up in hot ethanol (250 mL) and hot-filtered to remove unreacted $NH_3{\cdot}BH_3$. The major parts of the ethanol were evaporated off under reduced pressure and the residue was washed with warm water (10 mL). The mixture was concentrated (to less than 1 mL) so that compound (*S*)-**1a** recrystallised as a white solid (82% yield, 340 mg).

12.2.2.3 Analytical Method

The enantiomeric excess was determined by chiral HPLC analysis: CHIRALPAK ZWIX® (−) column (150 × 4 mm; DAICEL Chiral Technologies), (50 mM FA + 25 mM DEA) in MeOH, flow rate = 0.4 mL.min^{-1}, 220 nm UV detector, t_R = 9.02 min (*R*) and t_R = 13.34 min (*S*), column temperature = 25 °C.

Melting point: 263–266 °C; $[\alpha]_D{}^{20}$ = +68 [~0.5, 1 M HCl]; ^{1}H NMR (500 MHz, D_2O): δ 7.48–7.39 (m, 1H), 7.36–7.07 (m, 3H), 4.87 (s, 1H), 3.56–3.46 (m, 1H), 3.42–3.33 (m, 1H), 3.04–2.94 (m, 2H); ^{13}C NMR (125 MHz, D_2O): δ171.9, 131.7, 128.7, 128.3, 128.1, 127.9, 126.9, 58.6, 39.8, 24.5; HRMS (ESI-TOF) m/z: calcd for $C_{10}H_{12}NO_2$ $[M+H]^+$: 178.0863; found: 178.0887.

12.2.3 Procedure 3: One-Pot Chemo-Enzymatic Deracemisation of *rac*-5a

12.2.3.1 Materials and Equipment

- Racemic 1,2,3,4-tetrahydroisoquinoline-3-carboxylic acid (*rac*-**5a**; J&K Scientific Ltd)
- Acetonitrile (MeCN) for HPLC, gradient grade, ≥99.9%
- Other materials and equipment as in Procedure 2

12.2.3.2 Procedure

1. In a 100 mL three-necked flask, substrate *rac*-**5a** (400 mg, 2.26 mmol) was dissolved in dH_2O (48 mL).
2. The pH was carefully adjusted to 8.0 by addition of aqueous ammonium hydroxide (10 M).
3. $NH_3{\cdot}BH_3$ (1395 mg, 45.2 mmol) was added to the solution.
4. In a 100 mL centrifuge tube, the lyophilised *E.coli* cells containing *Fs*DAAO from 400 mL of fermentation broth were resuspended in dH_2O (20 mL). The crude lysate was prepared by sonication and centrifugation as described in Procedure 2.
5. The supernatant (12 mL) and catalase (100 mg) were added into the substrate solution.
6. The flask was placed in a thermostatic water bath at 30 °C and magnetically stirred.
7. The reaction was monitored by chiral HPLC.
8. Upon completion of the reaction, the proteins were removed with an Amicon Ultra 15 mL centrifugal filter (3 kDa).
9. The solution was freeze-dried, taken up in hot ethanol (200 mL) and hot-filtered to remove unreacted $NH_3{\cdot}BH_3$. The major parts of the ethanol were evaporated off under reduced pressure and the residue was washed with warm water (10 mL). The mixture was concentrated (to less than 1 mL) so that compound (*S*)-**5a** recrystallised as a white solid (73% yield, 292 mg).

12.2.3.3 Analytical Method

The enantiomeric excess was determined by chiral HPLC analysis: CHIRALPAK ZWIX® (−) column (150 × 4 mm, DAICEL Chiral Technologies), (50 mM FA + 25 mM DEA) in MeOH/MeCN = 50 : 50 v/v, flow rate = 0.5 mL.min^{-1}, 210 nm UV detector, t_R = 8.81 min (*R*) and t_R = 11.28 min (*S*), column temperature = 25 °C.

Melting point: 294–296 °C; $[\alpha]_D^{20}$ = −177 [*c* 0.5, 1M NaOH]; ^{1}H NMR (500 MHz, D_2O) δ 7.36–7.09 (m, 4H), 4.41–4.23 (m, 2H), 4.01–3.88 (m, 1H), 3.31 (dd, *J* = 17.2, 5.3 Hz, 1H), 3.07 (dd, *J* = 17.2, 11.2 Hz, 1H); ^{13}C NMR (125 MHz, D_2O): 173.8, 131.4, 128.9, 128.1, 127.6, 127.2, 126.5, 56.1, 44.1, 28.8; HRMS (ESI-TOF) m/z: calcd for $C_{10}H_{12}NO_2$ $[M+H]^+$: 178.0863; found: 178.0886.

12.2.4 Conclusion

We have developed a chemo-enzymatic method for one-pot synthesis of (*S*)-1,2,3,4-tetrahydroisoquinoline carboxylic acids. By using this method, a panel of racemic 1-carboxyl-substituted THIQs **2a–4a** with different substitutions (Cl, OH, OCH_3) on the phenyl ring was transformed to the corresponding (*S*)-enantiomers with conversion up to >96 and >99% ee (Table 12.3) [4]. Preparative-scale deracemisation of *rac*-**1a** and **5a** was also demonstrated, with good yield (82 and 73%) and excellent enantioselectivity (>99% ee). We anticipate that this method might provide access to other pharmaceutically important building blocks, such as chiral carboxyl-substituted 1,2,3,4-tetrahydroquinolines and 1,2,3,4-tetrahydro ß carbolines.

References

1. (a) Welsch, M.E., Snyder, S.A. and Stockwell, B.R. (2010) *Current Opinion in Chemical Biology*, **14**, 347–361; (b) Chrzanowska, M., Grajewska, A. and Rozwadowska, M.D. (2016) *Chemical Reviews*, **116**, 12 369–12 465.
2. (a) Foulkes, J.M., Malone, K.J., Coker, V.S. *et al.* (2011) *ACS Catalysis*, **1**, 1589–1594; (b) Ghislieri, D., Green, A.P., Pontini, M. *et al.* (2013) *Journal of the American Chemical Society*, **135**, 10 863–10 869; (c) Schrittwieser, J.H., Groenendaal, B., Willies, S.C. *et al.* (2014) *Catalysis Science & Technology*, **4**, 3657–3664; (d) Heath, R.S., Pontini, M., Bechi, B. and Turner, N.J. (2014) *ChemCatChem*, **6**, 996–1002; (e) Aleku, G.A., Mangas-Sanchez, J., Citoler, J. *et al.* (2018) *ChemCatChem*, **10**, 515–519.
3. Isogai, T., Ono, H., Ishitani, Y. *et al.* (1990) *Journal of Biochemistry*, **108**, 1063–1069.
4. Ju, S., Qian, M., Xu, G. *et al.* (2019) *Advanced Synthesis & Catalysis*, **361**, 3191–3199.

Table 12.3 Chemo-enzymatic deracemisation of rac-**2a–4a** employing *Fs*DAAO.

Entry	Substrate (mM)	*Fs*DAAO (mg.mL^{-1})	$NH_3 \cdot BH_3$ (equiv)	Time (h)	Conversion (C) (%)	ee_s (%)/ configuration
1	**2a** (10)	0.1	4	5	>94	>99 (*S*)
2	**3a** (10)	0.1	4	10	>94	>99 (*S*)
3	**4a** (5)	0.1	4	10	>96	>99 (*S*)

12.3 Amine Oxidase-Catalysed Deracemisation of (*R*,*S*)-4-Cl-Benzhydrylamine into the (*R*)-Enantiomer in the Presence of a Chemical Reductant

Kazuyuki Yasukawa[1,2] and Yasuhisa Asano*[1]
[1] *Biotechnology Research Center and Department of Biotechnology, Toyama Prefectural University, Imizu, Toyama, Japan*
[2] *Toyama Institute of Health, Imizu, Toyama, Japan*

Organic and enzyme chemists are collaborating to develop efficient and ecofriendly synthetic methods of creating a sustainable chemical industry. Projects to synthesise an enantiopure chiral compound in high yield have proved challenging [1]. The deracemisation reaction, in which one of the enantiomers of the racemate is converted into the opposite enantiomer, is an attractive solution to this problem [2]. In particular, deracemisation using an oxidase is a simple synthetic method that can be carried out under mild conditions without the need for a protecting group in the substrate.

The porcine kidney D-amino acid oxidase (pkDAAO, EC; 1.4.3.3) catalyses the oxidation of D-amino acids into the corresponding α-keto acids, hydrogen peroxide and ammonia. DAAO and its variants have been widely utilised for chiral synthesis and integration into biosensors [3]. Soda et al. [4] reported the synthesis of L-proline from the racemate using the deracemisation method with a DAAO in combination with a chemical reducing agent. Asano et al. [5] succeeded in generating the (*R*)-stereoselective amine oxidase from pkDAAO by removing a critical residue in the active site that recognised the carboxylic acid of the substrate and applied it to the production of (*S*)-methylbenzylamine from its racemate using the deracemisation method.

In this section, we describe the deracemisation reaction using a newly designed I230A/R283G pkDAAO variant, which catalyses (*S*)-4-Cl-benzhydrylamine oxidation (Scheme 12.4) [6].

The enzymatic deracemisation of (*R*,*S*)-4-Cl-benzhydrylamine involves the following steps: (i) a stereoselective oxidation step: the I230A/R283G variant enzyme catalyses the oxidation of (*S*)-4-Cl-benzhydrylamine to form its imine product; and (ii) a non-stereoselective reduction step: a chemical reducing agent ($NaBH_4$) acts on the imine to reform the starting material, (*R*,*S*)-4-Cl-benzhydrylamine. These two reactions are

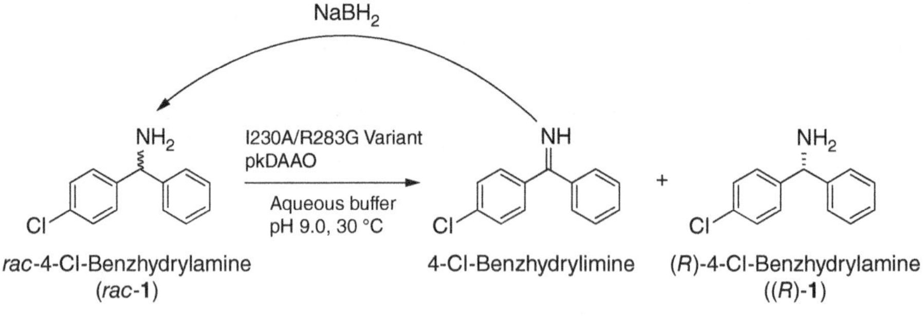

Scheme 12.4 *Overview of the amine oxidase-catalysed deracemisation of* rac-***1*** *to form (R)-**1**. pkDAAO, porcine kidney D-amino acid oxidase.*

run until the unwanted (*S*)-enantiomer concentration becomes negligible. Finally, the (*R*)-enantiomer is achieved in close to 100% yield, without undesired byproducts.

12.3.1 Procedure 1: Amine Oxidase Activity Assay

12.3.1.1 *Materials and Equipment*

- K_2HPO_4
- KH_2PO_4
- (*S*)-4-Cl-Benzhydrylamine·HCl (Combi-Blocks)
- 50 mM (*S*)-4-Cl-Benzhydrylamine stock solution (12.7 $mg.mL^{-1}$ in DMSO)
- Distilled water (dH_2O)
- Dimethyl sulfoxide (DMSO)
- 20 mM Phenol stock solution (1.8 $mg.mL^{-1}$ in dH_2O)
- 15 mM 4-Aminoantipyrine stock solution (3.05 $mg.mL^{-1}$ in dH_2O)
- 200 $U.mL^{-1}$ Horseradish peroxidase (HRP; 100 $U.mg^{-1}$) stock solution (2 $mg.mL^{-1}$ in dH_2O)
- pH meter
- Spectrophotometer (JASCO Co.)

12.3.1.2 *Procedure*

1. 1 M potassium phosphate buffer (KPi), pH 8.0 stock solution was prepared as follows: K_2HPO_4 (69.7 g) was dissolved in dH_2O (400 mL) and the pH was adjusted to 8.0 with 1 M KH_2PO_4 solution (0.136 $g.mL^{-1}$ in dH_2O).
2. Amine oxidase activity was assayed via spectrophotometer by measuring the absorbance at 505 nm and the formation of 4-*N*-(*p*-benzoquinoneimine)-antipyrine (molecular extinction coefficient; 13 600 $M^{-1}.cm^{-1}$). The assay mixture (total 1 mL) consisted of the following: (*S*)-4-Cl-Benzhydrylamine (0.1 mL of 50 mM), 1 M KPi pH 8.0 (0.1 mL), 20 mM phenol (0.1 mL), 15 mM 4-aminoantipyrine (0.1 mL), 200 $U.mL^{-1}$ HRP solution (0.01 mL) and an appropriate amount of the enzyme solution at 30 °C. 1 U of oxidation activity was defined as the amount of enzyme that produces 1 μmol of hydrogen peroxide per minute.

12.3.2 Procedure 2: Selection of Porcine Kidney D-Amino Acid Oxidase Variant with (*S*)-4-Cl-Benzhydrylamine Oxidase Activity

12.3.2.1 *Materials and Equipment*

- 1 M KPi, pH 8.0 from Procedure 1
- (*R,S*)-4-Cl-Benzhydrylamine·HCl (Sigma-Aldrich)
- (*S*)-4-Cl-Benzhydrylamine·HCl (Combi-Blocks)
- (*R*)-4-Cl-Benzhydrylamine·HCl (Astatech Inc.)
- Peptone
- Yeast extract
- NaCl
- Lysogenic broth (LB) (1% w/v peptone, 0.5% w/v yeast extract and 1% w/v NaCl in dH_2O, autoclaved at 121 °C for 15 min)

- Distilled water (dH_2O)
- 80 mg.mL^{-1} ampicillin stock solution (0.8 g in 10 mL dH_2O, filter-sterilised)
- 1 M isopropyl-β-D-1-thiogalactopyranoside (IPTG) stock solution (0.238 g in dH_2O, filter-sterilised)
- Dimethyl sulfoxide (DMSO)
- *Escherichia coli* JM109 competent cells (prepared by $CaCl_2$ method)
- QuikChange Lightning Site-Directed Mutagenesis Kit (Stratagene)
- Terminator v3.1 Cycle Sequencing Kit (Applied Biosystems)
- ABI 3500 Genetic analyser (Applied Biosystems)
- 50 mM (*R,S*)-4-Cl-Benzhydrylamine stock solution (12.7 mg.mL^{-1} in DMSO)
- 20 mM Phenol stock solution (1.8 mg.mL^{-1} in dH_2O)
- 15 mM 4-Aminoantipyrine stock solution (3.05 mg.mL^{-1} in dH_2O)
- 200 U.mL^{-1} HRP (100 U.mg^{-1}) stock solution (2 mg.mL^{-1} in dH_2O)

12.3.2.2 Procedure

1. Site-specific saturation mutagenesis at residues Lue51, Ile215 and Ile230 of the R283G and Y228L/R283G variant pkDAAO-encoding gene (as *Bam*HI-TAAGGAGGACTAGCTC-*gene-Hind*III; Shine–Dalgarno sequence underlined) in plasmid pUC18 vector was carried out using the QuikChange Lightning Site-Directed Mutagenesis Kit according to the manufacturer's protocol [5, 6].
2. The gene sequences of R283G and Y228L/R283G are shown in Figures 12.2 and 12.3, respectively. These mutation sites were carefully selected by analysis of the crystal structure with the Y228L/R283G variant-(*R*)-methylbenzylamine complex (PDB: 3WGT) [5–7].
3. The engineered plasmids were transformed into *E. coli* JM109 competent cells. These mutants were cultivated at 37 °C for 24 hr on an LB agar plate containing 80 μg.mL^{-1} ampicillin, and then single colonies were cultivated at 37 °C for 24 hr in 100 μL LB media containing 80 μg.mL^{-1} ampicillin and 1 mM IPTG in a Nunc™ DeepWell™ plate.
4. After cultivation, cells were collected by centrifugation at 10 000× g and 4 °C for 5 min and suspended in 0.1 mL of 100 mM KPi (pH 8.0).
5. The cell suspension (10 μL) was added to the oxidase activity assay mixture (0.1 mL), consisting of 100 mM KPi (pH 8.0), 2 mM phenol, 1.5 mM 4-aminoantipyrine, 2 U HRP and 5 mM (*R,S*)-4-Cl-benzhydrylamine.
6. Variants were selected if they showed high activity. The stereoselectivity of mutants was assayed under the same conditions with 5 mM (*R*)- or (*S*)-4-Cl-benzhydrylamine as a substrate.
7. Finally, the I230A/R283G variant with (*S*)-4-Cl-benzhydrylamine oxidase activity was obtained. The gene sequence is shown in Figure 12.4.
8. The mutation site of the encoding gene was determined using the dideoxynucleotide chain termination sequencing method. Sequencing reactions were carried out with a BigDye® Terminator v3.1 Cycle Sequencing Kit using M13 primers, and the reaction mixture was electrophoresed on an ABI 3500 genetic analyser. The sequence of the M13 primer set is as follows: M13 forward primer, 5'-CAGGAAACAGCTATGAC-3'; M13 reverse primer, 5'-GTAAAACGACGGCCAGT-3'

```
1    atgcgtgtggtggtgattggagcaggcgtcattgggctgtccactgccctctgcatccacgagcgctaccactctgtcctgcagcccctc
      M  R  V  V  V  I  G  A  G  V  I  G  L  S  T  A  L  C  I  H  E  R  Y  H  S  V  L  Q  P  L

91   gatgtgaaggtctacgcagaccgcttcaccccgttcaccaccactgatgtagctgctggcctctggcagccctacacctctgagcccagc
      D  V  K  V  Y  A  D  R  F  T  P  F  T  T  T  D  V  A  A  G  L  W  Q  P  Y  T  S  E  P  S

181  aacccacaggaggcgaactggaaccaacagaccttcaactatctcctgagtcacatcggttctcccaatgctgcaaacatgggtctgacc
      N  P  Q  E  A  N  W  N  Q  Q  T  F  N  Y  L  L  S  H  I  G  S  P  N  A  A  N  M  G  L  T

271  ccagtctcaggctacaacctcttccgtgaagctgttccggacccttactggaaagacatggtcctgggattccgaaagctgactccaga
      P  V  S  G  Y  N  L  F  R  E  A  V  P  D  P  Y  W  K  D  M  V  L  G  F  R  K  L  T  P  R

361  gaactggacatgtttcctgattatagatatggctggttcaacacatccctgattctggagggaaggaagtacctacagtggctgacagaa
      E  L  D  M  F  P  D  Y  R  Y  G  W  F  N  T  S  L  I  L  E  G  R  K  Y  L  Q  W  L  T  E

451  aggttaactgagaggggagtgaaattcttcctgaggaaggtggaatcttttgaggaggtggcaagaggtggcgctgatgtgattatcaac
      R  L  T  E  R  G  V  K  F  F  L  R  K  V  E  S  F  E  E  V  A  R  G  G  A  D  V  I  I  N

541  tgcactggggtgtgggctggggtgctgcaaccggatcccctgctgcagccaggccgggggcagatcattaaggtggatgccccttggctg
      C  T  G  V  W  A  G  V  L  Q  P  D  P  L  L  Q  P  G  R  G  Q  I  I  K  V  D  A  P  W  L

631  aagaacttcattatcacccatgacctagagagaggcatctacaactctccatacatcattccagggctgcaggcagtgacacttggaggc
      K  N  F  I  I  T  H  D  L  E  R  G  I  Y  N  S  P  Y  I  I  P  G  L  Q  A  V  T  L  G  G

721  accttccaggtggggaactggaatgagataaataatatccaggaccacaacaccatctgggaaggctgctgcagactggagcccacactg
      T  F  Q  V  G  N  W  N  E  I  N  N  I  Q  D  H  N  T  I  W  E  G  C  C  R  L  E  P  T  L

811  aaggatgcaaaaattgttggtgaatatactggcttcggaccagtacgcccccaggttcggctagaaagagaacagcttcgctttggatct
      K  D  A  K  I  V  G  E  Y  T  G  F  G  P  V  R  P  Q  V  R  L  E  R  E  Q  L  R  F  G  S

901  tcaaacacagaggtcatccacaactatggccatggaggctatgggctcaccatccactggggctgtgccctagaggtggccaagctcttt
      S  N  T  E  V  I  H  N  Y  G  H  G  G  Y  G  L  T  I  H  W  G  C  A  L  E  V  A  K  L  F

991  gggaaagtcctggaagaaaggaatttgctcacaatgccaccatcccacctctga
      G  K  V  L  E  E  R  N  L  L  T  M  P  P  S  H  L  *
```

Figure 12.2 *Gene sequence of the R283G variant. The mutated residue is underlined.*

12.3.3 Procedure 3: Expression of the I230A/R283G Variant pkDAAO

12.3.3.1 Materials and Equipment

- Peptone
- Yeast extract
- NaCl
- LB (1% w/v peptone, 0.5% w/v yeast extract and 1% w/v NaCl in 5 L dH_2O, autoclaved at 121 °C for 15 min)
- K_2HPO_4
- KH_2PO_4
- Phosphate-buffered saline (PBS; 0.8 g NaCl in 100 mL of 10 mM KPi, pH 7)
- dH_2O
- Ampicillin stock solution (80 mg.mL^{-1}, 0.8 g in 10 mL dH_2O, filter-sterilised)
- 1 M IPTG stock solution (0.238 g in dH_2O, filter-sterilised)
- *E. coli* JM109 harbouring the expression vector pUC18 with the gene encoding the engineered porcine kidney D-amino acid oxidase (I230A/R283G) [6]
- 10× 0.5 mL test tubes with cotton caps

```
1    atgcgtgtggtggtgattggagcaggcgtcattgggctgtccactgccctctgcatccacgagcgctaccactctgtcctgcagcccctc
     M  R  V  V  V  I  G  A  G  V  I  G  L  S  T  A  L  C  I  H  E  R  Y  H  S  V  L  Q  P  L

91   gatgtgaaggtctacgcagaccgcttcaccccgttcaccaccactgatgtagctgctggcctctggcagccctacacctctgagcccagc
     D  V  K  V  Y  A  D  R  F  T  P  F  T  T  T  D  V  A  A  G  L  W  Q  P  Y  T  S  E  P  S

181  aacccacaggaggcgaactggaaccaacagaccttcaactatctcctgagtcacatcggttctcccaatgctgcaaacatgggtctgacc
     N  P  Q  E  A  N  W  N  Q  Q  T  F  N  Y  L  L  S  H  I  G  S  P  N  A  A  N  M  G  L  T

271  ccagtctcaggctacaacctcttccgtgaagctgttccggacccttactggaaagacatggtcctgggattccgaaagctgactcccaga
     P  V  S  G  Y  N  L  F  R  E  A  V  P  D  P  Y  W  K  D  M  V  L  G  F  R  K  L  T  P  R

361  gaactggacatgtttcctgattatagatatggctggttcaacacatccctgattctggagggaaggaagtacctacagtggctgacagaa
     E  L  D  M  F  P  D  Y  R  Y  G  W  F  N  T  S  L  I  L  E  G  R  K  Y  L  Q  W  L  T  E

451  aggttaactgagaggggagtgaaattcttcctgaggaaggtggaatcttttgaggaggtggcaagaggtggcgctgatgtgattatcaac
     R  L  T  E  R  G  V  K  F  F  L  R  K  V  E  S  F  E  E  V  A  R  G  G  A  D  V  I  I  N

541  tgcactggggtgtgggctggggtgctgcaaccggatcccctgctgcagccaggccgggggcagatcattaaggtggatgccccttggctg
     C  T  G  V  W  A  G  V  L  Q  P  D  P  L  L  Q  P  G  R  G  Q  I  I  K  V  D  A  P  W  L

631  aagaacttcattatcacccatgacctagagagaggcatctacaactctccattgatcattccagggctgcaggcagtgacacttggaggc
     K  N  F  I  I  T  H  D  L  E  R  G  I  Y  N  S  P  L  I  I  P  G  L  Q  A  V  T  L  G  G

721  accttccaggtggggaactggaatgagataaataatatccaggaccacaacaccatctgggaaggctgctgcagactggagcccacactg
     T  F  Q  V  G  N  W  N  E  I  N  N  I  Q  D  H  N  T  I  W  E  G  C  C  R  L  E  P  T  L

811  aaggatgcaaaaattgttggtgaatatactggcttcggaccagtacgcccccaggttcggctagaaagagaacagcttcgctttggatct
     K  D  A  K  I  V  G  E  Y  T  G  F  G  P  V  R  P  Q  V  R  L  E  R  E  Q  L  R  F  G  S

901  tcaaacacagaggtcatccacaactatggccatggaggctatgggctcaccatccactggggctgtgccctagaggtggccaagctcttt
     S  N  T  E  V  I  H  N  Y  G  H  G  G  Y  G  L  T  I  H  W  G  C  A  L  E  V  A  K  L  F

991  gggaaagtcctggaagaaaggaatttgctcacaatgccaccatcccacctctga
     G  K  V  L  E  E  R  N  L  L  T  M  P  P  S  H  L  *
```

Figure 12.3 *Gene sequence of the Y228L/R283G variant. The mutated residues are underlined.*

- 10× 2 L Erlenmeyer flask (baffled) with cotton caps
- Orbital shaker (Innova44, New Brunswick Scientific Co.)
- Autoclave
- Refrigerated centrifuge

12.3.3.2 Procedure

1. Recombinant *E. coli* cells were cultivated at 37 °C for 24 hr on an LB agar plate (3% w/v agarose) containing 80 μg.mL^{-1} ampicillin, and then single colonies were precultivated at 37 °C with shaking for 12 hr in LB medium containing 80 μg.mL^{-1} ampicillin (10 × 5 mL test tubes).
2. After 12 hr, 5 mL of pre-culture medium was placed into a 2 L Erlenmeyer flask containing 400 mL LB medium, 80 μg.mL^{-1} ampicillin and 1 mM IPTG and the culture was incubated at 37 °C for 24 hr using an orbital shaker (200 rpm).
3. The cells were collected by centrifugation at 10 000× g and 4 °C for 5 min.
4. The collected cells were resuspended in PBS (5 mL.$g_{wet\ weight}{}^{-1}$) and centrifuged at 10 000× g and 4 °C for 5 min (this procedure was repeated twice).
5. Finally, the cell pellet (25 g wet weight) was stored frozen at −20 °C for later use.

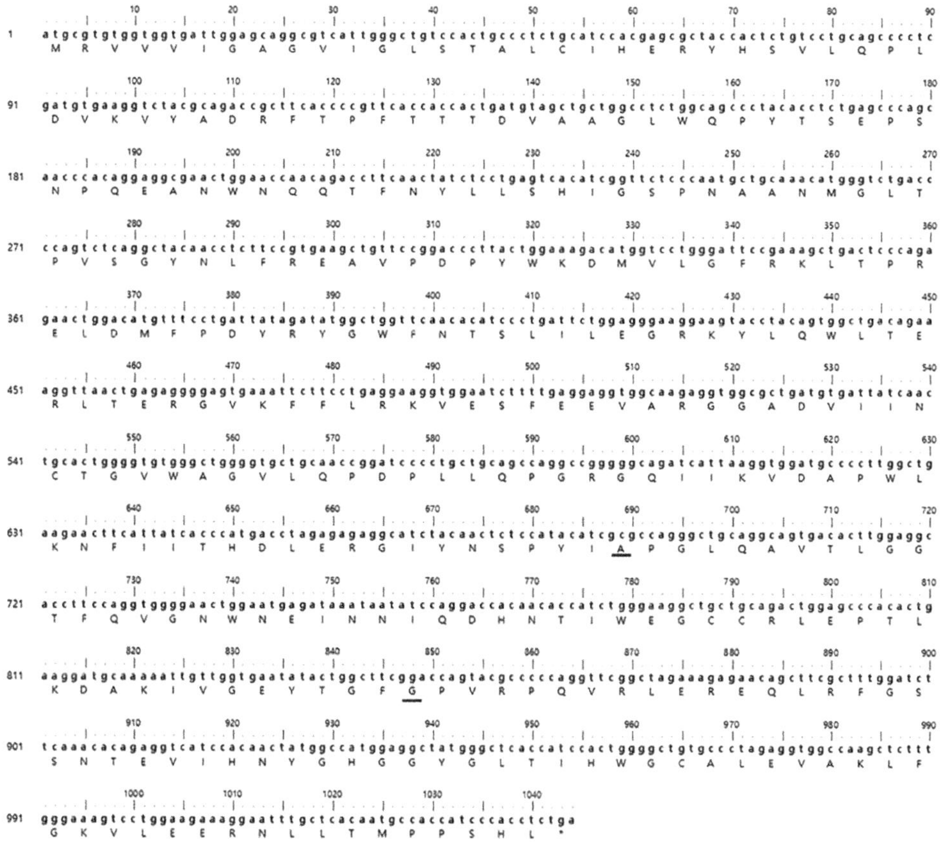

Figure 12.4 *Gene sequence of the I230A/R283G variant. The mutated residues are underlined.*

12.3.4 Procedure 4: Preparation of Purified Enzyme

12.3.4.1 Materials and Equipment

- 1 M KPi stock solution from Procedure 1
- NaCl
- dH_2O
- Ammonium sulfate
- 2-Mercaptoethanol
- TOYOPEARL DEAE-650M (Tosho Bioscience)
- TOYOPEARL Butyl-650M (Tosho Bioscience)
- Protein assay dye reagent (Bio-Rad Laboratories)
- Bovine serum albumin (BSA; 1 mg.mL^{-1} in dH_2O)
- Insonator 201M (Kubota Co.)

12.3.4.2 Procedure

1. *E. coli* cells were suspended in 5 volumes (5 mL.$g_{wet\ weight}^{-1}$) of 10 mM KPi, pH 8.0 containing 0.1% v/v 2-mercaptoethanol.

2. The cell suspension was disrupted by sonication for 20 min at 180 W using the Insonator 201M.
3. The cell debris was removed by centrifugation at 10 000× g and 4 °C for 30 min and the enzyme activity in the cell-free extract was determined through Procedure 1.
4. The cell-free extract was fractionated by 20% ammonium sulfate precipitation. Ammonium sulfate (114 g.L^{-1}) was added to the cell-free extract and gently stirred on a magnetic stirrer at 4 °C for 1 hr. The precipitant of the 20% saturation fraction was then removed by centrifugation at 10 000× g and 4 °C for 30 min. The supernatant of the 20% saturation fraction was used for the next step and fractionated with 33% ammonium sulfate (114 g.L^{-1}). The precipitant of the 33% saturation fraction was suspended in a small amount of 10 mM KPi (pH 8.0) containing 0.1% v/v 2-mercaptoethanol and dialysed against 5 L of the same buffer.
5. The dialysed crude extract was centrifuged at 10 000× g and 4 °C for 30 min and the debris was removed. The supernatant was applied to a DEAE-Toyopearl column (ϕ 6.0 × 13 cm) equilibrated with 5 column volumes (CV) of 10 mM KPi containing 0.1% v/v 2-mercaptoethanol. The column was then washed with the same buffer, and the enzyme was eluted using a linear gradient of 0–0.5 M NaCl. The active fraction was collected and ammonium sulfate was added up to 20% saturation.
6. The 20% ammonium sulfate saturated enzyme solution was applied to a Butyl-Toyopearl column (ϕ 3.0 × 22.0 cm) equilibrated with 5 CV of 10 mM KPi containing 0.1% 2-mercaptoethanol and 20% ammonium sulfate. The column was then washed with 5 CV of the same buffer, and the enzyme was eluted by stepwise elution with 10 and 0% ammonium sulfate in 10 mM KPi containing 0.1% v/v 2-mercaptoethanol (100 mL each).
7. Finally, active fractions were pooled as purified enzyme and dialysed against 5 L of 10 mM KPi containing 0.1% v/v 2-mercaptoethanol.
8. Purified enzyme (40 mL, protein concentration 3 mg.mL^{-1}) was obtained. The specific activity of the purified enzyme was shown to be 15.5 U.mg^{-1} with 5 mM (*S*)-4-Cl-benzhydrylamine as substrate. The activity assay is described in Procedure 1. Protein concentration was determined using the Bradford method with the protein-assay dye reagent. BSA was used as standard.
9. Purified enzyme was stored at 4 °C for later use.

12.3.5 Procedure 5: Enzymatic Deracemisation Reaction of (*S*)-4-Cl-Benzhydrylamine

12.3.5.1 Materials and Equipment

- (*R,S*)-4-Cl-Benzhydrylamine·HCl (Sigma-Aldrich)
- (*S*)-4-Cl-Benzhydrylamine·HCl (Combi-Blocks)
- (*R*)-4-Cl-Benzhydrylamine·HCl (Astatech Inc.)
- $NaBH_4$
- DMSO
- Glycine
- Aqueous NaOH (6 M)
- dH_2O
- *n*-Hexane

- 2-Propanol
- Anhydrous sodium sulfate (Na_2SO_4)
- Purified I230/R283 enzyme from Procedure 3 (substrate specificity as follows: (*S*)-4-Cl-benzhydrylamine (5 mM), 100%; (*R,S*)-4-Cl-benzhydrylamine (5 mM), 53% (*S*-selective); (*R,S*)-4-F-benzhydrylamine (5 mM), 47% (*S*-selective); (*R,S*)-4-Br-benzhydrylamine (1 mM), 10% (*S*-selective); benzhydrylamine (5 mM), 13%; (*R*)-methylbenzylamine (10 mM); 3% [6])
- 100 mL recovery flask
- 300 mL separatory funnel
- pH meter
- Vortex mixer
- Desiccator
- Rotary evaporator
- High-performance liquid chromatography (HPLC) system (Alliance HPLC System, Waters)
- Chiral HPLC column (OD-H column: ϕ 0.46 × 25 cm; Daicel Chiral Technologies Co. Ltd)
- Bruker Biospin AVANCE II 400 system (Bruker Biospin)
- ATAGO AP-300 polarimeter (Atago Co.)

12.3.5.2 *Procedure*

1. 1 M glycine-NaOH buffer (pH 9.0) stock solution was prepared as follows: glycine (37.5 g) was dissolved in dH_2O (400 mL) and the pH was adjusted to 9.0 with 6 M NaOH (aq). The solution was brought up to 500 mL with dH_2O.
2. The deracemisation reaction mixture (40 mL in 100 mL recovery flask) contained (*R,S*)-4-Cl-benzhydrylamine (50 mg; 0.197 mmol in 4 mL DMSO), 1 M glycine-NaOH buffer pH 9.0 (4 mL), $NaBH_4$ (0.15 g) and 4 U purified enzyme (~0.3 mg protein).
3. The progress of the reaction was monitored by HPLC with a chiral OD-H column (ϕ 0.46 × 25 cm) using a solvent system composed of hexane/2-propanol 9 : 1 (absorbance 220 nm, flow rate 1 mL.min^{-1}, column temperature 30 °C). Analyte samples were prepared as follow: 0.1 mL of reaction sample was added to hexane/2-propanol 9 : 1 (0.9 mL) in a 1.5 mL tube and mixed vigorously with a vortex mixer. The mixture was centrifuged at 10 000× g for 5 min and the organic layer was analysed by HPLC. The retention time is shown in Table 12.4.
4. After 1 hr, the reaction was quenched by the addition of 1 M NaOH (aq) solution (10 mL).

Table 12.4 *Retention times for LC analysis.*

Substance	Retention (min)
(*R*)-4-Cl-benzhydrylamine	11.8
(*S*)-4-Cl-benzhydrylamine	10.0
DMSO	12.7

5. The reaction mixture was extracted with hexane (3 × 40 mL) in a 300 mL separatory funnel. The organic phases were collected in Erlenmeyer flasks, dried over anhydrous Na_2SO_4 and filtered into a 500 mL recovery flask. The mixture was concentrated by distillation under reduced pressure.
6. The remaining oil product was further dried using a desiccator with a rotary pump for 1 day to afford (*R*)-4-Cl-benzhydrylamine 19.8 mg (46.2% yield, 95.7% ee, $[\alpha]_D^{27} = -9.99°$ (c 1.0, EtOH)).

12.3.6 Analytical Method

The product was identified by nuclear magnetic resonance (^{1}H-NMR) in DMSO-d_6 solvent and the configuration of the product in ethanol (1 mg.mL^{-1}) was determined by an ATAGO AP-300 polarimeter with a 10.1 mm cell. The ^{1}H-NMR spectra were as follows: (400 MHz; DMSO-d_6) δ_H (ppm) 7.18–7.43 (m, 9H), 5.09 (s, 1H), 2.33 (s, 2H). The enantiomeric excess value of the product was determined by HPLC with an OD-H column. HPLC conditions were described in Procedure 3. Working standard solutions of 7.5, 5, 2.5, 1, 0.5 and 0.2 mM were prepared by diluting 50 mM (*R*,*S*)-4-Cl-benzhydrylamine stock solution (12.7 mg.mL^{-1} in DMSO) with dH_2O. 0.1 mL of standard solutions were added to hexane/2-propanol 9 : 1 (0.9 mL) in a 1.5 mL tube and mixed vigorously with a vortex mixer. The mixture was centrifuged at 10 000× g for 5 min and the organic layer was analysed.

12.3.7 Conclusion

This section provided a basic procedure for the production of (*R*)-4-Cl-benzhydrylamine from the racemate using an amine oxidase-catalysed deracemisation method with easy purification of the product. In recent years, successful improvement of enzyme substrate specificity, stereoselectivity and catalytic efficiency has been reported [5–8]. Imine reductases and an artificial transfer hydrogenase are attracting interest as alternatives for the chemical reduction step in the deracemisation reaction because they offer relative safety and economy [9]. The choice of a suitable combination of enzyme and reducing agent is critical for the development of successful oxidative deracemisation reactions.

Acknowledgements

This work was supported by JST ERATO Asano Active Enzyme Molecule Project (Grant Number JPMJRER1102), Japan and by a Grant-in-Aid for Scientific Research (S) from the Japan Society for the Promotion of Sciences (Grant No. 17H06169), awarded to Y. Asano.

References

1. (a) Jesus, A.V., Daniel, G.M. and Vicente, G.F. (2017) *Biocatalysis and Biotransformation*, **36**, 102–130; (b) Bezborodov, A.M. and Zagustina, N.A. (2016) *Applied Biochemistry and Microbiology*, **52**, 237–249.
2. (a) Musa, M.M., Frank, H. and Francesco, G.M. (2019) *Catalysis Science & Technology*, **9**, 5487–5503; (b) Asano, Y. and Hölsch, K. (2012) Isomerisation, in *Catalysis in Organic Synthesis* (eds K. Drauz, H. Groger and O. May), Springer Verlag, pp. 1607–1684; (c) Asano, Y. (2019)

Bioscience, Biotechnology, and Biochemistry, **83**, 1402–1416; (d) Asano, Y. and Yasukawa, K.Y. (2019) *Current Opinion in Chemical Biology*, **49**, 76–83.
3. Pollegioni, L. and Molla, G. (2011) *Trends in Biotechnology*, **29**, 276–283.
4. Huh, J.W., Yokoigawa, K., Esaki, N. and Soda, K. (1992) *Journal of Fermentation and Bioengineering*, **74**, 189–190.
5. Yasukawa, K., Nakano, S. and Asano, Y. (2014) *Angewandte Chemie International Edition*, **53**, 4428–4431.
6. Yasukawa, K., Motojima, F., Ono, A. and Asano, Y. (2018) *ChemCatChem*, **10**, 3500–3505.
7. Nakano, S., Yasukawa, K., Tokiwa, T. *et al.* (2016) *Journal of Physical Chemistry B*, **120**, 10 736–10 743.
8. (a) Alexeeva, M., Enright, A., Dawson, M.J. *et al.* (2022) *Angewandte Chemie International Edition*, **41**, 3177–3180; (b) Ghislieri, D., Green, A.P., Willies, S.C. *et al.* (2013) *Journal of the American Chemical Society*, **135**, 10 863–10 869; (c) Herter, S., Medina, F., Wagschal, S. *et al.* (2017) *Bioorganic & Medicinal Chemistry*, **26**, 1338–1346; (d) Li, G., Yao, P., Cong, P. *et al.* (2016) *Scientific Reports*, **6**, 2497.
9. (a) Heath, R.S., Pontini, M., Hussain, S. and Turner, N.J. (2016) *ChemCatChem*, **8**, 117–120; (b) Köhler, V., Wilson, Y.M., Dürrenberger, M. *et al.* (2013) *Nature Chemistry*, **5**, 93–99.

12.4 Asymmetric Synthesis of 1-Phenylpropan-2-amine from Allylbenzene through a Sequential Strategy Involving a Wacker–Tsuji Oxidation and a Stereoselective Biotransamination

Daniel González-Martínez, Vicente Gotor and Vicente Gotor-Fernández*
Organic and Inorganic Chemistry Department, University of Oviedo, Oviedo, Spain

Optically active amines are valuable key building blocks in the synthesis of pharmaceuticals and drug intermediates [1]. Within this broad group of nitrogen-containing compounds, 1-arylpropan-2-amines, also known as amphetamines, are considered privileged motifs in commercialised pharmaceuticals such as Dextroamphetamine, Lisdexamphetamine, Benzphetamine, Selegiline, Tamsulosin and (*R*,*R*)-Formoterol, whose effectiveness has been largely demonstrated in the treatment of several diseases related to the central nervous system.

Various enzyme families have been efficiently employed in the stereoselective preparation of 1-arylpropan-2-amines, including amine transaminases (ATA) [2], amine dehydrogenases [3], imine reductases [4], reductive aminases [5] and lipases [6]. Herein, we focus on the search for an efficient match between metal and enzyme catalysis enabling the transformation of allylbenzenes **1** into optically active 1-arylpropan-2-amines **3** (Scheme 12.5).

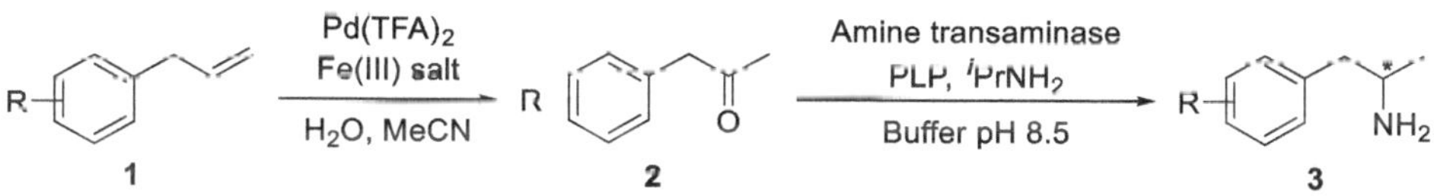

Scheme 12.5 *Transformation of allylbenzenes **1** into optically active 1-arylpropan-2-amines **3** through the synthesis and subsequent biotransamination of the corresponding 1-arylpropan-2-one intermediates **2**.*

With this aim, a sequential and selective approach has been recently described involving a metal-catalysed Wacker–Tsuji oxidation of allylbenzenes **1** followed by the ATA-catalysed biotransamination of the resulting 1-arylpropan-2-ones **2** [7]. After optimisation of the individual Wacker–Tsuji oxidation of allylbenzene and screening of different ATAs for the biotransamination reaction of the resulting 1-phenylpropan-2-one (R= H), the stereoselective preparation of 1-phenylpropan-2-amine is here disclosed, starting from commercially available and inexpensive allylbenzene and using commercially available selective ATAs for the production of the requested (*S*)- or (*R*)-amine enantiomer.

12.4.1 Procedure 1: Wacker–Tsuji Oxidation of Allylbenzene

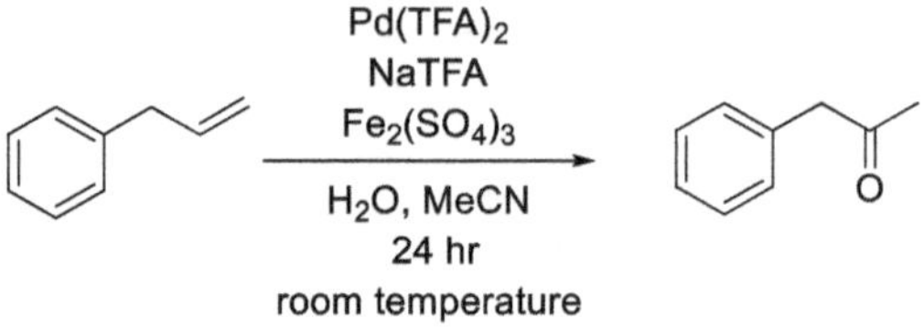

Scheme 12.6 *Wacker–Tsuji oxidation of allylbenzene.*

12.4.1.1 Materials and Equipment

- Palladium(II) trifluoroacetate ($Pd(TFA)_2$, 2.0 mg, 0.006 mmol)
- Sodium trifluoroacetate (NaTFA, 6.8 mg, 0.050 mmol)
- Iron(III) sulfate hydrate ($Fe(SO_4)_3 \cdot nH_2O$, 150.0 mg, 0.375 mmol)
- Water (9.5 mL)
- Allylbenzene (33.1 μL, 0.25 mmol)
- Acetonitrile (MeCN, 0.5 mL)
- Sodium sulfate (Na_2SO_4)
- Ethyl acetate (EtOAc, 20 mL) as solvent for liquid–liquid extraction and column chromatography purification
- Hexane as part of the eluent (10% EtOAc/hexane) for column chromatography purification
- 50 mL Schlenk flask sealed with a glass stopper
- IKA® WERKE RCT basic control stirrer hotplate
- Stirring bar
- 5 mL glass vial to prepare the allylbenzene solution
- 100 mL separating funnel
- 100 mL round-bottom flask
- Conical funnel and fluted filter paper
- Rotary evaporator
- Agilent 6860 gas chromatograph (GC) with flame ionisation detector (FID); Agilent HP-1 column (30 m × 0.32 mm × 0.25 μm) or Agilent HP-5 column (30 m × 0.32 mm × 0.25 μm) for the measurement of product ratios
- Silica gel thin-layer chromatography (TLC) plates (silica gel 60 F254)
- Silica gel for column chromatography purification

12.4.1.2 Procedure

1. $Pd(TFA)_2$ (2.0 mg), NaTFA (7.0 mg) and $Fe_2(SO_4)_3 \cdot nH_2O$ (150 mg) were added to a Schlenk flask and stored under nitrogen atmosphere sealed with a glass stopper.

2. Distilled water (9.5 mL) was added, forming a yellowish solution.
3. A solution of allylbenzene (33.1 μL, 0.25 mmol) in acetonitrile (0.5 mL) was prepared in an Eppendorf tube and added to the solution, maintaining the nitrogen atmosphere in the Schlenk flask.
4. The reaction was stirred at 30 °C for 24 hr in the absence of light.
5. The mixture was extracted with EtOAc (2 × 10 mL) and the organic phases were combined and dried over Na_2SO_4.
6. An aliquot was taken for the measurement of product ratios by GC (Tables 12.5 and 12.6).
7. After filtering off the Na_2SO_4, the solvent was removed by distillation under reduced pressure.
8. The resulting crude reaction mixture was purified by column chromatography over silica gel using 10% EtOAc/hexane as eluent to yield 1-phenylpropan-2-one (28.5 mg, 85% isolated yield) (Scheme 12.6).

Table 12.5 *GC method for the measurement of product ratios.*

Temperature program (r = °C.min^{-1})	Duration
90 °C 2 min – 10 r → 160 °C 0 min– 40 r → 240 °C 0 min	17 min

Table 12.6 *Retention times for GC analysis using an Agilent HP-1 or HP-5 column.*

Substance	Retention (min) HP-1 column	Retention (min) HP-5 column
Allylbenzene	4.3	4.5
1-Phenylpropan-2-one	8.6	9.3
trans-β-Methylstyrene	6.5	6.8
Propiophenone	9.4	10.0
Benzaldehyde	4.3	4.9
3-Phenylpropanal	9.3	10.9
Cinnamaldehyde	10.5	12.0

12.4.2 Procedure 2: Biotransamination of 1-Phenylpropan-2-One Using (*R*)- or (*S*)-Selective ATAs

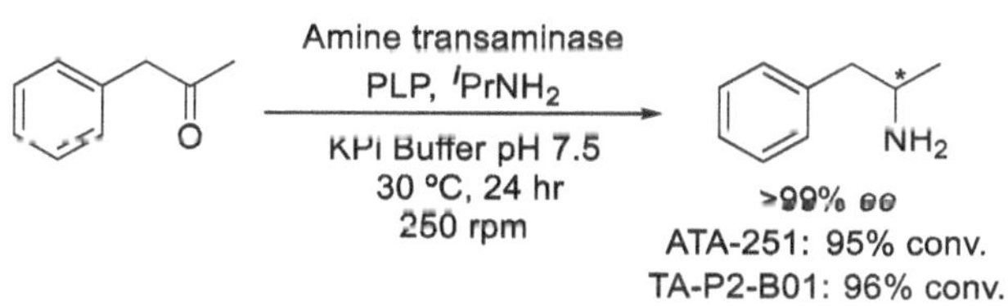

Scheme 12.7 Biotransamination of 1-phenylpropan-2-one using (*R*)- or (*S*)-selective ATAs.

12.4.2.1 Materials and Equipment

- 100 mM phosphate buffer pH 7.5 (500 µL) containing isopropylamine ($^{i}PrNH_2$, 1.0 M) and pyridoxal-5-phosphate (PLP, 1 mM)
- 1-Phenylpropan-2-one (1.34 mg, 0.01 mmol)
- Codex Transaminase ATA Screening Kit (ATASK-000250) and PLP (Codexis Inc.); other in-house ATAs overexpressed in *Escherichia coli* were employed for screening purposes, but lower conversion values were found [7]
- ATA-251 from Codexis Inc. (2.0 mg) for the synthesis of (*S*)-1-phenylpropan-2-amine
- TA-P2-B01 from Codexis Inc. (2.0 mg) for the synthesis of (*R*)-1-phenylpropan-2-amine
- NaOH 4 M aqueous solution (200 µL)
- EtOAc (1.5 mL) as solvent for liquid–liquid extraction
- Water for liquid–liquid extraction
- Na_2SO_4
- Acetic anhydride (1 drop) for the determination of amine enantiomeric excess
- Hexane and isopropanol of high-performance liquid chromatography (HPLC) quality for chiral HPLC analyses
- 1.5 mL Eppendorf tube
- IKA® KS 4000 ic controlled orbital shaker
- BIOSAN TS-100 thermo shaker
- Agilent 6860 GC with FID detector; Agilent HP-1 column (30 m × 0.32 mm × 0.25 µm)
- Hewlett-Packard 1100 HPLC chromatograph with UV detector at 210, 215 and 254 nm; Chiralcel OD column (25 cm × 4.6 mm × 5 µm) using hexane/isopropanol (97 : 3 v/v) as mobile phase and isocratic 0.8 $mL.min^{-1}$ flow

12.4.2.2 Procedure

1. 1-Phenylpropan-2-one (1.34 mg, 0.01 mmol) was added to a 100 mM phosphate buffer, pH 7.5 (500 µL) containing $^{i}PrNH_2$ (1.0 M) and PLP (1 mM) in a 1.5 mL Eppendorf tube.
2. A commercially available ATA (2.0 mg) was added, and the mixture shaken at 30 °C and 250 rpm for 24 hr in an orbital shaker.
3. The reaction was quenched by addition of a 4 M NaOH aqueous solution (200 µL) and extracted with EtOAc (3 × 500 µL) by shaking in a thermo shaker.
4. The combined organic phases were washed with water (500 µL), dried over Na_2SO_4 and filtered.
5. The resulting solution was analysed by GC for conversion measurement (Tables 12.7 and 12.8): (a) 95% with the ATA-251 for the formation of (*S*)-1-phenylpropan-2-amine; (b) 96% with the TA-P2-B01 for the formation of (*R*)-1-phenylpropan-2-amine.

Table 12.7 *GC method for the measurement of reaction conversion.*

Temperature program (r = $°C.min^{-1}$)	Duration
60 °C 5 min – 10 r → 100 °C 0 min – 40 r → 200 °C 0 min	11.5 min

Table 12.8 Retention times for GC analysis using the Agilent HP-1 column.

Substance	Retention (min)
1-Phenylpropan-2-one	4.3
1-Phenylpropan-2-amine	8.8

6. Derivatisation of the resulting solution was performed by addition of acetic anhydride (10 μL) and shaken at 250 rpm and 30 °C for 1 min using a thermo shaker, obtaining the corresponding acetamide for the measurement of amine enantiomeric excess via HPLC (Tables 12.9 and 12.10; Scheme 12.7).

Table 12.9 HPLC method for the measurement of amine enantiomeric excess.

HPLC conditions	Duration
Chiralcel OD column (25 cm × 4.6 mm × 5 μm) using hexane/isopropanol 97 : 3 v/v as mobile phase and an isocratic 0.8 mL.min^{-1} flow	33 min

Table 12.10 Retention times for HPLC analysis using Chiralcel OD column.

Substance	Retention (min)
(*S*)-1-Phenylpropan-2-amine	25.7
(*R*)-1-Phenylpropan-2-amine	28.3

12.4.3 Procedure 3: Synthesis of 1-Phenylpropan-2-Amine Enantiomers from Allylbenzene through a Chemoenzymatic Cascade Sequence Using Isopropylamine as Amine Donor

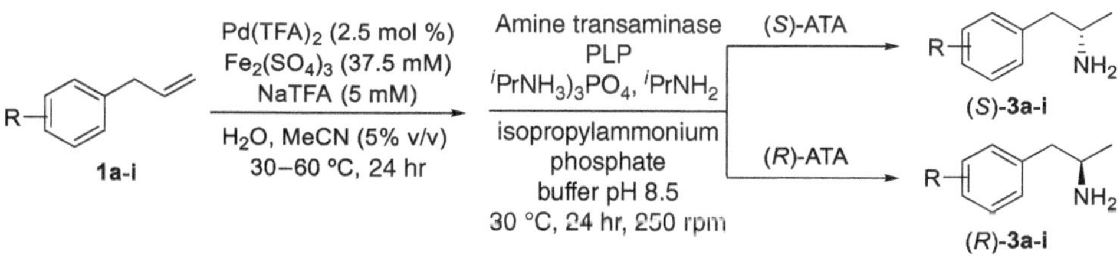

Scheme 12.8 Synthesis of 1-phenylpropan-2-amine enantiomers from allylbenzene through a chemoenzymatic cascade sequence using isopropylamine as amine donor.

12.4.3.1 Materials and Equipment

- Palladium(II) trifluoroacetate (Pd(TFA)$_2$, 0.5 mg, 0.0015 mmol)
- Sodium trifluoroacetate (NaTFA, 1.7 mg, 0.0125 mmol)

- Iron(III) sulfate hydrate ($Fe(SO_4)_3$, 37.5 mg, 0.0938 mmol)
- Distilled water (2.4 mL)
- Allylbenzene (8.3 μL, 0.0625 mmol)
- Acetonitrile (MeCN, 125 μL)
- Isopropylammonium phosphate (172 mg, 0.625 mmol)
- Isopropylamine (32 μL, 0.375 mmol)
- Pyridoxal-5-phosphate (0.6 mg, 0.0025 mmol)
- ATA-251 from Codexis Inc. (7.4 mg) for the synthesis of (*S*)-1-phenylpropan-2-amine
- TA-P2-B01from Codexis Inc. (7.4 mg) for the synthesis of (*R*)-1-phenylpropan-2-amine
- HCl 0.5 M aqueous solution (to bring solution to acidic pH)
- NaOH 4 M aqueous solution (1 mL)
- EtOAc (12.5 mL) as solvent for liquid–liquid extraction
- Na_2SO_4
- 1.5 mL Eppendorf tube for preparation of allylbenzene solution
- 20 mL Schlenk flask sealed with glass stopper
- IKA® WERKE RCT basic control stirrer hotplate
- IKA® KS 4000 ic controlled orbital shaker
- 10 mL Falcon tube for liquid–liquid extraction
- Thermo Scientific Heraeus Megafuge 16R Centrifuge
- Agilent 6860 GC with FID detector; conversion values calculated as described in Procedures 1 and 2
- Hewlett-Packard 1100 HPLC chromatograph; enantiomeric excess values calculated as described in Procedure 2

12.4.3.2 Procedure

1. $Pd(TFA)_2$ (0.5 mg), NaTFA (1.7 mg) and $Fe_2(SO_4)_3{\cdot}H_2O$ (37.5 mg) were added to a Schlenk flask and stored under nitrogen atmosphere sealed with a glass stopper.
2. Distilled water (2.4 mL) was added, forming a yellowish solution.
3. A solution of allylbenzene (8.3 μL) in acetonitrile (125 μL) was prepared in an Eppendorf tube and added to the solution, maintaining the nitrogen atmosphere in the Schenk flask.
4. The reaction was stirred at 30 °C for 24 hr in the absence of light within the closed flask.
5. Isopropyl ammonium phosphate (172 mg) and isopropylamine (31 μL) were added for a final amine donor concentration of 0.9 M and pH 8.5. PLP (0.6 mg, 1 mM) and the corresponding ATA (7.4 mg) were successively added.
6. The obtained suspension was shaken at 250 rpm and 30 °C for 24 hr in an orbital shaker. The reaction was then quenched by acidification with a 0.5 M HCl aqueous solution (2 mL).
7. The resulting solution was transferred to a Falcon tube and extracted with EtOAc (2 × 2.5 mL) using a centrifugue. The aqueous phase was collected, basified with a 4 M NaOH aqueous solution (1 mL) and extracted with EtOAc (3 × 2.5 mL) in the same centrifuge.

8. The organic phases were combined, dried over Na_2SO_4 and filtered, and the solvent was removed by distillation under reduced pressure, yielding enantiopure 1-phenylpropan-2-amine as a colourless oil: (a) 83% conversion and 78% isolated yield (6.6 mg) after acid–basic liquid–liquid extraction for the (*S*)-amine; (b) 83% conversion for the (*R*)-amine (Scheme 12.8).

Table 12.11 *Sequential cascade process to transform allylbenzenes into optically active 1-arylpropan-2-amines.*

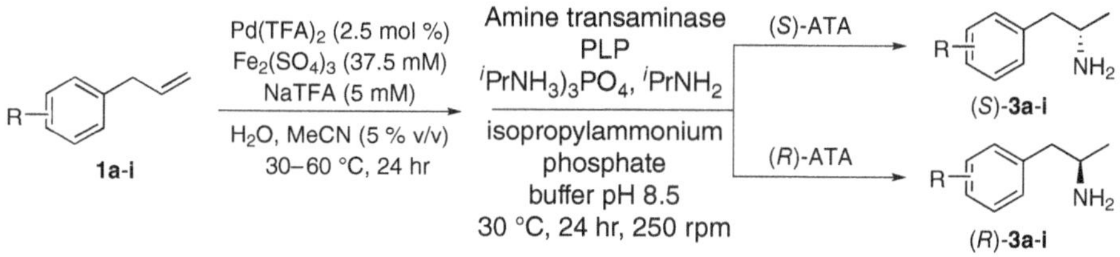

Entry	R	T (°C)[a]	ATA	Amine **3a–i** yield (%)	Amine **3a–i** ee (%)
1	H (**a**)	30	TA-P1-G06	83 (78)	>99 (*S*)
2	H (**a**)	30	TA-P2-B01	83	>99 (*R*)
3	2-CH_3 (**b**)	30	ATA-251	86 (85)	>99 (*S*)
4	2-CH_3 (**b**)	30	ATA-412	79	>99 (*R*)
5	3-CH_3 (**c**)	30	TA-P1-G06	87	>99 (*S*)
6	3-CH_3 (**c**)	30	TA-P2-B01	92 (92)	>99 (*R*)
7	4-CH_3 (**d**)	30	TA-P1-G06	84	99 (*S*)
8	4-CH_3 (**d**)	30	TA-P2-B01	92 (90)	>99 (*R*)
9	4-CF_3 (**e**)	45	TA-P1-F03	74	>99 (*S*)
10	4-CF_3 (**e**)	45	ATA-412	80 (70)	>99 (*R*)
11	2-OCH_3 (**f**)	60	ATA-251	89 (87)	>99 (*S*)
12	2-OCH_3 (**f**)	60	ATA-412	87 (82)	>99 (*R*)
13	4-OCH_3 (**g**)	30	TA-P1-G06	87 (84)	>99 (*S*)
14	4-OCH_3 (**g**)	30	TA-P2-B01	88 (86)	>99 (*R*)
15	3,4-OCH_2O- (**h**)	60	TA-P1-G05	81 (76)	>99 (*S*)
16	3,4-OCH_2O- (**h**)	60	TA-P2-B01	86 (85)	>99 (*R*)
17	3,4-$(OCH_3)_2$ (**i**)	45	TA-P1-F03	75	>99 (*S*)
18	3,4-$(OCH_3)_2$ (**i**)	45	ATA-412	79 (79)	>99 (*R*)

Reaction conditions: allylbenzene **1a–i** (25 mM), $Pd(TFA)_2$ (2.5 mol %), Fe(III) (37.5 mM, $Fe_2(SO_4)_3$ for **1a–e** or $FeCl_3$ for **1f–i**), NaTFA (5 mM), MeCN (5% v/v) in water were stirred at 30–60 °C for 24 hr under inert atmosphere. Then isopropylammonium phosphate (0.25 M), isopropylamine (0.15 M), PLP (0.5 mM) and the corresponding ATA (1 : 1 w/w enzyme/substrate) were successively added. The mixture was shaken at 250 rpm and 30 °C for 24 hr.

[a]Temperature of the Wacker–Tsuji oxidation.

12.4.4 Conclusion

The described chemoenzymatic approach allows the production of 1-phenylpropan-2-amine enantiomers via Wacker–Tsuji oxidation of allylbenzene, which smoothly leads to 1-phenylpropan-2-one and subsequent stereoselective biotransamination of the ketone intermediate by the appropriate selection of a commercially available ATA (Scheme 12.5). The strategy occurs through a sequential approach in aqueous medium, and has been successfully extended to a series of nine optically active 1-arylpropan-2-amines, which can be obtained with good to very high conversions (74–92%) and excellent selectivities (>99% ee) after optimisation of the Wacker and biotransamination reactions (Table 12.11) [7].

References

1. (a) Höhne, M. and Bornscheuer, U.T. (2009) *ChemCatChem*, **1**, 42–51; (b) Nugent, T.C. (2010) *Chiral Amine Synthesis: Methods,* Developments and Applications, Wiley-VCH.
2. (a) Martínez-Montero, L., Gotor, V., Gotor-Fernández, V. and Lavandera, I. (2016) *Advanced Synthesis & Catalysis*, **358**, 1618–1624; (b) Gomm, A., Lewis, W., Green, A.P. and O'Reilly, E. (2016) *Chemistry: A European Journal*, **22**, 12 692–12 695; (c) Gomm, A., Grigoriou, S., Peel, C. *et al.* (2018) *European Journal of Organic Chemistry*, **2018**, 5282–5284.
3. (a) Pushpanath, A., Siirola, E., Bornadel, A. *et al.* (2017) *ACS Catalysis*, **7**, 3204–3209; (b) Liu, J., Pang, B.Q.W., Adams, J.P. *et al.* (2017) *ChemCatChem*, **9**, 425–431; (c) Knaus, T., Böhmer, W. and Mutti, F.G. (2017) *Green Chemistry*, **19**, 453–463; (d) Knaus, T., Cariati, L., Masman, M.F. and Mutti, F.G. (2017) *Organic and Biomolecular Chemistry*, **15**, 8313–8325.
4. (a) Roiban, G.-D., Kern, M., Liu, Z. *et al.* (2017) *ChemCatChem*, **9**, 4475–4479; (b) Matzel, P., Gand, M. and Höhne, M. (2017) *Green Chemistry*, **19**, 385–389.
5. (a) Aleku, G.A., France, S.P., Man, H. *et al.* (2017) *Nature Chemistry*, **9**, 961–969; (b) Aleku, G.A., Mangas-Sanchez, J., Citoler, J. *et al.* (2018) *ChemCatChem*, **10**, 515–519.
6. (a) Muñoz, L., Rodriguez, A.M., Rosell, G. *et al.* (2011) *Organic and Biomolecular Chemistry*, **9**, 8171–8177; (b) Rodríguez-Mata, M., Gotor-Fernández, V., González-Sabín, J. *et al.* (2011) *Organic and Biomolecular Chemistry*, **9**, 2274–2278.
7. González-Martínez, D., Vicente Gotor, V. and Gotor-Fernández, V. (2019) *Advanced Synthesis & Catalysis*, **361**, 2582–2593.

12.5 Chemoenzymatic Synthesis of (2*S*,3*S*)-2-Methylpyrrolidin-3-ol

Raquel Roldan,[1] Karel Hernandez,[1] Jesús Joglar,[1] Jordi Bujons,[1] Teodor Parella,[2] Wolf-Dieter Fessner[3] and Pere Clapés*[1]

[1]*Instituto de Química Avanzada de Cataluña IQAC-CSIC, Barcelona, Spain*
[2]*Servei de Ressonància Magnètica Nuclear, Universitat Autònoma de Barcelona, Bellaterra, (Cerdanyola del Vallès), Spain*
[3]*Institut für Organische Chemie und Biochemie, Technische Universität Darmstadt, Darmstadt, Germany*

Nitrogen heterocycles are important compounds found in many naturally occurring biologically active compounds and a structural moiety extensively utilised in pharmaceutical drug development [1]. We have investigated the chemoenzymatic preparation of *N*-heterocyclic

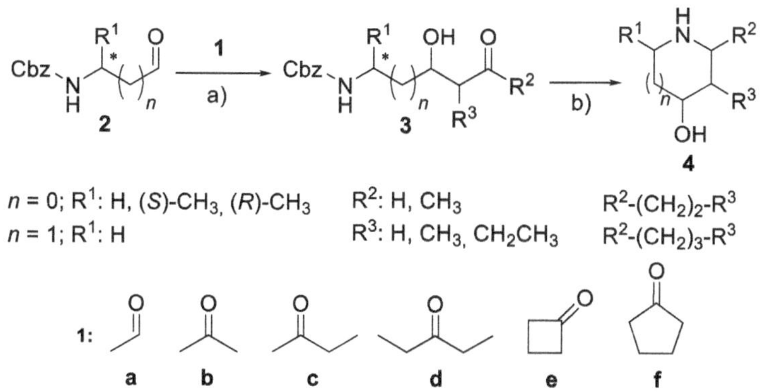

Scheme 12.9 Two-step chemo-enzymatic synthesis of iminocyclitols. (a) Aldolase-catalysed aldol reaction of D-fructose-6-phosphate aldolase from *Escherichia coli* (FSA) and variants or of 2-deoxy-D-ribose-5-phosphate aldolase from *Thermotoga maritima* ($DERA_{Tma}$). (b) hydrogenolysis of Cbz and reductive amination using H_2 in the presence of palladium over charcoal.

compounds with low hydroxylation levels **4** in a two-step strategy (Scheme 12.9). Step 1 consists of an aldol addition of ethanal and simple linear or cyclic aliphatic ketone **1** to a *N*-Cbz-protected amino aldehyde **2** catalysed by D-fructose-6-phosphate aldolase from *Escherichia coli* (FSA) and variants or 2-deoxy-D-ribose-5-phosphate aldolase from *Thermotoga maritima* ($DERA_{Tma}$) as catalyst [2]. Step 2 sees the aldol adducts transformed into the corresponding *N*-heterocycles via intramolecular reductive amination using H_2-palladium over charcoal, yielding *N*-heterocyclic derivatives of piperidine, pyrrolidine and *N*-bicyclic structures with stereoselectivities of 96–98% ee and 97 : 3 dr in isolated yields ranging from 35 to 79% [2]. This two-step strategy starting with *N*-Cbz-protected amino aldehyde was used to produce (2*S*,3*S*)-2-methylpyrrolidin-3-ol following the detailed procedure described herein (Scheme 12.10).

12.5.1 Procedure: 1 Preparation of (*S*)-*N*-Cbz-2-Aminopropanal 5

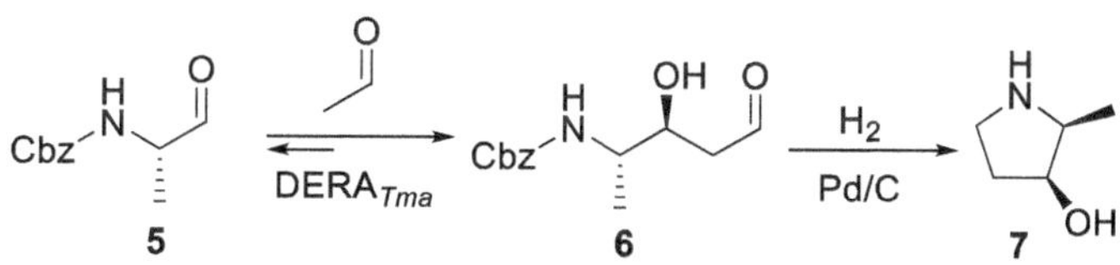

Scheme 12.10 Two step chemoenzymatic synthesis of (2*S*,3*S*)-2-methylpyrrolidin-3-ol using $DERA_{Tma}$: 2-deoxy-D-ribose-5-phosphate aldolase from *Thermotoga maritima*.

12.5.1.1 Materials and Equipment

- *S*-2-Amino-1-propanol (3.15 g, 42.0 mmol)
- *N*-(Benzyloxycarbonyloxy)succinimide (Cbz-OSu; 10.5 g, 42 mmol)
- *o*-Iodoxybenzoic acid (IBX, 5.36 g, 9.56 mmol)
- Ethyl acetate (EtOAc)

- Hexane
- Dioxane
- Anhydrous sodium sulfate
- Sodium hydrogen carbonate
- Citric acid
- Filter paper
- Thin-layer chromatography (TLC) plates: ALUGRAM® Xtra SIL G aluminium sheets (Silica Gel 60)
- AriumTM Pro Ultrapure Water
- 500 mL round-bottom flask connected to a water-cooled Liebig or Vigreux condenser
- 150 mL Buchner funnel with 60 mm-diameter coarse-porosity (4) fritted disc
- 500 mL Buchner flask
- 500 mL Erlenmeyer flasks
- Magnetic stirrer
- Magnetic bar

12.5.1.2 Procedure

1. (*S*)-2-Amino-1-propanol (3.15 g, 42.0 mmol) was dissolved in dioxane/water 4 : 1 (100 mL).
2. Cbz-OSu (10.5 g, 42.0 mmol) in dioxane/water 4 : 1 (10–50 mL) was added dropwise at 25 °C.
3. After stirring for 24 hr, the mixture was evaporated to dryness under reduced pressure.
4. The residue was dissolved with EtOAc (150 mL) and washed successively with citric acid 5% w/v (3 × 50 mL), $NaHCO_3$ 10% w/v (3 × 50 mL) and brine (2 × 50 mL). After drying over anhydrous Na_2SO_4, the organic layer was evaporated under reduced pressure to afford a white solid of (*S*)-*N*-Cbz-2-aminopropanol (7.90 g, 90% yield; 98% purity by HPLC, t_R = 15.5 min).
5. IBX (5.36 g, 9.56 mmol) was added to a solution of (*S*)-*N*-Cbz-2-aminopropanol (2.0 g, 9.56 mmol) in EtOAc (250 mL) and the mixture was heated at reflux for 6 hr, open to the atmosphere [3].
6. The mixture was cooled to room temperature and filtered, the filtrate was washed with brine (3 × 100 mL) and the organic phase was dried with anhydrous sodium sulfate. The solvent was removed by distillation under reduced pressure to afford (*S*)-*N*-Cbz-2-aminopropanal **5** (1.98 g, quantitative yield) as a pale yellow oil ready to be used in the enzymatic reaction without any further purification step. $[\alpha]^{20}{}_D = -16.3$ (c=1 in MeOH); ^1H NMR (300 MHz, $[D_6]$DMSO, 25 °C): δ 9.5 (1H, s, CHO), 5.0 (2H, s; CH_2O), 4.0 (1H, m; CH), 1.2 ppm (3H, d; CH_3); ^{13}C NMR (75 MHz, $[D_6]$DMSO, 25 °C): δ 201 (CHO), 156.1 (CONH), 65.7 (CH_2O), 55.1 (CH), 13.8 ppm (CH_3).
7. HPLC analyses were performed on an X-Bridge™ C18, 5 μm, 4.6 × 250 mm column (Waters). Samples (30 μL) were injected and eluted under the following conditions: solvent system (a) aqueous trifluoroacetic acid (TFA; 0.1% v/v) and (b) TFA (0.095% v/v) in CH_3CN/H_2O (4 : 1), gradient elution from 10 to 100% (b) in 30 min, flow rate 1 mL.min^{-1}, detection at 215 nm, column temperature 30 °C. t_R ((*S*)-*N*-Cbz-2-aminopropanol) = 15.5 min; t_R **5** = 15.9 min. TLC analysis: eluent hexane/EtOAc 1 : 1 v/v; Rf ((*S*)-*N*-Cbz-2-aminopropanol) = 0.2; Rf **5** = 0.5.

12.5.2 Procedure: 2 Preparation of (4*S*,3*S*)-*N*-Cbz-4-Amino-3-Hydroxypentanal 6

12.5.2.1 Materials and Equipment

- Ethanal (120 µL, 2 mmol)
- (*S*)-*N*-Cbz-2-aminopropanal (414.0 mg, 2.0 mmol; prepared in Procedure 1)
- AriumTM Pro Ultrapure Water, used for analytical and preparative high-performance liquid chromatography (HPLC), buffer preparations and other assay solutions
- 2-Deoxy-D-ribose-5-phosphate aldolase from *T. maritima* ($DERA_{Tma}$) as cell-free crude extract from Prozomix Ltd (activity 0.056 $U.mg^{-1}$ lyophilised powder); 1 U is defined as the amount of protein that catalyses the cleavage of 1 µmol of 2-deoxy-D-ribose-5-phosphate (Dr5P) per minute at [Dr5P] = 0.7 µM, 50 mM triethanolamine buffer pH 7.5 and 30 °C
- 50 mL Falcon tubes
- Shaker (1000 rpm)
- 150 mL Buchner funnel with 60 mm-diameter coarse-porosity (4) fritted disc
- 500 mL Buchner flask
- Analytical HPLC
- Column chromatography glass column (47 × 4.5) packed with silica gel (100 g, 35–70 µm, 200–500 mesh)
- Silica gel (100 g, 35–70 µm, 200–500 mesh)
- Celite™ filter aid

12.5.2.2 Procedure

1. $DERA_{Tma}$ (60 mg, 3 $mg.mL^{-1}$) was dissolved in water (18.0 mL), then 1 M buffer triethanolamine, pH 8 solution (1 mL) was added.
2. (*S*)-*N*-Cbz-2-aminopropanal (414.0 mg, 2.0 mmol, 100 mM in the reaction) and ethanal (120 µL, 2 mmol, 100 mM in the reaction) were added.
3. The reaction mixture was shaken at 1000 rpm and 25 °C for 24 hr (98% aldol product formed as determined by HPLC). HPLC analyses were performed on an X-Bridge™ C18, 5 µm, 4.6 × 250 mm column (Waters). Samples (30 µL) were injected and eluted under the following conditions: solvent system (a) aqueous trifluoroacetic acid (TFA; 0.1% v/v) and (b) TFA (0.095% v/v) in CH_3CN/H_2O 4 : 1, gradient elution from 10 to 100% (b) in 30 min, flow rate 1 $mL.min^{-1}$, detection at 215 nm, column temperature 30 °C. t_R **5** = 15.9 min; t_R **6** = 15.1 min. The amount of aldol adduct product was quantified from the peak areas using an external standard methodology. Samples were withdrawn (50 µL) from the reaction medium, centrifuged, diluted with methanol (450 µL) and analysed with HPLC.
4. The reaction was stopped by addition of MeOH (30 mL) to precipitate the enzyme, and the mixture was filtered through Celite™ (~10–15 g).
5. MeOH was evaporated and the reaction was lyophilised.
6. (4*S*,3*S*)-*N*-Cbz-4-Amino-3-hydroxypentanal was purified by silica gel column chromatography and eluted with a step gradient of hexane/EtOAc 100 : 0 (200 mL) 80 : 20 (200 mL), 70 : 30 (200 mL), 60 : 40 (200 mL) and 50 : 50 (700 mL), yielding 171.2 mg (34% yield) of aldol adduct.

12.5.3 Procedure 3: Preparation of (2*S*,3*S*)-2-Methylpyrrolidin-3-ol: Reductive Amination

12.5.3.1 Materials and Equipment

- (4*S*,3*S*)-*N*-Cbz-4-Amino-3-hydroxypentanal (171.2 mg, 0.68 mmol, prepared in Procedure 2)
- Palladium over charcoal 10% Pd
- Hydrogen (50 psi)
- Pressure-vessel, heavy-wall, round-bottom flask equivalent, equipped with an inlet cap
- Magnetic stirrer
- Stirrer bar
- 150 mL Buchner funnel with 60 mm-diameter coarse-porosity (4) fritted disc
- 500 mL Buchner flask
- Celite™ filter aid

12.5.3.2 Procedure

1. (4*S*,3*S*)-*N*-Cbz-4-Amino-3-hydroxypentanal (171.2 mg, 0.68 mmol) was dissolved in MeOH (400 mL).
2. The mixture was flushed with N_2 for 20 min.
3. 10% Pd on C with 50% humidity (200 mg) was added.
4. The mixture was shaken under H_2 (2.5 atm) at room temperature overnight.
5. The reaction was filtered through Celite™ and the solvent was evaporated. Product (51.2 mg, 74% yield) was characterised without any further purification.

^{1}H NMR (400 MHz, CD_3OD) δ 4.25 (t, $J = 3.1$ Hz, 1H), 3.48 (tt, $J = 2 \times 6.7, 2 \times 3.4$ Hz, 1H), 3.46 – 3.34 (m, 1H), 2.26 – 2.13 (m, 1H), 2.09 – 2.00 (m, 1H), 1.38 (d, $J = 6.7$ Hz, 2H). ^{13}C NMR (101 MHz, CD_3OD) δ 72.1, 61.6, 44.0, 34.1, 11.7.

12.5.4 Conclusion

The procedure presented here provides a straightforward way to prepare (2*S*,3*S*)-2-methylpyrrolidin-3-ol in high yield and with high stereoselectivity using standard chemical laboratory and an aldolase DERA widely employed in aldol addition reactions. This procedure can be adapted to similar substrates as described in [2].

References

1. (a) Marcantoni, E. and Petrini, M. (2016) *Advanced Synthesis & Catalysis*, **358**, 3657–3682; (b) Lazzara, P.R., Fitzpatrick, K.P. and Eichman, C.C. (2016) *Chemistry: A European Journal*, **22**, 16 779–16 782; (c) Lovering, F., Bikker, J. and Humblet, C. (2009) *Journal of Medicinal Chemistry*, **52**, 6752–6756; (d) Rubiralta, M., Giralt, E. and Diez, A. (1992) *Piperidine: Structure, Preparation, Reactivity, and Synthetic Applications of Piperidine and its Derivatives*, Vol. **40**, Wiley-VCH Verlag GmbH; (e) Schneider, M.J. (1996) Pyridine and piperidine alkaloids: an update, in *Alkaloids: Chemical and Biological Perspectives*, Vol. **10** (ed. S.W. Pelletier), Pergamon, pp. 155–299.
2. Roldán, R., Hernández, K., Joglar, J. *et al.* (2019) *Advanced Synthesis & Catalysis*, **361**, 2673–2687.
3. Ocejo, M., Vicario, J.L., Badia, D. *et al.* (2005) *Synlett*, **2005**, 2110–2112.

13

Whole-Cell Procedures

13.1 Semipreparative Biocatalytic Synthesis of (*S*)-1-Amino-1-(3′-Pyridyl)methylphosphonic Acid

Ewa Żymańczyk-Duda,*[1] Natalia Dunal,[1] Małgorzata Brzezińska-Rodak,[1] Tomasz K. Olszewski,[2] Magdalena Klimek-Ochab[1] and Monika Serafin-Lewańczuk[1]

[1]*Department of Bioorganic Chemistry, Faculty of Chemistry, Wrocław University of Science and Technology, Wrocław, Poland*

[2]*Department of Organic Chemistry, Faculty of Chemistry, Wrocław University of Science and Technology, Wrocław, Poland*

α-Aminophosphonates, analogues of the natural amino acids, can act as antibiotics, crop-protection agents, herbicides or peptide mimetics. Such biological activities are characteristic of single-enantiomer chiral compounds of defined absolute configuration [1]. Phosphonate derivatives with a heteroatom incorporated into the aryl ring have attracted attention because they are considered compounds of low toxicity and good bioavailability compared to previously reported organophosphonates. Bioconversion of chemically synthesised racemic mixtures of 1-amino-1-(3′-pyridyl)methylphosphonic acid hydrochloride **3** (Scheme 13.1) with immobilised mycelium of *Penicillium funiculosum* has allowed the optically pure (*S*)-isomer to be produced (Scheme 13.2).

13.1.1 Procedure 1: Synthesis of Racemic 1-Amino-1-(3′-Pyridyl)methylphosphonic Acid Hydrochloride 3

13.1.1.1 Materials and Equipment

- Benzhydrylamine (8.1 mL, 47.0 mmol)
- 3-Pyridinecarboxaldehyde (4.38 mL, 47.0 mmol)
- Dichloromethane (DCM; 50 mL)

Applied Biocatalysis: The Chemist's Enzyme Toolbox, First Edition. Edited by John Whittall and Peter W. Sutton.
© 2021 John Wiley & Sons Ltd. Published 2021 by John Wiley & Sons Ltd.

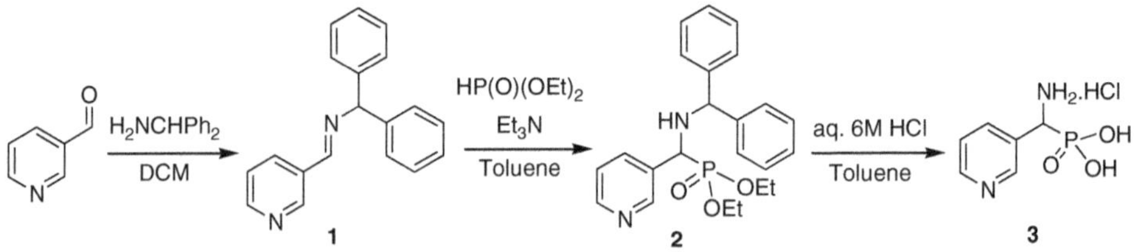

Scheme 13.1 Preparation of racemic mixture of 1-amino-1-(3′-pyridyl)methylphosphonic acid hydrochloride **3**.

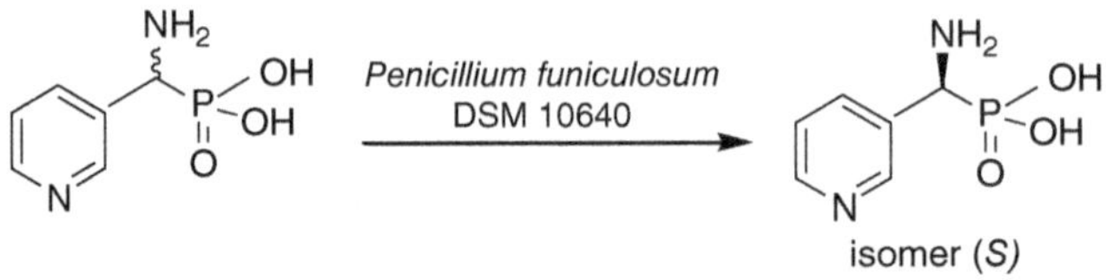

Scheme 13.2 *Bioconversion synthesis of 1-amino-1-(3′-pyridyl)methylphosphonic acid.*

- Anhydrous sodium sulfate (anh. Na_2SO_4; 5 g)
- Diethyl phosphite [$HP(O)(OEt)_2$] (4.76 mL, 37.0 mmol)
- Toluene (2 × 50 mL)
- Triethylamine (Et_3N; 5.1 mL)
- Hydrochloric acid 6 M (aq. HCl; 15 mL)
- Distilled water (dH_2O; 20 mL)
- Ethanol (10 mL)
- 3× 100 mL one-necked round-bottom flasks
- Glass stopcock
- 2× 50 mL glass sintered disc filter funnels, porosity G4
- 2× Erlenmeyer-shape 100 mL filtering flasks for vacuum filtrations, with glass hose connector
- Liebig condenser
- 250 mL separator funnel, cone shape
- Magnetic stirrer with heating (Heidolph MR Hei-Standard)
- 250 mL oil bath
- Vacuum rotary evaporator (Buchi R100 equipped with vacuum pump V-100 and vacuum controller I-100)

13.1.1.2 Procedure

1. In a 100 mL one-necked round-bottom flask placed on a magnetic stirrer, neat benzhydrylamine (8.1 mL, 47.0 mmol) was added to a solution of 3-pyridinecarboxaldehyde (4.38 mL, 47.0 mmol) in DCM (50 mL) and the reaction mixture was stirred at room temperature overnight lightly stoppered with a glass stopper to avoid evaporative losses. The next day, anh. Na_2SO_4 (5 g) was added to the reaction in order to remove the formed water and stirring was continued for an additional 30 min.
2. The drying agent was filtered off under vacuum over a glass sinter.

3. The resulting filtrate was concentrated by distillation under reduced pressure to yield 10 g of crude imine **1** as white solid, usable without further purification.
4. Crude imine **1** (10 g) was placed in a 100 mL one-necked round-bottom flask and dissolved in toluene (50 mL). $HP(O)(OEt)_2$ (4.76 mL, 37.0 mmol) and then Et_3N (5.1 mL) were added. The mixture was stirred at 120 °C for 8 hr.
5. The reaction mixture was cooled to room temperature and concentrated by distillation under reduced pressure to yield 10 g of crude aminophosphonate diethyl ester **2** as a yellowish solid, which was used directly in the final synthetic step.
6. Crude aminophosphonate ester **2** (10.0 g) was placed in a 100 mL one-necked round-bottom flask and dissolved in toluene (50 mL). 6M HCl (aq) (15 mL) was added and the mixture was stirred at 120 °C for 4 hr.
7. The reaction was cooled to room temperature and transferred to a separating funnel and the layers were separated.
8. The organic layer was discarded and the aqueous phase was evaporated to dryness by distillation under reduced pressure to crude 1-amino-1-(3′-pyridyl)methylphosphonic acid hydrochloride **3** (5.0 g), which was further purified by recrystallisation from a mixture of dH_2O/EtOH (20 : 10 v/v). After crystallisation, the desired aminophosphonic acid hydrochloride **3** was filtered off under vacuum over a glass sinter to afford 1-amino-1-(3′-pyridyl)methylphosphonic acid hydrochloride **3** as a white solid in 85% yield (4.7 g).

^{1}H NMR (600.58 MHz; D_2O) δ 8.81(s, 1 H, py-2), 8.73 (d, 1 H, py-6, $J-5.8$ Hz), 8.61 (d, 1H, py-4, $J=6.8$ Hz), 8.08–8.04 (m, 1H, py-5), 4.69 (d, 1H, CHP, $J=16.3$ Hz). ^{31}P NMR (243.12 MHz; D_2O) δ 8.42 (s).

13.1.2 Procedure 2: Preparation and Immobilisation of *P. funiculosum* DSM 10640

13.1.2.1 Materials and Equipment

- Potato dextrose broth (PDb; Difco, cat. nr: 90003-494)
- Potato dextrose agar (PDa; Difco, cat. nr: 90000-758)
- Slant with stock culture of *P. funiculosum* DSM 10640 (DSMZ)
- Sterile microbiological loop (Roth)
- 2× 250 mL Erlenmeyer flask with cotton cap
- Laminar flow unit (BIOAIR Instruments Ins. GFL 3019)
- Orbital shaker with adjustable temperature (New Brunswick Scientific Innova 4000)
- Autoclave (Classic Prestige Medical)
- Petri dish (diam. 10 cm)
- Sterile aqueous solution containing 0.05% Triton X-100 (10 mL)
- Automatic pipette with adjustable volume (1–10 mL)
- Sterile tips for automatic pipette with adjustable volume (1–10 mL)
- 30× Polyurethane foams: FILTREN TM25133, pore size μm (1060–1600), 1 × 1 × 1 cm cube (Bogmar)
- Funnel (diam. 15 cm)
- Paper filter (diam. 20 cm, 79–85 g.m^{-2})
- dH_2O

13.1.2.2 Procedure

1. Pre-culture preparation: *P. funiculosum* DSM 10640 spores were transferred from the stock culture (stored as slant on PDa, 4 °C) with a sterile microbiological loop into a 250 mL flask containing 100 mL PDb under sterile conditions (laminar). The mixture was incubated at 135 rpm and 25 °C for 4 days.
2. 100 μL of the resultant culture was transferred to a Petri dish (containing 20 mL PDa). After 7 days (stationary cultivation, 25 °C), 10 mL Triton X-100 was added to the Petri dish and 1 mL of the spore suspension was transferred into a 250 mL flask filled with PDb medium (100 mL) and 30 cubes of polyurethane foams. The mixture was shaken at 135 rpm and 25 °C for 4 days, after which mycelia had overgrown the foams.
3. The supernatant was separated from the foams by gravity filtration using a filter paper. The immobilised fungal biomass was washed with dH_2O to afford the immobilised biocatalyst ready for immediate use (Figure 13.1).

Figure 13.1 *Mycelia after immobilisation.*

13.1.3 Procedure 3: Bioconversion of Racemic Mixture of 1-Amino-1-(3′-Pyridyl)methylphosphonic Acid

13.1.3.1 Materials and Equipment

- Column bioreactor – plastic chromatographic tube without sinter (13.5 cm length and 2.5 cm internal diameter)
- Peristaltic pump (Ecoline, Ismatec)

- Reservoir with biotransformation medium containing substrate (200 mL, 11.2 mM solution in water in a 300 mL Erlenmeyer flask with a ground joint with adaptor for bottle gas washing; Drechsler, SJ 29/32)
- Silicone tubes (diam. 6 mm)
- Magnetic stirrer
- Round-bottom flask with sinter (250 mL)
- Rotary evaporator
- Nuclear magnetic resonance (NMR) spectrometer (Bruker Advance™ 600)
- D_2O (99.9% of atom D; Merck)
- α-Cyclodextrin (Merck)
- NMR tubes
- pH meter (HANNA Instruments pH 210)
- 5 M NaOD solution in D_2O
- Polarimeter (polAAr 31; Index Instruments Inc.)

13.1.3.2 Procedure

1. The column bioreactor was packed with immobilised biocatalyst (30 pieces of foam overgrown by fungus).
2. The biotransformation was carried out in a continuous-operation system consisting of filled-column bioreactor and peristaltic pump forcing the flow of the water substrate solution (200 mL, 11.2 mM) (Figure 13.2). 200 mL of the substrate solution was pumped (5 $mL.min^{-1}$) from the reservoir into the bioreactor. The process was conducted for 5 days.

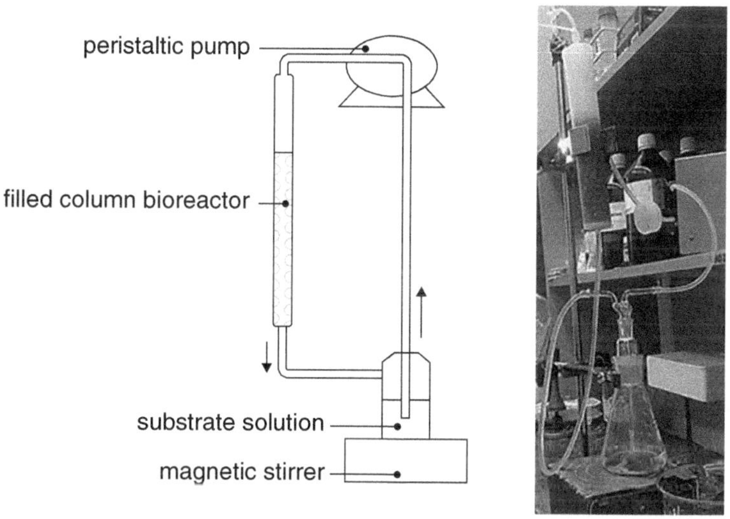

Figure 13.2 *System for continuous-flow column experiments.*

13.1.4 Procedure 4: Product Isolation and Analysis

13.1.4.1 Materials and Equipment

- Round-bottom flask with sinter (250 mL)
- Rotary evaporator

- NMR spectrometer (Bruker Advance™ 600)
- D_2O (99.9% of atom D; Merck)
- α-Cyclodextrin (Merck)
- NMR tubes
- pH meter (HANNA Instruments pH 210)
- 5 M NaOD (solution in D_2O)
- Polarimeter (polAAr 31; Index Instruments Inc.)

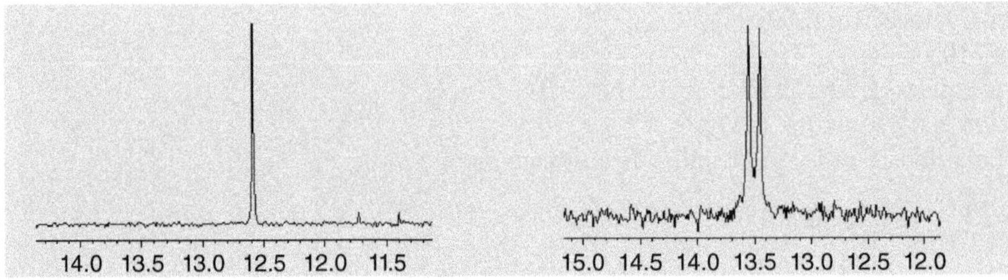

Figure 13.3 *^{31}P NMR spectra of 1-amino-1-(3′-pyridyl)methylphosphonic acid with the addition of cyclodextrin. Left: after bioconversion. Right: before bioconversion.*

13.1.4.2 Procedure

1. After the system was stopped, the whole post-biotransformation medium, collected in reservoir, was evaporated in the round-bottom flask with sinter and the residues were analysed.
2. Enantiomeric excess was evaluated by ^{31}P NMR technique. Spectra were measured at 243.12 MHz in D_2O (99.9% of atom D) and chemical shifts (δ) were reported in ppm. NMR sample was prepared as follows: 5.8 mg of product was dissolved in 600 μL of D_2O with the addition of 58 mg of α-cyclodextrin as a chiral solvating agent. The pH value of the NMR sample was adjusted with NaOD to pD ~10–11.
3. Absolute configuration was established tentatively using the model-similar compound 1-aminophenylmethanephosphonic acid, assigned as (*S*)-1-amino-1-(3′-pyridyl)methylphosphonic acid by optical rotation measurement and comparison to the literature data (optical rotation $[\alpha]_D = -2$ (2M NaOH)) (Figure 13.3) [2].

13.1.5 Conclusion

The described procedure allowed the synthesis of a chiral building block in optically pure form with 100% yield. This method can be considered a low-cost protocol and is not harmful to the environment.

References

1. Wolinska, E., Haldys, K., Gora, J. *et al.* (2019) *Chemistry & Biodiversity*, **16** (7), e1900167
2. Rudzinska, E., Dziedzioła, G., Berlicki, Ł. and Kafarski, P. (2010) *Chirality*, **22**, 63–68.

13.2 Practical and User-Friendly Procedure for the Regio- and Stereoselective Hydration of Oleic, Linoleic and Linolenic Acids, Using Probiotic *Lactobacillus* Strains as Whole-Cell Biocatalysts

Stefano Serra* and Davide De Simeis

CNR Istituto di Chimica del Riconoscimento Molecolare (ICRM), Milan, Italy

Hydroxy fatty acids (HFAs) are important chemicals widely used for a number of industrial applications [1]. The chemical synthesis of HFAs is usually achieved by hydration of the unsaturated fatty acids (UFAs) straightforwardly available from natural sources. Unfortunately, this kind of reaction lacks stereochemical control, and complex mixtures of isomers are usually formed. On the other hand, high regio- and stereoselective transformation of UFAs into HFAs can be conveniently accomplished by taking advantage of the catalytic activity of the enzyme group of fatty acid hydratases (EC 4.2.1.53) [2]. Different studies have reported the 10-hydratase activity of some *Lactobacillus* species [3]. These microorganisms are currently used as probiotics [4], and their use in industrial processes does not involve any safety concerns [5].

For these reasons, we have established a preparative, whole-cell-based procedure that exploits the hydratase activity of bacteria belonging to the *Lactobacillus* genus [6]. *L. rhamnosus* LGG (ATCC 53103) and *L. plantarum* 299V are suitable biocatalysts for the hydration reaction of the most common UFAs, namely oleic acid **1a**, linoleic acid **1b** and linolenic acid **1c** (Scheme 13.3). We have discovered that the addition of the latter fatty acids to an anaerobic culture of the latter bacteria, during the early part of their exponential phase of growth, allows the regiospecific production of the corresponding 10-hydroxy derivatives. Accordingly, the biotransformation of acids **1a–c** affords (*R*)-10-hydroxystearic acid **2a**, (*S*)-(12*Z*)-10-hydroxy-octadecenoic acid **2b** and (*S*)-(12*Z*,15*Z*)-10-hydroxy-octadecadienoic acid **2c**, respectively, in very high stereoselectivity. The hydration reactions are conveniently performed with an acid concentration up to 6 g.L^{-1} to produce suitable HFAs with up to 55% yield.

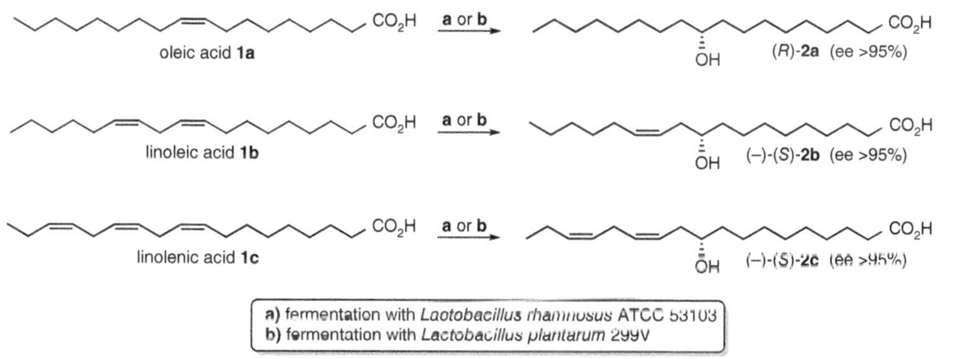

Scheme 13.3 *Lactobacillus* strain-mediated biotransformation of oleic acid **1a**, linoleic acid **1b** and linolenic acid **1c** into (*R*)-10-hydroxystearic acid **2a**, (*S*)-(12*Z*)-10-hydroxy-octadecenoic acid **2b** and (*S*)-(12*Z*,15*Z*)-10-hydroxy-octadecadienoic acid **2c**, respectively.

13.2.1 Procedure 1: Preparation of *Lactobacillus* Inoculum Using Anaerobic Flasks

13.2.1.1 Materials and Equipment

- MRS broth (powder, Millipore, 5.1 g)
- Cysteine (100 mg)
- Sodium thioglycolate (100 mg)
- Resazurine sodium salt (1 mg)
- Distilled water (dH_2O)
- Tween 80 (0.1 mL)
- 2× silicone rubber septa
- 2× conical vacuum flasks
- Skimmed milk (sterilised, 4 mL)
- *L. rhamnosus* ATCC 53103, lyophilised probiotic powder (trade name Kaleidon 60; 4.5×10^9 cfu; Malesci Spa)
- *L. plantarum* 299V, lyophilised probiotic powder (trade name Smebiocta LP299V; 5×10^9 cfu; Ipsen Pharma)
- Incubator shaker (Eppendorf, Innova 42R)
- Autoclave (Fedegari FVS/3)
- Cooling centrifuge (min. 3500× g)

13.2.1.2 Procedure

1. MRS broth powder (5.1 g), cysteine (100 mg), sodium thioglycolate (100 mg), tween 80 (0.1 mL) and resazurine sodium salt (1 mg) were dissolved in dH_2O (100 mL) and the resulting pink solution was dispensed in two 100 mL conical vacuum flasks. The flasks were flushed with nitrogen until complete removal of the oxygen content, as indicated by a change in the colour solution from pink to pale yellow.
2. The flasks were sealed with the silicone rubber septa then sterilised at 121 °C for 15 min.
3. The lyophilised probiotic powder containing the bacteria *L. rhamnosus* ATCC 53103 (4.5×10^9 cfu) was rehydrated for 15 min using sterilised skimmed milk (2 mL). The obtained suspension was inoculated via syringe into the anaerobic flask, which was incubated at 37 °C and 130 rpm for 12 hr. The culture was centrifuged at 3500× g and 4 °C for 5 min, the supernatant removed and the cells suspended in 5 mL of sterilised skimmed milk. The obtained suspension was used as inoculum for the fatty acid biotransformation.
4. The lyophilised probiotic powder containing the bacteria *L. plantarum* 299V (5×10^9 cfu) was rehydrated for 15 min using sterilised skimmed milk (2 mL). The obtained suspension was inoculated via syringe into the anaerobic flask, which was incubated at 37 °C and 130 rpm for 12 hr. The culture was centrifuged at 3500× g and 4 °C for 5 min, the supernatant removed and the cells suspended in 5 mL of sterilised skimmed milk. The obtained suspension was used as inoculum for the fatty acid biotransformation.

13.2.2 Procedure 2: Whole-Cell-Catalysed Hydration of Oleic Acid, Linoleic Acid and Linolenic Acid

13.2.2.1 Materials and Equipment

- MRS broth powder (Millipore, 51 g)
- Cysteine (1 g)

- Tween 80 (1 mL)
- 5 L Fermenter (Biostat A BB-8822000, Sartorius-Stedim)
- dH_2O
- Glucose (25% w/v in dH_2O, sterilised)
- Aqueous NH_3 (10% w/v, sterilised)
- Aqueous acetic acid (10% w/v, sterilised)
- Ethyl acetate (EtOAc)
- Hexane
- NaCl-saturated solution (brine)
- Anhydrous sodium sulfate (anh. Na_2SO_4)
- Celite
- Thin-layer chromatography (TLC) Silica Gel 60 F_{254} plates (plastic foil; Merck)
- TLC visualisation reagent (25 g phosphomolybdic acid hydrate, 10 g $CeSO_4{\cdot}4H_2O$, 60 mL H_2SO_4 96%, dH_2O to prepare 1L of reagent)
- Silica gel for column chromatography (70–230 mesh; Sigma-Aldrich)
- Rotary evaporator
- Inoculum of *L. rhamnosus* ATCC 53103 from Procedure 1
- Inoculum of *L. plantarum* 299V from Procedure 1
- Oleic acid (94%, 4 g; Sigma-Aldrich)
- Linoleic acid (94%, 4 g; Sigma-Aldrich)
- Linolenic acid (85%, 4 g; Nissan – Nippon Oil and Fats Co.)
- Autoclave (Fedegari FVS/3)

13.2.2.2 Procedure

1. MRS broth powder (51 g), cysteine (1 g) and tween 80 (1 mL) were dissolved in dH_2O (1 L) and the resulting solution was introduced into the 5 L fermenter. The apparatus was sealed and sterilised at 121 °C for 15 min.
2. The fermenter vessel was flushed with nitrogen until complete removal of the oxygen content. The temperature, stirring speed and pH were set to 37 °C, 170 rpm and 6.2, respectively.
3. The inoculum of the suitable *Lactobacillus* strain from procedure 1 was added via syringe. The pH was controlled by dropwise addition of sterilised aqueous solutions (10% w/w in water) of either acetic acid or ammonia. After 2–8 hr, the fermentation showed an exponential phase of growth as indicated by starting of the continuous addition of base, required to neutralise the lactic acid produced by the glucose bacterial catabolism.
4. The solution of the suitable fatty acid (4 g) in ethanol (4 mL) was added when 60 mmol of ammonia had been introduced (which corresponds to the start of the exponential growth phase) and a sterilised solution of glucose (60 mL, 250 g L^{-1} in water) was added when overall 220 mmol of ammonia had been used (end of the exponential phase).
5. The hydration reaction was monitored by TLC analysis, eluting with *n*-hexane/AcOEt 7 : 3 and visualising the spots by means of a cerium/phosphomolybdic reagent. The 10-hydroxyacids formed and the starting fatty acids showed Rf values of 0.21 and 0.40, respectively. The fermentation was stopped when the biotransformation ceased to proceed (48–94 hr after the introduction of inoculum). At that time, the reaction mixture was acidified at pH 4 by addition of diluted HCl aq. (3% w/v) and then filtered over celite.

6. The aqueous phase was extracted three times with EtOAc (3 × 200 mL) and the combined organic layers were washed with brine and dried over Na_2SO_4 and the solvent was removed by distillation under reduced pressure.
7. The residue was purified by column chromatography using a gradient of 9 : 1 to 7 : 3 *n*-hexane/AcOEt as eluent to afford the unreacted fatty acid (first eluted fractions) followed by the hydroxy acid derivative.
8. The *L. rhamnosus* ATCC 53103-mediated biotransformations afforded: (*R*)-10-hydroxystearic acid **2a** (1.90 g, 45% yield) as a colourless solid showing mp = 82–84 °C, 94% chemical purity and >95% ee; (*S*)-(12*Z*)-10-hydroxy-octadecenoic acid **2b** (2.0 g, 47% yield) as a pale yellow oil showing 95% chemical purity and >95% ee; and (*S*)-(12*Z*,15*Z*)-10-hydroxy-octadecadienoic acid **2c** (1.50 g, 36% yield) as a pale yellow oil showing 83% chemical purity and >95% ee. Chemical and enantiomeric purity were determined as described in Procedure 3.
9. The *L. plantarum* 299V-mediated biotransformations afforded: (*R*)-10-hydroxystearic acid **2a** (2.17 g, 51% yield) as a colourless solid showing mpt = 83–84 °C, 95% chemical purity and >95% ee; (*S*)-(12*Z*)-10-hydroxy-octadecenoic acid **2b** (1.95 g, 46% yield) as a pale yellow oil showing 94% chemical purity and >95% ee; and (*S*)-(12*Z*,15*Z*)-10-hydroxy-octadecadienoic acid **2c** (2.13 g, 50% yield) as a pale yellow oil showing 82% chemical purity and >95% ee. Chemical and enantiomeric purity were determined as described in Procedure 3.

13.2.3 Procedure 3: Determination of the Chemical and Enantiomeric Purity of the Obtained (*R*)-10-Hydroxystearic Acid, (*S*)-(12*Z*)-10-Hydroxy-Octadecenoic Acid and (*S*)-(12*Z*,15*Z*)-10-Hydroxy-Octadecadienoic Acid

13.2.3.1 Materials and Equipment

- Diazomethane (ethereal solution)
- Acetic anhydride
- Pyridine
- Dichloromethane (DCM)
- 4-(Dimethylamino)pyridine (DMAP)
- (*R*)-10-Hydroxystearic acid, (*S*)-(12*Z*)-10-hydroxy-octadecenoic acid and (*S*)-(12*Z*,15*Z*)-10-hydroxy-octadecadienoic acid from Procedure 2
- *N,N′*-Dicyclohexylcarbodiimide (DCC)
- (*S*)-*O*-Acetylmandelic acid
- Celite
- Anh. Na_2SO_4
- NaCl-saturated solution (brine)
- $NaHCO_3$-saturated solution
- Diethyl ether (Et_2O)
- Silica gel for column chromatography (70–230 mesh; Sigma-Aldrich)
- Hexane
- EtOAc
- Rotary evaporator
- TLC Silica Gel 60 F_{254} plates (plastic foil; Merck)

- TLC visualisation reagent (25 g phosphomolybdic acid hydrate, 10 g $CeSO_4 \cdot 4H_2O$, 60 mL H_2SO_4 96%, dH_2O to prepare 1 L of reagent)
- Gas chromatography/mass spectrometry (GC-MS) analysis: HP-6890 GC equipped with a 5973 mass detector (Hewlett-Packard)
- GC column: HP-5MS (30 m × 0.25 mm, 0.25 μm film thickness, Hewlett-Packard)
- Nuclear magnetic resonance (NMR) analysis: Bruker-AC-400 spectrometer (400 MHz), $CDCl_3$ solutions at rt

13.2.3.2 Procedure

1. The hydroxy acid sample (100 mg, 0.33 mmol) was treated with an excess of the ethereal solution of freshly prepared diazomethane (0.2 M, 5 mL). As soon as the evolution of nitrogen ceased (1 min), the solvent was removed under reduced pressure and the resulting methyl ester was treated at room temperature with a 1 : 1 mixture of pyridine/acetic anhydride (4 mL) and DMAP (10 mg). After 5 hr, the excess of reagents was removed under reduced pressure and the residue was analysed by GC-MS in order to determine the chemical purity of the HFA. The GC temperature programme and the retention times of the hydroxy acid derivatives are reported in Tables 13.1 and 13.2, respectively.
2. The hydroxy acid sample (100 mg, 0.33 mmol) was treated with an excess of the ethereal solution of freshly prepared diazomethane (0.2 M, 5 mL). As soon as the evolution of nitrogen ceased (1 min), the solvent was removed under reduced pressure and the resulting methyl ester was dissolved in dry CH_2Cl_2 (5 mL), treated with (*S*)-*O*-acetylmandelic acid (130 mg, 0.67 mmol), DCC (140 mg, 0.68 mmol) and DMAP (10 mg) and stirred at room temperature for 6 hr. The reaction was then quenched by the addition of water and diethyl ether (60 mL). The dicyclohexylurea formed was removed by filtration on celite and the organic phase was washed with aq. $NaHCO_3$ and brine and dried on Na_2SO_4. The solvent was then removed under reduced pressure and the residue was purified by column chromatography, collecting every fraction containing the fatty acid mandelates (TLC analysis using *n*-hexane/AcOEt 8 : 2 and visualising the spots by means of a cerium/phosphomolybdic reagent). The solvents were removed under reduced pressure

Table 13.1 *GC method.*

Temperature program	Duration
120 °C 3 min; 12 °C.min^{-1}; 195 °C 10 min; 12 °C.min^{-1}; 300 °C 10 min Carrier gas: He; constant flow 1 mL.min^{-1}; split ratio 1 : 30	38 min

Table 13.2 *Retention times for GC analysis.*

Substance	Retention (min)
Methyl 10-acetoxystearate	24.47
Methyl (12*Z*)-10-acetoxy-octadecenoate	24.28
Methyl (12*Z*,15*Z*)-10-acetoxy-octadecadienoate	24.33

Table 13.3 ^{1}H-NMR analysis of the (*S*)-*O*-acetylmandelate esters deriving from 10-hydroxystearic acid, (12*Z*)-10-hydroxy-octadecenoic acid and (12*Z*,15*Z*)-10-hydroxy-octadecadienoic acid methyl esters ($CDCl_3$, 400 MHz).

Starting hydroxyl acid	^{1}H chemical shift of the signals used in enantiomeric purity determination
(*R*)-10-hydroxystearic acid	3.665
(*S*)-10-hydroxystearic acid	3.669
(*S*)-(12*Z*)-10-hydroxy-octadecenoic acid	5.878
(*R*)-(12*Z*)-10-hydroxy-octadecenoic acid	5.889
(*S*)-(12*Z*,15*Z*)-10-hydroxy-octadecadienoic acid	5.877
(*R*)-(12*Z*,15*Z*)-10-hydroxy-octadecadienoic acid	5.890

and the residue was analysed by ^{1}H-NMR in order to determine the enantiomeric purity of the starting HFA. The chemical shifts of the signal used for this measurement are reported in Table 13.3.

3. The reference standards of the racemic 10-hydroxystearic acid, (12*Z*)-10-hydroxy-octadecenoic acid and (12*Z*,15*Z*)-10-hydroxy-octadecadienoic acid were synthesised as reported in [6, 7].
4. (*S*)-*O*-acetylmandelic acid was synthesised as reported in [8].

13.2.4 Conclusion

The probiotic bacteria *L. rhamnosus* (ATCC 53103) and *L. plantarum* 299V can be used as whole-cell biocatalysts for the hydration of oleic acid, linoleic acid and linolenic acid to produce (*R*)-10-hydroxystearic acid, (*S*)-(12*Z*)-10-hydroxy-octadecenoic acid and (*S*)-(12*Z*,15*Z*)-10-hydroxy-octadecadienoic acid, respectively. The presented biotransformation protocol holds preparative significance, is scalable, makes use of standard chemical and microbiological laboratory equipment and allows synthesis of the HFAs with high regio- and stereoselectivity.

References

1. (a) Kim, K.-R. and Oh, D.-K. (2013) *Biotechnology Advances*, **31**, 1473–1485; (b) Lu, W., Ness, J.E., Xie, W. *et al.* (2010) *Journal of the American Chemical Society*, **132**, 15 451–15 455; (c) Serra, S., Fuganti, C. and Brenna, E. (2005) *Trends in Biotechnology*, **23**, 193–198.
2. Engleder, M. and Pichler, H. (2018) *Applied Microbiology and Biotechnology*, **102**, 5841–5858.
3. Cao, Y. and Zhang, X. (2013) *Applied Microbiology and Biotechnology*, **97**, 3323–3331.
4. George Kerry, R., Patra, J.K., Gouda, S. *et al.* (2018) *Journal of Food and Drug Analysis*, **26**, 927–939.
5. Bourdichon, F., Casaregola, S., Farrokh, C. *et al.* (2012) *International Journal of Food Microbiology*, **154**, 87–97.
6. Serra, S. and De Simeis, D. (2018) *Catalysts*, **8**, 109.
7. Serra, S. and De Simeis, D. (2018) *Journal of Applied Microbiology*, **124**, 719–729.
8. Ebbers, E.J., Ariaans, G.J.A., Bruggink, A. and Zwanenburg, B. (1999) *Tetrahedron Asymmetry*, **10**, 3701–3718.

13.3 Clean Enzymatic Oxidation of 12α-Hydroxysteroids to 12-Oxo-Derivatives Catalysed by Hydroxysteroid Dehydrogenase

Fabio Tonin,[1] Natália Alvarenga,[1] Jia Zheng Ye,[1] Isabel W.C.E. Arends[1,2] and Ulf Hanefeld*[1]

[1]*Department of Biotechnology, Delft University of Technology, Delft, The Netherlands*

[2]*Faculty of Science, Utrecht University, Utrecht, The Netherlands*

Ursodeoxycholic acid **3b** (UDCA) and chenodeoxycholic acid **3a** (CDCA) are widely used pharmaceutical ingredients. They can be obtained by transformation of cholic acid **1a** (CA) or ursocholic acid **1b** (UCA), the cheapest and most available natural hydroxysteroids [1]. The C_{12}-specific oxidation of hydroxysteroids is an essential reaction required for the preparation of 12-dehydroxy steroids **3**, which can be synthesised by Wolff–Kishner reduction of the biocatalytically prepared 12-oxo-hydroxysteroids **2** (Scheme 13.4).

The application of enzymes, particularly 12α-hydroxysteroid dehydrogenases (12α-HSDHs), allows the selective oxidation of the C_{12} hydroxyl group with high yields under mild conditions and without the need for the protection and deprotection steps required in chemical synthesis [2].

The recombinantly expressed NAD^+-dependent HSDH from *Eggerthella lenta* (*El*12α-HSDH) was employed for the preparation (gram-scale) of 12-oxo-hydroxysteroids [3].

In order to perform a viable and atom-efficient enzymatic hydroxysteroid oxidation, NAD(P)H oxidase (NOX) was employed as co-factor regeneration system. NOX uses oxygen (O_2) as a sacrificial substrate and produces only water as a side product. A batch method using L-malate dehydrogenase for recycling NAD^+ is also described.

The feasibility of a catalytic aerobic-NAD^+-dependent enzymatic oxidation was shown on a preparative scale (oxidation of CA **1a** into 12-oxo-CDCA **2a**), employing the *El*12α-HSDH-NOX system in a segmented-flow reactor, where 10 mM of CA was fully and selectively converted to 12-oxo-CDCA in 24 hr.

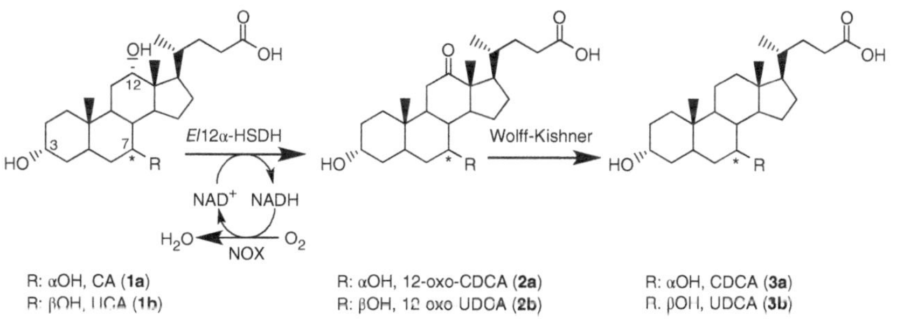

Scheme 13.4 Oxidation of 12α-OH group of hydroxysteroids catalysed by 12α-HSDH.

13.3.1 Procedure 1: Recombinant Expression of *El*12α-HSDH in *E. coli* BL21(DE3)

13.3.1.1 Materials and Equipment

- Lysogenic broth (LB; 1.5 L for main culture + 50 mL pre-culture – tryptone from casein 10 g.L^{-1}, yeast extract 5 g.L^{-1} and NaCl 10 g.L^{-1})

- Demineralised water (dH_2O)
- Washing buffer: 10 mM KPi buffer, pH 8.0 (100 mL)
- Kanamycin (30 mg.mL^{-1} in dH_2O, filter-sterilised)
- Isopropyl-β-D-1-thiogalactopyranoside (IPTG; 1 M in dH_2O, filter-sterilised)
- LB agar (+ 30 μg.mL^{-1} kanamycin) plate with colonies of *E. coli* BL21(DE3) harbouring the expression vector pET28b(+)-*El*12α-HSDH (GenBank: WP_114518444.1) [3]
- 0.3 L Erlenmeyer flask with cotton cap
- 5 L Erlenmeyer flasks with cotton caps
- Sterile syringe filters (cellulose acetate, 0.2 μm; GE Healthcare)
- UV-Vis spectrophotometer (Ultrospec 2100 *pro*, Pharmacia)
- Orbital shaker (Innova 44, New Brunswick Scientific)
- Autoclave (Getinge)
- Centrifuge with temperature control (17 000× g; Sorvall RC6+ Centrifuge, Thermo Fisher Scientific)

13.3.1.2 Procedure

1. Tryptone (15 g), yeast extract (7.5 g) and NaCl (15 g) were dissolved in dH_2O (1.5 L) and autoclaved (20 min, 121 °C) in 5 L Erlenmeyer flasks with cotton caps to give sterile LB medium.
2. To prepare the pre-culture, sterile LB medium (50 mL) was placed into a sterile 0.3 L Erlenmeyer flask. Afterwards, kanamycin stock solution (50 μL) was added to reach a final concentration of 30 μg.mL^{-1}. The solution was inoculated with a single colony of *E. coli* BL21(DE3) harbouring pET28b(+)-*El*12α-HSDH and shaken at 200 rpm and 37 °C overnight (17 hr).
3. The next day, the LB medium (1.5 L) was supplemented with kanamycin (1.5 mL) stock solution to a final concentration of 30 μg.mL^{-1} in a sterile 5 L Erlenmeyer flask with cotton cap. The flask was inoculated from the LB pre-culture to give an OD_{600} of 0.05 (spectrophotometrically measured).
4. The cells were grown at 37 °C and 200 rpm until an OD_{600} of 1 was reached (~3.5 hr). Then, the cultivation was cooled to 25 °C and expression of recombinant *El*12α-HSDH was induced by the addition of 375 μL IPTG stock solution (final conc. 0.25 mM). The expression was performed at 25 °C for 20 hr.
5. The cells were harvested by centrifugation at 17 000× g and 4 °C for 10 min, washed with 100 mL of 10 mM KPi buffer, pH 8.0 and frozen at −80 °C. 8.2 g of cells were obtained.

13.3.2 Procedure 2: Purification of *El*12α-HSDH

13.3.2.1 Materials and Equipment

- Loading buffer: 50 mM KPi, 500 mM NaCl, 10% glycerol, pH 8.0 (500 mL)
- Elution buffer: 50 mM KPi, 500 mM imidazole, 5% glycerol, pH 8.0 (250 mL)
- Storage buffer: 50 mM KPi, 5% glycerol, pH 8.0 (2 L)
- Protein purification system (NGC Chromatography system, Bio-Rad)
- Hi-Trap FF Crude column preloaded with Ni (5 mL volume; GE Healthcare)
- Ultrasonic wave generator (Branson Sonifier 250, Branson Ultrasonics)

- Centrifuge with temperature control (50 000× g; Sorvall RC6+ Centrifuge, Thermo Fisher Scientific)
- 2× 0.5 L Schott bottles with screw caps
- Bradford assay solution (Bio-Rad) used accordingly to manufacturer's instructions
- Plate reader (Synergy HTX, BioTek)
- Activity-measurement solution (1.0 mM CA, 1.0 mM NAD^+, in 50 mM KPi buffer, pH 8.0 and 10% v/v methanol)
- UV-transparent microplates in 96-well configuration (UV-Star®, Greiner bio-one, Alphen aan den Rijn)

13.3.2.2 Procedure

1. Cells (8.2 g) were resuspended in 35 mL of loading buffer. In order to break them, 5 cycles of sonication (30 sec, 30% power) interspersed with 1 min of resting in an ice-water bath were performed.
2. The lysate was centrifuged at 50 000× g and 4 °C for 30 min, yielding 42 mL of clear solution (crude extract) and a cell-debris pellet.
3. The lysate was loaded on to a Hi-Trap FF Crude chelating column (5 mL) with a flow rate of 3 mL.min^{-1}. After sample application, the column was washed with loading buffer till $Abs_{280nm} \approx 0$. Unspecifically bound proteins were eluted with 5 : 95 elution/loading buffer, then *El*12α-HSDH was eluted with 50 : 50 elution/loading buffer. The obtained preparation was dialysed overnight in storage buffer (Figure 13.4).
4. The concentration of the enzyme in the final preparation was assayed by Bradford assay using bovine serum albumin (BSA) as standard curve (range 2–50 μg.mL^{-1}).
5. 322 mg of pure *El*12α-HSDH was obtained (Table 13.4).
6. The activity of the pure *El*12α-HSDH preparation was assayed spectrophotometrically in a 96-well plate reader: to 200 μL of activity-measurement solution was added 10 μL of *El*12α-HSDH enzyme preparation (diluted 1 : 100). The increase in absorbance at 340 nm (NADH production, $\varepsilon_{340\,nm} = 6.2\,mM^{-1}.cm^{-1}$) was followed. 1 U was defined

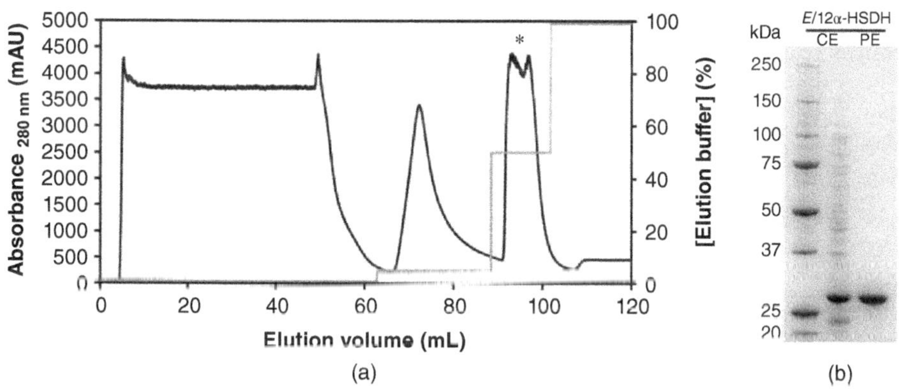

Figure 13.4 (a) Elution profile of El12α-HSDH. The black and grey lines correspond to the absorbance at 280 nm and to the percentage of elution buffer applied to the column during the purification, respectively. The peak * contains the El12α-HSDH. (b) SDS PAGE analysis of the crude extract (CE) and purified enzyme (PE)obtained in Procedure 2.

as the amount of enzyme producing 1 μmol of product per minute at 25 °C and pH 8.0. Under these conditions, the obtained enzymatic preparation showed a specific activity of 30.1 $U.mg^{-1}$ and a volumetric activity of 422 $U.mL^{-1}$.

Table 13.4 *Purification table for El12α-HSDH.*

*El*12α-HSDH: 1.5 L LB media (8.2 g cell)						
	Volume (mL)	[Protein][a] ($mg.mL^{-1}$)	Protein (mg)	Activity[b] ($U.mL^{-1}$)	U_{tot}	Specific activity ($U.mg^{-1}$)
Crude extract	42	37	1554	22.8	958	0.6
HiTrap-Chelating	23	14	322	422	9706	30.1

[a] Protein concentrations determined by Bradford assay.
[b] *El*12α-HSDH enzymatic activity in the crude extract and of the purified enzyme were determined at 25 °C using 1.0 mM CA, 1.0 mM NAD^+, in 50 mM KPi buffer, pH 8.0 and 10% v/v methanol. The production of NADH was followed at 340 nm.

13.3.3 Procedure 3: Biocatalytic Oxidation of CA 1a to 12-Oxo-CDCA 2a with NOX

13.3.3.1 Materials and Equipment

- CA stock solution (7.5 mL, 100 mM, prepared by dissolving 4085 mg in 10 mL high-purity methanol)
- Potassium phosphate buffer pH 8.0 (1 M)
- NAD^+ (potassium salt) stock solution (10 mM in water)
- Aqueous NaOH (3 M)
- *El*12α-HSDH from Procedure 2 (18 μL, 14 $mg.mL^{-1}$, 422 $U.mL^{-1}$ determined on CA as substrate)
- NAD(P)H oxidase (001) (PRO-NOX, Prozomix Ltd)
- 100 mL Schott bottle with screw cap
- Silicon reactor coil (ID 2 mm, volume 60 mL)
- Silicon connectors tube (ID 2 mm, volume 60 mL)
- 5 mL empty column (expansion vessel)
- Peristaltic pump (120 U; Watson Marlow)
- Mass flow controller (EL-FLOW, Bronkhorst)
- pH meter
- Benchtop centrifuge (min. 14 000 rpm)

13.3.3.2 Procedure

1. A flow reactor was set up as shown in Figure 13.5. O_2 flow rate was set at 5 $mL.min^{-1}$.
2. A solution (75 mL) of 10 mM CA **1a** (306 mg) and 0.5 mM NAD^+ (25 mg) in 10% v/v MeOH in 50 mM KPi buffer pH 8.0 was prepared by diluting the stock solutions. The pH value was adjusted to 8.0 by adding 1 M NaOH (aq).
3. To the solution were added *El*12α-HSDH (262 μg, 18 μL) and PRO-NOX (001) (37.5 mg) powder.

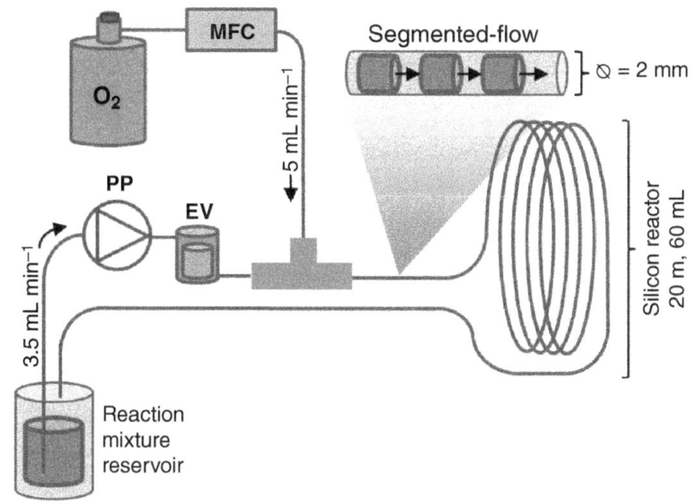

Figure 13.5 Scheme of the flow reactor employed in the preparative biotransformation of CA into 12-oxo-CDCA as catalysed by an *El*12α-HSDH-NOX system (Procedure 3). PP, peristaltic pump; MFC, mass flow controller; EV, expansion vessel.

4. The reaction mixture was pumped into a flow reactor by peristaltic pump at a flow rate of 3.5 mL.min^{-1} and recirculated at 21 °C for 24 hr.
5. For reaction monitoring by liquid chromatography (LC), 50 μL of reaction mixture was supplemented with 200 μL of mobile phase (H_2O/CH_3CN/TFA 70 : 30 : 0.1) and centrifuged at 14 000 rpm in a benchtop centrifuge for 2 min. 10 μL of sample was analysed by high-performance liquid chromatography (HPLC) to determine conversion (see Analytical Method).
6. After 98% conversion, the obtained 12-oxo-CDCA **2a** was isolated following Procedure 5.

13.3.4 Procedure 4 : Biocatalytic Oxidation of CA 1a into 12-Oxo-CDCA 2a with MDH

13.3.4.1 Materials and Equipment

- CA stock solution (5.0 mL, 100 mM, prepared by dissolving 4085 mg in 10 mL high-purity methanol)
- Oxaloacetate disodium salt (270 mg, 1.53 mmol, powder; Sigma-Aldrich)
- Potassium phosphate buffer pH 8.0 (1 M)
- NAD^+ (potassium salt) stock solution (10 mM in water)
- Aqueous NaOH (3 M)
- *El*12α-HSDH from Procedure 2 (12 μL, 14 mg.mL^{-1}, 422 U.mL^{-1} determined on CA as substrate)
- L-Malate dehydrogenase (25 μL, 1200 U.mg^{-1}, 10 mg.mL^{-1}; L-MDH, Sigma-Aldrich)
- 100 mL round-bottom flask
- pH meter
- Benchtop centrifuge (min. 14 000 rpm)

13.3.4.2 Procedure

1. In a 100 mL round-bottom flask, a 50 mL solution of CA **1a** (10 mM, 5 mL of 100 mL stock solution, 204 mg), 30 mM oxaloacetate (270 mg) and 0.5 mM NAD^+ (17 mg) in 10% v/v MeOH in 50 mM KPi buffer, pH 8.0 was prepared by diluting the stock solutions. The pH was adjusted to 8.0 by adding 3 M NaOH (aq).
2. *El*12α-HSDH (175 μg, 12 μL) and L-malate dehydrogenase (0.25 mg, 300 U) were added to the solution.
3. The reaction mixture was incubated in an orbital shaker at 25 °C for 4 hr.
4. For reaction monitoring by HPLC, 50 μL aliquots of reaction mixture were supplemented with 200 μL of mobile phase (H_2O/CH_3CN/TFA 70 : 30 : 0.1) and centrifuged at 14000 rpm in a benchtop centrifuge for 2 min. 10 μL of sample was injected (see Analytical Method).
5. After 99% conversion, the obtained 12-oxo-CDCA **2a** was isolated following Procedure 5.

13.3.5 Procedure 5: Workup of 12-Oxo-CDCA B1

13.3.5.1 Materials and Equipment

- Aqueous HCl (3 M)
- Diethyl ether (Et_2O) peroxide-free, control performed with peroxide test strips (Quantofix, Machery-Nagel)
- Dry Na_2SO_4
- 1 L separating funnel
- Sterile syringe filters (Teflon, 0.45 μm; GE Healthcare)
- pH meter
- Rotary evaporator (Rotavap)

13.3.5.2 Procedure

1. The reaction mixture obtained in Procedure 3 or 4 was acidified to pH ~3.0 with 3 M HCl (aq) and extracted with Et_2O (3 × 35 mL). The organic layers were collected, pooled and dried by adding Na_2SO_4 (~2 g).
2. The organic solvent was evaporated with the rotary evaporator.
3. The obtained dry residue (crude) was dissolved in absolute MeOH and filtered with a Teflon filter (0.45 μm).
4. MeOH was evaporated, giving 12-oxo-CDCA **2a** as a white powder. Conversions and isolated yields for the two different procedures are reported in Table 13.5.

Table 13.5 *Bioconversion tables for Procedures 3 and 4. In both cases, a reaction selectivity of 100% was observed.*

Entry	Reaction	Regeneration system	Scale (mg)	Conversion (%)	Yield (%)
Procedure 13.3.3	CA → 12-oxo-CDCA	NOX	306	98	90
Procedure 13.3.4	CA → 12-oxo-CDCA	L-MDH	204	>99	85

13.3.6 Analytical Method

HPLC analyses were performed on a Shimadzu apparatus equipped with a LC20AT pump and an ELSD-LTII detector and fitted with an XTerra RP C18 column (Waters) (length/internal diameter 150/4.6 mm, pore size 3.5 µm) under the following conditions: eluent $H_2O/CH_3CN/TFA$ (70 : 30 : 0.1), flow 1.0 mL.min^{-1}, temperature 35 °C (Figure 13.6). Calibration curves were obtained for CA and 12-oxo-CDCA by integrating the peak area obtained by seven injections of samples with known concentration of the two different compounds in the range of 0.1–15 mM. Retention times: 12-oxo-CDCA **2a**, 3.2 min; CA **1a**, 3.7 min.

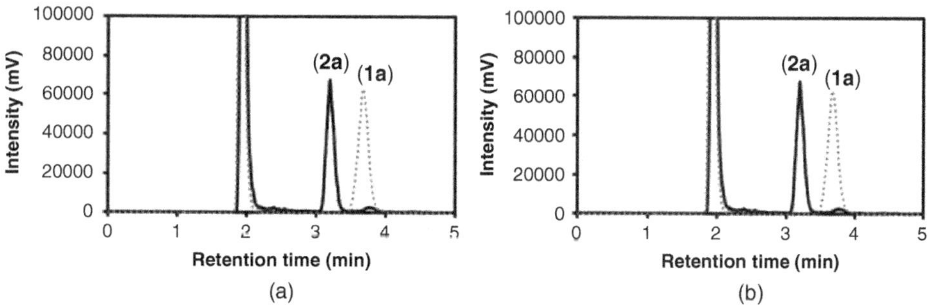

Figure 13.6 HPLC analyses of biocatalytic oxidation of CA **1a** into 12-oxo-CDCA **2a** with (a) NOX (Procedure 3) and (b) L-MDH (Procedure 4) as regeneration systems. In both cases, the grey dotted line corresponds to the reaction mixture before incubation with *E*/12α-HSDH, whilst the black continuous line is after the incubation.

13.3.7 Conclusion

The described procedures allowed the regioselective oxidation of CA in position C_{12} using *El*12α-HSDH enzyme and two different NAD^+-regeneration systems. These procedures can be applied for the preparation of a range of 12-oxo-hydroxysteroid derivatives.

References

1. Tonin, F. and Arends, I.W.C.E. (2018) *Beilstein Journal of Organic Chemistry*, **14**, 470–483.
2. Abdelraheem, E.M.M., Busch, H., Hanefeld, U. and Tonin, F. (2019) *Reaction Chemistry & Engineering*, **4**, 1878–1894.
3. Tonin, F., Alvarenga, N., Ye, J.Z. *et al.* (2019) *Advanced Synthesis & Catalysis*, **361**, 2448–2455.

13.4 Whole-Cell Biocatalysis Using PmlABCDEF Monooxygenase and Its Mutants: A Versatile Toolkit for Selective Synthesis of Aromatic *N*-Oxides

Vytautas Petkevičius,[1] Justas Vaitekūnas,[1] Daiva Tauraitė,[1] Jonita Stankevičiūtė,[1] Dovydas Vaitkus,[1] Jonas Šarlauskas,[2] Narimantas Čėnas[2] and Rolandas Meškys[1]
[1]Department of Molecular Microbiology and Biotechnology, Institute of Biochemistry, Life Sciences Center, Vilnius University, Vilnius, Lithuania
[2]Department of Xenobiotics Biochemistry Institute of Biochemistry, Life Sciences Center, Vilnius University, Vilnius, Lithuania

The current synthesis of aromatic *N*-oxides is exclusively based on chemical methods [1], some of which are hazardous and produce significant amounts of waste (oxidation by hydrogen peroxide [2] or peroxy acids [3]), lack selectivity (reagent $HOF{\cdot}CH_3CN$ [4]) or are relatively unspecific and complicated (condensation reactions, various rearrangements and cycloadditions [5]). The practicality of currently reported enzyme–based methods is limited by the low productivity and narrow range of available substrates [6]. We, however, have established the application of *Escherichia coli* whole cells harbouring PmlABCDEF monooxygenase as a productive, scalable, regio- and chemoselective enzyme-based method for the synthesis of aromatic *N*-oxides [7]. The application of both rational design and directed evolution enabled the development of a variety of PmlABCDEF mutants featuring expanded substrate scope, altered regioselectivity and the ability to produce certain di-*N*-oxides [8].

Here, we present detailed procedures for obtaining desired aromatic *N*-oxides via different biocatalytic pathways. The application of PmlABCDEF monooxygenase (biocatalyst A) and its mutant A113G (biocatalyst B) offers three distinct approaches (Scheme 13.5).

The biocatalytic production of 4–(pyridine–4–ylsulfanyl)pyridine-1-oxide (**1b**) from 4–(pyridine–4–ylsulfanyl)pyridine (**1a**) shows selectivity and novelty, as synthesis of the resulting compound has not been documented. The versatility is demonstrated by the synthesis of quinazoline–3-oxide (**2b**) and quinazoline–1–oxide (**2c**) from quinazoline (**2a**), employing different biocatalysts. Both enzyme variants were also applied in the conversions of quinoxaline (**3a**); biocatalyst A was used for the preparation of quinoxaline–1–oxide (**2b**), whilst biocatalyst B was employed in the synthesis of quinoxaline–1,4–dioxide (**2c**).

13.4.1 Procedure 1: Recombinant Expression of the PmlABCDEF Monooxygenase and Its Mutants in *E. coli* BL21(DE3)

13.4.1.1 Materials and Equipment

- Tryptone from casein (5 g)
- Yeast extract (2.5 g)
- NaCl (2.5 g)
- Distilled water (dH_2O)
- Glucose (40% in dH_2O) stock solution
- KOH (~1 M) stock solution
- Kanamycin ($40\,mg.mL^{-1}$ in dH_2O) stock solution

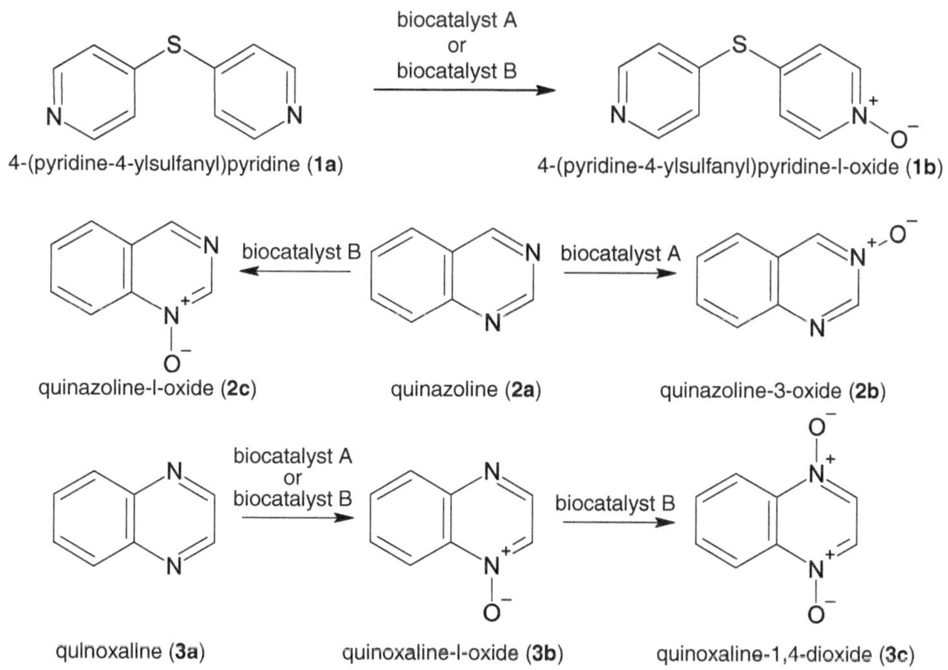

Scheme 13.5 Three different routes in the synthesis of aromatic *N*-oxides utilising PmlABCDEF.

- Isopropyl-β-D-1-thiogalactopyranoside (IPTG; 1 M in dH_2O)
- Biocatalyst A: lysogenic broth (LB) agar plate with colonies of *E. coli* BL21(DE3) harbouring the expression vector pET28b(+) with genes encoding wild-type PmlABCDEF monooxygenase (plasmid pET_pmlABCDEF) [7]
- Biocatalyst B: LB agar plate with colonies of *E. coli* BL21(DE3) harbouring the expression vector pET28b(+) with genes encoding mutant A113G of PmlABCDEF monooxygenase (plasmid pET_pmlABCDEF_A113G) [8]
- 100 mL plastic Erlenmeyer flasks with screw caps
- 2 L plastic Erlenmeyer flasks with screw caps
- Incubator shaker (Innova 44 R)
- Steam steriliser (Raypa)
- Refrigerated centrifuge (Eppendorf 5810R)
- 400 mL large-volume centrifuge cones

13.4.1.2 Procedure

1. Tryptone (5 g), yeast extract (2.5 g) and NaCl (2.5 g) were dissolved in dH_2O (0.5 L) and autoclaved (30 min, 121 °C) in two 2 L plastic Erlenmeyer flasks (250 mL in each) to give sterile LB medium.
2. Pre-culture was prepared in a sterile 100 mL plastic Erlenmeyer flask containing 20 mL of LB medium supplemented with kanamycin (20 μL) and glucose (0.5 mL) stock solutions to final concentrations of 40 mg.mL^{-1} and 50 mM, respectively. The

resulting solution was inoculated with a single colony of *E. coli* BL21(DE3) harbouring pET_pmlABCDEF (biocatalyst A) or pET_pmlABCDEF_A113G (biocatalyst B) and shaken at 180 rpm and 30 °C overnight.

3. The next day, each flask containing LB medium (250 mL) was supplemented with kanamycin (250 μL) and glucose (6.25 mL) stock solutions to final concentrations of 40 mg.mL^{-1}, and 2.5 mL of the LB pre-culture (1% v/v inoculum size) was added.
4. The cells were grown at 30 °C and 180 rpm until an OD_{600} of 0.8–1.0 was reached. Then, expression of the recombinant enzymes was induced by the addition of 125 μL IPTG stock solution (final conc. 0.5 mM) to every flask. The cell culture was placed in a 20 °C incubator and cultivation was continued for 12–16 hr.
5. After induction, cells were separated from the medium by centrifugation (4000× g, 4 °C for 5 min) in large-volume centrifuge cones (250 mL in each). The supernatant was discarded and the resulting cell paste (1.5–2.0 g) was handled further as described in Procedure 2.

13.4.2 Procedure 2: Preparation of Whole-Cell Biocatalyst A and B (*E. coli* BL21(DE3) Resting Cells Harbouring pET_pmlABCDEF or pET_pmlABCDEF_A113G

13.4.2.1 Materials and Equipment

- Potassium phosphate wash buffer (10 mM KH_2PO_4 (1.36 g) in dH_2O, volume adjusted to 1.0 L and pH to 7.0 (using 1 M KOH), transferred to a 1 L glass bottle and autoclaved (30 min, 121 °C))
- KOH (~1 M) stock solution
- dH_2O
- 1 L glass bottle with screw cap
- 400 mL large-volume centrifuge cones
- 50 mL centrifuge tubes
- Refrigerated centrifuge (Eppendorf 5810R)
- Ice bath

13.4.2.2 Procedure

1. The separated cell biomass was washed with ice-cold 10 mM potassium phosphate buffer. Wash buffer solution (10–20 mL) was poured in each large-volume centrifuge cone and cells were gently resuspended by pipetting.
2. The cell suspension was transferred to 50 mL centrifuge tubes and separated from the medium by centrifugation (4000× g, 4 °C for 5 min). The resulting cell culture was suspended in ice-cold 10 mM potassium phosphate buffer (50 mL) and kept on ice, ready for the bioconversions.

13.4.3 Procedure 3: Biocatalytic Conversions of 4–(Pyridin-4-ylsulfanyl)pyridine-1-Oxide, Quinazoline–3-Oxide, Quinazoline-1-Oxide, Quinoxaline-1-Oxide and Quinoxaline-1,4-Dioxide

13.4.3.1 Materials and Equipment

- 4-(Pyridin-4-ylsulfanyl)pyridine (47 mg, 0.25 mmol)
- Quinazoline (32.5 mg, 0.25 mmol)

- Quinoxaline (32.5 mg, 0.25 mmol)
- Glucose (40% in dH_2O) stock solution
- 10 mM potassium phosphate buffer (pH 7.0)
- *E. coli* BL21(DE3) resting cells harbouring pET_pmlABCDEF or pET_pmlABCDEF_A113G (from Procedure 2)
- 2 L plastic Erlenmeyer flasks with screw caps
- Incuabator shaker (Innova 44 R)
- Refrigerated centrifuge (Eppendorf 5810R)
- 400 mL large-volume centrifuge cones
- Thin-layer chromatography (TLC) silica gel aluminium sheets
- UV lamp, 254 nm
- Chloroform
- Methanol
- Rotary evaporator IKA RV 10
- 250 mL round-bottom flask
- 50 mL round-bottom flask
- Anhydrous sodium sulfate (anh. Na_2SO_4)
- Rectangular TLC jars
- 150 mL separation funnel
- Filter funnel
- Incubator shaker (Innova 44 R)

13.4.3.2 Procedure

1. The prepared biomass was transferred to a 2 L plastic Erlenmeyer flask and the reaction volume was adjusted to 250 mL with 10 mM potassium phosphate buffer. The reaction mixture was supplemented with appropriate substrate (final conc. 1 mM) and placed into an incubator shaker (30 °C, 180 rpm).
2. The bioconversion proceeded for 8 hr and was followed by addition of 6.25 mL glucose stock solution (final conc. 50 mM) and continued overnight, for a total of 24 hr.
3. The progress of the conversion was monitored by TLC. After 24 hr of incubation, 1 μL of reaction mixture was transferred on to a TLC silica gel sheet, adjacent to 1 μL of appropriate substrate. The mobile phase consisted of methanol and chloroform (1 : 9). The bioconversion was called complete when the reaction mixture did not contain any residual substrate.
4. The biomass was separated from the reaction mixture by centrifugation (4000× g, 4 °C for 1 hr). The resulting supernatant was processed for the product extraction.
5. The supernatant was evaporated under reduced pressure (8–10 torr) using a rotary evaporator (rotating 20 rpm at 40 °C). The reduction was performed in a 250 mL round-bottom flask, adding portions of the reaction mixture (~100 mL) until the total volume decreased to 15–20 mL.
6. The resulting mixture was transferred to the separation funnel, followed by 15–20 mL of chloroform. The separation funnel then was capped, thoroughly shaken, ventilated and put on the stand to allow the layers to separate. The organic phase (bottom) was collected, whilst the water phase was supplemented with 15–20 mL of chloroform. The extraction was repeated a total of three times.
7. Organic phases were combined and filtered/dried out through the filter funnel containing anh. Na_2SO_4. This solution was evaporated under reduced pressure until dry. In the case

of **1a** (using biocatalyst A) and **2a** (using biocatalyst A or B), the resulting compounds were the final products.

8. The synthesis of **3b** (using biocatalyst A) and **3c** (using biocatalyst B) was usually incomplete after 24 hr of incubation. In that case, the resulting mixture of compounds (Step 7) was dissolved in water and used as a substrate (Step 1) for another round of conversion using fresh whole cells. The rest of the procedure remained the same.

13.4.4 Analytical Method

Nuclear magnetic resonance (NMR) spectroscopy was performed on a Bruker Ascend 400. NMR spectra were recorded at ambient temperature using DMSO-d6 or CCl_3D as solvent, with proton and carbon resonances at 400 and 100 MHz, respectively. All NMR data are reported in ppm relative to the solvent signal as internal standard. The multiplicities are stated as follows: s, singlet; d, doublet; dd, doublet of doublets; t, triplet; q, quartet; m, multiplet.

4-(Pyridin-4-ylsulfanyl)pyridine-1-oxide **1b** was prepared in 81% yield using biocatalyst A. Brown crystals. ^{1}H-NMR (400 MHz, DMSO-d6): $\delta = 7.25$–7.28 (m, 2H, CH), 7.52–7.56 (m, 2H, CH), 8.26–8.30 (m, 2H, CH), 8.45–8.49 (m, 2H, CH). ^{13}C-NMR (100 MHz, DMSO-d6): δ = 122.6, 127.5, 131.3, 140.4, 146.5, 150.5.

Quinazoline-3-oxide **2b** was prepared in 67% yield using biocatalyst A. Yellow solid. ^{1}H-NMR (400 MHz, DMSO-d6): $\delta = 7.73$–7.79 (m, 1H, CH), 7.79–7.85 (m, 1H, CH), 7.93 (d, $J = 8.4$, 1H, CH), 7.99 (d, $J = 8.3$, 1H, CH), 9.06 (d, $J = 2.2$, 1H, CH), 9.28 (d, $J = 2.2$, 1H, CH). ^{13}C-NMR (100 MHz, DMSO-d6): δ = 124.7, 125.7, 128.2, 130.5, 131.8, 139.9, 140.1, 149.1.

Quinazoline-1-oxide **2c** was prepared in 73% yield using biocatalyst B. ^{1}H-NMR (400 MHz, DMSO-d6): δ = 7.89–7.95 (dd, J = 7.9, 7.2 Hz, 1H, CH), 8.07–8.14 (dd, J=8.1, 7.8 Hz, 1H, CH), 8.30 (d, J=8.2 Hz, 1H, CH), 8.47 (d, J=8.7, 1H, CH), 9.14 (s, 1H, CH), 9.31 (s, 1H, CH). ^{13}C-NMR (100 MHz, DMSO-d6): δ = 118.3, 128.6, 130.7, 135.4, 143.4, 147.4.

Quinoxaline-1-oxide **3b** was prepared in 66% yield using biocatalyst A. White solid. ^{1}H-NMR (400 MHz, DMSO-d6): δ = 7.84–7.90 (m, 1H, CH), 7.92–7.98 (m, 1H, CH), 8.15 (dd, J = 8.3, 1.4 Hz, 1H, CH), 8.45 (dd, J = 8.6, 1.5 Hz, 1H, CH), 8.6 (d, J = 3.6 Hz, 1H, CH), 8.78 (d, J=3.6 Hz, 1H, CH). ^{13}C-NMR (100 MHz, DMSO-d6): δ = 118.8, 130.1, 130.2, 131.1, 132.5, 137.0, 145.7, 147.5.

Quinoxaline-1,4-dioxide **3c** was prepared in 52% yield using biocatalyst B. Orange solid. ^{1}H-NMR (400 MHz, DMSO-d6): $\delta = 7.95$–8.05 (m, 2H, CH), 8.43–8.52 (m, 2H, CH), 8.54 (s, 2H, CH). ^{13}C-NMR (100 MHz, CCl_3D): δ = 120.6, 130.4, 132.2, 138.6.

13.4.5 Conclusion

The described method is a new biocatalytic approach to the production of aromatic *N*-oxides, offering a green alternative to existing chemical methods. Most importantly, we have demonstrated that tailoring PmlABCDEF monooxygenase offers versatility, as new enzyme variants could be applied in the advanced synthesis. The procedures exploit minimal substrate concentration (1 mM) and cover conversions of the most distinguishable examples. A detailed substrate scope and optimal working concentrations for a variety of substrates can be found in our previous publications [7, 8].

References

1. Mfuh, A.M. and Larionov, O.V. (2015) *Current Medicinal Chemistry*, **22**, 2819–2857.
2. Kokatha, H.P., Thomson, P.F., Bae, S. *et al.* (2011) *Journal of Organic Chemistry*, **76**, 7842–7848.
3. Mosher, H.S., Turner, L. and Carlsmith, A. (1963) *Organic Syntheses*, **4**, 828–830.
4. Rozen, S. (2014) *Accounts of Chemical Research*, **47**, 2378–2389.
5. (a) Chucholowski, A.W. and Uhlendorf, S. (1990) *Tetrahedron Letters*, **31**, 1949–1952; (b) Nesi, R., Giomi, D., Papaleo, S. and Turchi, S. (1992) *Journal of Organic Chemistry*, **57**, 3713–3716.
6. (a) Ullrich, R., Dolge, C., Kluge, M. and Hofrichter, M. (2008) *FEBS Letters*, **582**, 4100–4106; (b) Zhao, Y., Qian, G., Ye, Y. *et al.* (2016) *Organic Letters*, **18**, 2495–2498; (c) Mitsukura, K., Hayashi, M., Yoshida, T. and Nagasawa, T. (2013) *Journal of Bioscience and Bioengineering*, **115**, 651–653.
7. Petkevičius, V., Vaitekūnas, J., Tauraitė, D. *et al.* (2019) *Advanced Synthesis & Catalysis*, **361**, 2456–2465.
8. Petkevičius, V., Vaitekūnas, J., Vaitkus, D. *et al.* (2019) *Catalysts*, **9**, 356.

Index

Applied Biocatalysis: The Chemist's Enzyme Toolbox, First Edition. Edited by John Whittall and Peter W. Sutton.
© 2021 John Wiley & Sons Ltd. Published 2021 by John Wiley & Sons Ltd.